# Electronic Communications: Principles and Systems

# Electronic Communications: Principles and Systems

**William D. Stanley**

**John M. Jeffords**

**THOMSON**
™
**DELMAR LEARNING**

Australia   Canada   Mexico   Singapore   Spain   United Kingdom   United States

## THOMSON
™
## DELMAR LEARNING

Electronic Communications: Principles and Systems
William D. Stanley and John M. Jeffords

**Vice President, Technology and Trades SBU:**
Alar Elken

**Editorial Director:**
Sandy Clark

**Senior Acquisitions Editor:**
Stephen Helba

**Senior Development Editor:**
Michelle Ruelos Cannistraci

**Marketing Director:**
David Garza

**Senior Channel Manager:**
Dennis Williams

**Marketing Coordinator:**
Stacey Wiktorek

**Production Director:**
Mary Ellen Black

**Production Editor:**
Toni Hansen

**Art/Design Coordinator:**
Francis Hogan

**Senior Project Editor:**
Christopher Chien

**Senior Editorial Assistant:**
Dawn Daugherty

Library of Congress Cataloging-in-Publication Data

Stanley, William D.
    Electronic communications : principles and systems / William D. Stanley, John M. Jeffords.—1st ed.
        p.   cm.
    Includes bibliographical references and index.
    ISBN 1-4180-0003-5 (alk. paper)
    1. Telecommunication.   2. Digital communications.   I. Jeffords, John M. II. Title.
    TK5101.S6594 2006
    621.382—dc22                    2005019390

### NOTICE TO THE READER

# Table of Contents

# Lab Manual Contents

Lab Exercises to Accompany *Electronic Communications: Principles and Systems* is available on the accompanying Lab.*Source* CD.

# Preface

Electronic communications is arguably the broadest application subject area in the electrical field. Topics include traditional analog communication techniques such as AM and FM, modern digital systems, telecommunications, radar, microwaves, networking, and many other areas. No single textbook could do justice to all of these areas, so any broad-coverage text must necessarily limit its coverage to some extent. However, within the context of the previous terms, this book aims at a relatively broad coverage.

The mathematical techniques in this book have been kept to a relatively modest level, making it accessible to engineers and technologists who have only a moderate background in differential and integral calculus. Yet it provides exposure, from a practical point of view, to some of the sophisticated analytical techniques, such as Fourier transforms, spectral analysis, and the performance of systems in the presence of noise. In this coverage, the emphasis is directed toward the practical interpretation and application of the tools, rather than the formal mathematical derivations so characteristic of advanced communication texts.

There are many fine books with the broad level of coverage included in many of the books written at the two-year technology level. Yet many of these books assume such a modest level of mathematics on the part of the reader that only a superficial understanding of the subject matter can likely be achieved. This may be perfectly adequate for individuals who will be trained to operate communication equipment, but for the technologist or engineer who aspires to develop a deeper understanding of the "why," or to become involved in the design of such equipment, it is necessary to encounter more mathematical exposure of the concepts. This book has been written to provide a broad exposure employing mathematical concepts, but with the emphasis on the practical application and interpretation of the subject areas.

While this book has circuit diagrams where appropriate, the major emphasis throughout is on the signal processing at a block diagram or systems level. This approach will provide a better background for the student as the technology advances, and new integrated circuits or modules become available. For example, early phase-locked loop circuits were composed of discrete elements, first with tubes and later with transistors. Of course, they are now available as off-the-shelf integrated circuits. Yet the signal processing function of the circuit has remained the same throughout this evolution. When actual circuits are considered in this text, in most cases the approach will focus on the strategy of the circuit from a signal processing point of view, rather than on the actual internal details.

## Organization of Book

### 1. General Overview

A general overview of communication systems is presented to provide the reader with the "big picture" before dealing with the individual parts. Some of the earlier material is qualitative and descriptive and various terms are introduced and defined. The last few sections are devoted to decibel gain and loss definitions and computations including signal-to-noise ratios. Even though most readers will have been exposed to decibel forms, few are proficient in applying the somewhat confusing communication decibel forms that involve both dimensionless ratios and values referred to standard levels; e.g., the dBm.

## 2. Spectral Analysis I: Fourier Series

Spectral analysis is developed through an intuitive approach leading to the Fourier series. Emphasis is placed on the use of Fourier series pairs and their practical applications rather than on the derivations. The treatment classifies Fourier series in three categories: (a) quadrature (sine-cosine) form, (b) amplitude-phase form, and (c) exponential (two-sided) form.

## 3. Spectral Analysis II: Fourier Transforms and Pulse Spectra

The work of the preceding chapter is continued with the emphasis directed toward non-periodic functions and their spectral forms. As in the case of Fourier series, the emphasis is on the practical applications and interpretations of the mathematical forms. Properties of four types of pulses and their spectra will be emphasized: (a) baseband non-periodic, (b) baseband periodic, (c) RF non-periodic, and (d) RF periodic.

## 4. Communication Filters and Signal Transmission

The steady-state frequency response analysis of a circuit is developed and amplitude and phase functions are described. The criteria for distortionless transmission are developed and the various classes of filters are introduced. A survey of common filter types is presented and the analysis and design of resonant circuits are developed. Criteria for "coarse" and "fine" reproduction of pulses are provided.

## 5. Frequency Generation and Translation

The concept of oscillator design is introduced and various classical oscillator circuits are briefly surveyed. The phase-locked loop is introduced and its application in frequency synthesizers is studied. Mixer circuits are analyzed and both up-conversion and down-conversion strategies are developed. The properties of superheterodyne receivers are studied in some detail. Mixer specifications are introduced.

## 6. Amplitude Modulation Methods

The various forms of amplitude modulation are developed from a signal processing point of view. This includes double-sideband (DSB), single-sideband (SSB), and conventional amplitude modulation (AM). Comparisons are made with respect to bandwidth, spectral forms, power utilization, and other factors.

## 7. Angle Modulation Methods

The major forms of angle modulation are developed. This includes phase modulation (PM) and frequency modulation (FM). Various terms related to the FM process are defined and their meanings are discussed. Bandwidth estimates are presented utilizing Carson's rule. Some of the strategies for generating and detecting FM and PM signals will be considered. An introduction to frequency-division multiplexing (FDM) will be provided using IRIG standards as an example.

## 8. Pulse Modulation Methods and Time-Division Multiplexing

The process of sampling and the sampling theorem are developed. Both non-zero width sampling and impulse sampling are considered, the latter of which serves as the basis for digital signal processing (DSP). The concepts of pulse amplitude modulation (PAM), pulse-width modulation (PWM), and pulse-position modulation (PPM) are discussed. Time division multiplexing (TDM) is introduced.

## 9. Digital Communications I: Binary Systems

The concept of pulse-code modulation is developed. Strategies employed in the analog-to-digital and digital-to-analog conversion processes are discussed. Various forms of base-band encoding forms (line codes) are surveyed. Some of the basic RF digital modulation processes are described. This includes amplitude-shift keying (ASK), frequency-shift keying (FSK), binary phase-shift keying (BPSK), and differentially-encoded phase-shift keying (DPSK).

## 10. Digital Communications II: *M*-Ary Systems

The emphasis in this chapter is directed toward the newer areas of *M-ary* encoding. First, the concepts of bit rate and baud rate are introduced and compared. Next, the Shannon-Hartley theorem and the Shannon limit are explained. The concept of quadrature phase-shift keying (QPSK) is introduced as a basic method in which the bit rate exceeds the baud rate. Some of the other *M-ary* methods such as quadrature amplitude modulation (QAM) and the associated constellation diagrams are considered.

## 11. Computer Data Communications

The concepts of character encoding, particularly those associated with computer data processing, are developed. Various encoding schemes such as ASCII, EBCDIC, and Unicode are discussed. Numerous data processing standards and definitions are provided.

## 12. Noise in Communication Systems

The basic types of electrical noise are surveyed and attention is directed toward thermal noise models. The concepts of noise figure and effective noise temperature are developed. Complete analysis of the noise temperature and/or noise figure of a cascade system, including the effects of lossy elements, is performed.

## 13. Performance of Modulation Systems with Noise

Various performance measures for both analog and digital systems are provided. Two metrics are used in the comparison of analog systems: (a) baseband comparison gain and (b) receiver processing gain. The performance measure for digital systems involve (a) quantization noise and (b) decision noise. The basis for digital system performance is developed in terms of the bit energy to one-sided noise density ratio.

## 14. Transmission Lines and Waves

An overview of transmission line phenomena as it relates to practical communication systems is provided. The primary emphasis is on lossless or nearly lossless lines and the necessity for proper impedance matching. Reflection coefficient and standing-wave ratio definitions are considered. The concepts of electric and magnetic fields, the Poynting vector, and plane wave propagation are considered. Reflection and refraction at a boundary and Snell's law are introduced.

## 15. Introduction to Antennas

A general survey of antenna concepts is provided. The emphasis is directed toward the operational specifications, such as gain and beamwidth, rather than the detailed electromagnetic phenomena. Properties of some of the common antenna types are discussed and a major emphasis is directed toward parabolic reflector antennas due to their increasing importance in satellite systems.

## 16. Communication Link Analysis and Design

The chapter begins with a discussion of the Friis transmission link formula for direct ray propagation, including the various decibel forms widely used in industry. The radar or two-way link equation is also developed. Properties of both pulse radar and Doppler radar are discussed. After completion of this chapter, a reader should be able to deal with the analysis and design of a complete system in terms of transmitter power, antenna gains, path loss, receiver noise properties, and the resulting detected signal-to-noise ratio.

## 17. Satellite Communications

This chapter begins with a somewhat simplified coverage of some basic properties of orbital mechanics to provide understanding of a geostationary orbit. Emphasis is then directed toward the analysis and design of both the uplink and the downlink, along with the engineering tradeoffs inherent in the process.

## 18. Data Network Communication Basics

This chapter provides an overview of basic network concepts, architectures, protocols, and devices. The OSI seven-layer model is introduced and used as the basis for the consideration of protocols and devices. The importance of standards is discussed and the IEEE 802 family of standards is introduced. The operation of large and small networks is explored with an emphasis on the operation of TCP/IP networks for networks of all sizes, including the Internet. IP addressing and the application to the internet are described. Hubs, switches, bridges, routers, and other devices are discussed.

## 19. Wireless Network Communication

The basic concepts of wireless networking are introduced in this chapter. Advantages and disadvantages of different wireless networking methods are discussed. A comparison of the capabilities of the IEEE 802-11a, -11b and -11g standards are made and the basic operation of each method is described. Security risks involved in wireless networking are examined and current methods of countering the risks are discussed.

## 20. Optical Communications

The applications of optical fibers in communication systems are introduced. Single-mode and multimode fibers are described. Various components such as LEDs, lasers, and photodiodes are considered. Wavelength division multiplexing is described, and the link analysis of a fiber-optic system is performed.

## 21. Consumer Communication Systems

This chapter provides information on some of the common systems that are familiar to most consumers. This includes stereo FM, black and white television, color television, and the telephone system.

## Appendix A: MATLAB® FFT Programs for Spectral Analysis

Three programs written as MATLAB M-files are described for performing spectral analysis. The codes are provided in the appendix and are provided on the accompanying CD.

## Appendix B: Introduction to Multisim®

This appendix provides a brief introduction to Multisim for readers who are not familiar with the program.

## Appendix C: Introduction to MATLAB®

This appendix provides a brief introduction to MATLAB for readers who are not familiar with the program.

# Computer Software

This book contains numerous examples that were developed utilizing three distinctly different software programs, and the accompanying CD provides files to support the examples. These examples are optional, and may be omitted without loss of continuity of the text material.

## Multisim®

Multisim is a registered trademark of Electronics Workbench, a National Instruments Company. The primary use of this software within the text is to permit the actual simulation of circuits at the component level. Most chapters have a section devoted to Multisim, and Appendix B provides a brief introduction to the software. Other information is provided in the examples contained in the chapter sections.

For Multisim product information, please contact: Electronics Workbench, 111 Peter Street, Suite 801, Toronto, Ontario M5V 2H1; Tel: 800-263-5552; Fax: 416-977-1818; E-mail: ewb@electronicsworkbench.com; Web: www.electronicsworkbench.com.

## MATLAB®

MATLAB is a registered trademark of The MathWorks, Inc. The MATLAB program is a general purpose software package for technical computing and mathematical analysis. Its primary function within the text is to provide support in various mathematical operations. Most chapters have a section devoted to MATLAB Examples, and Appendix C provides a brief introduction to the software. Other information is provided in the examples contained in the chapter sections.

Other product or brand names are trademarks or registered trademarks of their respective holders.

MATLAB is a trademark of The MathWorks, Inc. and is used with permission. The MathWorks does not warrant the accuracy of the text or exercises in this book. This book's use or discussion of MATLAB software or related products does not constitute endorsement or sponsorship by The MathWorks of a particular pedagogical approach or particular use of the MATLAB software.

For MATLAB product information, please contact The MathWorks, Inc.: 3 Apple Hill Drive, Natick, MA, 01760-2098; Tel: 508-647-7000; Fax: 508-647-7001; E-mail: info@mathworks.com; Web: www.mathworks.com.

## SystemVue™

SystemVue by Eagleware-Elanix is a program for modeling communication and signal processing operations at the systems level, utilizing block diagrams for systems and subsystems. Many chapters begin and end with SystemVue examples. Further information on how the program is used in the text is provided in the section that follows.

For SystemVue product information, please contact Eagleware-Elanix Corporation: 3585 Engineering Drive, Suite 150, Norcross, GA 30092; Tel: 678-291-0995; Fax: 678-291-0971; E-mail: sales@eagleware.com; Web: www.eagleware.com.

### About SystemVue

The program SystemVue is used in the text to present communication applications at a level in which complete operations are performed through modular blocks (called tokens) that simulate actual components and subsystems. The use of this program is optional and is not a prerequisite to anything else in the book. SystemVue applications appear in short sections that open and close applicable chapters, beginning with Chapter 2.

Using Chapter 2 to discuss the format, a short section near the beginning has the title "Chapter 2 SystemVue Opening Application (Optional)." The reader will load the file 2-1 from the text CD and run it with the program. Little or no understanding of the system is expected at this point. However, the demonstration will provide an application of the types of concepts that will be covered in the chapter.

At the end of Chapter 2, a short section is included with the title "Chapter 2 SystemVue Closing Application (Optional)." The reader will reload file 2-1 from the CD. Instructions are now provided to make changes within the computer environment to reflect the learning objectives of the chapter. This exercise should help to solidify the concepts learned in the chapter. The reader may save the modified system file on a hard drive or disk, since the original file 2-1 remains the same on the disk.

The simulation parameters have been preset in the files and should initially run with no changes. As an additional measure to ensure success and to impart some understanding of the process, the CD contains files showing the windows for the tokens that were used to set the parameters. For Chapter 2, this file has the name "2-1 initial settings," and it begins with a display of the block diagram that should appear when the file is loaded. A window providing the time parameters for the simulation will follow on the same page. In later chapters, additional pages providing other parameters will also be given. As the reader develops more proficiency, only the block diagram and time parameters will be given.

The exercises at the ends of the chapters will require making changes for certain parameters. The instructions for making these changes will be provided as necessary.

### Token Numbering

An explanation of the token numbering system will be given in the first exercise in Chapter 2. Briefly, the smallest token numbers beginning with Token 0 will be used to monitor the various waveforms. These tokens (called Sink Tokens) will have the same numbers as the numerical portion of the graph windows, which begin with W0. Thus, the graph associated with Token 0 is W0; the graph for Token 1 is W1, and so on. This process should help the reader follow the results more easily.

### Obtaining SystemVue Software

Many academic institutions and industrial organizations have site licenses for SystemVue, so many readers will have access to the software through licensing agreements. However, Eagleware-Elanix has certain policies that will permit qualified users to obtain temporary access to a version of the software through their professor. Industry use of the software is also available through Eagleware-Elanix. The company's web site is http://www.eagleware.com. Professors wishing to pursue possible limited use options for students may call the main number at (678) 291-0995.

## CD

The CD enclosed with the text provides several important supplements.

### 1. Multisim® Circuits

Multisim files for all circuit examples used in the text are contained in a folder with the title Multisim Circuits. This folder contains three subfolders with the titles Version 6 (2001), Version 7, and Version 8. With a few exceptions, all circuits are available in all of the versions. They are coded according to the figure number. For example, the Multisim circuit shown in the text as Figure 1-8 is encoded as **01-08.ext**, with ext representing the extension for the particular version of Multisim.

## 2. MATLAB® Files

MATLAB software codes for all M-files are contained in a folder with the title MATLAB M-files. They have the names indicated in the MATLAB examples. In addition, the three M-files described in Appendix A are included.

## 3. SystemVue™ Systems

SystemVue files for beginning and ending chapter examples are contained in a folder with the title SystemVue Systems. Additional initial settings for the various demonstration are contained in a folder entitled SystemVue Initial Settings.

## 4. Lab.Source

Laboratory assignments on selected topics such as filters, modulation and demodulation methods, and analog-to-digital conversion, are included in the CD-ROM accompanying the book. These assignments are designed to use equipment that should be available in an electronics laboratory in most engineering schools and do not require equipment from specific vendors.

## Supplements

An e.resource CD (ISBN: 1-4180-0004-3) available to educators upon adoption of the text includes the following teaching tools:

- A Solutions Manual contains the solutions to all problems
- Several hundred Power Point Presentation Slides covering all chapters
- Over 200 questions in the Computerized Testbank
- The Lab.Builder template is designed to help create labs quickly by using pre-formatted text elements. Labs can be further customized by selecting images from the Image Library and importing them into each lab.
- An electronic Image Library contains all images from the textbook to customize handouts and tests.

## About the Authors

William D. Stanley is an Eminent Professor Emeritus at Old Dominion University and was a recipient of the State Council of Higher Education in Virginia Outstanding Faculty Award. He received his Bachelor of Science degree from the University of South Carolina and the Master of Science and Ph.D. degrees from North Carolina State University. All degrees were in Electrical Engineering. Dr. Stanley is a Registered Professional Engineer in the Commonwealth of Virginia. He is also a Senior Member of the IEEE and a Life Member of ASEE.

John Jeffords retired from the United States Navy after twenty-four years as an air crewman and aircraft maintenance officer. While attending law school at the College of William and Mary he started teaching electrical engineering technology courses at Old Dominion University as an adjunct faculty member and continued to do so after passing the bar and commencing his law practice. After several years he preferred teaching to practicing law, and taught electrical engineering technology and information technology courses as a full-time faculty member for fifteen years.

## Acknowledgments

The authors would like to express their deep appreciation to the following people for their important contributions in the development and production of this book:

From Thomson Delmar Learning

Steve Helba, *Senior Acquisitions Editor*

Michelle Ruelos Cannistraci, *Senior Development Editor*

Dave Garza, *Marketing Director*

Dennis Williams, *Senior Channel Manager*

Toni Hansen, *Production Editor*

Christopher Chien, *Senior Project Editor*

Francis Hogan, *Art & Design Coordinator*

Dawn Daugherty, *Senior Editorial Assistant*

From Interactive Composition Corporation

Michael Ryder, *Project Manager*

Finally, both the authors and Thomson Delmar Learning would like to thank the following reviewers for their valuable suggestions:

Vic Cerniglia, *Pearl River Community College, Polarville, MS*

David Loker, *Penn State Erie, The Behrend College, Erie, PA*

Greg Simmons, *DeVry University, DeVry University, Columbus, OH*

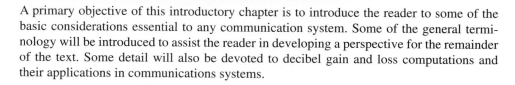

# Introduction

**1**

## OVERVIEW AND OBJECTIVES

A primary objective of this introductory chapter is to introduce the reader to some of the basic considerations essential to any communication system. Some of the general terminology will be introduced to assist the reader in developing a perspective for the remainder of the text. Some detail will also be devoted to decibel gain and loss computations and their applications in communications systems.

### Objectives

After completing this chapter, the reader should be able to:

1. Discuss on a simplified level the model of a communication system.
2. Convert between wavelength and frequency.
3. Discuss the radio-frequency spectrum.
4. Explain why electromagnetic signals can be more easily propagated at radio frequencies.
5. Discuss in general terms the role of bandwidth and spectrum.
6. Discuss the general concepts of modulation and demodulation (or detection).
7. Perform decibel gain and loss computations.
8. Explain decibel signal level standards and convert between some of the forms.
9. Perform system-level decibel gain and loss computations.
10. Define signal-to-noise ratio both in absolute and in decibel forms.

## 1-1 Simplified Communication System Model

A block diagram illustrating a simplified model of a communication system is shown in Figure 1–1. The block on the left represents a certain *source* of *information,* which is to be "delivered" or transmitted to a particular *destination* over a *channel.* The channel contains random *noise,* which adds to or otherwise disturbs the *signal* in some manner.

At one extreme (high power and/or close range), the effect of noise may be insignificant, and the signal reaches the destination without alteration. At the other extreme (low power and/or distant range), the signal may be so masked or overshadowed with noise that it is totally unintelligible. Most cases of practical importance fall somewhere between these extremes, and the choice of a given communication system can affect drastically the possible success of the desired objective.

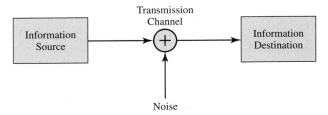

**FIGURE 1–1**
Simplified model of a communication system.

## Transmission Channel

The term "transmission channel" refers to the particular medium over which the signal is to be transmitted. The simplest medium is a pair of wires or a cable, and this type of medium is used in the telephone industry, for cable TV, and in computer links. When long distances are required, repeater amplifiers may be used along the path to maintain the signal integrity. In recent decades, fiber-optic cables have experienced increasing usage due to their low losses and high bandwidths. This medium will continue to grow, and find many new application areas in which it is feasible to provide a direct connection between source and destination. For sonar signals transmitted from a ship, water serves as the medium, and the signals are in the audio or sound range.

## Electromagnetic Wave Propagation

The most versatile type of communication channel employs electromagnetic wave propagation through the atmosphere or through free space. One of the greatest scientific discoveries of all time was the deduction and verification of the existence of electromagnetic waves, which led to the development of radio transmission. The early mathematical work of James Clerk Maxwell in England, followed by the experimental work of Heinrich Hertz in Germany, led Guglielmo Marconi in Italy to construct the first "wireless" system, and a new era of technology was initiated.

Electromagnetic fields are present in any circuit in which time-varying voltages and currents are present. However, in well-designed circuits in which radiation is not desired, the magnitudes of such fields can be controlled to acceptable levels by good component placement and shielding. When electromagnetic radiation is specifically desired, an interface between the circuit and the outside propagation medium is established in order to enhance the propagating field. The device that accomplishes this purpose is the *antenna*.

The *velocity of propagation* is the speed at which electrical waves travel. In many applications, the velocity of propagation is assumed to be the same as the speed of light, but in some cases it is less.

## Wavelength and Frequency

Three important parameters concerning wave propagation are the *frequency,* the *period,* and the *wavelength.* The *frequency* will be denoted as $f$ and the corresponding period will be denoted by $T$. The *period* is the time duration of one cycle and is related to the *frequency* by

$$T = \frac{1}{f} \tag{1-1}$$

Since the symbol $v$ will be reserved for voltage, the symbol $\nu$ (pronounced "noo") will be used to represent the *velocity of propagation* of the wave. The *wavelength* will be denoted as $\lambda$ and it is defined as the *distance a given point on a wave travels in one cycle*. The time duration of one cycle is the period, which means that the wavelength is

$$\lambda = vT = \frac{v}{f} \tag{1-2}$$

The velocity of propagation is then

$$v = f\lambda \tag{1-3}$$

The units in Equation 1-3 can be changed but they should be consistent on both sides of the equation. The frequency $f$ can be expressed in cycles per second or *hertz* (Hz). The velocity $v$ can be expressed in length units/second, and the units for $\lambda$ must be the same length units. For example, assume that $v$ is expressed in meters/second (m/s); in that case, the units for $\lambda$ will be meters (m).

The preceding result applies for any given velocity of wave propagation $v$. In many applications, the velocity of propagation is assumed to be that of free space. The symbol $c$ is commonly used to represent the free-space velocity. With that assumption, Equation 1-3 changes to

$$c = f\lambda \tag{1-4}$$

The values of $c$ to three significant figures in both meters/second (m/s) and miles/second (mi/s) are

$$c = 3 \times 10^8 \text{ m/s} = 186,000 \text{ mi/s} \tag{1-5}$$

The speed of propagation in the atmosphere is slightly smaller than that of free space, but in most routine applications, the free-space value can be assumed.

## Radio-Frequency Spectrum

A tabulation of terms, frequency ranges, and wavelength ranges commonly referred to as the "radio-frequency spectrum" is provided in Table 1–1. The frequency range of visible light is somewhat higher than the radio-frequency range shown in the table.

Contrary to popular belief, electromagnetic waves can theoretically be generated at any frequency, even in the audio range. The practical difficulty in implementation at low frequencies, however, is the cause of the widespread notion that radio propagation can be achieved only at much higher frequencies. The difficulty relates to the requirement that the antenna size must be an appreciable fraction of a wavelength to provide significant energy transfer to the wave. Note the values of the wavelengths at low frequencies. A few special radio communication systems have been developed for communicating with submerged submarines, but they require a huge amount of space for the shore antenna site. The reason that low frequencies are required in this case is that electromagnetic signal strength is greatly reduced in water as the frequency increases.

**Table 1–1** The Radio-Frequency Spectrum

| Band | Frequency Range | Wavelength Range | Sample Applications |
|---|---|---|---|
| Extremely Low Frequencies (ELF) | 30 to 300 Hz | 10,000 to 1000 km | Power transmission |
| Voice Frequencies (VF) | 300 Hz to 3 kHz | 1000 to 100 km | Audio; Sonar |
| Very Low Frequencies (VLF) | 3 to 30 kHz | 100 to 10 km | Submarine communications |
| Low Frequencies (LF) | 30 to 300 kHz | 10 to 1 km | Navigation |
| Medium Frequencies (MF) | 300 kHz to 3 MHz | 1 km to 100 m | AM broadcast |
| High Frequencies (HF) | 3 to 30 MHz | 100 to 10 m | Shortwave broadcast; Commercial |
| Very High Frequencies (VHF) | 30 to 300 MHz | 10 to 1 m | TV broadcast; FM broadcast |
| Ultra High Frequencies (UHF) | 300 MHz to 3 GHz | 1 m to 10 cm | Cellular telephones; Satellites |
| Super High Frequencies (SHF) | 3 to 30 GHz | 10 cm to 1 cm | Satellites; Radar |
| Extremely High Frequencies (EHF) | 30 to 300 GHz | 1 cm to 1 mm | Mostly experimental |

▌▌ **EXAMPLE 1-1**

A frequency of 1 MHz is close to the middle of the commercial AM band. Determine the wavelength.

**SOLUTION**   From Equation 1-4, the wavelength is

$$\lambda = \frac{c}{f} = \frac{3 \times 10^8}{1 \times 10^6} = 300 \, \text{m} \tag{1-6}$$

▌▌ **EXAMPLE 1-2**

A frequency of 100 MHz is close to the middle of the commercial FM band. Determine the wavelength.

**SOLUTION**   The wavelength is

$$\lambda = \frac{c}{f} = \frac{3 \times 10^8}{100 \times 10^6} = 3 \, \text{m} \tag{1-7}$$

Obviously, the wavelengths of commercial FM signals are considerably shorter than those of commercial AM signals.   ▌▌

## 1-2   Bandwidth and Spectrum

All signals that are processed in a communication system can be expressed in terms of sinusoidal components. The relative strength of the sinusoidal components that constitute the signal is called the *frequency spectrum*. For example, it is known that the human voice can be represented by sinusoidal components whose frequencies range from well under 100 hertz (Hz) to several kilohertz (kHz). In contrast, the video portion of a television signal has frequency components that extend above 4 megahertz (MHz).

Knowledge of the frequency content of the signal is very important in estimating the *bandwidth* required in the appropriate communications channel. Conversely, the bandwidth of a given channel can limit the bandwidth of possible signals that could be transmitted over the channel. For example, assume that a given channel is an unequalized telephone line having a frequency response from a few hundred hertz to several kilohertz. The channel might be perfectly adequate for simple voice data, but would hardly suffice for high-fidelity sound or for high-quality video. Thus, the frequency response of the channel must be adequate over the frequency range encompassing the spectrum of the signal, or severe distortion and degradation of the signal will result.

In some cases, it is possible to transmit information over a narrower channel by slowing down the information rate to a level compatible with the channel. For example, in some space communications systems, the bandwidth has been kept quite low to keep the total noise power to a minimum. In some cases, higher frequency data can be first recorded and then played back at a slower speed, which reduces the bandwidth. In this case, a longer processing time is being substituted for a wider bandwidth.

## 1-3   Modulation and Demodulation

Most information signals tend to have the most significant part of their spectra concentrated at relatively low frequencies. Such signals are referred to as *baseband signals*. As explained earlier, it is usually not practical to achieve wave propagation directly at baseband frequencies.

### Modulation

*Modulation* may be defined as the process of shifting the frequency spectrum (while possible changing the form) of a message signal to a frequency range in which more efficient transmission can be achieved. Numerous modulation methods have been developed over the years, and it is difficult to classify them in a simple, nonambiguous manner. On the broadest scale, modulation may be roughly classified in one of three categories as (1) *analog* (or *continuous-time*), (2) *pulse* (or *discrete-time*), and (3) *digital*. Many systems employ a combination of these methods.

The system that provides all of the signal processing at the source end of the communications link is called the *transmitter*. The transmitter contains a *modulator*, which is the circuit in which the message signal is applied to a *carrier*. The *carrier* is usually a single sinusoidal frequency in the range for which transmission is desired. The transmitter also contains numerous amplifier circuits.

### Analog Modulation

Analog modulation is the oldest type, and includes all methods in which some characteristic of the high-frequency signal represents the modulating signal on a continuous basis (i.e., an "analog" of the modulating signal). This includes such methods as *amplitude modulation* (including variations such as *double sideband* and *single sideband*) and *angle modulation* (which includes *phase* and *frequency modulation*).

### Pulse Modulation

*Pulse modulation* involves sampling of the analog signal in discrete samples and transmitting the signal in discrete samples. As will be seen later, it is absolutely necessary that a minimum sampling rate be employed if the signal is to be reconstructed. Among the methods of pulse modulation are *pulse amplitude modulation, pulse width modulation,* and *pulse position modulation*.

### Digital Modulation

Digital modulation is somewhat akin to pulse modulation, in that samples of the message signal are first taken at a minimum sampling rate. However, digital modulation takes the process one step farther, in that the amplitude levels are encoded into characters consisting of a finite number of levels. The most basic form is that of *binary encoding,* in which all characters consist of a combination of two levels. The binary words are then transmitted as a combination of *ones* and *zeros*. At one extreme, the binary words may be transmitted directly as a combination of positive and negative (or on and off) pulses at baseband. However, the more common case of interest for our discussion involves modulation. This may take the form of shifting between two different frequencies to represent the two different states or shifting between two different phases. The first method is called *frequency-shift keying* (FSK) and the second is called *phase-shift keying* (PSK). More recent digital methods involve a general approach referred to as *M-ary* encoding in which more than two levels are employed in the encoding process. This evolution has made possible the transmission of large amounts of data (e.g., the Internet).

## Demodulation

*Demodulation* is the process of shifting the spectrum back to the original baseband frequency range and reconstructing the original form. The term *detection* is widely used as an alternate term for *demodulation*. The device that performs all the signal processing at the destination source of the link is called a *receiver*. The receiver contains the demodulator, various amplifier stages, and other signal processing units.

## 1-4 Decibel Gain and Loss Ratios

Decibel measures are used extensively in the electronics field, but if there is one subject area in which the practice dominates, it is communications. Decibel forms date back to the early days of communications, and they are used extensively in equipment specifications and in signal processing. It is virtually impossible to deal with communication system analysis and design without dealing with decibel forms.

## Logarithm Identities

Prior to introducing decibel definitions, some properties of logarithms will be reviewed. Throughout this text, the convention will be used that *log* means "logarithm to the base 10." (This is indicated in many texts as $log_{10}$.) The following basic properties of logarithms will be used in the work that follows:

$$\log \frac{1}{x} = -\log x \qquad (1\text{-}8)$$

$$\log x^k = k \log x \qquad (1\text{-}9)$$

$$\log xy = \log x + \log y \qquad (1\text{-}10)$$

## Decibel Power Gain

The original concept of decibel measurement was based on a comparison of two power levels. Using signal gain as the basis for establishing the comparison, consider the block diagram shown in Figure 1–2, which could represent an amplifier. The input signal is assumed to deliver a power $P_1$ to the amplifier input, and the amplifier in turn delivers a power $P_2$ to the external load. The *absolute power gain G* is defined as

$$G = \frac{P_2}{P_1} \qquad (1\text{-}11)$$

The *decibel power gain* $G_{dB}$ in **decibels** (abbreviated as dB) is defined by

$$G_{dB} = 10 \log_{10} G = 10 \log_{10} \frac{P_2}{P_1} \qquad (1\text{-}12)$$

When the output power is greater than the input power, the absolute power gain is greater than one, and there is a true gain. In this case, the decibel gain is positive. However, when the output power is less than the input power, the absolute power gain is less than one, and this means there is actually a loss of power. In this case, the decibel power gain is negative.

## Decibel Power Loss

For systems in which the output power is usually less than the input power (e.g., attenuator circuits and many filters), it is often more convenient to deal with *loss* rather than *gain*. The *absolute power loss L* can be defined as

$$L = \frac{P_1}{P_2} = \frac{1}{G} \qquad (1\text{-}13)$$

The *decibel power loss* $L_{dB}$ can then be defined as

$$L_{dB} = 10 \log L = 10 \log \frac{P_1}{P_2} \qquad (1\text{-}14)$$

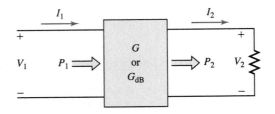

**FIGURE 1–2**
Block diagram of amplifier system used to define gain.

Application of Equation 1-8 to Equation 1-14 and a comparison with Equation 1-12 yield

$$L_{dB} = -G_{dB} \qquad (1\text{-}15)$$

Thus, a negative decibel gain is a positive decibel loss, and vice versa. For example, a cable having a loss of 6 dB could also be described as having a gain of −6 dB. In casual usage, the form having a positive value is more convenient. In a chain of calculations, however, one must be careful to maintain consistency. The following table should help the reader obtain an intuitive feeling for the concepts.

| Absolute Gain | Absolute Loss | Decibel Gain | Decibel Loss |
|---|---|---|---|
| >1 | <1 | + | − |
| <1 | >1 | − | + |
| 1 | 1 | 0 | 0 |

## Converting Decibel Values to Absolute Power Ratios

If the value of decibel gain is given, it is necessary to invert the form of Equation 1-12 to determine the corresponding absolute power gain. This is most readily done by first dividing both sides by 10, as follows:

$$\frac{G_{dB}}{10} = \log G \qquad (1\text{-}16)$$

Both sides of Equation 1-16 are then equated as powers of 10.

$$10^{G_{dB}/10} = 10^{\log G} \qquad (1\text{-}17)$$

The right-hand side of Equation 1-17 is simply $G$, since by definition, $\log G$ is a value such that when 10 is raised to that power, the result is $G$. Thus,

$$G = 10^{G_{dB}/10} \qquad (1\text{-}18)$$

A similar relationship can be deduced for $L$ in terms of $L_{dB}$.

$$L = 10^{L_{dB}/10} \qquad (1\text{-}19)$$

Most modern scientific calculators have a $y^x$ function, so operations of the forms of Equations 1-18 and 1-19 can be easily achieved.

## Decibel Measures of Voltage and Current

While the most basic decibel form is concerned with power ratios, it is also possible to convert voltage and current ratios to equivalent decibel forms if some care is exercised. Assume for the amplifier model previously discussed that the output power is delivered to a resistance $R$. Assume that the input resistance looking into the amplifier is also $R$. Let $V_2$ represent the output effective voltage (the equivalent power producing measure), and let $V_1$ represent the input effective voltage. The decibel power gain is then

$$G_{dB} = 10 \log \frac{P_2}{P_1} = 10 \log \frac{V_2^2/R}{V_1^2/R} = 10 \log \left(\frac{V_2}{V_1}\right)^2 = 10 \log A_v^2 \qquad (1\text{-}20)$$

where $A_v$ is the voltage gain as defined by

$$A_v = \frac{V_2}{V_1} \qquad (1\text{-}21)$$

Note that $R$ was canceled out in the process. Application of Equation 1-9 to Equation 1-20 with $k = 2$ results in

$$G_{dB} = 20 \log \frac{V_2}{V_1} = 20 \log A_v \qquad (1\text{-}22)$$

A similar development expressed in terms of the effective currents $I_2$ and $I_1$ for the case of *equal resistances* yields

$$G_{dB} = 20 \log \frac{I_2}{I_1} = 20 \log A_i \qquad (1\text{-}23)$$

where $A_i$ is the current gain as defined by

$$A_i = \frac{I_2}{I_1} \qquad (1\text{-}24)$$

Comparing Equations 1-12, 1-22, and 1-23, we see that **10** is the appropriate constant for computing decibels from power levels and **20** is the corresponding value for working with voltage or current.

## Converting Decibel Values to Voltage or Current Ratios

To convert decibel values to voltage or current ratios, we follow a process very similar to that employed in converting decibel values to power ratios. The only significant difference is that the first step in Equation 1-22 or 1-23 is to divide both sides by 20. Both sides can then be raised to be a power of 10, and the results are

$$A_v = \frac{V_2}{V_1} = 10^{G_{dB}/20} \qquad (1\text{-}25)$$

and

$$A_i = \frac{I_2}{I_1} = 10^{G_{dB}/20} \qquad (1\text{-}26)$$

It is recommended to the reader that the inversion formulas be worked out each time they are needed, rather than memorizing the different forms.

## Input and Output Resistances Not Equal

In developing the decibel forms for voltage and current, it was assumed that the input and load resistances were equal, and this is the proper assumption for decibel values derived from voltage and current ratios in order to represent a true power decibel measure. However, there has evolved a somewhat casual usage of decibel notation, and it is very common to find computations based on the forms of Equations 1-22 and 1-23 when the resistances are not equal. Many voltmeters, for example, have decibel scales that simplify measurements of relative voltage levels through the use of decibel forms without regard to the resistance levels. The intent here is neither to praise nor to condemn the practice, but merely to alert the reader to its usage so that it will be clear that true power measurements are not obtained in such cases. Since power measurements are so important in communications systems, the practice in this book will be to use voltage and current ratios in decibel computations only when the resistances are the same.

## Common Decibel Level Relationships

Many communications engineers and technicians learn to "think" in terms of certain common decibel levels in dealing with equipment specifications and measurements. To instill a feeling for many of these common levels, Table 1–2 is provided. Many of the useful levels

Table 1–2  Some Common Power, Voltage, and Current Ratios and Their Decibel Equivalents

| Power Ratio | Voltage or Current Ratio | Decibel Values |
|---|---|---|
| 2 | $\sqrt{2} = 1.414$ | 3 dB |
| 4 | 2 | 6 dB |
| 8 | $\sqrt{8} = 2.828$ | 9 dB |
| 10 | $\sqrt{10} = 3.162$ | 10 dB |
| 100 | 10 | 20 dB |
| $10^n$ | $10^{n/2}$ | $10n$ dB |
| $10^{2n}$ | $10^n$ | $20n$ dB |
| $1/2 = 0.5$ | $1/\sqrt{2} = 0.7071$ | $-3$ dB |
| $1/4 = 0.25$ | $1/2 = 0.5$ | $-6$ dB |
| $1/8 = 0.125$ | $1/\sqrt{8} = 0.3536$ | $-9$ dB |
| $1/10 = 0.1$ | $1/\sqrt{10} = 0.3162$ | $-10$ dB |
| $1/100 = 0.01$ | $1/10 = 0.1$ | $-20$ dB |
| $1/10^n = 10^{-n}$ | $1/10^{n/2} = 10^{-n/2}$ | $-10n$ dB |
| $1/10^{2n} = 10^{-2n}$ | $1/10^n = 10^{-n}$ | $-20n$ dB |

are illustrated, and by working with a combination of these, many other forms can be developed (as will be illustrated in the examples that follow this section).

All of the decibel values that are derived from power levels that are integer powers of 10 are exact. In accordance with widespread practice, however, the values that are derived from power levels that are integer powers of 2 have been rounded slightly. For example, the exact number of decibels corresponding to a power level of 2 is 3.01030 dB to five decimal places. This value is usually rounded to 3 dB, and this will be the practice in many places throughout the book. In some cases, the values will be carried out more accurately in order to minimize roundoff and to correlate two different approaches for the same analysis.

## Decibel Not an Absolute Measure

The reader should carefully note that a decibel is *not* an *absolute* unit; rather it compares one level with another. For example, a power level of 8 W is about 3 dB higher than a power level of 4 W. However, a power level of 4 mW is also about 3 dB higher than a power level of 2 mW. In one case, we are comparing watts and in the other case, we are comparing milliwatts. Yet the relative power level differences in decibels in the two cases are the same. Thus, to say that a certain signal level is 3 dB is meaningless unless a reference level is known.

---

**▌▌ EXAMPLE 1-3**

A communications amplifier has an absolute power gain of 175. Determine the decibel power gain.

**SOLUTION**  The decibel power gain is determined as follows:

$$G_{dB} = 10 \log G = 10 \log 175 = 10 \times 2.243 = 22.43 \, dB \tag{1-27}$$

---

**▌▌ EXAMPLE 1-4**

An amplifier gain is specified as 28 dB. Determine the absolute power gain.

**SOLUTION**  Although we could immediately use Equation 1-18, it is recommended that we start with the basic formula for decibels and work backwards. Thus, we begin with

$$G_{dB} = 10 \log G \tag{1-28}$$

or

$$28 = 10 \log G \tag{1-29}$$

We then divide both sides by 10 and raise both sides as a power of 10. This yields

$$G = 10^{2.8} = 631.0 \tag{1-30}$$

---

**▮▮ EXAMPLE 1-5**

Assuming equal input and load resistances, determine the voltage gain corresponding to the power gain of Example 1-4.

**SOLUTION**  Since the power gain is simply the square of the voltage gain or the current gain in the case of equal resistances, we have

$$A_v = \sqrt{G} = \sqrt{631.0} = 25.12 \tag{1-31}$$

---

**▮▮ EXAMPLE 1-6**

In a certain lossy line, only 28% of the power input to the line reaches the load. Determine the decibel gain and decibel loss.

**SOLUTION**  In terms of the standard definition of gain, we should say that $G = 0.28$. The decibel "gain" is then

$$G_{dB} = 10 \log G = 10 \log 0.28 = 10 \times (-0.5528) = -5.528 \, \text{dB} \tag{1-32}$$

Of course, this negative value of decibel gain is expected since the output power is less than the input power. The decibel loss is simply

$$L_{dB} = 10 \log \frac{P_1}{P_2} = 10 \log \frac{1}{0.28} = 10 \times 0.5528 = 5.528 \, \text{dB} \tag{1-33}$$

This calculation was made simply to illuminate the much simpler interpretation that

$$L_{dB} = -G_{dB} = -(-5.528) = 5.528 \, \text{dB} \tag{1-34}$$

It is usually more convenient, whenever possible, to work with absolute ratios greater than one and with positive decibel values. The strategy is to remember if the situation truly represents a gain or a loss and to make a final adjustment based on the true outcome.    ▮▮

---

## 1-5   Decibel Signal Level Reference Standards

A number of standard reference levels for decibel measures have been developed over the years, and many measurements are made in reference to these standard levels. In most cases, the unit incorporates a modifier that gives it a unique meaning. For example, a very common reference level is 1 milliwatt (mW), and all decibel measurements relative to that level are assigned the unit dBm. As a general rule, any decibel unit in which the dB abbreviation has a modifier means that there is some standard reference, and "pure" power ratios will be indicated simply by dB as a unit. Be aware, however, that there is some misuse of this convention in actual practice.

The reference levels that will be used at this point are 1 W, 1 mW, and 1 fW (femtowatt). The corresponding decibel measures are denoted as dBW, dBm, and dBf. Each will be defined in the next several paragraphs.

### dBW

$$\text{power level (dBW)} = 10 \log \frac{\text{power level (W)}}{1 \, \text{W}} \tag{1-35}$$

This is the simplest form to deal with if the power is initially expressed in watts, since division by one is a trivial process. The subtle presence of the 1 W reference, however, should be carefully understood, since there is a ratio involved. The dBW measure is convenient in working with transmitter output power levels, which are often well above 1 W.

## dBm

$$\text{power level (dBm)} = 10 \log \frac{\text{power level (mW)}}{1\,\text{mW}} = 10 \log \frac{\text{power level (W)}}{1 \times 10^{-3}\,\text{W}} \quad (1\text{-}36)$$

If the power level is given directly in mW, then division is by 1 in the same manner as for dBW. However, if the power level is given in watts, one can divide by $10^{-3}$ as indicated in the last form. This is arithmetically equivalent to multiplying the numerator by $10^3$, which amounts to converting watts to milliwatts. The dBm measure is very convenient in working with the typical signal levels representative of many amplifiers and signal processing equipment.

## dBf

$$\text{power level (dBf)} = 10 \log \frac{\text{power level (fW)}}{1\,\text{fW}} = 10 \log \frac{\text{power level (W)}}{1 \times 10^{-15}\,\text{W}} \quad (1\text{-}37)$$

The logic in dealing with the dBf is similar to that of the dBm discussed in the preceding paragraph, except that the dBf is a much smaller level (1 fW = $10^{-15}$ W). The dBf is convenient for working with the extremely small power levels encountered at the output terminals of an antenna. Many receiver input specifications are given in terms of dBf levels at the receiver input.

## Conversion between Various dB Reference Levels

To convert between the various dB reference levels just discussed, either add or subtract an integer multiple of 10. The integer multiple is the difference between the powers of 10 associated with the two reference level in watts, and whether it is plus or minus depends on the direction of conversion. Just remember that the smaller the unit, the higher the number on an algebraic scale required to represent a given level. The following conversions can be deduced from the preceding concept:

$$\text{Level in dBm} = \text{Level in dBW} + 30 \quad (1\text{-}38)$$

$$\text{Level in dBf} = \text{Level in dBW} + 150 \quad (1\text{-}39)$$

$$\text{Level in dBf} = \text{Level in dBm} + 120 \quad (1\text{-}40)$$

---

**▐▌ EXAMPLE 1-7**

A certain signal has a power level of 100 mW. Express the signal level in (a) dBm, (b) dBW, and (c) dBf.

**SOLUTION**

(a) Since the power level is given in milliwatts, the most direct form to determine is that of the dBm level. Hence,

$$P(\text{dBm}) = 10 \log \frac{P(\text{mW})}{1\,\text{mW}} = 10 \log \frac{100\,\text{mW}}{1\,\text{mW}} = 10 \times 2 = 20\,\text{dBm} \quad (1\text{-}41)$$

(b) While we could first convert the power to watts and use the definition of Equation 1-35, let's try a more intuitive approach. The difference in the integer powers of 10 between a watt and a milliwatt is 3, and since the dBW level will be expressed as a less positive value than the dBm level, we must subtract $3 \times 10 = 30$ from dBm to get dBW. Thus,

$$P(\text{dBW}) = P(\text{dBm}) - 30 = 20 - 30 = -10\,\text{dBW} \quad (1\text{-}42)$$

(c) Following the same approach as in (b), the difference in the integer powers of 10 between a milliwatt and a femtowatt is $15 - 3 = 12$. In this case, however, the level expressed in dBf will be expressed as a more positive level than the dBm level, and we add $12 \times 10 = 120$ to get dBf. Hence,

$$P(\text{dBf}) = P(\text{dBm}) + 120 = 20 + 120 = 140\,\text{dBf} \tag{1-43}$$

## 1-6 System Level Decibel Analysis

Having considered both decibel gain analysis and decibel signal levels based on a reference, we are ready to combine these concepts in a form to deal with signal behavior from input to output. Assume that a system has an absolute power gain of $G$ and a corresponding decibel gain of $G_{\text{dB}}$. Assume that the power input received from the source is $P_{\text{s}}$. The output power $P_{\text{o}}$ is simply

$$P_{\text{o}} = G P_{\text{s}} \tag{1-44}$$

Assume that both sides of Equation 1-44 are divided by some reference level (e.g., 1 mW). The results that follow apply to any reference *as long as both sides are divided by the same reference*. To generalize the results, we will use the modifier $x$ to refer to any arbitrary reference level. After taking the logarithm to the base 10 of both sides of Equation 1-44 with the reference level included, and multiplying both sides by 10, the following form results:

$$P_{\text{o}}(\text{dBx}) = P_{\text{s}}(\text{dBx}) + G_{\text{dB}} \tag{1-45}$$

where dBx is the decibel level referred to any reference level. Again, $P_{\text{s}}$ and $P_{\text{o}}$ must be based on the same reference level.

Please note the following point that often causes confusion with beginners: *While the input and output power levels are expressed in dB levels referred to specific but equal reference levels, the system gain is expressed in decibels (no reference)*. It may seem as if we're adding things together that have different units, but this illusion is one of the quirks of the logarithmic conversion. Gain is dimensionless, so the only way it can be expressed is in decibels. However, both input and output signal levels are referred to specific reference levels, so the modified decibel units are required. In a system level input-output equation, therefore, there will typically be only two variables (input and output power levels) that are measured in the dBx form, while all gain blocks will simply be measured in decibels.

### Cascade Systems

One of the most useful aspects of working with decibel forms is in analyzing a cascade of many successive stages of amplification and/or attenuation, as illustrated by the block diagram of Figure 1–3. Assume that all the blocks are represented by "equivalent gains" (meaning that true loss blocks are represented as gain values less than one). The overall power gain $G$ is

$$G = G_1 G_2 G_3 \cdots G_n \tag{1-46}$$

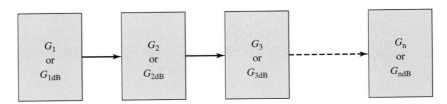

**FIGURE 1–3**
Block diagram of cascade system.

The logarithm of both sides can now be taken, and the property of Equation 1-10 can be used to expand the right-hand side. After multiplying both sides by 10, the following form is deduced:

$$G_{dB} = G_{1dB} + G_{2dB} + G_{3dB} + \cdots + G_{ndB} \qquad (1\text{-}47)$$

where each of the terms on the right represents the decibel gain of the given stage. Stated in words: *The decibel gain of a cascade of units is the sum of the individual decibel gains.* A block that has an actual loss must be treated as a negative decibel gain in the equation.

---

**▍▌ EXAMPLE 1-8**

The cascade system shown in Figure 1–4 has three components: (1) input line amplifier with power gain $G_1 = 5000$, (2) long transmission line with a power loss factor $L = 2000$, and (3) load amplifier with an absolute power gain $G_2 = 400$. Impedances are matched at all junctions. Determine (a) net system absolute gain, (b) system decibel gain using the result of (a), and (c) system decibel gain from individual decibel values.

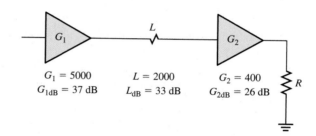

$G_1 = 5000$ $\qquad$ $L = 2000$ $\qquad$ $G_2 = 400$
$G_{1dB} = 37$ dB $\quad$ $L_{dB} = 33$ dB $\quad$ $G_{2dB} = 26$ dB

**FIGURE 1–4**
Cascade system of Examples 1-8 and 1-9.

**SOLUTION**

(a)  Note that an absolute power loss factor of $L$ can be considered as an absolute power gain factor of $1/L$. The net absolute gain can then be expressed as the product of the three "gains."

$$G = G_1 \times \left(\frac{1}{L}\right) \times G_2 = 5000 \times \left(\frac{1}{2000}\right) \times 400 = 1000 \qquad (1\text{-}48)$$

(b)  The system decibel gain is

$$G_{dB} = 10 \log 1000 = 10 \times 3 = 30 \text{ dB} \qquad (1\text{-}49)$$

(c)  The individual stage decibel values are

$$G_{1dB} = 10 \log G_1 = 10 \log 5000 = 10 \times 3.7 = 37 \text{ dB} \qquad (1\text{-}50)$$

$$L_{dB} = 10 \log L = 10 \log 2000 = 10 \times 3.3 = 33 \text{ dB} \qquad (1\text{-}51)$$

$$G_{2dB} = 10 \log G_2 = 10 \log 400 = 10 \times 2.6 = 26 \text{ dB} \qquad (1\text{-}52)$$

The net decibel gain is

$$G_{dB} = G_{1dB} - L_{dB} + G_{2dB} = 37 - 33 + 26 = 30 \text{ dB} \qquad (1\text{-}53)$$

Note that the process of *subtracting* the *positive loss* is equivalent to adding a *negative gain*.

---

**▍▌ EXAMPLE 1-9**

Assume that a signal source is connected to the input of the system of Example 1-8 as shown in Figure 1–5. The signal source delivers a power $P_s = 0.1$ mW into the line amplifier. Determine (a) the absolute power levels at each of the junctions, and (b) the power in

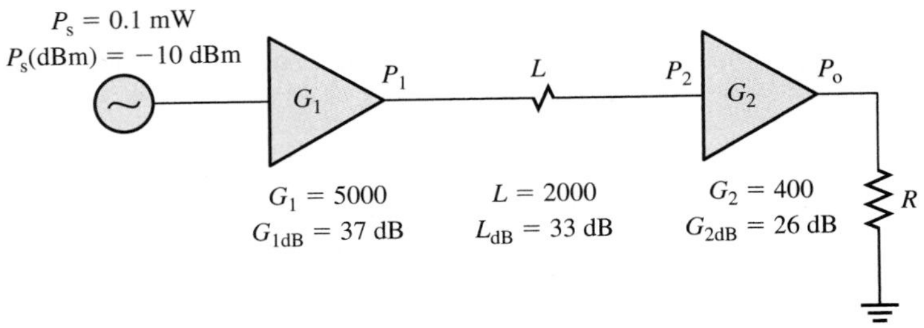

**FIGURE 1–5**
Cascade system of Examples 1-8 and 1-9 with source connected.

dBm at each of the junctions. (c) Convert the system output of (a) to a dBm level and compare with (b).

**SOLUTION**

(a) The absolute power level $P_1$ at the line amplifier output is

$$P_1 = G_1 P_s = 5000 \times 0.1 \text{ mW} = 500 \text{ mW} = 0.5 \text{ W} \qquad (1\text{-}54)$$

This power is also the input power to the line, and the line output power $P_2$ is then

$$P_2 = \frac{P_1}{L} = \frac{500 \text{ mW}}{2000} = 0.25 \text{ mW} \qquad (1\text{-}55)$$

The system output power to the load is

$$P_o = G_2 P_2 = 400 \times 0.25 \text{ mW} = 100 \text{ mW} \qquad (1\text{-}56)$$

(b) To utilize a decibel basis, we first convert the source power of 0.1 mW to the dBm level.

$$P_s(\text{dBm}) = 10 \log \left( \frac{P_s(\text{mW})}{1 \text{ mW}} \right) = 10 \log \left( \frac{0.1 \text{ mW}}{1 \text{ mW}} \right)$$

$$= 10 \log (0.1) = -10 \text{ dBm} \qquad (1\text{-}57)$$

The output level of the line amplifier is

$$P_1(\text{dBm}) = P_s(\text{dBm}) + G_{1\text{dB}} = -10 + 37 = 27 \text{ dBm} \qquad (1\text{-}58)$$

As explained earlier, while the input and output *signal levels* are expressed in dBm, the *gain* must be expressed in dB.
    The output level of the line is

$$P_2(\text{dBm}) = P_1(\text{dBm}) - L_{\text{dB}} = 27 - 33 = -6 \text{ dBm} \qquad (1\text{-}59)$$

The system output level is

$$P_o(\text{dBm}) = P_2(\text{dBm}) + G_{2\text{dB}} = -6 + 26 = 20 \text{ dBm} \qquad (1\text{-}60)$$

(c) It can be readily verified that the output power of 100 mW in (a) corresponds to 20 dBm, which is the result of (b). It can also be readily shown that when the net system gain of 30 dB is added to the input signal level of $-10$ dBm, the output level of 20 dBm is obtained in one step.

**▐▌ EXAMPLE 1-10**    A certain cable signal transmission system is shown in Figure 1–6. The nominal level at the signal source output is 10 dBm. The signal is amplified before transmission by the line amplifier whose gain is 13 dB. The first section of cable (section A) has a loss of 26 dB,

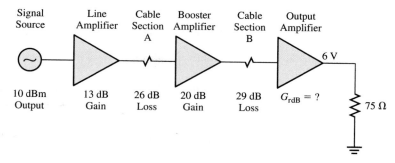

**FIGURE 1–6**
System of Example 1-10.

and a booster amplifier with a gain of 20 dB is placed at the output of that section. The second section of cable (section B) has a loss of 29 dB. The system is operated on a matched impedance basis with all effective sources and terminations having a resistive impedance of 75 Ω, and the characteristic impedance of the cable (to be defined in Chapter 14) is the same. (a) Determine the signal levels both in dBm and in volts at various points in the system. (b) Determine the gain of the receiver amplifier such that its nominal output level is 6 V across a 75-Ω termination.

**SOLUTION**

(a) To assist in keeping track of the computations, Table 1–3 provides a summary of the gains and levels at various points. Note that the two cable sections are shown with negative decibel gains. To determine the output level in dBm for any particular stage, we simply add the decibel gain in dB to the input level in dBm for that stage. For example, the line amplifier gain of 13 dB is added to the source level of 10 dBm to yield the amplifier output level of 23 dBm. The reader should check this process at all of the other stages.

　　To determine the voltage level at a particular point, we must first convert the power level in dBm to an absolute power level in watts. We start with

$$P(\text{dBm}) = 10\log\left(\frac{P(\text{mW})}{1\text{ mW}}\right) = 10\log\left(\frac{P(\text{W})}{1 \times 10^{-3}\text{ W}}\right) \qquad (1\text{-}61)$$

Since the final value for power is desired in watts, it was desirable to switch both numerator and denominator in the last form to watts. Since the reference for dBm is a milliwatt, it was necessary to express the reference level as $1 \times 10^{-3}$ W.

　　Solving for $P(\text{W})$, we obtain

$$P(\text{W}) = 1 \times 10^{-3} \times 10^{P(\text{dBm})/10} \qquad (1\text{-}62)$$

This expression can be used at any point in the system to convert from the dBm level to the absolute power level in watts.

Table 1–3   Tabulated Data for Example 1-10

|  | Gain | Output Level | Voltage |
|---|---|---|---|
| Signal Source | | 10 dBm | 0.8660 V |
| Line Amplifier | 13 dB | 23 dBm | 3.868 V |
| Cable Section A | −26 dB | −3 dBm | 0.1939 V |
| Booster Amplifier | 20 dB | 17 dBm | 1.939 V |
| Cable Section B | −29 dB | −12 dBm | 68.79 mV |

Since the voltages at various points are desired, the relationship between the effective (equivalent power producing) voltage value $V$, resistance $R$, and average power $P$ is

$$P = \frac{V^2}{R} = \frac{V^2}{75} \tag{1-63}$$

where the value of resistance at all junctions is used. Solving for $V$, we obtain

$$V = \sqrt{75P} \tag{1-64}$$

The reader is invited to verify the various voltage levels given in Table 1–3.

(b) The required output effective signal voltage level is 6 V. The corresponding output power $P_o$ in a 75-$\Omega$ load is

$$P_o = \frac{(6)^2}{75} = 480 \text{ mW} \tag{1-65}$$

The corresponding power in dBm is

$$P_o(\text{dBm}) = 10 \log \left( \frac{480 \text{ mW}}{1 \text{ mW}} \right) = 26.81 \text{ dBm} \tag{1-66}$$

The required gain $G_{\text{rdB}}$ for the receiver amplifier is

$$G_{\text{rdB}} = 26.81 \text{ dBm} - (-12 \text{ dBm}) = 38.81 \text{ dB} \tag{1-67}$$

An alternate way to achieve this result is to first observe from Table 1–3 that the output of the line is 68.79 mV, and the required amplifier output is 6 V. The voltage gain $A_v$ of the amplifier is

$$A_v = \frac{6 \text{ V}}{68.79 \times 10^{-3} \text{ V}} = 87.22 \tag{1-68}$$

The decibel gain of the receiving amplifier should then be

$$G_{\text{rdB}} = 20 \log 87.22 = 38.81 \text{ dB} \tag{1-69}$$

which agrees with the result of Equation 1-67. ▮

## 1-7 Signal-to-Noise Ratios

A detailed study of noise and its effects on various types of communications systems will be studied later in the text. However, while decibel computations are fresh in the reader's mind, and in anticipation of some additional concepts later in the chapter, some noise measures will be introduced.

Many measurements of communications signals are specified in terms of the average signal power at a certain point in the system as measured over a specific bandwidth. However, various types of noise will also be present. In general, noise is a random process that cannot be adequately described on an instantaneous basis. However, some of the statistical properties can be described, and it is possible to determine the average noise power over the same bandwidth as the signal.

### Absolute Signal-to-Noise Ratio

Assume that the average signal power is $P$ and that the average noise power is $N$, both measured in the same units. The absolute signal power to noise power ratio is often expressed symbolically as $S/N$, and it can be defined as

$$S/N = \frac{P}{N} \tag{1-70}$$

### Decibel Signal-to-Noise Ratio

The signal-to-noise ratio is often specified in decibels. This quantity can be defined as

$$(S/N)_{dB} = 10 \log (S/N) \tag{1-71}$$

The basic logarithmic properties can be used to expand this definition in the form

$$(S/N)_{dB} = 10 \log P - 10 \log N \tag{1-72}$$

This last result is not quite satisfying, since there is no absolute reference built into the logarithmic arguments—although we could think of an implied reference of 1 W in order to make the equation valid. In general, however, we can start with any arbitrary power reference, take its logarithm, multiply by 10, and add and subtract this quantity to Equation 1-72. After some basic logarithmic manipulations, the following result is obtained:

$$(S/N)_{dB} = P(dBx) - N(dBx) \tag{1-73}$$

Stated in words, *the signal-to-noise ratio in dB is equal to the signal power expressed in any arbitrary decibel signal level minus the noise power level expressed in the **same** units.*

---

**III EXAMPLE 1-11**

At a certain point in a communications system, the signal power is 5 mW, and the noise power is 100 nW. Determine (a) the absolute signal-to-noise ratio, and (b) the decibel signal-to-noise ratio. (c) Convert both the signal and noise power levels to dBm values and determine the decibel signal-to-noise ratio from these values.

**SOLUTION**

(a) Before the signal-to-noise ratio can be determined, both power levels must be expressed in the same units. The noise power of 100 nW corresponds to $10^{-4}$ mW. The absolute signal-to-noise ratio is then

$$S/N = \frac{5 \text{ mW}}{10^{-4} \text{ mW}} = 5 \times 10^4 \tag{1-74}$$

(b) The decibel signal-to-noise ratio is

$$(S/N)_{dB} = 10 \log(S/N) = 10 \log 5 \times 10^4 = 47 \text{ dB} \tag{1-75}$$

(c) The signal power in dBm is

$$P(dBm) = 10 \log \frac{P(\text{mW})}{1 \text{ mW}} = 10 \log \frac{5 \text{ mW}}{1 \text{ mW}} = 7 \text{ dBm} \tag{1-76}$$

The noise power in dBm is

$$N(dBm) = 10 \log \frac{N(\text{mW})}{1 \text{ mW}} = 10 \log \frac{10^{-4} \text{ mW}}{1 \text{ mW}} = -40 \text{ dBm} \tag{1-77}$$

The decibel signal-to-noise ratio is then

$$(S/N)_{dB} = P(dBm) - N(dBm) = 7 \text{ dBm} - (-40 \text{ dBm}) = 47 \text{ dB} \tag{1-78}$$

Note the somewhat strange looking pattern of units in Equation 1-78. Even though the signal and noise are both expressed in dBm, the difference is expressed in dB. As long as both the signal and noise decibel measures are referred to the same reference level, the logarithmic ratio will be in decibels.

III

---

## 1-8 Multisim® Example (Optional)

This chapter has introduced some of the basic terminology associated with communication systems and provided a general treatment of decibel forms. This section will introduce the use of Multisim to assist in the modeling and analysis of communication circuits and systems. In particular, the example provided should enhance the understanding of the decibel computations considered in the chapter.

### Reference to Appendix B

In the example of this section and in subsequent chapters, some introductory familiarity with Multisim will be assumed. Readers not having this background should first read Appendix B carefully and refer back to it when necessary. That information, along with further details provided in each example, should be sufficient to deal with the examples in the text.

The accompanying CD contains all the circuits in the text in three formats: Version 8, Version 7, and Version 6 (2001). Since most software programs undergo periodic updates, the reader may occasionally find a difference or change to a component or procedure as provided in the text. When in doubt, check the Multisim Help file.

### Notational Conventions

Before displaying Multisim schematics, some notational differences should be discussed. In mathematical equations and on most text schematic diagrams, variables will be given in italicized forms with subscripts and superscripts as appropriate. However, on Multisim schematic diagrams, the default patterns for variables are nonitalicized forms and all elements in a name will have the same level. For example, a resistance in a text equation or figure might be $R_1$. The corresponding symbol on a Multisim circuit schematic would be R1. This should not cause any difficulty in interpretation, but be aware that this is simply a reflection of different formats rather than careless editing. Notice also that, in some cases, there may not be a space between a component value and its units on the Multisim schematics.

### Standard Components versus Virtual Components

Figure B-1 in Appendix B provides a typical view of the Multisim screen. As discussed in that appendix, the two bars on the left represent the **Component Toolbars**. The one farthest to the left represents *standard components* and the one to its right represents *virtual components*. For detailed design work in which standard values are required, there is an obvious need in many cases to utilize standard values. However, for much of the work in this book, the virtual components are somewhat easier to use because they can be set to any arbitrary values, and they will be employed as appropriate.

### Reference ID

As components are placed on a schematic, they will each be labeled by an appropriate name. For example, the first resistor pulled from a resistor bin will be denoted **R1**, and the second one will be denoted **R2**, and so on. However, the names can be changed by the procedure that follows.

Double left-click on the device and the properties window for the device will open. Then left-click on the **Label** tab. Type the desired name in the **RefDes** slot and left-click **OK**.

*One caution:* You cannot use a name that is currently assigned to another component. Suppose you wish to swap the names of two resistors **R1** and **R2**. If you tried to change either one as they are, the program will generate an error message since the other one is already assigned the desired name. Therefore, it will be necessary to assign one temporarily to a name not currently on the schematic, and then you can perform the desired steps. For that reason, it is convenient to extract components to be partially identified by numbers in the order for desired labeling when possible.

### Node Labels

The procedure on node labels follows a pattern similar to that of component names. The first node connection will be **1**, followed by **2**, etc. The one exception is that of a grounded node, which will always be assigned the value **0**. Nodes can also be assigned names, such

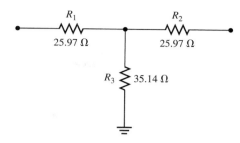

**FIGURE 1–7**
Precision 10-dB attenuator circuit.

as **INPUT** and **OUTPUT**, and so on. The same caution about duplication mentioned in the preceding paragraph applies to node numbers as well.

## Precision Attenuator Circuit

The circuit of Figure 1–7 is a precision constant-impedance T-pad attenuator circuit that will be analyzed shortly with Multisim. Such circuits are used in various communication modules to control the level of a signal and to provide some degree of isolation between input and output. It has been designed around a constant resistive impedance level of 50 Ω. This means that if the output is terminated in a 50-Ω resistance, the resistance "looking into" the input will also be 50 Ω. It is also designed to be driven by a source having an internal resistance of 50 Ω, and this means that the equivalent Thevenin resistance "looking back" from the output terminals will also be 50 Ω. The attenuation desired for this particular circuit is 10 dB, and this is based on the difference in levels between the voltage across the input terminals (not the source voltage) and the voltage across the output with the circuit properly terminated.

**▌▌ MULTISIM EXAMPLE 1-1**

Assume that a dc source of 2 V with an internal resistance of 50 Ω is connected to the input of the circuit of Figure 1–7, and assume that the output is terminated with 50 Ω. Model the circuit with Multisim and use the **DC Operating Point Analysis** to determine various voltages and the input current.

**SOLUTION**   The **DC Operating Point Analysis** is the simplest one to implement, but the data obtained are limited to (1) voltages at all nodes with respect to a reference ground, and (2) currents through voltage sources. Thus, a ground node must be established, as is the case with all Multisim circuits.

The three resistors in the attenuator were denoted $R_1$, $R_2$, and $R_3$ on Figure 1–7. The **Reference ID** names on the schematic will automatically be labeled as **R1**, **R2**, and **R3** provided that they are taken from the parts bin in that order. However, the source internal resistance label is changed to **Rs** and the load resistance is changed to **RL**, so it is recommended that they be the last ones placed on the circuit schematic. The initial label of the voltage source will be **V1**, but it is changed to **Vs**.

Using the procedure outlined in Appendix B, the circuit schematic is created as shown in Figure 1–8. The dc source is obtained by left-clicking below the **Show Power Source Family Bar** at the top, and is identified in that bin by the battery symbol and the designation **Place DC Voltage Source.** Left-click on that button and a source will attach to the arrow. Drag it to the approximate location where it is to be placed and left-click again. It will then be released and is available for modification. If it were necessary to change the orientation (which is not the case in this example), you would select it by left-clicking on it, then left-click on **Edit** on the **Menu Row.** You would then follow the procedure discussed in Appendix B.

FIGURE 1–8
Attenuator with matched
source and load.

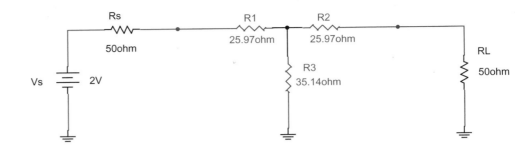

A circuit ground is also available in the same parts bin as the dc source. The title appearing on the screen when the mouse is placed over the ground symbol is **Place Ground.** Left-click on this button and it will attach to the arrow. Drag it to the desired location and left-click to detach it.

The initial value of the voltage source is set to **12 V.** To set any other value, double left-click on the source and a window with the title **POWER_SOURCES** will open. The tab will likely be in the default **Value** position, but if not, left-click on that tab. The value is then set in the slot provided. Note that there is a second slot that may be set for **V, mV,** etc. In this example, the voltage is set to **2 V** for reasons that will be clearer later. Leave the other slots at the default values of **0** since they are not used in this example. Left-click on **OK** and the new value will be retained. You can then change the label from **V1** to **Vs** using the procedure discussed earlier.

We won't discuss all of the options underneath the **Label** and **Display** tabs at this point since they are more "cosmetic" than operational, but the reader may wish to experiment with them as more experience with the program is acquired.

The resistors are obtained from the **Show Basic Family Bar,** which is the third one from the top. Left-click on the **Place Virtual Resistor** button and a resistor will attach to the arrow. Drag it to the approximate location where it is to be placed and left-click again. It will then be released and is available for further modification. If it is necessary to change the orientation, select it by left-clicking on it and then left-click on **Edit** on the **Menu Row** and follow the procedures discussed in Appendix B.

The initial value of the virtual resistance comes out as **1 kΩ.** To set any other value, double left-click on the resistor and a window with the title **BASIC_VIRTUAL** will open. The tab will likely be in the **Value** position, but if not, left-click on that tab. The value is then set in the slot provided. Note that there is a second slot that may be set for **Ohm, kOhm, MOhm,** etc. Leave all the blocks from **Tolerance** on down at their default values. Left-click on **OK** and the value will be set.

As in the case of the dc source window, some other tabs on the resistor window are the **Label** and **Display** tabs. Many other components will have similar options on the properties windows, and once you begin developing experience using the program, you may wish to alter the labels and displays as desired. For this circuit, the source resistance was changed to **Rs** and the load resistance was changed to **RL.**

The dc analysis is performed by first left-clicking on the **Analyses** button, which is next to the button resembling a triangular wave. A window providing different analysis options will then open, and a dc analysis is performed by left-clicking on the **DC Operating Point** button. An alternate path to get to the same point is to first left-click on **Simulate** on the top row and then hold the arrow over the **Analyses** option. The **Analyses** window will then open and you can left-click on **DC Operating Point.**

When the **DC Operating Point** button is clicked, a new window with the title **DC Operating Point Analysis** opens, from which a number of choices can be made. The default tab position is with the upper tab set to **Output,** and the possible variables that can be tabulated are the node voltages and the current through the voltage source. For this circuit, there are five variables. Each node voltage name or number is preceded by the dollar sign **$.** Thus, the four node designations in this example are **$1, $2, $3,** and **$s.** The additional variable represents the current through the voltage source. The desired variables must be

**FIGURE 1–9**
Data for Multisim Example 1-1.

**Operating Point**

| DC Operating Point | | |
|---|---|---|
| 3 | 316.27419m | |
| s | 2.00000 | |
| 1 | 999.96512m | |
| 2 | 480.54701m | |
| vvs#branch | -20.00070m | |

highlighted. Since more than one is to be read, hold the keyboard **Ctrl** key down while left-clicking on each of the five quantities. Next, left-click on the block entitled **Add.** The numbers will then appear in the right-hand block. Finally, left-click on **Simulate** and a window similar to Figure 1–9 appears. This is a compilation of the output data. The data may not necessarily appear in the same order as in Figure 1–9, but the values of the variables should be essentially the same.

The entry labeled **vvs#branch** is the current in amperes flowing through the voltage source, and is indicated as **−20 mA**. The negative value for the current may seem strange and needs some explanation. In the original SPICE program, currents could only be measured directly through voltage sources. The reference direction of conventional current flow for any SPICE component is *into the positive terminal and out of the negative terminal*. Since the actual current in this circuit flows out of the positive terminal, it appears with a negative value. Thus, the current is **20 mA** in a clockwise direction. Although we do not need to measure any other currents in the circuit, one can always place a voltage source of value 0 V in a branch for which a current measurement is desired.

Now we will investigate the various voltages within the circuit. The voltage at node 1 for all practical purposes is 1000 mV or 1 V. This tells us that the input impedance of the attenuator is 50 Ω since the 2-V value of the source is equally divided between its internal resistance and that of the circuit. The reason that 2 V was chosen for the source was that it would provide 1 V across the input to the attenuator. Another way to verify the input resistance of the attenuator is to divide the input voltage of 1 V (not the source open-circuit voltage) by the current of 20 mA, which yields 1 V/0.02 A = 50 Ω.

Next, we need to check the attenuation of the circuit. The voltage across the input is 1 V and the voltage across the output (node 3) is about 0.3163 V. Thus, the attenuation is $20 \log(1/0.3163) = 20 \log(3.162) = 10.0$ dB.

It should be noted that there is an additional loss of 6 dB between the open-circuit source voltage of 2 V and the input voltage to the attenuator of 1 V. However, this is an inherent property of circuits operated on a matched impedance basis.

## 1-9 MATLAB® Examples (Optional)*

The major benefits of MATLAB® will best be demonstrated with the more advanced applications in subsequent chapters. However, it is perfectly logical to use MATLAB in the same manner as one might use a scientific calculator if it is accessible. Indeed, the **Command Window** can be employed in an interactive dialog mode, and that will be the basis of many of the examples in the text, including those of this section.

*MATLAB® is a registered trademark of The MathWorks, Inc., 3 Apple Hill Drive, Natick, MA, 01760-2098 USA. Tel: 508-647-7000, Fax: 508-647-7001, E-mail: info@mathworks.com, Web: www.mathworks.com.

### Reference to Appendix C

Readers with little or no background with MATLAB may first wish to refer to Appendix C (entitled "MATLAB PRIMER"). That appendix provides a somewhat detailed introduction to MATLAB, and should enable the reader to follow the developments that employ the program throughout the text.

### Professional Version versus Student Version

Either the Professional Version or the Student Version may be employed to support the text. With either program, the optional **Signal Processing Toolbox** is necessary for some of the examples. Students may acquire that toolbox for a modest additional cost.

The main identifier for the Student Version is that each instruction line in the **Command Window** begins with the prompt **EDU>>**, which is automatically inserted by MATLAB. The corresponding prompt for the Professional Version is simply >>, and it will be assumed for simplicity. Following the typing of a command in the **Command Window**, it is activated by depressing **Enter** on the computer. This process will be assumed with all commands, whether stated in the procedure or not.

### Notational Conventions

As discussed in Appendix C, we will usually display the results exactly as they appear on the screen. This means that the type style may vary somewhat from the standard mathematical style in the text, but the symbols should be perfectly clear.

### Command Window

The **Command Window** will allow you to work in a manner similar to that of a calculator. You can enter a command and see the numerical results immediately. You can also enter code in the **Command Window** that will plot curves. The important thing is that it is interactive, in contrast to the writing of *M-files,* which is performed in a different window. An M-file is actually a computer program that can be written, saved, and activated as many times as desired. M-files are discussed in Appendix C.

---

**‖ MATLAB EXAMPLE 1-1**

Use the MATLAB Command Window to solve Example 1-3; that is, to determine the decibel gain corresponding to an absolute power gain of 175.

**SOLUTION**   The logarithm to the base 10 of any value x in MATLAB is denoted **log10(x)**. This value must be multiplied by 10 and the result is denoted here as **GdB**. The Command Window dialogue follows.

```
>> GdB = 10*log10(175)
GdB =
 22.4304
```

The result, after rounding, is equivalent to that of Equation 1-27. MATLAB does not provide units, but we obviously interpret the result to be in decibels.

Incidentally, the logarithm to the base e in MATLAB is denoted simply as **log(x)**.

---

**‖ MATLAB EXAMPLE 1-2**

Use the MATLAB Command Window to solve Example 1-4; that is, to determine the absolute power gain corresponding to 28 dB.

**SOLUTION**   Without showing the development again, Equation 1-30 is used. The MATLAB command is

```
>> G = 10^(28/10)
G =
  630.9573
```

The presence of the parentheses affirms the fact that 28 should be divided by 10 as the first step. The result then serves as the exponent of 10.

---

**MATLAB EXAMPLE 1-3**    Use the MATLAB Command Window to solve Example 1-5; that is, to determine the voltage gain corresponding to the power gain of Example 1-4.

**SOLUTION**   While we could take the square root of the preceding example result, as was done back in Example 1-5, it is instructive to show how several of the preceding operations can be performed in one step. For any value of x, the command for the square root of x in MATLAB is **sqrt(x)**. Combining the steps of the preceding MATLAB example with this one, we have

```
>> A = sqrt(10^(28/10))
A =
  25.1189
```

Note the nesting within the expression. Starting with the innermost parentheses, 28 is first divided by 10. That result is the exponent of 10. After the exponentiation process, the square root of the result is taken. It is possible to nest complex expressions with MATLAB as long as the hierarchy from innermost to outermost is established. Ultimately, there must be the same number of left parentheses as right parentheses.

Of course, it perfectly permissible to expand the operation into several steps, and this should be done when there is a possibility of misinterpretation.

---

## PROBLEMS

**1-1**    The lower frequency limit of the commercial AM band is about 540 kHz. Determine the wavelength.

**1-2**    The upper frequency limit of the commercial AM band is about 1600 kHz. Determine the wavelength.

**1-3**    The lower frequency limit of the commercial FM band is about 88 MHz. Determine the wavelength.

**1-4**    The upper frequency limit of the commercial FM band is about 108 MHz. Determine the wavelength.

**1-5**    One of the amateur radio bands is referred to as the "20-meter band." Determine the frequency corresponding to this wavelength.

**1-6**    The so-called Citizens Band was originally referred to as the "11-meter" band. Determine the frequency corresponding to this wavelength.

**1-7**    An amplifier has an absolute power gain of 750. Determine the decibel power gain.

**1-8**    An amplifier has an absolute power gain of 28,000. Determine the decibel power gain.

**1-9**    An amplifier has a gain of 16 dB. Determine the absolute power gain.

**1-10**    An amplifier has a gain of 38 dB. Determine the absolute power gain.

**1-11**    The output power from a long cable connection is only 4% of the input power. Determine the decibel loss.

**1-12**    The output power from an attenuator circuit is 20 mW and the input power is 0.15 W. Determine the decibel loss.

**1-13**    The signal transmitted from a satellite to Earth experiences a loss of 204 dB. Determine the absolute loss factor.

**1-14**    The output level of a circuit is 18 dB below the input level. Determine the output power if the input power is 2 W.

**1-15**    A microwave amplifier is specified as having a gain of 20 dB ± 0.5 dB. Determine the range of possible absolute power gains.

**1-16**    The absolute power gain of an amplifier can vary from 400 to 600. Determine a decibel power gain specification in the same format as in the preceding problem.

**1-17**    The voltage gain of an amplifier is 160. Assuming equal impedance levels at both input and output, determine the decibel gain.

**1-18**    The output current of a lossy circuit is 17% of the input current. Assuming equal impedance levels at both input and output, determine the decibel loss.

**1-19**    A signal has a power level of 85 W. Determine the level in (a) dBm, (b) dBW, and (c) dBf.

**1-20**  A signal has a power level of 0.24 mW. Determine the level in (a) dBm, (b) dBW, and (c) dBf.

**1-21**  Express the following power levels in dBm: (a) 1 kW, (b) 1 W, (c) 1 mW, (d) 1 μW, (e) 1 nW.

**1-22**  Express the following power levels in dBf: (a) 1 mW, (b) 1 μW, (c) 1 nW, (d) 1 pW, (e) 1 fW.

**1-23**  Consider the system of the following figure, with parameters shown and with impedances matched at all junctions. Determine (a) net system absolute gain, (b) net system decibel gain from the result of (a), and (c) net system decibel gain from individual decibel values.

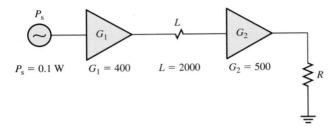

$P_s = 0.1$ W    $G_1 = 400$    $L = 2000$    $G_2 = 500$

**1-24**  Consider the system of the following figure, with parameters shown and with impedances matched at all junctions. Determine (a) net system decibel gain, (b) net system absolute gain from the result of (a), and (c) net system absolute gain from individual gain values.

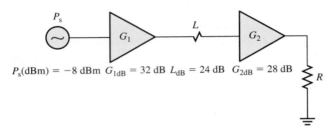

$P_s(dBm) = -8$ dBm  $G_{1dB} = 32$ dB  $L_{dB} = 24$ dB  $G_{2dB} = 28$ dB

**1-25**  In the system of Problem 1-23, the signal source delivers an input power of 0.1 W, as shown in the accompanying figure. Determine (a) the absolute power levels at each one of the junctions, and (b) the power in dBm at each one of the junctions. (c) Convert the system output of (a) to dBm and compare with (b).

**1-26**  In the system of Problem 1-24, the signal source delivers an input power of −8 dBm, as shown in the accompanying figure. Determine (a) the power levels in dBm at each one of the junctions, and (b) the absolute power level at each one of the junctions. (c) Convert the system output of (a) to an absolute power level and compare with the result of (b).

**1-27**  For the system of Problems 1-23 and 1-25, and with an impedance level at all junctions of 50 Ω, determine (a) the voltage across the input to the first amplifier, and (b) the voltage across the output load resistance. (c) Use the results of (a) and (b) to determine the system decibel gain and compare with the result of Problem 1-23.

**1-28**  For the system of Problems 1-24 and 1-26, and with an impedance level at all junctions of 50 Ω, determine (a) the current input to the first amplifier, and (b) the current in the output load resistance. (c) Use the results of (a) and (b) to determine the system decibel gain and compare with the result of Problem 1-24.

**1-29**  Perform all the computations of Example 1-10 if the system parameters are as follows: signal source output = −3 dBm, line amplifier gain = 18 dB, cable section A loss = 22 dB, booster amplifier gain = 26 dB, cable section B loss = 25 dB.

**1-30**  Perform all the computations of Example 1-10 if the system parameters are as follows: signal source output = 2 dBm, line amplifier gain = 15 dB, cable section A loss = 24 dB, booster amplifier gain = 18 dB, cable section B loss = 22 dB.

**1-31**  A satellite communication system is illustrated in the following figure. The *uplink* portion will be analyzed in this problem. The output power of the ground transmitter is 200 W. This power level is effectively increased in the direction of transmission by the ground antenna, which has a gain of 40 dB. The transmitted signal is attenuated by a path loss of 200 dB. The receiver antenna gain is 20 dB. (a) Determine the received signal power in femtowatts and in dBf. (b) If the total noise level at the input to the satellite receiver is 8 dBf, determine the received signal-to-noise ratio in dB.

Geostatic Communication Satellite

Transmitting Earth Station          Receiving Earth Station

**1-32**  Some computations involving the *downlink* portion of the satellite system in the figure in Problem 1-31 will be made in this problem. The output power of the satellite transmitter is 0.5 W. This power level is effectively increased in the direction of transmission by the transmitter antenna, which has a gain of 20 dB. The transmitted signal is attenuated by a path loss of 200 dB. The total noise level at the ground receiver input is estimated to be 2 dBf. The desired signal-to-noise ratio at the receiver input is 15 dB. Determine the minimum ground receiving antenna gain in dB that will meet the specifications.

# Spectral Analysis I: Fourier Series    2

The great French mathematician Jean Baptiste Fourier (1768–1830) made an outstanding contribution to scientific analysis utilizing the mathematical properties of sine waves. Fourier developed a method for representing arbitrary waveforms in terms of sine and/or cosine functions. This concept led to an area of applied mathematics known as *Fourier analysis,* which is utilized today in a wide range of engineering and scientific disciplines. The concept as applied to electrical signal analysis is commonly referred to as *spectral analysis.*

The concept of the spectrum is very important in communications systems. The spectrum information for a given signal identifies those frequencies present and their relative importance. The system is designed to process these frequencies and to reject those outside the range. Channel allocations are based on specific frequencies in order that different signals can be processed without interference.

Both this chapter and the next are devoted to a development of the basis for spectral analysis and the means for determining the spectrum of a given signal. The primary emphasis in this chapter will be on *Fourier series* as applied to *periodic signals;* in the next chapter, the emphasis will be directed toward the *Fourier transform,* which can be applied to *nonperiodic signals.*

## Objectives

After completing this chapter, the reader should be able to:

1. Classify signals as *periodic* or *nonperiodic* and as *deterministic* or *random.*
2. Perform simple calculations on periodic waveforms to predict the frequencies contained in the spectrum.
3. State the quadrature form of the Fourier series and explain the properties.
4. State the amplitude-phase forms of the Fourier series and explain the properties.
5. Convert between the quadrature and amplitude-phase forms of the Fourier series.
6. Apply symmetry conditions to deduce various spectral properties.
7. Determine the spectral rolloff rate for piecewise linear waveforms.
8. Utilize tables of Fourier series to analyze the spectra of common waveforms.
9. Calculate average power in the time domain and in the frequency domain.
10. State the complex exponential form of the Fourier series and explain its significance.
11. Plot one-sided and two-sided linear amplitude spectra.
12. Plot one-sided and two-sided power spectra.

## SystemVue™ Opening Application (Optional)

Insert the text CD in a computer having SystemVue™ installed and activate the program. Open the CD folder entitled **SystemVue Systems** and open the file entitled 2-1.

### Tokens

Various signal generator and processing blocks in SystemVue™ are referred to as **Tokens.** The specific token blocks used to view waveforms and obtain output data are referred to as **Sink Tokens.** We will refer to all tokens other than Sink Tokens as **Operational Tokens.**

### Token and Sink Token Numbers

In general, tokens are assigned numbers in the order in which they are placed in the **System Design Area** on the screen. However, there is a procedure for changing the token numbers with the mouse. Left-click on the **Edit** button on the **Menu Bar,** and then left-click on **Assign Token Numbers** followed by a left-click on **Use the Mouse.** A window will open providing the procedure, which amounts to assigning an initial value and a subsequent click on each token in the desired order.

### Format for Assignments in this Text

The various window tokens have fixed numbers beginning with 0, although names can be assigned as well. In general, window numbers will be different than sink token numbers, depending on the order in which they were entered. However, using the procedure discussed in the preceding paragraph, *all sink token numbers start from a value of 0 and match the number of the particular window used to monitor that waveform.* The authors have established this system because we believe that it will minimize confusion on the part of the reader. Thus, the waveform for Sink Token 0 will be monitored by Window W0; the waveform created by Sink Token 1 will be monitored by Window W1, and so on. This practice is followed throughout the book, and we recommend it as a good procedure to keep track of the relationship between tokens and sinks.

Each opening application will begin with a tabulated listing for the tokens used in that particular application. The example that follows is based on Chapter 2, and this will be the format in subsequent chapters. In this case, there is only one sink token.

### Sink Token

| Number | Name | Token Monitored |
| --- | --- | --- |
| 0 | Output | 6 |

### Operational Tokens

| Number | Name | Function |
| --- | --- | --- |
| 1 | F1 | sinusoidal source with frequency of 1 Hz |
| 2 | F2 | sinusoidal source with frequency of 3 Hz |
| 3 | F3 | sinusoidal source with frequency of 5 Hz |
| 4 | F4 | sinusoidal source with frequency of 7 Hz |
| 5 | F5 | sinusoidal source with frequency of 9 Hz |
| 6 | Sum | provides sum of previous 5 sinusoids |

## Further Description of System

The system has five sinusoidal source tokens as shown on the left. The source at the top has a frequency of 1 Hz and the additional sources have frequencies that are odd integer multiples of 1 Hz; that is, 3 Hz, 5 Hz, 7 Hz, and 9 Hz.

The sources are all connected to the adder token near the middle. This token forms the instantaneous sum of the sources connected to the input. Several of the lines connecting to the token may appear to be connected at some points, but this is simply the layout process generated by the program; each source is independently connected to an input of the adder token, and the apparent "short" is an illusion of the layout.

The block to the right of the adder is a sink token that is used to obtain output data. The larger block on its right is a graph to display the sink output as a plot. This particular plot is not large enough to show any detail, but it gives a small glimpse of the result. The process to be discussed shortly enlarges the graph.

## Some Important Buttons

Three of the most important buttons used in performing a simulation appear on the right-hand side of the **Tool Bar.** The button displaying a picture of a clock is the **Define System Time** control, in which various time parameters are set. The button to its right is the **Analysis Window** control, which opens various windows for observing waveforms. The button with a green triangle is the **Run System** control, which generates a simulation run.

## Activating the Simulation

We will assume initially that all the parameter settings have their initial values. Left-click on the **Run System** button and the simulation should run. You will see the small plot to the right of the sink, but its function is primarily to verify that something has happened. To see the plot on a better scale, left-click on the **Analysis Window** button. The graph will then occupy much of the screen, and the **Maximize** button can be used to make it occupy the entire screen if desired.

One other button that will be introduced now is the **Load New Sink Data** control. It is on the extreme left, just above the plot. If you return to the system to make a new run, the plot that was first generated will remain on the graph until the **Load New Sink Data** button is depressed. You can always tell when that control needs to be used because it will blink when new data are available.

## What You See

The waveform that appears on the graph resembles a square wave oscillating from $-1$ V to 1 V. You are seeing a demonstration of the concept of *Fourier series,* in which a periodic waveform can be represented as the sum of sinusoidal functions. In this case, we are taking some of the sine functions in the series and adding them together to produce an approximate square wave.

You will learn in this chapter how to determine the values of the sine and/or cosine functions that could be summed together to create arbitrary periodic waveforms. You will learn that functions having discontinuities, such as the square wave, require many more sinusoidal terms to achieve a very accurate representation of the waveform. Thus, with only five components in this demonstration, it is clear that there are significant discrepancies between the ideal square wave and the approximation shown. However, the trend is clearly leading in the direction of a square wave.

The ability to reproduce accurately a waveform with sinusoidal functions is important in determining the *bandwidth* required to transmit a signal. Thus, a square wave or rectangular pulse train would require a very large bandwidth for faithful reproduction.

Since digital information is often transmitted in the form of pulses, this has led to the development of certain types of rounded pulses, for which the Fourier series has fewer terms for faithful reproduction. This means that the resulting bandwidth is smaller. Thus, the determination of the bandwidth for a given signal and the determination of signals with minimum bandwidths are very important concepts in the design of communication systems.

After you have studied the material of this chapter, you will return to this same system and adjust the components to create a different periodic waveform, but one in which the practical bandwidth required will be much less.

## Checking the Settings

All of the parameter values were preset for this system, but you may wish now to verify some of the settings. Moreover, for the exercises at the ends of chapters, you will need to reset some of the parameters, so it is a good idea now to investigate how this is done.

There is a folder on the disk with the title **SystemVue Initial Settings.** The various files have been prepared in Microsoft Word format and show the initial settings of the parameters. For this system, the file is called **2-1 initial settings.** Some of the settings are calculated automatically when others are entered. For example, the program computed the **Frequency Resolution** value in the **Define System Time** procedure.

One point to note is that data values may be entered using MATLAB® mathematical operations. For example, the amplitude of the top source was entered as 4/pi. However, when you double-click on that particular oscillator in the actual system, the value will have been evaluated to a large number of decimal places.

## 2-1  Signal Terminology

The term "signal" will appear many times throughout this book. In a collective sense, the term will be used to represent any of the different kinds of waveforms that are encountered in communication system analysis. This will include both information or message waveforms and the various waveforms generated directly in the communication equipment. The term could refer to a voltage, a current, or some other variable, depending on the frame of reference.

### Notation and Conventions

The measurable variables that describe the behavior of various communication circuits and systems are usually voltages, currents, power, and energy. Voltages will be represented by $v$ or $V$, currents by $i$ or $I$, power by $p$ or $P$, and energy by $w$ or $W$. Various subscripts and/or superscripts may be added when there are two or more of these variables in the same analysis.

By convention, lowercase symbols such as $v$, $i$, $p$, or $w$ are generally used to represent instantaneous or time-varying variables. To emphasize this point, such variables are frequently written in functional forms as $v(t)$ and $i(t)$, where $t$ represents time in seconds (s). The quantity in parentheses, $t$ in this case, is called the *argument* of the function. To simplify notation, arguments will frequently be omitted, but remember that when lowercase symbols for variables are used, the functions *could* vary with time.

Uppercase symbols such as $V$ and $I$ are used to represent either dc values or some fixed property of the waveform such as the peak value or the effective value. Uppercase symbols are also used to represent *transform functions,* which will be introduced later.

For special reasons there may be an occasional deviation from the preceding rules, so they should be interpreted with some flexibility.

## Instantaneous Power

Let $p(t)$ represent the instantaneous power in watts (W) dissipated in a resistance $R$. For a voltage $v(t)$ in volts across the resistance, the instantaneous power is

$$p(t) = \frac{v^2(t)}{R} \tag{2-1}$$

For a current $i(t)$ in amperes flowing through the resistance, the instantaneous power is

$$p(t) = R i^2(t) \tag{2-2}$$

## Use of Voltage as Dominating Variable

In studying and tracing through the operation of many circuits, more often than not it is the voltage that is the major variable under observation. For that reason, and because it is usually the more convenient variable to measure, we will use *voltage* as the theme for much of the analysis throughout the book. Many of the results will apply equally well if current behavior is of interest. The one main area of difference is in the formulation of power. From Equations 2-1 and 2-2, we understand that to compute power with voltage, we must divide by $R$, but if we wish to compute power from current, we must multiply by $R$. Thus, if we desire to modify any power equations developed with voltages for use with currents, the necessity to make this change should be self-evident.

We will now investigate some classifications of signals.

## Periodic versus Nonperiodic

The first classification is to identify signals as either (1) *periodic* or (2) *nonperiodic*. A *periodic* signal is one that repeats itself in a predictable fashion as illustrated in Figure 2–1(a). The *period T* (in seconds) is the smallest value of time such that

$$v(t + T) = v(t) \quad \text{for all } t \tag{2-3}$$

An interval of one period is also referred to as *one cycle* of the signal.

A *nonperiodic* signal is one in which there is no finite value of $T$ that satisfies the condition given in the preceding paragraph. It can also be said that a nonperiodic signal is a limiting case of a periodic signal in which the period approaches infinity. (This concept will be useful in the next chapter.) An example of a nonperiodic signal is shown in Figure 2–1(b).

## Deterministic versus Random

The second type of classification is to identify signals as being either (1) *deterministic* or (2) *random*. A *deterministic* signal is one whose instantaneous value as a function of time

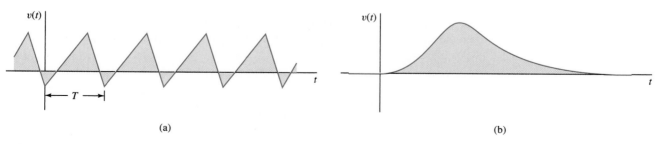

(a)                                                          (b)

**FIGURE 2–1**
Examples of (a) a periodic signal and (b) a nonperiodic signal.

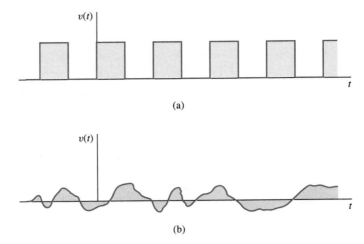

(a)

(b)

**FIGURE 2–2**
Examples of (a) a deterministic signal and (b) a random signal.

can be totally predicted. An example of a deterministic signal (a square wave) is shown in Figure 2–2(a). (This particular waveform is also periodic.)

A *random* signal is one whose instantaneous value cannot be predicted at any given time. While the exact value at any given time is not known, many random signals encountered in communications systems have properties that can be described in statistical or probabilistic terms. Much of the treatment of random signals will be delayed until later in the text. An example of a random signal is shown in Figure 2–2(b).

## Sinusoidal Functions

Because of the importance of sinusoidal functions in communication theory, some of their basic properties will be reviewed here. Both the sine function and the cosine function are considered sinusoidal functions. For the purposes of communication theory, these two functions can be expressed in either of the following forms:

$$v(t) = A \cos \omega t = A \cos 2\pi f t = A \cos \frac{2\pi t}{T} \tag{2-4}$$

and

$$v(t) = B \sin \omega t = B \sin 2\pi f t = B \sin \frac{2\pi t}{T} \tag{2-5}$$

The following quantities are defined:

$A$ or $B$ = peak value or amplitude, expressed in volts if the function is a voltage

$\omega = 2\pi f = \frac{2\pi}{T}$ = *angular* frequency in radians per second (rad/s)

$f = \frac{1}{T}$ = *cyclic* frequency in hertz (Hz)

$T$ = period in seconds (s)

The forms of the cosine and sine functions for $t \geq 0$ are shown in Figure 2–3. Remember that the cosine has a positive peak at the origin and the sine passes through the origin with a positive slope.

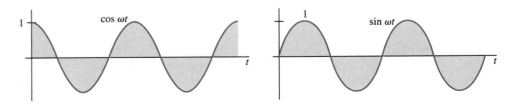

**FIGURE 2–3**
Cosine and sine functions.

## Combining Sine and Cosine Functions

Frequently, communication signals of the following form will arise:

$$v(t) = A \cos \omega t + B \sin \omega t \tag{2-6}$$

A signal of this type can always be expressed in either of the following forms:

$$v(t) = C \cos(\omega t + \theta) \tag{2-7}$$

or

$$v(t) = C \sin(\omega t + \phi) \tag{2-8}$$

Note that for this process to be possible, *the frequency must be the same in the cosine and sine terms.* As a byproduct, an expression of the form of either Equation 2-7 or 2-8 can be resolved into the sum of a cosine and a sine as given by Equation 2-6.

While various trigonometric identities could be employed in the conversion process, the simplest way is to utilize a semi-graphical approach coupled with rectangular-to-polar conversion, or vice versa. Consider the coordinate system depicted in Figure 2–4. The relative phase sequence of the cosine and sine functions is shown in this figure. The positive direction of rotation is counter-clockwise. When cosine and sine terms at a specific frequency are given, they can be interpreted as vectors (or phasors) along the $x$ and $y$ axes according to their signs. By the Pythagorean theorem, the length of the resulting combined vector is the value of $C$, which can be expressed as

$$C = \sqrt{A^2 + B^2} \tag{2-9}$$

The angle $\theta$ or $\phi$ is determined from the angle with respect to the appropriate axis. The process will be illustrated in Example 2-2.

Whenever it is desired to resolve a function of the form of either Equation 2-7 or 2-8 into a sum of a cosine and a sine function, draw a vector of length $C$ rotated by the angle $\theta$ or $\phi$ from the cosine or sine axis, depending on which form is given. Then, resolve the vector into horizontal and vertical components, and these components will be the values of $A$ and $B$, respectively. The signs of $A$ and $B$ will depend on the signs of the pertinent axes according to Figure 2–4.

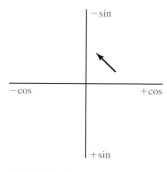

**FIGURE 2–4**
Relative phase sequence of sinusoids.

---

**▌ EXAMPLE 2-1**

A voltage (in volts) is given by

$$v(t) = 12 \cos 2\pi \times 2000t \tag{2-10}$$

Determine (a) peak value, (b) radian frequency, (c) cyclic frequency, and (d) period.

**SOLUTION**

(a) The peak value or amplitude is easily identified as

$$A = 12 \text{ V} \tag{2-11}$$

(b) The radian frequency is

$$\omega = 2\pi \times 2000 = 12.57 \times 10^3 \text{ rad/s} = 12.57 \text{ krad/s} \tag{2-12}$$

(c) The cyclic frequency is

$$f = \frac{\omega}{2\pi} = \frac{2\pi \times 2000}{2\pi} = 2000 \text{ Hz} = 2 \text{ kHz} \tag{2-13}$$

(d) The period is

$$T = \frac{1}{f} = \frac{1}{2000} = 0.5 \times 10^{-3} \text{ s} = 0.5 \text{ ms} \tag{2-14}$$

It should be noted in Equation 2-10 that the cyclic frequency in Hz could have been obtained by inspection, since the $2\pi$ factor is explicitly provided. This form is a very common way of expressing a sinusoid, and it makes the cyclic frequency immediately evident.

■■ **EXAMPLE 2-2**

A current (in amperes) is given by

$$i(t) = 4\cos 50t + 3\sin 50t \qquad (2\text{-}15)$$

Express the function as (a) a cosine function and (b) a sine function.

**SOLUTION**   Observe that both terms have a frequency of 50 rad/s or $50/2\pi = 7.958$ Hz. Otherwise, it would not be possible to combine the two terms into a single term. The form of the relative phase diagram of the two functions is shown in Figure 2–5. The first term is represented as a vector of length 4 along the +cosine axis, and the second term is represented as a vector of length 3 along the +sine axis (which is down). The hypotenuse is equal to the peak value of the net sinusoid (whether cosine or sine), and is the length of the hypotenuse that could be formed from a right triangle; thus, it is

$$C = \sqrt{A^2 + B^2} = \sqrt{4^2 + 3^2} = 5 \text{ A} \qquad (2\text{-}16)$$

(a) To express in terms of the cosine function, the angle is $\theta$ and it is

$$\theta = -\tan^{-1}\frac{3}{4} = -36.87° = -0.6435 \text{ rad} \qquad (2\text{-}17)$$

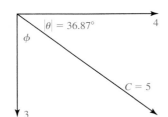

**FIGURE 2–5**
Relative phase sequence diagram of Example 2-2.

The negative sign is required since the angle *lags* (is *more negative* than) the +cosine axis. The current can then be expressed as

$$i(t) = 5\cos(50t - 36.87°) = 5\cos(50t - 0.6435) \qquad (2\text{-}18)$$

The form using degrees for the phase angle is commonly used in engineering applications and will be used freely in this text. However, it should be emphasized that the product $\omega t$ has the dimension of radians, so the phase angle must be expressed in radians if the quantities are actually to be combined at specific values of time.

(b) To express in terms of the sine function, the angle is $\phi$ as shown, and it is

$$\phi = 90 - |\theta| = 90 - 36.87 = 53.13° \qquad (2\text{-}19)$$

Alternately, the angle could be determined by taking the inverse tangent of 4/3. In this case, the angle *leads* (is more *positive* than) the +sine axis. Thus, the current can be expressed as

$$i(t) = 5\sin(50t + 53.13°) \qquad (2\text{-}20)$$

This example should demonstrate that a sinusoidal function with any arbitrary phase angle can be expressed either as a sine function plus a cosine function, a cosine function with a certain phase angle, or a sine function with a different phase angle.   ■■

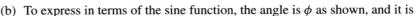

## 2-2  Simplified Initial Approach to Fourier Analysis

We will illustrate the concept of Fourier analysis by studying the effect of adding a number of sinusoids together. For simplicity, we will limit the consideration to three sinusoids, each of which is assumed to be a basic sine function. While we can show only a few cycles of each sinusoid, it will be assumed that they exist for all time, both positive and negative.

First, consider the sine function $v_1(t)$ of Figure 2–6(a), having a peak value of 12 V and a period of 0.1 s. The corresponding frequency $f_1$ is

$$f_1 = \frac{1}{T} = \frac{1}{0.1} = 10 \text{ Hz} \qquad (2\text{-}21)$$

The equation for this function is

$$v_1(t) = 12\sin 2\pi \times 10t = 12\sin 20\pi t \qquad (2\text{-}22)$$

This voltage is expressed as a function of time $t$ and is said to be a representation in the *time domain*.

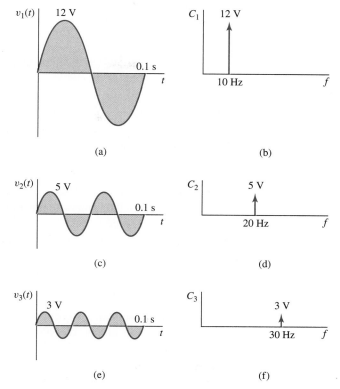

**FIGURE 2–6**
Three time domain sinusoids and their frequency domain spectra.

Next, look at Figure 2–6(b) and observe the graph shown there. In this case, the independent variable is *frequency f,* and the dependent variable is the *amplitude.* For this simple function, the graph consists of a single line of amplitude 12 V at a frequency of 10 Hz. This representation is said to be in the *frequency domain,* and the plot is called the *spectrum* of the signal. Thus, for a time function consisting of a single sinusoid, the spectrum is a single line whose abscissa represents the frequency and whose ordinate represents the amplitude.

Either of the preceding plots describes the major properties of the signal. The time domain plot of Figure 2–6(a) shows us what we would see on an *oscilloscope.* The plot of Figure 2–6(b), however, shows us what we would see on the screen of an ideal *spectrum analyzer,* which is an instrument that displays the frequency components or spectrum of the signal.

To continue this process, observe the function $v_2(t)$ of Figure 2–6(c), in which a sinusoid having an amplitude of 5 V and a period of 0.05 s is shown. The frequency $f_2$ is

$$f_2 = \frac{1}{0.05} = 20 \text{ Hz} \tag{2-23}$$

The time function is

$$v_2(t) = 5\sin(2\pi \times 20t) = 5\sin(40\pi t) \tag{2-24}$$

Next, observe the spectrum of $v_2(t)$, which is shown in Figure 2–6(d). In this case, a single line of amplitude 5 V appears at a frequency of 20 Hz. Compared with the previous case, this component has a higher frequency and a smaller amplitude.

Finally, consider $v_3(t)$ as shown in Figure 2–6(e), having an amplitude of 3 V and a period of 0.0333 s. The frequency is

$$f_3 = \frac{1}{0.0333} = 30 \text{ Hz} \tag{2-25}$$

This time function is

$$v_3(t) = 3\sin(2\pi \times 30t) = 3\sin(60\pi t) \tag{2-26}$$

The spectrum of the third component is shown in Figure 2–6(f). Once again a single line appears, but it is higher in frequency and smaller in amplitude than the preceding two components.

## Forming the Sum of the Components

Now suppose we add the three signals together to form a composite function $v(t)$, as follows:

$$v(t) = 12\sin(2\pi \times 10t) + 5\sin(2\pi \times 20t) + 3\sin(2\pi \times 30t) \tag{2-27}$$

The resulting function as generated by MATLAB over *two cycles* is shown in Figure 2–7.

The spectrum of the composite function is shown in Figure 2–8, and it is simply the sum of the three earlier spectra; that is, it consists of the three lines in the frequency domain.

Suppose that we had started with Figure 2–7, and had not observed the process of adding the components. Without some mystical power, it would have been impossible to determine by a simple inspection that the signal was composed of the three sinusoids with the amplitudes given. However, if the mathematical processes to be provided in subsequent sections were used to analyze the signal, the three components in Figure 2–8 would result. Also, if the waveform were applied to a spectrum analyzer, the screen would display the three components at their respective frequencies.

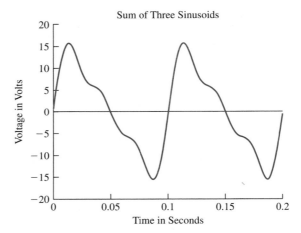

**FIGURE 2–7**
Sum of three sine functions used to illustrate Fourier series.

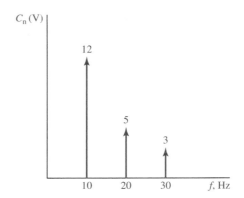

**FIGURE 2–8**
Spectrum of composite signal of Figure 2–7.

## Spectral Analysis

The process of analyzing a complex waveform to determine the frequency components is called *Fourier analysis* or *spectral analysis*. The information determined from a spectral analysis consists of the amplitudes and frequencies of the various components. An additional quantity that can be determined for each component is a *phase angle,* which has not been considered thus far because all the components in our example were basic sine functions.

It can be observed from our special function that the period of the composite function is that of the longest period (0.1 s). The corresponding frequency $f_1 = 10$ Hz is thus the frequency at which the complete waveform repeats itself, and this quantity is called the *fundamental frequency.* The other frequencies are called the *harmonics* of the fundamental frequency, and they are identified by their integer relationship to the fundamental. Thus, the component at $2f_1$ is called the *second harmonic,* and the component at $3f_1$ is called the *third harmonic.*

A word of caution about terminology: some references use the term *first harmonic* to refer to the fundamental since the frequency is $1 \times f_1$. However, this is a contradiction in terms since harmonic refers to a higher frequency. A few references even refer to $n = 2$ as the first harmonic, $n = 3$ as the second harmonic, and so on, but this is very confusing. The majority of communications references tend to avoid reference to a "first harmonic" and go directly from the "fundamental" $(n = 1)$ to the "second harmonic" $(n = 2)$, and that will be the practice in this book.

## Addition of DC Component

The function that was employed in the preceding development consisted only of three sinusoids, and the resulting waveform did not possess a dc level. A practical implication of this fact is that an ideal dc voltmeter connected to the waveform would read zero.

Now suppose we add a constant or dc voltage $v_0$ of value 4 V, that is,

$$v_0(t) = 4 \text{ V} \tag{2-28}$$

The function in the time domain is simply a level of 4 V as shown in Figure 2–9(a), and it is assumed to remain at that level forever. The corresponding spectrum is shown in Figure 2–9(b), and it is simply a line of height 4 V at $f = 0$. Note that a frequency of zero corresponds to dc.

Next, the dc component will be added to the function $v(t)$ to form a new function $v_A(t) = v(t) + 4$. This function is shown in Figure 2–10, and the new spectrum is shown in Figure 2–11.

The shape of $v_A$ is the same as that of $v$, but it is shifted upward by the amount of the dc component. When connected to an ideal dc voltmeter, the reading would be 4 V. The difference in the spectrum is now the fact that a line appears at $f = 0$ of magnitude 4 V.

The fundamental frequency is still considered to be 10 Hz since the waveform repeats at that rate. In other words, the presence of a dc component does not alter the basic definition of the fundamental frequency or the harmonic content.

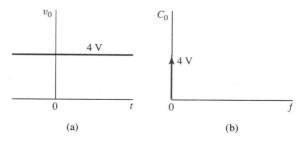

(a)          (b)

**FIGURE 2–9**

Constant (dc) function in time domain and its frequency domain spectrum.

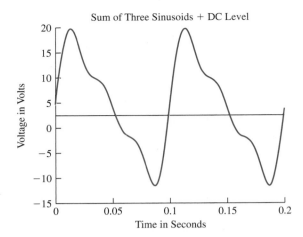

**FIGURE 2–10**
Sum of three sine functions plus a dc component.

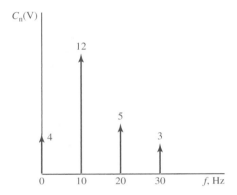

**FIGURE 2–11**
Linear amplitude spectrum of the function of Figure 2–10.

It will be shown later that a necessary and sufficient condition that the dc component be zero is that the net area over one complete cycle of the waveform is zero. (In this context, area above the time axis is positive and area below is negative.)

## Simple Spectral Terms by Inspection

Before getting into the more mathematical process of spectral analysis, we already have the tools to perform a simple spectral form analysis by inspection. This process should always be performed before resorting to the more complex processes to be discussed in the next section. The simple spectral form analysis proceeds as follows:

1. Measure the period $T$ of the waveform, which is the time duration of one cycle. The fundamental frequency $f_1$ of the spectrum is

$$f_1 = \frac{1}{T} \tag{2-29}$$

2. Make a simple inspection to see whether it is obvious if a dc is component present. If all the area is *either* positive *or* negative, there will definitely be a dc component. However, if there is area both above and below the time axis, the situation may not be obvious. In some cases, the best that can be said is there *may* be a dc component.

3. All other frequencies present in the spectrum will be *integer* multiples of the fundamental frequency. Thus, in general, there will be a second harmonic with frequency $2f_1$, a third harmonic with frequency $3f_1$, a fourth harmonic at frequency $4f_1$, and

so on. Certain symmetry conditions may force some of these components to be zero, as we will see later, but the best that can be said at this point is that there *may* be a second harmonic, a third harmonic, and so on.

The preceding process is simple to apply, but it does not tell us the relative magnitudes of the various spectral terms. For that purpose, we must resort to either an analytical formulation using Fourier analysis, a computer-based analysis process, or a spectrum analyzer for laboratory use.

## Bandwidth

A major reason for studying the spectrum of a given signal is to estimate the amount of bandwidth required for transmitting or processing the signal. The bandwidth refers to the range of frequencies over which the signal can be processed without significant alteration of its shape.

From the preceding work, it might appear that an infinite amount of bandwidth would be required if all harmonic terms are present in a signal. However, most real-life waveforms have the property that the amplitudes of the harmonic terms eventually get smaller and smaller as the frequency increases. Thus, in a practical sense, the effect of eliminating higher-numbered harmonics should not drastically alter the shape of the signal.

To illustrate this concept, refer to the two spectral plots shown in Figure 2–12. The fundamental frequency for both plots is 1 kHz. In the plot of 2–12(a), the highest visible component is the fifth harmonic (5 kHz). Unless there is some hidden component off the scale, we might conclude that the signal could be reproduced with a bandwidth extending from dc to a little above 5 kHz.

The plot of Figure 2–12(b) reveals a different story. Over the entire scale provided, all components have significant levels. Unless one knew that the spectrum terminated abruptly above 10 kHz, it is unlikely that we could estimate the bandwidth from the information provided.

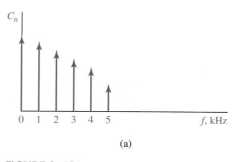    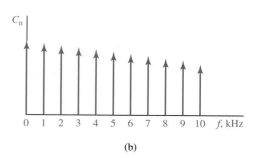

(a)                                    (b)

**FIGURE 2–12**
Two spectral plots illustrating bandwidth considerations.

---

**❙❙ EXAMPLE 2-3**

The waveform of Figure 2–13 represents one cycle of a periodic waveform. By a simple inspection process, list the lowest five frequencies in the spectrum (including dc if present).

**SOLUTION**   Since the positive area in one cycle is clearly much greater than the magnitude of the negative area, there will be a dc component, which will be positive in this case. The period is $T = 12.5 \ \mu s$, and the fundamental frequency is

$$f_1 = \frac{1}{T} = \frac{1}{12.5 \times 10^{-6}} = 80 \times 10^3 \text{ Hz} = 80 \text{ kHz} \qquad (2\text{-}30)$$

All other frequencies will be integer multiples of this frequency. Thus, the lowest five frequencies are 0 (dc), 80 kHz, 160 kHz, 240 kHz, and 320 kHz.

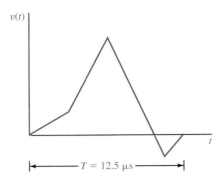

**FIGURE 2–13**
One cycle of the waveform of Example 2-3.

## 2-3  Fourier Series

Now that we have established an intuitive basis for spectral analysis, we are ready to consider the mathematical forms for the Fourier series. In this section, some closely related forms will be considered. They will be denoted as (1) the *quadrature form* and (2) the *amplitude-phase* forms (two variations). At this point, only *periodic* functions of a *deterministic* nature will be considered. Thus, assume a periodic function $v(t)$ having a period $T$.

### Quadrature Form

The *quadrature* form is the one most commonly presented first in circuits and mathematics texts. This form represents a periodic signal $v(t)$ as follows:

$$v(t) = A_0 + \sum_{n=1}^{\infty} (A_n \cos n\omega_1 t + B_n \sin n\omega_1 t) \tag{2-31}$$

where

$$\omega_1 = 2\pi f_1 = \frac{2\pi}{T} \tag{2-32}$$

The $n$th harmonic cyclic frequency is $nf_1$ and the $n$th harmonic radian frequency is $n\omega_1$. As a result of the equalities of Equation 2-32, the arguments of either the sine or the cosine functions may be expressed in either of the following equivalent forms:

$$n\omega_1 t = 2\pi n f_1 t = \frac{2\pi n t}{T} \tag{2-33}$$

The constant term $A_0$ in the series is the *dc term,* and it is the average value of the function over one cycle, that is,

$$A_0 = \frac{\text{net area under curve in one cycle}}{T} = \frac{1}{T} \int_0^T v(t) \, dt \tag{2-34}$$

Formulas for determining $A_n$ and $B_n$ are derived in applied mathematics texts and are summarized here as follows:

$$A_n = \frac{2}{T} \int_0^T v(t) \cos n\omega_1 t \, dt \tag{2-35}$$

for $n \geq 1$, but not for $n = 0$.

$$B_n = \frac{2}{T} \int_0^T v(t) \sin n\omega_1 t \, dt \tag{2-36}$$

for $n \geq 1$. (There is no term corresponding to $n = 0$ for the sine series.)

In all three of the preceding integration formulas, the limits may be changed provided that the integration is performed over one complete cycle. A common alternative range is from $-T/2$ to $T/2$.

The logic behind the designation *quadrature* is the fact that the cosine and sine functions have a phase difference of $90°$. This condition is referred to as one of *phase quadrature*. There are numerous applications in communication systems involving quadrature sinusoidal functions, and some of these will be encountered later in the text.

## Amplitude-Phase Forms

There are two variations on the *amplitude-phase* representation, summarized as follows:

**Cosine Amplitude-Phase Form**

$$v(t) = C_0 + \sum_{n=1}^{\infty} C_n \cos(n\omega_1 t + \theta_n) \tag{2-37}$$

**Sine Amplitude-Phase Form**

$$v(t) = C_0 + \sum_{n=1}^{\infty} C_n \sin(n\omega_1 t + \phi_n) \tag{2-38}$$

The first term $C_0$ is the dc value and is the same as given by Equation 2-34; that is, $C_0 = A_0$. It has been redefined here as $C_0$ in order to maintain a consistent form of notation.

Note that the amplitude $C_n$ at a given frequency is the same for either form, and is always related to the coefficients in the *quadrature* representation by

$$C_n = \sqrt{A_n^2 + B_n^2} \tag{2-39}$$

The only parameter that is different is the phase angle. For the cosine form, the phase angle is

$$\theta_n = \tan^{-1} \frac{-B_n}{A_n} \tag{2-40}$$

For the sine form, the phase angle is

$$\phi_n = \tan^{-1} \frac{A_n}{B_n} \tag{2-41}$$

It should be stressed that the signs of both the numerator and denominator in Equations 2-40 and 2-41 must be carefully noted in order that the angle can be located in the proper quadrant.

## Comments on the Calculation of Fourier Coefficients

For all but the most common waveforms, hand computation of the Fourier coefficients can be a laborious chore. Most of the common waveforms (such as square waves, triangular waves, and the like) have been worked out and are well tabulated. For more complex waveforms, computer techniques can be used. The major approach in this text will be the use of Fourier series forms that are well tabulated, or of MATLAB programs that will be introduced later.

## When Forms Are Equivalent

Many of the common waveforms (such as square waves, triangular waves, and other basic forms) can be placed around the time origin so that the resulting Fourier series will have *either* cosine terms only *or* sine terms only. For either possibility, *the quadrature form and one of the two amplitude-phase forms become equivalent.* Depending on whether it is in cosine terms or sine terms, $C_n = A_n$ or $C_n = B_n$.

## Plot of the Spectrum

Early in this chapter, we examined a spectral plot of a periodic function and observed how it can be used to convey the nature of the spectrum in the frequency domain. From a knowledge of the Fourier series of a given periodic function, the spectrum can be plotted easily. Normally, one would not plot the spectrum from the quadrature form since two amplitude components would be required at each frequency. Rather, one of the two amplitude-phase forms would normally be used.

The most useful spectral plot is that of the *amplitude spectrum,* which is a plot of $C_n$ versus frequency. One could also create a *phase spectrum,* which would be a plot of the phase (either $\theta_n$ or $\phi_n$) as a function of frequency, but this is usually less important than the amplitude spectrum.

Since the amplitude plot of $C_n$ is made only for positive frequencies (plus dc), it is called a *one-sided* spectrum. Later, we will study the concept of a *two-sided* spectrum, which is made for both positive and negative frequencies.

The *amplitude spectrum* can be also be characterized as a *linear amplitude spectrum* since it is normally a plot of voltage or current values. This is in contrast to a *power spectrum,* in which the power in watts associated with each frequency is plotted, and that concept will be considered in the next section. When the term *amplitude spectrum* is used without a modifier, it will be understood to be a *linear* plot of the voltage or current spectrum instead of the power.

When the terms *linear amplitude spectrum* (or simply *amplitude spectrum*) or *power spectrum* are used, they will be interpreted as *one-sided* in form. Whenever spectra defined for both positive and negative frequencies are employed later in the text, the modifier *two-sided* will usually precede the descriptions for clarity.

---

**▌▌ EXAMPLE 2-4**

A certain bandlimited voltage (in volts) has a Fourier series given by

$$v(t) = 5 + 8\cos(2\pi \times 500t + 30°) + 6\cos(2\pi \times 1000t - 45°)$$
$$+ 3\cos(2\pi \times 1500t + 80°) \tag{2-42}$$

(a) List the frequencies (in Hz) contained in the signal. (b) Plot the linear amplitude spectrum. (c) Determine the bandwidth required to transmit the signal.

**SOLUTION**

(a) The signal contains a constant or dc value, so $f = 0$ is a component. The three sinusoidal components each have a $2\pi$ factor included in the arguments, and the corresponding frequencies may be determined by inspection. The three frequencies are 500 Hz, 1000 Hz, and 1500 Hz. The *fundamental* is 500 Hz, and the signal thus contains both a *second* and a *third* harmonic.

(b) The linear amplitude spectrum is shown in Figure 2–14.

(c) From the spectral plot, it is seen that the bandwidth required for signal processing must extend from dc to slightly more than 1500 Hz. To simplify subsequent references to similar situations, we will assume the "slightly more" condition when referring to the highest spectral line, in which case we state for this example that the required bandwidth is 1500 Hz.

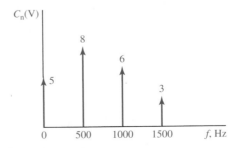

**FIGURE 2–14**
Linear amplitude spectrum of Example 2-4.

## 2-4 Power in a Fourier Series

It is frequently necessary to compute the power associated with a periodic signal. The power contained in the total signal could be of interest, or there may be an interest in the power associated with certain components of the Fourier series.

The forms for *instantaneous power* were given in Equations 2-1 and 2-2, and these results apply to all types of signals. For periodic functions, it is usually the *average power* that is of primary interest.

### Average Power in Time Domain

The definition of average power $P$ for a periodic waveform is determined by integrating the instantaneous power over one cycle and dividing by the period; that is,

$$P = \frac{1}{T} \int_0^T p(t)\, dt \tag{2-43}$$

In practice, the concept of the *effective* or *root-mean-square* (*rms*) value of the waveform is widely used in determining the average power of a periodic waveform. This value is a number that can be used in the same manner as a pure dc value for computing average power produced by the given waveform. The rms voltage $V_{rms}$ for any periodic voltage $v(t)$ can be determined from the following equation:

$$V_{rms} = \sqrt{\frac{1}{T} \int_0^T v^2(t)\, dt} \tag{2-44}$$

For a periodic current, replace $v(t)$ in Equation 2-44 by $i(t)$ and the result will be the rms current $I_{rms}$.

The average power $P$ dissipated in a resistance $R$ by a voltage or current is given by

$$P = \frac{V_{rms}^2}{R} \tag{2-45}$$

or

$$P = I_{rms}^2 R \tag{2-46}$$

### RMS Values from Terms in Fourier Series

First, consider a dc function defined as

$$v_0(t) = C_0 \tag{2-47}$$

The rms value in this case is simply

$$V_{0,rms} = C_0 \tag{2-48}$$

Stated in words, *the rms value of a dc voltage or current is the value of the voltage or current.*

Next, consider any one of the sinusoidal functions in the Fourier series having the form

$$v_n(t) = C_n \cos(n\omega_1 t + \theta) \tag{2-49}$$

The rms value for the sinusoidal function is

$$V_{n,rms} = \frac{C_n}{\sqrt{2}} = 0.7071 C_n \tag{2-50}$$

Finally, consider a complete Fourier series of the form

$$v(t) = C_0 + \sum_{n=1}^{\infty} C_n \cos(n\omega_1 t + \theta_n) \tag{2-51}$$

It can be shown by integration that the rms value of the total waveform is given by

$$V_{rms} = \sqrt{V_{0,rms}^2 + V_{1,rms}^2 + V_{2,rms}^2 + V_{3,rms}^2 + \cdots} = \sqrt{\sum_{n=0}^{\infty} V_{n,rms}^2} \qquad (2\text{-}52)$$

This equation could also be expressed as

$$V_{rms} = \sqrt{C_0^2 + \frac{C_1^2}{2} + \frac{C_2^2}{2} + \frac{C_3^2}{2} + \cdots} = \sqrt{C_0^2 + \frac{1}{2}\sum_{n=1}^{\infty} C_n^2} \qquad (2\text{-}53)$$

A few comments are in order. First, note that we do not simply add the rms values of the components at different frequencies. Rather, we add the *squares* of the respective components and then take the *square-root* of the result. For example, suppose that we have the sum of a dc value of 3 V and a single-frequency sinusoid with an rms value of 4 V. The rms value of the sum is *not* 7 V, but rather it is 5 V. Second, if we don't want to go through the trouble of converting to rms values, we can work with peak values of sinusoids *provided that we divide all peak values squared by* 2, as indicated by Equation 2-53. Of course, the square of the dc value is *not* divided by 2.

## Determining Power Directly from Fourier Series

The perspective thus far has been that of determining an equivalent rms value for the series and then using it to calculate the power according to Equation 2-45 or 2-46. An equally important point of view is to consider directly the power associated with each frequency, and then add the respective powers to determine the net power.

In general, power is a nonlinear variable, and superposition does not apply. However, for certain classes of functions called *orthogonal functions*, the net power associated with the sum is simply the sum of the individual power values. A Fourier series is an example of a set of orthogonal functions, and the net power is the sum of the individual power values.

Assuming a voltage for reference and a resistance $R$, the dc power $P_0$ is

$$P_0 = \frac{V_{0,rms}^2}{R} = \frac{C_0^2}{R} \qquad (2\text{-}54)$$

The average power $P_1$ in the *fundamental* is

$$P_1 = \frac{V_{1,rms}^2}{R} = \frac{(C_1/\sqrt{2})^2}{R} = \frac{C_1^2}{2R} \qquad (2\text{-}55)$$

In general, the normalized power $P_n$ in the $n$th harmonic is

$$P_n = \frac{V_{n,rms}^2}{R} = \frac{(C_n/\sqrt{2})^2}{R} = \frac{C_n^2}{2R} \qquad (2\text{-}56)$$

The total power is the sum of the power values associated with all the components; that is,

$$P = P_0 + P_1 + P_2 + P_3 + \cdots = \sum_{n=0}^{\infty} P_n \qquad (2\text{-}57)$$

## Parseval's Theorem

Parseval's theorem represents a consolidation of some of the preceding results. The average power is given by either of the following forms for a periodic voltage waveform:

$$P = \frac{V_{rms}^2}{R} = \frac{C_0^2}{R} + \frac{1}{2}\sum_{n=1}^{\infty} \frac{C_n^2}{R} \qquad (2\text{-}58)$$

Stated in words, this says that the average power can be computed by working directly with the total rms value, usually obtained in the time domain, or by summing the power values in the frequency domain.

Once again, we will remind ourselves that to adapt the results of Equation 2-58 to a current waveform, all terms would be multiplied by $R$ rather than divided by $R$.

## Power Spectrum

Along with the linear amplitude spectrum discussed in some detail earlier, it is also useful to display the power in the frequency domain. The *power spectrum* is the name associated with a plot of the power versus frequency. The symbol $P_n$ will be used to represent the discrete power associated with the $n$th component of the series.

---

**▐▌ EXAMPLE 2-5**

The voltage waveform of Example 2-4 is applied across a 5-$\Omega$ resistance. Determine (a) the rms value of the voltage, (b) the average power dissipated in the resistance, and (c) the average power associated with each frequency. (d) Plot the power spectrum.

**SOLUTION**    For convenience, the waveform is repeated here and is

$$v(t) = 5 + 8\cos(2\pi \times 500t + 30°) + 6\cos(2\pi \times 1000t - 45°)$$
$$+ 3\cos(2\pi \times 1500t + 80°) \tag{2-59}$$

(a)  The rms value of the dc component is the dc value and is 5 V. Although we could take the peak value of each of the sinusoidal components and divide by $\sqrt{2}$, it is simpler to use the logic of Equation 2-53; that is, take the sum of the squares of the peak values and divide by 2. Hence,

$$V_{rms} = \sqrt{C_0^2 + \frac{1}{2}(C_1^2 + C_2^2 + C_3^2)} = \sqrt{5^2 + \frac{1}{2}(8^2 + 6^2 + 3^2)}$$

$$= \sqrt{25 + 54.5} = \sqrt{79.5} = 8.916 \text{ V} \tag{2-60}$$

(b)  The average power dissipated in the resistance is

$$P = \frac{V_{rms}^2}{R} = \frac{79.5}{5} = 15.9 \text{ W} \tag{2-61}$$

Note that it was not necessary to square 8.916 since we already had its square.

(c)  The dc power is

$$P_0 = \frac{C_0^2}{R} = \frac{5^2}{5} = 5 \text{ W} \tag{2-62}$$

The power in the fundamental is

$$P_1 = \frac{C_1^2}{2R} = \frac{8^2}{2 \times 5} = 6.4 \text{ W} \tag{2-63}$$

The power values for the second and third harmonics are

$$P_2 = \frac{C_2^2}{2R} = \frac{6^2}{2 \times 5} = 3.6 \text{ W} \tag{2-64}$$

$$P_3 = \frac{C_3^2}{2R} = \frac{3^2}{2 \times 5} = 0.9 \text{ W} \tag{2-65}$$

It can be readily verified that the sum of the four individual power values at the four frequencies $(5 + 6.4 + 3.6 + 0.9)$ is equal to the net power of 15.9 W.

(d)  The power spectrum is obtained by simply plotting a line at each of the frequencies, with the height of the line representing the power level associated with that frequency. The plot is shown in Figure 2–15.

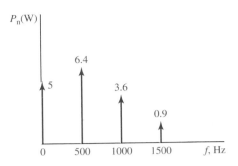

**FIGURE 2–15**
Power spectrum of Example 2-5.

## 2-5   Fourier Series Symmetry Conditions and Convergence

Many common waveforms have certain symmetry conditions that may be used to simplify the determination of the spectra. In fact, these properties may be exploited in certain applications to provide desired outcomes. A compilation of these conditions is provided in Table 2–1, and their practical significance will be discussed in this section. Proofs are given in various applied mathematics texts.

The equations at the top of the table are the various forms of the Fourier series that were discussed in the preceding section, along with the relationships for converting from one form to another. The first row of the table provides the general relationships, which can be used in all cases. Subsequent cases apply whenever the waveform possesses one or more symmetry conditions. For each of the symmetry conditions, the pertinent Fourier coefficients can be determined by integrating over one-half of a cycle and doubling the result. This property can simplify greatly the computation or interpretation of the spectrum in many cases. Each type will now be discussed.

**Table 2–1**   Fourier Series Properties

**Quadrature Form:** $v(t) = A_0 + \sum\limits_{n=1}^{\infty} (A_n \cos n\omega_1 t + B_n \sin n\omega_1 t)$

**Amplitude-Phase Form:** $v(t) = C_0 + \sum\limits_{n=1}^{\infty} C_n \cos(n\omega_1 t + \theta_n) = C_0 + \sum\limits_{n=1}^{\infty} C_n \sin(n\omega_1 t + \phi_n)$

$$\omega_1 = 2\pi f_1 = \frac{2\pi}{T} \qquad C_n = \sqrt{A_n^2 + B_n^2} \qquad \theta_n = \tan^{-1}\frac{-B_n}{A_n} \qquad \phi_n = \tan^{-1}\frac{A_n}{B_n}$$

| Condition | $A_n$ (except $n = 0$) | $B_n$ | Comments |
|---|---|---|---|
| General | $\dfrac{2}{T}\displaystyle\int_0^T v(t)\cos n\omega_1 t\, dt$ | $\dfrac{2}{T}\displaystyle\int_0^T v(t)\sin n\omega_1 t\, dt$ | |
| Even Function $v(-t) = v(t)$ | $\dfrac{4}{T}\displaystyle\int_0^{T/2} v(t)\cos n\omega_1 t\, dt$ | $0$ | Cosine terms only |
| Odd Function $v(-t) = -v(t)$ | $0$ | $\dfrac{4}{T}\displaystyle\int_0^{T/2} v(t)\sin n\omega_1 t\, dt$ | Sine terms only |
| Half-Wave Symmetry $v\left(t + \dfrac{T}{2}\right) = -v(t)$ | $\dfrac{4}{T}\displaystyle\int_0^{T/2} v(t)\cos n\omega_1 t\, dt$ | $\dfrac{4}{T}\displaystyle\int_0^{T/2} v(t)\sin n\omega_1 t\, dt$ | Odd-numbered harmonics only |
| Full-Wave Symmetry $v\left(t + \dfrac{T}{2}\right) = v(t)$ | $\dfrac{4}{T}\displaystyle\int_0^{T/2} v(t)\cos n\omega_1 t\, dt$ | $\dfrac{4}{T}\displaystyle\int_0^{T/2} v(t)\sin n\omega_1 t\, dt$ | Even-numbered harmonics only |

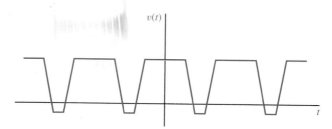

**FIGURE 2–16**
Example of an even function.

## Even Function

A function $v(t)$ is said to be *even* if

$$v(-t) = v(t) \tag{2-66}$$

This means that a function projected for negative time is like a "mirror image" of the function for positive time. An example of an even function is shown in Figure 2–16.

*The Fourier series of an even function has only cosine terms in the quadrature representation.* A summary of the properties follows:

$$B_n = 0$$
$$C_n = A_n \tag{2-67}$$
$$\theta_n = 0$$

## Odd Function

A function $v(t)$ is said to be *odd* if

$$v(-t) = -v(t) \tag{2-68}$$

This means that a function projected for negative time is like an "inverted mirror image" of the function for positive time. An example of an odd function is shown in Figure 2–17.

*The Fourier series of an odd function has only sine terms in the quadrature representation.* A summary of the properties follows:

$$A_n = 0$$
$$C_n = B_n \tag{2-69}$$
$$\phi_n = 0$$

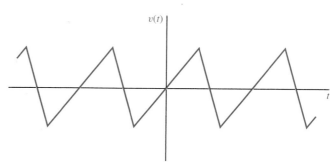

**FIGURE 2–17**
Example of an odd function.

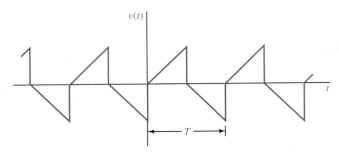

**FIGURE 2–18**
Example of a function with half-wave symmetry.

## Half-Wave Symmetry

A function $v(t)$ is said to possess *half-wave symmetry* if

$$v\left(t + \frac{T}{2}\right) = -v(t) \tag{2-70}$$

This means that if the function is shifted by one-half cycle and flipped over, it will coincide with the original function. An example of a function with half-wave symmetry is shown in Figure 2–18.

*The Fourier series of a function having half-wave symmetry has only odd-numbered frequency components.* This means that, in general, there will be a fundamental ($n = 1$), a third harmonic ($n = 3$), a fifth harmonic ($n = 5$), and so on. However, there will be *no second harmonic, no fourth harmonic,* and so on.

## Full-Wave Symmetry

A function $v(t)$ is said to possess *full-wave symmetry* if

$$v\left(t + \frac{T}{2}\right) = v(t) \tag{2-71}$$

This means that if the function is shifted by one-half cycle, it will coincide exactly with the original function. An example of a function with full-wave symmetry is shown in Figure 2–19.

*The Fourier series of a function having full-wave symmetry has only even-numbered frequency components.* This means that, in general, there will be a dc component and components corresponding to *even* values of $n$.

The idea that there are no odd-numbered components in the spectrum can pose an interesting dilemma, since the fundamental corresponds to $n = 1$. Is there no fundamental component in the series? The answer is based on an interpretation of the meaning of the symmetry statement. If the function repeats itself after an interval of $T/2$ as stipulated by Equation 2-71, the period is not $T$, but is actually $T/2$. Thus, the "true fundamental" is the

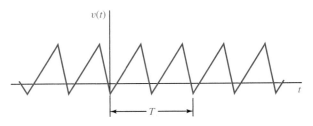

**FIGURE 2–19**
Example of a function with full-wave symmetry.

reciprocal of $T/2$, which is $2/T$ or twice the frequency originally assumed for the fundamental. The reasoning behind this dilemma is that this type of waveform frequently arises in certain processes in which it is desired to maintain an original reference frequency and period. With respect to that reference frequency, if the condition of Equation 2-71 occurs, then there are no odd-numbered components. Examples are a full-wave rectifier circuit and a frequency doubler circuit.

## Even and Odd Points of Confusion

There have been two different uses of the words "even" and "odd" in the preceding discussion, and these uses could be confusing. First, we have *even* and *odd functions,* which are well-established mathematical definitions. Next, we have *even-* and *odd-numbered* frequency components. The two uses are completely independent. For example, we can have an *even function* that has only *odd-numbered* spectral components.

## Effect of DC Component on Symmetry

The presence of a dc component can obscure symmetry and make it harder to detect. For example, consider the waveform of Figure 2–20. It is actually the waveform of Figure 2–17 with a dc component added. The function is no longer an odd function. However, if the dc component is subtracted from the signal, the resulting function will be odd and will have only sine terms.

Another case is shown in Figure 2–21, which is the waveform of Figure 2–18 with a dc component added. The function no longer has half-wave symmetry. However, if the dc component is subtracted from the signal, the resulting function will have half-wave symmetry and will have only odd-numbered spectral components.

When inspecting a waveform for possible symmetry, a useful rule is to mentally shift the waveform up or down to see if symmetry can be established. If so, note the presence of a dc component, and then use the shifted waveform to infer any possible properties associated with the symmetry condition established.

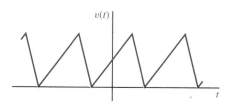

**FIGURE 2–20**
Function of Figure 2–17 with dc component added.

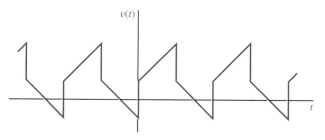

**FIGURE 2–21**
Function of Figure 2–18 with dc component added.

## Spectral Rolloff

Spectral rolloff or convergence is a measure of how fast an amplitude spectrum approaches zero above the dominant frequency range of the spectrum. For many common waveforms, it can be expressed as $1/n^k$, where $n$ is the harmonic number and $k$ is an integer. It can also be expressed as a rolloff rate (or change) of $-6k$ dB/octave, where an octave is a doubling of the frequency.

In general, the smoother a time function, the higher the value of $k$, and the more rapidly the amplitude spectrum approaches zero. The extreme case, of course, is a sinusoidal function, which is arguably the smoothest of all functions and which becomes a single spectral line. For other waveforms, rounded edges and slower transitions between levels take less bandwidth than sudden jumps and jagged edges.

For certain of the very common waveforms, particularly those composed of straight-line segments, the spectral rolloff rate can be predicted by visual inspection. The cases that can be visually determined are summarized as follows: (1) Time domain function has a finite discontinuity or "jump"; (2) time domain function does not have a discontinuity but there is a discontinuity in the slope; and (3) neither the function nor its slope has a finite discontinuity. The convergence or rolloff rates for these three cases are summarized in the table that follows.

| Time Domain Function | Amplitude Spectrum Rolloff |
| --- | --- |
| Function has finite discontinuity | $1/n$ or $-6$ dB/octave |
| Function is continuous but slope has discontinuity | $1/n^2$ or $-12$ dB/octave |
| Neither function nor slope has discontinuity | At least $1/n^3$ or $-18$ dB/octave |

If one could predict the order of the derivative that has a discontinuity, it would be possible to predict the rolloff rate for higher orders, but a visual determination is usually possible only for the cases shown in the table. For the last case, all we can say is that if both the function and its slope (first derivative) have no discontinuities, the rolloff rate will be at least $-18$ dB/octave.

It should be stressed that the *worst-case* condition is the dominating factor. For example, a function may be quite smooth over most of a cycle, but if there is one finite jump, the rolloff rate will be $-6$ dB/octave.

In the next section, a number of common waveforms will be tabulated, and observations concerning the rolloff rates will be made.

---

**▌▌ EXAMPLE 2-6**

Consider the waveform of Figure 2–22. (a) Determine possible symmetry conditions, and list the five lowest frequencies in the spectrum. (b) Determine the spectral rolloff rate.

**SOLUTION**

(a) Over an interval of one cycle, the area above the time axis is equal to the area below the time axis. Hence, the net area is zero, and there is no dc component. The period is 4 ms so the fundamental frequency is

$$f_1 = \frac{1}{T} = \frac{1}{4 \times 10^{-3}} = 250 \text{ Hz} \tag{2-72}$$

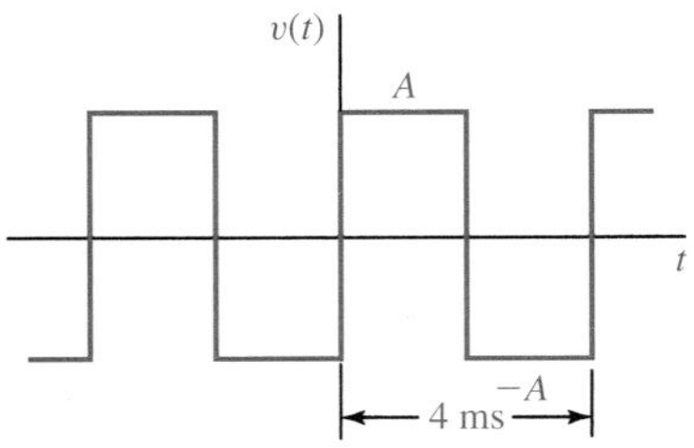

**FIGURE 2–22**
Waveform of Example 2-6.

To test for symmetry, we first ask whether the waveform is an *even* function, an *odd* function, or *neither.* For any negative value of time, it can be seen that the value of the function is an *inverted mirror image* of the function for the corresponding positive value of time. Hence $v(-t) = -v(t)$ and the function is *odd.* Thus, the quadrature form of the Fourier series will have only *sine* terms.

Next, we check for either *half-wave* or *full-wave* symmetry. It can be seen that if we start at any value of time and move over a half-cycle, the function will have the negative of the value at the starting point. Hence, $v(t + T/2) = -v(t)$ and the function has *half-wave symmetry.* Thus, the function will have only *odd-numbered* components.

Putting together the previous deductions, the Fourier series of the function will have *no dc component* and will consist only of *sine terms.* The five lowest frequencies are

250 Hz

750 Hz

1250 Hz

1750 Hz

2250 Hz

(b) The function "jumps" between levels so there is a finite discontinuity. Hence, the amplitude spectrum rolloff rate is $1/n$ or $-6$ dB/octave.

## 2-6   Table of Fourier Series

The focus on Fourier series in this text will be that of interpreting their meaning, and of using tabulated results in the practical analysis and design of communication systems. To that goal, a compilation of the Fourier series for some of the most common periodic waveforms is provided in Table 2–2.

For all of these basic waveforms, it has been possible to set the origin so that the function is either even or odd. However, the magnitudes of spectral components are independent of the time origin, so moving signals to the left and right will affect only the phase and will not affect $C_n$. Likewise, the addition of a dc level to any of these signals will not affect any components except $C_0$.

Some useful data concerning the dc and rms values are provided for each waveform. The rms value provided is the *total* value based on the level as shown. If the level is shifted up or down from the form given in the table, it is necessary to provide some alteration in the values. Assume that the dc values in the table are $V_{dc}$ and $V_{rms}$, respectively. Assume that the waveform has been shifted by an amount $V_{shift}$ (which could be either positive or negative). The modified dc and rms values will be denoted with primes, and they are

$$V'_{dc} = V_{dc} + V_{shift} \tag{2-73}$$

and

$$V'_{rms} = \sqrt{(V_{dc} + V_{shift})^2 + V_{rms}^2 - V_{dc}^2} \tag{2-74}$$

When the dc component of the initial waveform is zero, Equation 2-74 reduces to

$$V'_{rms} = \sqrt{V_{shift}^2 + V_{rms}^2} \quad \text{for } V_{dc} = 0 \tag{2-75}$$

The spectral convergence properties discussed in the last section may be readily observed in the Fourier series pairs shown in Table 2–2. The top entry is a square wave, which has finite discontinuities, and the spectral coefficients follow a $1/n$ form. (The absence of even-numbered harmonics does not change the basic pattern. Those that exist follow the predicted pattern.)

**Table 2–2** Some Common Periodic Waveforms and their Fourier Series

| Signal $v(t)$ | Fourier Series | DC | RMS |
|---|---|---|---|
| **Square Wave** | $\dfrac{4A}{\pi}\left(\cos \omega_1 t - \dfrac{1}{3}\cos 3\omega_1 t \right.$ $\left. + \dfrac{1}{5}\cos 5\omega_1 t - \dfrac{1}{7}\cos 7\omega_1 t \cdots\right)$ | 0 | $A$ |
| **Triangular Wave** | $\dfrac{8A}{\pi^2}\left(\cos \omega_1 t + \dfrac{1}{9}\cos 3\omega_1 t \right.$ $\left. + \dfrac{1}{25}\cos 5\omega_1 t + \cdots\right)$ | 0 | $\dfrac{A}{\sqrt{3}}$ |
| **Sawtooth Wave** | $\dfrac{2A}{\pi}\left(\sin \omega_1 t - \dfrac{1}{2}\sin 2\omega_1 t \right.$ $\left. + \dfrac{1}{3}\sin 3\omega_1 t - \dfrac{1}{4}\sin 4\omega_1 t + \cdots\right)$ | 0 | $\dfrac{A}{\sqrt{3}}$ |
| **Half-Wave Rectified Cosine** | $\dfrac{A}{\pi}\left(1 + \dfrac{\pi}{2}\cos \omega_1 t + \dfrac{2}{3}\cos 2\omega_1 t \right.$ $- \dfrac{2}{15}\cos 4\omega_1 t + \dfrac{2}{35}\cos 6\omega_1 t$ $\left. \cdots (-1)^{\frac{n}{2}+1}\dfrac{2}{n^2-1}\cos n\omega_1 t + \cdots\right)$ $n$ even | $\dfrac{A}{\pi}$ | $\dfrac{A}{2}$ |
| **Full-Wave Rectified Cosine** | $\dfrac{2A}{\pi}\left(1 + \dfrac{2}{3}\cos 2\omega_1 t - \dfrac{2}{15}\cos 4\omega_1 t \right.$ $+ \dfrac{2}{35}\cos 6\omega_1 t - \cdots (-1)^{\frac{n}{2}+1}\dfrac{2}{n^2-1}\cos n\omega_1 t$ $\left. + \cdots\right)$ $n$ even | $\dfrac{2A}{\pi}$ | $\dfrac{A}{\sqrt{2}}$ |
| | $Ad\left[1 + 2\left(\dfrac{\sin \pi d}{\pi d}\cos \omega_1 t \right.\right.$ $+ \dfrac{\sin 2\pi d}{2\pi d}\cos 2\omega_1 t + \dfrac{\sin 3\pi d}{3\pi d}\cos 3\omega_1 t$ $\left.\left. + \cdots\right)\right] \quad d = \tau/T$ | $dA$ | $\sqrt{d}\,A$ |

The triangular wave does not have a discontinuity, but there are abrupt changes in the slope (first derivative); the spectral terms, therefore, converge as $1/n^2$.

Two of the functions have a coefficient pattern of $2/(n^2 - 1)$. For large $n$, this approaches $2/n^2$. The factor "2" is simply a multiplier but the pattern can still be said to be a $1/n^2$ form.

---

**▮▮ EXAMPLE 2-7**

A symmetrical square-wave voltage has no dc component, a peak-to-peak value of 40 V, and a period of 0.2 ms. Assume that the voltage exists across a 50-Ω resistance. (a) Using the results of Table 2–2, tabulate the frequencies, peak values (amplitudes), and average powers of the first three nonzero components in the spectrum. (b) Plot both the linear amplitude spectrum and the power spectrum.

**SOLUTION**

(a) No information is given or requested about the origin or relative time shift of the waveform. Therefore, the basic symmetrical square-wave form given in the table may be assumed. (After all, the frequencies, amplitudes, and power levels of the components are independent of the origin.) The fundamental frequency is

$$f_1 = \frac{1}{T} = \frac{1}{0.2 \times 10^{-3}} = 5 \text{ kHz} \tag{2-76}$$

From the form of the series given, or by observing that the function has half-wave symmetry, it is seen that the spectrum has only odd-numbered harmonics; that is, the harmonics are 15 kHz, 25 kHz, and so on. Since the peak-to-peak value is 40 V, the peak value is 20 V. The magnitude of the fundamental is

$$C_1 = \frac{4A}{\pi} = \frac{4 \times 20}{\pi} = 25.465 \text{ V} \tag{2-77}$$

The third-harmonic term has the value

$$C_3 = \frac{4A}{3\pi} = \frac{4 \times 20}{3\pi} = 8.488 \text{ V} \tag{2-78}$$

The fifth-harmonic term has the value

$$C_5 = \frac{4A}{5\pi} = \frac{4 \times 20}{5\pi} = 5.093 \text{ V} \tag{2-79}$$

For any one of the components, the power $P_n$ is

$$P_n = \frac{C_n^2}{2R} = \frac{C_n^2}{2 \times 50} \tag{2-80}$$

We should remind ourselves that if there had been a dc component, the factor 2 in the denominator of Equation 2-80 would have been eliminated for that term.

The three amplitudes can now be substituted in Equation 2-80 to determine the three power values. All results of this problem are summarized in the table that follows.

| Frequency | Linear Amplitude | Power |
| --- | --- | --- |
| 5 kHz | 25.465 V | 6.485 W |
| 15 kHz | 8.488 V | 0.721 W |
| 25 kHz | 5.093 V | 0.259 W |

(b) Plots of the linear and power spectra are shown in Figures 2–23 and 2–24, respectively.

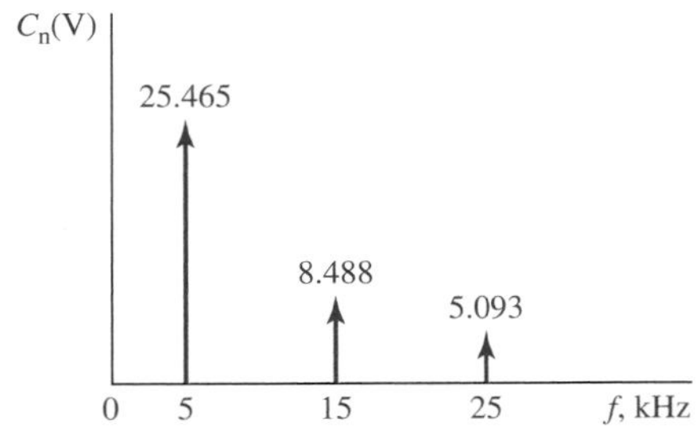

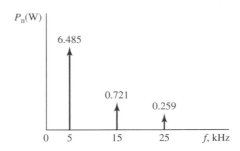

**FIGURE 2–23**
Linear amplitude spectrum of Example 2-7.

**FIGURE 2–24**
Power spectrum of Example 2-7.

---

**▌▌ EXAMPLE 2-8**

For the square wave of Example 2-7, determine the fraction of the total average power contained in the first three spectral components.

**SOLUTION**  The total power can be obtained directly from the rms value provided in Table 2–2. This value is noted from the table to be

$$V_{rms} = A = 20 \text{ V} \tag{2-81}$$

The total power is then

$$P = \frac{V_{rms}^2}{R} = \frac{20^2}{50} = 8 \text{ W} \tag{2-82}$$

Let $P_{135}$ represent the power contained in the fundamental, the third harmonic, and the fifth harmonic, that is, in the first three spectral components. This power is readily determined by summing the three values contained in the table in Example 2-7 or from the power spectrum plot of Figure 2–24.

$$P_{135} = 6.485 + 0.721 + 0.259 = 7.465 \text{ W} \tag{2-83}$$

This sum corresponds to the first three frequency domain terms on the right-hand side of Equation 2-58. In order to determine the total power in the frequency domain, we would have to theoretically sum an infinite number of terms. In practice, however, most of the power in baseband functions is contained in the dc component and a modest number of spectral components.

The percentage of power in the three components of interest is

$$\text{percentage of power} = \frac{P_{135}}{P} \times 100\% = \frac{7.465}{8} \times 100\% = 93.3\% \tag{2-84}$$

Thus, more than 93% of the total power is contained in the first three spectral components.

▌▌

## 2-7 Two-Sided Spectral Form: The Exponential Fourier Series

Both the quadrature and amplitude-phase forms of the Fourier series are *one-sided spectral forms,* meaning that the spectrum is considered to exist only for positive frequencies (and dc, of course). For reasons that will become clearer later, it is desirable in many theoretical developments to consider the concept of a *two-sided spectrum;* that is, one that exists for *both positive and negative frequencies.* Understand that the idea of a *negative frequency* is a sort of mathematical fiction, but it actually serves a useful purpose.

The form of the Fourier series used for the two-sided spectrum is the *exponential form.* This form is primarily used as a bridge to the Fourier transform for nonperiodic waveforms, which will be considered in the next chapter. The treatment here will be somewhat abbreviated, and will focus on interpretation and meaning rather than the detailed mathematical computations.

First, we consider *Euler's formula,* which can be stated as follows:

$$e^{j\omega t} = \cos \omega t + j \sin \omega t \tag{2-85}$$

where $j = \sqrt{-1}$ is the basis for the complex number system widely used in electrical theory.

When the sign of the argument is changed, the function becomes

$$e^{-j\omega t} = \cos \omega t - j \sin \omega t \tag{2-86}$$

If the two preceding equations are alternately added and subtracted, expressions for both cosine and sine functions expressed in terms of complex exponentials can be derived. These expressions are

$$\cos \omega t = \frac{e^{j\omega t} + e^{-j\omega t}}{2} \tag{2-87}$$

and

$$\sin \omega t = \frac{e^{j\omega t} - e^{-j\omega t}}{2j} \tag{2-88}$$

This means that the cosine and/or sine terms in a Fourier series could be regrouped and expressed in terms of complex exponentials. Those exponentials with positive arguments $(j\omega t)$ are considered *positive frequencies,* and the exponentials with negative arguments $(-j\omega t)$ are considered *negative frequencies.* Thus, each spectral component is considered to be composed of a positive frequency component and a corresponding negative frequency component. For example, if the Fourier series has a component at 1 kHz, the exponential form would require a component at 1 kHz and one at $-1$ kHz.

The mathematical form of the exponential Fourier series is

$$v(t) = \sum_{n=-\infty}^{\infty} \overline{V}_n e^{jn\omega_1 t} \tag{2-89}$$

where $\overline{V}_n$ represents a spectral term associated with the $n$th harmonic. Since the series is summed over both positive and negative values of $n$, except for dc, there will always be two terms. One will correspond to a positive value of $n$ and one will correspond to a negative value of $n$. The bar on $\overline{V}_n$ indicates that, in general, it is a complex value having both a magnitude and an angle, or a real and an imaginary part. The equation for determining $\overline{V}_n$ is

$$\overline{V}_n = \frac{1}{T} \int_0^T v(t) e^{-jn\omega_1 t} \, dt \tag{2-90}$$

It is readily apparent that the exponential form of the Fourier series is more compact than the other forms, and that is one of its virtues when dealing with many theoretical developments.

## Simplified Interpretation

For our purposes, we will provide a simplified interpretation. At this point, we will only be concerned with the magnitude of the spectral terms and not the phase. Let $V_n = |\overline{V}_n|$ represent the magnitude of the complex coefficient. The following relationships exist between the magnitudes of the two-sided forms and the one-sided forms:

$$V_n = V_{-n} = \frac{C_n}{2} \quad \text{for } n \neq 0 \tag{2-91}$$

and

$$V_0 = C_0 \tag{2-92}$$

The approach at this level is quite straightforward. With the exception of the dc term, we can think of the process as dividing the one-sided coefficients $C_n$ into two equal parts, placing one at the positive frequency and the other at the corresponding negative frequency. Since the dc term appears only once, it is not altered.

## Two-Sided Power Spectrum

The concept of the two-sided spectral representation can be applied to the power spectrum as well as to the linear amplitude spectrum. With the exception of dc, the power associated with any given frequency is simply divided in half, with one part being assigned to the positive frequency and the other to the negative frequency. The dc power is, of course, not altered.

For the power spectrum, there is a very simple relationship between the magnitude of the exponential coefficient and the power associated with a component in the two-sided power spectrum. Let $P_{2n}$ represent the average power associated with the frequency $nf_1$. This value is

$$P_{2n} = \frac{V_n^2}{R} \tag{2-93}$$

There is no need in this process to modify the magnitude to form an rms value, as in the case of the one-sided form, since the "mathematical bookkeeping" takes care of the process. One simply squares the magnitude of the exponential Fourier coefficient and the result is the contribution of the two-sided power spectrum at the positive frequency $nf_1$. With the exception of dc, however, *the total power at the real frequency $nf_1$ is the sum of the two-sided component at $nf_1$ and the corresponding component at $-nf_1$*. Hence, except for dc,

$$\text{total power at real frequency } nf_1 = \frac{V_n^2 + V_{-n}^2}{R} = \frac{2V_n^2}{R} \quad \text{for } n \neq 0 \tag{2-94}$$

The dc power is

$$\text{dc power} = \frac{V_0^2}{R} \tag{2-95}$$

---

**▌█ EXAMPLE 2-9**

For the functions of Examples 2-4 and 2-5, plot (a) the two-sided linear spectrum, and (b) the two-sided power spectrum.

**SOLUTION** The one-sided linear spectrum was given in Figure 2–14, and the one-sided power spectrum was given in Figure 2–15. All that is required to create the two-sided forms is to divide each positive frequency component in half and display the two resulting values at both the original positive frequency and the image negative frequency. The dc component is not altered. The two-sided linear spectrum is shown in Figure 2–25, and the two-sided power spectrum is shown in Figure 2–26.

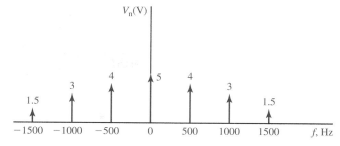

**FIGURE 2–25**
Two-sided linear amplitude spectrum of Example 2-9.

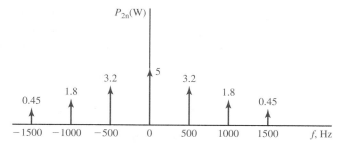

**FIGURE 2–26**
Two-sided power spectrum of Example 2-9.

It can be readily observed that Equation 2-94 is satisfied at each of the components in the figures. Remember, however, that (except for dc) the power associated with a given real frequency is the sum of the positive frequency value and the negative frequency value. For example, the power at 500 Hz is $3.2 + 3.2 = 6.4$ W.

## 2-8  Multisim® Examples (Optional)

Multisim can be used to determine the Fourier spectrum of a waveform appearing in a circuit simulation. In this section we will investigate some of the capabilities of this feature. However, prior to utilizing the Fourier analysis capabilities of Multisim, it is desirable to understand some of the properties of **Transient Analysis.** This important feature of Multisim permits the display of the instantaneous response of a circuit as a function of time. In a sense, the waveform can be considered in much the same manner as an oscilloscopic display. The **Fourier Analysis** option can then be applied to the waveform.

In Multisim Example 2-1, a waveform used early in the chapter to explain Fourier analysis will be generated on an instantaneous basis using **Transient Analysis.** In Multisim Example 2-2, the spectral content of the signal will be displayed using **Fourier Analysis.**

**▌▌ MULTISIM EXAMPLE 2-1**    Consider the voltage of Equation 2-27, which is repeated here for convenience:

$$v(t) = 12 \sin 2\pi \times 10t + 5 \sin 2\pi \times 20t + 3 \sin 2\pi \times 30t \qquad (2\text{-}96)$$

Generate the voltage over two cycles using **Transient Analysis.**

**SOLUTION**    In most applications of **Transient Analysis,** the objective is to determine the unknown response within a circuit excited by a given input waveform, so this particular application is a little unusual in that we know the functional form of the waveform in advance. However, the transient capability can be used to view a waveform on an instantaneous basis, and that is the goal here.

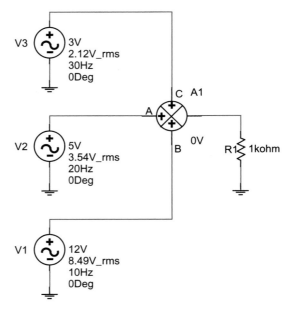

**FIGURE 2–27**
Circuit of Multisim Examples 2-1 and 2-2.

This circuit is the first of many that will utilize **Transient Analysis,** so a detailed explanation is in order. First, any voltage or current sources used must describe the voltage or current on an instantaneous basis as a function of time.

The first step is to create a circuit that will generate the desired waveform, and this process is that of adding three sinusoidal sources together. This process could be achieved by simply connecting the sources in series, but in a practical summing circuit, sources will typically have a common ground. Therefore, the summing module will be used. Refer to Figure 2–27 to support the discussion that follows.

To obtain a summing module, left-click on the **Place Source** icon at the top of the standard components bar. Then left-click on the **CONTROL_FUNCTIONS** option on the left, and highlight the **VOLTAGE_SUMMER** option. Click **OK** and a summing block will appear on the screen.

A double left-click on the summer will open a window with the title **CONTROL_FUNCTION_BLOCKS.** All of the entries may be left at their default positions for this example, but the various available parameters may be changed if necessary.

Left-click below the **Show Signal Source Family Bar.** (This parts bin contains a variety of different types of sources, and many will be utilized throughout the text.) To obtain a sinusoidal source, left-click on the **Place AC Voltage Source** and a sinusoidal source will attach to the arrow. It can then be moved to any desired location and an additional left-click will detach it from the arrow.

Three different sinusoidal voltage sources are obtained by the procedure previously discussed. Of course, it is necessary to set them to the three frequencies and amplitudes required. For each source, double left-click on it and a window with the title **SIGNAL_VOLTAGE_SOURCES** will open. The tab will likely be set at **Value,** but if not, left-click on that tab. Various parameters may then be set in the slots provided. Assuming that the first source is that of the fundamental frequency, set the **Voltage (Pk)** to **12** with the units of **V.**

The **Frequency (F)** slot should be set to **10** with the units of **Hz** for the fundamental. All other variables should be left at their default values.

The preceding procedure is repeated for the other two sinusoidal sources. One is set to a peak value of **5 V** at a frequency of **20 Hz**, and the last one is set to a peak value of **3 V** at a frequency of **30 Hz**. The lower terminal of each source is grounded as shown. To obtain a ground, left-click below the **Show Power Source Family Bar** and then left-click on **Place Ground.** The ground is then carried to the circuit and released by an additional left-click.

As in the original SPICE program, Multisim uses the positive sine function as the basis for any phase angle reference, which represents a 90° phase rotation with respect to the convention of this chapter. Therefore, all three sources in this example have phase angles of 0° since they are all positive sine functions. In the Multisim version used in preparing this text, the phase angles measured from the positive sine axis appear to be defined in a negative sense with respect to that axis. Therefore, to establish the +cosine function, the phase angle should be set to 270°, and to set the −cosine function, the phase angle should be set to 90°. This convention may be reversed by the company in the future, so a simple test should be used to ascertain the reference direction in future updates. In general, the phase angle is set in the **AC Analysis Phase** slot with the units in degrees as indicated by **Deg** to the right of the slot.

The voltage at the summer output (node 4) represents the sum of the three voltages with respect to ground. In order to establish this point, it is necessary to place a junction at this node. Left-click on the **Place** button on the top row, and when the menu opens, left-click on **Junction.** The result is a dot that can be moved to the desired location in the same manner as any other component. It must be connected *by wire* to the upper source output. As always, a ground reference must be established.

To activate a transient analysis, first left-click on **Analyses,** and when the menu window opens, left-click on **Transient Analysis.** A **Transient Analysis** window will then open and the default condition should initially be set to the **Analysis Parameters** tab. The default for the **Initial Conditions** block will probably be set at **Automatically determine initial conditions.** This option is satisfactory for the present problem, since all of the sinusoidal sources start with a value of zero. One must be careful, however, when there are sources in the circuit that could establish initial voltages on capacitors and/or initial currents in inductors. Other options for this condition will be discussed as the need arises.

The **Start time** and the **End time** data blanks are self-explanatory and represent, respectively, the values of time at which observation of the analysis will begin and end. In many situations, the starting time is set simply as 0, as will be the practice in many cases in the text. Other than 0, time values must be entered in the basic units of seconds. In this case the ending time could be expressed as **0.2**. For simulation problems involving very small time values (for example, nanoseconds or microseconds), the floating point notation is much more convenient. Thus, a time value of 2 ns could be entered as **2e-9**.

The next several blanks provide different options for the time steps involved in the numerical mathematical analysis and the associated output. The simplest choice for the beginner is the **Generate time steps automatically** option. For functions that don't change too quickly, this option is usually satisfactory. However, for rapidly changing functions with steep slopes and sharp curves, this option may not always produce the best results. One can always start with this option, but if the waveforms appear to be jagged, one of the other options may be best.

The other two time step options are the **Minimum number of time points** and the **Maximum time step.** Both of these forms are useful when fine resolution is desired, and they will be employed in many examples. Since the time interval is two cycles, the option **Minimum number of time points** was set to **200**, which represents 100 points per cycle at the fundamental frequency. (Later, we will see that this choice is not always practical with modulated waveforms.)

The next step is to left-click on the **Output** tab, and a new window will open for the purpose of selecting the desired variables for viewing on a graph. Several variables can be plotted on the same graph, but to keep the situation as clear as possible, we will

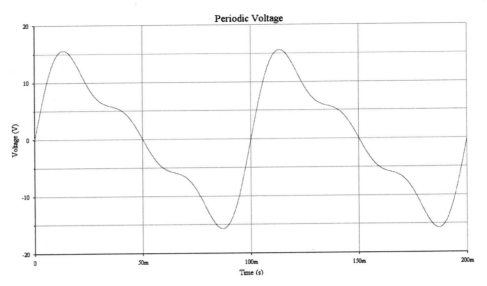

**FIGURE 2–28**
Periodic voltage of Multisim Examples 2-1 and 2-2.

initially choose to plot only one variable on the graph. In this case, only the output voltage will be plotted. Under **Variables in circuit,** left-click on **$4** (representing the voltage at node 4) and then left-click on **Add.** The number will be transferred to the right-hand column under **Selected variables for analysis.** Finally, left-click on **Simulate** and an **Analyses Graphs** window will appear on the screen providing a plot of the voltage at node 1. Many options for altering the form and labeling of the graph are provided.

The waveform of the voltage at node 4 after some alteration is shown in Figure 2–28. Two title lines initially appeared at the top of the figure. The top line is the *page title* and the second line is the *figure title*. The *page title* provides the name (without the extension) of the circuit file that was run to create the waveform.

To modify the *page title,* left-click on the **Page Properties** button and a window with the same name will open. The upper slot is the **Tab Name** and it reflects the type of analysis, which in this case is **Transient Analysis.** The lower slot is the **Title** and it was initially indicated as **02-27,** which was the author's name for the circuit file. In order to avoid confusion with the figure number, it was eliminated and the **Tab Name** was not changed. A left-click on **OK** completes the change to the *page title.*

To change the *figure title,* left-click on the **Properties** button and a window with the title **Graph Properties** will open. Under the **General** tab, the initial **Title** is **Transient Analysis,** and it is changed to **Periodic Voltage.** Various additional options are available. In this case, the **Left Axis** tab is selected and the axis is labeled **Voltage (V).** This version of Multisim uses upper-case **S** as the abbreviation for *seconds,* but it should be lower-case **s.** This is changed by left-clicking on the tab for **Bottom Axis** and making the change. Finally, left-click on **OK** and the changes will take effect. This last change will be made throughout the text, since upper-case **S** is the standard abbreviation for *siemens,* the unit for conductance.

Another useful option is the button with the title **Show/Hide Grid,** and this button allows a grid to be added as was done here.

The process of selecting and deselecting variables for viewing will appear many times throughout the text, and the procedure is generally the same as described here. If more than one variable is to be selected or deselected for a run, hold down the **Ctrl** key on the computer while making the selections.

We end this example with two comments. First, transient analysis is among the most complex forms of analysis with Multisim, since it involves numerical approximation to integration and is more prone to error than many of the other forms of analysis. Therefore, one needs to scrutinize the approach taken on any problem very carefully to ensure that the results can be trusted. Second, all of the options for labeling figures may seem a bit unwieldy, and some have several different ways to activate the operations. Therefore, don't become discouraged if you initially get confused with the different procedures. Simply experiment with the different options and/or use the **Help** file until you understand the steps involved.

---

**▎▎ MULTISIM EXAMPLE 2-2**

The Fourier analysis capability of Multisim will be demonstrated in this example by analyzing the function developed in the preceding example and considered early in the chapter. If necessary, review the form of the function in Equation 2-96, and the resulting circuit for generating the function shown in Figure 2–27. Use the Multisim **Fourier Analysis** operation to determine the spectrum.

**SOLUTION**  It should be clear from the outset that the function in Equation 2-96 is already in a basic Fourier series form, and by inspection, we see that $A_n = 0$ for all $n$, $B_1 = 12$, $B_2 = 5$, $B_3 = 3$, and $B_n = 0$ for $n > 3$. The fundamental frequency is 10 Hz, and the other two nonzero components are the second and third harmonics with frequencies of 20 Hz and 30, respectively. Our purpose is to demonstrate how Multisim analyzes the spectral content.

In general, the first step with Multisim is to develop a circuit that will generate the desired waveform, and that has already been achieved in Multism Example 2-1. To perform the spectral analysis, left-click on **Analyses** and then left-click on **Fourier Analysis.** A window with the title **Fourier Analysis** will then open. If the tab is not already in the **Analysis Parameters** position, left-click on that tab. The **Frequency resolution (Fundamental frequency)** is the spacing between successive frequency components, and we know that this value should be set at **10 Hz.** In general, this value should be set to the fundamental frequency, which is the reciprocal of the period.

The default value for the **Number of harmonics** is **9**. While it is expected that the given waveform only has 3 components, we will leave the number at **9** to see what happens.

The **Stop time for sampling** is set to 0.2 s. This value must be *at least the duration of one cycle.* If there is a transient interval that is different than the steady-state response, there should be at least one cycle in the steady-state form. The program will use the last full cycle at the fundamental frequency to perform the spectral analysis. While there is no transient or settling interval for this function, the stopping time corresponds to two cycles. It is recommended that an integer number of cycles be used when it is feasible.

Under **Results,** the default value on the left is a check for **Display as bar graph.** On the right under **Display,** the default is **Chart and Graph,** and under **Vertical scale,** it is **Linear.** These are the settings for the run that will be shown, but the reader may wish to experiment with different options.

Next, left-click on the **Output** tab and set the output on the right-hand side to node **$4** using the same procedure utilized in other examples. Left-click on **Simulate** and both a table and a graph will be displayed, as shown in Figure 2–29. Some alterations to the display were made using procedures discussed in the preceding example.

The graph at the bottom clearly depicts the spectrum as we would expect to see it. The values in the table must be interpreted properly. Since numerical integration is performed, there are slight variances from the actual values of the fundamental, second harmonic, and third harmonic, but they are insignificant. Note, however, that although the table provides values for all the components, the magnitudes are negligible above the third harmonic. The

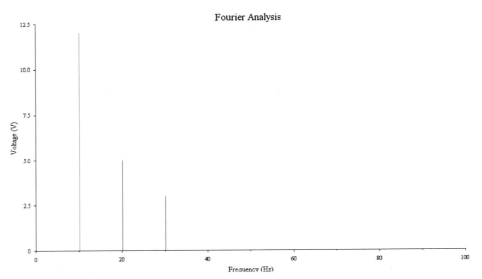

```
Fourier analysis for V(4):
DC component: -8.7087e-007

  No. Harmonics: 9, THD: 48.5222 %, Gridsize: 256, Interpolation Degree: 1

Harmonic Frequency    Magnitude   Phase        Norm. Mag    Norm. Phase
-------- ---------    ---------   -----        ---------    -----------
   1       10         11.9961     -1.7019e-005 1            0
   2       20         4.99343     -3.5492e-005 0.416256     -1.8473e-005
   3       30         2.99113     -4.8872e-005 0.249343     -3.1853e-005
   4       40         1.74175e-006 -84.375     1.45193e-007 -84.375
   5       50         1.23566e-005 170.393     1.03006e-006 170.393
   6       60         1.18281e-005 171.04      9.86002e-007 171.04
   7       70         1.06725e-005 172.238     8.89663e-007 172.238
   8       80         1.74175e-006 -78.75      1.45193e-007 -78.75
   9       90         0.0001131    -35.812     9.42807e-006 -35.812
```

**FIGURE 2–29**
Fourier analysis data of Multisim Example 2-2.

phase values of these components are, therefore, meaningless. In addition, the dc component is negligible.

The quantity THD needs some explanation. This quantity represents *total harmonic distortion,* and is the ratio of the rms sum of all harmonics divided by the fundamental component expressed as a percentage. This measure is primarily a figure of merit for an amplifier and is used to estimate the amount of distortion. For a function having a significant amount of harmonic content, it is simply a measure of the relative size of the harmonics.

## 2-9  MATLAB® Examples (Optional)

MATLAB® contains many features that can be employed in the determination of Fourier series, and in plotting both time and frequency domain functions. We will explore a few of these features.

The first example (MATLAB Example 2-1) will demonstrate how a function can be readily plotted using basic MATLAB operations. In the particular example employed, it will be a voltage expressed as a function of time, consisting of a finite number of terms in a Fourier series. However, the concept may be used to plot any functions that can be described on a point-by-point basis or by standard equations.

MATLAB is matrix oriented, and some of the terminology and references utilize matrix forms. One need not have a deep understanding of matrix theory to use MATLAB, but a modest amount of matrix terminology is helpful in relating to the various operations that will be used throughout the text.

## Matrix Definitions

A matrix of order $m$, $n$ will be momentarily denoted $[A]_{m,n}$, and is a rectangular array of numbers containing $m$ rows and $n$ columns. The general form can be expressed as

$$[A]_{m,n} = \begin{bmatrix} a_{11} & a_{12} & a_{13} & \cdots & a_{1n} \\ a_{21} & a_{22} & a_{23} & \cdots & a_{2n} \\ \vdots & \vdots & \vdots & & \vdots \\ a_{m1} & a_{m2} & a_{m3} & \cdots & a_{mn} \end{bmatrix} \tag{2-97}$$

If $m = n$, the matrix is said to be a *square matrix*.

If $m = 1$, the matrix is said to be a *row matrix* or *row vector*, and could be expressed as

$$[A]_{1,n} = [a_{11} \quad a_{12} \quad a_{13} \quad \cdots \quad a_{1n}] \tag{2-98}$$

If $n = 1$, the matrix is said to be a *column matrix* or *column vector*, and could be expressed as

$$[A]_{m,1} = \begin{bmatrix} a_{11} \\ a_{21} \\ \cdots \\ a_{m1} \end{bmatrix} \tag{2-99}$$

For the applications contained in this text, virtually all arrays of numbers will be either row vectors or column vectors, with the former being the most common. Therefore, the matrix terminology will be simplified, and the form of each array will be noted as it arises.

## Entering Matrices and Vectors in MATLAB

When a matrix is to be entered into MATLAB, the values are entered on a row-by-row basis. Elements on each row are separated by either a *space* or by a *comma*. Rows are separated by a *semicolon*.

Assume, for example, that a row vector **t** is to be created with the following values: 0, 1, 3, 6. The MATLAB command is

```
>> t = [0 1 3 6]
```

where a space has been used between entries (alternately, a comma could have been used). If entered, MATLAB memory would now contain a $1 \times 4$ matrix **t** with the four values expressed as a row.

Suppose we had desired to establish the preceding values as a column vector. This could have been achieved by the command

```
>> t = [0; 1; 3; 6]
```

If entered, MATLAB would now contain a $4 \times 1$ matrix with the four values expressed as a column.

## Transpose

A row vector may be converted to a column vector by the use of the *transpose*, which is achieved in MATLAB by the apostrophe ('). In general, the transpose of a matrix is a new matrix obtained by interchanging the rows with the columns. Thus, consider the expression

```
>> t = t'
```

If entered, the new value of **t** is redefined as the transpose of the previous value, meaning that a row vector will be converted to a column vector and vice versa.

## Developing an Equally Spaced Vector

We have seen that vectors can be created by placing the values between brackets. Consider now the possibility of generating a large number of values of time at which a function is to be evaluated, or a similar process with frequency values. For linear spacing, there are two possible simple commands that can be used to generate a large number of values in one step. We will illustrate with a time vector **t**. Let

t1 = starting time

t2 = ending time

tstep = increment between successive times

The following command will generate a row vector containing all the values:

```
>> t = t1:tstep:t2
```

The number of points $n$ generated by this process is

$$n = \frac{t_2 - t_1}{t_{\text{step}}} + 1 \tag{2-100}$$

Note the additional "+1". This is necessary to ensure that both the beginning and end times are covered. For example, if we step from 0 to 2 in steps of 0.5, the values are 0, 0.5, 1, 1.5, and 2. Thus, 5 values are generated and this is readily predicted from Equation 2-100.

An alternate command that will achieve the same objective is the **linspace** (linear spacing) operation. The format for that operation is

```
>> t = linspace(t1,t2,n)
```

---

**▐▌ MATLAB EXAMPLE 2-1**

The Fourier series of a certain bandlimited voltage was considered in Example 2-4 and given by Equation 2-42. Use MATLAB to plot the function over two cycles.

**SOLUTION**   For convenience, the function in the standard mathematical form of Equation 2-42 is repeated, as follows:

$$v(t) = 5 + 8\cos(2\pi \times 500t + 30°) + 6\cos(2\pi \times 1000t - 45°)$$

$$+ 3\cos(2\pi \times 1500t + 80°) \tag{2-101}$$

The voltage consists of a dc component, a fundamental at 500 Hz, a second harmonic at 1000 Hz, and a third harmonic at 1500 Hz. This information, of course, has already been deduced in Example 2-4.

The form of Equation 2-101, in which radians and degrees are mixed in the arguments of the sinusoidal functions for clarity, is very common in engineering texts. However, the first term in each of the arguments—that is the $\omega t$ term—is expressed in radians, and the second term is expressed in degrees. Therefore, before performing any numerical computations with the functions, it is necessary to convert the angles in degrees to radians. In general, this can be accomplished by the following conversion:

$$\text{angle in radians} = \left(\frac{\pi}{180}\right) \times \text{angle in degrees} \tag{2-102}$$

After conversion of the three phase angles, the function may be rewritten as

$$v(t) = 5 + 8\cos(2\pi \times 500t + \pi/6) + 6\cos(2\pi \times 1000t - \pi/4)$$
$$+ 3\cos(2\pi \times 1500t + 4\pi/9) \tag{2-103}$$

Now let us adapt the equation for MATLAB processing. The first step is to define a row vector that establishes the values of time at which the function is to be evaluated. The fundamental frequency is $f_1 = 500$ Hz, and the period is $T = 1/f_1 = 1/500 = 0.002$ s $= 2$ ms. Two cycles would then correspond to 4 ms.

As an arbitrary but reasonable assumption, assume a time step of 10 μs. A command to generate the row vector is

```
>> t=0:10e-6:4e-3;
```

It can be readily shown that this operation results in 401 values of time.

An alternate approach is to use the **linspace** command, and this operation is

```
>> t=linspace(0,4e-3,401);
```

Both forms produce identical results so it is a matter of choice. Note that the semicolon has been placed at the end of both of the preceding commands. Without it, 401 values of time would sweep down the screen.

Evaluation of the voltage could be resolved into several steps if desired. However, we will choose to carry out the operation in one big sweep. Within MATLAB, the symbol for $\pi$ is **pi.** Refer back to Equation 2-103 and note how the form is adapted to MATLAB in the command that follows.

```
>> v = 5+8*cos(2*pi*500*t+pi/6)+6*cos(2*pi*1000*t-pi/4)
       +3*cos(2*pi*1500*t+4*pi/9);
```

Since **t** in the argument is a row vector containing 401 points, the function **v** is computed at all points and will result in a 401-point row vector. Again, the semicolon suppresses the screen display of the 401 values.

Next, we will plot the function. A linear plot is generated by the command

```
>> plot(t,v)
```

where the first variable is the independent (horizontal) variable and the second is the dependent (vertical) variable. MATLAB will then generate a plot of **v** versus **t** with automatic scaling. Many different adjustments and labels may be added to the figure. At this point we will consider a few of the most basic.

First, a grid will be added by the command

```
>> grid
```

Next, the horizontal axis will be labeled by the command

```
>> xlabel('Time, seconds')
```

The vertical axis will be labeled by the command

```
>> ylabel('Voltage, volts')
```

Note that it doesn't matter what the actual horizontal and vertical variables are for the labeling commands. The horizontal labeling is achieved with **xlabel** and the vertical labeling is achieved with **ylabel.**

Finally, a title is provided with the command

```
>> title('Voltage of MATLAB Example 2-1')
```

The resulting function over two cycles is shown in Figure 2–30.

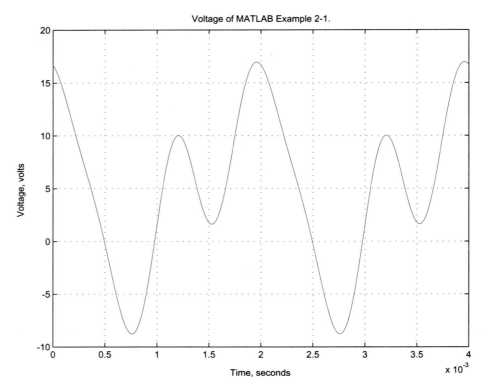

Voltage of MATLAB Example 2-1.

**FIGURE 2–30**
Voltage of MATLAB Example 2-1.

---

**MATLAB EXAMPLE 2-2**

Use the M-file program **fourier_series_1** to determine the spectrum of the square wave of Example 2-7.

**SOLUTION**   Before using the program, the reader should refer to Appendix A for a general discussion. Prior to executing the program, the time function must be entered as a row vector. Although the function is shown in the table as an even function of time, for the purpose of determining the amplitude spectrum, this is unnecessary, and it is easier to think of the square wave as having a level of 20 V for the first half-cycle and a level of –20 V for the second half-cycle. An arbitrary but reasonable decision was made to employ 200 points in the complete cycle. The fundamental frequency is 5 kHz, and the period is $1/5 \times 10^3 = 200$ μs. With 200 points, the time step is 200 μs/200 = 1 μs = 1e-6 s.

The easiest way to generate the time function is to use the "ones" command. A matrix containing m rows and n columns of ones can be generated by the instruction

```
ones(m,n);
```

Although we don't need it for this example, a similar process can be used to generate a matrix of zeros with the command

```
zeros(m,n)
```

First, we will create 100 values of a row matrix **first_half** by the command

```
>> first_half=20*ones(1,100);
```

The resulting matrix is a row vector containing 100 points with the value 20 at each point. Next we will create 100 values of a row matrix **second_half** by the command

```
>> second_half=-20*ones(1,100);
```

MATLAB permits the combining of matrices to form a larger matrix if the combination makes sense. We will thus form a 200-point row matrix v by the operation

```
>> v=[first_half second_half]
```

Alternately, the entire time function could have been created in one step by the command

```
>> v=[20*ones(1,100) -20*ones(1,100)]
```

With the 200-point row vector for **v** in memory, the M-file is run by the command

```
>> fourier_series1
```

The following dialogue now appears on the screen.

**The function v must already be in memory.**
**It should be a row vector defined over one cycle.**

**The number of points in one cycle is 200.**

**Enter the time step between points in seconds.**

The step of **1e-6** is typed on the screen and entered. The screen dialogue follows.

**The fundamental frequency is 5000 Hz.**

**The highest unambiguous frequency is 500000 Hz.**

**Enter integers requested as multiples of the fundamental.**
**For example, dc would be entered as 0.**
**The fundamental would be entered as 1.**
**The 2nd harmonic would be entered as 2, etc.**
**The highest integer that should be entered is 99.**

**Enter the lowest integer for plotting.**

Although the dc component is zero, we will choose to have the plot begin at dc, so the value entered should be 0. The next dialogue is

**Enter the highest integer for plotting.**

While we could enter a value as high as 99, corresponding to the 99th harmonic, the resulting plot would be very crowded. We will choose to enter the value 10. Following this step, the program computes the approximate amplitude spectrum desired, and a figure displaying the components requested appears on the screen. The result in this case is shown in Figure 2–31. Note the absence of the dc component and of the even harmonics, as expected.

It is of interest to compare the exact values of Example 2-7 with the approximate values obtained using the FFT algorithm. To determine the computed values more accurately than can be seen on the curve, it is necessary to request specific values of C for this purpose. The indices previously entered on the computer were set up so that dc corresponds to 0, the fundamental corresponds to 1, and so on. However, if we want to index the variables in memory, 1 is the lowest value and it corresponds to dc, 2 is the fundamental, 3 is the second harmonic, and so on. In Example 2-7, we computed values up to the fifth harmonic. To have the corresponding values from the FFT program listed on the screen, we can enter the statement

```
>> C(1:6)

ans =

  Columns 1 through 5

     0   25.4658      0   8.4914      0

Column 6

  5.0982
```

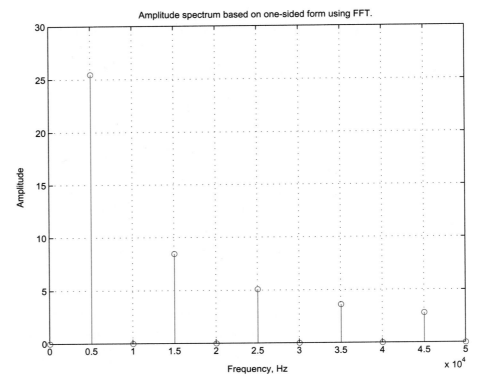

**FIGURE 2–31**
Amplitude spectrum of MATLAB Example 2-2.

A short table follows, providing a comparison between the exact values of Example 2-7 and the FFT-based values of this example.

| Frequency, kHz | Exact Value, V | Value from FFT, V |
|---|---|---|
| 0 | 0 | 0 |
| 5 | 25.465 | 25.466 |
| 10 | 0 | 0 |
| 15 | 8.488 | 8.491 |
| 20 | 0 | 0 |
| 25 | 5.093 | 5.098 |

It is obvious that the values obtained using the FFT-based MATLAB program are very close to the exact values. It should be noted in general that the results tend to be closer as the sampling time interval decreases. With the scaling provided in the Fourier analysis program, the FFT can be thought of, in some sense, as an approximation to numerical integration.

The program **fourier_series_2.m** is very similar to **fourier_series_1.m** and has a similar dialogue. The only difference is that **fourier_series_2.m** provides magnitude coefficients for dc and positive frequencies based on the two-sided exponential Fourier series, rather than for the amplitude-phase form. Only the positive frequency range is plotted, but the negative frequency range is simply a mirror image of the positive frequency range.

In subsequent examples using **fourier_series_1** or **fourier_series_2,** the dialogue expanded in this example will usually be omitted, and only the major parameters required for the program and the results will be provided.

Insert the text CD in a computer having SystemVue™ installed and activate the program. Open the CD folder entitled **SystemVue Systems** and load the file entitled 2-1.

## Description of System

The system shown on the screen is the one used for the chapter opening application. However, you will be modifying this file to create a different waveform, and you may choose to save the new file on a hard drive or a different disc. (Of course, the original file 2-1 will remain on the text CD.) Again, there are five sinusoidal sources shown on the left. The source at the top has a frequency of 1 Hz, and the additional sources have frequencies that are odd integer multiples of 1 Hz: 3 Hz, 5 Hz, and so on.

The sources are all connected to the adder token in the middle. This block creates the sum of the values for all the sources connected to it. The sink displays the result.

Now that you have completed the chapter, you may wish to obtain an initial run to observe how the five components provide an approximation to a square wave. This is achieved by a click on the **Run System** button on the toolbar. The approximation shows some ripple, and there is a significant rise time.

## Creating a Triangular Waveform

You will now modify the sinusoidal coefficients to create an entirely different function. Refer to Table 2–2 and note the Fourier series of the triangular wave. The function in the table is *even,* which would require that cosine functions be used if the origin as given were to be maintained. While we could modify the sources for this purpose, we will instead shift the origin to the left by a quarter of a cycle so that the function becomes *odd.* We will then be able to synthesize the waveform using only sine functions.

There is one nuance that must be recognized with the origin shift. It turns out the signs of the third and seventh harmonics must be changed from positive to negative. (If more harmonics were used, the sign change would be required at the eleventh, the fifteenth, etc.)

The coefficients of each of the components must be changed. Based on the pattern shown in Table 2–2, choose $A = 1$. Then compute the values of all harmonics up through the ninth, with the understanding that the signs of the third and seventh must be negative. In case there is any uncertainly concerning components above the fifth, the denominator factors are based on $n^2$ as observed in Table 2–2. You may enter the values in MATLAB format if you desire. For example, the third harmonic could be entered as -8/(9*pi^2). After you click **OK,** the value will be calculated to a large number of significant digits.

Click on the **Run System** button and observe the waveform. Maximize the waveform so that the graph occupies the full screen. If all procedures have been correctly followed, the function obtained should closely approximate a triangular waveform over two cycles. The triangular wave does not have a discontinuity, so the spectrum converges much more rapidly than that of a square wave.

Starting with the ninth harmonic, successively remove each higher-order harmonic, and observe the resulting waveform after each change. The easiest way to disconnect when there is more than one input to a token is to first left-click on the button with the name **Disconnect Tokens** near the left-end of the **Tool Bar.** A window will open providing the procedure. First, you left-click on the **From Token** and then you left-click on the **To Token.**

As you start to make a run with one or more tokens disconnected, a window will open alerting you to that fact. Just left-click on **Run System** in the window that opens, and the simulation will be performed anyway.

Even when you get to the point where the waveform has only a fundamental and a third harmonic, the resulting waveform is still reasonably close to the triangular form.

## PROBLEMS

**2-1**   A voltage (in volts) is given by

$$v(t) = 160 \cos 2\pi \times 500t$$

Determine (a) peak value, (b) radian frequency, (c) cyclic frequency, and (d) period.

**2-2**   A current (in amperes) is given by

$$i(t) = 7 \sin(2000t + 30°)$$

Determine (a) peak value, (b) radian frequency, (c) cyclic frequency, and (d) period.

**2-3**   A current (in amperes) is given by

$$i(t) = 0.01 \sin(10^6 t - 120°)$$

Determine (a) peak value, (b) radian frequency, (c) cyclic frequency, and (d) period.

**2-4**   A voltage (in volts) is given by

$$v(t) = 50 \cos(2\pi \times 2 \times 10^4 t + 150°)$$

Determine (a) peak value, (b) radian frequency, (c) cyclic frequency, and (d) period.

**2-5**   A voltage (in volts) is given by

$$v(t) = 12 \cos 200t - 5 \sin 200t$$

Express the voltage as (a) a cosine function and (b) a sine function.

**2-6**   A current (in amperes) is given by

$$i(t) = -4 \cos 50t - 3 \sin 50t$$

Express the current as (a) a cosine function and (b) a sine function.

**2-7**   A current (in amperes) is given by

$$i(t) = 12 \cos(3000t + 30°)$$

Express the current as the sum of a cosine function and a sine function.

**2-8**   A voltage (in volts) is given by

$$v(t) = 75 \sin(10^6 t - 60°)$$

Express the voltage as the sum of a cosine function and a sine function.

**2-9**   The waveform in the following figure represents *one cycle* of a periodic waveform. By a simple inspection process, list the lowest five frequencies (including dc if present) in the spectrum.

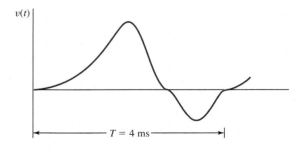

**2-10**   The waveform in the following figure represents *one cycle* of a periodic waveform. By a simple inspection process, list the lowest five frequencies (including dc if present) in the spectrum.

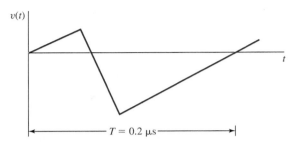

**2-11**   A periodic bandlimited signal has a Fourier series given by

$$v(t) = 12 + 20 \cos(200\pi t - 30°) + 18 \cos(400\pi t + 60°)$$

(a)  List the frequencies contained in the signal. (b) Plot the linear amplitude spectrum.

**2-12**   A periodic bandlimited signal has a Fourier series given by

$$v(t) = 20 + 16 \sin(10^4 t - 120°) + 12 \sin(2 \times 10^4 t + 45°)$$

(a) List the frequencies contained in the signal. (b) Plot the linear amplitude spectrum.

**2-13**   For the function of Problem 2-11, and a 50-Ω resistance, determine the (a) rms value, (b) total average power, and (c) average power associated with each frequency. (d) Plot the power spectrum.

**2-14**   For the function of Problem 2-12, and a 50-Ω resistance, determine the (a) rms value, (b) total average power, and (c) average power associated with each frequency. (d) Plot the power spectrum.

In Problems 2-15 through 2-20, you are not asked to compute the Fourier coefficients, but rather to determine as much information as possible from the symmetry conditions and other information provided by the figures. For each waveform, determine the following information: (a) whether there is a dc component, (b) whether the quadrature form of the series has *only cosine terms, only sine terms,* or *both,* (c) whether the series has *only odd-numbered harmonics, only even-numbered harmonics,* or *both,* (d) the spectral rolloff rate in decibels/octave, and (e) the lowest frequency (other than dc) that does not have an amplitude of zero. Assume that all waveforms continue indefinitely with the pattern shown.

**2-15**   See the figure that follows.

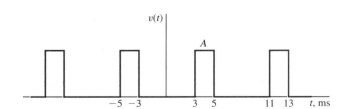

**2-16**  See the figure that follows.

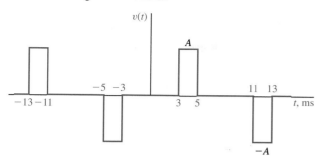

**2-17**  See the figure that follows.

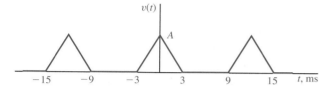

**2-18**  See the figure that follows.

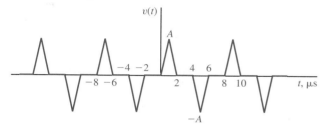

**2-19**  See the figure that follows.

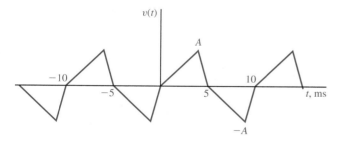

**2-20**  See the figure that follows.

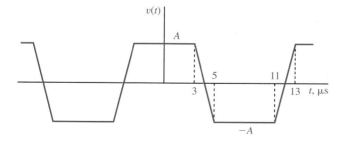

**2-21**  A symmetrical triangular voltage has no dc component, a peak-to-peak voltage of 40 V, and a period of 1 ms. Assume the voltage is applied across a 50-Ω resistance.

(a) Using the results of Table 2–2, tabulate the frequencies, peak values (amplitudes), and average powers of the first three nonzero components in the spectrum. (b) Plot both the linear amplitude spectrum and the power spectrum.

**2-22**  A symmetrical sawtooth voltage has no dc component, a peak-to-peak voltage of 40 V, and a period of 4 ms. Assume the voltage is applied across a 50-Ω resistance. (a) Using the results of Table 2–2, tabulate the frequencies, peak values (amplitudes), and average powers of the first three nonzero components in the spectrum. (b) Plot both the linear amplitude spectrum and the power spectrum.

**2-23**  A half-wave rectified current has a peak value of 20 A and a period of 2 ms. Assume the current is flowing in a 50-Ω resistance. (a) Using the results of Table 2–2, tabulate the frequencies, peak values (amplitudes), and average powers of the first four nonzero components in the spectrum. (b) Plot both the linear amplitude spectrum and the power spectrum.

**2-24**  A full-wave rectified current has a peak value of 20 A and a period of 2 ms. Assume the current is flowing in a 50-Ω resistance. (a) Using the results of Table 2–2, tabulate the frequencies, peak values (amplitudes), and average powers of the first four nonzero components in the spectrum. (b) Plot both the linear amplitude spectrum and the power spectrum.

**2-25**  Consider the square-wave voltage shown below, and assume that it is applied across a 50-Ω resistance. (a) Using the results of Table 2–2, tabulate the frequencies, peak values (amplitudes), and average powers of the first four nonzero components in the spectrum. (b) Plot both the linear amplitude spectrum and the power spectrum. (*Hint:* Separate the signal into a dc component and a function of a form given in Table 2–2.)

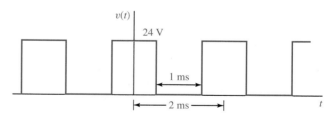

**2-26**  Consider the triangular-wave voltage shown below, and assume that it is applied across a 50-Ω resistance. (a) Using the results of Table 2–2, tabulate the frequencies, peak values (amplitudes), and average powers of the first four nonzero components in the spectrum. (b) Plot both the linear amplitude spectrum and the power spectrum. (*Hint:* Separate the signal into a dc component and a function of a form given in Table 2–2.)

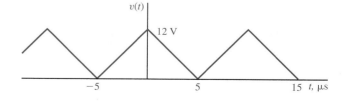

**2-27** Consider the square-wave voltage shown below, and assume that it is applied across a 50-$\Omega$ resistance. (a) Using the results of Table 2–2, tabulate the frequencies, peak values (amplitudes), and average powers of the first four nonzero components in the spectrum. (b) Plot both the linear amplitude spectrum and the power spectrum. (*Hint:* Separate the signal into a dc component and one without a dc component. Since none of the computations involve the signs of the amplitude terms or the phase spectrum, the desired results can be deduced from one of the functions in Table 2–2.)

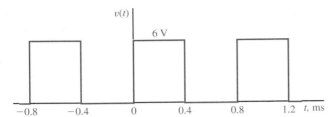

**2-28** Consider the triangular voltage shown as follows, and assume that it is applied across a 50-$\Omega$ resistance. (a) Using the results of Table 2–2, tabulate the frequencies, peak values (amplitudes), and average powers of the first four nonzero components in the spectrum. (b) Plot both the linear amplitude spectrum and the power spectrum. (*Hint:* Separate the signal into a dc component and one without a dc component. Since none of the computations involve the signs of the amplitude terms or the phase spectrum, the desired results can be deduced from one of the functions in Table 2–2.)

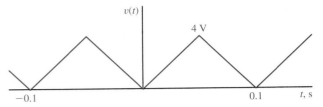

**2-29** For the triangular wave of Problem 2-21, determine the fraction of the total power contained in the first three spectral components.

**2-30** For the sawtooth wave of Problem 2-22, determine the fraction of the total power contained in the first three spectral components.

**2-31** For the square wave of Problem 2-25, determine the fraction of the total power contained in the first four spectral components.

**2-32** For the triangular wave of Problem 2-26, determine the fraction of the total power contained in the first four spectral components.

**2-33** For the function of Problems 2-11 and 2-13, plot (a) the two-sided linear amplitude spectrum, and (b) the two-sided power spectrum.

**2-34** For the function of Problems 2-12 and 2-14, plot (a) the two-sided linear amplitude spectrum, and (b) the two-sided power spectrum.

# Spectral Analysis II: Fourier Transforms and Pulse Spectra

**3**

The Fourier series approach developed in the last chapter permits the determination of the spectrum of a periodic function. However, most signals encountered in communications systems are nonperiodic in nature. The Fourier transform is the mathematical tool used in developing a spectrum for a nonperiodic function.

The differences in properties of the Fourier series and Fourier transform are delineated as follows:

| Time Domain Signal | Spectral Form | Property |
| --- | --- | --- |
| Periodic | Fourier series | Defined at discrete frequencies |
| Nonperiodic | Fourier transform | Defined at continuous frequencies |

## Objectives

After completing this chapter, the reader should be able to:

1. State the definitions of the Fourier and inverse Fourier transforms and compare with the Fourier series.
2. Discuss the various Fourier transform operations and their significance.
3. Determine the spectral rolloff rates for piecewise linear waveforms.
4. Utilize tables of Fourier transforms to analyze the spectra of common waveforms.
5. Discuss and sketch the spectrum of a single baseband pulse function.
6. Discuss and sketch the spectrum of a baseband periodic pulse train.
7. Discuss and sketch the spectrum of an RF pulse function.
8. Discuss and sketch the spectrum of an RF periodic pulse train.

## SystemVue™ Opening Application (Optional)

Insert the text CD in a computer having SystemVue™ installed and activate the program. Open the CD folder entitled **SystemVue Systems** and open the file entitled 3-1.

### Sink Token

| Number | Name | Token Monitored |
|--------|------|-----------------|
| 0 | Signal | 1 |

### Operational Token

| Number | Name | Function |
|--------|------|----------|
| 1 | Signal | generates different types of waveforms |

### Description of System

This is about as simple a system as could be devised; the intent is to familiarize the reader with the various waveforms that can be generated with a **Source Token.** Prior to studying the chapter, you will observe one of the many waveforms that can be generated.

### Activating the Simulation

We will assume initially that all the parameter settings have their initial values. Left-click on the **Run System** button and the simulation should run. To see the plot, left-click on the **Analysis Window** button. The graph will then occupy much of the screen, but the **Maximize** button can be used to make it occupy the entire screen if desired.

### What You See

The waveform that appears on the graph is a square wave with a dc component. The square wave should have a positive value for one-half of the cycle and zero for the other half. You will return to this system after the chapter has been studied in order to observe many different types of waveforms.

### Checking the Settings

All of the parameter values were preset for this system, but you may wish now to investigate some of the settings. Moreover, for the exercises at the ends of chapters, you will need to reset some of the parameters, so it is a good idea now to investigate how this is done.

You may wish to consult the Word file on the disk with the title **3-1 initial settings.** It will show the various settings of the parameters.

## 3-1   Fourier Transform

The starting point for the Fourier transform is the exponential form of the Fourier series and its two-sided spectrum, as developed in Section 2-7. Consider the periodic function $v(t)$ shown in Figure 3–1(a), and an assumed spectrum shown in Figure 3–1(b). Since the

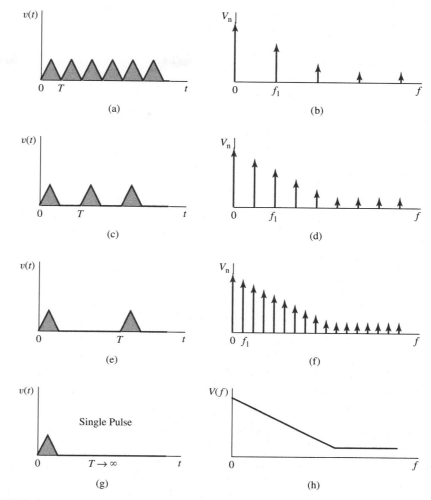

**FIGURE 3–1**

Development of the Fourier transform as a limiting case of the exponential Fourier series, as the period becomes infinite.

time function is periodic with period $T$, the spectrum is discrete, and it appears only at integer multiples of the fundamental frequency $f_1 = 1/T$.

Next, assume that the period is artificially doubled while retaining the same function shape as shown in Figure 3–1(c). This can be accomplished by leaving the signal blank during the second half of the period. The function integrated to determine the spectral coefficients has the same mathematical form as before, but since the period is twice as great, the fundamental frequency is half as great as before. The *relative shape* of the spectrum is shown in Figure 3–1(d). (Strictly speaking, all coefficients are half as large as before, but it is their *relative* sizes that are shown, and that is the significant property of concern.) Note that the spectral lines are half as far apart as in the first case.

Once again the period is doubled as shown in Figure 3–1(e). The spectral coefficients get closer together as shown in Figure 3–1(f).

Now consider the extreme limiting situation shown in Figure 3–1(g) as the period increases without limit. The spacing between the spectral lines now approaches zero. It is no longer desirable to plot the spectrum as a series of lines, since they all merge together. Instead, they could be represented as a series of points, which now merge into a continuous curve as shown in Figure 3–1(h). Thus, the frequency content is now a *continuous* function of frequency.

The *Fourier transform* is the commonly used name for the mathematical function that provides the frequency spectrum of a nonperiodic signal. For a nonperiodic time function $v(t)$, the Fourier transform is indicated by $\overline{V}(f)$. The overbar indicates that the Fourier transform is, in general, a complex function having a magnitude and an angle, or a real part and an imaginary part.

## Fourier Transform and Inverse Transform Definitions

The process of Fourier transformation of a time function is designated symbolically as

$$\overline{V}(f) = \mathbf{F}[v(t)] \tag{3-1}$$

The process of inverse transformation is designated symbolically as

$$v(t) = \mathbf{F}^{-1}[\overline{V}(f)] \tag{3-2}$$

The actual mathematical processes involved in these operations are as follows:

$$\overline{V}(f) = \int_{-\infty}^{\infty} v(t)e^{-j\omega t}\,dt \tag{3-3}$$

$$v(t) = \int_{-\infty}^{\infty} \overline{V}(f)e^{j\omega t}\,df \tag{3-4}$$

Note that the argument of $\overline{V}(f)$ and the differential of Equation 3-4 are both expressed in terms of the cyclic frequency $f$ (in hertz), but the arguments of the exponentials in both (3-3) and (3-4) are expressed in terms of the radian frequency $\omega$, where $\omega = 2\pi f$. These are the most convenient forms for manipulating and expressing the given functions. Most spectral displays are made in terms of $f$, while the analytical expressions are usually more convenient to deal with in terms of $\omega$. This should present no difficulty, as long as the $2\pi$ factor relating the two variables is understood.

## Amplitude and Phase Spectra

The Fourier transform $\overline{V}(f)$ is, in general, a complex function having both a magnitude (or amplitude) and a phase angle. It can thus be expressed as

$$\overline{V}(f) = V(f)\angle\theta(f) \tag{3-5}$$

The function $V(f)$ is the *amplitude* or *magnitude spectrum,* and the function $\theta(f)$ is the *phase spectrum.*

Unless the inverse transform is to be computed or a special phase-sensitive operation is to be performed, the amplitude spectrum is usually more significant than the phase spectrum. In many of the applications that follow, only the amplitude spectrum $V(f)$ will be studied. A representative amplitude response of a function having most of its energy at relatively low frequencies is shown in Figure 3–2.

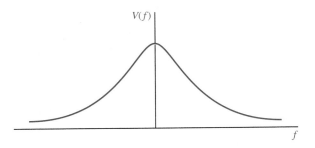

**FIGURE 3–2**
Typical amplitude spectrum of nonperiodic signal.

**EXAMPLE 3-1**

Determine the Fourier transform of the single nonperiodic pulse shown in Figure 3–3(a).

**SOLUTION** We should always start with the basic definition as given in Equation 3-3, which is repeated here for convenience.

$$\overline{V}(f) = \int_{-\infty}^{\infty} v(t)e^{-j\omega t}\, dt \tag{3-6}$$

The infinite integration range simply means that $v(t)$ must be integrated over all time, but since it is zero everywhere but from $-\tau/2$ to $\tau/2$, the actual final limits will be adjusted to reflect that range. Within that range, $v(t) = A$, and the transform can thus be expressed as

$$\overline{V}(f) = \int_{-\tau/2}^{\tau/2} Ae^{-j\omega t}\, dt \tag{3-7}$$

The integral is

$$\overline{V}(f) = \left[\frac{Ae^{-j\omega t}}{-j\omega}\right]_{-\tau/2}^{\tau/2} = \frac{Ae^{-j\omega\tau/2} - Ae^{j\omega\tau/2}}{-j\omega} = A\left(\frac{e^{j\omega\tau/2} - e^{-j\omega\tau/2}}{j\omega}\right) \tag{3-8}$$

The function in parentheses on the right-hand side of Equation 3-8 has a form very similar to that of Equation 2-88, in which the sine function is expressed in terms of the difference between two complex exponentials. We need only multiply numerator and denominator by 2 and bring out the $\omega$ factor. The result is

$$\overline{V}(f) = \frac{2A \sin(\omega\tau/2)}{\omega} \tag{3-9}$$

While this is a perfectly fine function as it is, for reasons that will be clearer in a later section, a few additional adjustments will be made. First, the angular frequency will be expressed as $\omega = 2\pi f$ both in the sine function and in the denominator. Next, the numerator and denominator will be multiplied by $\tau/2$. The final form then is

$$\overline{V}(f) = A\tau\frac{\sin \pi f\tau}{\pi f\tau} \tag{3-10}$$

Since this function is real, it is convenient to define the amplitude and phase functions as

$$V(f) = A\tau\frac{\sin \pi f\tau}{\pi f\tau} \tag{3-11}$$

and

$$\theta(f) = 0 \tag{3-12}$$

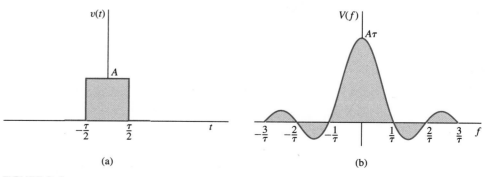

(a)　　　　　　　　　　　　　　　　　　(b)

**FIGURE 3–3**
Single nonperiodic pulse and its amplitude spectrum (based on the natural form of the function), with both positive and negative values permitted.

Strictly speaking, the amplitude response of Equation 3-11 assumes both positive and negative values, which is contrary to the pure definition of a magnitude, but this form is the most convenient to use. This function will be referred to as the "$\sin x/x$" function, and it will be discussed in great detail later in the chapter. It is also denoted elsewhere in the literature as the *sinc* function. The form of the amplitude spectrum with negative values permitted is shown in Figure 3–3(b). The properties of this function will be discussed in great detail in later sections.

---

**▮▮ EXAMPLE 3-2**

The unit *impulse* function $\delta(t)$ is a fictitious mathematical function that has many engineering applications in dealing with waveforms that have "spike-like" characteristics (i.e., a sudden very short burst with a large amplitude and narrow width). The model is represented graphically as shown in Figure 3–4(a) and is assumed to occur at $t = 0$. In the theoretical limit, it approaches a width of zero, an infinite height, and an area of 1. It is assumed to obey the following *sifting property*:

$$\int_{-\infty}^{\infty} g(t)\delta(t)\, dt = g(0) \tag{3-13}$$

where $g(t)$ is any continuous function. Determine the Fourier transform of the impulse function.

**SOLUTION**    Once again we start with the basic definition, which is

$$\overline{V}(f) = \int_{-\infty}^{\infty} v(t)e^{-j\omega t}\, dt \tag{3-14}$$

Substituting $v(t) = \delta(t)$ in Equation 3-14, we have

$$\overline{V}(f) = \int_{-\infty}^{\infty} \delta(t)e^{-j\omega t}\, dt \tag{3-15}$$

Observing the property of Equation 3-13, the integral is simply the exponential evaluated at $t = 0$. Thus,

$$\overline{V}(f) = 1 \tag{3-16}$$

The amplitude and phase functions are

$$V(f) = 1 \tag{3-17}$$

and

$$\theta(f) = 0 \tag{3-18}$$

The amplitude response is shown in Figure 3–4(b). While there is a certain "mystery" surrounding the mathematical impulse function, its Fourier transform is the simplest possible function: unity for a unit impulse. Of course, for a weighted impulse $K\delta(t)$, the amplitude spectrum would be $K$.

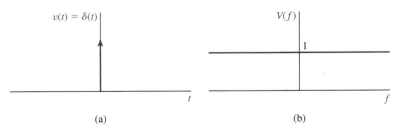

(a)    (b)

**FIGURE 3–4**
Impulse function and its spectrum, as developed in Example 3-2.

Let's get away from the math briefly and explain the physical significance of this result. Basically, it says that a fictitious impulse would generate an infinitely wide frequency spectrum. In practical terms, it says that a very narrow impulsive type of burst would generate frequency terms over a very wide frequency range. Anyone who has had a radio receiver turned on in the vicinity of various electrical appliances, unfiltered automobile ignition systems, and the like will likely have experienced this phenomenon irrespective of where the dial may have been tuned. Such impulsive type sources will typically have a very broad spectrum.

## 3-2 Fourier Transform Operations

Communications signals are altered as they pass through various stages of a signal processing system or a signal transmission channel. These operations may be used to deliberately change the form of the signal, or may arise as a natural property of the system.

The primary Fourier transform operation pairs of interest are summarized in Table 3–1. The practical significance of these operations will be discussed in this section. The following notational form will be used here and in some subsequent sections:

$$v(t) \Leftrightarrow \overline{V}(f) \tag{3-19}$$

This notation indicates that $v(t)$ and $\overline{V}(f)$ are a corresponding transform pair. In each case, whatever affects the time function in the manner shown on the left affects the transform function on the right in the manner indicated. The equation numbers of the following operation pairs correspond to those in Table 3–1.

### Superposition Principle

$$av_1(t) + bv_2(t) \Leftrightarrow a\overline{V}_1(f) + b\overline{V}_2(f) \tag{O-1}$$

This statement means that the Fourier transform obeys the principle of superposition, as far as the level of a signal and the combination of several signals is concerned. Thus, the Fourier transform of a constant times a function is the constant times the Fourier transform of the function, and the Fourier transform of the sum of two or more functions is the sum of the individual Fourier transforms.

### Differentiation

$$\frac{dv(t)}{dt} \Leftrightarrow j2\pi f \overline{V}(f) \tag{O-2}$$

**Table 3–1** Fourier Transform Operation Pairs

| $v(t)$ | $\overline{V}(f) = \mathbf{F}[v(t)]$ | |
|---|---|---|
| $av_1(t) + bv_2(t)$ | $a\overline{V}_1(f) + b\overline{V}_2(f)$ | (O-1) |
| $\dfrac{dv(t)}{dt}$ | $j2\pi f \overline{V}(f)$ | (O-2) |
| $\displaystyle\int_{-\infty}^{t} v(t)\, dt$ | $\dfrac{\overline{V}(f)}{j2\pi f}$ | (O-3) |
| $v(t - \tau)$ | $e^{-j2\pi f\tau}\overline{V}(f)$ | (O-4) |
| $e^{j2\pi f_0 t}v(t)$ | $\overline{V}(f - f_0)$ | (O-5) |
| $v(at)$ | $\dfrac{1}{a}\overline{V}\left(\dfrac{f}{a}\right)$ | (O-6) |

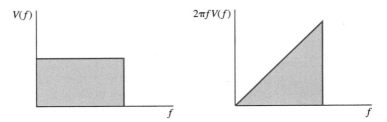

**FIGURE 3–5**
Effect on the spectrum of differentiating a time signal.

This theorem states that when a time function is differentiated, the spectrum is multiplied by $j2\pi f$. Multiplication by $j2\pi f$ has the effect of decreasing the relative level of the spectrum at low frequencies and increasing the relative level at high frequencies. The phase shift of all spectral components is increased by 90° as a result of the j factor. A sketch illustrating the general effect on the amplitude spectrum is shown in Figure 3–5. Note than any dc component in the spectrum is removed by differentiation.

## Integration

$$\int_{-\infty}^{t} v(t)\, dt \Leftrightarrow \frac{\overline{V}(f)}{j2\pi f} \tag{O-3}$$

This theorem, which is the reverse of Equation O-2, states that when a signal is integrated, the spectrum is divided by $j2\pi f$. Division by $j2\pi f$ has the effect of increasing the relative level of the spectrum at low frequencies and decreasing the relative level at high frequencies. The phase shift of all spectral components is decreased by 90° as a result of the j factor in the denominator (equivalent to $-j$ in the numerator). A sketch illustrating the general effect on the amplitude spectrum is shown in Figure 3–6.

## Time Delay

$$v(t - \tau) \Leftrightarrow e^{-j2\pi f\tau}\,\overline{V}(f) \tag{O-4}$$

The function $v(t - \tau)$ represents the delayed version of a signal $v(t)$, as illustrated in Figure 3–7. This operation could occur as a result of passing an analog signal through a delay line, or a digital signal through a shift register. It can be readily shown that the amplitude spectrum is not changed by the time delay, but the phase spectrum is shifted by $-2\pi f\tau$ radians. Certainly, one would expect the amplitude spectrum of a given signal to be independent of the time at which the signal occurs, but the phase shifts of all the components are increased to reflect the result of the time delay.

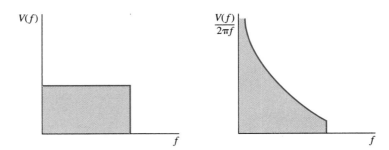

**FIGURE 3–6**
Effect on the spectrum of integrating a time signal.

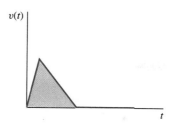
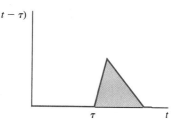

**FIGURE 3–7**
A time signal and its delayed form.

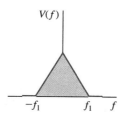
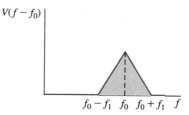

**FIGURE 3–8**
Effect on the spectrum of the modulation operation.

## Modulation

$$e^{j2\pi f_0 t} v(t) \Leftrightarrow \overline{V}(f - f_0) \qquad \text{(O-5)}$$

This operation bears the same relationship to the frequency domain as the time delay theorem does to the time domain. It is an important result in the study of modulation, and will be encountered many times throughout the text in different forms. If a time signal is multiplied by a complex exponential, the spectrum is translated or shifted to the right by the frequency of the exponential, as illustrated in Figure 3–8. In practical cases, complex exponentials appear in pairs, with a term of the form of the left-hand side of Equation O-5 along with its conjugate (as will be seen later).

## Time Scaling

$$v(at) \Leftrightarrow \frac{1}{a}\overline{V}\left(\frac{f}{a}\right) \qquad \text{(O-6)}$$

If $a > 1$, $v(at)$ represents a "faster" version of the original signal, while if $a < 1$, $v(at)$ represents a "slower" version. In the first case, the spectrum is broadened, while in the second case, the spectrum is narrowed. These concepts are illustrated in Figure 3–9. Basically, the theorem says that time and bandwidth have something of an inverse relationship to one another. To transmit a signal in a shorter amount of time, more bandwidth is required, and vice versa.

## Spectral Rolloff

As in the case of Fourier series, the spectral rolloff or convergence rate of the Fourier transform can be predicted for some of the most common piecewise linear waveforms. In the case of the Fourier transform, the convergence of the amplitude spectrum is specified as $1/f^k$ or $-6k$ dB/octave.

In general, the smoother the time function, the higher the value of $k$, and the more rapidly the amplitude spectrum approaches zero. As in the case of periodic functions, the cases that can be visually determined are as follows: (1) Time domain function has a

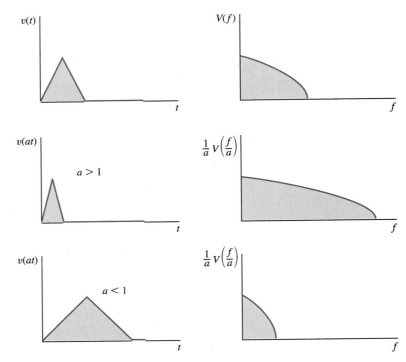

**FIGURE 3–9**
Effect on the spectrum of the time scaling operation.

finite discontinuity or "jump." (2) Time domain function does not have a discontinuity but there is a discontinuity in the slope. (3) Neither the function nor its slope has a finite discontinuity. The rolloff rates for these three cases are summarized in the table that follows.

| Time Domain Function | Amplitude Spectrum Rolloff |
|---|---|
| Function has finite discontinuity. | $1/f$ or $-6$ dB/octave |
| Function is continuous but slope has discontinuity. | $1/f^2$ or $-12$ dB/octave |
| Neither function nor slope has discontinuity. | At least $1/f^3$ or $-18$ dB/octave |

The basic continuity rules are essentially the same for nonperiodic signals as for periodic signals, but the harmonic number $n$ is replaced by the frequency $f$.

## 3-3  Table of Fourier Transforms

As with Fourier series, the treatment of Fourier transforms in this text will focus on interpreting their meaning and using tabulated results in the practical analysis and design of communication systems. Fourier transforms of some of the most common nonperiodic waveforms are provided in Table 3–2.

For all but one case, the origin has been established so that the function is even. (The one exception is the sawtooth pulse, which is neither even nor odd.) The magnitudes of spectral components are independent of the time origin, so moving signals to the left and right will affect only the phase and will not affect $V(f)$.

The spectral convergence properties discussed in the last section may be readily observed in the Fourier transform pairs of Table 3–2. The top entry is the rectangular pulse previously discussed, and it has a finite discontinuity. As expected, it has a $1/f$ form. (The presence of $f$ as part of the sine function argument does not affect the envelope of the function. The denominator $f$ dominates the rolloff behavior.)

Table 3–2    Some Common Waveforms and Their Fourier Transforms

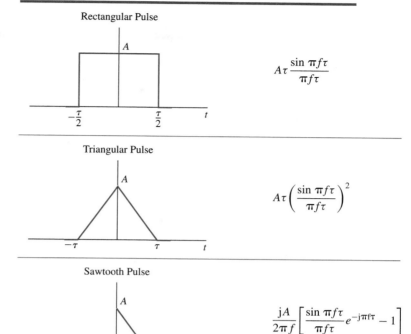

Rectangular Pulse

$$A\tau \frac{\sin \pi f\tau}{\pi f\tau}$$

Triangular Pulse

$$A\tau \left(\frac{\sin \pi f\tau}{\pi f\tau}\right)^2$$

Sawtooth Pulse

$$\frac{jA}{2\pi f}\left[\frac{\sin \pi f\tau}{\pi f\tau}e^{-j\pi f\tau} - 1\right]$$

Cosine Pulse

$$\frac{2A\tau}{\pi}\frac{\cos \pi f\tau}{1 - 4f^2\tau^2}$$

Neither the triangular pulse nor the cosine pulse has discontinuities, but their first derivatives or slopes do have discontinuities. The spectrum of the triangular pulse follows a $1/f^2$ form, and for large $f$, the spectrum of the cosine pulse approaches a $1/f^2$ form.

The spectrum of the sawtooth pulse has both a $1/f^2$ and a $1/f$ term. However, the first term approaches zero much faster than the second term, so for large $f$ the $1/f$ term dominates.

---

**■■ EXAMPLE 3-3**

A nonperiodic voltage pulse waveform $v(t)$ (as shown in Figure 3–10) has the following parameters: $A = 20$ V and $\tau = 1$ ms. (a) Using Table 3–2, write an equation for the Fourier transform $\overline{V}(f)$. (b) Determine the amplitude spectrum at the following values of $f$: 0, 500 Hz, 1 kHz, 1.5 kHz. (c) Determine the amplitude spectrum in decibels at each frequency *relative* to the value at dc.

**SOLUTION**

(a)  The form of the Fourier transform (from Table 3–2) is

$$\overline{V}(f) = A\tau \frac{\sin \pi f\tau}{\pi f\tau} = V(f) \tag{3-20}$$

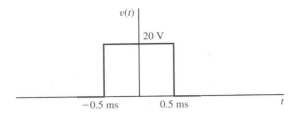

**FIGURE 3–10**
Nonperiodic pulse waveform of Example 3-3.

where the latter term is based on permitting the amplitude spectrum to assume both positive and negative real values (as discussed in Example 3-1).

Substituting the values $A = 20$ V and $\tau = 10^{-3}$ s, we obtain the amplitude spectrum.

$$V(f) = 0.02 \left( \frac{\sin 0.001\pi f}{0.001\pi f} \right) \tag{3-21}$$

The parentheses have been added to better clarify the separation between the basic $\sin x/x$ form and the multiplicative constant.

(b) If one attempts to substitute $f = 0$ in Equation 3-21, an indeterminate form of 0/0 will be obtained. However, by a theorem known as L'Hospital's rule, it can be shown that the limiting value of $\sin x/x$ is 1 as $x$ approaches 0. Thus,

$$V(0) = 0.02 \times 1 = 0.02 \tag{3-22}$$

(Strictly speaking, the units of the Fourier transform of a voltage are volts $\times$ seconds, but we will momentarily sidestep that property and leave the units undefined.)

The other three values of frequency may now be substituted in Equation 3-21 to determine the amplitude spectrum values. It should be emphasized that in the basic $\sin x/x$ function, **the angle of the sine function must be expressed in radians.** Thus, to verify the numbers that follow, the reader should set a calculator to radian mode if available. Alternately, the argument of the sine function calculated at each frequency can be multiplied by $180/\pi$ and the resulting value will be in degrees. However, the denominator factor must be retained in the form given.

To illustrate the process, the value at 500 Hz will be calculated. We have

$$V(500) = 0.02 \times \frac{\sin 0.001 \times 500 \times \pi}{0.001 \times 500 \times \pi} = 0.02 \times \frac{\sin 0.5\pi}{0.5\pi}$$

$$= 0.02 \times \frac{1}{0.5\pi} = 0.01273 \tag{3-23}$$

This process is then repeated at the two additional frequencies and the results are

$$V(1000) = 0 \tag{3-24}$$

and

$$V(1500) = -0.004244 \tag{3-25}$$

This negative value is acceptable at this point because our definition of amplitude spectrum permitted both positive and negative values. Strictly speaking, however, the magnitude spectrum would be 0.004244 and we would then have to say that the phase spectrum at 1500 Hz is 180° or $\pi$ radians.

(c) The decibel magnitude response at any frequency *relative* to that at dc will be defined as $R_{dB}$, which can be expressed as

$$R_{dB} = 20 \log \left| \frac{V(f)}{V(0)} \right| \tag{3-26}$$

Note the use of magnitude bars in the equation, since we do not want negative values in the argument of the logarithmic function. We can now substitute in Equation 3-26 the various frequencies and obtain the relative decibel values. As an additional point of interest, however, suppose we first substitute Equations 3-21 and 3-22 in Equation 3-26, in which case we obtain

$$R_{dB} = 20 \log \left| \frac{\sin 0.001 \pi f}{0.001 \pi f} \right| \tag{3-27}$$

The $A\tau$ factor has cancelled out in the process. In effect, this result indicates that the *relative response* is independent of the actual pulse amplitude, and can be determined strictly from the $\sin x / x$ function. Within the literature, there are variations in the amplitude scaling of the Fourier transform, but it is the *relative* magnitude of the spectrum that is usually of significance.

The results are provided in the table that follows.

| $f$ | $V(f)$ | $R_{dB}$ |
|---|---|---|
| 0 | 0.02 V | 0 dB |
| 500 Hz | 0.01273 | −3.924 dB |
| 1000 Hz | 0 | −∞ dB |
| 1500 Hz | −0.004244 | −13.47 dB |

## 3-4 Baseband Pulse Functions

A signal that finds extensive application in communications systems is the rectangular pulse function. Its Fourier transform was derived in Example 3-1, and some sample computations on its spectrum were made in Example 3-3. Due to its importance, both this section and the next are devoted to discussing many of the significant features of various pulse waveforms and their spectra. In this section, we will concentrate on *baseband* pulse functions. *Baseband signals* are those whose spectra are concentrated at lower frequencies, and which have not been shifted in frequency by modulation.

Consider first the single nonperiodic pulse $v(t)$, shown in Figure 3–11(a). The spectrum of this function $\overline{V}(f)$ was given in Equation 3-10, and is repeated here for convenience:

$$\overline{V}(f) = V(f) = A\tau \frac{\sin \pi f \tau}{\pi f \tau} \tag{3-28}$$

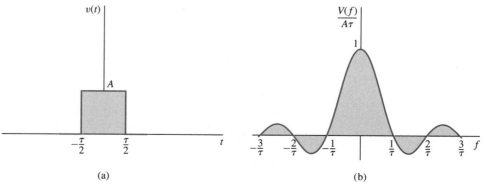

(a)    (b)

FIGURE 3–11
Single nonperiodic pulse and its amplitude spectrum.

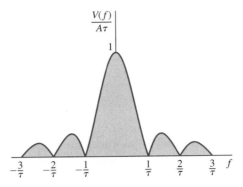

**FIGURE 3–12**
Amplitude spectrum of single pulse, displaying true magnitude.

where we are permitting the amplitude spectrum $V(f)$ to assume both positive and negative values (as was done earlier). With this assumption, the phase spectrum is zero.

A sketch of the amplitude spectrum allowing both positive and negative values is shown in Figure 3–11(b). For convenience, it is normalized with respect to $A\tau$.

When the strict definition of the magnitude (which can assume only positive values) is used, the form of the spectrum will be as shown in Figure 3–12. This is the form that would actually appear on an ideal spectrum analyzer. Both forms will be used in the text.

## Important Properties of Pulse Spectrum

Some important properties of the pulse function and its spectrum are the following:

1. The spectrum is continuous, as is to be expected for a nonperiodic function. The spectral range theoretically extends over an infinite frequency range. Practical estimates of the minimum bandwidth will be developed in the next chapter.

2. The spectrum is bounded by a $1/f$ or a $-6$ dB/octave rolloff rate as a result of the finite discontinuity.

3. The major lobe of the spectrum is located in the lower frequency range, with the maximum value at dc.

4. The first zero crossing of the spectrum occurs at a frequency $f_0 = 1/\tau$. Subsequent zero crossings occur at integer multiples of the frequency $f_0$.

5. As the pulse width narrows, the width of the main lobe of the spectrum out to the first zero crossing widens. Hence, for any practical pulse reproduction, the required bandwidth increases as the pulse width decreases, and vice versa.

## Periodic Pulse Train

Next, consider the case of a periodic pulse train, as shown in Figure 3–13(a). The pulse width is $\tau$ and the period is $T$. The *duty cycle d* of the pulse train is defined as

$$d = \frac{\tau}{T} \tag{3-29}$$

Stated in words, the duty cycle is the ratio of the time interval in which the pulse is on to the total period. Frequently, the duty cycle is expressed as a percentage by multiplying the preceding definition by 100%. For example, a duty cycle of 0.2 could be expressed as 20%.

To correlate well with the nonperiodic pulse, the exponential form of the Fourier series is the most convenient to use for the periodic pulse train. It can be shown that the coefficients $\overline{V}_n$ are given by

$$\overline{V}_n = V_n = Ad\frac{\sin \pi dn}{\pi dn} \tag{3-30}$$

with the amplitude spectrum $V_n$ allowed to assume both positive and negative values.

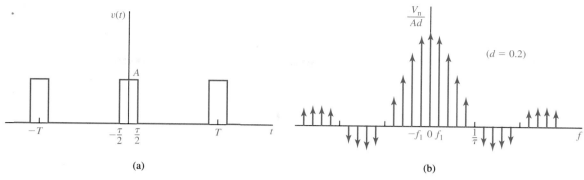

FIGURE 3–13
Periodic pulse train and its amplitude spectrum.

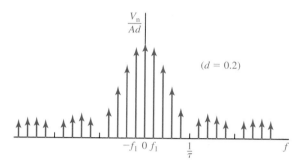

FIGURE 3–14
Amplitude spectrum of periodic pulse train, displaying true magnitude.

A comparison of Equations 3-28 and 3-30 reveals a very close similarity between the spectrum of a pulse train and a single pulse. However, aside from slightly different notation and different levels, the most important difference is that the spectrum of the pulse train exists only at frequencies that are integer multiples of the fundamental frequency, as expected for a periodic function. The form of the spectrum for a duty cycle of 0.2 is shown in Figure 3–13(b). The corresponding true magnitude spectrum is shown in Figure 3–14.

The similarity between the spectra suggests a simplified approach for sketching the spectrum of a periodic pulse train, which can be stated as follows:

1. Momentarily assume a single nonperiodic pulse of width $\tau$, and lightly sketch the $\sin x/x$ spectrum, carefully noting the zero crossings at frequencies that are integer multiples of $1/\tau$. This curve becomes the *envelope* of the spectrum.

2. Starting at dc ($f = 0$), draw vertical lines representing the spectral terms to the envelope curve at all frequencies that are integer multiples of $f_1 = 1/T$.

The fundamental frequency $f_1$ is often referred to as the *pulse repetition frequency* (*prf*).

---

**▌▌ EXAMPLE 3-4**

For the pulse waveform of Example 3-3, sketch the spectrum out to the third zero crossing for both positive and negative frequencies.

**SOLUTION**   As determined in Example 3-3, the value of the spectrum at dc is $A\tau = 0.02$. The first zero crossing is

$$f_0 = \frac{1}{\tau} = \frac{1}{0.001} = 1000\,\text{Hz} = 1\,\text{kHz} \tag{3-31}$$

The second zero crossing is $2 \times 1\,\text{kHz} = 2\,\text{kHz}$, and the third zero crossing is $3 \times 1\,\text{kHz} = 3\,\text{kHz}$. A sketch of the spectrum is shown in Figure 3–15.

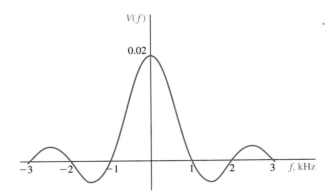

**FIGURE 3–15**
Amplitude spectrum of Example 3-4.

---

**▌▌ EXAMPLE 3-5**

The periodic pulse train shown in Figure 3–16(a) is based on the 1-ms pulse of Examples 3-3 and 3-4, but with the pulse repeated at intervals of 5 ms. Sketch the spectrum for positive and negative frequencies out to the third zero crossing.

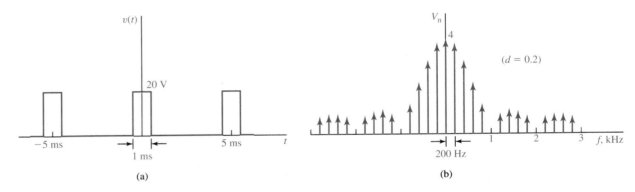

(a)                                              (b)

**FIGURE 3–16**
Periodic pulse train of Example 3-5 and its amplitude spectrum (showing true magnitudes of components).

**SOLUTION** The envelope of the spectrum will have the same form as the spectrum of Example 3-4; that is, it will have zero crossings at integer multiples of 1 kHz. However, components of the spectrum will appear only at integer multiples of the fundamental frequency $f_1$, which is

$$f_1 = \frac{1}{T} = \frac{1}{5 \times 10^{-3}} = 200 \, \text{Hz} \tag{3-32}$$

The result is shown in Figure 3–16(b). The true magnitudes of the components are shown in this case.                                                                            ▌▌

## 3-5 Radio Frequency Pulse Functions

With baseband pulses as considered in the last section, the most significant part of the spectrum is concentrated at low frequencies. In contrast, the pulses considered in this section will have the most significant part of their spectrum concentrated within some frequency range other than near dc. To relate the concept to applications in communication theory, the term *radio frequency* (*RF*) pulse will be used. However, the mathematical results are applicable in the audio range or any other frequency range in which the major lobe of the spectrum is shifted away from dc.

Consider first the single nonperiodic pulse function $v_{rf}(t)$ shown in Figure 3–17(c). This function can be considered as the product of the baseband pulse function $v_{bb}(t)$ shown in Figure 3–17(a) and the infinitely long sinusoid shown in Figure 3–17(b). Based

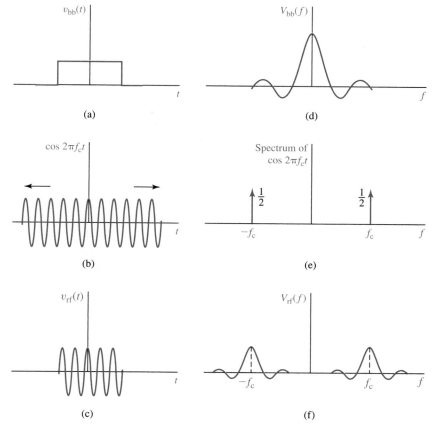

**FIGURE 3–17**
Development of the spectrum of a nonperiodic RF pulse.

on the assumption that the sinusoid is the basic cosine function with a frequency $f_c$, the RF pulse can be expressed as

$$v_{rf}(t) = v_{bb}(t) \cos 2\pi f_c t = \frac{v_{bb}(t)}{2} e^{j2\pi f_c t} + \frac{v_{bb}(t)}{2} e^{-j2\pi f_c t} \tag{3-33}$$

where the definition of the cosine function in terms of complex exponentials has been used. The spectrum of the signal is determined by taking the Fourier transform, and the modulation theorem of Equation O-5 is applicable. The result is

$$\overline{V}_{rf}(f) = \frac{1}{2}\overline{V}_{bb}(f - f_c) + \frac{1}{2}\overline{V}_{bb}(f + f_c) \tag{3-34}$$

This result is best explained by parts (d), (e), and (f) of Figure 3–17. The amplitude spectrum of the baseband spectrum is shown in (d), and two spectral lines, representing the spectrum of the sinusoid, are shown in (e). The final RF amplitude spectrum shown in (f) is generated by shifting the baseband spectrum both to the right and to the left by $f_c$.

For clarity in visualizing this result, a portion of the amplitude spectrum of the RF pulse is expanded in Figure 3–18. The portion of the spectrum shown is the center of the first spectral term in Equation 3-34. The second term in Equation 3-34 is the mirror image shifted to the left, and its main lobe will be in the negative frequency range. In most practical applications, the effects of any foldover of this component into the positive frequency range will be negligible, and that will be the assumption here.

From the display of Figure 3–18, the following properties of the spectrum of the RF pulse can be deduced:

1. The spectrum of the RF pulse has the same form as the baseband pulse; that is, the $\sin x / x$ function. However, while the center of the spectrum for the baseband pulse is

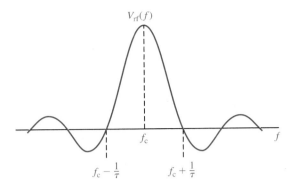

**FIGURE 3–18**
Expanded amplitude spectrum of RF pulse near center frequency.

   dc, the center of the spectrum for the RF pulse is $f_c$. Thus, the spectrum is centered at
   the frequency of the sinusoidal oscillation.

2. The zero-crossing frequencies are located at integer multiples of $1/\tau$ on either side
   of the center frequency $f_c$.

3. The width of the main lobe of the spectrum as measured between corresponding zero
   crossings is $2/\tau$.

4. For any type of bandwidth approximation for pulse reproduction, the RF pulse
   requires twice the bandwidth as for a baseband pulse since the spectrum extends
   in two directions from the positive frequency spectrum center.

## Periodic RF Pulse Train

The last and most complex case for consideration is the RF periodic pulse train. An example of such a function is shown in Figure 3–19(a). There are three parameters to consider

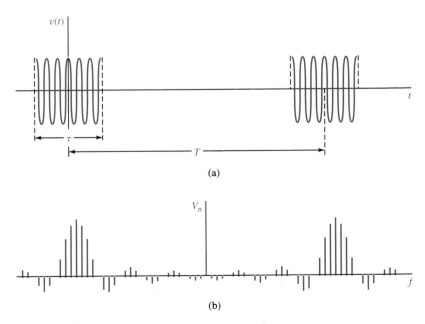

(a)

(b)

**FIGURE 3–19**
Periodic RF pulse train and its spectrum.

in this case: (1) the pulse width $\tau$, (2) the period $T$ (or the pulse repetition rate $f_1 = 1/T$), and (3) the RF frequency $f_c$ of the sinusoidal oscillation. We will not develop the mathematical details in this case, but will show the results from a qualitative point of view.

As in the case of the nonperiodic RF train, the spectrum is now centered at the RF frequency $f_c$, as shown in Figure 3–19(b). Likewise, zero crossings occur at frequencies that are integer multiples of $1/\tau$ on either side of the center frequency. However, as in the case of the baseband periodic pulse train, the spectrum is defined only at frequencies that are integer multiples of the fundamental frequency $f_1 = 1/T$. Thus, an envelope can be sketched to determine the basic form of the spectrum, and lines can be drawn to the envelope at those frequencies that are integer multiples of the fundamental.

**EXAMPLE 3-6**

An RF oscillator produces a 30-MHz sine wave voltage. It is gated on for an interval of 1 ms and then turned off. For the single RF pulse produced, sketch the amplitude spectrum for positive frequencies in the vicinity of the major lobe.

**SOLUTION** The function will be of the form of Figure 3–17(c) with a duration $\tau = 1$ ms. The frequency of the RF sinusoid is 30 MHz, which means there will be many cycles of the sinusoid within the pulse interval. Specifically, there will be $\tau/(1/f_c) = \tau f_c = 1 \times 10^{-3} \times 30 \times 10^6 = 30{,}000$ cycles. Hence, it is not too practical to sketch the time function!

The amplitude spectrum in the neighborhood of the main lobe can be sketched from the data given. First, the main lobe is centered at 30 MHz. The first zero crossing on either side of the center frequency is located $1/\tau = 1/10^{-3} = 1000 \text{ Hz} = 1 \text{ kHz}$ away from the peak component on both sides. The resulting amplitude spectrum is shown in Figure 3–20.

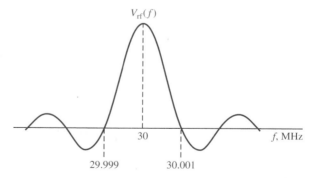

**FIGURE 3–20**
Amplitude spectrum of Example 3-6.

**EXAMPLE 3-7**

Assume that the RF pulse of Example 3-6 is gated on and off periodically. Assume that it is turned on for 1 ms as in Example 3-6 but that it is then turned off for 4 ms to complete a cycle. Sketch the amplitude spectrum in the vicinity of the main lobe.

**SOLUTION** The envelope of the amplitude spectrum will have the same form as the spectrum of Example 3-6. However, the spectrum will now be discrete. The period is $1 \text{ ms} + 4 \text{ ms} = 5 \text{ ms}$, and the fundamental frequency is $1/5 \times 10^{-3} = 200 \text{ Hz}$. Hence, components of the spectrum will appear only at integer multiples of 200 Hz on both sides of the center frequency. A sketch of the spectrum is shown in Figure 3–21.

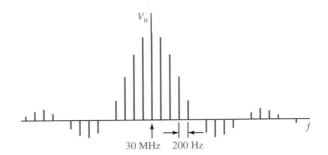

**FIGURE 3–21**
Amplitude spectrum
of Example 3-7.

---

**EXAMPLE 3-8**

The spectrum analyzer display of a periodic RF pulse train is shown in Figure 3–22. Determine (a) the RF burst frequency, (b) the pulse repetition rate, and (c) the width of the RF pulses.

**SOLUTION**

(a) The RF burst frequency is the center of the spectrum and is readily identified as 500 MHz.

(b) The pulse repetition rate is the separation between lines and is 1 MHz.

(c) The first zero crossings occur 10 MHz above and below the center frequency of the spectrum. Thus, the width of the pulses is

$$\tau = \frac{1}{f_0} = \frac{1}{10 \times 10^6} = 0.1 \ \mu s \tag{3-35}$$

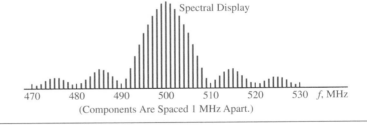

**FIGURE 3–22**
Spectrum analyzer display
of Example 3-8.

---

## 3-6  Multisim® Example (Optional)

We have explored the **Fourier Analysis** option of Multisim in the preceding chapter by observing the spectrum of a periodic signal. While this operation is primarily applicable to periodic functions, it may be used to estimate the **form** of the spectrum of a nonperiodic function under certain conditions. The function under observation must be interpreted as a periodic function, and the values obtained represent the coefficients of the Fourier series. However, by a simple adjustment of the amplitude values, the results may be used to infer the form of the spectrum of a nonperiodic signal.

---

**MULTISIM EXAMPLE 3-1**

Use the periodic pulse function of Multisim to approximate a single pulse of width 0.1 ms and a level of 1 V. Use the **Fourier Analysis** function to approximate the spectrum.

**SOLUTION**   To obtain the periodic pulse function, left-click on the **Show Signal Source Family Bar** and the source is obtained from the **Place Pulse Voltage Source** button. Based on procedures covered earlier, obtain a **Junction** and connect it to the reference

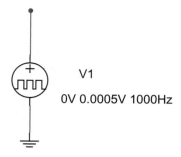

**FIGURE 3–23**
Circuit of Multisim Example 3-1.

positive output terminal, and obtain a **Ground** and connect it to the lower terminal. The circuit is shown in Figure 3–23.

Left-click on the source and a properties window will open. The default position for the tab should be at **Value,** but if not, left-click on this tab. Various pulse parameters must now be set.

Some of the pulse parameters can probably be left at their initial values, but all should be noted. This source allows two constant values within a cycle, and these are designated **Initial Value** and **Pulsed Value.** To approximate a nonperiodic positive pulse, the **Initial Value** should be set at **0**.

In theory, the final value could be set to any value, and the approximate *form* (but not the absolute level) of the spectrum could be obtained. However, there is a difference in the levels based on the common definitions most widely used. In the case of the Fourier transform as employed in this text, there is no constant in front. However, based on the one-sided form of the Fourier series—and this is compatible with Multisim—there is a $2/T$ factor in front of all but the dc term. Therefore, the Fourier series may be made to "look like" a Fourier transform if the series is multiplied by $T/2$. We will arbitrarily set the period to be 1 ms, and this would mean that the amplitude should be multiplied by $1\,\text{ms}/2 = 0.5 \times 10^{-3}$. Thus, the **Pulsed Value** is set to **0.5e-3.** The **Tolerance** block is left unchecked.

The definitions of the additional quantities are as follows:

**Delay Time:** Initial time delay before **Pulsed Value** is first set.

**Rise Time:** Time for leading edge of pulse to increase from initial value to pulsed value.

**Fall Time:** Time for trailing edge of pulse to decrease from pulsed value to initial value.

**Pulse Width:** Time width of pulse.

**Period:** Time width of period (reciprocal of frequency).

Set the **Delay Time** to 0. This will result in the **Pulsed Value** being generated first. The default values of **Rise Time** and **Fall Time** are 1 ns, and these should be acceptable for the present simulation since the period is 10 ms. In general, these values must be small compared with both the period and pulse width if a close approximation to a pulse train is desired. Set the **Pulse Width** to 0.1 ms = 1e-4 s.

The next step in the process is to left-click on **Analyses** and then left-click on **Fourier Analysis.** When the **Fourier Analysis** window opens, the **Frequency resolution** is set to **1000 Hz.** In this case the **Number of harmonics** was set to 39. The **Stop time** was set to 1e-3 s (one cycle) and the simulation is activated by a left-click on **Simulate.**

The resulting spectrum is shown in Figure 3–24. Obviously, this is a line spectrum, but it should closely approximate a continuous spectrum of a nonperiodic pulse at the points on the curve. In fact, the value at $f = 0$ is 1e-4, which is the value of $A\tau$ for the specific function under consideration.

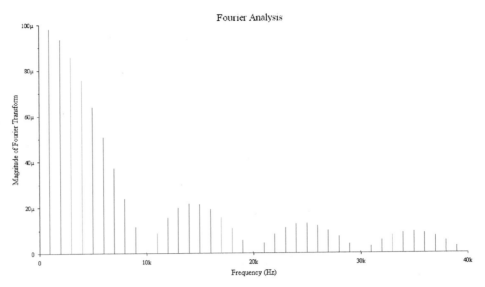

**FIGURE 3–24**
Magnitude spectrum of Multisim Example 3-1.

## 3-7  MATLAB® Examples (Optional)

As in the case of Chapter 2, MATLAB will be employed in two different ways: (1) to plot a spectral function whose mathematical form is given, and (2) to plot the spectrum of a given time function using a fast Fourier transform (FFT)–based program to closely approximate the analytical Fourier transform.

**MATLAB EXAMPLE 3-1**    Use MATLAB to plot the amplitude spectrum of the voltage pulse waveform of Example 3-3.

**SOLUTION**    The Fourier transform was determined directly from Table 3–2, with the appropriate values substituted. The result was given in Equation 3-21, and is repeated here for convenience.

$$V(f) = 0.02 \left( \frac{\sin 0.001\pi f}{0.001\pi f} \right) \tag{3-36}$$

In order to plot the function, a row vector containing values of frequency for plotting will be created. For reasons that will be clearer in MATLAB Example 3-2, a choice will be made to use 50 values of frequency with a spacing of 100 Hz between components. Since the first zero crossing occurs at $f = 1\,\text{kHz}$, this will provide 10 frequency values before the crossing. A command to accomplish this task is

```
>> f = 0:100:4900;
```

An alternate command would be

```
>> f = linspace(0, 4900, 50);
```

There is a problem with MATLAB in trying to evaluate Equation 3-36 at $f = 0$, since the result will appear as 0/0 and will result in a reading of **NaN** ("not a number"). There are two ways to get around this problem. One way is to add an extremely small increment to $f$ so that it never quite reaches 0. In fact MATLAB contains a number **"eps"** for this purpose and the approximate value is eps $= 2.22 \times 10^{-16}$. The second way, and the one we will choose to use, is to simply write a separate statement for the first value, making use of the fact that $\sin x/x$ approaches unity as $x$ approaches 0. Let Vexact represent the exact Fourier transform as given by Equation 3-36. (An approximate form will be developed in the next example.) We have

```
>> Vexact(1) = 0.02;
>> Vexact(2:50) = 0.02*sin(0.001*pi*f(2:50))./(0.001*pi*f(2:50));
```

The first statement defines the dc value of Vexact as 0.02, a value we can easily determine from Equation 3-36. The second statement defines the second through the 50th value from the equation. Note that it is necessary to index the frequency with the same values as for the function.

Another important quantity in the expression for Vexact(2:50) is the period at the end of the numerator. In effect, we are dividing all values of the numerator row vector by all values of the denominator row vector on a point-by-point or **array** basis. Such an operation requires the period (.) at the end of the first vector.

We will choose to deal with the magnitudes of the various spectral terms, and these can be generated by the command

```
>> Vexact = abs(Vexact);
```

The preceding operation forms the absolute value or magnitude of each of the values of Vexact. The results could have been assigned to a new variable name, but the simplest choice is to retain the same name as before. A single variable or a dimensioned variable corresponds to one or more memory locations, and a command such as the preceding one means that the modified variables are stored in the same locations as before.

The function may now be plotted by the command

```
>> plot(f, Vexact);
```

The result is shown in Figure 3–25, with additional labeling provided as discussed earlier.

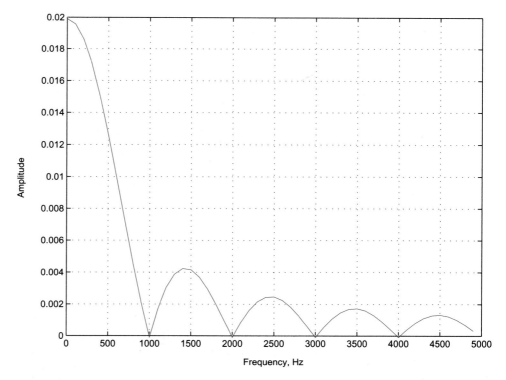

**FIGURE 3–25**
Amplitude spectrum of MATLAB Example 3-1.

**▍▌ MATLAB EXAMPLE 3-2**

Use the M-file program **fourier_transform** to determine the approximate amplitude spectrum of the voltage waveform of Example 3-3 and MATLAB Example 3-1.

**SOLUTION**    Before using the program, the reader should refer to Appendix A for a general discussion. Prior to executing the program, the time function must be entered as a row vector. This example is considered to be an extension of MATLAB Example 3-1, and some of the parameters of that example will be used here. Since the frequency scale in MATLAB Example 3-1 was defined at 50 points, ranging from dc to 4900 Hz, the function used for evaluating the FFT should be defined at 100 points. The frequency step in MATLAB Example 3-1 was 100 Hz, and this would imply a total duration of $1/100 = 0.01$ s $= 10$ ms. The pulse width in Example 3-3 was 1 ms. The voltage v will then be defined as

```
>> v = 20*[ones(1,10) zeros(1,90)];
```

Thus, the row vector of voltage has been defined at a level of 20 V for the first 10 points and a value of 0 for the last 90 points. This time interval is sufficiently long that the pulse may be considered nonperiodic for many practical purposes.

    With the 100-point row vector for v in memory, the M-file is initiated by the command

```
>> fourier_transform
```

The following dialogue now appears on the screen:

**The function v must already be in memory.**
**It should be a row vector defined over the time interval.**

**The total number of points is 100.**

**Enter the time step between points in seconds.**

Since 10 steps are being used to represent a pulse width of 1ms, the time step is entered as 1e-4. The screen dialogue follows.

**The frequency increment between points is 100 Hz.**
**The highest unambiguous frequency is 5000 Hz.**

**Enter integers as multiples of the frequency increment.**
**For example, dc would be entered as 0.**
**The lowest non-zero frequency would be entered as 1.**
**The next frequency would be entered as 2, etc.**
**The highest integer that should be entered is 49.**
**Enter the lowest integer for plotting.**

We wish to have the plot begin at dc so the value **0** is entered. The next dialogue is

**Enter the highest integer for plotting.**

In this case the value **49** is entered. The resulting plot is shown in Figure 3–26. The labeling is automatically provided as a part of the program.

    To compare the exact Fourier transform of the previous example with the FFT approximation of this example, the two plots can be shown on the same graph. We will choose to show the exact function as a solid curve and the approximate function as a series of x's. Assuming that Vexact is still in memory, the following command will generate both functions:

```
>> plot(f, Vexact, f, V, 'x')
```

The two functions are shown in Figure 3–27 with the FFT approximation displayed as the x symbols. In the frequency range up to the first zero crossing, there is no discernible difference in the curves. As the highest possible frequency is approached, the approximate spectrum differs slightly from the true spectrum, but the approximation is quite good for most practical purposes.

    In subsequent examples using **fourier_transform,** the dialogue expanded in this example will usually be omitted, and only the major parameters required for the program and the results will be provided.

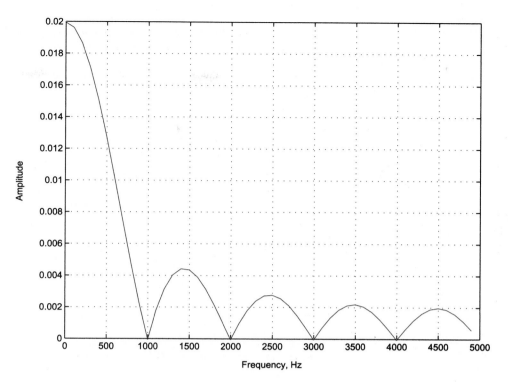

**FIGURE 3–26**
Amplitude spectrum of Fourier transform using FFT.

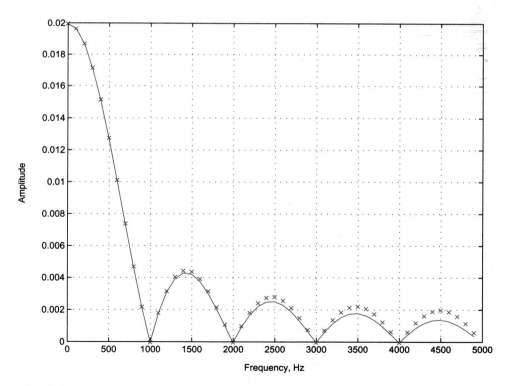

**FIGURE 3–27**
Comparison of exact and approximate spectra.

## SystemVue™ Closing Application (Optional)

Insert the text CD in a computer having SystemVue™ installed and activate the program. Open the CD folder entitled **SystemVue Systems** and load the file entitled 3-1.

### Description of System

The system shown on the screen is the one used for the chapter opening application. However, you will be modifying this file to create different waveforms, and you may choose to save any new files on a hard drive or a different disk.

You may wish to obtain an initial run to observe the square wave again. This is achieved by a click on the **Run System** button on the **Tool Bar.** The square wave has a period of 1 s. The value for half of the cycle is 1 and the value for the other half is 0.

### Changing Parameters of the Pulse Train

The initial square wave has a duty cycle of 0.5. The first change will be to set a duty cycle of 0.1. Double click on the **Source Token** and then left-click on **Parameters.** You will see several slots that may be used to provide parameters for the pulse train. Perform the following steps:

1. Change the duty cycle to 0.1 while maintaining the **Frequency** at 1 Hz. This process may be achieved by changing the **Pulse Width.** Perform a new run and observe the waveform. (As a review point, remember that it is necessary to left-click on the **Load New Sink Data** button to load a new curve on the screen.) Measure the pulse width on the screen to ensure that the objective has been met.

2. While maintaining the pulse width set in the previous step, change the **Frequency** to 5 Hz. Perform a new run and observe the waveform. Verify that the frequency has the correct value.

3. Return to the **Parameters** window for the source and left-click on **Square Wave.** Left-click **OK** on that window, then left-click **OK** on the source window to return to the system. Perform a new run and compare the properties of the square wave with those of the pulse train previously viewed.

4. To reduce the number of cycles on the display, open the **System Time Specification** window and set the **Stop Time** to a value that will result in two cycles on the screen. Note that it is necessary to left-click on **Update** after entering changes to this window.

5. Experiment with different settings on both the **Pulse Train** and **System Time Specification** windows so that you will feel comfortable in making such changes in future assignments. Note again that after making changes in the **System Time Specification** window that it is necessary to left-click on **Update.** After you have completed this set of assignments, you should realize that the process involved is somewhat similar to adjusting the controls on a signal generator and tracking the changes with an oscilloscope to observe the resulting waveforms.

### Other Waveforms

1. Double left-click on the **Source** token and note the various waveforms that may be used.

2. Experiment with several of the waveforms and adjust the **System Time Specification** window to enhance the viewing. You will probably not understand the meaning of all of the forms at this point, but many will be employed throughout the text in demonstrations.

# PROBLEMS

**3-1**  Refer to the nonperiodic voltage pulse waveform $v(t)$ shown below. (a) Using Table 3–2, write an equation for the Fourier transform $\overline{V}(f)$. (b) Determine the amplitude spectrum at the following values of $f$: 0, 2 Hz, 4 Hz, 6 Hz. (c) Determine the amplitude spectrum in decibels at each frequency *relative* to the value at dc.

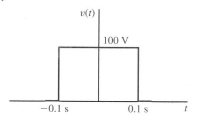

**3-2**  Refer to the nonperiodic current pulse waveform $i(t)$ shown below. (a) Using Table 3–2, write an equation for the Fourier transform $\overline{I}(f)$. (b) Determine the amplitude spectrum at the following values of $f$: 0, 40 kHz, 80 kHz, 160 kHz. (c) Determine the amplitude spectrum in decibels at each frequency *relative* to the value at dc.

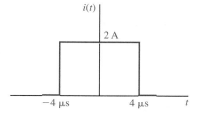

**3-3**  Refer to the nonperiodic voltage triangular waveform $v(t)$ shown below. (a) Using Table 3–2, write an equation for the Fourier transform $\overline{V}(f)$. (b) Determine the amplitude spectrum at the following values of $f$: 0, 2 Hz, 4 Hz, 6 Hz. (c) Determine the amplitude spectrum in decibels at each frequency *relative* to the value at dc. (d) How does the spectral rolloff for this function compare with that of the rectangular pulse of Problem 3-1?

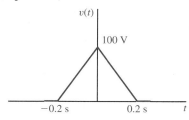

**3-4**  Refer to the nonperiodic current triangular waveform $v(t)$ shown below. (a) Using Table 3–2, write an equation for

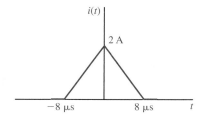

the Fourier transform $\overline{V}(f)$. (b) Determine the amplitude spectrum at the following values of $f$: 0, 40 kHz, 80 kHz, 160 kHz. (c) Determine the amplitude spectrum in decibels at each frequency *relative* to the value at dc. (d) How does the spectral rolloff for this function compare with that of the rectangular pulse of Problem 3-2?

**3-5**  For the pulse waveform of Problem 3-1, sketch the spectrum out to the third zero crossing for both positive and negative frequencies.

**3-6**  For the pulse waveform of Problem 3-2, sketch the spectrum out to the third zero crossing for both positive and negative frequencies.

**3-7**  The pulse of Problem 3-1 is repeated at intervals of 0.8 s. Sketch the spectrum for positive and negative frequencies out to the third zero crossing.

**3-8**  The pulse of Problem 3-2 is repeated at intervals of 80 μs. Sketch the spectrum for positive and negative frequencies out to the third zero crossing.

**3-9**  Consider the periodic pulse train shown below. Sketch the amplitude spectrum for both positive and negative frequencies out to the first zero crossing.

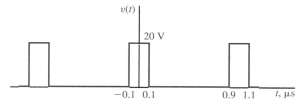

**3-10**  Consider the periodic pulse train shown below. Sketch the amplitude spectrum for both positive and negative frequencies out to the first zero crossing.

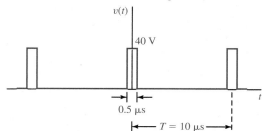

**3-11**  The waveform shown below represents a single nonperiodic burst of a sinusoidal voltage. Sketch the form of the amplitude spectrum for positive frequencies in the vicinity of the major lobe, and label important frequencies.

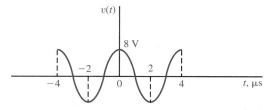

**3-12**  The nonperiodic waveform shown as follows represents a single burst of a sinusoidal voltage. Sketch the form of the amplitude spectrum for positive frequencies in the vicinity of the major lobe, and label important frequencies.

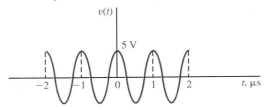

**3-13** The periodic waveform shown below represents a gated sinusoid. Sketch the form of the amplitude spectrum for positive frequencies in the vicinity of the major lobe, and label important frequencies.

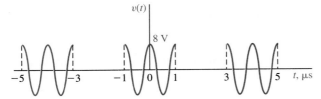

**3-14** The periodic waveform shown below represents a gated sinusoid. Sketch the form of the amplitude spectrum for positive frequencies in the vicinity of the major lobe, and label important frequencies.

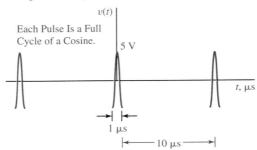

**3-15** An RF sinusoidal oscillator operating at 10 MHz is gated on for an interval of 1 μs by a single baseband pulse. Sketch the form of the amplitude spectrum in the vicinity of the major positive frequency lobe, and label significant frequencies.

**3-16** An audio oscillator operating at 1 kHz is gated on for an interval of 5 ms by a single baseband pulse. Sketch the form of the amplitude spectrum in the vicinity of the major positive frequency lobe, and label significant frequencies.

**3-17** Repeat Problem 3-15 if the gating signal is a periodic pulse train that turns on the oscillator for 1 μs, but in which the period is 5 μs.

**3-18** Repeat Problem 3-16 if the gating signal is a periodic pulse train that turns on the oscillator for 5 ms, but in which the pulse repetition rate is 20 Hz.

**3-19** A technologist is performing some measurements on a baseband periodic pulse train, but the only instrument available is a *frequency-selective voltmeter* (a form of a

spectrum analyzer in which each component of the spectrum is individually tuned and measured). The instrument does not measure dc. Spectral components are measured at the following frequencies (in kHz): 2, 4, 6, 10, 12, 14, 18, 20, 22, 26, and so on. (a) Determine the *pulse repetition rate,* (b) the width of each pulse, and (c) the duty cycle.

**3-20** A baseband pulse train with a fixed period of 200 ns is operating in the vicinity of a narrowband receiver tuned to 100 MHz. What is the smallest duty cycle at which the pulse train can be set in order to ensure that no interference will be heard on the receiver?

**3-21** The spectrum analyzer display of a gated RF periodic pulse train is shown below. (A spectrum analyzer displays true magnitude values of positive frequency components.) Determine (a) the RF burst frequency, (b) the pulse repetition rate, and (c) the width of the RF pulses.

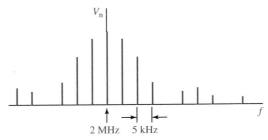

**3-22** Repeat the analysis of Problem 3-21 for the display shown below.

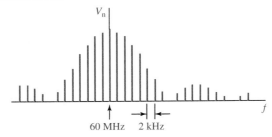

**3-23** An instrumentation signal of duration 5 minutes has an estimated bandwidth of 4 MHz. It must be transmitted over a baseband channel having a bandwidth of only 1 MHz. To accomplish this task, a technologist proposes to first record the signal and then transmit it at a slower rate. (a) Explain how this can be accomplished using one of the theorems of Table 3–1. (b) What is the minimum time interval over which the signal must be transmitted?

**3-24** A recorded data signal of one hour duration has an estimated bandwidth of 50 kHz. It is to be transmitted over a channel having a bandwidth of 750 kHz. If it were desired to take advantage of the available bandwidth to shorten the transmission time, by what factor could speed of transmission be increased, and what would be the resulting transmission time?

# Communication Filters and Signal Transmission

# 4

## OVERVIEW AND OBJECTIVES

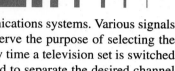

Electrical filters play a very important role in all communications systems. Various signals are characterized by their frequency spectra and filters serve the purpose of selecting the desired signal while rejecting all unwanted signals. Every time a television set is switched from one channel to another, one or more filters are tuned to separate the desired channel from all the others. Likewise, when a radio is tuned from one station to another, a filter is used in the tuning process. In this chapter, the concept of filter response will be explored, and some of the most common types of filters will be discussed.

## Objectives

After completing this chapter, the reader should be able to:

1. Explain the concept of the *steady-state transfer function* and explain its significance.
2. Determine the steady-state transfer function for a circuit.
3. Determine the *amplitude* and *phase response* functions for a circuit.
4. State and explain the criteria for a distortionless filter.
5. Describe the ideal amplitude response functions for *low-pass, high-pass, band-pass,* and *band-rejection* filters.
6. Define *passband, stopband,* and *transition band* for a filter.
7. Define *phase delay* and *group delay* and explain their significance.
8. Discuss the strategy of operation for a ladder filter structure.
9. Sketch circuit diagrams for typical low-pass, high-pass, band-pass, and band-rejection filters.
10. Discuss the response forms for *Butterworth, Chebyshev,* and *maximally flat time delay* filters.
11. Analyze and/or design a one pole-pair band-pass filter using either a series or a parallel resonant circuit.
12. State the pulse bandwidth criteria for coarse and fine reproduction.

**SystemVue™ Opening Application (Optional)**

Insert the text CD in a computer having SystemVue™ installed and activate the program. Open the CD folder entitled **SystemVue Systems** and load the file entitled 4-1.

## Sink Tokens

| Number | Name | Token Monitored |
|--------|------|-----------------|
| 0 | Input | 2 |
| 1 | Output | 3 |

## Operational Tokens

| Number | Name | Function |
|--------|------|----------|
| 2 | Input | generates input pulse train |
| 3 | Filter | filters the pulse train |

## Description of System

The system shown on the screen consists of a **Pulse Train Source** followed by a **Butterworth Low-Pass Filter.** The **Pulse Width** of the pulse train has been set at 1 second and the **Frequency** has been set at 0.1 Hz. This results in a duty cycle of 0.1, and the **System Time Specification** window has been set for a duration of 10 seconds. Thus, one cycle of the pulse train will be observed.

## Activating the Simulation

Left-click on the **Run System** button and the simulation should run. To see the plots, left-click on the **Analysis Window** button. The two graphs will then be available and the waveforms can be maximized.

## What You See

The waveform that appears on the graph of Token 0 is the input pulse train shown for one cycle. The waveform that appears on the graph of Token 1 is the output of a Butterworth filter having a 3-dB cutoff frequency of 0.5 Hz. The pulse has been rounded considerably and has lost a slight amount of amplitude, but it is still distinguishable as a pulse.

## How This Demonstration Relates to the Chapter Learning Objectives

In Chapter 4, you will learn about different classifications of filters and their specifications. Filters are among the most important components of a communication system. You will learn about the bandwidth required to process pulse waveforms. After this chapter is completed, you will return to this system and study the effects of changing bandwidth or pulse width.

## 4-1 Steady-State Transfer Function

The means by which the frequency response of a filter (or any circuit) is determined is through the *steady-state transfer function*. The transfer function concept may be applied to any linear circuit containing fixed values of inductance, capacitance, and resistance. The circuit is converted to the steady-state ac phasor form, the details of which are covered in basic circuit texts dealing with ac circuits.

The instantaneous relationship between the input and the output in the time domain involves a differential equation. However, by tranforming the variables to their steady-state ac or phasor forms, the input-output relationship can be reduced to an algebraic form. The result applies to any sinusoidal signal with a radian frequency $\omega$.

### Phasor Domain Forms

In the frequency response analysis of interest, it is necessary to define an *input* variable and an *output* variable. While these variables may be any combination of voltages and currents, for convenience in this particular formulation we will assume an input voltage $v_i(t)$ and an output voltage $v_o(t)$. In the various problems that follow, the notation will be modified as appropriate. In all cases, however, it is necessary to define the input and output variables before the transfer function can be defined.

The process of determining the transfer function is outlined in several steps:

1. Convert the circuit parameters to their steady-state ac impedance forms, but leave the frequency $\omega$ as a variable in the impedances (or admittances). The various steady-state impedances are delineated in Figure 4–1, and the conversion process is outlined as follows:

   (a) Each resistance $R$ remains as a real value $R$.
   (b) Each inductance $L$ is replaced by an impedance $j\omega L$ ($j = \sqrt{-1}$).
   (c) Each capacitance $C$ is replaced by an impedance $1/j\omega C$ (or $-j/\omega C$).

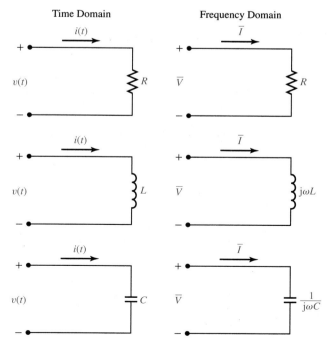

**FIGURE 4–1**
Time domain and frequency domain models of passive circuit parameters.

2. The input is represented as a phasor $\overline{V}_i$.

3. The output is represented as a phasor $\overline{V}_o$.

4. The output $\overline{V}_o$ is determined by basic circuit analysis and algebraic methods in terms of the input $\overline{V}_i$.

## Steady-State Transfer Function Relationships

Once the output is expressed in terms of the input, the transfer function is determined as

$$H(f) = \frac{\overline{V}_o}{\overline{V}_i} \tag{4-1}$$

The input–output relationship is now algebraic in form, and the output phasor at any given frequency may be determined from the input phasor and the transfer function as

$$\overline{V}_o = H(f)\overline{V}_i \tag{4-2}$$

In basic ac circuit theory, the frequency is usually considered to be fixed at a specific value. In frequency response analysis, however, the frequency is considered as a variable, which permits a determination of the response for any possible frequency.

The Fourier transform of a signal can be considered a sort of "extended phasor," in that it represents all frequency components constituting the signal. Therefore, for any frequency domain analysis, a phasor may be replaced with a Fourier transform. The main difference is that a phasor is usually considered at a single frequency, while the Fourier transform is based on an ensemble of frequencies representing the spectrum.

Based on the preceding discussion, any phasor $\overline{V}$ can be interchanged with a Fourier transform $\overline{V}(f)$, or vice versa, as the need arises. For example, Equation 4-2 can be expressed in terms of the input and output Fourier transforms as

$$\overline{V}_o(f) = H(f)\overline{V}_i(f) \tag{4-3}$$

Once the transfer function of a circuit or system is known, it is often convenient to represent it in a *block diagram form,* as shown in Figure 4–2. The input variable is simply a line shown entering the box, and the output variable is a line exiting the box. This type of diagram is often used at the systems level.

## System Response with Fourier Transform

In theory, the response of a linear system to any input can be determined through the use of the Fourier transform and the transfer function. The procedure can be stated as follows:

1. Determine the Fourier transform of the input.

2. Multiply the Fourier transform of the input by the transfer function.

3. Determine the inverse Fourier transform of the output.

Step 3 can be expressed as follows:

$$v_o(t) = \int_{-\infty}^{\infty} \overline{V}_o(f)e^{j\omega t}\, df = \int_{-\infty}^{\infty} H(f)\overline{V}_i(f)e^{j\omega t}\, df \tag{4-4}$$

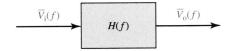

**FIGURE 4–2**
Block diagram representation of linear circuit or system.

**EXAMPLE 4-1**

The circuit shown in Figure 4–3(a) is one of the simplest forms of a *low-pass filter*. The input is $v_1$ and the output is $v_2$. Determine the transfer function $H(f) = \overline{V}_2/\overline{V}_1$.

**SOLUTION**   The circuit is converted to the steady-state frequency domain form by the procedure described in this section, and the resulting circuit is shown in Figure 4–3(b). The input and output time domain variables are replaced by their phasor forms, and the components are replaced by phasor impedances.

To determine the transfer function, the output must be determined in terms of the input, and this is most easily achieved by a simple application of the voltage divider rule. We have

$$\overline{V}_2 = \frac{1/j\omega C}{R + 1/j\omega C} \times \overline{V}_1 = \frac{1}{1 + j\omega RC} \times \overline{V}_1 \tag{4-5}$$

Once the output is determined in terms of the input without other variables present, the transfer function is obtained by forming the ratio of the output to the input. In this case, the result is simply

$$H(f) = \frac{\overline{V}_2}{\overline{V}_1} = \frac{1}{1 + j\omega RC} = \frac{1}{1 + j2\pi fRC} \tag{4-6}$$

A block diagram model of this transfer function is shown in Figure 4–3(c). The analysis of this circuit will be extended in Example 4-3 at the end of the next section.

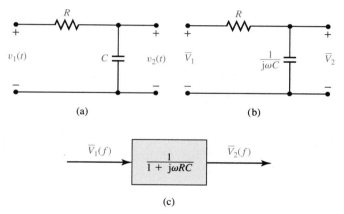

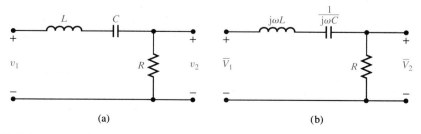

**FIGURE 4–3**
Circuit models for Example 4-1.

**EXAMPLE 4-2**

The circuit of Figure 4–4(a) is a *series resonant circuit,* which can function as one of the most simplified forms of a band-pass filter. The input is $v_1$ and the output is $v_2$. Determine the transfer function $H(f) = \overline{V}_2/\overline{V}_1$.

**FIGURE 4–4**
Circuit models for Example 4-2.

**SOLUTION** The circuit is first converted to the frequency domain, as shown in Figure 4–4(b). As in the preceding example, the simplest way to determine the output in terms of the input is to use the voltage divider. In this case, it will be formulated directly as

$$\frac{\overline{V}_2}{\overline{V}_1} = H(f) = \frac{R}{R + j\omega L + 1/j\omega C} \tag{4-7}$$

The expression on the right can be simplified by multiplying numerator and denominator by $j\omega C$. With some rearrangement, we have

$$H(f) = \frac{j\omega RC}{1 - LC\omega^2 + j\omega RC} \tag{4-8}$$

Further analysis of this function will be made in Example 4-4 at the end of the next section.

---

## 4-2 Frequency Response

The most common practical interpretation of the steady-state transfer function is that of the *amplitude* and *phase* response functions. Since $H(f)$ is a complex function, it may be expressed in polar form in terms of a magnitude and an angle. Each will be a function of the frequency $f$, and they have very important practical applications.

### Amplitude or Magnitude Response

The *amplitude response* function will be denoted $A(f)$, and it is the magnitude of the complex function $H(f)$; that is,

$$A(f) = |H(f)| \tag{4-9}$$

where the vertical bars mean "magnitude of." The *amplitude response* is also called the *magnitude response*.

### Phase Response

The *phase response* function will be denoted $\beta(f)$, and it is the angle of the complex function $H(f)$; that is,

$$\beta(f) = \text{ang}\,[H(f)] \tag{4-10}$$

where "ang" is used to represent "angle of."

### Effect of Amplitude and Phase

At any frequency $f$, the effect of the amplitude response is to alter the amplitude of the spectral component, and the effect of the phase response is to alter the phase. Assume that the input component $v_i(t)$ at a frequency $f$ is of the form

$$v_i(t) = V_{pi}\cos(\omega t + \theta) = V_{pi}\cos(2\pi ft + \theta) \tag{4-11}$$

The steady-state output $v_o(t)$ will then be of the form

$$v_o(t) = A(f)V_{pi}\cos\,[\omega t + \theta + \beta(f)] = V_{po}\cos\,[2\pi ft + \phi] \tag{4-12}$$

If we consider the frequency domain phasor forms, we can express input and output phasors as

$$\overline{V}_i = V_{pi}\angle\theta \tag{4-13}$$

and

$$\overline{V}_o = V_{po}\angle\phi \tag{4-14}$$

Then

$$V_{po} = A(f)V_{pi} \tag{4-15}$$

and

$$\phi = \beta(f) + \theta \tag{4-16}$$

The results of the preceding equations can be summarized in practical terms by the following:

1. Multiply the amplitude of the input phasor or sinusoid at a frequency $f$ by the amplitude response of the circuit evaluated at that particular frequency to determine the amplitude of the output phasor or sinusoid.

2. Add the phase response of the circuit at the particular frequency to the input phase shift to determine the output phase.

### Decibel Amplitude Response

As in the case of gain and loss, a common practice is to specify the amplitude response of a circuit in decibel form. Since the amplitude response $A(f)$ is usually a linear amplitude response based on voltage and/or current variables (as opposed to power), the constant **20** provides the correct scaling factor for decibels. Let $A_{dB}(f)$ represent the decibel amplitude response, which is defined as

$$A_{dB}(f) = 20 \log \left( \frac{A(f)}{A_0} \right) \tag{4-17}$$

where $A_0$ is some convenient reference level. When $A(f) = A_0$, $A_{dB}(f) = 0$ dB, so the reference level is often chosen to normalize the amplitude response to a relative level of 0 dB at some convenient frequency.

In most of our subsequent work, we will choose the simple level $A_0 = 1$ for convenience. In this case, the decibel amplitude response will be expressed simply as

$$A_{dB}(f) = 20 \log A(f) \tag{4-18}$$

With this simplified form, the decibel amplitude response will be positive when the linear response is greater than 1 and negative when the linear amplitude response is less than 1. When the linear response is unity, the decibel response will have a level of 0 dB.

### Frequency Response Measurement

The manner in which the frequency response of a circuit can be measured is illustrated in Figure 4–5. A sinusoidal signal generator with variable frequency is applied to the input, and the frequency is incremented in steps. At each frequency, the input and output voltages (or other variables) are measured, and the phase shift is measured. The ratio of the output

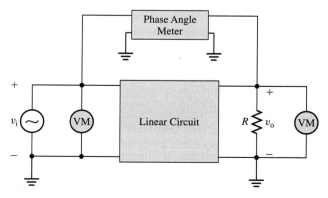

**FIGURE 4–5**

Measurement of amplitude and phase response functions for a circuit.

amplitude to the input amplitude at each frequency is $A(f)$ at that particular frequency, and the difference in phase is $\beta(f)$. Many instruments are capable of measuring the decibel amplitude response directly.

---

**▌▌ EXAMPLE 4-3**

For the simple low-pass filter of Example 4-1, determine (a) the amplitude response (including the 3-dB down frequency), (b) the phase response, and (c) the decibel amplitude response.

**SOLUTION**    The steady-state transfer function was determined in Example 4-1, and is repeated here for convenience.

$$H(f) = \frac{\overline{V}_2}{\overline{V}_1} = \frac{1}{1 + j\omega RC} = \frac{1}{1 + j2\pi fRC} \tag{4-19}$$

(a)  The amplitude response is the magnitude of the transfer function; the magnitude of the transfer function is the magnitude of the numerator divided by the magnitude of the denominator. Hence,

$$A(f) = |H(f)| = \left| \frac{1}{1 + j2\pi fRC} \right| = \frac{|1|}{|1 + j2\pi fRC|} = \frac{1}{\sqrt{1 + (2\pi fRC)^2}} \tag{4-20}$$

It is readily determined that the dc response is $A(0) = 1$. It is convenient to define a reference "cutoff frequency" $f_c$ as the frequency at which the response is $1/\sqrt{2} = 0.7071$, corresponding to 3 dB down from the dc level. By setting the quantity underneath the square root in Equation 4-20 to 2, the 3-dB frequency is readily determined as

$$f_c = \frac{1}{2\pi RC} \tag{4-21}$$

Substitution of this definition in the amplitude response results in the following form:

$$A(f) = \frac{1}{\sqrt{1 + \left(\frac{f}{f_c}\right)^2}} \tag{4-22}$$

(b)  The phase response is the angle of the numerator minus the angle of the denominator. The angle of the numerator is $0°$, and thus

$$\beta(f) = 0° - \tan^{-1}\frac{2\pi fRC}{1} = -\tan^{-1}\left(\frac{f}{f_c}\right) \tag{4-23}$$

At the frequency $f = f_c$, the phase response is readily determined as $-45°$ or $-\pi/4$ radians.

(c)  The decibel amplitude response is determined by taking the logarithms of both sides of Equation 4-22 and multiplying by 20. Making use of the fact that $\log(1/x) = -\log x$ and $\log x^k = k\log x$, the result is

$$A_{dB}(f) = -20\log\sqrt{1 + \left(\frac{f}{f_c}\right)^2} = -10\log\left[1 + \left(\frac{f}{f_c}\right)^2\right] \tag{4-24}$$

The decibel amplitude response and the phase response functions are shown in Figure 4–6. The shapes of the curves shown are based on semilog plots with the frequency scale logarithmic. (Readers familiar with Bode plot analysis will recognize these basic forms.) At the frequency $f = f_c$, the response is down 3 dB from the response at dc, and at frequencies above this value, the response approaches a $-6$ dB/octave rolloff. (An octave represents a doubling of the frequency.)

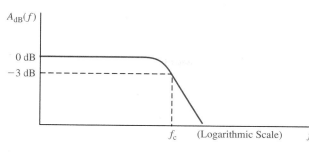
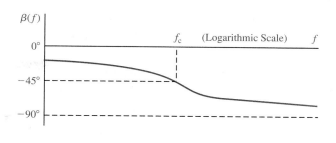

**FIGURE 4–6**
Amplitude and phase responses for the one-pole low-pass filter of Examples 4-1 and 4-3.

This simple filter is not intended to portray the degree of sharpness required in demanding communications filtering applications, but it is provided as a basic exercise in evaluating amplitude and phase functions. It does have wide applications in situations in which a gradual rolloff is adequate to eliminate frequency components well above the desired frequency range, or when the undesired components are already very small in amplitude.

**▌▌ EXAMPLE 4-4**

For the series resonant circuit of Example 4-2, determine (a) the amplitude response, and (b) the phase response.

**SOLUTION**    The transfer function was given in Equation 4-8, and is repeated here for convenience.

$$H(f) = \frac{j\omega RC}{1 - LC\omega^2 + j\omega RC} \tag{4-25}$$

(a) The amplitude response is the magnitude of the transfer function, which can be expressed as the magnitude of the numerator divided by the magnitude of the denominator.

$$A(f) = \frac{\omega RC}{\sqrt{(1 - LC\omega^2)^2 + (\omega RC)^2}} \tag{4-26}$$

(b) The phase response is the angle of the numerator minus the angle of the denominator. The angle of the numerator is 90° or $\pi/2$ radians, and the angle of the denominator can be expressed as an inverse tangent function. Hence,

$$\beta(f) = 90° - \tan^{-1}\left(\frac{\omega RC}{1 - LC\omega^2}\right) \tag{4-27}$$

These functions could be simplified further, but we will revisit band-pass functions and resonant circuits at some length later in the chapter, so we will defer further consideration at this point. The purpose here was to practice determining a transfer function, as well as determining amplitude and phase functions.

## 4-3    Ideal Frequency Domain Filter Models

There are certain characteristics that define the ideal filter model, and before investigating realistic filters, it is desirable to understand these limiting cases. The first consideration is the manner in which an ideal filter would process an input signal.

### Distortionless Filter

Consider the block diagram shown in Figure 4–7. The input to the block is the *desired* signal $v_i(t)$ plus any *undesired* signal or group of signals, denoted here as $v_u(t)$. The output is

**FIGURE 4–7**
Block diagram of filter used to separate desired signal from undesired signal.

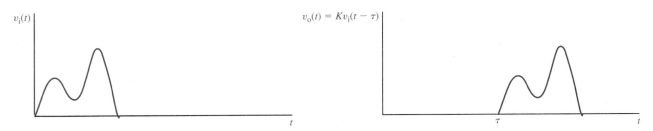

**FIGURE 4–8**
Input and output of distortionless filter.

$v_o(t)$. The purpose of the filter is to eliminate the undesired signal while preserving the form of the desired signal as close as possible to the original form.

The process of filtering results in a certain amount of delay produced by the reactive elements of the filter, and a possible change in the signal level. Therefore, the best we can hope for is that the output signal will be a delayed version of the desired signal, with a possible change in the amplitude level. If the shape of the desired signal is preserved, the output can be expressed as

$$v_o(t) = K v_i(t - \tau) \tag{4-28}$$

where $K$ represents a change in level and $\tau$ is the delay. This concept is illustrated in Figure 4–8 for a hypothetical signal. In this particular case $K = 1$, but it could be either greater or less than 1. The undesired part of the input is not shown, since it does not appear in the output. A filter that can produce this effect would be called a *distortionless filter.*

## Constant Amplitude and Linear Phase

The frequency domain behavior of this ideal filter will now be investigated. Taking the Fourier transform of both sides of Equation 4-28, and applying operation O-4 of Table 3–1, we obtain

$$\overline{V}_o(f) = Ke^{-j\omega\tau}\overline{V}_i(f) \tag{4-29}$$

The steady-state transfer function is obtained as

$$H(f) = \frac{\overline{V}_o(f)}{\overline{V}_i(f)} = Ke^{-j\omega\tau} = K\angle - \omega\tau \tag{4-30}$$

The amplitude and phase functions are

$$A(f) = K \tag{4-31}$$

and

$$\beta(f) = -\omega\tau = -2\pi\tau f \tag{4-32}$$

From these results, the following appear to be the desired properties of an ideal filter:

1. The amplitude response should be *constant* over the passband of the desired signal.
2. The phase response should be a *linear* function of frequency; that is, a *constant times frequency* over the passband of the desired signal.

Obviously, the amplitude response cannnot be constant over all frequencies or the desired and undesired signals would be treated in the same way. The process of frequency domain filtering depends on the assumption that the spectrum of the undesired signals occupies a frequency range different than that of the desired signal. With this assumption, the amplitude response should be zero or close to it in the frequency range of the components to be rejected.

## Frequency Regions

An ideal frequency domain filter amplitude responses has two primary regions of interest: (1) *passband* and (2) *stopband.*

### Passband

The passband of an ideal filter is the frequency range in which the amplitude response is a constant nonzero value. Often, $A(f) = 1$ or $A_{dB}(f) = 0$ dB in this range. However, as long as $A(f)$ has a constant nonzero value in this range, any change in level can be adjusted by external components. The important consideration is that the amplitudes of all frequency components in the passband be treated alike.

### Stopband

The stopband of an ideal filter is the frequency range in which the amplitude response is zero, meaning that the decibel response would have an infinite negative value.

## Four Ideal Amplitude Response Models

In general, there are four ideal frequency domain models for the ideal amplitude response characteristics. These four forms are shown in Figure 4–9.

### Low-Pass

A low-pass filter is characterized by a passband at low frequencies and a stopband at high frequencies, as shown in Figure 4–9(a). The theoretical ideal model would pass all frequency components below the cutoff frequency $f_c$ and reject all frequencies above it.

### High-Pass

A high-pass filter is characterized by a stopband at low frequencies and a passband at high frequencies, as shown in Figure 4–9(b). The nature of the high-pass filter is exactly opposite to that of the low-pass filter, and the cutoff frequency $f_c$ is at the lower edge of the passband.

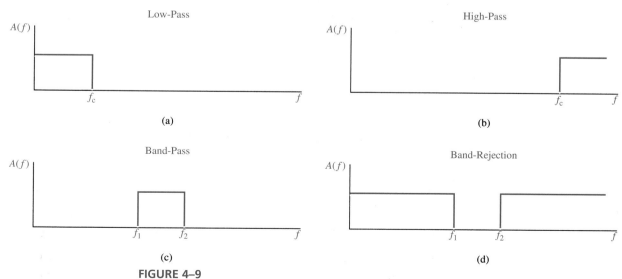

**FIGURE 4–9**

Ideal frequency domain amplitude response models.

**Band-Pass**

A band-pass filter has a passband that lies between two frequencies $f_1$ and $f_2$, as shown in Figure 4–9(c). Thus, it will pass only frequencies lying between two band-edge frequencies, and the stopband consists of two regions, one below and the other above the passband.

**Band-Rejection**

A band-rejection filter has a stopband that lies between two frequencies $f_1$ and $f_2$, as shown in Figure 4–9(d). This filters rejects all frequencies over a certain range, and the passband consists of two regions, one below and the other above the stopband.

## Practical Filter Characteristics

No real filter exhibits either a perfectly constant amplitude response or a perfectly linear phase response over any reasonable frequency range. A more realistic goal is that of a nearly constant amplitude response over a prescribed frequency range and a response close to zero outside of this range. Likewise, the phase response should be close to linear over the frequency range of the desired signal (the phase response in the range in which the amplitude response is near zero is usually not important). These more realistic characteristics are illustrated for an arbitrary band-pass filter in Figure 4–10. In the case of a realistic filter, there is a third region called the *transition band,* which is defined as follows.

**Transition Band**

The transition band is the range of frequencies between the passband and the stopband. The definition is somewhat arbitrary and depends on the nature of the filter. For example, the passband may be defined as the range of frequencies in which the relative decibel amplitude response is between 0 dB and $-3$ dB, and the stopband may be defined as the range of frequencies in which the relative decibel amplitude response is below $-50$ dB. The transition band would then be the range in which the response changes from a level of $-3$ dB to $-50$ dB. Values vary considerably from one filter type to another, and these should be considered as examples only. For a band-pass or band-rejection filter, there are two transition bands. (See Figure 4–10.)

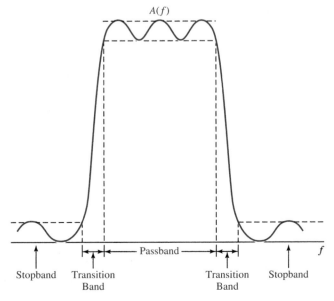

**FIGURE 4–10**
Regions of a practical filter.

## Time Delay Characteristics

In some applications, the time delay of a signal passing through a filter is more significant than the phase shift. Two definitions of certain time delay characteristics will now be given: the *phase delay* $T_p$ and the *group delay* (or *envelope delay*) $T_g$. These functions are defined as follows:

$$T_p(f) = \frac{-\beta(f)}{\omega} \tag{4-33}$$

$$T_g(f) = \frac{-d\beta(f)}{d\omega} \tag{4-34}$$

The graphical significance of these operations is illustrated in Figure 4–11. The phase delay at a given frequency is proportional to the slope of the *secant line* from dc to the particular frequency, and is a type of overall delay parameter. The group delay is proportional to the slope of the *tangent line* at a particular frequency, and represents a narrow-range delay parameter.

Consider now the case of a filter with a constant amplitude response and a linear phase response as described by Equations 4-31 and 4-32, respectively. It is readily verified that

$$T_p(f) = T_g(f) = \tau \tag{4-35}$$

For the ideal filter, the phase and group delay parameters are identical and represent the exact time delay of the signal, which has not been distorted in this case. In the more general case, where the amplitude response is not constant in the passband and the phase response is not linear, it is more difficult to precisely define the delay, since a signal will undergo some distortion in passing through the filter.

## Interpreting Phase and Group Delay

The following statements can be used as guidelines for interpreting the significance of phase and group delay functions when the phase does not deviate markedly from the ideal linear case:

1. The phase delay computed over the range of frequencies constituting the spectrum of a wideband signal may represent a reasonable approximation to the delay of the signal.

2. When a narrow-band modulated signal is passed through a filter, the group delay computed at the center of the spectrum is a good approximation to the delay of the intelligence or envelope of the signal.

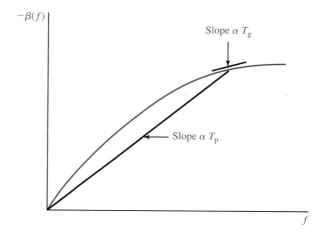

**FIGURE 4–11**
Graphical significance of phase delay and group delay.

## 4-4   Implementation of Filters

A practical consideration for any filter is its design and implementation. The practical design of filters is an outgrowth of a classical area of electrical theory called *network synthesis*. Much research has been done in this area since the early days of the communications industry, and many books and journal articles have been written on the subject. Here we will acquaint the reader with some of the major properties of filters, especially in regard to their general performance and specifications.

While there are a variety of ways of implementing filters, one of the most common approaches in the frequency ranges encountered in communications is through the use of *ladder networks*. A typical ladder structure is shown in Figure 4–12, and it consists of successive *series* and *shunt* blocks. A block may consist of a single element or a special combination such as a resonant circuit.

In the most common situation there is a resistance at either the source end or the load end, or both, but with all other elements *reactive*—that is, inductance and capacitance. When the resistance appears at only one end, the filter is said to be *single-terminated,* and when resistance appears at both ends, it is said to be *double-terminated.* Often a terminating resistance may be the source resistance or the load resistance, so it may not be part of the filter itself, but its presence is required to produce the proper response. If the filter is designed to operate with an ideal voltage source input, the first element is a series reactance, and if the filter is designed to operate with an ideal current source input, the first element is a shunt reactance. In both of these cases, a terminating resistance is required at the output. If the filter is designed with a terminating resistance on the input, any practical source having the required output resistance will work.

The basic strategy is as follows: The elements in the series blocks should present a low impedance in the passband and a high impedance in the stopband. Conversely, the elements in the shunt blocks should present a high impedance in the passband and a low impedance in the stopband. This combination results in very little opposition to current flow along the top and very little current shunted to the common terminal in the passband. However, in the stopband, there is large opposition to current flow along the top and much of the current is shunted to the common terminal. Along with the reactive elements of inductance and/or capacitance, there will usually be one or more terminating resistances, which are required as part of the overall system response. Of course, the impedance of an ideal resistance does not vary with frequency.

### Poles

All filters are characterized by a parameter known as the *number of poles*. This is a term relating to the mathematical properties of the transfer function. In general, the complexity of the filter increases with the number of poles. In the most basic filter types, *the number of poles is equal to the number of reactive elements (L and C) in the circuit,* and that will be the property of most filters considered in this text. However, some filter types have more reactive elements than the actual number of poles in the transfer function.

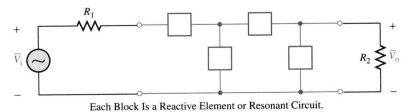

Each Block Is a Reactive Element or Resonant Circuit.

**FIGURE 4–12**
Ladder structure used for many filter circuits.

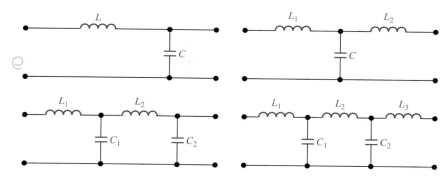

**FIGURE 4–13**
Typical low-pass filter structures (terminating resistances not shown).

## Low-Pass Filter

For low-pass filters designed with ladder structures, the *series* elements are *inductors* and the *shunt* elements are *capacitors*. This produces the expected results, since the impedance of an inductor increases with frequency while that of a capacitor decreases with frequency. Some typical low-pass filters with poles ranging from two to five are shown in Figure 4–13.

## High-Pass Filter

For high-pass filters designed with ladder structures, the *series* elements are *capacitors* and the *shunt* elements are *inductors*. This arrangement works since the impedance of the series path is lowest at high frequencies and highest at low frequencies, and the opposite property is true of the shunt components. Some typical high-pass filters with poles ranging from two to four are shown in Figure 4–14.

## Band-Pass Filter

A circuit that produces a *low impedance* over a specific frequency range while providing a *high impedance* above and below that frequency range is a *series resonant circuit*. In contrast, a circuit that produces a *high impedance* over a frequency range while providing a *low impedance* above and below that frequency range is a *parallel resonant* circuit. This means that a band-pass response can be created by employing *series resonant circuits* as the *series* elements and *parallel resonant circuits* as the *shunt* elements. Technically, each

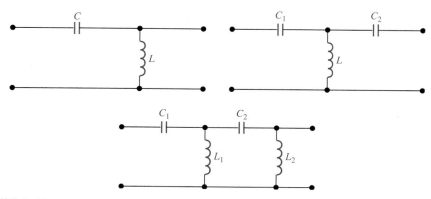

**FIGURE 4–14**
Typical high-pass filter structures (terminating resistances not shown).

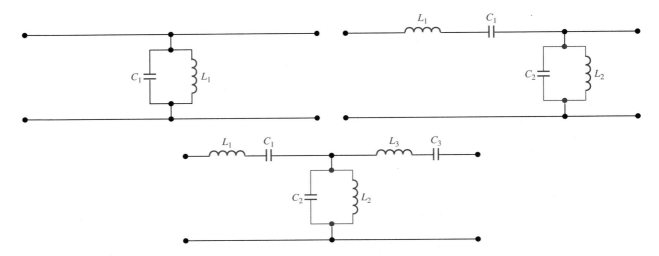

**FIGURE 4–15**
Typical band-pass filter structures (terminating resistances not shown).

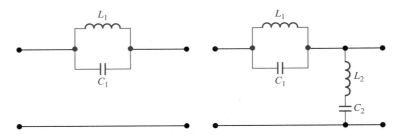

**FIGURE 4–16**
Typical band-rejection filter structures (terminating resistances not shown).

resonant circuit produces two poles, but it is a common practice in specifying band-pass filter characteristics to use the term *pole-pair*. Thus, each resonant circuit or its equivalent produces a *pole-pair*. Some typical band-pass filters with the number of pole-pairs ranging from one to three are shown in Figure 4–15.

## Band-Rejection Filter

A band-rejection filter is obtained by using the opposite strategy to that of a band-pass filter; that is, *parallel resonant circuits* are used as the *series* elements and *series resonant circuits* are used as the *shunt* elements. Some typical band-rejection filters with one and two pole-pairs are shown in Figure 4–16.

## Other Filter Implementations

While the ladder structures previously discussed are probably the most widely used methods for implementing communication filters, a few other types will be briefly mentioned. *Active filters* have become widely used in the past few decades, particularly in relatively low frequency applications. Active filters have the advantage that they can achieve the same type of characteristics as passive *RLC* filters without using inductors. At low frequencies, very large inductances are required with *RLC* filters, and the inductors required are very large and lossy, so *active RC filters* have significant advantages in that range. However, these filters require active devices, such as one or more operational amplifiers, and feedback is required. While there are no doubt exceptions to the rule, active filters are

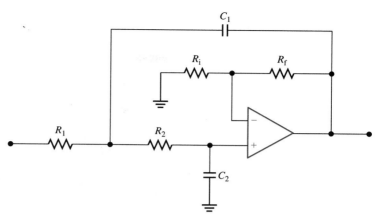

**FIGURE 4–17**
Typical active filter circuit.

generally most practical in the frequency range well below 1 MHz or so. A typical low-pass active filter (two poles) is shown in Figure 4–17.

In the frequency range of several hundred megahertz and above, *transmission line filters* are practical. These filters make use of the property that a transmission line terminated in a short or an open appears as a reactance. Depending on the length, the reactance may be either positive or negative; that is, the line may act as either an inductance or a capacitance. Thus, the filter is implemented with sections of transmission line chosen to have the required reactances at the cutoff frequency. As the frequency range is increased in the microwave range above about 1 GHz or so, *cavity resonator filters* find application.

Finally, no discussion of modern filter implementation would be complete without some mention of the concept of a *digital filter*. A digital filter is a numerical algorithm that transforms an input data signal into an output data signal, with a desired filtering objective accomplished through data processing. Most digital filters employ *constant coefficient difference equations* that relate the output data stream to the input data stream. If the coefficients of the difference equations are chosen properly, the digital filter accomplishes the same operation on the data as an *RLC* filter.

A possible system employing a digital filter is illustrated in Figure 4–18. The input analog signal is applied to an *analog-to-digital (A/D) converter,* which converts the analog signal into a sequence of digital numbers (or *words*). The signal is then processed by the digital filter, which might be a general purpose computer programmed as a digital filter, or it might be a dedicated microprocessor-based signal processing IC chip. The output is then converted back to an analog signal by means of a *digital-to-analog (D/A) converter*. With some systems, the input might be in digital form initially, and/or the output might be desired in a digital form, in which case one or more of the conversions could be eliminated. Using the accuracy attainable with modern digital circuitry, it is possible to realize responses not physically possible with conventional analog circuitry. It is also possible to take advantage of multiplexing techniques, in which a large number of signals are processed by one digital signal processing unit.

**FIGURE 4–18**
Illustration of digital filter with analog input and output.

## 4-5  Some Common Filter Characteristics

The ideal block characteristics considered in earlier sections cannot be attained with "real-life" filters. It turns out also that some of the filter types having the best amplitude response tend to have the poorest phase response, and vice versa. Therefore, all filter design is a compromise, in which some deviation from the ideal amplitude and phase characteristics must be accepted.

The form of the amplitude or phase response for any filter constructed from lumped elements of resistance, capacitance, and inductance must satisfy certain mathematical forms. From these possible mathematical forms, a number of useful and widely employed filter characteristics have been developed. Some of the most widely used types will be surveyed here.

### Frequency Variables

It turns out that the same mathematical forms may be used for all four of the basic filter types by altering the form of the frequency variable. For that purpose, we will employ a general normalized frequency variable $u$ that can be used in all cases. First, the following definitions should be noted:

$f$ = frequency variable

$f_c$ = cutoff frequency for low-pass and high-pass filters

$f_0$ = geometric center frequency for band-pass and band-rejection filters

The geometric center frequency $f_0$ for the band-pass and band-rejection filters is related to the band-edge frequencies $f_1$ and $f_2$ by

$$f_0 = \sqrt{f_1 f_2} \tag{4-36}$$

It is convenient to define the filter bandwidth $B$ and $Q$ as

$$B = f_2 - f_1 \tag{4-37}$$

$$Q = \frac{f_0}{B} \tag{4-38}$$

The variable $u$ is then defined according to filter type, as follows:

low-pass:    $$u = \frac{f}{f_c} \tag{4-39}$$

high-pass:    $$u = \frac{f_c}{f} \tag{4-40}$$

band-pass:    $$u = Q\left(\frac{f}{f_0} - \frac{f_0}{f}\right) \tag{4-41}$$

band-rejection:    $$u = \frac{1}{Q\left(\frac{f}{f_0} - \frac{f_0}{f}\right)} \tag{4-42}$$

### Butterworth Response

The Butterworth amplitude response of order $m$ is given by

$$A(u) = \frac{1}{\sqrt{1 + u^{2m}}} \tag{4-43}$$

The low-pass function is illustrated for several values of $m$ in Figure 4–19. The value $u = 1$ in the normalized response corresponds to the cutoff frequency for low-pass and high-pass filters, and to the band-edge frequencies for the band-pass and band-rejection filters. At

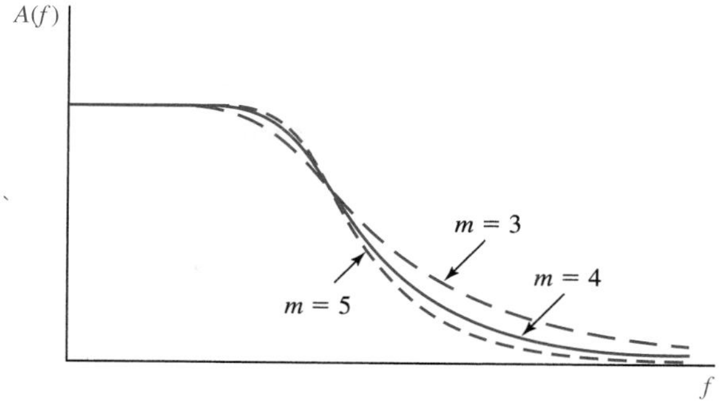


FIGURE 4–19
Amplitude response functions for several Butterworth filters.

these frequencies, the amplitude response is $1/\sqrt{2} = 0.7071$, which corresponds to a decibel response of $-3$ dB.

It should be noted that while $m$ in Equation 4-43 is the actual number of poles for low-pass and high-pass filters, it is the number of pole-pairs for band-pass and band-rejection filters. Thus, for band-pass and band-rejection filters, the resulting number of poles is actually $2m$.

The Butterworth amplitude response can be shown to be optimum near dc in the *maximally flat sense.* This means that as many lower-order derivatives of the approximation as possible are equated to zero at $f = 0$. The Butterworth function is a *monotonically decreasing* function of frequency, which means that the amplitude response always continues to decrease as the band-edge is approached; that is, the slope of the response never changes sign.

As the order $m$ increases, the response becomes "flatter" in the passband, and becomes "sharper" in the stopband. Above cutoff, the Butterworth amplitude response of order $m$ approaches a high-frequency asymptote having a slope of $-6m$ dB/octave.

## Chebyshev Response

The Chebyshev or *equiripple* amplitude response is derived from a set of mathematical functions called the *Chebyshev polynomials.* The Chebyshev polynomial of order $m$ is denoted $C_{\text{m}}(u)$, and can be determined from the relationship

$$C_{\text{m}}(u) = \cos(m \cos^{-1} u) \tag{4-44}$$

Believe it or not, this function can be simplified to a polynomial function for any integer value of $m$.

The basic Chebyshev amplitude response is defined as

$$A(u) = \frac{K}{\sqrt{1 + \varepsilon^2 C_{\text{m}}^2(u)}} \tag{4-45}$$

where $K$ is a constant used to adjust the overall level and $\varepsilon^2$ is a parameter used to adjust the ripple level. The parameter $m$ is both the order of the Chebyshev polynomial and the number of poles (low-pass and high-pass) or pole-pairs (band-pass and band-rejection) r the filters.

Since both the order of the filter and the ripple level are parameters, a variety of different Chebyshev filter responses can be created. The general forms for several orders are illustrated in Figure 4–20. The amplitude response exhibits a ripple effect in the passband, with the ripple maxima and minima exhibiting a constant ratio. For example, a ripple level in the passband of 1 dB means that the decibel ratio of all maxima is 1 dB above all minima. The number of maxima and minima in the passband is equal to the order of the filter.

Most references define the cutoff frequency of Chebyshev filters as the point at which the amplitude response drops below the specified passband ripple bound. This definition is

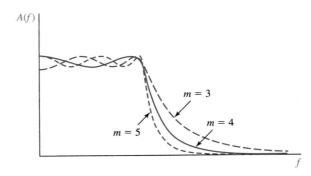

**FIGURE 4–20**
Amplitude response functions for several Chebyshev filters.

consistent with that of the Butterworth filter only in the case of a passband ripple bound of 3 dB. However, some filter references use the 3 dB point as the definition of cutoff for all ripple values. In using filter design data, therefore, check the manner in which the cutoff frequency is defined.

As the order of the filter is increased, the attenuation in the stopband increases for a given passband ripple. For a given order, the stopband attenuation increases as the passband ripple is allowed to increase. Thus, there is a direct trade-off between the allowable passband ripple and the stopband attenuation. As in the case of the Butterworth response, the Chebyshev response of order $m$ approaches a high frequency rolloff rate with a slope of $-6m$ dB/octave. However, for a given order, the Chebyshev filter will usually have a larger attenuation than the Butterworth filter.

## Maximally Flat Time Delay Filter

Both the Butterworth and Chebyshev responses were obtained from approximations involving only the amplitude response, and no attention was paid to the phase response in either case. In some applications, the phase (or time delay) is more important than the amplitude response.

One approximation in which linear phase (or constant time delay) is optimized is the *Gaussian* or *maximally flat time delay* (MFTD) response. The amplitude response that results from the MFTD approximation has a low-pass shape with a monotonically decreasing behavior as the frequency is increased, as illustrated in Figure 4–21. However, compared with the Butterworth amplitude response, the MFTD response is not nearly as flat in the passband, and the attenuation in the stopband is not as great as either the Butterworth or Chebyshev responses. It does, however, eventually provide a rolloff rate of $-6$ dB/octave.

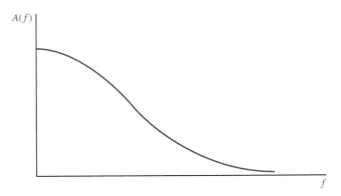

**FIGURE 4–21**
Form of amplitude response for Gaussian (maximally flat time delay) filter.

The MFTD filter is used where excellent phase shift (or time delay) characteristics are required, but where the amplitude response need not display a rapid attenuation increase just above cutoff.

## Comparison

Of the three filter types considered thus far, a brief and somewhat simplified comparison is provided in the table that follows. The comparison assumes the same number of poles (or pole-pairs) for the three cases.

| Filter | Amplitude | Phase |
|---|---|---|
| Butterworth | medium | medium |
| Chebyshev | best | worst |
| MFTD | worst | best |

The Chebyshev filter offers the "best" amplitude response in the sense that it displays a more rapid cutoff rate just above the cutoff frequency, while the MFTD filter is "worst" in this respect. In contrast, the Chebyshev filter has the "worst" phase response (greatest deviation from linear phase), and the MFTD filter has the "best" phase response; that is, the most linear phase or most constant time delay. The Butterworth filter falls between these extremes, and offers a reasonable compromise between amplitude and phase.

## Other Filter Types

A few other filter types will be briefly mentioned here. The *inverted Chebyshev response* is characterized by a maximally flat passband response and an equiripple stopband response, as illustrated in Figure 4–22. The *Cauer* or *elliptic function filter* has both an equiripple passband and an equiripple stopband, as illustrated in Figure 4–23.

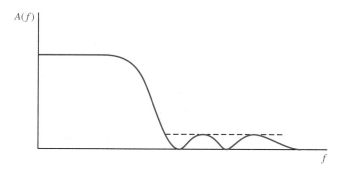

**FIGURE 4–22**
Form of amplitude response for inverted Chebyshev filter.

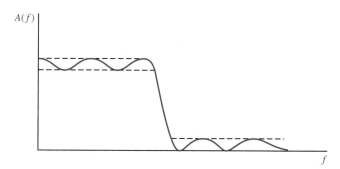

**FIGURE 4–23**
Form of amplitude response for Cauer (or elliptic function) filter.

An earlier form of filter technology is that of *image parameter* filter design, which employs so-called *constant-k* and *m-derived* filter sections. This technique is based in part on principles similar to transmission line theory, and involves matching the impedances between successive sections. Most early communication system filters were designed around this concept. There are still situations in which the image parameter approach yields the most satisfactory solution, particularly in systems involving sections of transmission lines.

---

**▐▌ EXAMPLE 4-5**

A four-pole low-pass Butterworth filter has a 3-dB cutoff frequency of 1 kHz. Calculate the linear amplitude response and the decibel amplitude response at each of the following frequencies: 500 Hz, 2 kHz, and 10 kHz.

**SOLUTION**  The basic form of the Butterworth response was given in Equation 4-43, and when the definition of $u$ from Equation 4-39 and other parameters are substituted, the low-pass four-pole response is

$$A(f) = \frac{1}{\sqrt{1 + \left(\frac{f}{1000}\right)^8}} \tag{4-46}$$

For convenience, the functional form has been redefined as $A(f)$. The decibel amplitude response is

$$A_{dB}(f) = 20 \log A(f) = -10 \log \left[1 + \left(\frac{f}{1000}\right)^8\right] \tag{4-47}$$

where the latter form utilized some basic logarithmic identities. We can now substitute specific values of frequency. At 500 Hz, we have

$$A(500) = \frac{1}{\sqrt{1 + \left(\frac{500}{1000}\right)^8}} = \frac{1}{\sqrt{1 + 0.003906}} = 0.9981 \tag{4-48}$$

The decibel amplitude response at this frequency is

$$A_{dB}(500) = 20 \log 0.9981 = 20 \times (-0.0008259) = -0.016 \text{ dB} \tag{4-49}$$

Obviously, the response has not deviated very much at this frequency from the dc level of 0 dB.

The preceding process is repeated at the other frequencies, and the results are provided in the table that follows.

| $f$ | $A(f)$ | $A_{dB}(f)$ |
|---|---|---|
| 500 Hz | 0.9981 | −0.016 dB |
| 2 kHz | 0.06238 | −24.1 dB |
| 10 kHz | 0.0001 | −80 dB |

Above the cutoff frequency, the response drops very rapidly, as expected.    ▐▌

---

## 4-6  One Pole-Pair Resonant Circuits

One pole-pair resonant circuits are among the most common band-pass filter types utilized in communications systems. While they don't exhibit the nearly "brick wall" characteristics possible with higher-order precision filters, they are very useful in many applications. They are also relatively straight-forward to design and tune, and one does not have to be a filter specialist to implement them.

The general form of the transfer function of a one pole-pair band-pass filter is given by

$$H(f) = \frac{1}{1 + jQ\left(\frac{f}{f_0} - \frac{f_0}{f}\right)} \tag{4-50}$$

The various parameters are given by

$f_0$ = geometric center frequency

$B$ = 3-dB bandwidth = $f_2 - f_1$

$f_1$ and $f_2$ = lower and upper 3-dB frequencies

The following relationships apply:

$$f_0 = \sqrt{f_1 f_2} \tag{4-51}$$

$$Q = \frac{f_0}{B} \tag{4-52}$$

In a sense, the one pole-pair band-pass filter can be considered a limiting case of a Butterworth band-pass function for $m = 1$. (Recall that since this is a band-pass filter, it is actually a second-order circuit.) In some cases, the numerator in Equation 4-50 may have some value other than unity, but the *relative* response will be the same.

The amplitude response of the one pole-pair filter may be expressed as

$$A(f) = \frac{1}{\sqrt{1 + Q^2\left(\frac{f}{f_0} - \frac{f_0}{f}\right)^2}} \tag{4-53}$$

The decibel amplitude response can be expressed as

$$A_{\mathrm{dB}}(f) = -10 \log\left[1 + Q^2\left(\frac{f}{f_0} - \frac{f_0}{f}\right)^2\right] \tag{4-54}$$

Plots for several values of $Q$ are shown in Figure 4–24. Note that the width of the band-pass response decreases as the $Q$ increases.

## Series Resonant Circuit

The basic form of a series resonant circuit arranged as a one pole-pair band-pass filter is shown in Figure 4–25. This circuit was partially analyzed in Examples 4-2 and 4-4. The

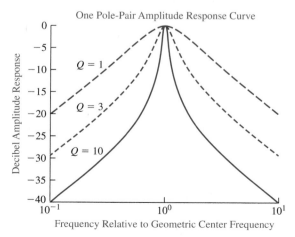

**FIGURE 4–24**
One pole-pair band-pass filter curves.

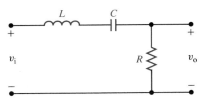

**FIGURE 4–25**
Series resonant circuit for implementing one pole-pair band-pass filter.

inquisitive reader may wish to try to verify the results that follow. The various relationships for some of the parameters are

$$f_0 = \frac{1}{2\pi\sqrt{LC}}$$

(4-55)

and

$$Q = \frac{X}{R} = \frac{2\pi f_0 L}{R} = \frac{1}{2\pi f_0 RC} = \frac{1}{R}\sqrt{\frac{L}{C}}$$

(4-56)

where $X$ is either the inductive reactance or the magnitude of the capacitive reactance at resonance. (They are equal at that frequency.)

## Parallel Resonant Circuit

The most common form in which a parallel resonant circuit occurs is shown in Figure 4–26. The best way to analyze this circuit is to first perform a source transformation on the voltage source with series resistance, and convert the combination to a current source in parallel with the resistance. (One can also apply Norton's theorem to achieve the same result.)

For the parallel circuit, the relationships are

$$f_0 = \frac{1}{2\pi\sqrt{LC}}$$

(4-57)

and

$$Q = \frac{R}{X} = \frac{R}{2\pi f_0 L} = 2\pi f_0 RC = R\sqrt{\frac{C}{L}}$$

(4-58)

where again $X$ is either the inductive reactance or the magnitude of the capacitive reactance. Note that the expression for resonant frequency is the same for parallel as for series resonance. However, *the expressions for Q in a parallel resonant circuit are the reciprocals of the relationships for a series resonant circuit.* Thus, a combination of components that would give a high $Q$ for a series circuit would give a very low $Q$ if rearranged in parallel form. This property often establishes the practicability of one or the other circuit in a given situation. Because of the usual presence of fairly high equivalent resistances at the input and output circuits of active devices, parallel circuits are probably more often used in communications circuits in order to achieve high $Q$ values.

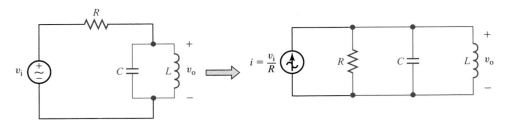

**FIGURE 4–26**
Parallel resonant circuit for implementing one pole-pair band-pass filter.

<table>
<tr><td>■■ EXAMPLE 4-6</td><td>A parallel resonant circuit has $L = 4\ \mu H$, $C = 9$ pF, and $R = 20$ kΩ. Determine (a) the resonant frequency, (b) $Q$, and (c) 3-dB bandwidth.</td></tr>
</table>

**SOLUTION**

(a)  The resonant frequency is

$$f_0 = \frac{1}{2\pi\sqrt{LC}} = \frac{1}{2\pi\sqrt{4 \times 10^{-6} \times 9 \times 10^{-12}}} = 26.53 \text{ MHz} \tag{4-59}$$

(b)  The $Q$ is

$$Q = R\sqrt{\frac{C}{L}} = 20 \times 10^3 \times \sqrt{\frac{9 \times 10^{-12}}{4 \times 10^{-6}}} = 30 \tag{4-60}$$

(c)  The 3-dB bandwidth is

$$B = \frac{f_0}{Q} = \frac{26.53 \text{ MHz}}{30} = 884 \text{ kHz} \tag{4-61}$$

## 4-7  Pulse Transmission Approximations

From the preceding material, we have seen that the effect of a transmission system with finite bandwidth is to reject some of the Fourier spectral components of a wideband signal. The resulting output signal will then display some distortion. The limited bandwidth can be deliberately introduced in some cases (e.g., a channel separation filter), or it may occur as a natural undesirable property of many components.

Extensive and detailed system analysis at a fundamental level necessarily requires the use of Fourier series and transform theory to predict accurately the results of finite band limiting. This level of analysis is often not feasible in dealing with many practical, day-to-day hardware applications. Fortunately, there are some simple "rules of thumb" or estimates that permit the applied communications technologist or engineer to predict approximate bounds for certain bandwidth requirements, and our focus will be directed toward these estimates.

It must be stated at the outset that the formulas to be given are *not* intended as *exact* formulas. They are simply *estimates* that predict the approximate range of bandwidth requirements, and their use must be tempered by good judgment and allowances for worst-case conditions. Indeed, there are variations in these formulas in the literature. Therefore, accept the results that follow as general guidelines.

The approximations will be applied to pulse transmission. A pulse is a worst-case condition as far as bandwidth is concerned, and the results are quite appropriate for dealing with digital data transmitted as pulses.

We will consider two distinctly separate types of criteria: (1) approximate pulse reproduction, which will be denoted as *coarse* reproduction, and (2) reasonably exact reproduction, which will be denoted as *fine* reproduction.

### Coarse and Fine Reproduction

Consider for purposes of discussion the pulse shown in Figure 4–27(a). The *coarse* condition is common in situations in which the presence of the pulse must be recognized, and in which the peak level must be approximately preserved, but in which some smearing or rounding of the beginning and ending of the pulse is acceptable. This condition applies to many forms of digital data transmission and radar pulses. In digital data transmission, for example, the transmitted "ones" and "zeros" may be restored perfectly at the receiver as

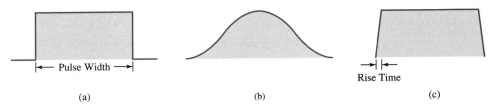

**FIGURE 4–27**
(a) Pulse waveform. (b) Filtered pulse corresponding to "coarse" reproduction. (c) Filtered pulse corresponding to "fine" reproduction.

long as the receiver can recognize which was transmitted. The coarse criterion of pulse reproduction is illustrated in Figure 4–27(b).

The *fine* condition applies to systems in which a more exact pulse shape must be preserved. This condition is required in modulation systems, in which the beginning and ending of pulses are important and in which negligible smearing or spreading should occur. The fine criterion is illustrated in Figure 4–27(c).

## Bandwidth Requirements

Extensive studies have been made of the bandwidth requirements for these conditions, with the following conclusions.

### Coarse Reproduction

The bandwidth $B$ for coarse reproduction is inversely proportional to the width of a given pulse. In other words,

$$B = \frac{K_1}{\text{pulse width}} \tag{4-62}$$

where $K_1$ is a constant that will be discussed shortly.

### Fine Reproduction

The bandwidth required for accurate reproduction of a pulse is proportional to the reciprocal of the rise (or fall) time, and can be expressed as

$$B = \frac{K_2}{\text{rise time}} \tag{4-63}$$

where $K_2$ will also be discussed shortly. For Equation 4-63 to be valid, the rise time of the pulse prior to filtering must be much less than the allowable rise time.

To summarize, *the bandwidth required for coarse reproduction is inversely proportional to the pulse width, and the bandwidth required for fine reproduction is inversely proportional to the allowable rise time.*

## Determining the Coarse Reproduction Constant

Consider first a baseband pulse of width $\tau$. There are several approaches that can be used to approximate the coarse bandwidth criterion, but one of the simplest is to assume a worst-case condition of a periodic pulse train oscillating between 0 and an amplitude $A$ with a 50% duty cycle. The period is $2\tau$ and, hence, the fundamental frequency is $f_1 = 1/T = 1/2\tau = 0.5/\tau$. The dc component plus the fundamental component would provide a vertically shifted sinusoidal pulse having sufficient amplitude, and this criterion would suggest that $K_1 = 0.5$. For a nonperiodic pulse, this choice results in the inclusion of all spectral components up to one-half of the first zero-crossing frequency.

For RF pulses, the spectrum appears on both sides of the center frequency. Hence, the same type of criterion would result in a choice of $K_1 = 1$.

### Determining the Fine Reproduction Constant

The reader may be familiar with a formula from basic electronics that states that the rise time of a basic amplifier circuit is 0.35/(3-dB bandwidth). If we used this formula, we would select $K_2 = 0.35$. However, that formula is derived on the basis of a one-pole low-pass −6 dB/octave rolloff model, and it uses the standard IEEE definition of rise time as the time between the 10% and 90% points. For filters exhibiting a sharper cutoff rate, the constant may be higher.

To be conservative and to make the result easier to remember, we will choose $K_2 = 0.5$ for the baseband case and $K_2 = 1$ for the RF case. These reasonable choices result in the same numerator constants for fine reproduction as for coarse reproduction.

### Summary of Formulas for Baseband Pulses

The formulas for baseband pulse transmission are as follows:

$$\text{coarse reproduction:} \quad B = \frac{0.5}{\text{pulse width}} \tag{4-64}$$

$$\text{fine reproduction:} \quad B = \frac{0.5}{\text{rise time}} \tag{4-65}$$

### Summary of Formulas for RF Pulses

$$\text{coarse reproduction:} \quad B = \frac{1}{\text{pulse width}} \tag{4-66}$$

$$\text{fine reproduction:} \quad B = \frac{1}{\text{rise time}} \tag{4-67}$$

---

**▌▌ EXAMPLE 4-6**

A baseband pulse generator produces pulses having widths of 2 μs. Determine the approximate bandwidth for (a) coarse reproduction and (b) fine reproduction, if the allowable rise and fall times are 10 ns. (Assume that the initial rise and fall times of the pulse are much smaller than 10 ns.)

**SOLUTION**

(a) The approximate bandwidth for coarse reproduction is

$$B = \frac{0.5}{\text{pulse width}} = \frac{0.5}{2 \times 10^{-6}} = 250 \text{ kHz} \tag{4-68}$$

(b) The approximate bandwidth for fine reproduction is

$$B = \frac{0.5}{\text{rise time}} = \frac{0.5}{10 \times 10^{-9}} = 50 \text{ MHz} \tag{4-69}$$

▌▌

---

## 4-8 Multisim® Examples (Optional)

The first few examples presented in this section will focus on the concept of steady-state ac analysis with the frequency swept over a specific range. Multisim can thus generate the frequency response for the circuit under consideration. Plots may be generated for the *linear amplitude response,* the *decibel amplitude response,* and the *phase response.*

Following the consideration of frequency response, some of the pulse reproduction criteria will be investigated using a realistic filter. It should be understood that the criteria discussed in the chapter represent reasonable approximations rather than exact formulas. Therefore, the results will vary somewhat with different filter characteristics.

**MULTISIM EXAMPLE 4-1**    Use Multisim to plot the decibel amplitude and phase response curves for the one-pole low-pass filter whose schematic is shown in Figure 4–28. The 3-dB down frequency is about 1 kHz.

SOLUTION    The schematic diagram of Figure 4–28 utilizes the **AC Voltage Source** obtained from the **Signal Source Family Bar.** The voltage source frequency will be swept over a range of frequencies to be specified, and the ac steady-state output voltage/input voltage ratio, along with the phase shift, will be determined at each frequency. Thus, the ac source will assume an entirely different character than in the transient analysis discussed earlier.

As a check, double left-click on the source and the **SIGNAL_VOLTAGE_ SOURCES** window will open. However, the entries for transient analysis at the top of the window are not used in this case, and may be left at their default values. Instead, the frequency range will be specified in an analysis window to be discussed shortly.

The only entries of significance are the **AC Analysis Magnitude,** which should be set at **1 V,** and the **AC Analysis Phase,** which should be set at **0 Deg.** All of the other parameters are unimportant for the work of this chapter. The preceding parameters are automatically activated for an ac frequency response analysis, and they override any parameters that may have been given to the source voltage for other types of analysis.

Just so the schematic diagram is not confusing to the reader, the **Display** tab properties window was opened by left-clicking on it, and the **Use Schematic Option Global Setting** was deactivated by removing the check mark. The **Show values** option was then deactivated by removing the check mark in that block. While not strictly necessary, this practice will be followed in all frequency response examples.

To perform a frequency response analysis, left-click on **Analyses** and then left-click on **AC Analysis.** The **AC Analysis** window then opens, and various types of sweep parameters must be entered. If not already open, left-click on the **Frequency Parameters** tab.

The **Start frequency** and **Stop frequency** values are self-explanatory and define the sweep range. There are three **Sweep types: Decade, Octave,** and **Linear.** Decade sweeps are usually the best for a broad range of frequencies, and they work best if the start and stop frequencies are integer powers of 10. In this example, the two values were selected as **100 Hz** and **10 kHz,** which represent two decades.

The linear sweep is convenient for a narrow range of frequencies, and the octave sweep is convenient when the range of frequencies desired is an integer power of two.

The **Number of points** specifies the number of frequencies **per decade** for a decade sweep, the number of points **per octave** for an octave sweep, and the total number of points for a linear sweep. For this example, the **Sweep type** was set to **Decade** and the **Number of points per decade** was set to **100.** The **Vertical scale** has four options: **Linear, Logarithmic, Decibel,** and **Octave** The most common for frequency response is **Decibel,** and that option was chosen for this example.

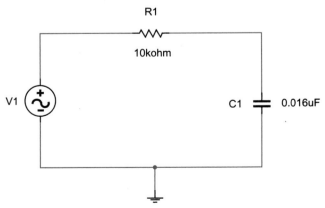

**FIGURE 4–28**
Circuit of Multisim Example 4-1.

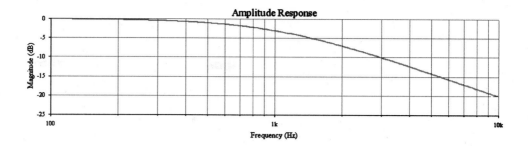

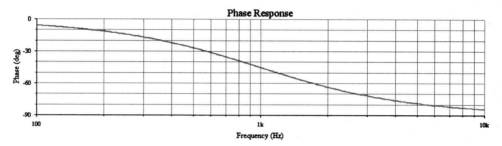

**FIGURE 4–29**
Amplitude and phase response functions of Multisim Example 4-1.

Left-click on the **Output** tab, and the desired output node can be set. This window works like others encountered throughout the text, and the output selected was node **2**. Left-click on **Add** and the node number transfers to the right. Left-click on **Simulate** and both amplitude and phase curves are obtained. The resulting response curves (after some relabeling and adjustment of the scales) are shown in Figure 4–29. The process for changing labels is essentially the same as employed in earlier chapters for **Transient Analysis,** and the reader may wish to review Multisim Example 2-1 for more details or experiment with the controls. Since there are two plots, it is necessary to "select" either one by left-clicking on it, and a marker appears on the left axis to identify that it is active for labeling.

---

**⦚⦚ MULTISIM EXAMPLE 4-2**

The filter of Figure 4–30 is a five-pole low-pass Chebyshev filter with a 1-dB bandwidth of 100 kHz. Use Multisim to plot the decibel amplitude and phase response curves.

**SOLUTION**   The procedure is basically the same as for the simple filter of the preceding example, except for some changes to enhance the presentation. First, note that this filter has a resistive termination of 50 Ω at each end (this is a very common situation in communication

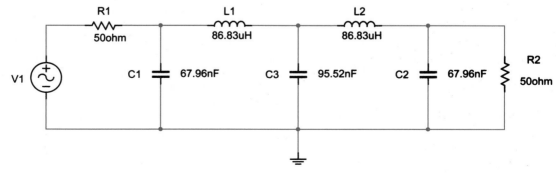

**FIGURE 4–30**
Chebyshev filter of Multisim Example 4-2.

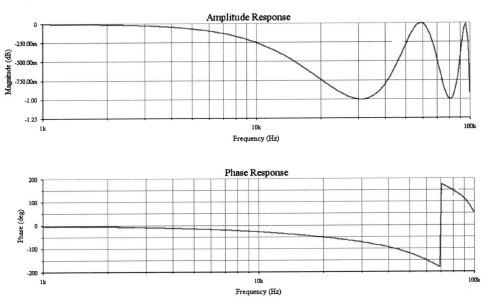

**FIGURE 4–31**
Amplitude and phase response functions in passband for filter of Multisim Example 4-2.

circuits). The resistance on the left is part of the source, and the resistance on the right is the load termination, which could represent the input resistance to the next circuit. If the source voltage remains at the default value of 1 V, the effects of the two resistances will cause the output voltage at dc and very low frequencies to be 0.5 V, and this will be displayed on the graph as a 6-dB drop in amplitude. While this is certainly a real reduction, it is a *flat loss* in that it affects all frequencies by the same amount, and for most filter responses, it is the *relative* effect that is most significant. The amplitude response will look "cleaner" if we make the source voltage 2 V. This will result in 1 V across the load at dc and very low frequencies, and will provide a 0-dB reference level.

The other change will be to make two sweeps instead of one. The first will be from 1 kHz to 100 kHz to observe the passband behavior; the second will then be from 10 kHz to 1 MHz to see both the passband and the stopband.

Each of the sweeps was implemented by the procedure discussed in Multisim Example 4-1. The amplitude and phase response curves obtained in the first sweep (1 kHz to 10 kHz) are shown in Figure 4–31 (after some relabeling). Note the passband amplitude ripple of 1 dB. At a frequency of 100 kHz, the response function is passing through the last point at which the response is down by 1 dB.

For this example and many other frequency response plots, the apparent discontinuities in the phase should be explained. Multisim always provides a value of phase based on the smallest magnitude of the angle. The phase angle for a low-pass filter typically rotates in a negative or clockwise direction. Therefore, at low frequencies, the phase assumes a negative sign; that is, a phase lag. However, once the phase passes −180°, the smallest magnitude is represented by a positive angle measured in a positive or counter-clockwise direction. For example, a phase shift of −190° will be represented as +170°. Mathematically, this is perfectly correct, but the appearance of these apparent discontinuities can be misleading if the process involved is not understood. Note that to the right of the discontinuity, the phase continues rotating clockwise, but this is represented by a decreasing positive angle as the frequency increases.

The frequency sweep from 10 kHz to 1 MHz is shown in Figure 4–32 (after some relabeling). The pronounced drop in the amplitude response above the passband is very obvious in this case. Indeed, the attenuation at 1 MHz is well above 100 dB. The passband ripple can barely be detected because of the scale of this presentation. Note the appearance of another apparent discontinuity in phase.

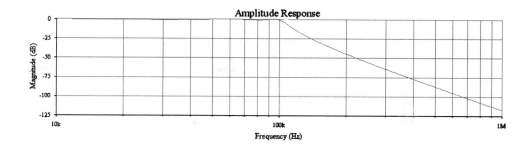

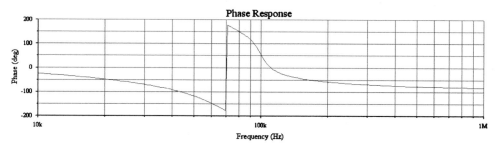

**FIGURE 4–32**
Overall amplitude and phase response functions for filter of Multisim Example 4-2.

**▍▍ MULTISIM EXAMPLE 4-3**

The filter of Figure 4–33 is a three pole-pair band-pass Chebyshev filter with a center frequency of 10 MHz and a 1-dB bandwidth of 1 MHz. Use Multisim to plot the decibel amplitude and phase response curves over the range from 9 MHz to 11 MHz.

**SOLUTION** Most of the steps involved follow those of the preceding two examples. Since the filter has equal resistive termination at both ends, the source voltage was set to 2 V as in Example 4-2.

One difference in this example as compared with the previous two is that a **Linear** sweep was employed. While a logarithmic sweep would certainly be fine, in most cases the linear sweep provides a better way of viewing a fairly narrow range.

The resulting amplitude and phase response curves are shown in Figure 4–34. Although there is geometric symmetry in the amplitude response, the points at which the response begins to drop rapidly are approximately 9.5 MHz and 10.5 MHz.

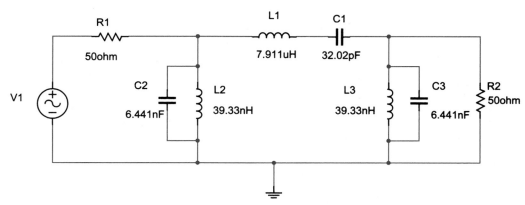

**FIGURE 4–33**
Band-pass filter of Multisim Example 4-3.

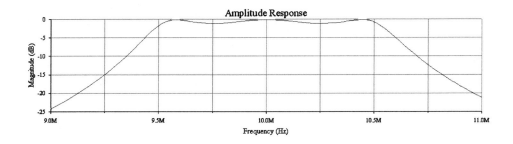

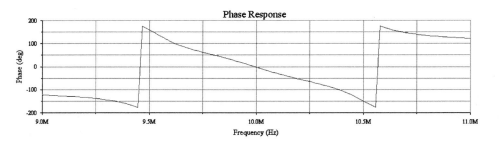

**FIGURE 4–34**
Amplitude and phase response functions of Multisim Example 4-3.

**▐▐ MULTISIM EXAMPLE 4-4**

The Multisim schematic of Figure 4–35 is a three-pole Butterworth filter designed to be driven by an ideal voltage source, and the 3-dB cutoff frequency is 1 kHz. This circuit will be used in this example and the next two to investigate some of the approximations developed in this chapter for pulse transmission. Excite the circuit with a pulse that satisfies the "coarse" criteria for pulse reproduction. Use Multisim to show both the input and output pulses for the conditions required.

**SOLUTION** To run a test of this type for different conditions, one of two possible approaches could be used: (a) the pulse width of the input could be fixed and the cutoff frequency of the filter could be varied, or (b) the cutoff frequency of the filter could be fixed and the width of the pulse could be varied. With software simulation, there is little difference in the labor involved with either approach, but in a practical laboratory situation, it is much easier to use fixed filter characteristics and vary the width of the pulse. To be more realistic, we will take that approach in this problem.

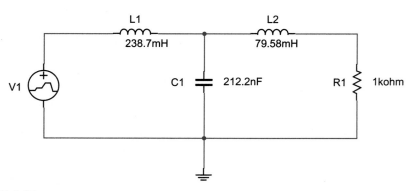

**FIGURE 4–35**
Circuit of Multisim Examples 4-4, 4-5, and 4-6.

The low-pass criterion indicates that the bandwidth $B$ required for "coarse" pulse reproduction is about

$$B = \frac{0.5}{\tau} \qquad (4\text{-}70)$$

where $\tau$ is the bandwidth. Since $B$ is to be fixed, the pulse width based on this approximation is

$$\tau = \frac{0.5}{B} \qquad (4\text{-}71)$$

Based on this criterion, the pulse width for coarse reproduction is

$$\tau = \frac{0.5}{10^3} = 0.5 \text{ ms} \qquad (4\text{-}72)$$

The voltage source that can be used to create this particular waveform (and many others) is the **Piecewise Linear Voltage Source,** which will hereafter be referred to as the **PWL** voltage source. This source can be used to create any waveform that is composed of straight-line segments. It is obtained from near the bottom of the **Signal Source Family Bar.**

This source is first dragged to the **Circuit Window.** To construct the desired voltage, double left-click on the source and a window with the title **PWL Voltage** will open. The default tab should be **Value,** but if not, left-click on that tab. The dot should be in the **Enter Points** circle. Appropriate values of time and voltage can now be entered on the right-hand side. The two columns read **Time** and **Voltage.**

Assume that the waveform starts at time $t_1$ with value $v_1$, then moves in a straight line to a value $v_2$ at time $t_2$, and so on. The format for the data is as follows:

| | |
|---|---|
| $t_1$ | $v_1$ |
| $t_2$ | $v_2$ |
| $t_3$ | $v_3$ |
| etc. | etc. |

It is recommended that the first and last voltage values be set to zero. If a nearly perfect pulse is desired, very small values of rise and fall times may be used to allow realistic level changes as needed. Since the pulse width in this case is to be 0.5 ms, a reasonable value of 1 $\mu$s will be assumed for both the rise and fall times in this example (and in the next two).

Referring back to Figure 4–35, the description of the source is listed in the file as

| | |
|---|---|
| 0 | 0 |
| 1e-6 | 1 |
| 0.5e-3 | 1 |
| 0.501e-3 | 0 |

Incidentally, the block on the **PWL** source only provides a limited number of slots. However, one can establish a longer pattern by using a text editor such as **Notepad** and checking the slot with the title **Open Data File.** The particular file containing the PWL source can then be indicated as the source for the voltage.

To activate a transient analysis, first left-click on **Analyses,** and when the menu window opens, left-click on **Transient Analysis.** A **Transient Analysis** window will then open. The options chosen for this example were to perform an analysis from **0** to **2.5 ms** with the **Maximum time step (TMAX)** set to **1e-6;** that is, 1 $\mu$s. It was also desired in this example to show both the input voltage (node 1) and the output voltage (node 3).

A graph showing both the input and output pulses is provided in Figure 4–36. While there is definite rounding and some delay, the output pulse is degraded very little in amplitude. However, for digital data for which regeneration of the pulses can be performed, this bandwidth could be adequate. Different filter characteristics result in variations of the shape, but this condition can serve as a very useful approximation for minimal reproduction.

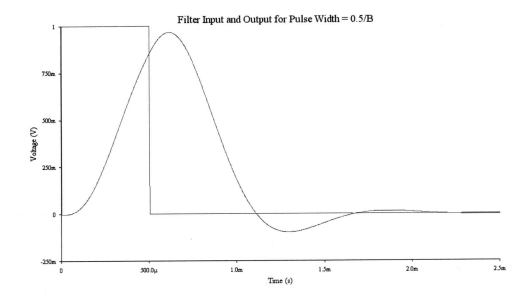

**FIGURE 4–36**
Input and output waveforms
of Multisim Example 4-4.

---

**▌▌ MULTISIM EXAMPLE 4-5**

For the circuit of Example 4-4, change the pulse width to 20% of the minimum for coarse reproduction, and repeat the analysis.

**SOLUTION**    The choice of 20% means that the pulse width is to be only 0.1 ms. The data for the PWL source are changed to the following:

| | |
|---|---|
| 0 | 0 |
| 1e-6 | 1 |
| 0.1e-3 | 1 |
| 0.101e-3 | 0 |

All other parameters in the analysis were maintained with the same values as in the preceding example.

The result in this case is shown in Figure 4–37. Obviously, the pulse has been severely degraded in amplitude and the width has been broadened considerably. Therefore, the bandwidth is inadequate for the narrow pulse used.

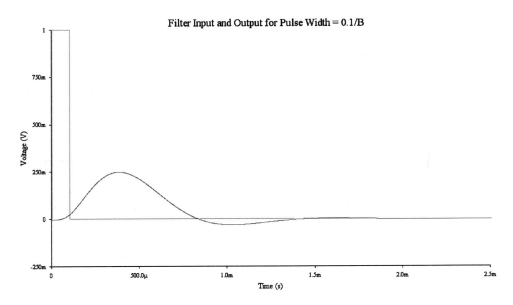

**FIGURE 4–37**
Input and output waveforms
of Multisim Example 4-5.

**▌▌ MULTISIM EXAMPLE 4-6**

For the circuit of Example 4-4, change the pulse width to five times the minimum for coarse reproduction, and repeat the analysis.

**SOLUTION** The choice of five times the minimum means that the pulse width is to 2.5 ms. The data for the PWL source are changed to the following:

| | |
|---|---|
| 0 | 0 |
| 1e-6 | 1 |
| 2.5e-3 | 1 |
| 2.501e-3 | 0 |

Since the pulse width is now equal to the upper time limit of the previous two examples, the value of **End Time** was changed to 5 ms for this example. The result is shown in Figure 4–38.

While the output pulse is not perfect, as a result of both rise time and overshoot, it is starting to look more like the input pulse than the previous two cases.

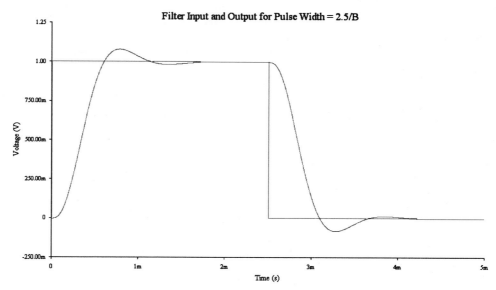

**FIGURE 4–38**
Input and output waveforms of Multisim Example 4-6.

## 4-9 MATLAB® Examples

MATLAB may be used to evaluate both the time response and the frequency response of various communication filters. Some of these operations will be demonstrated in this section.

**▌▌ MATLAB EXAMPLE 4-1**

Use MATLAB to plot the decibel amplitude response on a semilog scale for the simple low-pass filter of Example 4-3, based on a normalized independent variable $f/f_c$.

**SOLUTION** The decibel amplitude response of the one-pole low-pass filter was given by Equation 4-24, and will be repeated here for convenience. Note that the normalized independent variable is $f/f_c$, which will be denoted $u$ in the equation that follows.

$$A_{dB}(f) = -10\log\left[1 + \left(\frac{f}{f_c}\right)^2\right] = -10\log(1 + u^2) \qquad (4\text{-}73)$$

The command for generating a *logarithmic spacing* in MATLAB has the form

```
u = logspace(a, b, Np)
```

The quantities **a** and **b** are the powers of 10 defining the beginning and ending frequencies, and **Np** is the number of points in the range. Thus, $10^a$ is the beginning frequency and $10^b$ is the ending frequency. It is very desirable to assume integer values for **a** and **b**, so that the beginning and end of the plot occur at integer powers of 10.

We will choose a logarithmic scale for **u** based on the range from one decade below cutoff to one decade above cutoff. The command is

```
>> u = logspace(-1,1,100);
```

The row vector **u** will thus contain 100 points on a logarithmic scale ranging from $10^{-1} = 0.1$ to $10^1 = 10$. This corresponds to two decades.

Recall that the MATLAB form for the logarithm to the base 10 is denoted **log10( )**. The vector containing the values of the amplitude response is then given by

```
>> AdB = -10*log10(1 + u.^2);
```

Note that **u** is followed by a period (.) This tells MATLAB to square the entries in **u** on a point-by-point or **array** basis. Without the period, MATLAB would start to interpret the result as a matrix multiplication, and this is not possible for two-row matrices.

A plot may now be generated by the command

```
>> semilogx(u, AdB)
```

This command provides a logarithmic horizontal axis and a linear vertical axis. Similarly, the command **semilogy( )** would provide a logarithmic vertical scale and a linear horizontal scale. Finally, the command **loglog( )** would provide logarithmic scales in both x and y directions.

The resulting decibel amplitude response is shown in Figure 4–39. Additional labeling was provided on the figure in accordance with procedures developed earlier.

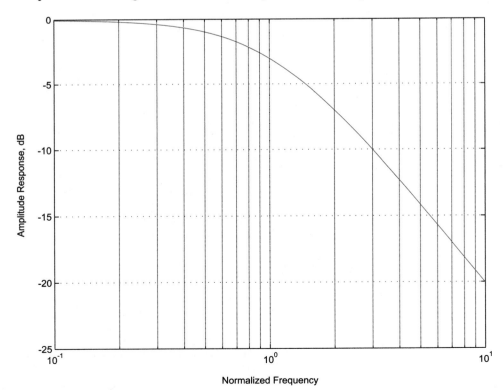

**FIGURE 4–39**
Amplitude response of one-pole low-pass filter.

**▐▌ MATLAB EXAMPLE 4-2**

Use MATLAB to plot the decibel amplitude response of a five-pole low-pass Butterworth filter over the frequency range from $0.1 f_c$ to $10 f_c$.

**SOLUTION**  The form of the linear amplitude response of the Butterworth filter characteristic was given in Equation 4-43, and is repeated here for convenience.

$$A(u) = \frac{1}{\sqrt{1 + u^{2m}}} = \frac{1}{\sqrt{1 + u^{10}}} \tag{4-74}$$

The decibel form can be determined by taking the logarithm of both sides of Equation 4-74 and multiplying by 20. Making use of some logarithmic identities, the result can be simplified to

$$A_{dB} = -10\log(1 + u^{10}) \tag{4-75}$$

In the same spirit as in MATLAB Example 4-1, the normalized variable u will be defined over the interval from 0.1 to 10. The command is

```
>> u = logspace(-1, 1, 100);
```

The decibel response can then be determined by the command

```
>> AdB = -10*log10(1 + u.^10);
```

Note the necessary period after the variable u, since the exponentiation is to be performed on each term in the array.

The function can be plotted by

```
>> semilogx(u,AdB)
```

The plot is shown in Figure 4–40, with additional labeling provided as discussed in earlier chapters.

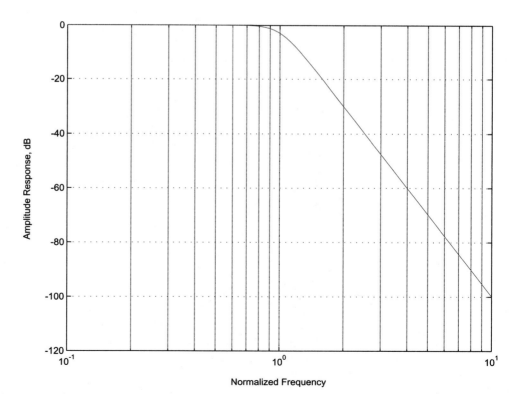

**FIGURE 4–40**
Amplitude response of five-pole low-pass Butterworth filter.

**▌▌ MATLAB EXAMPLE 4-3**

Consider a band-pass Butterworth filter with five pole-pairs and $Q = 10$. Extend the results of MATLAB Example 4-2 to plot the band-pass amplitude response over the frequency range from $0.9 f_c$ to $(1/0.9) f_c = 1.111 f_c$.

**SOLUTION**   This seemingly odd choice for the two frequencies will result in an amplitude response that is the same at the two endpoints. Most wideband amplitude response functions are best achieved with a logarithmic scale, but the linear scale is probably best for a relatively narrowband region near the center frequency.

First we define a variable **x** as

```
>> x = linspace(0.9, 1/0.9, 100);
```

The variable **u** required in the Butterworth equation is then defined as

```
>> u =10*(x - 1./x);
```

Note the period after the 1 in the last term in parentheses; this forces the division to be performed on a term-by-term basis.

We may now use the same Butterworth command as in MATLAB Example 4-2. We have

```
>> AdB = -10*log(1+ u.^10);
```

A plot is shown in Figure 4–41. The frequency scale was adjusted for convenience.

It should be noted that the **Signal Processing Toolbox** of MATLAB contains built-in filter functions of many different types, and some will be used in later work in the text. The emphasis in this chapter was on the computation of the response functions based on the mathematical forms.

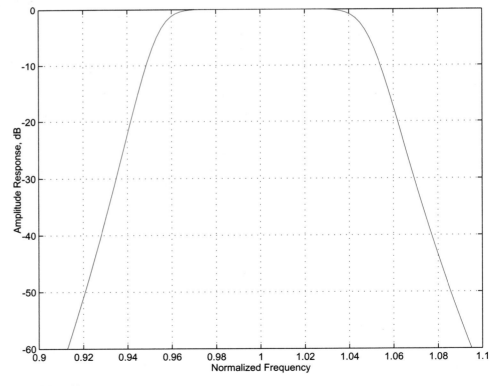

**FIGURE 4–41**
Amplitude response of Butterworth band-pass filter.

## SystemVue™ Closing Application (Optional)

Insert the text CD in a computer having SystemVue™ installed and activate the program. Open the CD folder entitled **SystemVue Systems** and load the file entitled 4-1.

### Description of System

The system shown on the screen is the one used for the chapter opening application. You will be modifying the system in this application to study how filter bandwidth affects pulse transmission.

The various properties to be studied could be implemented by either changing the pulse width or changing the filter characteristics. Since you studied how to make changes in pulse characteristics in Chapter 3, we will leave the pulse waveform fixed and perform changes on the filter characteristics.

### Filter Parameters

Token 3 was obtained from the **Operator Library** and can be used to implement a variety of different types of operations, some of which are well beyond the scope of this exercise. Therefore, don't worry if you don't understand some of the terminology and operations involved. Many of them will be introduced gradually throughout the text. For this exercise, double-click on token 3. The upper left-hand window should read **Filters/Systems,** and the **Linear Sys. Filters** button should be underlined. If not, left-click on that button.

Left-click on **Parameters** and a window with the title **SystemVue Linear System** should open. This window will likely be a little intimidating, but left-click on **Analog** near the upper right corner. You are telling the program that you want to observe the parameters for a Butterworth analog filter that has been preset.

A window with the title **SystemVue Analog Filter Library** will now open. It should show the value of **No. of Poles** to be set to 3, **Low Cutoff** to be set to 0.5, and **Filter Pass-Band** to be set to **Low-Pass.** Thus, the filter is a 3-pole Butterworth low-pass filter set to a cutoff frequency of 0.5 Hz, which is the approximate value for "coarse" reproduction of a baseband pulse, as established in the chapter.

### Changing the Filter Parameters

1. Change the cutoff frequency of the filter to 0.1 Hz and run the simulation over the given time interval. What is your conclusion about the results in this case?

2. Next, change the cutoff frequency of the filter to 10 Hz and run the simulation over the given time interval. What is your conclusion in this case?

## PROBLEMS

**4-1**  An $RL$ one-pole low-pass filter is shown below. The input is $v_1$ and the output is $v_2$. Determine the transfer function $H(f) = \overline{V}_2/\overline{V}_1$.

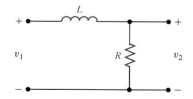

**4-2**  An $RC$ one-pole filter having some attenuation at dc is shown below. The input is $v_1$ and the output is $v_2$. Determine the transfer function $H(f) = \overline{V}_2/\overline{V}_1$.

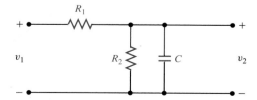

**4-3**  An *RC* one-pole high-pass filter is shown below. The input is $v_1$ and the output is $v_2$. Determine the transfer function $H(f) = \overline{V}_2/\overline{V}_1$.

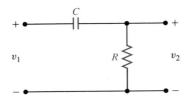

**4-4**  An *RL* one-pole high-pass filter is shown below. The input is $v_1$ and the output is $v_2$. Determine the transfer function $H(f) = \overline{V}_2/\overline{V}_1$.

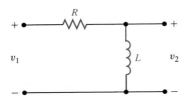

**4-5**  For the circuit of Problem 4-1, determine (a) the amplitude response (including the 3-dB down frequency), (b) the phase response, and (c) the decibel amplitude response.

**4-6**  For the circuit of Problem 4-2, determine (a) the amplitude response (including the 3-dB down frequency), (b) the phase response, and (c) the decibel amplitude response. (*Note:* 3 dB down from the dc value.)

**4-7**  For the circuit of Problem 4-3, determine (a) the amplitude response (including the 3-dB down frequency), (b) the phase response, and (c) the decibel amplitude response.

**4-8**  For the circuit of Problem 4-4, determine (a) the amplitude response (including the 3-dB down frequency), (b) the phase response, and (c) the decibel amplitude response.

**4-9**  The circuit shown below represents a normalized passive low-pass two-pole Butterworth filter designed to be driven with an ideal voltage source input. In the normalized form, the 3-dB frequency is 1 rad/s and a 1-$\Omega$ termination is used. (By procedures called *frequency* and *impedance scaling,* the circuit can be converted to any desired frequency and impedance levels.) (a) Determine the transfer function $H(f) = \overline{V}_2(f)/\overline{V}_1(f)$. (b) Determine the amplitude response $A(f)$ and the phase response $\beta(f)$. (c) Show that $A(f)$ is of the form of Equation 4-43, with $f_c = 1/2\pi$ Hz and $m = 2$.

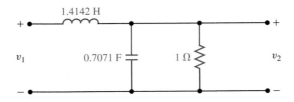

**4-10**  The circuit shown below represents a normalized passive low-pass two-pole Butterworth filter designed to be terminated at both ends. In the normalized form, the 3-dB frequency is 1 rad/s and a 1-$\Omega$ termination is used. (By procedures called frequency and impedance scaling, the circuit can be converted to any desired frequency and impedance levels.) (a) Determine the transfer function $H(f) = \overline{V}_2(f)/\overline{V}_1(f)$. (b) Determine the amplitude response $A(f)$ and the phase response $\beta(f)$. (c) Show that $A(f)$ is of the form of Equation 4-43 with $f_c = 1/2\pi$ Hz and $m = 2$. (*Note:* The numerator constant will be different than 1 in this case, but it is the denominator function that determines the form of the response.)

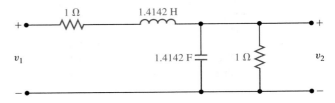

**4-11**  For the circuit of Problem 4-9, determine (a) the phase delay function and (b) the group delay function.

**4-12**  For the circuit of Problem 4-10, determine (a) the phase delay function and (b) the group delay function.

**4-13**  A baseband pulse generator produces pulses having widths of 0.5 μs. Determine the approximate baseband bandwidths for (a) coarse reproduction and (b) fine reproduction, based on maximum allowable rise and fall times of 40 ns.

**4-14**  A baseband pulse generator produces pulses having widths of 20 ms. Determine the approximate baseband bandwidths for (a) coarse reproduction and (b) fine reproduction, based on maximum allowable rise and fall times of 5 ns.

**4-15**  Assume that the pulses of Problem 4-13 modulate an RF carrier. Based on the same criteria as in Problem 4-13, determine the bandwidths for (a) coarse reproduction and (b) fine reproduction.

**4-16**  Assume that the pulses of Problem 4-14 modulate an RF carrier. Based on the same criteria as in Problem 4-14, determine the bandwidths for (a) coarse reproduction and (b) fine reproduction.

**4-17**  A cable link is to be used to process digital data, which consists of a combination of pulses and spaces, representing ones and zeros. The amplitude response is nearly flat from dc to 1 MHz, but it drops off rapidly above this frequency. Determine the approximate maximum data rate (in bits per second) that can be transmitted over this link. The bits are assumed to be transmitted in succession with no extra spaces inserted. (*Note:* For digital data, it is only necessary that the two levels be recognized at the receiver.)

**4-18**  For the cable link of Problem 4-17, assume that some *pulse position modulated* signals are to be transmitted over the system. Assume that the rise (or fall) time of each pulse cannot exceed 0.1% of the total time allocated to the pulse. Determine the approximate maximum data rate in pulses per second that would be permissible in this case. (*Note:* For pulse position modulated signals, it is necessary that the exact beginning and ending of each pulse be determined.)

**4-19**  An oscilloscope is to be selected to accurately measure some logic signals in a high-speed data system. The rise and fall times of the pulses are in the neighborhood of 10 ns. Suppose the measurement criteria are specified so that the rise time introduced by the oscilloscope should not exceed 10% of the rise time to be measured. Determine the approximate bandwidth required for the oscilloscope.

**4-20**  A circuit that can be used to partially compensate for the undesirable effects of input shunt capacitance (such as lead and vertical amplifier input capacitance in an oscilloscope) is shown as follows. The capacitance $C_2$ is the fixed circuit input capacitance and $C_1$ is the adjustable probe capacitance. The resistances $R_1$ and $R_2$ are selected to form a fixed attenuation ratio. (a) Determine the transfer function $H(f) = \overline{V}_2(f)/\overline{V}_1(f)$, in terms of arbitrary parameters. (b) Show that if $R_1 C_1 = R_2 C_2$, the amplitude response $A(f)$ is a constant value independent of frequency. For the condition of (b), assume that the resistors are selected such that a 10-to-1 attenuation results; that is, $A(f) = 0.1$. Determine the relationship between $C_1$ and $C_2$ for this case.

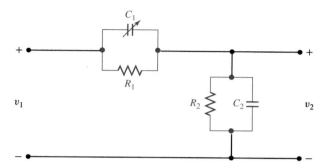

# Frequency Generation and Translation

# 5

## OVERVIEW AND OBJECTIVES

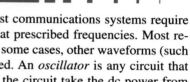

To assist in the processing of information signals, most communications systems require various standard periodic waveforms to be generated at prescribed frequencies. Most required waveforms are sinusoidal in nature, although in some cases, other waveforms (such as square waves and triangular waves) may be required. An *oscillator* is any circuit that generates the required waveform. The components of the circuit take the dc power from the power supply and convert a portion of it to output power in the form of an oscillation. Most oscillator circuits require the use of an active device such as a transistor or integrated circuit.

Classical oscillator circuits of the past were tuned by varying one or more reactive elements in the circuit. In recent years, however, the use of *frequency synthesis* technology has become the dominant approach in both transmitters and receivers. A *frequency synthesizer* utilizes one or more fixed frequency stable oscillators coupled with *phase-locked loops, frequency dividers,* and *mixers* to create a wide variety of output waveforms at highly controllable frequencies.

*Mixers* are devices that permit signals to be shifted in frequency while retaining their primary spectral properties. Mixers are used in both transmitters and receivers.

## Objectives

After completing this chapter, the reader should be able to:

1. Discuss the Barkhausen criterion and define the conditions for oscillations to occur.

2. Discuss some of the common oscillator circuits.

3. Describe a *phase-locked loop (PLL)* circuit.

4. Describe the operation of a basic *frequency synthesizer* circuit.

5. Describe the operation of an ideal *mixer* circuit, and show the mathematical form of the output.

6. Define *frequency conversion, up-conversion,* and *down-conversion.*

7. Predict the output frequency components of an ideal mixer for both the local oscillator below the signal frequency and above the signal frequency.

8. Discuss the operation of a *tuned radio frequency (TRF)* receiver.

9. Discuss the operation of a *superheterodyne* receiver and discuss its advantages.

10. Define *image frequency* and predict its values for a receiver.

11. Discuss practical mixer operation and specifications.

## SystemVue™ Opening Application (Optional)

Insert the text CD in a computer having SystemVue™ installed and activate the program. Open the CD folder entitled **SystemVue Systems** and load the file entitled 5-1.

### Sink Tokens

| Number | Name | Token Monitored |
|---|---|---|
| 0 | Input | 3 |
| 1 | LO | 4 |
| 2 | Output | 6 |

### Operational Tokens

| Number | Name | Function |
|---|---|---|
| 3 | Input | input source whose frequency is to be shifted |
| 4 | LO | source that will mix with input (often called a "local oscillator") |
| 5 | Mixer | generates product of input signal and LO signal |
| 6 | Filter | passes desired sideband while rejecting the superfluous one |

### Zooming in on the Graphs

It will be quite easy to verify that the waveform at Sink 0 has a frequency of 1 kHz, since only 5 cycles will appear on the screen. However, it will be difficult to make measurements at Sinks 1 and 2 due to the large number of cycles present.

Any portion of a plot may be enlarged by the procedure that follows. Pick a starting point and press the left mouse button. While holding it down, move vertically and then horizontally around a few cycles of the waveform. Be sure to enclose both positive and negative peaks. A box composed of dashed line segments will appear. After the box has been closed, release the button. The area that was framed will now occupy the entire screen, which will allow you to perform time measurements on the waveform more accurately.

You may want to experiment with this feature several times to obtain a good display for Sinks 1 and 2. At any point, you can return to the original scale by a left-click on the **Rescale** button located near the left end of the **Tool Bar.** This button is labeled with perpendicular sets of arrows.

### What You See

Using the zooming procedure when necessary, measure the frequencies of the two input sources to the multiplier (Sinks 0 and 1), and verify that they are 1 kHz and 9 kHz. Next, observe the output of the band-pass filter, which is observed by Sink 2. There is a transient response while the filter is settling, so move over to the last few cycles and zoom in on the waveform. Measure the frequency, which should be 10 kHz.

### How This Demonstration Relates to the Chapter Learning Objectives

In Chapter 5, you will learn about frequency translation and shifting. You will learn that when two sinusoids are multiplied, the output consists of the difference frequency and the sum frequency. These are referred to as the *lower sideband* and the *upper sideband,* respectively. A band-pass filter can be used to extract the desired sideband while rejecting the other one. The frequency shifting property permits communication signals to be shifted up and down in frequency while retaining any intelligence (i.e., information) that has been imparted to the waveform. After this chapter is completed, you will return to this system and study the effects of changing input and output frequencies.

## 5-1    Oscillator Circuits

There are many classical oscillator circuits that have been used since the beginning of the electronics industry. In the early days, vacuum tubes were used as the active elements, but beginning in the 1950s, most of these circuits were converted to transistorized versions. Still later, of course, integrated circuits were developed to produce oscillations from the output of a single chip.

### Barkhausen Criterion

Most classical oscillator circuits used in communications circuits utilize the *Barkhausen criterion* as the basis for oscillation. This concept is based on creating a loop in which positive feedback from the output is fed back to the input. The concept is illustrated in Figure 5–1. The block labeled $A$ normally represents the gain of an amplifier circuit, and $\beta$ normally represents the transfer function of some frequency-dependent passive network.

Assume some kind of signal $e$ existing at the input of the amplifier. It will be multiplied by the gain $A$, and the output will thus be $Ae$. This signal will in turn be multiplied by the transfer function $\beta$ of the feedback network, and that will produce a signal at its output given by $A\beta e$. To sustain oscillations, this signal must have the same level and phase as the assumed input $e$. Thus, $A\beta e = e$ or $A\beta = 1$. The product $A\beta$ is called the *loop gain*.

The preceding development is a simplified explanation of the Barkhausen criterion. It states that for oscillations to occur at a specific frequency, the loop gain must be unity at that frequency. This means that the loop gain magnitude must be unity, and the phase shift must be $0°$ or an integer multiple of $360°$. A frequency-dependent network is chosen for the feedback block that will produce the correct phase shift at the desired frequency, and this will force the oscillations to occur at that frequency.

The inquisitive reader may wonder what created the assumed input signal $e$ at the beginning. In most oscillators, random noise in the circuit will actually initiate the process. In practice, however, most oscillators have a loop gain greater than unity at small signal

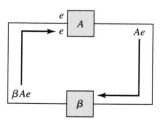

**FIGURE 5–1**

Loop illustrating Barkhausen criterion.

levels in order to initiate the process. As the signal starts to grow, some limiting feature of the circuit will eventually take over and force the oscillations to settle at the level of unity loop gain.

## Oscillator Survey

Several of the classical circuits used in communications will be surveyed in this section. Since these circuits can be implemented with vacuum tubes, bipolar junction transistors (BJTs), field effect transistors (FETs), and even operational amplifiers, our approach will be to show the active device simply as a block. These circuits are thoroughly analyzed in many basic electronic devices texts and communication circuit texts. While some of these circuits may still be found in older equipment, most newer equipment utilizes crystal controlled oscillators (the last type to be considered in this section) and frequency synthesizer technology (to be considered in the next section).

Most of the circuits to be considered can be implemented either with inverting gain or noninverting gain, depending on the way in which the feedback is connected. Because common cathode (for vacuum tubes), common emitter (for BJTs), and common source (for FETs) are inverting, and because they have probably been used more often over the years in the classical oscillator circuits, all of the configurations to be shown utilize inverting amplifiers. The gain value $A$ shown on the circuit diagrams will have a negative (inverting) value.

### Hartley Oscillator

The inverting amplifier configuration for a Hartley oscillator is shown in Figure 5–2. An $LC$ circuit is used to establish the frequency of oscillation, and feedback is established by either utilizing two inductors or by providing a tap on one inductor. The voltage across $L_2$ represents the feedback voltage. If the amplifier input current is small compared to the current in the branches of the tuning circuit, and if the resonant circuit has a relatively high $Q$, the feedback fraction $\beta$ is approximately

$$\beta \approx \frac{-L_2}{L_1} \tag{5-1}$$

The minus sign is a result of the fact that the voltage fed back to the active device is inverted with respect to the voltage across the combination.

The frequency of oscillation is approximately given by

$$f_0 \approx \frac{1}{2\pi\sqrt{(L_1 + L_2)C}} \tag{5-2}$$

### Colpitts Oscillator

The inverting amplifier configuration for a Colpitts oscillator is shown in Figure 5–3. As in the case of the Hartley oscillator, an $LC$ circuit is used to establish the frequency of

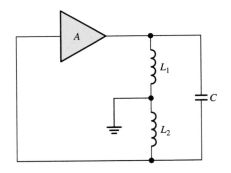

**FIGURE 5–2**

Hartley oscillator for inverting amplifier. ($A$ is negative.)

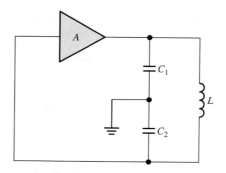

**FIGURE 5–3**
Colpitts oscillator for inverting amplifier. (*A* is negative.)

oscillation. In this case, however, the capacitive circuit shown establishes the feedback. The voltage across $C_2$ represents the feedback voltage. If the amplifier input current is small compared to the current in the branches of the tuning circuit, and if the resonant circuit has a relatively high $Q$, the feedback fraction $\beta$ is approximately

$$\beta \approx \frac{-C_1}{C_2}$$

(5-3)

As in the Hartley oscillator, the minus sign is a result of the fact that the voltage fed back to the active device is inverted with respect to the voltage across the combination. However, the capacitive ratio is opposite to the inductive ratio of the Hartley circuit.

The approximate frequency of oscillation is

$$f_o \approx \frac{1}{2\pi\sqrt{LC_{eq}}}$$

(5-4)

where $C_{eq}$ is the equivalent series capacitance, determined from Equation 5-5:

$$\frac{1}{C_{eq}} = \frac{1}{C_1} + \frac{1}{C_2}$$

(5-5)

### Clapp Oscillator

The form of the Clapp oscillator is shown in Figure 5–4. It differs from the Colpitts oscillator in that a third capacitor is inserted in series with the inductor. If this capacitor is small relative to the other two, the oscillation frequency is almost completely dependent on $C_3$ and $L$, and relatively insensitive to the other two capacitance values. This has the desirable effect of making the oscillation frequency relatively independent of active device input and output capacitance. Based on the same assumptions as made for the Colpitts oscillator, the value of $\beta$ is given approximately by Equation 5-3.

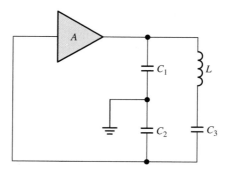

**FIGURE 5–4**
Clapp oscillator for inverting amplifier. (*A* is negative.)

A reasonable approximation of the theoretical frequency of oscillation is given by

$$f_o \approx \frac{1}{2\pi\sqrt{LC_{eq}}} \tag{5-6}$$

where $C_{eq}$ is determined from the equation that follows.

$$\frac{1}{C_{eq}} = \frac{1}{C_1} + \frac{1}{C_2} + \frac{1}{C_3} \tag{5-7}$$

In practice, $C_3$ is chosen to be much smaller than the other two capacitors (as previously noted), so the equivalent capacitance can be approximated as

$$C_{eq} \approx C_3 \tag{5-8}$$

### Crystal Oscillator

A crystal oscillator is a circuit that employs a *crystal* to establish the oscillation frequency. The most common type of crystal is the *quartz crystal. Quartz* is a substance that exhibits a property called the *piezoelectric effect*. When a mechanical stress is applied across the crystal, a voltage is developed at the frequency of the mechanical vibration. Conversely, when a voltage is applied across the crystal, it vibrates.

The schematic symbol of a crystal is shown in Figure 5–5(a), and an equivalent circuit model is shown in Figure 5–5(b). The crystal exhibits both a series resonant frequency and a parallel resonant frequency. The parallel resonant frequency is higher than the series resonant frequency.

The operating frequency of a crystal is typically accurate to well within ±0.01%. However, the frequency will vary somewhat with temperature. Crystals are specified with a temperature coefficient, which can be either positive or negative, depending on the manner in which the crystal is cut. The coefficient can be as high as 100 parts per million (ppm) per degree Celsius. The frequency $f_T$ at a temperature $T$ can be determined from the formula

$$f_T = f_0 + kf_0(T - T_0) \tag{5-9}$$

where $f_0$ is the frequency at the reference temperature $T_0$, and $k$ is the temperature coefficient converted from parts per million to decimal form. The signs of $k$ and the temperature difference must be carefully noted in the formula.

An oscillator circuit employing a crystal is shown in Figure 5–6. This particular circuit is a modified Colpitts oscillator, and the parallel mode of the crystal is utilized. The assumptions made earlier for the Colpitts oscillator apply here as well.

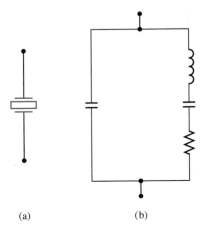

(a)                    (b)

**FIGURE 5–5**
Schematic diagram for quartz crystal and its equivalent circuit.

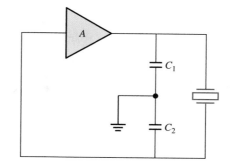

**FIGURE 5–6**

Typical crystal oscillator circuit with inverting amplifier. (*A* is negative.)

The primary advantage of a crystal oscillator is the excellent frequency stability provided. Commercial broadcast transmitters utilize crystal oscillators with the crystal kept in a temperature-stabilized oven to provide additional stability in the operating frequency.

**▌▌ EXAMPLE 5-1**

The circuit of Figure 5–7 represents a Clapp oscillator circuit implemented with a common source FET amplifier. The inductance label "RFC" is an old term referring to "radio frequency choke." It is essentially a short for dc, but it represents nearly an open circuit at the oscillation frequency. The components $R_1$, $R_2$, and $R_s$ form the bias circuit and $C_s$ is the emitter bypass capacitor. In the analysis that follows utilize the approximations provided in the text. (a) Determine the more "exact" frequency of oscillation based on the effects of $C_1$, $C_2$, and $C_3$. (b) Determine the approximate frequency of oscillation based on the assumption that $C_3$ provides complete control of the frequency. (c) Determine the required gain of the amplifier to sustain oscillations.

**SOLUTION**

(a) The equivalent capacitance is determined from Equation 5-7. For convenience, all units at this point will be expressed in pF.

$$\frac{1}{C_{eq}} = \frac{1}{C_1} + \frac{1}{C_2} + \frac{1}{C_3} = \frac{1}{200} + \frac{1}{2400} + \frac{1}{8} \tag{5-10}$$

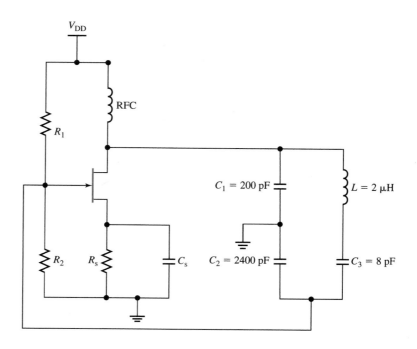

**FIGURE 5–7**

Clapp oscillator of Example 5-1.

This leads to

$$C_{eq} = 7.668 \text{ pF} \qquad (5\text{-}11)$$

Note that $C_{eq}$ is only slightly smaller than $C_3$.

The frequency of oscillation is determined from Equation 5-6 as

$$f_o = \frac{1}{2\pi\sqrt{LC_{eq}}} = \frac{1}{2\pi\sqrt{2 \times 10^{-6} \times 7.668 \times 10^{-12}}} = 40.64 \text{ MHz} \qquad (5\text{-}12)$$

(b) With the assumption that $C_{eq} \approx C_3$, the frequency of oscillation is determined as

$$f_o \approx \frac{1}{2\pi\sqrt{LC_3}} = \frac{1}{2\pi\sqrt{2 \times 10^{-6} \times 8 \times 10^{-12}}} = 39.79 \text{ MHz} \qquad (5\text{-}13)$$

(c) The feedback fraction $\beta$ can be determined from Equation 5-3, based on the assumptions made at that point.

$$\beta = \frac{-C_1}{C_2} = \frac{-200}{2400} = -0.08333 \qquad (5\text{-}14)$$

Since $A\beta = 1$ for oscillation, the gain must be

$$A = \frac{1}{\beta} = \frac{1}{-0.08333} = -12 \qquad (5\text{-}15)$$

Because of stray circuit parameters, loading, and other higher-order effects, these results must be considered reasonable approximations. Setting the amplifier gain to the value indicated will depend in part on the equivalent resistance exhibited by the resonant circuit at the oscillation frequency.

---

**▌▌ EXAMPLE 5-2**

A crystal-controlled portable transmitter is required to operate over a temperature range from $-8°C$ to $32°C$. The signal is derived from a crystal with a temperature coefficient of 40 ppm/degree C, and the transmitter frequency is exactly 148 MHz at 20°C. Determine maximum and minimum possible frequency values.

**SOLUTION** The pertinent relationship is Equation 5-9, and it is repeated here for convenience.

$$f' = f_0 + kf_0(T - T_0) \qquad (5\text{-}16)$$

The value of $k$ as given in ppm must be arranged as a decimal fraction. The value is $40 \times 10^{-6}$ and it is positive. Substituting values in Equation 5-16, the maximum possible frequency $f'_{max}$ is

$$f'_{max} = 148 + 40 \times 10^{-6} \times 148 \times (32 - 20) = 148.071 \text{ MHz} \qquad (5\text{-}17)$$

The minimum possible frequency $f'_{min}$ is

$$f'_{max} = 148 + 40 \times 10^{-6} \times 148 \times (-8 - 20) = 147.834 \text{ MHz} \qquad (5\text{-}18)$$

The reader should note that in a problem of this nature, we are required to carry out the answers to a reasonable number of digits for the results to make any sense. ▌▌

## 5-2 Frequency Synthesizer Concepts

Unless a transmitter or receiver is designed to operate at one specific frequency, it is necessary to provide some means for changing the frequency. Early transmitters and receivers utilized oscillator circuits of the type considered earlier, and tuning was often accomplished by varying one of the reactive elements. Equipment utilizing this approach is still around, and there are still some applications where such technology is useful.

Most modern communication equipment utilizes *frequency synthesizers* to generate the various frequencies required. Frequency synthesizers utilize one or more crystal oscillators operating at very stable frequencies, and accomplish tuning by the use of *phase-locked loop* circuits and digital *divider* or *counter* circuits.

There are several books entirely devoted to the subject of frequency synthesizers, and several books devoted to phase-locked loops, so only a limited description is possible in one section. However, the basic concept will be established so the reader can develop some insight into the process involved.

## Phase-Locked Loops

A fundamental element in the tuning process of a frequency synthesizer is a *phase-locked loop (PLL)*. A PLL is a closed-loop feedback circuit in which a generated signal can be locked in phase or frequency with an input signal. Moreover, its frequency can be made to be an integer multiple of the input frequency, an integer divisor of the input frequency, or a ratio of two integers times the input frequency. Various types of PLL circuits are readily available as integrated circuit (IC) chips.

We will start the discussion with a PLL block diagram in which no frequency multiplication or division takes place; the layout is shown in Figure 5–8. The first block on the left is a *phase comparator,* which compares the phase of the input and feedback signals. Its output voltage, which is called the *error voltage,* is proportional to the difference in phase between input and feedback signals. The loop filter provides smoothing of the error signal and ensures that the loop response meets certain design criteria. The output voltage of the loop is applied to a *voltage-controlled oscillator (VCO)*, whose details will be studied in more depth in Chapter 7 as related to *frequency modulation (FM)*. For the moment, think of the VCO as an oscillator whose frequency is established by an input voltage. When the voltage is zero, the output frequency is some reference frequency called the *free-running frequency,* but as the input voltage changes, the frequency increases or decreases in accordance with the sign of the control voltage.

Assume initially that the loop is locked at the center reference frequency of the VCO; that is, the input frequency is the same as the VCO free-running frequency. Next, assume that the input frequency changes by a certain amount. This will cause an error voltage to appear at the output of the phase comparator. After filtering, this error voltage is applied to the input of the VCO. If the loop has been designed correctly, this voltage will drive the VCO frequency in the direction of the input frequency, such as to reduce the frequency difference between the two signals. Eventually, lock will be established at the new frequency, and the output frequency $f_o$ will be the same as the input frequency $f_i$; that is,

$$f_o = f_i \tag{5-19}$$

The basic circuit as given has applications in which a noise-free signal is to be locked in frequency with a "noisy" reference, and it also has wide application in FM demodulation (which will be considered in Chapter 7). In the latter application, the control voltage of the VCO will be the desired output. For frequency synthesis, however, we need to provide a variety of possible output frequencies.

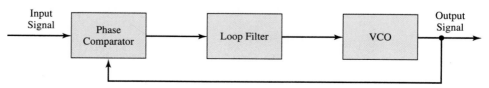

**FIGURE 5–8**
Block diagram of basic phase-locked loop.

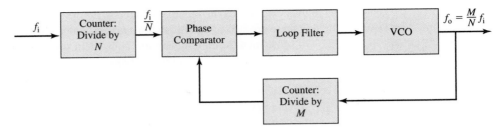

**FIGURE 5–9**
Block diagram showing frequency multiplication and/or division with PLL.

## Frequency Multiplication and Division

Operation of the PLL as a frequency multiplier and/or divider is illustrated in Figure 5–9. This circuit contains a "divide by $N$" counter ahead of the loop and a "divide by $M$" counter in the feedback path within the loop.

Assume that the input frequency is $f_i$. This frequency is divided by $N$ in the first counter, so one input to the phase comparator is a signal with frequency $f_i/N$. If the VCO output were applied directly as the other input to the phase comparator, a VCO output frequency of $f_i/N$ would result in zero frequency error. However, since the divider in the feedback path divides the VCO frequency by $M$, the VCO must provide an output at $M$ times the input frequency in order to bring the frequency error to zero at the point of comparison. Said differently, the division by $M$ in the feedback path forces the oscillator to operate at a frequency $M$ times as large, so the frequency after division will be the same as the input frequency. Thus, the output frequency $f_o$ is

$$f_o = \frac{M}{N} f_i \tag{5-20}$$

Using this concept, it is theoretically possible to generate any output frequency that can be expressed as a ratio of integers times an input frequency. By employing programmable divider circuits, a wide range of possible output frequencies can be generated. There are numerous other "tricks" (such as frequency offsets in the loop) than can be used to enhance the process.

## Capture and Lock Ranges

Two important parameters in the operation of a phase-locked loop are the *capture range* and the *lock range*. The *capture range* is the range of frequencies about the center frequency at which the PLL can initially establish synchronization. The *lock range* is the range of frequencies about the center frequency at which the PLL can hold lock, once it is initially established.

Although the lock and capture ranges may be the same in some cases, as a general rule, the lock range is larger than the capture range. This means that the PLL can hold lock, once established, over a broader frequency range than the range in which it could initially establish lock.

---

**▌▌ EXAMPLE 5-3**

Consider the PLL loop shown in Figure 5–10. (a) Determine the output frequency. (b) Determine the frequencies $f_1$ and $f_2$. What can be concluded?

**SOLUTION**

(a)  The input division is $N = 5$, and the feedback division is $M = 8$. The output frequency is thus

$$f_o = \frac{M}{N} f_i = \frac{8}{5} \times 4 \text{ MHz} = 6.4 \text{ MHz} \tag{5-21}$$

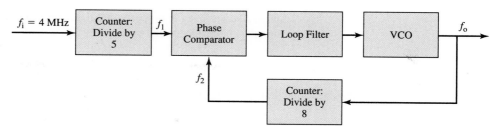

**FIGURE 5–10**
Frequency synthesizer circuit of Example 5-3.

(b) The frequency $f_1$ is the input frequency divided by 5.

$$f_1 = \frac{f_i}{5} = \frac{4\ \text{MHz}}{5} = 0.8\ \text{MHz} \tag{5-22}$$

The frequency $f_2$ is the output frequency divided by 8.

$$f_2 = \frac{f_o}{8} = \frac{6.4\ \text{MHz}}{8} = 0.8\ \text{MHz} \tag{5-23}$$

The two frequencies are the same, as required for lock to be established.

---

▌▌ **EXAMPLE 5-4**

It is desired to generate a 2 MHz signal locked to a given 2.4 MHz signal. Specify the smallest frequency division ratios for a PLL that can be used to achieve the objective.

**SOLUTION**  In order for an exact solution to exist for the basic circuit of Figure 5–9, the ratio of the two frequencies must be expressible as a ratio of integers. The desired ratio in this case is 2 MHz/2.4 MHz = 20/24. Thus, one solution would be $M = 20$ and $N = 24$. However, the smallest integer ratio is obtained by simplifying 20/24 to its most reduced form with integers, which is 20/24 = 5/6. The input frequency is thus divided by 6 and the VCO output frequency is divided by 5.

▐▌

## 5-3  Ideal Mixers

Before studying some of the methods for frequency shifting, we need to understand the operation of a mixer, which is one of the most basic of communication components. We will begin with the assumption of an *ideal mixer*. Once its operation is understood, some of the practical nuances will be considered.

### Ideal Mixer or Multiplier

Consider the block diagram shown in Figure 5–11. The block with the "×" represents a common schematic symbol for a mixer. The device has two inputs, $v_i(t)$ and $v_{LO}(t)$, and one output, $v_o(t)$. The subscript "LO" refers to "local oscillator," a classical term used for many years in communication circuits.

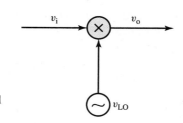

**FIGURE 5–11**
Ideal mixer block, with signal and local oscillator inputs.

An *ideal mixer* has the property that it performs a multiplication operation on the two input signals, and produces an output given by

$$v_o = K_1 v_i v_{LO} \tag{5-24}$$

where $K_1$ is a constant. Assume that $v_i$ is some arbitrary bandlimited input signal, and assume that $v_{LO}$ is a single frequency sinusoid of the form

$$v_{LO} = K_2 \cos \omega_0 t \tag{5-25}$$

Substitution of Equation 5-25 in Equation 5-24 results in

$$v_o(t) = K v_i(t) \cos \omega_0 t \tag{5-26}$$

where $K = K_1 K_2$ is the net constant.

## Spectrum of Mixer Output

The output spectrum of the mixer is determined by taking the Fourier transforms of both sides of Equation 5-26. The cosine function is first expanded into the sum of two complex exponential functions, and the modulation theorem is applied to each of the two resulting product terms. We obtain

$$\overline{V}_o(f) = \frac{K}{2} \overline{V}_i(f - f_0) + \frac{K}{2} \overline{V}_i(f + f_0) \tag{5-27}$$

The two-sided spectral interpretation of this result is shown in Figure 5–12. The spectrum shown in part (a) represents that of the input signal, and the output signal spectrum is shown in part (b). The output spectrum contains two components. One is obtained by shifting the spectrum to the right by $f_0$, and the other is obtained by shifting the spectrum to the left by $f_0$. This is a situation in which the concept of the two-sided spectrum starts to become useful. The results of mixing operations are very much enhanced by considering the input spectra as two-sided. One of the final positive frequency components can be considered to arise from the original negative frequency terms. Once the final shift is completed, we need to concentrate only on the positive frequencies. (Incidentally, if the input signal were a pulse waveform, the output would have the form of an RF pulse, as considered in Chapter 3.)

The various types of mixer outputs and their applications will be considered in the next section.

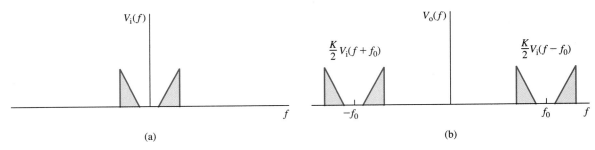

**FIGURE 5–12**
Input spectrum and output spectrum of ideal mixer.

**▌▌ EXAMPLE 5-5**

Consider the mixer block diagram of Figure 5–11, and assume that the local oscillator frequency is 100 kHz. Assume as a simple case that the input signal is also a single frequency sinusoid with a frequency of 20 kHz. Sketch the mixer input and output spectra.

**SOLUTION**    The two-sided input spectrum is shown in Figure 5–13(a), and it consists of a single line at 20 kHz and the corresponding negative frequency line at −20 kHz. From the results of Equation 5-27, multiplication of the input signal by the local oscillator will result in a positive frequency shift of 100 kHz and a negative frequency shift of the same

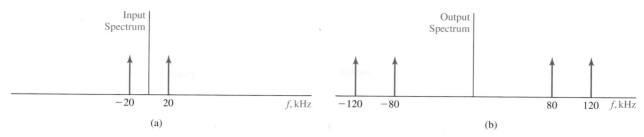

**FIGURE 5–13**
Spectra of Example 3-5.

amount. Since we don't know the value of the mixer constant $K$, the output level will be assumed to be the same as the input level.

The resulting spectrum is shown in Figure 5–13(b). The positive shift moves the component at 20 kHz to $20 + 100 = 120$ kHz and the component at $-20$ kHz to $-20 + 100 = 80$ kHz. The negative shift moves the component at 20 kHz to $20 - 100 = -80$ kHz and the component at $-20$ kHz to $-20 - 100 = -120$ kHz. The resulting spectrum must necessarily be symmetrical about the origin.

There is another thought process that should be mentioned here. For a simple situation such as this, a commonly used shortcut is the statement that the output frequencies of the mixer will be the sum and difference frequencies of the input and local oscillator frequencies. Using this rule, the output frequencies would simply be $100 + 20$ and $100 - 20$, or 120 kHz and 80 kHz, both expressed as positive frequencies. This rule works fine for a simple input signal such as considered here, and this rule will be utilized at various places in the text. For more complex input signals, and for certain combinations of local oscillator and input frequencies, however, this seemingly simple approach can get a little tricky and can lead to misleading results. For all but the simplest cases, it is recommended that a two-sided spectral layout be employed to represent the input to a mixer, and that the form of Equation 5-27 be used to predict the output.  ▌▊

## 5-4  Frequency Conversion

*Frequency conversion* is the process in which the spectrum of a given signal is shifted to a different frequency range, but in which *the shape of the spectrum remains unchanged*. The process is illustrated for positive frequencies in Figure 5–14(a). An arbitrary signal spectrum is shown in part (a). If the signal is shifted to a higher frequency as shown in Figure 5–14(b), the process is referred to as *up-conversion*.

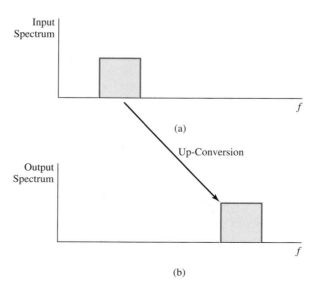

**FIGURE 5–14**
Illustration of frequency
up-conversion.

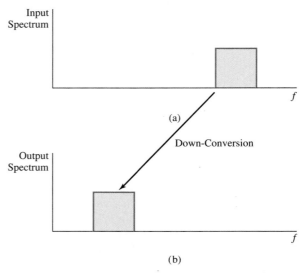

**FIGURE 5–15**
Illustration of frequency down-conversion.

Consider the same original spectrum again in Figure 5–15(a). If the signal is shifted to a lower frequency range as shown in Figure 5–15(b), the process is referred to as *down-conversion*. Other terms that are used synonymously with *frequency conversion* are *frequency translation* and *frequency shifting*.

A brief analysis of an ideal mixer was made in the last section. It was shown there that when a signal is multiplied with a sinusoid, two components are generated. Each has the shape of the original spectrum, although the spectral sense of one is turned around. Therefore, to perform a frequency conversion, we need only filter out one of the components. As will be seen later in the chapter, however, there may be other undesirable components that appear at the output of the mixer, but if the system is designed properly, the filter can remove those components as well. At this point, we will continue to assume an ideal mixer, since the major job is to remove the undesired sideband. We thus conclude that frequency conversion can be achieved with the combination of a mixer, an oscillator, and a filter. The block diagram of a typical system is shown in Figure 5–16.

## Predicting Frequency Conversion Output

The process of predicting the major components in the output of a mixer involves simple arithmetic, but the analysis can sometimes be a little tricky, especially when the spectral sense is inverted or there is some possible spectral overlap. The following steps will help to clarify the process.

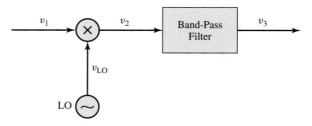

**FIGURE 5–16**
Frequency conversion system.

Assume that the spectrum consists of a bandlimited process extending from $f_1$ to $f_2$, where $f_2 > f_1$. Assume that the mixing oscillator frequency is $f_0$. This oscillator is referred to as the *local oscillator* (LO) in certain applications. The following steps are suggested:

1. Draw a two-sided spectrum, in which a continuous spectrum is defined from $f_1$ to $f_2$ for positive frequencies and from $-f_1$ to $-f_2$ for negative frequencies.

2. Shift both components to the *right* by $f_0$. Label the frequencies at the two edges of each spectral block.

3. Shift both components to the *left* by $f_0$. Again, label the frequencies at the edges.

4. The results of steps 2 and 3 are the outputs of an ideal mixer, and it will have four segments and be an even function of frequency. Depending on the desired objective, two of these components (one at positive frequencies and its image at negative frequencies) may be eliminated by a filter, provided that there is no overlap between desired and undesired components. The filter amplitude response is an even function of frequency so whenever a positive frequency component is removed, the corresponding negative frequency component will also be removed.

5. Redraw the spectrum after the undesired components are eliminated, and the result will be an even two-sided spectrum that has been either up-converted or down-converted.

The preceding processes will be illustrated for two particular cases: a local oscillator at lower frequency than any component within the spectrum, and a local oscillator at higher frequency than any component within the spectrum.

### LO Frequency below the Signal Spectrum

Consider the situation depicted in Figure 5–17(a). A band-pass spectrum exists over a frequency range from $f_1$ to $f_2$. (To enhance the presentations that follow, including spectral folding, it has been given a distinct shape that identifies clearly the lower and upper limits of the passband. The spectrum could have any form, however, including a symmetrical one about its center, a characteristic found in many modulated signals.) The LO mixing frequency $f_0$ is shown as a dashed line, and is lower than the lowest frequency $f_1$ in the spectrum. The positive frequency component is denoted A, and the negative frequency component is denoted B.

The process of multiplying the input signal by the mixing oscillator output results in a shift to the right and a shift to the left of the original spectrum. The shift to the right produces the components A' and B', and the shift to the left produces the components A'' and B'', as shown in Figure 5–17(b).

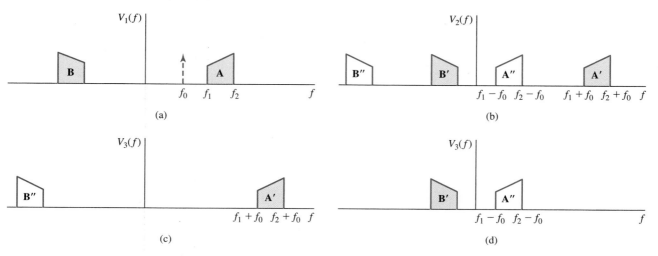

**FIGURE 5–17**

Frequency conversion based on oscillator frequency below spectrum.

The frequency bounds for A′ are the original frequency bounds plus $f_0$. Thus, $f_1$ moves to $f_1 + f_0$ and $f_2$ moves to $f_2 + f_0$.

The frequency bounds for A″ are the original frequency bounds minus $f_0$. Thus, $f_1$ moves to $f_1 - f_0$ and $f_2$ moves to $f_2 - f_0$.

The negative frequency components B′ and B″ are simply the mirror images of the positive frequency components.

If up-conversion is desired, a filter is used to accept A′ and reject A″, as shown in Figure 5–17(c). If down-conversion is desired, a filter is used to accept A″ and reject A′, as shown in Figure 5–17(d). In practice, a band-pass filter having a bandwidth just adequate to pass the desired spectrum is almost always used. Aside from the large sideband being eliminated, there will be other spurious components resulting from nonidealities in the mixing process as well as interference and noise. Both A′ and A″ have the same spectral sense as A; that is, higher frequency components of the original spectrum appear as higher frequency components of the shifted spectrum, and vice versa for the lower frequency components.

A conclusion from the analysis is that *for frequency conversion with the mixing frequency lower than the spectrum, one component will always be up-converted and the other component will always be down-converted.* Moreover, *both components will have the same spectral sense as the original signal.*

### LO Frequency above the Signal Spectrum

Consider next the situation depicted in Figure 5–18(a). The mixing frequency in this case is higher in frequency than the original spectrum. For the situation shown, it is only moderately higher, which will be a point of significance later.

The shift of the spectrum to the right now produces the components A′ and B′, and the shift to the left produces the components A″ and B″, as shown in Figure 5–18(b). However, unlike the previous case, A′ and B′ become the positive frequency components, while A″ and B″ become the negative frequency components. (Compare Figures 5–17(b) and 5–18(b) to see the difference.) While A′ has the same spectral sense as the original spectrum, B′ is *folded* or *reversed in frequency* with respect to the original component. Thus, the higher frequency components of the original spectrum become the lower frequency components of B′, and vice versa.

A short pause is in order at this point, to note that reversing the sense of the spectrum does not necessarily make that component useless. For certain types of symmetrical spectra that arise in some modulation systems, the spectral order is immaterial. Even when the spectral sense is important, however, an additional shifting operation of a similar nature

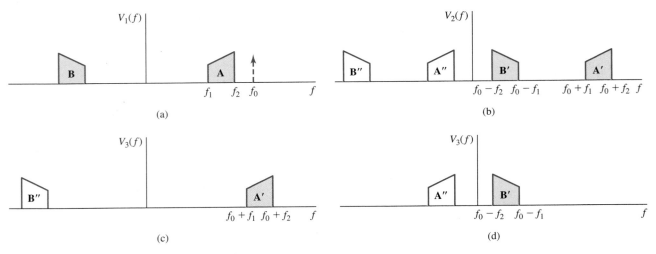

**FIGURE 5–18**

Frequency conversion based on oscillator frequency above spectrum.

would return the spectrum to its original sense. Thus, when the sense is important, it is necessary that there be an even number of reversals.

Back to the present development, the frequency bounds for A′ are the original frequency bounds plus $f_0$, as for the other case. Thus $f_1$ moves to $f_0 + f_1$ and $f_2$ moves to $f_0 + f_2$. (Note that we have written $f_0$ first in this case, for reasons that should be clear shortly.)

The frequency bounds for B′ are determined by subtracting $f_1$ and $f_2$ from $f_0$. Thus $f_1$ moves to $f_0 - f_1$ and $f_2$ moves to $f_0 - f_2$.

Figures 5–18(c) and (d) show the results of up-conversion and down-conversion after appropriate filtering. The component of (c) will always represent an up-conversion. In this case, the component of (d) is a down-conversion, since the mixing frequency was only moderately higher than the signal frequencies. Strictly speaking, however, this component could also be up-converted if the mixing frequency were high enough. It can be shown that if $f_0 > f_1 + f_2$, this component will also be up-converted.

A summary of the major points of the past few paragraphs follows:

1. If the LO frequency is lower than the signal frequency range, one component at the output of the mixer will be up-converted and the other will be down-converted. Both will have the same spectral sense as the original spectrum.

2. If the LO frequency is higher than the original frequency range, one component will be up-converted and the other component may be either up-converted or down-converted, depending on the frequency values. The component that is always up-converted will have the same spectral sense as the original spectrum, but the second component will have the spectral sense reversed or *folded*.

---

**▍▍ EXAMPLE 5-6**

A frequency conversion system is shown in Figure 5–19, and the spectrum of the input signal is shown in Figure 5–20(a). For an LO frequency of 110 MHz, sketch the spectra at the output of the mixer, and the output of the band-pass filter if the lower sideband is selected. Assume an ideal mixer.

**SOLUTION**  This is a case in which the LO frequency is lower than the frequency range of the input signal. The two-sided input amplitude spectrum $V_1(f)$ has positive frequencies ranging from 120 to 121 MHz and negative frequencies ranging from $-120$ to $-121$ MHz, as shown in Figure 5–20(a). To determine the output of the mixer, we first move the spectrum to the right by 110 MHz. The original positive frequency component moves to the range of $120 + 110 = 230$ MHz to $121 + 110 = 231$ MHz, as shown in 5–20 (b). However, the original negative frequency image moves to the range of $-121 + 110 = -11$ MHz to $-120 + 110 = -10$ MHz (not shown).

Next, the spectrum is shifted to the left by 110 MHz. The positive portion of the original spectrum now moves to the range of $120 - 110 = 10$ MHz to $121 - 110 = 11$ MHz, as also shown in Figure 5–20(b). The negative portion of the original spectrum moves to the range of $-120 - 110 = -230$ to $-121 - 110 = -231$ MHz (not shown). In the final analysis, the spectrum should be symmetrical about the origin; that is, the negative frequency components should be the mirror images of the positive frequency components.

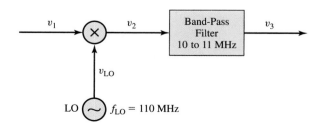

**FIGURE 5–19**
Frequency conversion system of Example 5-6.

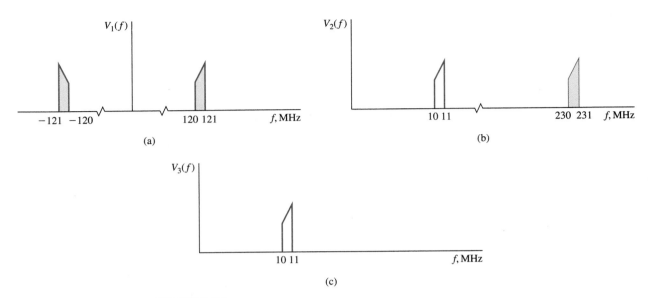

**(a)**

**(b)**

**(c)**

**FIGURE 5–20**
Spectral diagrams of Example 5-6.

When the signal is passed through a band-pass filter having a passband of 10 to 11 MHz, the output spectrum is as shown in Figure 5–20(c). Note that the spectral sense is the same as that of the original signal.

Because the lower sideband was selected in this example, the result was a down-conversion. Suppose, however, that the desired output frequency range were 230 to 231 MHz. In that case, the upper sideband could have been selected.

**▌▌ EXAMPLE 5-7**

Consider the frequency conversion system shown in Figure 5–21 with the input spectrum shown in Figure 5–22(a). For an LO frequency of 40 MHz, sketch the spectrum at the output of the mixer and at the output of the band-pass filter, based on selecting the upper side-band. Assume an ideal mixer.

**SOLUTION**  The two-sided input spectrum $V_1(f)$ has a spectrum ranging from 6 to 8 MHz for positive frequencies and from $-6$ to $-8$ MHz for negative frequencies. First, the spectrum is shifted to the right by 40 MHz and then shifted to the left by 40 MHz. On the right shift, the components in the range from $-8$ to $-6$ MHz move to the range of 32 to 34 MHz. The components in the range from 6 to 8 MHz move to the range of 46 to 48 MHz. The shift to the left would produce the negative frequency images of these spectral terms, which are not shown. The positive frequency results are shown in Figure 5–22(b).

When the output of the mixer is passed through a band-pass filter having a passband from 46 to 48 MHz, the output spectrum is as shown in Figure 5–22(c). Since this component is the upper sideband of the mixer output spectrum, it has the same spectral sense as the original signal.

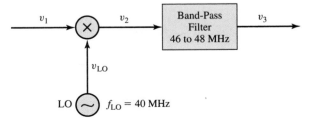

**FIGURE 5–21**
Frequency conversion system of Example 5-7.

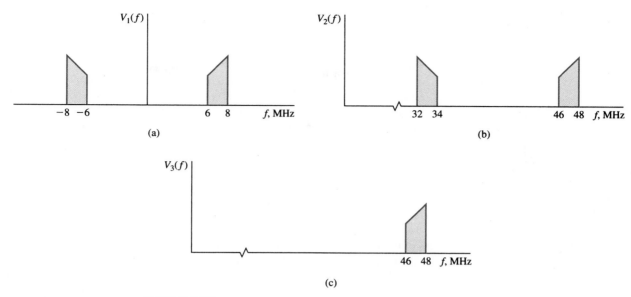

**FIGURE 5–22**
Spectral diagrams of Example 5-7.

**|‖ EXAMPLE 5-8**

In a certain receiver, it is desired to down-convert the RF signal to a lower frequency range for processing. The RF signal is centered at 200 MHz, and the center frequency of the output is desired to be 10 MHz. (a) Determine two possible LO frequencies that could be employed. (b) Sketch the forms of the output spectra in both cases.

**SOLUTION**

(a) This part is very simple and amounts to nothing more than the sum and difference of the two frequencies; that is, $200 - 10 = 190$ MHz and $200 + 10 = 210$ MHz. In other words, an LO frequency of either 190 MHz or 210 MHz will produce one component centered at 10 MHz when mixed with a signal centered at 200 MHz.

(b) A sketch of the original spectrum with an assumed distinct shape for identification is shown in Figure 5–23(a). When the signal is mixed with 190 MHz, the output

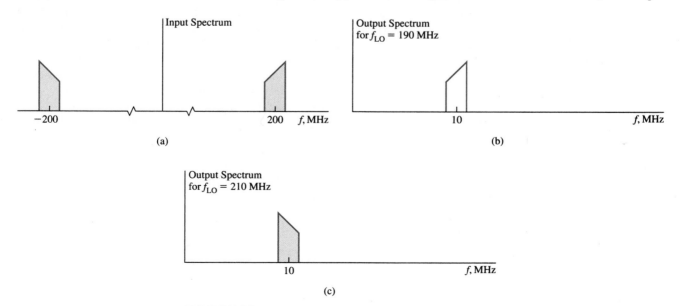

**FIGURE 5–23**
Spectral diagrams for Example 5-8.

spectrum centered at 10 MHz arises from the left shift of the positive portion of the original spectrum as shown in Figure 5–23(b). Only the left-shifted positive frequency component is shown. The component shifted to the right would be centered at 390 MHz and is easily eliminated.

When the signal is mixed with 210 MHz, the output spectrum centered at 10 MHz arises from the right shift of the negative portion of the original spectrum, as shown in Figure 5–23(c). The other component arising from the right shift would be centered at 410 MHz and is easily eliminated.

For the first case, the spectrum retains the same spectral sense as the original, but in the second case, it is reversed or folded.

---

**▐█ EXAMPLE 5-9**

A certain modulated signal has a center frequency of 1 MHz and a bandwidth of 100 kHz. It is desired to shift the center frequency to 400 MHz while retaining the exact form and sense of the spectrum. To ease the filtering requirements, a somewhat arbitrary but reasonable constraint is imposed on the output of each mixer, to the effect that all sideband spectral components to be eliminated should be displaced by no less than 5% from the adjacent band edge of the desired sideband. Plan a tentative design for an appropriate frequency translation system, specify mixing frequencies, and sketch the various spectra.

**SOLUTION**    A problem such as this has a number of possible solutions, and some trial and error is usually involved. First, we will determine if it would be practical to make the desired frequency shift in one step. An assumed spectrum in which the sense is easily recognized is shown in Figure 5–24(a). (In this part, both positive and negative frequencies are shown, but in subsequent steps, only the positive frequencies will be shown.) To shift the center of the spectrum all the way to 400 MHz and retain the same spectral sense, a mixing frequency of 399 MHz would be required. (A mixing frequency of 401 MHz would also shift the center to 400 MHz, but the spectral sense would be reversed.)

The two sidebands at the output of the mixer corresponding to a mixing frequency of 399 MHz are shown in Figure 5–24(b). The lowest frequency of the desired sideband is about 399.95 MHz, and the highest frequency of the sideband that must be removed is about 398.05 MHz. The second frequency is slightly greater than 0.995 times the first frequency, or less than 0.5% removed from the desired sideband. Adding to this is the fact that in any practical mixer, a small component of the mixing oscillator (399 MHz) appears at the output, and this component would be less than 0.24% removed from the desired sideband. Thus, the filtering requirements in this case cannot be achieved with the original constraints as stated, so the possibility of translation in one step is eliminated.

Next, the possibility of employing two steps of frequency shifting will be considered. Since the center frequency must eventually be translated by a factor of 400, a reasonable approach is to shift by a factor of 20 or so in each step.

The two-step or *double-conversion* system proposed is shown in Figure 5–25. The spectra at various points are shown in Figure 5–26, so refer to both figures in the discussion that follows.

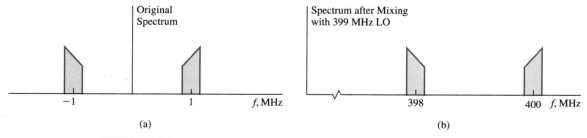

(a)                                                    (b)

**FIGURE 5–24**
Input and output spectra with LO frequency of 399 MHz in Example 5-9.

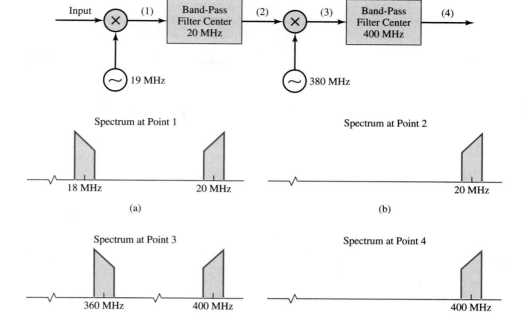

**FIGURE 5–25**
Proposed double conversion system of Example 5-9.

**FIGURE 5–26**
Spectral diagrams for double conversion system in Example 5-9.

The first LO frequency is chosen as 19 MHz. When the original two-sided spectrum centered at 1 MHz is shifted both to the right and left by 19 MHz, the resulting spectrum (point 1) on the positive frequency side is as shown in Figure 5–26(a). The desired component in this case is the one at 20 MHz. The lowest frequency of the desired sideband is 19.95 MHz, and the highest frequency of the undesired sideband is 18.05 MHz. The second frequency is about 0.905 times the first frequency or about 9.5% removed from the desired sideband. The mixing frequency at 19 MHz is displaced from the desired sideband by slightly less than 5%, but since any component appearing at the output at that frequency will be very small anyway (typically 40 dB or more down from the sideband levels), it should not cause any special filtering requirements. Thus, the first band-pass filter eliminates the lower sideband, and the spectrum at point 2 is shown in Figure 5–26(b).

The second mixer employs an LO frequency of 380 MHz. Although we have shown only the positive portion of the spectrum at point 2, we must consider both it and its negative frequency image in performing the subsequent shift. The positive frequency portion at 20 MHz moves to a center at $380 + 20 = 400$ MHz, and the negative portion moves to a center at $-20 + 380 = 360$ MHz. The resulting spectrum at point 3, displayed for positive frequencies, is shown in Figure 5–26(d). The lowest frequency of the desired component is 399.95 MHz, and the highest frequency of the component to be removed is 360.05 MHz. The second frequency is about 0.9 times the first frequency or about 10% removed from the desired sideband, so the filtering constraint is again satisfied. The mixing frequency of 380 MHz is displaced from the desired sideband by less than 5%, but for the same reason as before, this should not pose any serious problem. This system will be proposed as a possible solution, and the output spectrum at point 4 is shown in Figure 5–26(d).

## 5-5 Receivers

A *receiver* is the portion of the overall communcation system that processes the incoming signal and converts it back to the message or intelligence for which transmission is desired. The emphasis here will be on receivers designed to deal with radiated electromagnetic signals, although the term may also be applied to the destination processing in wire-carrier and fiber-optic systems as well. We study receivers before considering all of the different modulation methods, because all systems have some type of receiver and there are many common features.

## Antennas

For receivers operating in the RF electromagnetic frequency range, an *antenna* is required. An *antenna* is a physical device that "captures" some of the energy from electromagnetic waves that are present, and converts it into small voltages and currents that can be processed by the receiver circuits. In most cases, the signals at the antenna output are so small that normal voltmeters and/or ammeters cannot measure them. Hence, they must be amplified by a tremendous amount before useful output can be obtained. For this reason, any noise present can play a major role in the quality of the signal.

In many applications, the antenna is located physically well above the surface of the earth in order to obtain maximum signal strength. In other less critical applications, such as common household radios operated close to transmitters with high output power, an antenna inside the receiver may suffice. More detailed material on antennas will be considered in Chapter 15.

Some general definitions and properties important to all receivers will now be considered.

## Selectivity

*Selectivity* is a measure of how well a receiver separates the desired signal from the undesired signals in adjacent frequency ranges. The higher the selectivity, the more capability the receiver has in rejecting interfering components very close to the frequency of the desired signal. Selectivity is often specified by two bandwidth parameters, one of which indicates the passband bandwidth, and the other of which indicates the bandwidth at which attenuation is some minimum stopband value. Thus, a selectivity specification is similar to a filter specification, and in many receivers, it is based on one particular filter that establishes the major level of selectivity for the whole receiver.

Consider now the task of separating the desired signal from other signals present, in the case in which the receiver must tune over a wide frequency range. Good selectivity demands a band-pass filter having a flat passband characteristic and a very pronounced attenuation rate in the stopband. Such filters are relatively straightforward to build for operation near a limited frequency range, but tuning such filters over a very wide frequency range is something else entirely! Even with single series or parallel resonant circuits, which represent some of the simplest band-pass structures, maintaining a constant bandwidth over a wide frequency range is very difficult.

The concept of $Q$ was introduced in Chapter 4 for band-pass and band-rejection filters, and the term may be applied equally well to the selectivity measure of a receiver. When the filter or filters are connected to various input and output circuits, the term *loaded Q* is often used, and it can be defined as

$$Q = \frac{\text{center frequency}}{\text{bandwidth}} \tag{5-28}$$

In the basic resonant circuits considered in Chapter 4, the bandwidth was defined as the frequency difference between the 3-dB points. In more complex receiver characteristics, it may be defined as the frequency difference between other points (e.g., 1 dB or 6 dB).

One problem of tuning over a wide frequency range is the possible extreme range of $Q$ that may be required. If the required $Q$ is too low, there is difficulty in establishing proper stopband attenuation while simultaneously maintaining constant passband amplitude. Conversely, if the required $Q$ is too high, the resulting filter is very sensitive to element value tolerances and variations, and the range of component values may be very large. It is difficult to establish an exact range for $Q$ that is always feasible, due to the many forms of technology available and the wide variation of component characteristics over the frequency spectrum. For example, in the microwave range, a $Q$ of several thousand can be obtained with a cavity resonator, but such a value is virtually impossible in the lumped parameter frequency range. Realizing that there may be many exceptions to this pattern, the range $10 \leq Q \leq 100$ appears to be the most common approximate realistic working range for frequencies well below the microwave region.

## Sensitivity

*Sensitivity* is a measure of how well the receiver can respond to very weak signals. Theoretically, a receiver can be made to respond to an arbitrarily small signal by the addition of more amplifier stages. However, the important criterion is how well it can respond to small signals without masking the desired signal by the low-level noise introduced in the early stages. The major treatment of noise effects will be found in later chapters, but it should be noted here that it is a difficult design task to produce a highly sensitive receiver that can maintain a signal level well above the background noise level.

Sensitivity has been specified in several ways. One common receiver sensitivity specification is the level in microvolts required to produce a certain signal-to-noise ratio at the output. A more recent standard is to specify the signal power in dBf (decibels above 1 femtowatt, where 1 femtowatt $= 10^{-15}$ W) required to produce a certain signal-to-noise ratio at the output. The latter standard has the advantage that it is independent of the exact impedance level at the receiver input. With both of these standards, a *smaller* value for the sensitivity would indicate a *more sensitive* receiver. Thus, in qualitative references, one might casually say that the receiver with the smaller value of sensitivity is the more sensitive receiver.

A number of receiver types have been developed over the years. We will consider two types here: the *tuned-radio frequency* type and the *superheterodyne* type.

## Tuned Radio Frequency Receiver

The *tuned radio frequency* (*TRF*) receiver is the simplest and oldest type of receiver. (Refer to Figure 5–27 in the discussion that follows.) Other than the antenna (which may or may not be part of the actual receiver), the major units composing the TRF receiver are a *radio frequency* (*RF*) amplifier, a *detector* or *demodulator*, and a *post-detector* amplifier.

The RF amplifier is tuned to the center of the incoming signal spectrum, and its bandwidth should be sufficient to pass the desired signal spectrum while rejecting others. The detector demodulates or recovers the original intelligence from the modulated input signal. Depending on the nature of the modulating signal, the post-detection amplifier could be a baseband amplifier such as an *audio frequency* (*AF*) amplifier (for sound), a *video amplifier* when the output contains video information, or some other data.

TRF receivers are still used for some limited applications in which little or no tuning of the passband is required. However, when continual tuning over a wide frequency range

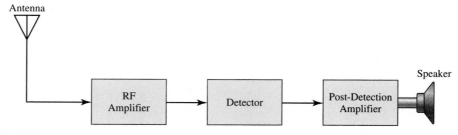

**FIGURE 5–27**
Block diagram of tuned radio frequency (TRF) receiver.

is required, a TRF receiver is very limited. The basic problem (as stated earlier) is that it is very difficult to vary the frequency of a band-pass filter over a wide range and still maintain high selectivity, particularly at the higher frequencies.

## Superheterodyne Receivers

Most receivers used in modern communications systems employ a concept known as the *superheterodyne principle*. Basically, the concept involves down-converting the *radio frequency* signals to one or more fixed *intermediate frequency* (*IF*) stages, in which the major portion of the filtering is achieved. Said differently, all of the necessary filtering requirements are translated from a higher frequency variable range to one or more lower frequency fixed ranges, in which precision fixed filters establish the selectivity.

Superheterodyne receivers are classified as *single-conversion* when there is only one IF frequency range, *double-conversion* when there are two IF frequency ranges, and so on. Single-conversion receivers are the simplest and most common in the consumer field, and that form will be used in the discussion that follows.

The block diagram of a single conversion superheterodyne receiver is shown in Figure 5–28(a), and a corresponding spectral diagram is shown in Figure 5–28(b). Observe the break in the frequency scale, since it is difficult to show on a single graph the typical frequencies to scale. In the discussion that follows, the reference will be shifted frequently back and forth from the block diagram to the spectral diagram.

The center frequencies of various portions of the receiver are indicated above the block diagram in Figure 5–28(a), and the corresponding bandwidths are indicated below the block diagram. The incoming signal is amplified first by an RF amplifier. The center frequency of this stage is $f_c$. A typical RF amplifier has a band-pass characteristic sufficiently

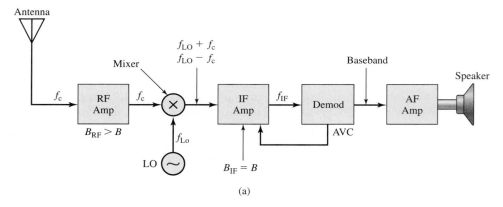

(a)

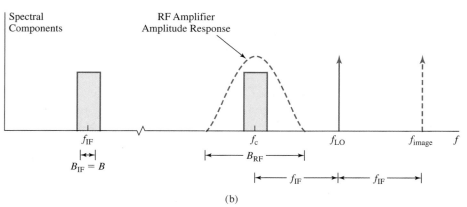

(b)

**FIGURE 5–28**
Block diagram of single-conversion superheterodyne receiver and associated spectral components.

broad to pass the highest bandwidth signal required. As we will see later, some selectivity is desired in the RF stage, but it need not possess the sharp stopband attenuation rate eventually needed to reject adjacent interfering signals. Indeed, if it did, all the filtering could be done in this stage and we would not need the IF stage! In some inexpensive consumer receivers and in some special remote-sensing receivers, the RF stage has no band-pass filter at all, but most receivers will have some RF stage filtering. A typical broadband characteristic is shown over the top of the other components about $f_c$ in Figure 5–28(b).

In the mixer stage, the incoming RF signal centered at $f_c$ is mixed with a local oscillator (LO) sinusoid with frequency $f_{LO}$. The LO tuning circuit is coupled with the RF amplifier tuning circuit so that the frequency difference between $f_c$ and $f_{LO}$ is a constant, and this constant frequency difference is the intermediate frequency, which will be denoted $f_{IF}$. Coupling of the tuning was traditionally achieved with a two-section variable capacitance in which both values change as the tuning knob is turned. However, modern frequency synthesizer technology has provided newer techniques to achieve the necessary frequency difference.

The LO frequency may be higher or lower than the RF frequency. When the LO frequency is higher, we have

$$f_{IF} = f_{LO} - f_c \qquad (5\text{-}29)$$

When the LO frequency is lower, the relationship is

$$f_{IF} = f_c - f_{LO} \qquad (5\text{-}30)$$

The spectral diagram assumes a higher LO frequency, and this assumption will be made in this development.

The major components at the output of the mixer are the sum and difference frequency components centered at $f_{LO} + f_c$ and $f_{LO} - f_c$. The sum frequency is off the scale of the figure, and only the difference frequency is of interest. Thus, the IF frequency is centered at the value given by Equation 5-29.

### IF Amplifier

The IF amplifier is a tuned amplifier stage having an amplitude characteristic that effectively establishes the selectivity of the receiver. The desired frequency component is shifted to the fixed passband of this stage, while undesired components should fall outside the passband. (We will shortly see that there may be some undesired components that get into the passband as well.) Ideally, the bandwidth of the IF stage should match the bandwidth of the desired signal. Many consumer receivers have fixed passband characteristics, but some general purpose communications receivers designed to accommodate a variety of signal types have means provided to vary this bandwidth about the fixed IF center frequency.

The undesired mixer output component at $f_{LO} + f_c$ will normally be well outside of the IF passband and is easily rejected by the IF stage. Other undesirable products of the mixer will be rejected by the IF amplifier unless they happen to occur in the IF bandwidth.

The strategy of the superheterodyne receiver can now be summarized. Instead of the virtually impossible chore of attempting to tune across a wide RF spectrum with a highly selective variable-center frequency filter, a fixed-center frequency filter is designed and optimized. The various desired signals are then shifted to the frequency range of the fixed IF filter by the mixer and LO, and are then separated from other components in that frequency range. The selectivity of the receiver is then determined by the IF stage. Incidentally, the superheterodyne principle is also employed in many spectrum analyzers and other frequency selective instruments.

Following amplification and filtering in the IF amplifier, the signal is applied to the demodulator or detector stage, where the desired demodulation is performed. The diagram of Figure 5–28(a) indicates an audio frequency (AF) amplifier and a speaker, which would suggest a consumer household radio.

**Automatic Volume Control**

One additional item in Figure 5–28(a) is the *automatic volume control* (*AVC*) line. Signals appearing at the antenna terminals of receivers vary considerably in their levels. The AVC signal is a rectified bias voltage or current proportional to the level of the demodulated signal. This voltage is applied to an earlier stage in the receiver in which the gain can be reduced by application of the particular voltage. Thus, stronger signals cause a larger AVC voltage to appear, and this in turn reduces the gain. A typical AVC circuit does not completely eliminate the variation of signal strengths, but it does reduce the effective dynamic range between the strongest and weakest signals.

## Image Frequency

We have just seen that the superheterodyne receiver offers significant advantages in establishing a fixed selectivity over a broad frequency range. As is often the case with engineering design, however, there is a tradeoff. A potential problem called *image frequency interference* suddenly appears. Refer again to the spectral diagram in Figure 5–28(b). Observe the component $f_{\text{image}}$ shown as a dashed line. This component is called the *image frequency,* and the value of this frequency is

$$f_{\text{image}} = f_{\text{LO}} + f_{\text{IF}} = f_{\text{c}} + 2f_{\text{IF}} \qquad (5\text{-}31)$$

Suppose a signal at the image frequency is passed by the RF amplifier. It will then mix with the LO signal, and one of its output components will appear at $f_{\text{IF}}$ along with the desired signal. Once it appears in the IF passband at the same frequency as the desired signal, it cannot be easily separated from the desired output.

The technique for minimizing image interference is to ensure that it does not get through the RF amplifier. This can be achieved by having a moderate degree of selectivity in that stage. If $B$ is the bandwidth of the desired signal, the frequency range from $f_{\text{image}} - B/2$ to $f_{\text{image}} + B/2$ represents the range that when mixed with $f_{\text{LO}}$ would fall in the IF bandpass. Consequently, this range should be well into the stopband portion of the RF amplifier response curve.

As the IF frequency is increased, the moderate filtering on the RF amplifier is reduced and the corresponding image frequency is farther away from $f_{\text{c}}$. Conversely, however, it is often easier to design and implement an IF amplifier with high selectivity at a relative low frequency. The selection of an IF frequency is then an engineering compromise.

## LO Frequency Lower Than IF Frequency

Throughout the analysis, we have assumed that $f_{\text{LO}} > f_{\text{c}}$. When $f_{\text{LO}} < f_{\text{c}}$, the same general principles apply except that the image component is now

$$f_{\text{image}} = f_{\text{LO}} - f_{\text{IF}} = f_{\text{c}} - 2f_{\text{IF}} \qquad (5\text{-}32)$$

For either case, the image component is on the opposite side of the LO frequency from the desired signal by a frequency increment equal to the IF frequency.

## Multiple Conversion

Because of the conflicting requirements of high IF selectivity at lower IF frequencies and ease of image rejection at high IF frequencies, complex receivers employing more than one level of conversion are used in many applications. Such receivers are described as *double-conversion, triple-conversion,* and so on, depending on the number of conversions.

A block diagram of a double-conversion receiver is shown in Figure 5–29. The first mixer stage down-converts the signal to an IF frequency $f_{\text{IF}}^{(1)}$. This particular IF frequency is relatively high and eases the burden on the RF amplifier. The second mixer stage down-converts the signal to a final IF frequency range $f_{\text{IF}}^{(2)}$, which is usually much lower than the first IF frequency. This stage has been optimized to produce a very nearly rectangular band-pass characteristic, so the selectivity of the receiver is established at this point.

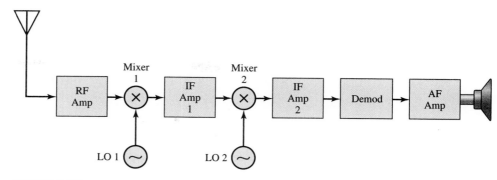

**FIGURE 5–29**
Block diagram of double-conversion superheterodyne receiver.

**EXAMPLE 5-10**

A single-conversion superheterodyne receiver is tuned to an RF frequency of 40 MHz, and the IF frequency is 5 MHz. If the LO frequency is higher than the RF frequency, determine (a) the LO frequency and (b) the image frequency.

**SOLUTION**

(a) The LO frequency must be 5 MHz higher than the input RF frequency; that is,

$$f_{LO} = f_c + f_{IF} = 40 + 5 = 45 \text{ MHz} \qquad (5\text{-}33)$$

(b) The image frequency is 5 MHz higher than the LO frequency; that is,

$$f_{image} = f_{LO} + f_{IF} = 45 + 5 = 50 \text{ MHz} \qquad (5\text{-}34)$$

**EXAMPLE 5-11**

Repeat the analysis of Example 5-10 if the LO frequency is lower than the RF frequency.

**SOLUTION**

(a) The LO frequency must be 5 MHz lower than the input RF frequency:

$$f_{LO} = f_c - f_{IF} = 40 - 5 = 35 \text{ MHz} \qquad (5\text{-}35)$$

(b) The image frequency is 5 MHz lower than the LO frequency:

$$f_{image} = f_{LO} - f_{IF} = 35 - 5 = 30 \text{ MHz} \qquad (5\text{-}36)$$

**EXAMPLE 5-12**

Most commercial FM receivers are single-conversion superheterodyne types, and are capable of tuning over the FM broadcast band from approximately 88 to 108 MHz. The IF frequency in these receivers is usually 10.7 MHz. The bandwidth allocated by the FCC for each station is 200 kHz. (a) Suppose that very selective filtering were performed in the RF stage (i.e., the TRF receiver concept). Determine the range of approximate loaded $Q$'s that would be required. (b) Determine the approximate $Q$ required for the IF filter in the superheterodyne receiver. (c) Determine the range of LO and image frequencies if the LO oscillator frequency is higher than the signal frequency. (d) Repeat part (c) if the LO frequency is lower than the signal frequency.

**SOLUTION**

(a) For this part and for part (b), a bandwidth of 200 kHz will be used. This may or may not be the optimum bandwidth to use for the filter, depending on the exact passband shape and how sharply it attenuates in the stopband. However, it is sufficiently close for the desired purpose, since approximate values were requested.

At the low end of the FM band, the $Q$ required would be

$$Q = \frac{88 \times 10^6}{200 \times 10^3} = 440 \tag{5-37}$$

At the high end of the FM band, the $Q$ required would be

$$Q = \frac{108 \times 10^6}{200 \times 10^3} = 540 \tag{5-38}$$

Both of these $Q$ values are rather difficult to obtain in the frequency range involved.

(b) At the IF frequency for the superheterodyne receiver, the required $Q$ is

$$Q = \frac{10.7 \times 10^6}{200 \times 10^3} = 53.5 \tag{5-39}$$

This value is much more reasonable. Moreover, the IF amplifier frequency is fixed, while the TRF tuning circuit would require maintaining a high $Q$ over a 20-MHz tuning range.

(c) If the LO frequency is higher than the signal frequency, it must be 10.7 MHz higher at all frequencies. Thus, if the signal varies from 88 to 108 MHz, the LO oscillator must tune from $88 + 10.7 = 98.7$ MHz to $108 + 10.7 = 118.7$ MHz.

The image frequency is 10.7 MHz above the LO oscillator frequency, so at the low end of the dial, the image is $98.7 + 10.7 = 109.4$ MHz, and at the high end of the dial, the image is $118.7 + 10.7 = 129.4$ MHz.

The ranges involved are summarized as follows:

Signal frequency = 88 to 108 MHz

LO frequency = 98.7 to 118.7 MHz

Image frequency = 109.4 to 129.4 MHz

At a given frequency, the RF amplifier passband should be sufficiently selective to reject the corresponding image frequency, but since the image frequency is 21.4 MHz away, a much more moderate $Q$ for the RF tuning circuit would suffice. The minimum $Q$ of the RF amplifier would depend on the level of the image rejection desired (and other factors), but a realistic design can be achieved, as evidenced by the millions of commercial FM receivers in operation.

It is interesting to note that the lowest possible image frequency (109.4 MHz) is above the high end of the FM band by a margin of 1.4 MHz. This eliminates the possibility of another FM station being the interfering image component, and this constraint was no doubt considered in the selection of the common IF frequency.

(d) The analysis proceeds in essentially the same fashion as in part (c) except that the LO frequency is always lower than the signal frequency by 10.7 MHz, and the image frequency is always lower than the LO frequency by 10.7 MHz. The reader is invited to perform the calculations, and the results are summarized as follows:

Signal frequency = 88 to 108 MHz

LO frequency = 77.3 to 97.3 MHz

Image frequency = 66.6 to 86.6 MHz

Note that the highest image frequency is 1.4 MHz below the low end of the FM band.

## 5-6  Realistic Mixers

Thus far in the chapter, each mixer under consideration has been assumed to be an ideal multiplier in which the instantaneous output is proportional to the product of two inputs, one of which is the signal and the other of which is usually a local oscillator. Precision

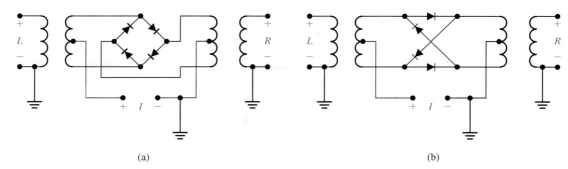

**FIGURE 5–30**
Double-balanced mixer, drawn two ways.

analog multiplier modules are available, and they could theoretically be used for some mixer applications. However, they are more suited to specialized control and instrumentation functions at low frequencies. Mixers designed specifically for RF communication circuits deviate somewhat from the ideal multiplier concepts, as we will see shortly. The most widely employed mixer utilized in communication applications is the *double-balanced mixer,* and that circuit will be our primary area of focus.

Two separate ways to show the circuit diagram of a *double-balanced mixer* are provided in Figure 5–30. Both forms appear in the literature, but after a bit of rearranging, it can be seen that they are the same circuit. A mixer is an example of a *three-port* device, and the labels shown in Figure 5–30 are the forms provided in manufacturers' labels. These labels are provided in relationship to down-conversion terminology, and they are described in the sections that follow.

## L-Port

The *L-Port* represents the *local oscillator* input port. The local oscillator may be lower or higher in frequency than the signal to be down-converted. The LO signal should normally be much larger in magnitude than the RF signal to ensure that switching is controlled primarily by the LO signal. This minimizes the creation of undesired products in the output. In a typical situation, the LO should have a level at least 20 dB above the RF level.

## R-Port

The *R-Port* represents the *RF* input port. This is the signal to be down-converted.

## I-Port

The *I-Port* is the *intermediate frequency* port. The output signal at the IF frequency is taken at this port.

## Single Sideband Conversion Loss

In a passive mixer of the type considered here, the output at the IF frequency will have a power level lower than the RF input signal. Losses occur as a result of spurious components at integer multiples of the frequencies involved, diode losses, impedance mismatches at the ports, and the creation of other undesired mixing components. Typical SSB (single sideband) conversion losses range from about 6 to 9 dB.

## Intermodulation Products

Assume in the discussion that follows that the RF signal is a sinusoid of frequency $f_R$, and the LO signal is a sinusoid of frequency $f_L$. In general, the output will consist of many

**Table 5–1**    Typical Mixer Intermodulation Specification Chart

|         | $n = 0$ | $n = 1$ | $n = 2$ | $n = 3$ | $n = 4$ | $n = 5$ |
|---------|---------|---------|---------|---------|---------|---------|
| $m = 0$ | NA      | 36      | 45      | 52      | 63      | 45      |
| $m = 1$ | 25      | 0       | 39      | 13      | 45      | 22      |
| $m = 2$ | 69      | 72      | 79      | 67      | 75      | 66      |
| $m = 3$ | 51      | 49      | 53      | 51      | 55      | 48      |
| $m = 4$ | 80      | 79      | 82      | 77      | 82      | 76      |
| $m = 5$ | 72      | 70      | 71      | 52      | 77      | 46      |

The integer $m$ is the multiple of $f_R$, and the integer $n$ is the multiple of $f_L$.

The tabulated values are the decibel levels *down* from a principal sideband level; that is, from the level corresponding to $m = n = 1$.

frequencies, representing both sum and difference frequencies as well as harmonics of both $f_R$ and $f_L$. The sum and difference frequencies are referred to as *intermodulation products*.

Let $f_o$ represent any one of the various output frequencies. Such a frequency satisfies the following relationship:

$$f_o = \pm m f_R \pm n f_L \tag{5-40}$$

where $m$ and $n$ are positive integers. An ideal mixer with $f_R > f_L$ would have only $f_R + f_L$ and $f_R - f_L$ corresponding to $m = 1$ and $n = 1$. However, the existence of the other components must be understood, and their levels may need to be estimated in order that adequate filtering may be employed.

Companies producing mixers publish tables providing typical levels of harmonic and intermodulation products, for which a sample is shown in Table 5–1. The horizontal axis represents the harmonics of the LO frequency $f_L$, and the vertical axis represents the harmonics of the RF frequency $f_R$. Intermodulation products are determined from the intersection of $n$ (the harmonic integer of $f_L$) and $m$ (the harmonic integer of $f_R$). All values in the table represent decibel levels *down* from the reference. In other words, these numbers should all be interpreted as *negative* levels with respect to the reference.

The reference level is based on the intersection of $n = 1$ and $m = 1$, which is indicated as 0 dB. This means that all other values are **down** from the levels of the components at $f_R \pm f_L$ by the number of decibels indicated.

## Isolation

In an ideal mixer, all of the output would be at the IF frequency, and no RF or LO components would appear at the IF port. In practice, some portion of the IF and RF signals will appear at the IF port. The degree of undesirable coupling is specified by an *isolation* specification in decibels. Typical values of L-to-I and R-to-I isolation are in the range from 20 to 30 dB. These numbers represent the amount by which the components are *down* from the L or R level, respectively. The most critical of these is the L-to-I specification, since the level of the LO is much greater than that of the RF signal.

## Dynamic Range

*Dynamic range* is the RF input power range (in decibels) over which the mixer should be operated. The lower level of the range is limited by the noise of the system. The upper level is usually defined as the RF power level in dBm at which the conversion loss increases by 1 dB.

At signal levels below the dynamic range upper level, the slope of the output–input curve (both measured in dB) is $+1$. However, as the input signal approaches the upper level of the dynamic range, the slope of the curve decreases and a compression occurs. In general, the 1-dB compression point is typically about 5 to 10 dB lower than the LO input power.

## 5-7 Multisim® Examples (Optional)

Two examples dealing with the topics covered in this chapter will be modeled with Multisim in this section. The first example is that of the junction field-effect transistor (JFET) oscillator of Example 5-1. The second example is that of an up-conversion frequency translation circuit typical of some of the chapter examples. In both cases, **Transient Analysis** will be employed.

**▐▌ MULTISIM EXAMPLE 5-1**

Model the Clapp oscillator circuit of Example 5-1 (Figure 5–7) with Multisim, and perform a **Transient Analysis** to verify the operating conditions for the circuit.

**SOLUTION**    The circuit is constructed using the procedures of Appendix B, and is shown in Figure 5–31. Although a battery could have been employed, instead the **VDD** bus in the **Power Source Family Bar** was employed and set to the value of 12 V. This source is convenient in that its ground terminal is automatically established. Of course, the remainder of the overall circuit must have a ground terminal.

The FET **Q1** is obtained from the **Transistors Family Bar.** Instead of employing a specific part number, the one designated as **JFET_N_Virtual** was employed. This is an n-channel JFET having characteristics that are somewhat typical of many in the same category.

Oscillator circuits can be a bit tricky to simulate in some cases, since they are totally dependent on circuit feedback within a loop. (The author has encountered some examples that actually required a "jump-start" in the form of some initial conditions set within the circuit.) In this case, the oscillator started on its own, but as is true with many oscillator circuits, some settling time was required for steady-state conditions to be reached. Therefore, some trial and error was involved in establishing appropriate simulation parameters. As

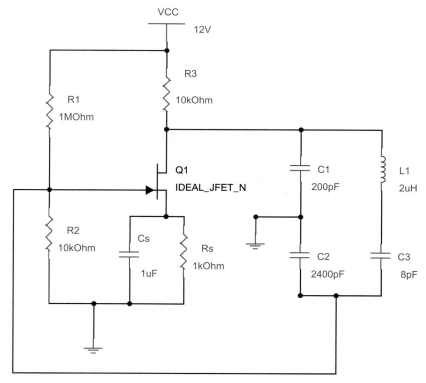

**FIGURE 5–31**
Oscillator circuit of Multisim Example 5-1.

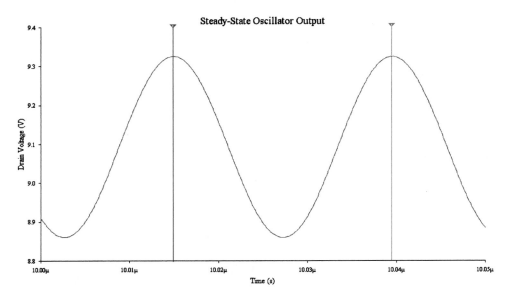

**FIGURE 5–32**
Waveform of Multisim Example 5-1.

noted in earlier work, **Transient Analysis** may require more than a casual effort to ensure that the results are reasonable.

If a view of the simulation were initiated at $t = 0$, it would be necessary to employ a time scale that would obscure a good view of the oscillations (as a result of the settling time). After some experimentation, it was decided to begin observation at 10 μs. Therefore, the **Start time** was set to **10e-6**. Understand that the actual internal simulation begins at $t = 0$, but we skip over the settling time and begin observation after steady-state conditions are reached. In view of the approximate period expected, the **End time** was set to **10.05e-6**. This should allow close to two cycles of observation at the expected frequency.

Because of the high frequency involved, it is also necessary to use a very small time step. The value selected for the **Maximum Time Step** was **0.1 ns (1e-10)**. This proved to yield good results. However, with this small time step and the fact that the first portion of the response is not shown, the simulation will take a little time before the results begin to display.

The output response measured at the drain terminal of the FET (node **2**) is shown in Figure 5–32. As expected, there is a dc level, which could be removed with a capacitor, but our interest is centered on the oscillation. In order to measure the frequency, the **Cursors** were used to determine the period. The two times at which the peak response occurred were determined by the cursors to be at 10.0396 μs and 10.0150 μs. The period is then estimated as $10.0396 - 10.0150 = 0.0246$ μs. This value leads to a frequency of $1/(24.6\text{e-}9) = 40.65$ MHz. This compares with the projected frequency of 40.64 MHz!

To be perfectly honest, the author is surprised that it came out that close to the calculated value, and there is probably a bit of luck involved. We are dealing with a numerical approximation in which a relatively high frequency is being generated with very small time steps. Moreover, the measurement with the cursors is subject to quantization errors, and we have to form a small difference between two much larger numbers to estimate the period. Don't expect simulation results of this nature to always be so close.

---

**▮▮ MULTISIM EXAMPLE 5-2**   Model an up-conversion circuit with Multisim that will shift a 1 MHz signal to 10 MHz, and verify that the circuit works as intended.

**SOLUTION**   Refer to the circuit diagram shown in Figure 5–33 for the discussion that follows. An ideal mixer can be modeled with the aid of the **MULTIPLIER** module, which is

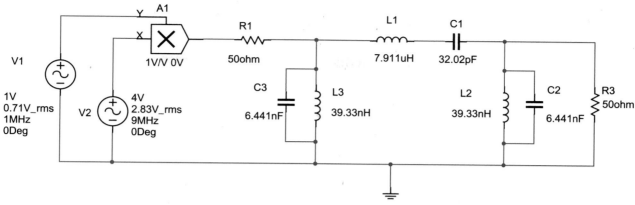

**FIGURE 5–33**
Circuit of Multisim Example 5-2.

located in the **CONTROL_FUNCTION** bin under the **Sources** toolbar. This module performs the product of the signal at the **Y** port with the signal at the **X** port. For this ideal module, either could be used for the input signal or the local oscillator. We will arbitrarily select the **Y** input for the 1-MHz signal, in which a sinusoidal source is used. The LO could either be 9 or 11 MHz, and since no reference to spectral sense is given, we will use a 9-MHz sinusoid.

Along with the desired component at 10 MHz, there will also be a component at 8 MHz, which must be eliminated. Recall from Multisim Example 4-3 that a band-pass filter with three pole-pairs, a center frequency of 10 MHz, and a 1-dB bandwidth of 1 MHz was presented. The circuit was presented in Figure 4–33, but is shown again at the output of the mixer in Figure 5–33. Let's give it a try.

The instantaneous output of the filter beginning at $t = 0$ and continuing to about 8 μs is shown in Figure 5–34. We see the initial transient response of the filter during the settling time, but the scale is too broad to adequately observe the oscillations. A magnified scale after the settling interval is shown in Figure 5–35, based on a time interval from 5 μs to 5.2 μs. There are exactly two cycles in this interval, so the period is 0.1 μs. The frequency is thus $1/(0.1e-6) = 10$ MHz, so the circuit is behaving as desired.

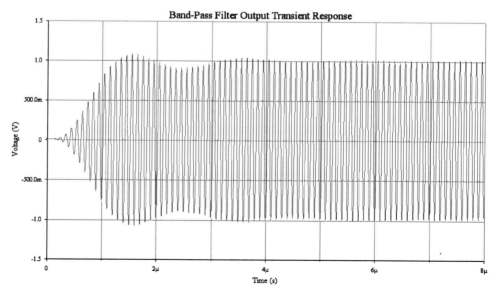

**FIGURE 5–34**
Response in Multisim Example 5-2 during settling interval.

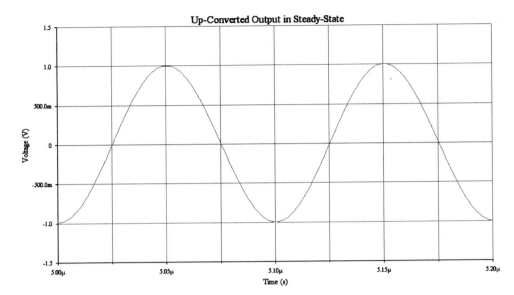

**FIGURE 5–35**
Steady-state output in Multisim Example 5-2.

## 5-8   MATLAB® Example

MATLAB will be used in this section to demonstrate the output of a frequency-shifting circuit. In particular, a down-conversion will be utilized, but the same process could just as easily be used for up-conversion. In the example that follows, some operations from the **Signal Processing Toolbox** will be utilized. Therefore, to achieve the results provided, it is necessary that this toolbox be installed in the version of MATLAB utilized.

**▌▌▌ MATLAB EXAMPLE 5-1**

Consider a mixer circuit with an input sinusoidal signal of 100 kHz and an LO frequency of 90 kHz. (a) Use MATLAB to determine the instantaneous output of the mixer and plot it. (b) Use MATLAB to implement a 4 pole-pair Butterworth band-pass filter to select the lower frequency component and plot it.

**SOLUTION**   There is no particular time step or time length that is optimum for the purpose of this simulation, so we will arbitrarily select a time step of 1 μs (1e-6 s) and a time duration of 2 ms (2e-3 s). The time vector is thus generated as

```
>> t = 0:1e-6:2e-3;
```

The input sinusoid **v1** has a frequency of 100 kHz, and can be generated by the command

```
>> v1 = sin(2*pi*1e5*t);
```

The LO **vlo** can be generated by the command

```
>> vlo = sin(2*pi*90e3*t);
```

Utilizing a gain constant of unity, the output **v2** of the balanced modulator is

```
>> v2 = v1.*vlo;
```

Note the necessity to put a period after the first variable, which tells MATLAB to multiply the corresponding elements of the two row vectors on a point-by-point or array basis.

The spectrum of **v2** consists of a component at $100 - 90 = 10$ kHz and a component at $100 + 90 = 190$ kHz. Although a low-pass filter could accomplish the task in this idealized case, we will choose instead to use a band-pass filter. Let us arbitrarily select a four pole-pair Butterworth filter with the two 3-dB frequencies at 8 and 12 kHz. Assuming that the **Signal Processing Toolbox** is installed in MATLAB, the numerator polynomial coefficients **n** and denominator polynomial coefficients **d** can be determined from the command

```
>> [n d] = butter(4, 2*pi*[8e3 12e3], 's');
```

where **butter** is the code for Butterworth and the value of **4** represents four pole-pairs. The quantity 's' represents the code for an analog filter based on the Laplace transform or s-domain form from linear system theory. Note that radian frequencies are required in the command, and this is the reason why both **8e3** and **12e3** in brackets are multiplied by **2*pi**.

The time-domain output can then be determined from the **lsim** command (for *linear system simulation*). The command based on the preceding functions, with the output denoted **v3**, is

```
>> v3 = lsim(n, d, v2 ,t);
```

A plot of the output over the entire time interval can be generated by the command

```
>> plot(t, v3)
```

This function is shown in Figure 5–36. Note the transient or settling interval required for the filter to reach steady-state conditions. To demonstrate that the output after settling has the appropriate form, we will next plot the function over two cycles between 1 ms and 1.2 ms. This operation can be achieved by the command

```
>> plot( t(1000:1200), v3(1000:1200))
```

This result is shown in Figure 5–37, and it can be readily verified that this function is a sinusoid with a frequency of 10 kHz.

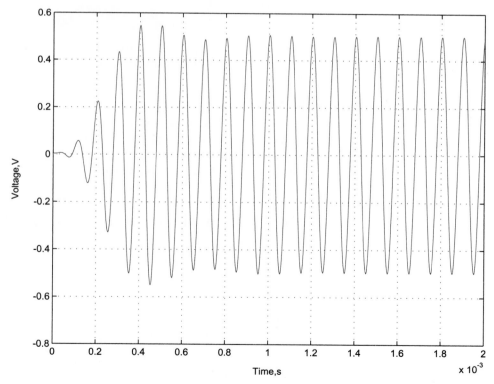

**FIGURE 5–36**
Output of filter beginning at $t = 0$.

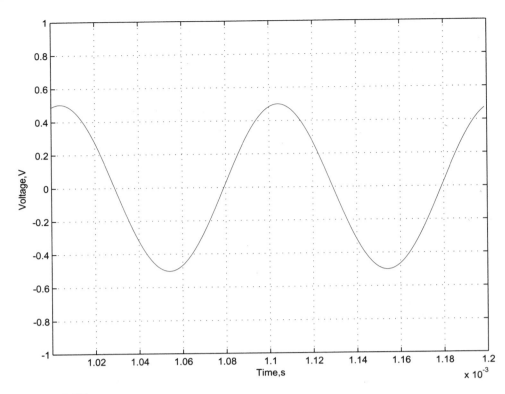

**FIGURE 5–37**
Output of filter over two cycles in steady state.

## SystemVue™ Closing Application (Optional)

Insert the text CD in a computer having SystemVue™ installed and activate the program. Open the CD folder entitled **SystemVue Systems** and load the file entitled 5-1.

### Review of System

The system shown on the screen was described in the chapter opening application, and you may wish to review the discussion there. Recall that the input was a 1-kHz signal, and it was shifted to 10 kHz by mixing it in an ideal multiplier with a 9-kHz oscillator followed by a filter. In that case, an up-conversion was performed with the LO on the lower side of the desired output frequency.

### Up-Conversion with LO above the Desired Output Frequency

1. Set the LO to a frequency above 10 kHz that will produce one component at 10 kHz. Since the output filter is centered at 10 kHz, it should suffice for this purpose.

2. Perform measurements of the signals to verify that the up-conversion as desired has been performed.

### Down-Conversion with LO below the Input Frequency

1. Change the input frequency (Token 3) to 10 kHz.

2. Change the LO frequency (Token 4) to a frequency lower than 10 kHz that will produce an output at 1 kHz.

3. Change the output filter to a low-pass filter having a cutoff frequency slightly above 1 kHz.

4. Perform a run, measure the frequencies, and verify that the objective has been met.

## Down-Conversion with LO above the Input Frequency

Repeat the preceding steps with the LO frequency set to a higher value than the 10-kHz input.

## PROBLEMS

**5-1** A Hartley oscillator of the form shown in Figure 5–2 has $L_1 = 19$ μH, $L_2 = 1$ μH, and $C = 100$ pF. Determine (a) the approximate frequency of oscillation and (b) the required gain for oscillation.

**5-2** A Colpitts oscillator of the form shown in Figure 5–3 has $C_1 = 20$ pF, $C_2 = 150$ pF, and $L = 50$ μH. Determine (a) the approximate frequency of oscillation and (b) the required gain for oscillation.

**5-3** For the Hartley oscillator of Problem 5-1 with the inductance values given, determine the approximate capacitance value to produce oscillations at 5 MHz.

**5-4** For the Colpitts oscillator of Problem 5-2 with the capacitance values given, determine the approximate inductance value to produce oscillations at 3 MHz.

**5-5** A crystal has a termperature coefficient of 50 ppm/°C, and the frequency is 4 MHz at 20°. Determine the frequency at (a) 35°C and (b) −5°C.

**5-6** A crystal has a termperature coefficient of −10 ppm/°C, and the frequency is 7 MHz at 20°. Determine the frequency at (a) 35°C and (b) −5°C.

**5-7** For the circuit below, $f_i = 4$ MHz, $N = 8$, and $M = 9$. Determine (a) the loop output frequency $f_o$ and (b) the frequency at which the phase detector operates.

**5-8** For the circuit above, $f_i = 3.5$ MHz, $N = 12$, and $M = 7$. Determine (a) the loop output frequency $f_o$ and (b) the frequency at which the phase detector operates.

**5-9** It is desired to generate a 5.1-MHz signal locked in phase to a given 3.4-MHz signal. Specify the smallest integer values of $N$ and $M$ in the circuit shown in Problem 5-7 that will achieve the objective.

**5-10** It is desired to generate an 8-MHz signal locked in phase to a given 5.6-MHz signal. Specify the smallest integer values of $N$ and $M$ in the circuit shown in Problem 5-7 that will achieve the objective.

**5-11** An ideal mixer circuit has a single frequency input of 8 kHz and an LO frequency of 40 kHz. Specify the positive output frequencies.

**5-12** An ideal mixer circuit has a single frequency input of 5 kHz and an LO frequency of 120 kHz. Specify the positive output frequencies.

**5-13** An ideal mixer circuit has an LO frequency of 120 kHz. List the positive frequencies in the output if the input signal consists of two spectral components: 3 kHz and 5 kHz.

**5-14** An ideal mixer circuit has an LO frequency of 2 MHz. List the positive frequencies in the output if the input signal consists of three spectral components: 2 kHz, 4 kHz, and 8 kHz.

**5-15** A frequency conversion system is shown in part (a) below, and the spectrum of the input signal is shown in part (b). For an LO frequency of 75 MHz, sketch the spectra at the output of the mixer and at the output of the band-pass filter, if the lower sideband is selected.

$$f_o = \frac{M}{N} f_i$$

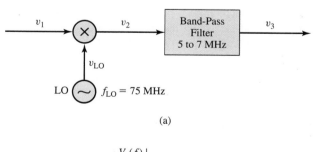

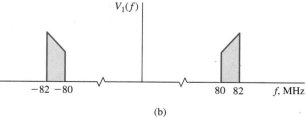

(a)

(b)

**5-16**   A frequency conversion system is shown in part (a) below, and the spectrum of the input signal is shown in part (b). For an LO frequency of 200 MHz, sketch the spectra at the output of the mixer and at the output of the band-pass filter, if the lower sideband is selected.

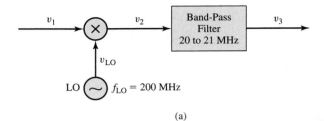

(a)

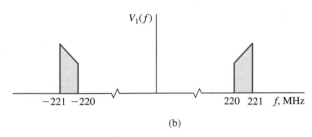

(b)

**5-17**   Repeat Problem 5-15 if the LO frequency is changed to 87 MHz.

**5-18**   Repeat Problem 5-16 if the LO frequency is changed to 241 MHz.

**5-19**   Suppose in Problem 5-15 that the desired output is the upper sideband of the mixer output. Specify the passband range of the band-pass filter, and sketch the spectra at different points.

**5-20**   Suppose in Problem 5-16 that the desired output is the upper sideband of the mixer output. Specify the passband range of the band-pass filter, and sketch the spectra at different points.

**5-21**   Suppose in Problem 5-17 that the desired output is the upper sideband of the mixer output. Specify the passband range of the band-pass filter, and sketch the spectra at different points.

**5-22**   Suppose in Problem 5-18 that the desired output is the upper sideband of the mixer output. Specify the passband range of the band-pass filter, and sketch the spectra at different points.

**5-23**   A signal occupies the frequency range from 20 to 21 MHz. It is desired to shift this spectrum in one step so that it occupies the range from 150 to 151 MHz. (a) Determine the LO frequency required, if the original spectral sense must be preserved. (b) Draw spectral diagrams of the signals at each point in the system.

**5-24**   Repeat Problem 5-23 if the desired output from 150 to 151 MHz is to have the spectral sense reversed from the input spectrum.

**5-25**   A signal occupies the frequency range from 400 to 403 MHz. It is desired to shift this spectrum in one step

so that it occupies the range from 30 to 33 MHz. (a) Determine the LO frequency required, if the original spectral sense must be preserved. (b) Draw spectral diagrams of the signals at each point in the system.

**5-26**   Repeat Problem 5-25 if the desired output from 30 to 33 MHz is to have the spectral sense reversed from the input spectrum.

**5-27**   A special-purpose receiver tunes from 3.5 to 5 MHz, and the IF frequency is 800 kHz. Determine the range of LO oscillator and image frequencies if the LO frequency is higher than the signal frequency.

**5-28**   A special-purpose receiver tunes from 200 MHz to 212 MHz, and the IF frequency is 30 MHz. Determine the range of LO oscillator and image frequencies if the LO frequency is higher than the signal frequency.

**5-29**   Repeat Problem 5-27 if the LO frequency is lower than the signal frequency.

**5-30**   Repeat Problem 5-28 if the LO frequency is lower than the signal frequency.

**5-31**   In the system of Example 5-9, the LO frequency was selected in each step such that the sense of the original spectrum was preserved all the way through. Redesign the system with the LO oscillator frequencies such that the sense of the spectrum is reversed after the first translation, but restored in the second translation. The center frequency at the end of the first step can be 20 MHz as before. Sketch the spectral layout and specify all LO frequencies.

**5-32**   The system shown below a simplified *speech scrambler* used to ensure communication privacy and to foil wiretapping. By sketching the spectrum at each stage, demonstrate that the original spectrum is reversed in frequency. How would you unscramble the signal? Assume ideal filters. (From A. B. Carlson, *Communications Systems,* 2nd Ed., McGraw-Hill Book Co., New York, 1975.)

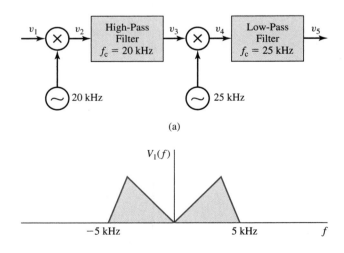

(a)

(b)

# Amplitude Modulation Methods  6

## OVERVIEW AND OBJECTIVES

As discussed briefly in Chapter 1, modulation is the process of transferring a baseband message or intelligence signal to a high-frequency carrier in order to provide a means of transmitting the message. The oldest form of modulation is that of amplitude modulation. *Amplitude modulation will be defined as any process in which the instantaneous amplitude of a high frequency carrier is varied in accordance with the amplitude of the message signal.* There are several variations of amplitude modulation with significant differences in generation, detection, efficiency, and performance. While the various forms of amplitude modulation are important in themselves, they also occur frequently in conjunction with other modulation methods. For example, some forms of digital modulation utilize amplitude modulation operations.

The various modulation methods that can be properly classified under the heading of amplitude modulation include the following:

1. Conventional amplitude modulation (usually referred to simply as AM)
2. Double sideband (DSB)
3. Single sideband (SSB)
4. Vestigial sideband (VSB)

Note that the term *amplitude modulation* is being used to refer to both the general collection of methods as well as one particular method within the collection. It will usually be clear in which context the term is being used.

## Objectives

After completing this chapter, the reader should be able to:

1. Define *amplitude modulation* and list the various forms.
2. Discuss *double sideband modulation,* including its generation and detection, and sketch waveforms and the form of the spectrum.
3. Discuss *single sideband modulation,* including its generation and detection, and sketch waveforms and the form of the spectrum.
4. Discuss *conventional amplitude modulation,* including its generation and detection, and sketch waveforms and the form of the spectrum.
5. Discuss advantages and disadvantages of the forms of amplitude modulation with respect to generation, ease of detection, bandwidths, and relative power requirements.
6. Compare the modulation methods with single-tone modulation.

7. Calculate the total power for each of the modulation methods with single-tone modulation.

8. Discuss the operation of a class C modulated amplifier.

## SystemVue™ Opening Application (Optional)

Insert the text CD in a computer having SystemVue™ installed and activate the program. Open the CD folder entitled **SystemVue Systems** and load the file entitled 6-1.

### Sink Tokens

| Number | Name | Token Monitored |
|--------|------|-----------------|
| 0 | Modulation | 4 |
| 1 | Carrier | 5 |
| 2 | DSB Signal | 6 |
| 3 | SSB Signal | 7 |

### Operational Tokens

| Number | Name | Function |
|--------|------|----------|
| 4 | Modulation | generates single-tone modulating signal at 1 kHz |
| 5 | Carrier | generates carrier at 10 kHz |
| 6 | Modulator | generates DSB signal |
| 7 | Filter | removes one sideband to form SSB signal |

The preset values for the run are from a starting time of 0 to a final time of 10 ms, with a time step of 1 µs. Run the simulation and observe the waveforms at the four sinks. You may need to review the zooming procedure discussed in the opening application for the preceding chapter.

### What You See

Using the zooming procedure when necessary, measure the frequencies of the two input sources to the multiplier (Sinks 1 and 2) and verify that they are 1 kHz and 10 kHz, respectively. Next, observe the output of the multiplier with Sink 3. This is a *double sideband* (DSB) signal, and the view should be over several cycles of the modulating signal period. The filter (Token 7) eliminates one of the sidebands; the signal seen with Sink 4 is a *single sideband* (SSB) signal, and has the form of the up-converted signal of the preceding chapter. As in that case, there is a transient response while the filter is settling, but move over to the last few cycles and zoom in on the waveform. The frequency of this signal should be 11 kHz.

### How This Demonstration Relates to the Chapter Learning Objectives

In Chapter 6, you will learn about the various amplitude modulation (AM) methods. The three that will be studied are *double sideband* (DSB), *single sideband* (SSB), and *conventional AM*. The present demonstrations have shown the first two forms based on a single frequency modulating signal. At the end of this chapter, you will use the layout with a few changes to create a conventional AM signal.

## 6-1 Approach and Terminology

Virtually all signals of interest in communications systems are nonperiodic ;
dom. However, in studying the operation of communications circuits and :
tests on equipment, it is often convenient to employ common simple waveforms, or which
the sinusoidal function is the most common. In the developments that follow (both in this
chapter and in later chapters), we will frequently alternate between complex waveforms and
the sinusoidal function in order to enhance the explanations. When a sinusoidal function is
assumed as a modulating signal, it will be referred to as a *single-tone **modulating** signal.*
The resulting high-frequency signal will be referred to as a *single-tone **modulated** signal.*

### Time Domain versus Frequency Domain

Each form of modulation will be described both in the *time domain* and in the *frequency
domain.* The time domain form illustrates the nature of the signal as it would be observed
on an oscilloscope, and the frequency domain form illustrates the nature of the spectrum as
it would be observed on a spectrum analyzer.

### Essential Trigonometric Identities

In studying the various modulation operations both in this chapter and in later chapters,
several basic trigonometric identities are essential. The identities that follow should be
carefully noted for frequent reference throughout the book.

$$\cos A \cos B = \frac{1}{2}\cos(A - B) + \frac{1}{2}\cos(A + B) \tag{6-1}$$

$$\sin A \sin B = \frac{1}{2}\cos(A - B) - \frac{1}{2}\cos(A + B) \tag{6-2}$$

$$\sin A \cos B = \frac{1}{2}\sin(A - B) + \frac{1}{2}\sin(A + B) \tag{6-3}$$

$$\cos A \sin B = -\frac{1}{2}\sin(A - B) + \frac{1}{2}\sin(A + B) \tag{6-4}$$

Actually, Equation 6-4 is an alternate way of expressing Equation 6-3, but both forms are
useful because of the sign difference of one term. The sign difference is due to the fact that
the sine function is an odd function, and when the order of the argument terms is reversed,
a negative sign arises in the process.

Some properties of these identities are the following:

1. The *product* of two sinusoidal functions can be expressed as the *sum* of two sinusoidal
   functions, but with different arguments.

2. The arguments of the two separate sinusoids in the sum consist of, respectively, the
   difference and the sum of the arguments of the sinusoids being multiplied.

3. When the two sinusoids being multiplied are either *both cosine* or *both sine* functions,
   the sum consists of *two cosine* functions.

4. When one of the functions being multiplied is a *sine* function and the other is a *cosine*
   function, the sum consists of *two sine* functions.

When applying these identities, the argument of a sine or a cosine function may initially
appear negative. When this occurs, convert the function to one with a positive argument by
using the even and odd properties of cosine and sine functions, which are

$$\cos(-A) = \cos A \tag{6-5}$$

$$\sin(-A) = -\sin A \tag{6-6}$$

## Notation

It is helpful in studying various modulation and demodulation techniques to have a standard form of notation. As in earlier chapters, voltage will be assumed as the primary variable in studying the various operations. Since we will now need to study both modulating and modulated signals, as well as carriers, it is necessary to add various subscripts. To the extent possible, the following terminology will be employed:

$v_m(t) =$ modulating or message signal (usually at baseband)

$v_c(t) =$ carrier signal or function (usually a sinusoid)

$v_o(t) =$ output of modulator

$v_r(t) =$ received signal

$v_d(t) =$ detected or demodulated signal

$f_c =$ carrier frequency in hertz

$\omega_c =$ carrier frequency in radians per second $= 2\pi f_c$

$W =$ baseband bandwidth of modulating signal in hertz

$B_T =$ transmission bandwidth of the modulated signal in hertz

In the ideal distortion-free and noise-free case, the following properties would be true:

1. The actual transmitted signal would have the same form as the modulator output $v_o(t)$, and that function can be used as the basis for studying the transmitted signal.

2. The received signal $v_r(t)$ would have the same form as the transmitted signal or as $v_o(t)$, but typically at a much lower level due to transmission losses.

3. The detected signal $v_d(t)$ would have the same form as the modulating or message signal $v_m(t)$ but possibly at a different level.

As has been the case in earlier chapters, the arguments $(t)$ and $(f)$ will frequently be omitted from the voltage variables in order to simplify the equations.

## Continuous and Discrete Spectra

Some modulation operations are best explained with the Fourier transform, and some are best explained with the Fourier series. When the Fourier transform is to be used, the modulating signal $v_m(t)$ will be assumed to be *nonperiodic,* and the spectral magnitude $V_m(f)$ will be a *continuous* function of frequency, as illustrated in Figure 6–1. A signal having this property will be referred to as a *continuous-spectrum signal.*

When the Fourier series is to be used, the modulating signal $v_m(t)$ will be assumed to be *periodic,* and the one-sided spectral magnitude $C_n$ will be employed. In this case, a *discrete* function of frequency will be assumed, as illustrated in Figure 6–2. A signal having this property will be referred to as a *discrete-spectrum signal.*

Note that for the discrete-spectrum case, the bandwidth $W$ corresponds to the highest frequency in the baseband spectrum; that is, $W = Nf_1$, where $N$ is an integer representing the order of the highest harmonic.

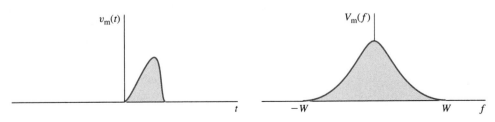

**FIGURE 6–1**
Example of a signal with a continuous spectrum.

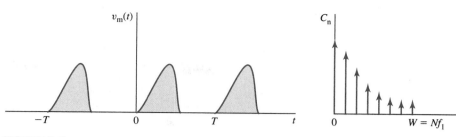

**FIGURE 6–2**
Example of a signal with a discrete spectrum.

In some developments, a discrete spectrum may be assumed in the mathematical development, but the spectrum may be shown as a continuous function of frequency. This will be purely a matter of convenience and should not cause any confusion.

## 6-2 Double Sideband Modulation

We will begin the study of AM processes with *double sideband modulation* (DSB) because it the easiest to analyze. This process is also called *double sideband–suppressed carrier* and denoted in some references as DSB-SC, but we will employ the shorter acronym DSB. A DSB signal can be generated by multiplying the baseband signal by a high-frequency sinusoid in the same manner as a mixer. In fact, a mixer circuit could be used to generate a DSB signal. In practice, however, active circuits providing some degree of amplification are more often used to generate a DSB signal. The name *balanced modulator* is used to describe a circuit used to generate a DSB signal. At the most theoretical level, however, a *balanced modulator* and a *mixer* perform essentially the same function.

### Balanced Modulator Operation

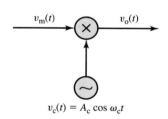

**FIGURE 6–3**
Block diagram form for a balanced modulator.

A block diagram showing the operation of a balanced modulator is shown in Figure 6–3. It is immediately obvious that the diagram is the same as for a mixer. The only differences will be in our notation and frame of reference. The input on the left is now assumed to be a baseband *modulating* or *message* signal $v_m(t)$, and the mixing oscillator is now denoted as the *carrier signal* $v_c(t)$. The form of the carrier is assumed to be

$$v_c = A_c \cos \omega_c t \tag{6-7}$$

where $A_c$ is the carrier amplitude, $\omega_c = 2\pi f_c$, and $f_c$ is the *carrier frequency*. The output of the balanced modulator is assumed to be proportional to an ideal multiplication of the two inputs, as given by

$$v_o(t) = K_b v_m(t) v_c(t) = K_b A_c v_m(t) \cos \omega_c t = K v_m(t) \cos \omega_c t \tag{6-8}$$

where $K_b$ is a modulator constant, and $K = K_b A_c$.

The function $v_o(t)$ is a DSB signal. The significance will be observed by first assuming that $v_m(t)$ is a continuous-spectrum signal and taking the Fourier transforms of both sides. Expressing the cosine as a combination of exponential terms and applying the modulation theorem of Chapter 3, we have

$$v_o(t) = \frac{K}{2} v_m(t) e^{j2\pi f_c t} + \frac{K}{2} v_m(t) e^{-j2\pi f_c t} \tag{6-9}$$

and

$$\overline{V}_o(f) = \frac{K}{2} \overline{V}_m(f - f_c) + \frac{K}{2} \overline{V}_m(f + f_c) \tag{6-10}$$

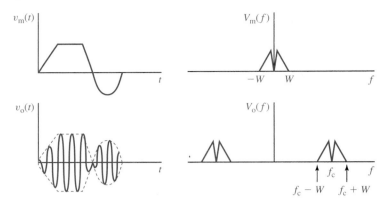

**FIGURE 6–4**
Generation of DSB signal from a baseband signal, and its spectrum.

The process involved is illustrated in Figure 6–4. An arbitrary modulating signal is shown at the top left in Figure 6–4, and its amplitude spectrum is assumed to be of the form shown to the right. Note that the baseband spectrum is assumed to be band-limited from near dc to $W$. It would certainly be proper to say from $-W$ to $W$, but the negative frequency range is automatically understood in the two-sided spectral form, so it is more customary to express the actual values in terms of the "real-world" positive frequency range.

The DSB signal corresponding to Equation 6-7 is shown on the bottom left in Figure 6–4, and its spectrum corresponding to Equation 6-10 is shown on the right. The first term on the right-hand side of Equation 6-10 represents a translation or shift of the original spectrum to the right by $f_c$ hertz, and is now centered at $f_c$. The second term represents a shift of the original spectrum to the left by $f_c$ hertz, and is now centered at $-f_c$ hertz.

## DSB Modulation versus Mixing

We are, in effect, revisiting the concept of mixing as discussed in Chapter 5. The main difference in this case is the fact that we are starting with a baseband message signal and transferring it to a center at high frequencies, whereas in Chapter 5 the emphasis was on an arbitrary shift to either a lower or a higher frequency while retaining the exact shape of the spectrum.

## Sidebands

The message signal that was previously bandlimited from dc to $W$ hertz for positive frequencies now appears on both sides of the carrier frequency $f_c$. Hence, the term *double sideband* is quite appropriate. The portion above $f_c$ is called the *upper sideband (USB)*, and the portion below $f_c$ is called the *lower sideband (LSB)*. The upper sideband has the same spectral sense as the modulating signal, but the lower sideband has spectral sense opposite to that of the modulating signal. Each sideband contains all the message information, however, if processed properly.

An interesting point is that *the carrier at $f_c$ does not appear in the spectrum at all!* This may seem a little strange considering that the signal is oscillating at a rate of $f_c$ hertz. However, the successive phase reversals at the carrier rate caused by zero crossings of the envelope, as may be observed in Figure 6–4, results in a complete cancellation of the carrier in the ideal case. In practical balanced modulators, a small carrier component can usually be measured, but it is typically 40 dB or more below the sideband levels in well-designed modulators. The cancellation of the carrier is the basis for the term *suppressed carrier* that is often used in the description of the DSB process.

## Possible Misinterpretation

One possible misinterpretation of the terms in Equation 6-10 will be noted. Frequently, students (and others) have a tendency to think that one of the terms in that equation is the lower sideband, and the other is the upper sideband. This interpretation is *not correct*. Each of the terms in Equation 6-10 contains a portion of both the upper and lower sidebands. An equation expressing only the lower or upper sideband is more difficult to formulate for the continuous-spectrum case, and will not be pursued at this time.

The result of the modulation operation is that the low-frequency baseband signal has been shifted upward in frequency and is now centered at $f_c$ instead of dc. Since $f_c$ is arbitrary, it could very well be in the radio-frequency range. Obviously, the scale of the figure does not permit us to illustrate typical frequency shifts, which could be, for example, several orders of magnitude times the baseband frequency range. By shifting the signal to the RF range, it is possible to utilize electromagnetic radiation to propagate the signal.

## Discrete-Spectrum Signal

It is of educational value to see how the system is analyzed when the input signal has a discrete-spectrum form. Assume that the signal $v_m(t)$ is of the form

$$v_m(t) = \sum_{n=1}^{N} C_n \cos(n\omega_1 t + \theta_n) \tag{6-11}$$

The form of the spectrum is shown in Figure 6–5(a). Note that the baseband bandwidth is

$$W = Nf_1 \tag{6-12}$$

The output of the balanced modulator is now

$$v_o(t) = K_b A_c \cos \omega_c t \sum_{n=1}^{N} C_n \cos(n\omega_1 t + \theta_n)$$

$$= K \sum_{n=1}^{N} C_n \cos \omega_c t \cos(n\omega_1 t + \theta_n) \tag{6-13}$$

where $K = K_b A_c$ as before. The identity of Equation 6-1 can now be applied to each of the terms in Equation 6-13, and the total number of terms is doubled. The result is

$$v_o(t) = \frac{K}{2} \sum_{n=1}^{N} C_n \cos\left[(\omega_c - n\omega_1)t - \theta_n\right] + \frac{K}{2} \sum_{n=1}^{N} C_n \cos\left[(\omega_c + n\omega_1)t + \theta_n\right] \tag{6-14}$$

This equation describes a DSB discrete-spectrum signal. *In the discrete-spectrum case, the first term is the lower sideband and the second term is the upper sideband.* This makes the discrete-spectrum case easier to work with in some mathematical developments than the continuous-spectrum case. The DSB spectrum is shown in Figure 6–5(b).

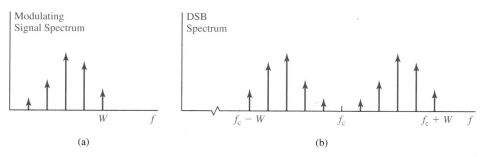

(a)                                                (b)

**FIGURE 6–5**
Discrete spectrum for a message signal, and its DSB modulated form.

### Transmission Bandwidth

The transmission bandwidth for the final DSB signal is simply

$$B_T = 2W \qquad (6\text{-}15)$$

Thus, the required transmission bandwidth is twice the baseband bandwidth.

---

**▌▌▌ EXAMPLE 6-1**

A continuous-spectrum baseband signal has a spectrum extending from near dc to 15 kHz. It is applied as one input to an ideal balanced modulator. The carrier input is a sinusoid having a frequency of 1 MHz. (a) Determine the range of frequencies contained in the DSB output spectrum. (b) Determine the required transmission bandwidth.

**SOLUTION**

(a) The spectrum of the DSB signal ranges from $f_c - W$ to $f_c + W$. The lowest frequency is thus

$$f_c - W = 1\,\text{MHz} - 15\,\text{kHz} = 1000\,\text{kHz} - 15\,\text{kHz} = 985\,\text{kHz} \qquad (6\text{-}16)$$

The highest frequency is

$$f_c + W = 1\,\text{MHz} + 15\,\text{kHz} = 1000\,\text{kHz} + 15\,\text{kHz} = 1015\,\text{kHz} \qquad (6\text{-}17)$$

The range of frequencies is thus from 985 kHz to 1015 kHz.

(b) Since $W = 15$ kHz, the transmission bandwidth is

$$B_T = 2W = 2 \times 15\,\text{kHz} = 30\,\text{kHz} \qquad (6\text{-}18)$$

---

**▌▌▌ EXAMPLE 6-2**

A discrete-spectrum baseband modulating signal has components at the following frequencies: 1 kHz, 3 kHz, and 5 kHz. It is applied as one input to an ideal balanced modulator, and the other input is a sinusoid having a frequency of 250 kHz. (a) List the frequencies appearing at the output of the balanced modulator. (b) Determine the required transmission bandwidth.

**SOLUTION**

(a) The various frequencies appearing at the output are the respective sum and difference frequencies between the carrier and the baseband signal. They can be divided into the lower sideband and upper sideband frequencies as follows:

$$
\begin{aligned}
\text{LSB:} \quad & 250\,\text{kHz} - 1\,\text{kHz} = 249\,\text{kHz} \\
& 250\,\text{kHz} - 3\,\text{kHz} = 247\,\text{kHz} \\
& 250\,\text{kHz} - 5\,\text{kHz} = 245\,\text{kHz} \\
\text{USB:} \quad & 250\,\text{kHz} + 1\,\text{kHz} = 251\,\text{kHz} \\
& 250\,\text{kHz} + 3\,\text{kHz} = 253\,\text{kHz} \\
& 250\,\text{kHz} + 5\,\text{kHz} = 255\,\text{kHz}
\end{aligned}
$$

(b) One could argue that the baseband bandwidth is the highest baseband frequency minus the lowest baseband frequency, and since there is no dc component, the result using that approach would be $5 - 1 = 4$ kHz. However, it is customary with baseband signals to consider the bandwidth from dc to the highest frequency, even if there is no dc component. From the point of view of the transmission of DSB, this certainly makes sense, so the baseband bandwidth may be considered to be $W = 5$ kHz. The net transmission bandwidth is then

$$B_T = 2W = 2 \times 5\,\text{kHz} = 10\,\text{kHz} \qquad (6\text{-}19)$$

▌▌

## 6-3    Single Sideband Modulation

We saw in the last section that when the product of a baseband signal and a higher frequency sinusoid is generated in a balanced modulator, a double sideband signal is obtained at the output. This process is reviewed in Figure 6–6(a) and (b). (For convenience, the translated components are shown at the same level as the baseband signal spectrum.) Due to the symmetry of the spectrum about $f_c$, either of the two sidebands contains all the spectral information associated with the message. Admittedly, the spectral sense of the lower sideband is turned around, but the process can be reversed again at the receiver to compensate. In some references, the longer term *single sideband–supressed carrier* (SSB-SC) is used, but we will use the shorter term SSB.

The basic strategy in *single sideband* is to eliminate one of the sidebands and transmit only one sideband. This concept is illustrated in Figure 6–6(d) for the case in which the upper sideband is used and in Figure 6–6(f) for the case in which the lower sideband is used. (The functions in Figure 6–6(c) and (e) will be discussed shortly.)

### Transmission Bandwidth

The transmission bandwidth for SSB is

$$B_T = W \qquad (6\text{-}20)$$

Unless some form of data compression is utilized, SSB requires the minimum bandwidth of all analog modulation systems and this bandwidth is that of the baseband signal.

From the work of Chapter 5 on mixers, it can be recognized that SSB generation is, in reality, another application of the frequency translation process. In the case of SSB generation, translation is achieved by moving a baseband spectrum up to some RF frequency range to enhance the transmission process. This is a case in which the oscillator frequency is higher than any frequencies in the original spectrum, so the LSB has the spectral sense reversed, while the USB has the same spectral sense as that of the input signal.

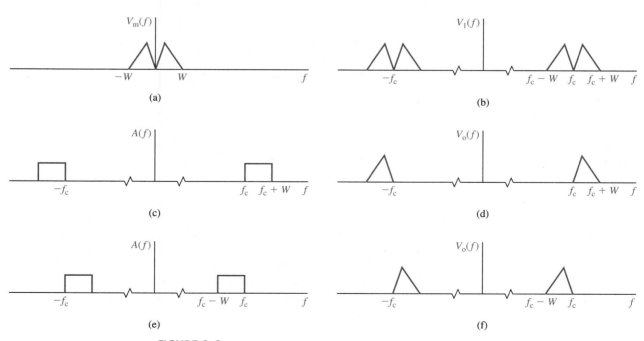

**FIGURE 6–6**
Development of SSB spectra for both lower and upper sidebands.

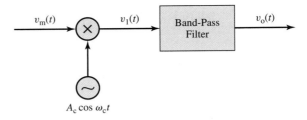

**FIGURE 6–7**
Block diagram of SSB generator using the filter method.

## Filter Method

While there are several methods of generating an SSB signal, the most common approach, and by far the easiest to explain, is the *filter method,* which is illustrated in Figure 6–7. The notation on this figure correlates with that of Figure 6–6. A DSB signal $v_1$ is first generated in a balanced modulator. A band-pass filter having a bandwidth $W$ and a passband aligned exactly with that of the desired sideband is used to pass the appropriate sideband and to reject the other sideband. Ideal block filter characteristics are shown in Figure 6–6(c) and (e).

The perceptive reader might wonder if a high-pass filter would be sufficient in the case of USB, and if a low-pass filter would be sufficient in the case of LSB. In theory, such filters would produce the desired result. However, in practice, a band-pass filter is usually employed in order to eliminate additional spurious components that might be produced by imperfections in the balanced modulator.

## Sideband Filter

The sideband filter is one of the most complex components in an SSB generator from a design point of view. The primary reason is that the two sidebands are very close together, and the filter must pass one while rejecting the other. For example, suppose it were desired to preserve frequencies in the message down to 50 Hz. The lowest frequency in the upper sideband would only be 100 Hz above the highest frequency in the lower sideband. It is difficult in the RF frequency range to design a filter that can reject a component that close to the passband.

The actual implementation of sideband filters is eased by two approaches: (1) The SSB signal is usually generated at a relatively low frequency below the RF range, where the necessary transition region represents a larger relative frequency band. The modulated signal is then translated by one or more additional frequency shifts to the final required frequency range. (2) Wherever practical, the lowest message signal frequency is set at as high a frequency as intelligibility and accuracy of reproduction will permit, thus easing the filtering requirement. For example, one of the most common applications of SSB is for voice transmission, and the lower frequency components can be sacrificed without significant loss of intelligibility.

## Discrete-Spectrum Case

The equation of an SSB signal modulated by a continuous-spectrum message signal can only be expressed through the use of an operation called the Hilbert transform, which is not within the intended scope of this book. However, the assumption of a discrete spectrum leads to an appropriate expression. In fact, we can make use of the results of the preceding section for the DSB signal. Assume that the message signal is of the form of Equation 6-11, which is repeated here for convenience.

$$v_m(t) = \sum_{n=1}^{N} C_n \cos(n\omega_1 t + \theta_n) \tag{6-21}$$

The output of the balanced modulator is of the form of Equation 6-14, and either of the two parts can be selected in accordance with the passband of the filter. The possible SSB output functions are

$$\text{LSB:} \qquad v_o(t) = \frac{K}{2} \sum_{n=1}^{N} C_n \cos\left[(\omega_c - n\omega_1)t - \theta_n\right] \qquad (6\text{-}22)$$

$$\text{USB:} \qquad v_o(t) = \frac{K}{2} \sum_{n=1}^{N} C_n \cos\left[(\omega_c + n\omega_1)t + \theta_n\right] \qquad (6\text{-}23)$$

In both Equations 6-22 and 6-23, the phase shifts and any amplitude variation produced by the filter on the spectrum have been ignored at this time since they do not contribute to the present discussion. However, such effects may have to be considered in a complete system analysis.

## Phase-Shift Method

The concept used in the phase-shift method appears in one form or another in many types of communication signal processing applications, and is worthy of some attention. Consider the block diagram shown in Figure 6–8. The system contains two balanced modulators, an oscillator, a summing circuit, and two $-90°$ phase-shift circuits. The phase-shift circuit above the lower balanced modulator need only impart a phase shift at the specific frequency $f_c$, and this operation is straightforward. However, the phase-shift circuit along the lower path must provide a constant $-90°$ phase shift for *all* message frequencies, and this has traditionally been a complex design challenge. Such a circuit is referred to in the literature as a *Hilbert transformer.* Modern implementations of this type employ digital signal processing.

Assume that the message signal is of the form

$$v_m(t) = \sum_{n=1}^{N} C_n \cos(n\omega_1 t + \theta_n) \qquad (6\text{-}24)$$

Let $\hat{v}_m(t)$ represent the signal at the output of the lower phase-shift network. It can be expressed as

$$\hat{v}_m(t) = \sum_{n=1}^{N} C_n \cos(n\omega_1 t + \theta_n - 90°) = \sum_{n=1}^{N} C_n \sin(n\omega_1 t + \theta_n) \qquad (6\text{-}25)$$

The direct output of the oscillator $v_c(t)$, which is applied as one input to the upper balanced modulator, is

$$v_c(t) = A_c \cos \omega_c t \qquad (6\text{-}26)$$

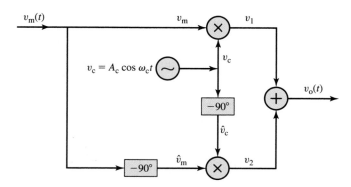

FIGURE 6–8
Block diagram of SSB generator using the phase-shift method.

This component is shifted in phase by $-90°$, and the resulting function $\hat{v}_c(t)$, which is applied as one input to the lower balanced modulator, is

$$\hat{v}_c(t) = A_c \cos(\omega_c t - 90°) = A_c \sin \omega_c t \qquad (6\text{-}27)$$

The output $v_1$ of the upper balanced modulator is

$$v_1 = K_b v_c v_m = K \cos \omega_c t \sum_{n=1}^{N} C_n \cos(n\omega_1 t + \theta_n)$$

$$= \frac{K}{2} \left\{ \sum_{n=1}^{N} C_n \cos\left[(\omega_c - n\omega_1 t) - \theta_n\right] + \sum_{n=1}^{N} C_n \cos\left[(\omega_c + n\omega_1 t) + \theta_n\right] \right\} \qquad (6\text{-}28)$$

where $K = K_b A_c$ and the identity of Equation 6-1 was employed in the second part of the equation. The output of the lower balanced modulator is

$$v_2 = K_b \hat{v}_m \hat{v}_c = K \sin \omega_c t \sum_{n=1}^{N} C_n \sin(n\omega_1 t + \theta_n)$$

$$= \frac{K}{2} \left\{ \sum_{n=1}^{N} C_n \cos\left[(\omega_c - n\omega_1 t) - \theta_n\right] - \sum_{n=1}^{N} C_n \cos\left[(\omega_c + n\omega_1 t) + \theta_n\right] \right\} \qquad (6\text{-}29)$$

where the identity of Equation 6-2 was employed in the second part of the equation.

The output summing circuit performs the operation

$$v_o = v_1 + v_2 \qquad (6\text{-}30)$$

When the function of Equations 6-28 and 6-29 are substituted in Equation 6-30, it can be readily seen that the first set of terms are in phase and add, while the second set of terms are out of phase and cancel. The output is

$$v_o = K \sum_{n=1}^{N} C_n \cos\left[(\omega_c - n\omega_1)t - \theta_n\right] \qquad (6\text{-}31)$$

The result is an SSB signal with the LSB component. In effect, one of the sidebands has cancelled out in the process. It can be shown that if the difference between the two components is formed in the output, the result will be the USB component.

---

**▌▌ EXAMPLE 6-3**

Refer back to the signal and system of Example 6-1. It is desired to convert the DSB signal to an SSB signal by passing the signal through an appropriate filter. Determine the range of frequencies contained in the output spectrum for (a) LSB and (b) USB. (c) Determine the required transmission bandwidth.

**SOLUTION**

(a)  For LSB, the range of frequencies is from 985 kHz to 1 MHz.

(b)  For USB, the range of frequencies is from 1 MHz to 1.015 MHz.

(c)  The required transmission bandwidth is

$$B_T = W = 15 \text{ kHz} \qquad (6\text{-}32)$$

---

**▌▌ EXAMPLE 6-4**

Refer back to the signal and system of Example 6-2. It is desired to convert the DSB signal to an SSB signal by appropriate filtering. List the frequencies appearing at the output of the filter for (a) LSB and (b) SSB. (c) Determine the required transmission bandwidth.

**SOLUTION**

(a)  For LSB transmission, the frequencies are 249, 247, and 245 kHz.

(b) For USB transmission, the frequencies are 251, 253, and 255 kHz.

(c) As explained in the solution to Example 6-2, it is appropriate to consider the baseband spectrum as ranging from dc to the upper frequency limit even though there is no dc component. From that perspective, the transmission bandwidth is

$$B_T = W = 5 \text{ kHz} \tag{6-33}$$

## 6-4  Product Detection of DSB and SSB

In the previous two sections, we have seen how DSB and SSB signals could be generated with a balanced modulator and associated circuitry. At the receiving end, it is necessary to perform an inverse process on the composite signal in order to recover the desired baseband message signal. The signal processing required to extract the original baseband signal from the composite modulated signal is called *detection* or *demodulation*. The circuit that performs this function is called a *detector* or a *demodulator*.

The detection process normally used for either DSB or SSB is called *product detection*. Product detection is achieved by mixing a carrier generated at the receiver with the incoming signal in a balanced modulator or mixer circuit, followed by low-pass filtering. If the carrier is locked exactly in phase and frequency with the carrier used for generating the signal, the product-detection process is also referred to as *synchronous* or *coherent detection*.

### DSB Synchronous Detection

As a first step, we will consider ideal synchronous or coherent detection. While the received signal will have undergone a time shift, it is convenient to redefine the time scale so that the received DSB signal $v_r(t)$ can be expressed as

$$v_r(t) = K_r v_m(t) \cos \omega_c t \tag{6-34}$$

where $K_r$ is a constant indicating that the level at the receiver is different than at the transmitter. Assume that the received signal is applied as one input to an ideal balanced modulator, as shown in Figure 6–9. Assume that the other input is a sinusoid of the form

$$v_c(t) = A_c \cos \omega_c t \tag{6-35}$$

Note that the sinusoid has been assumed to be *in phase* with the received signal. This is an important consideration, and is based on our assumption of synchronous detection.

The output of the balanced modulator is

$$v_1 = K_b v_r v_c = K_b K_r A_c v_m(t) \cos \omega_c t \cos \omega_c t = B v_m(t) \cos^2 \omega_c t \tag{6-36}$$

where $B = K_b K_r A_c$ is a constant. The cosine-squared function may be expanded by the identity

$$\cos^2 \theta = \frac{1}{2}(1 + \cos 2\theta) \tag{6-37}$$

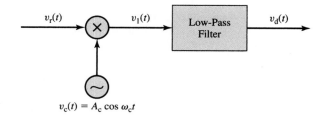

**FIGURE 6–9**
Block diagram of product detector employing coherent reference.

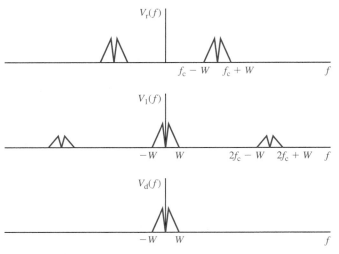

**FIGURE 6–10**
Spectral terms arising in the detection process of a DSB signal.

Thus, Equation 6-36 can be expressed as

$$v_1(t) = \frac{B}{2} v_m(t) + \frac{B}{2} v_m(t) \cos 2\omega_c t \qquad (6\text{-}38)$$

The first term on the right-hand side of Equation 6-38 is the baseband message signal multiplied by a constant, and is what we want. For the present discussion, the constant multiplier is of little importance, since the signal has probably undergone a significant change in level since it left the transmitter. The point is that the term is proportional to the message signal.

The amplitude spectra of the received signal and the output of the balanced detector are shown as the top two parts of Figure 6–10. Along with the desired signal, the spectrum $V_1(f)$ also contains a component representing the DSB product of the message signal modulating the second harmonic of the carrier. In most practical applications, this latter component will be much higher in frequency than that of the message signal, and can easily be eliminated with a low-pass filter. The detected output of the low-pass filter is then of the form

$$v_d(t) = \frac{B}{2} v_m(t) \qquad (6\text{-}39)$$

where any additional change in level or time shift resulting from the filter has been ignored. The form of the output spectrum $V_d(f)$ is shown as the lower part of Figure 6–10. The magnitudes of the various components were selected to illustrate the *relative* sizes. The exact levels will depend on the constant $B$.

## SSB Synchronous Detection

Due to the difficulty in expressing the equation of an SSB signal with a continuous-spectrum modulating signal, we will deal exclusively with the discrete-spectrum form. We will assume a USB signal as an arbitrary choice, although the result applies equally well to either case. Consider then a received USB signal of the form

$$v_r(t) = K_r \sum_{n=1}^{N} \frac{C_n}{2} \cos\left[(\omega_c + n\omega_1)t + \theta_n\right] \qquad (6\text{-}40)$$

The local carrier has the form as for DSB; that is,

$$v_c(t) = K_c \cos \omega_c t \qquad (6\text{-}41)$$

The output $v_1(t)$ of the balanced detector is then

$$v_1 = K_b v_r v_c = K_b \left[ K_r \sum_{n=1}^{N} \frac{C_n}{2} \cos\left[(\omega_c + n\omega_1)t + \theta_n\right] \right] [A_c \cos \omega_c t]$$

$$= B \sum_{n=1}^{N} \frac{C_n}{2} \cos\left[(\omega_c + n\omega_1)t + \theta_n\right] \cos \omega_c t \qquad (6\text{-}42)$$

where $B = K_b K_r A_c$. Each term in the summation can be expanded by the use of the identity of Equation 6-1. Following this expansion, the result is

$$v_1 = \frac{B}{4} \sum_{n=1}^{N} C_n \cos(n\omega_1 t + \theta_n) + \frac{B}{4} \sum_{n=1}^{N} \cos\left[(2\omega_c + n\omega_1)t + \theta_n\right] \qquad (6\text{-}43)$$

The first series in Equation 6-43 is a constant times the message signal. The second series represents an SSB signal located above $2f_c$. This last series is easily removed by filtering, provided that the carrier frequency is much higher than the baseband bandwidth. The detected signal $v_d$ is then of the form

$$v_d(t) = \frac{B}{4} v_m(t) \qquad (6\text{-}44)$$

where any time delay and amplitude variations in the low-pass filter response have been ignored. The message signal is thus recovered by the synchronous detector. The various steps involved in the detection process are illustrated in Figure 6–11 with a *continuous* spectrum.

## Relative Magnitudes of DSB and SSB

Comparing Figures 6–10 and 6–11, it is noted that the *relative* magnitudes of the recovered spectrum as compared to the input spectrum appear to be twice as large for DSB, as for SSB, if all other factors are equal. The reason for this is that with DSB, two translated sidebands overlap and add to each other coherently, while for SSB, there is only one component in each segment. Thus, for a given signal level, DSB produces a larger detected signal than SSB. However, as will be seen later in the text, the input noise power for the DSB signal is twice as great as for SSB, so the final signal-to-noise ratios for the two methods are the same.

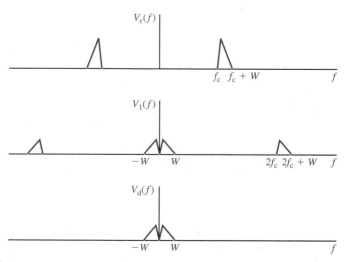

**FIGURE 6–11**

Spectral terms arising in the detection process of an SSB signal.

## Nonsynchronous Carrier at the Receiver

We have seen so far that synchronous or coherent detection can theoretically be used to recover the message from either a DSB or an SSB signal. Ignoring for the moment any differences in the levels of the detected signals, both modulation methods appear to provide similar results *if a coherent carrier reference is available at the receiver.*

Consider, however, the case in which the carrier reference is not locked exactly in phase and frequency with the one at the transmitter. All oscillators will have a tendency to drift over time, and unless there is some way of maintaining coherency between the carriers, the assumption we are about to make is a realistic one in many situations.

Assume that the carrier to be inserted at the receiver is of the form

$$v_c(t) = A_c \cos\left[(\omega_c + \Delta\omega)t + \Delta\theta\right] \tag{6-45}$$

where $\Delta\omega$ is a shift in frequency and $\Delta\theta$ is a shift in phase. The DSB detection process is easier to analyze with the continuous-spectrum signal, and the SSB detection process is more readily analyzed with the discrete-spectrum signal. The steps in the analysis follow a pattern similar to the previous developments in the text, but we will leave the exact details to the interested reader. The results after low-pass filtering are summarized as follows:

$$\text{DSB:} \qquad v_d(t) = \frac{B}{2}v_m(t)\cos(\Delta\omega t + \Delta\theta) \tag{6-46}$$

$$\text{SSB:} \qquad v_d(t) = \frac{B}{4}\sum_{n=1}^{N} C_n \cos\left[(n\omega_1 - \Delta\omega)t + \theta_n - \Delta\theta\right] \tag{6-47}$$

Inspection of Equations 6-46 and 6-47 reveals some significant differences between the ways in which the errors in frequency and phase affect the two different processes. The DSB signal is no longer simply the message signal $v_m(t)$, but it is the message signal modulated by a low-frequency sinusoid. Since both the frequency and phase will drift with time, it is virtually impossible to recover the desired signal under these conditions.

In contrast, the effect of the frequency difference on the SSB process does not affect the amplitudes of the spectral components, but rather causes a shift in the frequency and phase of each component. While this certainly represents a distortion of the original signal, it is not nearly as serious as the case of DSB, provided that the frequency shift is fairly small. Anyone who has tuned SSB voice signals on a communications receiver will be aware of this phenomenon. Depending on the side of the carrier one is tuning, it can result in a bass voice sounding like a soprano or vice versa. Nevertheless, the signal may be perfectly intelligible provided that the local oscillator is at least moderately stable over a period of time. Of course, the possible shifts in the frequencies could limit the types of data that one would transmit with SSB, but for noncritical applications (e.g., simple point-to-point voice transmission), SSB is very useful.

## Applications in Which DSB Modulation Is Usable

Some applications in which DSB may be used are the following:

1. It is possible to transmit a small pilot carrier that may be used to establish frequency and phase lock with a much larger carrier at the receiver for detection purposes. This is the strategy used for the so-called subcarrier in FM stereo transmission.

2. There are some signal processing schemes that may be used to generate a carrier directly from the sidebands, but their complexity limits their use to special applications. Some of the processes will be considered later in the text in relationship to digital communication systems.

3. Turning to the field of automatic control systems, there are various control components of the "ac-carrier" type that employ DSB signals for controlling system response. These devices utilize DSB signals based on relatively low carrier frequencies

(e.g., 60 or 400 Hz). The use of modulated techniques allows ac-coupled amplifiers and components to be employed where direct-coupled amplifiers would otherwise be needed for very low frequency error signals. Synchronization is no problem in this situation, since the reference carrier is readily available.

---

**▐▐ EXAMPLE 6-5**

Consider the lower SSB signal generated in Example 6-4(a). Assume that the signal is applied as one input to an ideal multiplier, and other input is a sinusoid having a frequency of 250 kHz. (a) List the positive frequencies at the output of the multiplier. (b) List the frequencies at the output of the detector low-pass filter.

**SOLUTION**

(a) Reviewing the work of Example 6-4(a), the following frequencies represent the input to the multiplier:

> 249, 247, 245 kHz

When the input signal is multiplied by the carrier having a frequency of 250 kHz, the output frequencies are the sum and difference frequencies of the receiver carrier at 250 kHz and all frequencies at the input. These frequencies are:

$$250 - 249 = 1 \text{ kHz} \qquad 250 + 249 = 499 \text{ kHz}$$
$$250 - 247 = 3 \text{ kHz} \qquad 250 + 247 = 497 \text{ kHz}$$
$$250 - 245 = 5 \text{ kHz} \qquad 250 + 245 = 495 \text{ kHz}$$

(b) After low-pass filtering, the actual output frequencies are

> 1, 3, and 5 kHz

These components represent the original modulating signal frequencies.

---

**▐▐ EXAMPLE 6-6**

Repeat the analysis of Example 6-5 for the upper SSB signal of Example 6-4(b).

**SOLUTION**

(a) Reviewing the work of Example 6-4(b), the following frequencies represent the input to the multiplier:

> 251, 253, 255 kHz

Once again, we form difference and sum frequencies between the receiver carrier frequency and the input frequencies. In this case, the input carrier frequency is subtracted from the input frequency. When working with a one-sided spectrum, the difference is always formulated in a manner that will produce positive frequencies. Thus, the frequencies at the output of the modulator are

$$251 - 250 = 1 \text{ kHz} \qquad 251 + 250 = 501 \text{ kHz}$$
$$253 - 250 = 3 \text{ kHz} \qquad 253 + 250 = 503 \text{ kHz}$$
$$255 - 250 = 5 \text{ kHz} \qquad 255 + 250 = 505 \text{ kHz}$$

(b) After low-pass filtering, the actual output frequencies are again

> 1, 3, and 5 kHz

As in the preceding example, these components represent the original modulating signal frequencies.

---

**▐▐ EXAMPLE 6-7**

Assume in the detection circuit of Example 6-5 that the receiver oscillator drifts by 100 Hz, so that the injected frequency is actually 250.1 kHz. Repeat the analysis of Example 6-5.

## SOLUTION

(a) The various difference and sum frequencies at the output of the multiplier are now

$$250.1 - 249 = 1.1 \text{ kHz} \qquad 250.1 + 249 = 499.1 \text{ kHz}$$
$$250.1 - 247 = 3.1 \text{ kHz} \qquad 250.1 + 247 = 497.1 \text{ kHz}$$
$$250.1 - 245 = 5.1 \text{ kHz} \qquad 250.1 + 245 = 495.1 \text{ kHz}$$

(b) After low-pass filtering, the output frequencies are now

1100, 3100, and 5100 Hz.

Thus, the spectral components of the recovered signal are all shifted by 100 Hz with respect to the modulating signal. Depending on the nature of the signal, this effect may or may not represent a severe degradation.

## 6-5 Conventional Amplitude Modulation

Within the general class of amplitude modulation methods being considered, the specific method that we have chosen to designate as *conventional amplitude modulation* will now be considered. In most usage, it is referred to simply as *amplitude modulation*. It is also sometimes referred to as *amplitude modulation–large carrier (AM-LC)*, but we will employ the shorter acronym AM.

Conventional amplitude modulation is the oldest form of modulation, and it served as the primary basis for commercial broadcasting for many years. Later, frequency modulation and television broadcasting systems were established, but conventional amplitude modulation still remains as a stable and entrenched sector of the commercial broadcasting industry.

A significant boon to the utilization of amplitude modulation occurred with the rapid increase in the number of citizen's band (CB) operators in the 1970s. The technology for implementing AM systems has been so well developed that it has been possible to mass-produce CB transceivers for all the "good buddies" at amazingly low costs. (A *transceiver* is a combination transmitter and receiver, in which some of the circuitry functions in both transmit and receive modes.)

As we will see shortly, conventional AM is not nearly as efficient in power utilization as either DSB or SSB, and it does not have the noise-reducing capabilities of frequency modulation. However, there are two factors that support its probable continuing use: (1) As already noted, it was the first commercial method, and it is now firmly established. Billions of dollars have been invested in AM systems over the years, and any attempt to drastically change such systems would be met with strong public resistance. (2) A conventional AM signal can be demodulated much more easily than most other forms of modulation. This fact has led to the mass production of small receivers that can be purchased for a few dollars.

### AM Generation by Addition of DC Bias

We will illustrate the generation of AM by a process that correlates with the developments of DSB and SSB. Refer both to the system shown in Figure 6–12(a) and the diagrams of

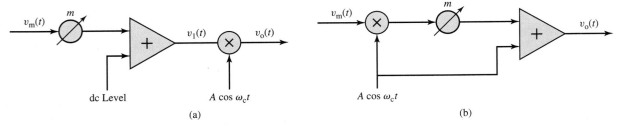

(a)

(b)

**FIGURE 6–12**
Block diagrams of possible AM generators.

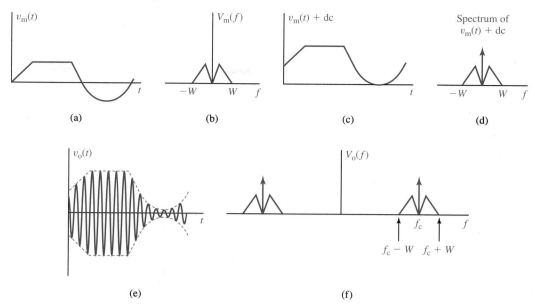

**FIGURE 6-13**
Generation of conventional AM signal from baseband signal, and spectra.

Figure 6–13. The modulating signal is assumed to vary both positively and negatively, as shown in Figure 6–13(a). The spectrum is shown in Figure 6–13(b). Assume that a dc level is added to the modulating signal $v_m(t)$ before it is applied to the balanced modulator. The dc level should be larger than any negative peaks so that the sum signal never goes negative, as shown in Figure 6–13(c). (The level control along the top in Figure 6–12(a) can be adjusted to ensure that this condition is met.) In the frequency domain, a discrete line now appears at the origin, as shown in Figure 6–13(d). The output of the balanced modulator now has the form shown in Figure 6–13(e). *Note that the modulating signal now appears as the envelope of the RF signal.* In fact, both the positive and negative envelopes of the modulated signal contain the original modulating signal, but the negative envelope is inverted. The spectrum now consists of two sidebands plus the carrier, as shown in Figure 6–13(f).

An alternate form of signal processing is shown in Figure 6–12(b), in which the carrier is added to a DSB signal. The carrier phase must be the same at the balanced modulator as at the point where it is added to the DSB signal. (In the next chapter, we will show that this process can be used to achieve angle modulation if the phase shift is different.)

The two methods shown in Figure 6–12 are both useful for explaining the form of AM, and these concepts could be employed in low-power signal generators. However, in practical applications involving moderate and large power levels, such as employed in commercial broadcasting, a nonlinear class C amplifier and a modulator are usually employed, and this process will be considered later in the chapter.

## AM Equation for Arbitrary Modulating Signal

In terms of a general modulating waveform, an AM signal can be described by an equation of the form

$$v_o(t) = A \cos \omega_c t + B v_m(t) \cos \omega_c t \qquad (6-48)$$

The first term corresponds to the carrier, and the second term, which by itself would represent a DSB signal, corresponds to the sidebands. It is necessary that the peak values (both positive and negative) of the second term be smaller than the first term, in order to avoid a condition known as *overmodulation*. This concept will become clearer shortly.

## AM with Single-Tone Modulation

Much insight into the AM process can be achieved by the assumption of a single frequency modulating signal. Indeed, transmitters are often tested with a single frequency sinusoid for the modulating signal. Continuing with $f_c$ as the carrier frequency, let $f_m$ represent the frequency of a sinusoidal modulating signal of the form $\cos \omega_m t$, with the modulating frequency assumed to be much smaller than the carrier frequency. The equation for AM in this case can be expressed as

$$v_0(t) = A \cos \omega_c t + Am \cos \omega_m t \cos \omega_c t$$
$$= A(1 + m \cos \omega_m t) \cos \omega_c t \tag{6-49}$$

The quantity $m$ is called the *modulation factor* or *index*. In order to produce AM without distortion, it is necessary that $0 \leq m \leq 1$. Often the modulation is expressed as a percentage, which is given by

$$\text{percentage modulation} = m \times 100\% \tag{6-50}$$

## Maximum and Minimum Envelope Levels

The factor $A(1 + m \cos \omega_m t)$ in the second part of Equation 6-49 is the equation for the envelope. Its maximum positive value or *crest* occurs when the cosine function is $+1$, and its minimum positive value or *trough* occurs when the cosine function is $-1$. It is convenient then to define a crest or maximum value $V_{max}$ and a trough or minimum value $V_{min}$ for the envelope as

$$V_{max} = A(1 + m) \tag{6-51}$$

$$V_{max} = A(1 - m) \tag{6-52}$$

As we will see shortly, the positive envelope varies between these limits.

## Tone-Modulated Form as Modulation Factor Is Varied

The frequency domain form for the AM signal with single-tone modulation is obtained by expanding the second term in the first form of Equation 6-49. The result is

$$v_0(t) = A \cos \omega_c t + \frac{mA}{2} \cos\left[(\omega_c - \omega_m)t\right] + \frac{mA}{2} \cos\left[(\omega_c + \omega_m)t\right] \tag{6-53}$$

Next, the time domain and frequency domain forms for several different modulation factors will be shown. Refer to Figure 6–14, beginning at the top. The first situation is for $m = 0$, in which case there is an unmodulated carrier. We will pause here to explain that this is one of the wasteful aspects of conventional AM. Even during pauses between modulation, the carrier power is transmitted but no intelligence is conveyed. The spectrum in this case is simply a line at the carrier frequency.

Moving downward in Figure 6–14, a modulation factor $m = 0.33$ is considered next. The envelope varies from $1.33A$ to $0.67A$. In the frequency domain, the two sidebands have amplitudes of $0.165A$. The next case corresponds to $m = 0.67$, in which case the envelope varies from $1.67A$ to $0.33A$, and the two sidebands have amplitudes of $0.335A$.

The last case corresponds to $m = 1$ or 100% modulation. In this pronounced case, the envelope amplitude varies from $2A$ to 0. The two sidebands now have amplitudes of $0.5A$.

The preceding equations are no longer valid when the modulation exceeds 100%, since most AM transmitters utilize a class C amplifier (to be discussed later), and severe degradation of the waveform occurs at that point. The spectrum now becomes more complex and extends well beyond the intended frequency range. The Federal Communications Commission (FCC) has stringent rules prohibiting commercial radio stations from *overmodulating* (i.e., exceeding 100%).

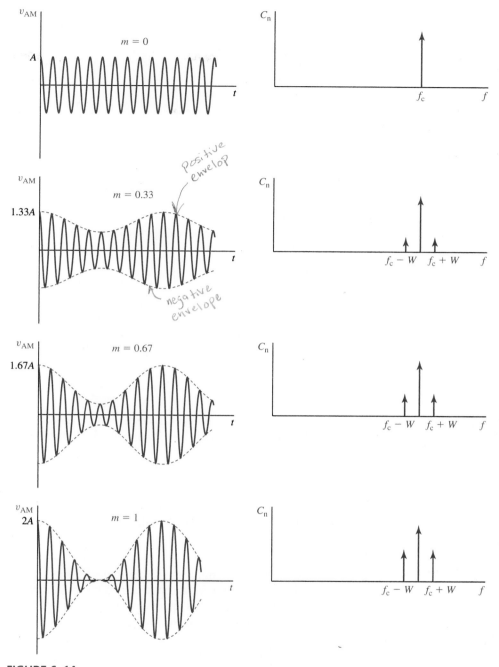

**FIGURE 6–14**
AM waveforms as modulation index is varied, and the spectra.

## Bandwidth

The transmission bandwidth of an AM signal is the same as for DSB, and is

$$B_T = 2W \qquad (6\text{-}54)$$

Thus, two forms of AM (conventional AM and DSB) have a bandwidth equal to twice the baseband bandwidth, and one form (SSB) has a bandwidth equal to the baseband bandwidth.

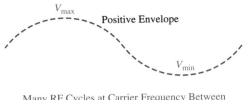

Many RF Cycles at Carrier Frequency Between
Positive and Negative Envelopes.

**FIGURE 6–15**
Determination of modulation index from maximum and minimum values of envelope.

## Measurement of Modulation Index

The results of Equation 6-51 and 6-52 may be used as a way of measuring the modulation index on an oscilloscope. The process is illustrated in Figure 6–15. The maximum and minimum values of the positive envelope are measured. By eliminating $A$ between the two equations, it may be readily shown that

$$m = \frac{V_{\max} - V_{\min}}{V_{\max} + V_{\min}} = \frac{1 - \frac{V_{\min}}{V_{\max}}}{1 + \frac{V_{\min}}{V_{\max}}} \tag{6-55}$$

In the latter form, it is not necessary that the actual maximum and minimum values be known; rather, the ratio may be measured as divisions on the oscilloscope.

## Two Observations Concerning 100% Modulation

When $m = 1$, the following two facts are readily observed for single-tone modulation: (1) The peak amplitude of the modulated signal is twice the value of the unmodulated carrier amplitude, and (2) each of the sidebands has one-half the amplitude of the carrier.

---

**III EXAMPLE 6-8**

The discrete-spectrum signal of Example 6-2 having components at 1, 3, and 5 kHz is applied as the input to a conventional AM transmitter having a carrier frequency of 250 kHz. (a) List the frequencies appearing at the output of the transmitter. (b) Determine the required transmission bandwidth.

**SOLUTION**

(a) The spectrum contains an LSB consisting of the difference frequencies, a USB consisting of the sum frequencies, *and the carrier frequency.* The list follows:

LSB:     249, 247, 245 kHz

Carrier:    250 kHz

USB:     251, 253, 255 kHz

(b) The transmission bandwidth is

$$B_T = 2W = 2 \times 5 \text{ kHz} = 10 \text{ kHz} \tag{6-56}$$

---

**III EXAMPLE 6-9**

An AM signal has an unmodulated peak carrier level of 100 V. Determine the maximum and minimum levels of the positive envelope for 60% modulation.

**SOLUTION**  The modulation factor is $m = 0.6$. The maximum value of the positive envelope is

$$V_{max} = A(1 + m) = 100(1 + 0.6) = 160 \text{ V} \tag{6-57}$$

The minimum value of the positive envelope is

$$V_{min} = A(1 - m) = 100(1 - 0.6) = 40 \text{ V} \tag{6-58}$$

---

**▐▌ EXAMPLE 6-10**

An oscilloscope is used to measure the modulation factor of an AM transmitter. The baseline of the oscilloscope is set at zero voltage. On the positive side of the envelope, the peak occurs at 2 divisions and the trough occurs at 0.5 divisions. Determine the modulation factor and the percentage modulation.

**SOLUTION**  Although we do not know the actual values of voltage, the ratio of the peak to the trough value is given by $0.5/2 = 0.25$. From Equation 6-55, we have

$$m = \frac{1 - \frac{V_{min}}{V_{max}}}{1 + \frac{V_{min}}{V_{max}}} = \frac{1 - 0.25}{1 + 0.25} = \frac{0.75}{1.25} = 0.6 \tag{6-59}$$

The percentage modulation is therefore 60%.

---

**▐▌ EXAMPLE 6-11**

In a conventional AM modulating system, a single-tone modulating signal is used to modulate a high-frequency carrier whose unmodulated amplitude is 10 V. Determine the amplitude of the sidebands for (a) $m = 0$, (b) $m = 0.5$, and (c) $m = 1$.

**SOLUTION**  From the expansion of Equation 6-53, it is noted that for a carrier amplitude of $A$, the amplitude of each sideband is $mA/2$. Thus, for $A = 10$, each sideband amplitude is $5m$ in volts.

(a) For $m = 0$, the sideband amplitudes are zero; in other words, there are no sidebands!
(b) For $m = 0.5$, each sideband has an amplitude of 2.5 V.
(c) For $m = 1$, each sideband has an amplitude of 5 V.

---

## 6-6  Envelope Detection of Conventional AM

It was shown in the last section that a conventional AM signal displays the form of the message signal directly on the envelope, which is not the case for the modulation forms considered earlier. This property permits the intelligence to be extracted through a process called *envelope detection,* which is one of the simplest concepts to implement in practice.

For envelope detection to be practical, it is necessary that the carrier frequency be much higher than the highest modulating frequency. Let $T_c = 1/f_c$ represent the period of the carrier, and let $T_m = 1/W$ represent the shortest possible period of a component of the modulating signal. (The shortest period occurs at the highest modulating frequency $W$, representing the baseband bandwidth.)

Consider the circuit shown in Figure 6–16 and the waveforms shown in Figure 6–17. For the purpose of this discussion, it will be assumed that $R_o \gg R$, in which case there is a negligible loading effect from the part of the circuit containing $R_o$ and $C_o$ on the remainder of the circuit. With this assumption, and with the diode reverse biased, the discharge path of the capacitor $C$ is primarily through the resistance $R$.

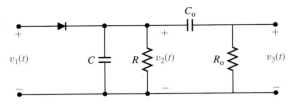

**FIGURE 6–16**
Envelope detector circuit.

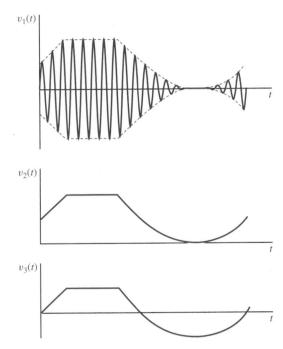

**FIGURE 6–17**
Waveforms occurring in envelope detector circuit.

Let $\tau = RC$ represent the time constant of the parallel $RC$ combination. Since the carrier frequency is several orders of magnitude higher than the highest message frequency, it is possible to select a time constant that will satisfy *both* of the following inequalities:

$$\tau \gg T_c \tag{6-60}$$

and

$$\tau \ll T_m \tag{6-61}$$

If both of these inequalities hold, the following conditions hold: (1) The voltage $v_2(t)$ cannot follow the rapid variation of the carrier due to the first inequality. (2) The voltage $v_2(t)$ will, however, be able to follow variations of the message signal due to the second inequality.

The result is that the voltage $v_2(t)$ represents very closely the positive envelope of the modulating signal. (Slight ripples due to capacitor discharge between peaks are not shown.) Recall that the envelope consisted of the message signal plus a dc bias component chosen to "lift" the baseband signal so that it could not go negative. The dc level can be readily removed by the simple high-pass filter consisting of $C_o$ and $R_o$ in Figure 6–16, and the voltage $v_3(t)$ in Figure 6–17 represents the message signal. Because of the need to add an extra dc level to the signal at the transmitter, conventional AM is usually not employed when dc data (i.e., very slowly varying data) are to be transmitted. Most applications of conventional AM involve signals having no spectral components at dc or a few hertz, in order that the dc component can be freely added and extracted without significantly affecting the signal.

One final point is that the negative envelope could just as easily have been used, and the only change required in the circuit would be to reverse the direction of the diode. The output signal extracted from the negative envelope is reversed in phase with respect to the signal obtained from the positive envelope, and this fact may or may not be important, depending on the application.

**▌▌ EXAMPLE 6-12**

An envelope detector of the form shown in Figure 6–16 is to be designed for an AM superheterodyne receiver in which the IF frequency is 455 kHz. Assume that the highest modulating frequency is 5 kHz. As an arbitrary design criterion, assume that the time constant is selected to be 10 times the period at the carrier frequency. (a) If $C$ is selected as 0.01 μF, determine $R$. (b) Show that $\tau \ll T_m$.

**SOLUTION**

(a) The carrier period is

$$T_c = \frac{1}{f_c} = \frac{1}{455 \times 10^3} = 2.20 \times 10^{-6} \text{ s} = 2.20 \text{ μs} \tag{6-62}$$

The time constant is selected as

$$\tau = 10T_c = 10 \times 2.20 \times 10^{-6} = 22.0 \times 10^{-6} \text{ s} \tag{6-63}$$

Since $\tau = RC$, $R$ is determined as

$$R = \frac{\tau}{C} = \frac{22.0 \times 10^{-6}}{0.01 \times 10^{-6}} = 2200 \text{ } \Omega \tag{6-64}$$

This value turns out to be a standard value.

(b) The shortest modulation period $T_m$ is based on the highest modulation frequency of 5 kHz. Thus,

$$T_m = \frac{1}{W} = \frac{1}{5 \times 10^3} = 0.2 \times 10^{-3} \text{ s} = 200 \text{ μs} \tag{6-65}$$

The shortest period is about $200/22 = 9.09$ times the time constant. Of course, at lower modulating frequencies (longer periods), the inequality is even greater.

▌▌

## 6-7 Comparison of Different Forms with Single-Tone Modulation

As a matter of special interest and practical importance, a comparison of the waveforms and associated spectra of the three different AM methods discussed thus far will be made in this section, based on a sinusoidal modulating signal. Assume that the modulating signal is of the form $\cos \omega_m t$ and the carrier is of the form $\cos \omega_c t$. As we have seen, the carrier frequency is usually much greater than the modulating frequency in practical systems, but for the sake of illustration simplicity, only a moderately higher carrier frequency will be assumed. The modulating signal is shown in Figure 6–18(a), and its one-sided spectrum is shown in (b).

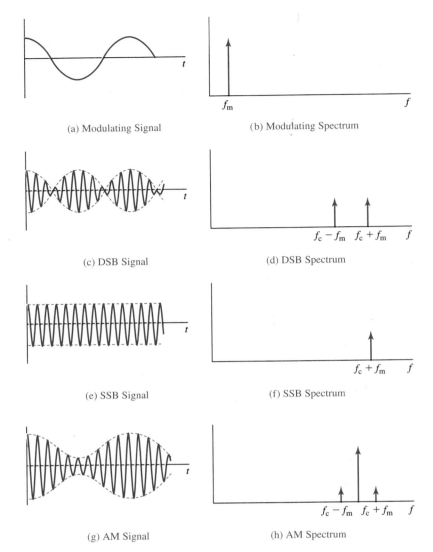

**FIGURE 6–18**
Tone-modulated AM signals and their spectra.

## DSB

The form of the single-tone modulated DSB signal is of the form

$$v(t) = A \cos \omega_m t \cos \omega_c t$$

$$= \frac{A}{2} \cos(\omega_c - \omega_m)t + \frac{A}{2} \cos(\omega_c + \omega_m)t \qquad (6\text{-}66)$$

where $A$ is an arbitrary level factor. The time domain form is shown in Figure 6–18(c) and the associated spectrum is shown in 6–18(d).

## SSB

An SSB signal for single-tone modulation can be obtained by selecting either of the two sideband components in the second form of Equation 6-66. However, in order to maintain the same peak amplitude as for the DSB component, the amplitude will be doubled. Arbitrarily selecting the upper sideband, we have

$$v(t) = A \cos(\omega_c + \omega_m)t \qquad (6\text{-}67)$$

This signal is shown in Figure 6–18(e) and its spectrum is shown in 6–18(f). For a sinusoidal modulating signal, an SSB signal is also a sinusoid, but with a much higher frequency.

## SSB with Two-Tone Modulating Signal

A test often performed with SSB transmitters is to modulate with two tones of equal amplitude. The resulting SSB signal can be assumed to be of the form

$$v(t) = \frac{A}{2}\cos(\omega_c + \omega_{m1})t + \frac{A}{2}\cos(\omega_c + \omega_{m2})t \qquad (6\text{-}68)$$

Both the time domain and frequency domain forms are similar to those of a DSB signal with single-tone modulation, and the results of Figure 6–18(c) and (d) provide appropriate descriptions.

## AM

The conventional AM signal with single-tone modulation was considered earlier, and has the form

$$v(t) = A(1 + m\cos\omega_m t)\cos\omega_c t$$

$$= A\cos\omega_c t + \frac{mA}{2}\cos(\omega_c - \omega_m)t + \frac{mA}{2}\cos(\omega_c + \omega_m)t \qquad (6\text{-}69)$$

This waveform is shown in Figure 6–18(g) for $m = 0.5$, and the spectrum is shown in 6–18(h). To get a sense of the difference in forms between DSB and AM, compare the time domain waveforms. The modulating envelopes cross with DSB, whereas with AM the complete modulating signal rides on the envelope. It is this property that makes conventional AM relatively easy to demodulate.

---

**▐▌ EXAMPLE 6-13**

For a single-tone modulating signal of 5 kHz and a carrier frequency of 200 kHz, list the frequencies appearing in the spectra of the following modulated signals: (a) DSB, (b) conventional AM, (c) SSB-LSB, (d) SSB-USB.

**SOLUTION**  We have performed this type of analysis quite a bit in previous sections devoted to the different methods, but this example is provided as one last comparison when the modulating signal is a single sinusoid.

(a)  For DSB, there are two frequencies: 195 and 205 kHz.

(b)  For conventional AM, there are three frequencies: 195, 200, and 205 kHz.

(c)  For SSB-LSB, there is one frequency: 195 kHz.

(d)  For SSB-USB, there is one frequency: 205 kHz.

▐▌

---

## 6-8  Power Relationships for Single-Tone Modulation

In this section, we will continue the analysis of single-tone modulation by investigating the relative magnitude and distribution of power for each of the three forms of AM previously considered. While the results based on tone modulation are not necessarily the same as for complex modulating signals, they serve to set representative bounds on power levels that are encountered in the general case. Moreover, many power measurements are made and equipment specifications are stated on the basis of tone modulation, so the results are very useful.

As has been the approach for most of the text thus far, waveforms will be assumed as voltages, and power equations are obtained from $v^2/R$ forms. The only change required to adapt the results to currents would be to use $i^2R$ forms instead.

## Power Definitions

We need to first review some earlier power equations, and then introduce a new one.

### Instantaneous Power $p(t)$

Instantaneous power was introduced very early in the text, and is given by

$$p(t) = \frac{v^2(t)}{R} \tag{6-70}$$

### Average Power $P$

Average power was also introduced early, and is given by

$$P = \frac{1}{T}\int_0^T p(t)\,dt = \frac{1}{RT}\int_0^T v^2(t)\,dt = \frac{V_{rms}^2}{R} \tag{6-71}$$

### Peak Envelope Power $P_p$

The peak envelope power, which is sometimes abbreviated as PEP, is a new definition and needs some explanation. It can be defined as the average power produced by a sinuosid whose peak voltage amplitude (or peak current amplitude) is the same as that of the given signal. Thus, if $V_p$ is the peak value of the envelope, the peak envelope power is

$$P_p = \frac{V_p^2}{2R} \tag{6-72}$$

This definition may seem a little strange and somewhat contradictory. However, it serves a useful purpose in dealing with modulated signals. Because the envelope of a modulated signal may be much greater than the level without modulation, a power level calculated on the basis of the peak amplitude level provides a measure of the power level generated during the peak intervals for the waveform. For example, suppose a transmitter produces an average power of 100 W and a peak envelope power of 200 W. Components in the power amplifier stage must be chosen to handle the requirements based on peak power considerations.

The average and peak envelope power levels will now be developed for the three basic AM forms, with the assumption of single-tone modulation.

## DSB

The instantaneous form of the signal was given in Equation 6-66, and is repeated here for convenience.

$$v(t) = A\cos\omega_m t\cos\omega_c t = \frac{A}{2}\cos(\omega_c - \omega_m)t + \frac{A}{2}\cos(\omega_c + \omega_m)t \tag{6-73}$$

The average power $P$ in a resistance $R$ is best determined from the second form by adding the powers for the two components.

$$P = \frac{\frac{1}{2}\left(\frac{A}{2}\right)^2}{R} + \frac{\frac{1}{2}\left(\frac{A}{2}\right)^2}{R} = \frac{A^2}{4R} \tag{6-74}$$

The peak amplitude of the signal is $A$, so the peak envelope power $P_p$ is

$$P_p = \frac{A^2}{2R} \tag{6-75}$$

The ratio of the peak envelope power to the average power for a single-tone modulating signal is

$$\frac{P_p}{P} = \frac{V_p^2/2R}{V_p^2/4R} = 2 \tag{6-76}$$

## SSB

The instantaneous form of the signal was given in Equation 6-67, and is repeated here for convenience.

$$v(t) = A\cos(\omega_c + \omega_m)t \tag{6-77}$$

The average power is

$$P = \frac{A^2}{2R} \tag{6-78}$$

However, since the function is a sine wave, the peak power is also

$$P_p = \frac{A^2}{2R} \tag{6-79}$$

The ratio of the peak envelope power to the average power for a single-tone modulating signal is

$$\frac{P_p}{P} = \frac{A^2/2R}{A^2/2R} = 1 \tag{6-80}$$

Thus, the average power and the peak envelope power are the same for SSB *when the modulating signal is a single tone.* It should be stressed that this relationship is not true for more complex modulating signals.

## SSB with Two-Tone Modulation

More realistic measures of the average and peak power ratings of SSB systems can be achieved from the two-tone test. Since the time and frequency domain forms in this case are the same as for DSB with single-tone modulation, the results of Equations 6-74, 6-75, and 6-76 are applicable.

## AM

The instantaneous form of the signal was given in Equation 6-69, and is repeated here for convenience.

$$v(t) = A(1 + m\cos\omega_m t)\cos\omega_c t$$
$$= A\cos\omega_c t + \frac{mA}{2}\cos(\omega_c - \omega_m)t + \frac{mA}{2}\cos(\omega_c + \omega_m)t \tag{6-81}$$

The average power $P$ is best obtained by adding the three powers associated with the three terms in the second form of Equation 6-81.

$$P = \frac{A^2}{2R} + \frac{\frac{1}{2}\left(\frac{mA}{2}\right)^2}{R} + \frac{\frac{1}{2}\left(\frac{mA}{2}\right)^2}{R} = \frac{A^2}{2R} + \frac{m^2 A^2}{4R} = \frac{A^2}{2R}\left(1 + \frac{m^2}{2}\right) \tag{6-82}$$

It is convenient to define the unmodulated average carrier power $P_c$ as

$$P_c = \frac{A^2}{2R} \tag{6-83}$$

The unmodulated carrier power is the power level specified in commercial broadcasting. The average total power can then be expressed as

$$P = P_c\left(1 + \frac{m^2}{2}\right) \tag{6-84}$$

The peak amplitude of the envelope of the composite signal was developed much earlier and is repeated here for convenience.

$$V_{max} = A(1 + m) \tag{6-85}$$

Table 6–1    Power Relationships for Single-Tone Modulation

| Method | $P$ | $P_p$ | $P_p/P$ |
|---|---|---|---|
| DSB | $\dfrac{A^2}{4R}$ | $\dfrac{A^2}{2R}$ | 2 |
| SSB | $\dfrac{A^2}{2R}$ | $\dfrac{A^2}{2R}$ | 1 |
| AM | $\left(1+\dfrac{m^2}{2}\right)P_c$ | $(1+m)^2\,P_c$ | $\dfrac{(1+m)^2}{1+m^2/2}$ |
| AM, $m=0.5$ | $1.125\,P_c$ | $2.25\,P_c$ | 2 |
| AM, $m=1$ | $1.5\,P_c$ | $4\,P_c$ | 2.667 |

The peak envelope power is then

$$P_p = \frac{V_{max}^2}{2R} = \frac{(1+m)^2 A^2}{2} = P_c(1+m)^2 \tag{6-86}$$

The ratio of peak envelope power to average power for a single-tone modulated AM signal is

$$\frac{P_p}{P} = \frac{(1+m)^2}{1+m^2/2} \tag{6-87}$$

The preceding quantities are tabulated in Table 6–1, with the AM data shown for several values of $m$. It is interesting to observe the manner in which the peak power varies with the modulation factor. For small values of the modulation factor, the average power increases only slightly, but the peak power increases substantially. For 100% single-tone modulation, *the average power is 1.5 times the unmodulated carrier power, and the peak power is 4 times the unmodulated carrier power.* As stated earlier, the power levels of most AM transmitters are specified in terms of the unmodulated carrier power. Thus, a 1000-W commercial transmitter would generate 1500 W of average power and 4000 W of peak power under conditions of 100% single-tone modulation. These limits must be considered in designing the RF power amplifier stage and the antenna system.

**III EXAMPLE 6-14**

A single-tone modulated DSB signal has a peak envelope voltage of 200 V across a 50-$\Omega$ resistive load. Determine (a) average power and (b) peak envelope power.

**SOLUTION**

(a)  From either Equation 6-74 or Table 6–1, the average power is

$$P = \frac{A^2}{4R} = \frac{200^2}{4 \times 50} = 200 \text{ W} \tag{6-88}$$

(b)  From either Equation 6-75 or Table 6–1, the peak envelope power is

$$P_p = \frac{A^2}{2R} = \frac{200^2}{2 \times 50} = 400 \text{ W} \tag{6-89}$$

While we have taken the easy route of simply "plugging in" the formulas, the reader should note that the values could have been determined by using the basic form of the DSB signal and the definitions of average and peak power.

**III EXAMPLE 6-15**

A commercial AM transmitter has an unmodulated carrier power level of 1 kW, and the antenna presents a resistive load of $R = 50\ \Omega$ to the transmitter at the operating frequency. Determine the unmodulated values of antenna rms voltage and antenna rms current.

SOLUTION   This problem reduces to one in basic circuit theory, but is used as a review and starting point for several other power problems to follow. The average power is related to the voltage by

$$P = \frac{V_{rms}^2}{R}$$

(6-90)

Solving for the unmodulated rms carrier voltage, we have

$$V_{rms} = \sqrt{RP} = \sqrt{50 \times 1000} = 223.6 \text{ V}$$

(6-91)

The average power is related to the current by

$$P = I_{rms}^2 R$$

(6-92)

Solving for the unmodulated rms carrier current, we have

$$I_{rms} = \sqrt{\frac{P}{R}} = \sqrt{\frac{1000}{50}} = 4.472 \text{ A}$$

(6-93)

Alternately, we could have first solved for either the voltage or the current and used Ohm's law to determine the other variable. The approach used here illustrates better how either voltage or current can be directly determined in terms of power and resistance.

---

**❚❙ EXAMPLE 6-16**

For the AM transmitter of Problem 6-15, determine, for both 50% and 100% modulation, (a) the total average power, (b) the peak envelope power, and (c) the sideband power.

**SOLUTION**

(a)  The total average power is given by

$$P = \left(1 + \frac{m^2}{2}\right) P_c$$

(6-94)

For $m = 0.5$, the total power is

$$P = \left(1 + \frac{(0.5)^2}{2}\right) \times 1000 = 1.125 \times 1000 = 1125 \text{ W}$$

(6-95)

For $m = 1$, the total power is

$$P = \left(1 + \frac{(1)^2}{2}\right) \times 1000 = 1.5 \times 1000 = 1500 \text{ W}$$

(6-96)

Note the nonlinear manner in which the power increases with modulation factor. When the modulation factor doubled, the *change* in average power increased by a factor of four (from 125 W to 500 W).

(b)  The peak envelope power is given by

$$P_p = (1 + m)^2 P_c$$

(6-97)

For $m = 0.5$, the peak envelope power is

$$P_p = (1 + 0.5)^2 P_c = 2.25 P_c = 2.25 \times 1000 = 2250 \text{ W}$$

(6-98)

For $m = 1$, the peak envelope power is

$$P_p = (1 + 1)^2 P_c = 4 P_c = 4 \times 1000 = 4000 \text{ W}$$

(6-99)

Table 6–2   Summary of Results of Example 6-16

| Modulation Factor | Average Power | Peak Power | Sideband Power |
|---|---|---|---|
| 0 | 1000 W | 1000 W | 0 |
| 50% | 1125 W | 2250 W | 125 W |
| 100% | 1500 W | 4000 W | 500 W |

(c) Let $P_{SB}$ represent the total sideband power. It can be determined by taking the difference between the total average power and the unmodulated carrier power.
For $m = 0.5$, the sideband power is

$$P_{SB} = 1125 - 1000 = 125 \text{ W} \tag{6-100}$$

For $m = 1$, the sideband power is

$$P_{SB} = 1500 - 1000 = 500 \text{ W} \tag{6-101}$$

The power in one sideband is half the total sideband power. Thus, for 50% modulation, the power in one sideband is 62.5 W, and for 100% modulation, the power in one sideband is 250 W.

The values computed in this example are summarized in Table 6–2.

**▎▎ EXAMPLE 6-17**

For the transmitter of Examples 6-15 and 6-16, determine the antenna rms voltage and current for (a) 50% modulation and (b) 100% modulation.

**SOLUTION**   The procedure is the same as in Example 6-15, and Equations 6-91 and 6-93 are applicable. The difference in this case is that the average power values from Example 6-16 will be used to determine the rms voltage and current for each modulation index.

(a) For $m = 0.5$, the average power was determined to be 1125 W. Hence,

$$V_{rms} = \sqrt{RP} = \sqrt{50 \times 1125} = 237.2 \text{ V} \tag{6-102}$$

and

$$I_{rms} = \sqrt{\frac{P}{R}} = \sqrt{\frac{1125}{50}} = 4.743 \text{ A} \tag{6-103}$$

(b) For $m = 1$, the average power was determined to be 1500 W. The rms voltage and current are then

$$V_{rms} = \sqrt{RP} = \sqrt{50 \times 1500} = 273.9 \text{ V} \tag{6-104}$$

and

$$I_{rms} = \sqrt{\frac{P}{R}} = \sqrt{\frac{1500}{50}} = 5.477 \text{ A} \tag{6-105}$$

It should be noted that the changes in rms voltage and current are much more moderate than the changes in the power.   ▎▎

## 6-9   Class C Modulated Amplifier Stage

Recall that the development of conventional AM was introduced through some simplified signal processing schemes illustrated in Figure 6–12. Techniques such as this are utilized in function generator chips and in very low power AM generators. However, when an AM

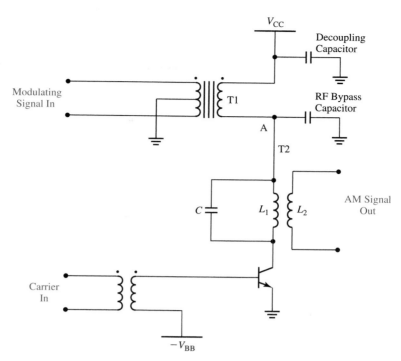

**FIGURE 6–19**
Typical collector-modulated class C amplifier circuit.

signal is generated at a very low level, all subsequent amplification must be achieved with *linear* amplifier stages (such as class A or push-pull class B), and this means that higher efficiency configurations such as class C amplifiers cannot be used in the remaining stages.

The process of *high-level modulation* is one in which the modulated signal is generated in the output power amplifier stage, usually with a class C amplifier. Both transistors and vacuum tubes are used in this application, but for the highest power levels, vacuum tubes still excel.

A typical layout of a class C collector-modulated transistor output stage is shown in Figure 6–19. With no signal applied, the transistor is at cutoff, and as a result of the negative voltage $-V_{BB}$, conduction will occur for somewhat less than a half-cycle at the input carrier frequency. This means that the collector current will flow for short pulses, with conduction angles less than 180°. Special power transistors with relatively large base-to-emitter reverse breakdown voltage ratings are required.

A complete quantitative analysis of a class C amplifier is rather difficult, due to the nonlinear nature of the circuit. Indeed, proper design of the circuit seems to be as much of an art as a science. However, by making certain simplifying assumptions, it is possible to obtain approximate operational bounds for the circuit.

## Analysis with No Modulation

First, we will consider the basic operation of the circuit with no modulation applied. With this assumption, the secondary of the modulation transformer T1 will be essentially a short circuit for dc. The RF bypass capacitor will prevent the voltage at point A from varying with the RF signal, so it must assume the dc level $V_{CC}$.

Refer to the waveforms shown in Figure 6–20. The carrier input signal is shown at the top, and the resulting RF current pulses are shown as the second waveform from the top.

The parallel combination of $C$ and $L_1$ is resonant at the carrier frequency. If the $Q$ of the circuit is sufficiently high to appear as a very low impedance for all harmonics of the

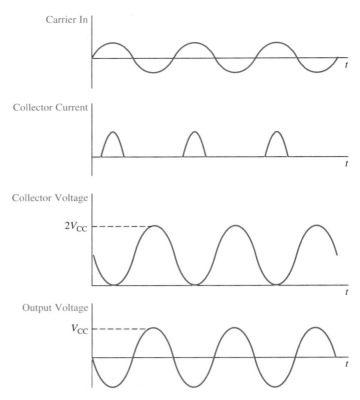

**FIGURE 6–20**
Waveforms in class C amplifier circuit with no modulation applied.

current pulse spectrum, the voltage across the resonant circuit should be sinusoidal at the carrier frequency.

Assume that the transistor is driven to saturation on the positive peaks of the input signal, which means that the collector voltage is nearly zero, as shown by the third waveform from the top in Figure 6–20. The so-called flywheel effect of the tuned circuit will then cause the voltage at the collector to reach a level of about $2V_{CC}$ at the positive peak. The output transformer T2 couples the signal to the external load, and is designed to provide an optimum impedance match for the load. The inductive coupling process eliminates the dc component, and the output voltage will have the form shown as the bottom waveform of Figure 6–20. For convenience, the output sinusoidal voltage is shown with the same peak-to-peak value as shown at the collector. In practice, the level will depend on the proper impedance match between the collector output and the load.

## Analysis with Modulation

Consider now the process of applying a modulating signal to the primary of transformer T1. In the lower modulating frequency range, the RF bypass capacitor is assumed to be an open circuit. This will allow the voltage at point A to vary at the modulating frequency rate, which causes the peak of the current pulses to vary in accordance with the modulating signal. Assuming that the bandwidth of the resonant circuit is sufficiently wide to accommodate the two sideband frequencies, the voltage across the resonant circuit and the voltage at the collector will become amplitude modulated forms.

## Power and Voltage Relationships

It was shown earlier that the sideband power for 100% single-tone modulation is 50% of the carrier power. The modulator supplies this power. Let $P_c$ represent the unmodulated carrier power. The modulator power must then be capable of delivering a power $P_m$, given by

$$P_m = 0.5P_c \qquad (6\text{-}106)$$

It has already been shown that the collector voltage is twice the power supply voltage when there is no modulation. To achieve 100% modulation, the collector must be able to swing all the way from zero to twice the voltage corresponding to no modulation. Thus, a very important design rating for the collector voltage is its maximum possible value $V_{c,max}$, which is given by

$$V_{c,max} = 4V_{CC} \qquad (6\text{-}107)$$

## 6-10  Multisim® Examples (Optional)

The Multisim examples of this section will focus on AM waveform generation and detection. The first example will investigate the **AM Source** in the **Sources** bin, and the second example will deal with a **Fourier Analysis** of its output. The third example will deal with a diode envelope detector and its ability to demodulate an AM signal.

**■ MULTISIM EXAMPLE 6-1**

Investigate the amplitude modulated voltage source and display the waveform for 100% modulation, based on a carrier frequency of 10 kHz and a modulation frequency of 1 kHz.

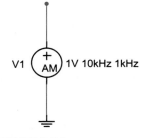

V1  AM  1V 10kHz 1kHz

**FIGURE 6–21**
Circuit of Multisim Example 6-1.

**SOLUTION**  The AM waveform is obtained from the **Signal Source Family Bar,** and it has the name **AM Source.** The simple circuit used to investigate the signal is shown in Figure 6–21. Along with the AM signal, a **Junction** is used to observe the signal and provide a valid node number.

A double left-click on the signal opens the properties window. The **Carrier Amplitude** is set to **1 V** and the **Carrier Frequency** is set to **10 kHz.** The modulation factor is referred to here as the **Modulation index** and is set to **1.** Finally, the **Intelligence Frequency** is set to **1 kHz.**

A **Transient Analysis** is run from $t = 0$ to $t = 2$ ms, which represents two cycles at the modulating frequency. The **Maximum time step** was set to 1 μs (1e-6 s). The result after some relabeling is shown in Figure 6–22.

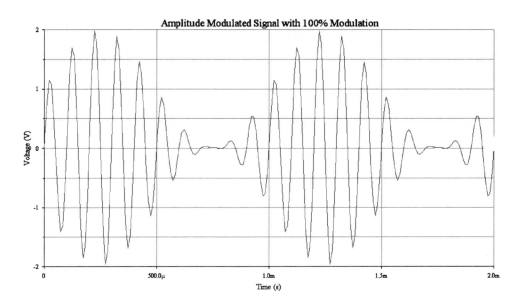

**FIGURE 6–22**
AM waveform of Multisim Example 6-1.

**MULTISIM EXAMPLE 6-2**     Perform a **Fourier Analysis** of the AM signal of Multisim Example 6-1.

SOLUTION     Although the spectrum is centered at 10 kHz, the fundamental is 1 kHz. In order to view the complete spectrum, the frequency range must extend at least to 11 kHz. Arbitrarily, the **Number of harmonics** was set to 19 and the **Stopping time for sampling** was set to **0.002 s**, which represents two cycles at the modulating frequency.

The resulting spectrum is shown in Figure 6–23. As in the earlier cases of spectral analysis, most of the tabulated data represent negligible components that arise only through the numerical approximation process. However, as expected, there is a carrier at 10 kHz, and sidebands at 9 kHz and 11 kHz. Moreover, the sidebands have 50% of the amplitude of the carrier, as expected.

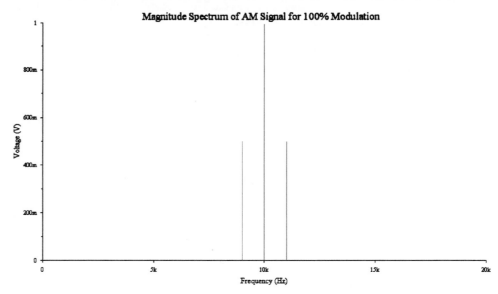

```
Fourier analysis for V(1):
DC component: 4.19071e-006

  No. Harmonics: 19, THD: 1.42474e+007 %, Gridsize: 512, Interpolation Degree: 1

Harmonic  Frequency    Magnitude     Phase       Norm. Mag    Norm. Phase
--------  ---------    ---------    -----        ---------    -----------
   1        1000     8.52546e-006  -64.531        1            0
   2        2000     1.05608e-005  -167.47        1.23873     -102.94
   3        3000     1.00525e-005  114.499        1.17911      179.03
   4        4000     1.80493e-005  -31.604        2.1171       32.9267
   5        5000     3.7221e-005   -125.8         4.36586     -61.266
   6        6000     3.61002e-005  169.22         4.2344       233.751
   7        7000     2.99264e-005  86.8751        3.51024      151.406
   8        8000     4.64765e-005  -54.216        5.4515       10.3143
   9        9000     0.496655      90.0057        58255.5      154.536
  10       10000     0.991792      -0.00048926    116333       64.5302
  11       11000     0.495047      -90.007        58066.8     -25.476
  12       12000     4.03496e-005  110.78         4.73283      175.311
  13       13000     8.83481e-005  59.3128        10.3628      123.843
  14       14000     0.000145727   -33.331        17.0931      31.1997
  15       15000     9.66736e-005  -122.94        11.3394     -58.405
  16       16000     4.5246e-005   166.849        5.30716      231.379
  17       17000     3.41216e-005  80.0897        4.00231      144.62
  18       18000     3.10882e-005  -27.51         3.64651      37.0203
  19       19000     2.88023e-005  -95.655        3.37839     -31.125
```

**FIGURE 6–23**
Spectral data for Multisim Example 6-2.

**MULTISIM EXAMPLE 6-3**     The circuit of Figure 6–24 represents a Multsim model of an AM envelope detector. Assuming a carrier frequency of 1 MHz, a modulating frequency of 1 kHz, and 50% modulation, plot the output of the detector.

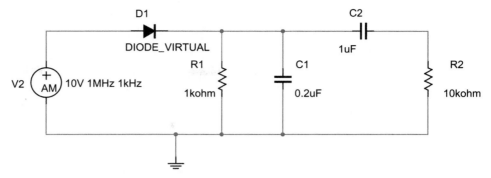

FIGURE 6–24
Circuit of Multisim Example 6-3.

SOLUTION   The only new component introduced in this circuit is a diode, which is obtained from the **Diode Family Bar.** The one employed is denoted **DIODE_ VIRTUAL.** The **AM Source** is used again, and the following parameters are set: **Carrier Amplitude = 10 V, Carrier Frequency = 1 MHz, Modulation Index = 0.5,** and **Intelligence Frequency = 1 kHz.**

Although some formal computations were made at the beginning, the parameter values in the circuit were adjusted experimentally to obtain the best demodulated waveform. It is appropriate here to discuss some of the values.

The period of the carrier is $1/(1e6) = 1$ μs, and the period of the modulating signal is $1/(1e3) = 1$ ms. The time constant of the $RC$ parallel combination to the right of the diode is $1e3 \times 0.2e\text{-}6 = 0.2$ ms. This value is large compared with the carrier period, and is about 20% of the length of the modulation period. Actually, this is a bit of an oversimplification, since the loading effect of the circuit to the right changes the situation somewhat, but its impedance level is relatively large compared to the $RC$ parallel combination.

The **Maximum time step** was set at 0.1 μs (1e-7 s). Following some changes in the labels, the waveform at the diode output is shown in Figure 6–25. This voltage has the correct shape, but it has the expected dc bias level. After removal of the dc component, the resulting waveform is shown in Figure 6–26.

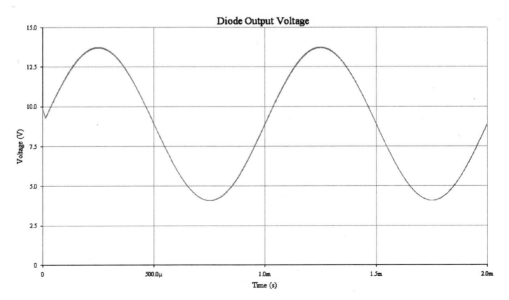

FIGURE 6–25
Detector output prior to removal of the dc component.

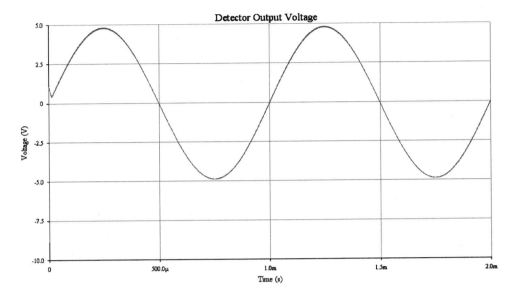

**FIGURE 6–26**
Detector output after removal of the dc component.

## 6-11   MATLAB® Examples (Optional)

The examples in this section will utilize MATLAB for plotting both the time function and the spectrum for modulated signals. The first two examples will be based on a single-tone modulated double sideband signal and its spectrum; the next two examples will be based on a conventional amplitude modulated signal and its spectrum.

**▌ MATLAB EXAMPLE 6-1**   Consider an ideal balanced modulator (multiplier) with a 1-kHz sinusoidal modulating signal and a 20-kHz carrier. Use MATLAB to generate the DSB output signal, and plot it over two cycles of the modulating frequency.

**SOLUTION**   We will somewhat arbitrarily select a time step of 1 μs (1e-6 s). In anticipation of the spectral analysis that will be performed in the next example, we will stop one point short of two complete cycles, so that the number of time points is an even number. The row vector for the time **t** is generated by the command

```
>> t = 0:1e-6:2e-3-1e-6;
```

Note that the ending time is the difference between 2 ms (2e-3) and the time step of 1 μs (1e-6).

Although cosine functions were used in the text to simplify the mathematical developments, a "cleaner" plot is generated with sine functions. The modulating signal **vm** is generated by the command

```
>> vm = sin(2*pi*1000*t);
```

The carrier **vc** is generated by the command

```
>> vc = sin(2*pi*20e3*t);
```

Multiplying the previous two functions on a point-by-point or array basis generates the DSB signal **vdsb.**

```
>> vdsb = vm.*vc;
```

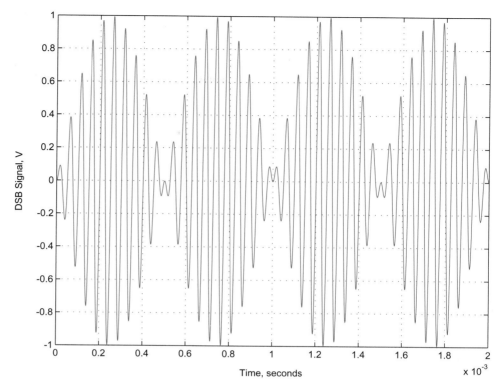

**FIGURE 6–27**
Plot of double sideband signal.

The function is plotted by the command

```
>> plot(t,vdsb)
```

The resulting DSB function, after appropriate labeling, is shown in Figure 6–27.

---

**▌ MATLAB EXAMPLE 6-2**

Keeping the function of the previous example in memory, use the program **fourier_series_1** to determine the amplitude spectrum.

**SOLUTION**   In order to use the program, the function must be defined as **v**. This process is about as easy as any command can be, and is simply

```
>> v = vdsb;
```

We then type the following command and enter it:

```
>> fourier_series_1
```

The dialogue that follows requires that we enter the time step (1e-6) and the harmonic numbers corresponding to the beginning and ending of the plot. The frequency step is 500 Hz, so we will choose to select the two integers as 20 and 60, respectively. This will provide a spectral plot between 10 kHz and 30 kHz. The resulting plot is shown in Figure 6–28.

---

**▌ MATLAB EXAMPLE 6-3**

Generate an AM signal with a carrier frequency of 20 kHz, a modulating frequency of 1 kHz, and 50% modulation.

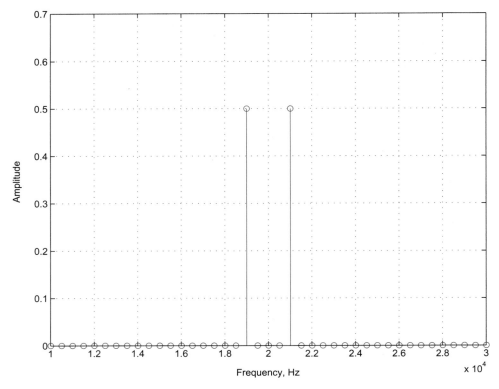

**FIGURE 6–28**
Amplitude spectrum based on one-sided form, using FFT.

**SOLUTION**   While we could start from the beginning, some of the parameters specified are those of MATLAB Example 6-1. If the vectors generated in that problem are still in memory, we need only add the carrier to the DSB sideband signal modified by the modulation factor ($m = 0.5$). Using the results of that example and defining the result as **vam**, we have

```
>> vam = vc + 0.5*vdsb;
```

The function is plotted by the command

```
>> plot(t,vam)
```

The resulting AM signal with 50% modulation is shown in Figure 6–29.

---

■ MATLAB EXAMPLE 6-4

Keeping the AM waveform of the previous example in memory, use the program **fourier_series_1** to determine the amplitude spectrum.

**SOLUTION**   First, we redefine the function to conform to the program.

```
>> v = vam
```

We then enter the command

```
>> fourier_series_1
```

The dialogue then requires that we enter the time step (1e-6). The resulting parameters are the same as for MATLAB Example 6-2 with a frequency step of 500 Hz. As in that example, we will choose 20 and 60 for the respective frequency integers. The resulting plot will then be generated from 10 kHz to 30 kHz, and is shown in Figure 6–30. Note the presence of the large carrier in this case. For 50% modulation, the two sidebands each have a height of one-fourth the level of the carrier.                                                                    ▌▌▌

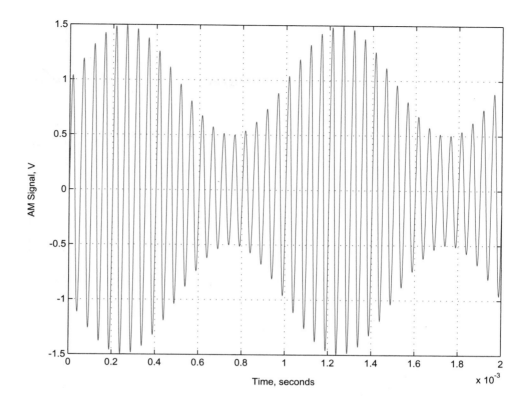

**FIGURE 6–29**
Plot of AM signal with 50% modulation.

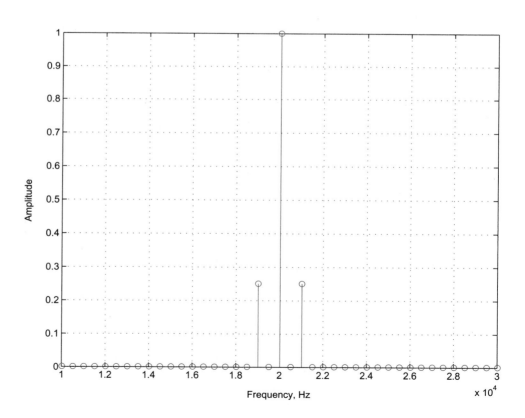

**FIGURE 6–30**
Amplitude spectrum based on one-sided form, using FFT.

## SystemVue™ Closing Application (Optional)

Insert the text CD in a computer having SystemVue™ installed and activate the program. Open the CD folder entitled **SystemVue Systems** and load the file entitled 6-1.

### Review of System

The system shown on the screen was described in the chapter opening application, and you may wish to review the discussion there. Recall that the input was a 1-kHz signal, and both a DSB and an SSB signal were created.

### Generation of a Conventional AM Signal

Delete the filter (Token 7). This can be achieved by a right-click on the token followed by a left-click on **Delete Token.** Next, drag an **Adder** from the token library on the left-hand side of the screen and place it in approximately the same location as the filter. Connect the output of the carrier source (Token 5) to the adder. As you connect it, a window with the title **Sinusoid Token 4** will open and the statement **Select Output** will appear. You have a choice between a sine function and a cosine function, but in this case the **Sine** should be selected in order to have the proper AM phase relationship. (It will likely already be in that position.) Also, connect the output of the multiplier (Token 6) as an input to the adder. Connect the output of the adder to Sink 3. Activate the circuit and observe the various waveforms.

### What You See

The signals at Sinks 0, 1, and 2 are the same as for the opening application. However, you should now observe a conventional AM signal at the output of the adder with Sink 4.

### Additional Design

For the parameter values as given, there will be a fixed modulation factor. Show how the modulation factor can be modified while holding the carrier level constant. Observe the output AM signal for several modulation factors, ranging from zero to 100% modulation.

## PROBLEMS

**6-1**   A continuous-spectrum baseband signal has a spectrum extending from near dc to 6 kHz. It is applied as one input to an ideal balanced modulator. The carrier input is a sinusoid having a frequency of 1.2 MHz. (a) Determine the range of frequencies contained in the DSB output spectrum. (b) Determine the required transmission bandwidth.

**6-2**   A continuous-spectrum baseband signal has a spectrum extending from near dc to 3 kHz. It is applied as one input to an ideal balanced modulator. The carrier input is a sinusoid having a frequency of 40 kHz. (a) Determine the range of frequencies contained in the DSB output spectrum. (b) Determine the required transmission bandwidth.

**6-3**   A discrete-spectrum baseband modulating signal has components at the following frequencies: 500 Hz, 1.5 kHz,

and 3.5 kHz. It is applied as one input to an ideal balanced modulator, and the other input is a sinusoid having a frequency of 1.2 MHz. (a) List the frequencies appearing at the output of the balanced modulator. (b) Determine the required transmission bandwidth.

**6-4**   A discrete-spectrum baseband modulating signal has components at the following frequencies: 1.2 kHz, 3.6 kHz, and 6 kHz. It is applied as one input to an ideal balanced modulator, and the other input is a sinusoid having a frequency of 1.5 MHz. (a) List the frequencies appearing at the output of the balanced modulator. (b) Determine the required transmission bandwidth.

**6-5**   Refer back to the signal and system of Problem 6-1. It is desired to convert the DSB signal to an SSB signal by passing the signal through an appropriate filter. Determine the range of frequencies contained in the output spectrum for (a) LSB and (b) USB. (c) Determine the required transmission bandwidth.

**6-6** Refer back to the signal and system of Problem 6-2. It is desired to convert the DSB signal to an SSB signal by passing the signal through an appropriate filter. Determine the range of frequencies contained in the output spectrum for (a) LSB and (b) USB. (c) Determine the required transmission bandwidth.

**6-7** Refer back to the signal and system of Problem 6-3. It is desired to convert the DSB signal to an SSB signal by appropriate filtering. List the frequencies appearing at the output of the filter for (a) LSB and (b) USB. (c) Determine the required transmission bandwidth.

**6-8** Refer back to the signal and system of Problem 6-4. It is desired to convert the DSB signal to an SSB signal by appropriate filtering. List the frequencies appearing at the output of the filter for (a) LSB and (b) USB. (c) Determine the required transmission bandwidth.

**6-9** Consider the lower SSB signal generated in Problem 6-7(a). Assume that the signal is applied as one input to an ideal multiplier, and the other input is a sinusoid having a frequency of 1.2 MHz. (a) List the positive frequencies at the output of the multiplier. (b) List the frequencies at the output of the detector low-pass filter.

**6-10** Consider the lower SSB signal generated in Problem 6-8(a). Assume that the signal is applied as one input to an ideal multiplier, and the other input is a sinusoid having a frequency of 1.5 MHz. (a) List the positive frequencies at the output of the multiplier. (b) List the frequencies at the output of the detector low-pass filter.

**6-11** Consider the upper SSB signal generated in Problem 6-7(b). Assume that the signal is applied as one input to an ideal multiplier, and the other input is a sinusoid having a frequency of 1.2 MHz. (a) List the positive frequencies at the output of the multiplier. (b) List the frequencies at the output of the detector low-pass filter.

**6-12** Consider the upper SSB signal generated in Problem 6-8(b). Assume that the signal is applied as one input to an ideal multiplier, and the other input is a sinusoid having a frequency of 1.5 MHz. (a) List the positive frequencies at the output of the multiplier. (b) List the frequencies at the output of the detector low-pass filter.

**6-13** Assume in the detection circuit of Problem 6-11 that the receiver oscillator drifts by 200 Hz, so that the injected frequency is actually 1.2002 MHz. Repeat the analysis of Problem 6-11.

**6-14** Assume in the detection circuit of Problem 6-12 that the receiver oscillator drifts by −500 Hz, so that the injected frequency is actually 1.4995 MHz. Repeat the analysis of Problem 6-12.

**6-15** The discrete-spectrum signal of Problem 6-3, having components at 500 Hz, 1.5 kHz, and 3.5 kHz, is applied as the input to a conventional AM transmitter having a carrier frequency of 1.2 MHz. (a) List the frequencies appearing at the output of the transmitter. (b) Determine the required transmission bandwidth.

**6-16** The discrete-spectrum signal of Problem 6-4, having components at 1.2 kHz, 3.6 kHz, and 6 kHz, is applied as the input to a conventional AM transmitter having a carrier frequency of 1.5 MHz. (a) List the frequencies appearing at the output of the transmitter. (b) Determine the required transmission bandwidth.

**6-17** An AM signal has an unmodulated peak carrier level of 200 V. Determine the maximum and minimum levels of the positive envelope for the following modulation percentages: (a) 25%, (b) 50%, and (c) 100%.

**6-18** An AM signal has an unmodulated peak carrier level of 80 V. Determine the maximum and minimum levels of the positive envelope for the following modulation percentages: (a) 25%, (b) 50%, and (c) 100%.

**6-19** An oscilloscope is used to measure the modulation index of an AM transmitter. The crest of the positive envelope is 180 V and the trough is 60 V. Determine the modulation index and the percentage modulation.

**6-20** An oscilloscope is used to measure the modulation index of an AM transmitter. The crest of the positive envelope is 60 V and the trough is 30 V. Determine the modulation index and the percentage modulation.

**6-21** In a conventional AM transmitter, a single-tone modulating signal is used to modulate a high-frequency carrier whose unmodulated amplitude is 120 V. Determine the amplitude of the sidebands for the following percentages of modulation: (a) 25%, (b) 50%, and (c) 100%.

**6-22** In a conventional AM transmitter, a single-tone modulating signal is used to modulate a high-frequency carrier whose unmodulated amplitude is 200 V. Determine the amplitude of the sidebands for the following percentages of modulation: (a) 40%, (b) 75%, and (c) 100%.

**6-23** For a single-tone modulating signal of 2 kHz and a carrier frequency of 100 kHz, list the frequencies appearing in the spectra of the following modulated signals: (a) DSB, (b) conventional AM, (c) SSB-LSB, and (d) SSB-USB.

**6-24** For a single-tone modulating signal of 5 kHz and a carrier frequency of 600 kHz, list the frequencies appearing in the spectra of the following modulated signals: (a) DSB, (b) conventional AM, (c) SSB-LSB, and (d) SSB-USB.

**6-25** A single-tone modulated DSB signal has a peak envelope voltage of 300 V across a 75-Ω load. Determine (a) average power and (b) peak envelope power.

**6-26** A single-tone modulated DSB signal has a peak envelope current of 4 A flowing in a 50-Ω load. Determine (a) average power and (b) peak envelope power.

**6-27** A two-tone test is run on an SSB transmitter. The peak envelope voltage across a 50-Ω load is 100 V. Determine (a) average power and (b) peak envelope power.

**6-28** A two-tone test is run on an SSB transmitter. The peak envelope voltage across a 75-Ω load is 200 V. Determine (a) average power and (b) peak envelope power.

**6-29** A commercial AM transmitter has an unmodulated carrier power level of 50 kW, and the antenna presents a resistive load of 50 Ω. Determine the unmodulated values of antenna rms voltage and rms current.

**6-30**    A commercial AM transmitter has an unmodulated carrier power level of 10 kW, and the antenna presents a resistive load of 50 Ω. Determine the unmodulated values of antenna rms voltage and rms current.

**6-31**    For the AM transmitter of Problem 6-29, determine for both 50% and 100% modulation, (a) the total average power, (b) the peak envelope power, and (c) the sideband power.

**6-32**    For the AM transmitter of Problem 6-30, determine for both 50% and 100% modulation, (a) the total average power, (b) the peak envelope power, and (c) the sideband power.

**6-33**    For the transmitter of Problems 6-29 and 6-31, determine the antenna rms voltage and current for (a) 50% modulation and (b) 100% modulation.

**6-34**    For the transmitter of Problems 6-30 and 6-32, determine the antenna rms voltage and current for (a) 50% modulation and (b) 100% modulation.

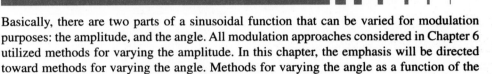

# Angle Modulation Methods

# 7

## OVERVIEW AND OBJECTIVES

Basically, there are two parts of a sinusoidal function that can be varied for modulation purposes: the amplitude, and the angle. All modulation approaches considered in Chapter 6 utilized methods for varying the amplitude. In this chapter, the emphasis will be directed toward methods for varying the angle. Methods for varying the angle as a function of the modulating signal are collectively referred to as *angle modulation*.

The primary methods that can be classified as angle modulation include phase modulation (PM) and frequency modulation (FM). Some AM waveforms can be identified by the presence of dominating features; this is usually not the case with angle modulation. As we will see later in the chapter, it is often impossible to tell whether a given angle-modulated waveform is a PM function or an FM function. In some cases, a signal is the combination of the two methods.

## Objectives

After completing this chapter, the reader should be able to:

1. State the general form of an *angle-modulated* signal.
2. Define *instantaneous signal phase* and *instantaneous signal frequency.*
3. State and apply the mathematical relationships between the two quantities mentioned in objective 2.
4. Discuss the differences between *phase modulation* and *frequency modulation.*
5. State the forms of a single-tone PM signal and a single-tone FM signal.
6. Define *frequency deviation* and *modulation index.*
7. Discuss the general nature of the spectrum of a tone-modulated FM signal.
8. State *Carson's rule,* and apply it to estimate the bandwidth required for FM or PM transmission.
9. Define *deviation ratio* and discuss its significance in estimating FM or PM bandwidth.
10. Define *frequency multiplication,* and compare its effects on modulated waveforms with those of frequency translation.
11. Discuss some of the circuit strategies used in generating FM and PM signals.
12. Discuss some of the circuit strategies used in detecting FM and PM signals.

## SystemVue™ Opening Application (Optional)

Insert the text CD in a computer having SystemVue™ installed and activate the program. Open the CD folder entitled **SystemVue Systems** and load the file entitled 7-1.

### Sink Tokens

| Number | Name | Token Monitored |
|--------|--------|-----------------|
| 0 | Input | 2 |
| 1 | Output | 3 |

### Operational Tokens

| Number | Name | Function |
|--------|------|----------|
| 2 | Input | provides input modulating signal at 1 Hz |
| 3 | FM Signal | generates FM signal centered at 10 Hz with modulation index of 5 |

The preset values for the run are from a starting time of 0 to a final time of 2 s, with a time step of 1 ms. Run the simulation and observe the waveforms at the two sinks. Locate a button on the **Tool Bar** with the title **Open All Windows** and left-click on it. Then locate another button on the **Tool Bar** with the title **Tile Horizontal** and left-click on that. These steps should align the two graph windows so that the display resembles that of a dual-trace oscilloscope. You may move them up or down with the mouse to change the vertical order if you desire.

### What You See

The 1-Hz sinusoidal source is frequency modulating the 10-Hz carrier of the FM generator. Carefully observe how the frequency of the FM generator increases as the modulating signal amplitude increases, and decreases as the modulating signal amplitude decreases. Note, however, that the peak amplitude of the modulated signal does not change. The modulation or "intelligence" is being transferred to the frequency of the FM signal.

### How This Demonstration Relates to the Chapter Learning Objectives

In Chapter 7, you will learn about the various angle modulation methods. The two that will be studied are *phase modulation* (PM) and *frequency modulation* (FM). In practice, FM usually has superior characteristics, but the interrelationships of the two forms necessitate the study of both. In fact, the waveforms are quite similar, and it is usually difficult to tell which form is present from a visual observation of a given modulated signal. At the end of the chapter, you will return to this system and study some properties with different types of waveforms.

# 7-1 Angle Modulation

Consider the basic sinusoidal function given by

$$v(t) = A \cos \omega_c t \qquad (7\text{-}1)$$

This is an "ordinary" sinusoidal function having a constant amplitude and a constant frequency. The subscript "c" for the frequency implies a carrier reference. For reasons that will be clear shortly, let us define an *instantaneous phase* $\theta_i(t)$ and an *instantaneous radian frequency* $\omega_i(t)$ as follows:

$$\theta_i(t) = \omega_c t \qquad (7\text{-}2)$$

$$\omega_i(t) = \frac{d\theta_i(t)}{dt} = \omega_c \qquad (7\text{-}3)$$

The question now is this: Do the preceding two definitions make sense? Certainly, the carrier phase progresses at a linear rate, and, at any time, is given by Equation 7-2. If we form the derivative of this carrier phase with respect to time, we obtain the carrier radian frequency. Therefore, we conclude that the definitions seem to make sense, at least as far as a constant-frequency sinusoid is concerned.

## Time-Varying Signal Phase Angle

Now that we have established certain concepts for the fixed carrier frequency portion of a sinusoid, we will leave it in its basic form and concentrate on the possibility of a modulated signal portion. Consider the function

$$v(t) = A \cos\left[\omega_c t + \theta_s(t)\right] \qquad (7\text{-}4)$$

We have a tendency to refer to this function as a sinusoid, and if $\theta_s(t)$ were a constant, the function would be a simple sinusoid nearly like Equation 7-1, but with an additional constant phase shift. However, because $\theta_s(t)$ is some arbitrary function of time, the complete function is no longer a simple sinusoid, even though it is expressed in terms of the sinusoidal function. An exaggerated form of an angle-modulated waveform is shown in Figure 7–1.

The function $\theta_s(t)$ will be called the *instantaneous signal phase function*. With angle modulation, the information or "intelligence" is somehow transferred to this function and the amplitude $A$ remains constant.

## Instantaneous Signal Frequency

The *instantaneous signal radian frequency* $\omega_s(t)$ is defined as the derivative of the instantaneous signal phase with respect to time. Thus,

$$\omega_s(t) = \frac{d\theta_s(t)}{dt} \qquad (7\text{-}5)$$

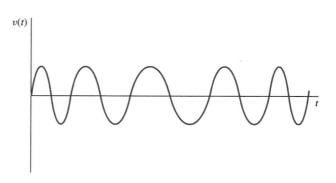

**FIGURE 7–1**

Illustration of an arbitrary angle-modulated "sinusoid."

It is also convenient to define an *instantaneous signal cyclic frequency* $f_s(t)$ as

$$f_s(t) = \frac{1}{2\pi}\omega_s(t) \tag{7-6}$$

If the instantaneous signal frequency is the derivative of the instantaneous signal phase, then it follows that the instantaneous signal phase is the integral of the instantaneous signal frequency. Thus,

$$\theta_s(t) = \int_0^t \omega_s(t)\,dt = 2\pi \int_0^t f_s(t)\,dt \tag{7-7}$$

where the framework of analysis is assumed to start at $t = 0$.

## Phase Modulation

Phase modulation (PM) is the process of creating an instantaneous signal phase function that is proportional to the modulating signal. Let $v_m(t)$ represent the modulating signal voltage. With phase modulation, the instantaneous signal phase is made to be proportional to the modulating signal:

$$\theta_s(t) = K_p v_m(t) \tag{7-8}$$

where $K_p$ is a modulation constant measured in radians/volt (rad/V).

The instantaneous signal frequency for a PM signal is obtained by differentiating Equation 7-8 with respect to time:

$$\omega_s(t) = \frac{d\theta_s(t)}{dt} = K_p \frac{dv_m(t)}{dt} \tag{7-9}$$

## Frequency Modulation

Frequency modulation (FM) is the process of creating an instantaneous signal frequency function that is proportional to the modulating signal. With FM, the strategy is

$$f_s(t) = K_f v_m(t) \tag{7-10}$$

where $K_f$ is a modulation constant measured in hertz/volt (Hz/V).

The result can also be expressed in terms of radian frequency as

$$\omega_s(t) = 2\pi K_f v_m(t) = 2\pi f_s(t) \tag{7-11}$$

In fact, a new constant equal to $2\pi K_f$ (and measured in radians per second/volt) could be defined if desired, but there is little reason to do so at this point.

The instantaneous signal phase for an FM signal is obtained by integrating Equation 7-11 with respect to time:

$$\theta_s(t) = \int_0^t \omega_s(t)\,dt = 2\pi \int_0^t K_f v_m(t)\,dt \tag{7-12}$$

## Brief Summary

It is likely that the reader is confused at this point, so a brief summary is in order.

1. With PM, the signal phase is directly proportional to the modulating signal.
2. With FM, the signal frequency is directly proportional to the modulating signal.
3. Since frequency is proportional to the derivative of phase with respect to time, the signal frequency for a PM signal is proportional to the derivative of the modulating signal.
4. Since phase is proportional to the integral of frequency, the signal phase for an FM signal is proportional to the integral of the modulating signal with respect to time.

Well, you are probably still confused, so hold on until we get all of these concepts fully explained. The example that follows will help to clarify them to some extent.

**▌▊ EXAMPLE 7-1**

A particular low-frequency function generator can be angle modulated with various waveforms. Based on a 100-Hz unmodulated sinusoid, the voltage with modulation applied can be expressed as

$$v(t) = 10\cos(200\pi t + 3\pi t^2) \tag{7-13}$$

(a) Is this signal PM or FM? Determine (b) instantaneous signal phase and (c) instantaneous signal frequency (radian and cyclic). (d) Sketch the forms of the instantaneous signal phase and frequency.

**SOLUTION**

(a) Since no statement is made concerning the nature of the modulating signal, it is impossible to state whether the signal is PM or FM without further information. All we can say is that it is angle modulated in some fashion.

(b) Note that the term $200\pi t$ within the argument represents the carrier phase, corresponding to a cyclic frequency of 100 Hz. The signal instantaneous phase is the additional term given by

$$\theta_s(t) = 3\pi t^2 \tag{7-14}$$

(c) The instantaneous radian signal frequency is

$$\omega_s(t) = \frac{d\theta_s(t)}{dt} = 6\pi t \tag{7-15}$$

The instantaneous cyclic signal frequency is

$$f_s(t) = \frac{1}{2\pi}\omega_s(t) = 3t \tag{7-16}$$

(d) Curves depicting the nature of the signal phase and frequency as functions are shown in Figure 7–2. If the modulating signal were a parabolic function, it would be appropriate to say that the given waveform is a PM signal, whereas if the modulating signal were a linear function, the waveform would be an FM signal.

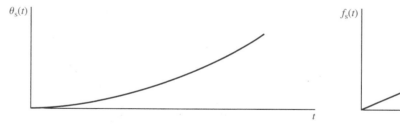
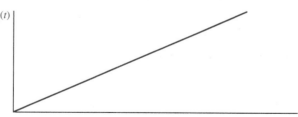

**FIGURE 7–2**
Curves for instantaneous signal phase and frequency in Example 7-1.    ▌▊

## 7-2 Single-Tone Angle Modulation

A complete analysis of an FM signal with an arbitrary modulating signal is somewhat difficult, especially in regard to the determination of the spectrum. Fortunately, operational bounds and the range of the spectrum may be determined from an analysis based on a single-tone modulating signal. The strategy is to determine the transmission bandwidth based on the assumption of a single tone having a frequency equal to the highest frequency in the modulating signal, which serves as a worst-case situation. Consequently, much attention will be paid from this point on to the determination of the signal properties based on a single sinusoidal modulating signal. As we will see, determination of the spectrum even for this simple modulating signal is not an easy chore.

In the steps that follow, a single sinusoid of the form $\cos \omega_m t$ will be assumed for both PM and FM. The level will be adjusted by a constant factor whose value will play a significant role in the analysis. A sine function could just as easily be assumed, and the resulting conclusions would be the same, but the signs of some of the terms would be different.

## Phase Modulation

Let $\Delta\theta$ represent the maximum phase deviation in radians from the carrier phase. For PM, the signal phase is directly proportional to the modulating signal, and is

$$\theta_s(t) = \Delta\theta \cos \omega_m t \tag{7-17}$$

The composite modulated signal is then given by

$$v(t) = A \cos(\omega_c t + \Delta\theta \cos \omega_m t) \tag{7-18}$$

Note that the maximum phase deviation is a function only of the phase deviation constant $\Delta\theta$ and is independent of the modulating frequency. This point is significant, and will be considered again later when comparing PM and FM.

The instantaneous signal frequency for a PM signal is obtained by applying Equation 7-5 to Equation 7-17. We have

$$\omega_s(t) = \frac{d\theta_s(t)}{dt} = -\omega_m \Delta\theta \sin \omega_m t \tag{7-19}$$

which indicates that the instantaneous frequency deviation for a PM signal is proportional to both the phase deviation and the modulating frequency. Thus, as the modulating frequency increases, the instantaneous frequency deviation increases.

## Frequency Modulation

Let $\Delta\omega$ represent the maximum frequency deviation (in radians per second) from the carrier frequency. For FM, the signal frequency is directly proportional to the modulating signal, and is

$$\omega_s(t) = \Delta\omega \cos \omega_m t \tag{7-20}$$

This function must be integrated according to Equation 7-7 to determine the signal phase, which appears in the composite signal. Hence,

$$\theta_s(t) = \int_0^t \omega_s(t)\,dt = \int_0^t \Delta\omega \cos \omega_m t\,dt = \frac{\Delta\omega}{\omega_m} \sin \omega_m t \tag{7-21}$$

This equation indicates that for FM, the instantaneous phase deviation is a function of both the frequency deviation and the modulating frequency. Specifically, the phase deviation increases linearly with the frequency deviation, but inversely with the modulating frequency. Thus, higher modulating frequencies result in smaller phase deviations.

The composite modulated signal is of the form

$$v(t) = A \cos\left(\omega_c t + \frac{\Delta\omega}{\omega_m} \sin \omega_m t\right) \tag{7-22}$$

## Modulation Index

An important parameter for FM is the *modulation index* $\beta$, which is defined as

$$\beta = \frac{\Delta\omega}{\omega_m} \tag{7-23}$$

The frequency deviation $\Delta\omega$ in radians per second can be expressed in terms of the frequency deviation $\Delta f$ in hertz by

$$\Delta\omega = 2\pi\Delta f \tag{7-24}$$

Likewise, the modulating frequency $\omega_m$ in radians per second can be expressed in terms of the modulating frequency in hertz by

$$\omega_m = 2\pi f_m \tag{7-25}$$

When these forms are substituted in Equation 7-23, the constant terms cancel, and the modulation index becomes

$$\beta = \frac{\Delta f}{f_m} \tag{7-26}$$

This is the most useful form, and it will be used in most subsequent work. Substitution in Equation 7-22 yields for the FM signal

$$v(t) = A\cos(\omega_c t + \beta \sin \omega_m t) \tag{7-27}$$

## Comparisons between PM and FM

For single-tone modulation, exaggerated forms for FM and PM are displayed in Figure 7–3. The frequency deviation $\Delta f$ is assumed to be positive, in which case the maximum instantaneous frequency occurs in FM on the positive peaks of the modulating signal, and the minimum frequency occurs on the negative peaks. There is a 90° shift of the peaks for PM. Without the presence of the modulating signal on the same time scale, it would not be possible to tell which is FM and which is PM.

Several important comparisons between PM and FM can be made from the equations developed in this section.

1. In PM, the *phase deviation* is a function only of the modulating signal amplitude and is independent of the frequency.
2. In FM, the *frequency deviation* is a function only of the modulating signal amplitude and is independent of frequency.
3. In PM, the *frequency deviation* increases linearly with both the modulating signal amplitude and the modulating frequency.
4. In FM, the *phase deviation* increases linearly with the modulating signal amplitude and inversely with the modulating frequency.

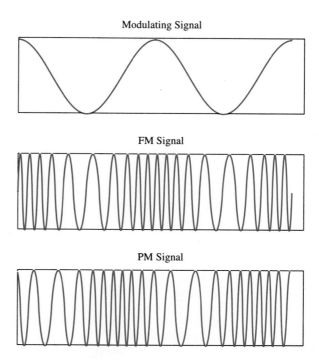

**FIGURE 7–3**
Sinusoidal modulating signal and the corresponding forms for FM and PM.

For any given modulating frequency, the peak phase deviation $\Delta\theta$ for PM has the same effect as the modulation index $\beta$ for FM. This similarity will be useful in dealing with bandwidth computations later.

In the preceding development, a cosine function was assumed for the modulating signal, which resulted in a cosine function within the argument for the PM signal and a sine function within the argument for the FM signal. However, this should not be interpreted as any type of general identification. We could just as easily have assumed a sine function for the modulating signal, which would have reversed the sine/cosine references in the PM and FM forms. Likewise, the carrier could have been assumed as a sine function. Therefore, the use of sine or cosine functions for carrier and modulation forms is arbitrary, but the choices that were made were based on desired mathematical forms.

## PM or FM Power

Although the angle-modulated waveform is not a pure sinusoid (due to the modulation within the angle), the power averaged over a period of the modulating signal turns out to be the same as that of a pure sinusoid. Thus, the average power for either a PM signal of the form of Equation 7-18 or an FM signal of the form of Equation 7-27 is given by

$$P = \frac{A^2}{2R} \tag{7-28}$$

---

**■ EXAMPLE 7-2**

A single-tone angle-modulated signal is given by

$$v(t) = 80\cos[2\pi \times 10^8 t + 20\sin(2\pi \times 10^3 t)] \tag{7-29}$$

Assume in this example that the signal is an FM signal. Determine the (a) unmodulated carrier frequency, (b) modulating frequency, (c) modulation index, and (d) maximum frequency deviation. Express all frequencies in hertz. (e) Determine the average power dissipated in a 50-$\Omega$ resistor.

**SOLUTION**  Much of this analysis involves comparing the given function with the general form of Equation 7-27. The only difference in the way they are expressed is that the $2\pi$ factor is provided for both the carrier and modulating frequencies in Equation 7-29.

(a)  The carrier cyclic frequency is identified as

$$f_c = 10^8 \text{ Hz} = 100 \text{ MHz} \tag{7-30}$$

(b)  The modulating frequency is identified as

$$f_m = 10^3 \text{ Hz} = 1 \text{ kHz} \tag{7-31}$$

(c)  The modulation index is identified as

$$\beta = 20 \tag{7-32}$$

(d)  Since $\beta = \Delta f / f_m$, and since both $\beta$ and $f_m$ are known, the frequency deviation is determined as

$$\Delta f = \beta f_m = 20 \times 1 \text{ kHz} = 20 \text{ kHz} \tag{7-33}$$

(e)  The average power is

$$P = \frac{A^2}{2R} = \frac{80^2}{2 \times 50} = 64 \text{ W} \tag{7-34}$$

---

**■ EXAMPLE 7-3**

Assume that the function in Example 7-2 is a PM signal. Determine the maximum phase deviation.

**SOLUTION**  The quantity that was identified as $\beta$ for FM becomes $\Delta\theta$ for PM. Hence, by a simple comparison,

$$\Delta\theta = 20 \text{ rad} \tag{7-35}$$

**EXAMPLE 7-4**

Let the modulating frequency in the function of Example 7-2 be changed to 2 kHz, and the signal remain an FM signal with the same frequency deviation. Write an equation for the signal.

**SOLUTION**   Since $\beta = \Delta f/f_m$ and $\Delta f$ remains constant, the modulation index is halved and becomes 10. The modulating frequency in the argument is also doubled, and the result is

$$v(t) = 80\cos[2\pi \times 10^8 t + 10\sin(4\pi \times 10^3 t)] \tag{7-36}$$

**EXAMPLE 7-5**

Let the modulating frequency in the function of Example 7-2 be changed to 2 kHz, and the signal be assumed to be PM. Write an equation for the signal.

**SOLUTION**   In this case, the maximum phase deviation of 20 rad remains the same, and only the modulating signal frequency changes. Hence,

$$v(t) = 80\cos[2\pi \times 10^8 t + 20\sin(4\pi \times 10^3 t)] \tag{7-37}$$

**EXAMPLE 7-6**

A 100-MHz carrier is to be frequency modulated by a 4-kHz tone. The maximum frequency deviation is to be adjusted to 12 kHz. Write an expression for the composite FM signal, assuming cosine functions for the modulating signal and the carrier.

**SOLUTION**   Reviewing the form of the tone-modulated FM signal, it is observed that a cosine modulating signal results in a sine term in the argument due to the integration of frequency to obtain phase. The modulation index is $\Delta f/f_m = 12\text{ kHz}/4\text{ kHz} = 3$. Assuming an arbitrary amplitude $A$ for the signal, we have

$$v(t) = A\cos[2\pi \times 10^8 t + 3\sin(8\pi \times 10^3 t)] \tag{7-38}$$

**EXAMPLE 7-7**

A 100-MHz carrier is to be phase modulated by a 5-kHz tone. The maximum phase deviation is to be 6 rad. Write an expression for the composite PM signal assuming cosine functions for the modulating signal and the carrier.   Δθ is the phase deviation

**SOLUTION**   Reviewing the form for a PM signal, it is observed that a cosine modulating signal results in a cosine term in the argument since the signal phase function is directly proportional to the modulating signal. The phase deviation constant is $\Delta\theta = 6$ rad and the composite function is

$$v(t) = A\cos[2\pi \times 10^8 t + 6\cos(10\pi \times 10^3 t)] \tag{7-39}$$

## 7-3   Spectrum of Tone-Modulated FM Signal

Thus far, angle modulation has been considered from a somewhat general point of view, and equal emphasis has been given to both PM and FM. In practice, FM has some advantages that make it more desirable than PM for most (but not all) applications. From this point forward, we will focus primarily on FM, but where necessary, the results can be adapted to PM.

The primary objective in this section is to investigate the form of the spectrum of a tone-modulated FM signal, and to use this result as a basis for determining the bandwidth requirements for complex modulating signals. With the various forms of AM, it was possible in most cases to derive complete expressions for the spectra even when the modulating signals were complex. Unfortunately, this is not the case with FM. It is possible to derive meaningful closed-form expressions for FM spectra only for a few special cases. Moreover, the bandwidth requirements for most useful FM systems are far greater than for AM. Therefore, much emphasis has been directed in FM analysis to the problem of estimating the transmission bandwidth for complex waveforms, by using approximate bounds determined from assuming sinusoidal modulating signals. Fortunately, this approach works quite well.

The analysis begins with the form of a single-tone modulated FM signal given by Equation 7-27, and repeated here for convenience:

$$v(t) = A \cos(\omega_c t + \beta \sin \omega_m t) \tag{7-40}$$

This function is periodic and, in theory, one could utilize the basic Fourier series equations to determine the one-sided spectral coefficients. In practice, this is very difficult, and the Fourier expansion is much easier to determine using an indirect approach.

The first step is to expand Equation 7-40 using a basic trigonometric identity for the cosine of the sum of two angles. The identity reads

$$\cos(x + y) = \cos x \cos y - \sin x \sin y \tag{7-41}$$

Application of this identity to Equation 7-40 yields

$$v(t) = A \cos \omega_c t \cos(\beta \sin \omega_m t) - A \sin \omega_c t \sin(\beta \sin \omega_m t) \tag{7-42}$$

The next step involves using appropriate closed-form expansions appearing in advanced applied mathematics texts to simplify the sinusoidal terms with sinusoidal arguments. These expansions make use of Bessel functions of the first kind, whose values are tabulated in mathematical handbooks. The specific expansions involved are the following:

$$\cos(\beta \sin \omega_m t) = J_0(\beta) + \sum_{\substack{n=2 \\ n \text{ even}}}^{\infty} 2J_n(\beta) \cos n\omega_m t \tag{7-43}$$

$$\sin(\beta \sin \omega_m t) = \sum_{\substack{n=1 \\ n \text{ odd}}}^{\infty} 2J_n(\beta) \sin n\omega_m t \tag{7-44}$$

The function $J_n(\beta)$ represents a Bessel function of the first kind, of order $n$ and argument $\beta$. For a particular value of $n$ and a particular value of $\beta$, the quantity $J_n(\beta)$ is a single value, so the presence of the Bessel functions should not obscure the trigonometric forms of the preceding expansions.

When the expressions of Equations 7-43 and 7-44 are substituted in Equation 7-42, two series involving sinusoidal products are obtained. In each series, the terms will involve the product of a sinusoid at the carrier frequency with one at the modulating frequency. The identities provided at the beginning of Chapter 6 may be used to simplify the expansions. The reader is invited to show that the result obtained after simplification is

$$
\begin{aligned}
\frac{v(t)}{A} =\ & J_0(\beta) \cos \omega_c t \\
& + J_1(\beta) \cos(\omega_c + \omega_m)t - J_1(\beta) \cos(\omega_c - \omega_m)t \\
& + J_2(\beta) \cos(\omega_c + 2\omega_m)t + J_2(\beta) \cos(\omega_c - 2\omega_m)t \\
& + J_3(\beta) \cos(\omega_c + 3\omega_m)t - J_3(\beta) \cos(\omega_c - 3\omega_m)t \\
& + \cdots
\end{aligned}
\tag{7-45}
$$

The form of Equation 7-45 suggests that there are an infinite number of components in the spectrum. However, it will be shown later that terms beyond a certain point in the spectrum are negligible in any practical situation.

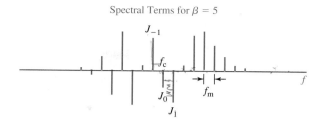

**FIGURE 7–4**

Representative spectrum of tone-modulated FM signal, exhibiting signs of components.

## Spectral Display

The form of the spectrum as deduced from Equation 7-45 is interesting, and it leads to some important properties. A component having the carrier frequency $f_c$ appears at the center of the spectrum. All other terms are collectively referred to as sidebands and, as previously noted, there are theoretically an infinite number of such components. Spectral frequencies greater than $f_c$ represent the upper sideband components, and frequencies less than $f_c$ represent the lower sideband components.

All spectral components are displaced apart by a fixed frequency increment equal to the modulating frequency $f_m$. The *magnitudes* of components an equal frequency increment on either side of the carrier are equal; that is, the magnitudes of spectral components have even symmetry about $f_c$. However, the signs of components an odd integer multiple of $f_m$ on one side of the carrier are the opposite of the corresponding components on the other side.

A representative spectrum is shown in Figure 7–4. The components having negative signs are actually inverted in this figure to reinforce the preceding discussion. In subsequent spectral diagrams, only the magnitudes will be shown.

## Instantaneous Frequency versus Spectral Frequency

A point that may be puzzling to the reader is the fact that the Fourier expansion of the FM signal results in a series of *fixed-frequency* sinusoids. This seems to contradict the basic assumption that the FM signal has a frequency that is *varying* with time. This paradox is resolved by recognizing that two separate concepts of the term *frequency* are at work here. The concept of *instantaneous frequency* refers to a dynamic parameter relating the variation of an angular function with respect to a modulating frequency. However, the concept of *spectral frequency* permits us to determine those spectral components that constitute the signal in a Fourier sense, and each of the components is a steady-state sinusoid with a fixed frequency. If all the fixed-frequency components were added together in the form described by the Fourier expansion, the result would be the FM signal, in which the instantaneous frequency varied with time.

## Bandwidth

An important question for consideration at this point is that of determining the approximate transmission bandwidth required for an FM signal. Expressed differently: How many terms in the expansion of Equation 7-45 are needed for reproduction, such that the resulting signal will have negligible distortion?

Most estimates for FM bandwidth have been developed from studies of the Bessel sideband components as a function of modulation index. The graphical forms of the Bessel functions of several orders as a function of the modulation index are illustrated in Figure 7–5. Since only limited accuracy can be obtained from the curves, several tabulated values are given in Table 7–1. Only values greater than 1% of the carrier level are shown, so it is possible to determine the number of sideband components in each case that have magnitudes greater than the 1% level. Incidentally, this particular choice is one that has been used as a criterion for bandwidth, but it is not as easy to work with as the ones that will be provided in this text.

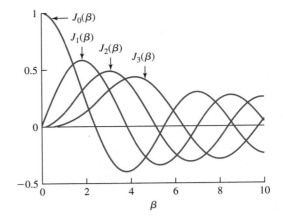

**FIGURE 7–5**
The Bessel functions $J_n(\beta)$ for several different values of $n$.

**Table 7–1**    Selected Values of the Bessel Functions*

| | $J_n(\beta)$ | | | | | |
|---|---|---|---|---|---|---|
| $n$ | $\beta = 0.1$ | $\beta = 0.5$ | $\beta = 1$ | $\beta = 2$ | $\beta = 5$ | $\beta = 10$ |
| 0 | 0.9975 | 0.9385 | 0.7652 | 0.2239 | −0.1776 | −0.2459 |
| 1 | 0.0499 | 0.2423 | 0.4401 | 0.5767 | −0.3276 | 0.04347 |
| 2 | | 0.03125 | 0.1149 | 0.3528 | 0.04657 | 0.2546 |
| 3 | | | 0.01956 | 0.1289 | 0.3648 | 0.05838 |
| 4 | | | | 0.03400 | 0.3912 | −0.2196 |
| 5 | | | | | 0.2611 | −0.2341 |
| 6 | | | | | 0.1310 | −0.01446 |
| 7 | | | | | 0.05338 | 0.2167 |
| 8 | | | | | 0.01841 | 0.3179 |
| 9 | | | | | | 0.2919 |
| 10 | | | | | | 0.2075 |
| 11 | | | | | | 0.1231 |
| 12 | | | | | | 0.06337 |
| 13 | | | | | | 0.02897 |
| 14 | | | | | | 0.01196 |

*For a given value of $\beta$, only components greater than 1% of the peak carrier level are shown.

Prior to developing a general rule for arbitrary values of modulation index, two limiting cases will be considered, due to their simplicity. They will be referred to as *narrowband frequency modulation (NBFM)* and *very wideband frequency modulation (VWBFM)*. The actual ranges of modulation index included in these definitions are somewhat arbitrary, but to simplify the process for our purposes, ranges have been assumed in the definitions that follow.

### Narrowband FM: $\beta \leq 0.25$

In this region of operation, the frequency deviation is much smaller than the modulating frequency. As it turns out, all sidebands are negligible except the ones adjacent to each side of the carrier. Therefore, the transmission bandwidth in this case is the same as for DSB and conventional AM, and is

$$B_T = 2f_m \tag{7-46}$$

This case corresponds to the smallest possible bandwidth for FM, and the various noise-reducing properties of FM do not apply to this limiting case. However, narrowband FM can be used as the first stage of a wider bandwidth system, as will be seen later.

### Very Wideband FM: $\beta \geq 50$

In this region of operation, virtually all the spectral energy lies in the frequency range over which the deviation occurs. Therefore, the transmission bandwidth in this case can be assumed to be

$$B_T = 2\Delta f \tag{7-47}$$

### Wideband FM

Values of $\beta$ between these extreme limits will be referred to simply as *wideband frequency modulation* (*WBFM*). Most practical applications of FM generally fall within the range of WBFM. The NBFM and very wideband FM (VWBFM) cases represent extreme cases for the purpose of bandwidth estimation.

---

**▌ EXAMPLE 7-8**

An FM tone-modulated signal source has a center frequency of 100 kHz, a modulating frequency of 2 kHz, and a frequency deviation of 400 Hz. Determine the transmission bandwidth.

**SOLUTION** The modulation index should be calculated first, in order to determine whether the signal is narrowband or wideband. The value is

$\beta$ = modulation Index

$$\beta = \frac{\Delta f}{f_m} = \frac{400 \text{ Hz}}{2000 \text{ Hz}} = 0.2 \tag{7-48}$$

This falls in the category of NBFM, and the bandwidth is

$$B_T = 2f_m = 2 \times 2 \text{ kHz} = 4 \text{ kHz} \tag{7-49}$$

Two points should be noted: (1) Even though the frequency swings only out to 400 Hz above and below the center frequency, the required transmission bandwidth is much greater. This illustrates the fact that spectral energy can exist far outside of the range of the frequency swing. (2) The carrier frequency does not enter into the calculation of the transmission bandwidth. However, the carrier frequency will be at the center of the composite spectrum. Thus, the spectrum of the composite signal lies between 98 and 102 kHz.

---

**▌ EXAMPLE 7-9**

An FM tone-modulated signal has a center frequency of 250 kHz, a modulating frequency of 100 Hz, and a frequency deviation of 8 kHz. Determine the transmission bandwidth.

**SOLUTION** The modulation index is

$$\beta = \frac{\Delta f}{f_m} = \frac{8000 \text{ Hz}}{100 \text{ Hz}} = 80 \tag{7-50}$$

This falls in the category of VWBFM, and the bandwidth is

$$B_T = 2\Delta f = 2 \times 8 \text{ kHz} = 16 \text{ kHz} \tag{7-51}$$

Unlike the previous example, essentially all the spectral energy in this case falls within the bounds of the frequency deviation. Thus, the bandwidth of the composite signal ranges from 242 kHz to 258 kHz. ▐

---

## 7-4 Carson's Rule for Estimating FM and PM Bandwidth

In the preceding section, simple formulas for estimating the transmission bandwidth for the limiting cases of NBFM ($\beta \leq 0.25$) and VWBFM ($\beta \geq 50$) were given. Most FM systems operate somewhere between these limits (WBFM), and an approximation is desired for that situation.

While there are several approaches provided in the literature, probably the simplest and easiest to apply is known as *Carson's rule*. It is based on the criterion that the number of sidebands selected should be the minimum number that will result in transmission of no less than 98% of the total power. This choice may seem arbitrary, but it results in a rather simple bandwidth estimation formula. In practice, the rule has proved to be quite adequate for initial estimates, and it can be modified if necessary.

The basis for Carson's rule is that if $\beta$ assumes integer values, it can be verified by computation that the number of sidebands on either side of the carrier required for 98% of the power is always $\beta + 1$. For example, if $\beta = 3$, at least 98% of the power is transmitted if 4 sidebands on either side of the carrier are selected, resulting in the need to transmit 8 sidebands (plus the carrier). The formula is then extended to noninteger values of $\beta$, and Carson's rule for transmission bandwidth can be stated as

$$B_T = 2(1 + \beta) f_m \tag{7-52}$$

where the center of the bandwidth is assumed to be at the carrier frequency.

An alternative (and equally useful) form of Carson's rule is obtained by expanding Equation 7-52 and recognizing that $\beta f_m = \Delta f$. This form reads

$$B_T = 2(\Delta f + f_m) \tag{7-53}$$

Both forms will be used freely throughout the book.

## Limiting Cases

In order to tie together the limiting cases of the preceding section with Carson's rule, let's see what happens with the latter when the modulation index is either very small or very large. For NBFM, $\beta << 1$, and Equation 7-52 can be approximated as

$$B_T \approx 2 f_m \quad \text{for NBFM} \tag{7-54}$$

For VWBFM, $\beta >> 1$, and Equation 7-52 can be approximated as

$$B_T \approx 2\beta f_m = 2\Delta f \quad \text{for VWBFM} \tag{7-55}$$

Thus, Carson's rule does reduce approximately to the limiting cases in the preceding section for either NBFM or VWBFM. One could, of course, use Carson's rule for all cases, and the results would still be reasonable, considering that all cases are approximations. However, within the text, we will use the simpler approximations of the last section for the limiting cases of NBFM and VWBFM.

## Some Spectral Comparisons

The variation of the spectra with different modulation parameters will be illustrated with some spectral plots. Consider the magnitude plots shown in Figure 7–6 in the vicinity of $f_c$. The modulating frequency is fixed at $f_m = 1$ kHz in all these plots, so the spacing between components is fixed at 1 kHz for all cases. In Figure 7–6(a), $\Delta f = 100$ Hz and $\beta = 100/1000 = 0.1$, so the result is NBFM. Only two sidebands are significant and $B_T = 2$ kHz. In Figure 7–6(b), $\Delta f = 1$ kHz and $\beta = 1000/1000 = 1$. Two sideband components on either side of the carrier are significant and $B_T = 4$ kHz. In Figure 7–6(c), $\Delta f = 10$ kHz and $\beta = 10/1 = 10$. In this case, it is assumed that there are $10 + 1 = 11$ significant components on either side of the carrier, and $B_T = 22$ kHz. Although not shown, if $\Delta f$ were increased to 100 kHz, in which case $\beta = 100/1 = 100$, the bandwidth would increase to approximately $B_T = 200$ kHz.

The plots shown in Figure 7–7 illustrate the effect of holding $\Delta f$ constant and varying $f_m$. In all these plots, the frequency deviation is fixed at $\Delta f = 10$ kHz. In Figure 7–7(a), the modulating frequency is $f_m = 40$ kHz, and $\beta = 10/40 = 0.25$. The result is NBFM and $B_T = 80$ kHz. In Figure 7–7(b), the modulating frequency is reduced to $f_m = 10$ kHz and $\beta = 10/10 = 1$. The bandwidth is $B_T = 40$ kHz. In Figure 7–7(c), the modulating frequency is assumed to be reduced all the way down to 100 Hz and $\beta = 10,000/100 = 100$. The bandwidth is $B_T = 20$ kHz, and all significant components lie in the range governed

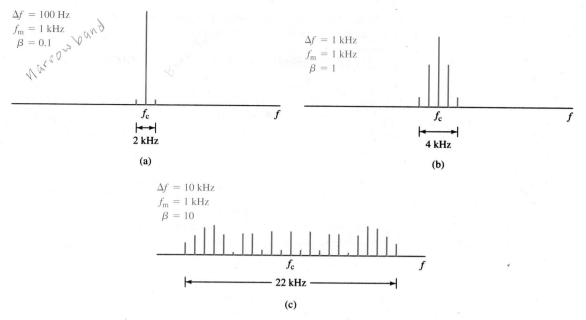

**FIGURE 7–6**

Magnitude spectra of tone-modulated FM signal for different values of frequency deviation, with the modulating frequency varied.

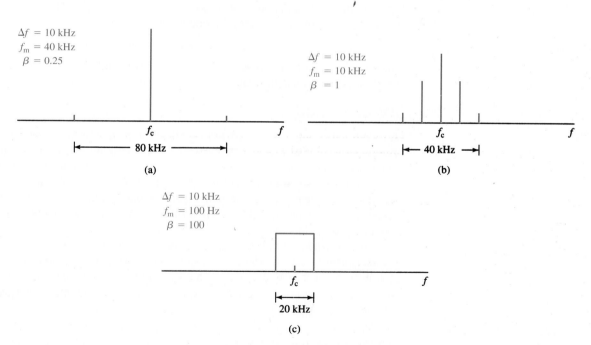

*Note:* This Block Shows Only the Frequency Limits of the 201 Significant Components.

**FIGURE 7–7**

Magnitude spectra of tone-modulated FM signal for different values of modulating frequency, with the frequency deviation fixed.

by the $\pm 10$-kHz frequency swing. In view of the large number of components in the spectrum, a rectangular block has been shown on the figure to define the proper range.

## Arbitrary Modulating Signal

The general estimates that have been developed were based on single-tone modulation, so a fundamental question is how to determine the bandwidth required for arbitrary modulating signals. The FM spectra resulting from most modulating signals are very difficult to determine analytically. Fortunately, the results for the tone-modulated signal can be used to estimate the bandwidth for more complex signals.

The strategy involved is to select the set of conditions for the complex modulating signal that would produce the largest transmission bandwidth when viewed from a single-tone modulation basis. Since it is assumed that we are dealing with the basic form of FM at this time, the maximum frequency deviation $\Delta f$ does not vary with modulating frequency. With this assumption, the highest modulating frequency determines the bandwidth. (This point will be demonstrated in Example 7-12.)

## Deviation Ratio

A particular parameter of interest when dealing with complex baseband modulating signals is the *deviation ratio D*. This quantity is defined as

$$D = \text{minimum value of } \beta = \frac{\Delta f}{W} \qquad (7\text{-}56)$$

where $W$ represents the highest possible modulating frequency. (It is also the baseband bandwidth.) If the deviation ratio and the highest modulating frequency are specified, the appropriate bandwidth may be determined from Carson's rule. Thus, Carson's rule may be restated for an arbitrary baseband modulating signal as

$$B_T = 2(1 + D)W = 2(\Delta f + W) \qquad (7\text{-}57)$$

Observe that the minimum value of $\beta$ determines the bandwidth for the case when $\Delta f$ is fixed (pure FM), since this value occurs at the highest frequency.

## PM Bandwidth

A final point of interest concerns the application of the preceding results to a PM signal. The development of the Fourier expansion for a tone-modulated PM signal follows the same general approach employed with the FM signal. The results differ only in minor sign and phase details and in the fact that $\beta$ is replaced by $\Delta\theta$ in the argument of the Bessel function coefficients. Thus, the magnitudes of the spectral coefficients are of the forms $J_n(\Delta\theta)$. This means that all bandwidth rules previously stated apply to PM, but with the role of $\beta$ changed to $\Delta\theta$.

For PM with a complex modulating signal, it should be stressed that $\Delta\theta$, unlike $\beta$, remains constant with modulating frequency. Thus, the parameter $D$ has no significance for PM. Instead, the maximum bandwidth is determined from the specified $\Delta\theta$ and the highest modulation frequency $W$. Carson's rule, or one of the simpler approximations from the last section, is then applied.

---

**▌▌ EXAMPLE 7-10**

A baseband signal having components from near dc to 2 kHz is used to frequency modulate a high-frequency carrier, and the maximum frequency deviation is set at $\pm 6$ kHz. Determine the approximate transmission bandwidth.

**SOLUTION**   First, the deviation ratio is calculated from the maximum frequency deviation and the highest modulating frequency:

$$D = \frac{\Delta f}{W} = \frac{6 \text{ kHz}}{2 \text{ kHz}} = 3 \qquad (7\text{-}58)$$

This value falls in the range in which Carson's rule is applicable. Hence,

$$B_T = 2(\Delta f + W) = 2(6 + 2) = 16 \text{ kHz} \qquad (7\text{-}59)$$

For conventional AM, the required transmission bandwidth would only be 4 kHz, so it is obvious that FM can require considerably more bandwidth. As will be seen later in the text, however, a considerable improvement of the signal-to-noise ratio can result from the FM process.

---

**▌ EXAMPLE 7-11**

The baseband signal of Example 7-10 is used to phase modulate a high-frequency carrier, and the maximum phase deviation is set at $\Delta\theta = \pm3$ rad. Determine the approximate transmission bandwidth.

**SOLUTION**   Carson's rule applied to the PM signal reads

$$B_T = 2(1 + \Delta\theta)W = 2(1 + 3)2 = 16 \text{ kHz} \qquad (7\text{-}60)$$

This problem was "rigged" to give the same answer as in the preceding example, in order to make a point. The value of the maximum phase deviation in this example is the same as the value of the deviation ratio in the preceding example. Therefore, for the same modulating signal, the bandwidth is the same. Of course, the modulation processes are different. In Example 7-10, the frequency deviation is independent of the modulating frequency, while in Example 7-11, the frequency deviation increases linearly with the modulating frequency.

---

**▌ EXAMPLE 7-12**

The commercial FM broadcasting system in the United States employs the following parameters for monaural transmission, as specified by the Federal Communications Commission: maximum frequency deviation $= \pm75$ kHz, and highest modulating frequency $= 15$ kHz. (a) For single-tone FM and a fixed deviation of $\pm75$ kHz , determine the approximate transmission bandwidth for each of the following modulating frequencies: 25 Hz, 75 Hz, 750 Hz, 1.5 kHz, 5 kHz, 10 kHz, 15 kHz. (b) Determine the approximate transmission bandwidth required when the FM transmitter is modulated by a complex audio signal whose highest frequency is 15 kHz.

**SOLUTION**

(a) For each of the specific modulating frequencies, the value of $\beta = \Delta f/f_m$ should be determined, and the bandwidth may be calculated either from Carson's rule or from one of the earlier limiting formulas. The results are provided in Table 7–2, and some of the trends will be briefly discussed.

For relatively low modulating frequencies, the values of $\beta$ are quite large and fall into the upper range of VWBFM. Thus, for the first three entries in the table, the bandwidth is $B_T = 2\Delta f = 150$ kHz. Eventually, $\beta$ drops to the range in which $\beta + 1$ sidebands on either side of the carrier are considered significant, and the remaining results were determined on that basis. Note that none of the results falls in the NBFM range.

(b) It is readily observed from Table 7–2 that the highest modulating frequency results in the largest bandwidth, as expected. Note that the deviation ratio is

$$D = \Delta f/W = 75 \text{ kHz}/15 \text{ kHz} = 5 \qquad (7\text{-}61)$$

This value of $\beta$ results in the largest bandwidth. The transmission bandwidth for the composite modulated signal is approximately

$$B_T = 180 \text{ kHz} \qquad (7\text{-}62)$$

Commercial FM stations are actually spaced 200 kHz apart, which provides some additional guard band and a transition region between channels. In addition, two stations in the same local area are not normally assigned to adjacent channels, so this serves as an additional safeguard.

**Table 7–2**  Data for Example 7-12

| $f_m$ | $\beta$ | $B_T$, kHz |
|-------|---------|------------|
| 25 Hz | 3000 | 150 |
| 75 Hz | 1000 | 150 |
| 750 Hz | 100 | 150 |
| 1.5 kHz | 50 | 153 |
| 5 kHz | 15 | 160 |
| 10 kHz | 7.5 | 170 |
| 15 kHz | $D = 5$ | 180 |

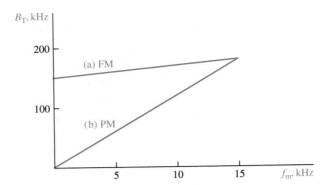

**FIGURE 7–8**
Transmission bandwidth as a function of modulating frequency, using conditions given in Example 7-12.

In actual practice, commercial FM stations employ preemphasis to the modulating signal to improve the signal-to-noise ratio. The result is that $\Delta f$ does not remain constant over the entire modulating frequency range as assumed here, and so commercial FM is not "pure FM" in the strictest sense. More will be said about this later, but the results of this problem would be directly applicable to commercial monaural FM without preemphasis, and the results with preemphasis do not differ significantly from those obtained here.

The results of Table 7–2 are shown graphically as curve (a) in Figure 7–8. An interesting observation is that the bandwidth does not vary widely with modulating frequency. As the modulating frequency varies over the very wide range of 25 Hz to 15 kHz, the transmission bandwidth varies only over the range of 150 kHz to 180 kHz.

**▌ EXAMPLE 7-13**

As a problem of hypothetical interest, suppose that commercial monaural FM were changed to pure PM. Assume that the maximum deviation of ±75 kHz and the highest modulating frequency of 15 kHz still applied. Determine the bandwidth as a function of the modulating frequency.

**SOLUTION**   From a review of Equation 7-19 and subsequent discussions, it can be deduced that the maximum frequency deviation for tone-modulated PM is a linear function of the modulating frequency. Since the maximum frequency deviation is ±75 kHz and the highest modulating frequency is 15 kHz, then $\Delta f/f_m = 75$ kHz$/15$ kHz $= 5$. Thus, $\Delta\theta$ in this example is the same as $D$ in the previous example. However, in this example, $\Delta\theta$ will remain at a fixed value of 5, while $\Delta f$ will change with frequency.

Since $\Delta\theta$ remains at 5, Carson's rule predicts that there will be 6 significant sidebands on either side of the carrier. The bandwidth is then

$$B_T \approx 12 f_m \qquad (7\text{-}63)$$

The result shows that the bandwidth for PM is *directly proportional to the modulating frequency*. The results for PM are shown as curve (b) of Figure 7–8. Because of the manner in which the two examples were linked together, the curves coincide at the highest modulating frequency.

From a comparison of the curves for FM and PM, it can be concluded that the variation of bandwidth with modulating frequency is far more dramatic for PM than for FM.

**▌**

## 7-5   Frequency Multiplication

Much of Chapter 5 was devoted to the concept of *frequency translation* or *shifting*, in which signals could be moved around from one frequency range to another without altering the basic spectral forms. As a general rule, frequency translation is applicable to all types of signals, both analog and digital, and is widely used in both transmission and reception.

There is another way to modify the operating frequency range by an integer multiple, which may be applied to certain types of signals, particularly those that are angle modulated.

This process is called *frequency multiplication,* and it may be used to advantage in angle-modulated signals.

The basis for frequency multiplication is that of a nonlinear circuit element. Circuits having nonlinear input–output characteristics generate various harmonic components of the input signal, with the amplitudes of the harmonic components being a function of the type of nonlinearity. Frequency multiplication is obtained by applying the input signal to a nonlinear circuit that has a strong component at the particular harmonic frequency desired. A filter is then used to pass the particular harmonic and to reject all other components.

## Frequency Doubler

Although there are many types of nonlinear circuits that will generate various types of harmonics, we will assume for the purpose of illustration an ideal square-law characteristic. The results should provide enough insight to generalize the concept without attempting a more formidable (and unwieldy) derivation. Assume an angle-modulated signal of the form

$$v_1(t) = A \cos [\omega_{c1} t + \theta_1(t)] \tag{7-64}$$

where $\omega_{c1}$ is the radian carrier frequency and $\theta_1(t)$ is the information. This could be either an FM or a PM signal, depending on the manner in which the instantaneous signal phase is established.

Refer to the block diagram shown in Figure 7–9. The output voltage of the square-law block on the left is assumed to be related to the input voltage by

$$v_2' = K_1 v_1^2 \tag{7-65}$$

where $K_1$ is a constant for the particular circuit.

When the voltage of Equation 7-64 is substituted in Equation 7-65, we obtain

$$v_2' = K_1 A^2 \cos^2 [\omega_{c1} t + \theta_1(t)] \tag{7-66}$$

By a basic trigonometric identity, this result can be expanded to

$$v_2' = \frac{K_1 A^2}{2} + \frac{K_1 A^2}{2} \cos [2\omega_{c1} t + 2\theta_1(t)] \tag{7-67}$$

The first term in Equation 7-67 is a dc term, and is easily eliminated by the filter in Figure 7–9. The multiplier of the second term is the same constant value, and can be simplified as

$$K = \frac{K_1 A^2}{2} \tag{7-68}$$

The output of the filter can then be written as

$$v_2 = K \cos [2\omega_{c1} t + 2\theta_1(t)] \tag{7-69}$$

This result may be expressed as

$$v_2 = K \cos [\omega_{c2} t + \theta_2(t)] \tag{7-70}$$

where $\omega_{c2}$ is the output carrier or center frequency and $\theta_2(t)$ is the output phase intelligence. By comparing Equations 7-69 and 7-70, the following deductions are made:

$$\omega_{c2} = 2\omega_{c1} \tag{7-71}$$

and

$$\theta_2(t) = 2\theta_1(t) \tag{7-72}$$

Both the center frequency and the signal phase function are doubled. However, the *form* of the signal phase function is not affected, which means that the resulting spectrum of the

**FIGURE 7–9**

Frequency doubler using a square-law characteristic.

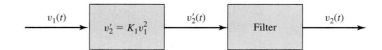

phase function remains the same. Therefore, the effective maximum phase deviation and the resulting maximum frequency deviation are *doubled*. This concept may be used to advantage for increasing the center frequency and the phase or frequency deviation of an angle-modulated signal.

## Effect on Amplitude-Modulated Signals

Unfortunately, the process of frequency multiplication cannot be applied to analog AM systems without introducing serious distortion. The reason is that the amplitude factor $A$ is squared. For angle modulation, this is not a problem, since it is a constant and the square of a constant is also a constant. For AM, however, the intelligence is contained in the amplitude, which is a time-varying function, and its square is a totally different function. For certain types of digital amplitude data, where it is only necessary to recognize the presence of a finite number of levels, squaring of the amplitude may be acceptable in some cases.

## Higher-Order Multiplication

The preceding development has been based on a frequency doubler, and it was shown that a square-law characteristic would be the ideal form for that purpose. For the moment, we will revert to basic mathematical notation for convenience. Assume an input variable $x$ and an output variable $y$. A general input–output nonlinear characteristic can be described by an equation of the form

$$y = a_0 + a_1 x + a_2 x^2 + a_3 x^3 + a_4 x^4 + \cdots \qquad (7\text{-}73)$$

The first term $a_0$ is a dc term, and the second term $a_1$ is the linear term. An ideal amplifier would have only the first two terms, and no new frequencies would be introduced. The square-law characteristic discussed earlier had only a second-degree term, and we saw that the frequency was doubled.

Extrapolating the concept to higher-order terms, it can be shown that to triple the frequency, the term $a_3$ is necessary (i.e., a cubic characteristic is required); to quadruple the frequency, the term $a_4$ is required, and so on. (These higher-order terms also introduce more components at lower frequencies as well.) The higher-order terms are generally much smaller than the lower-order terms, so most frequency multipliers of this type are limited in practice to doublers, triplers, and quadruplers. Higher-order multiplication can be achieved by cascading lower-order multipliers, and phase-locked loops can be used as discussed in Chapter 5.

## Summary

The effect of frequency multiplication by a factor $N$ on an angle-modulated signal can be summarized as follows: (1) The center frequency is multiplied by $N$. (2) The frequency deviation is multiplied by $N$. (3) The modulation index, deviation ratio, and phase deviation are multiplied by $N$.

---

**❚❚ EXAMPLE 7-14**

A baseband signal with spectral components from near dc to 5 kHz frequency modulates a 100-kHz carrier, and the frequency deviation is $\pm 2$ kHz. The signal is applied as the input to a frequency tripler. Determine (a) the output center frequency, (b) the output frequency deviation, and (c) the output deviation ratio.

**SOLUTION**  The subscripts "1" and "2" will refer to input and output, respectively.

(a) The output center frequency is

$$f_{c2} = 3 f_{c1} = 3 \times 100 \text{ kHz} = 300 \text{ kHz} \qquad (7\text{-}74)$$

(b) The output frequency deviation is

$$\Delta f_2 = 3 f_1 = 3 \times 2 = 6 \text{ kHz} \qquad (7\text{-}75)$$

(c) The input deviation ratio is $D_1 = 2\text{ kHz}/5\text{ kHz} = 0.4$, and the output deviation ratio is then

$$D_2 = 3D_1 = 3 \times 0.4 = 1.2 \qquad (7\text{-}76)$$

Alternately, the output deviation ratio could have been determined by dividing the output frequency deviation by the highest modulating frequency of 5 kHz.

---

**▌ EXAMPLE 7-15**

It is desired to multiply the frequency and deviation of an FM signal by a factor of 72. Using only frequency doublers and triplers, show how this can be done.

**SOLUTION** This involves a little trial and error, but the number 72 can be expressed as $2 \times 2 \times 2 \times 3 \times 3$. Hence, one solution is the use of three frequency doublers and two frequency triplers.

▐▌

---

## 7-6 FM Modulator Circuits

In this section, a survey of some representative techniques employed in circuits designed for generating FM and PM signals will be made. Although the technology used in implementing such circuits is constantly changing, some of the same strategies continue to be used. The focus, therefore, will be on these general strategies rather than on detailed circuit diagrams.

### VCO

While the term *FM modulator* is perfectly proper to use for any circuit designed for generating an FM signal, the term *voltage-controlled oscillator* (*VCO*) is widely used in various segments of the industry for such circuits. The term *voltage-to-frequency* (*V/F*) *converter* is also used.

**Ideal VCO Properties**

Any circuit used for generating an angle-modulated signal should produce a steady-state sinusoidal output at a fixed frequency when no modulating signal is present. When the data signal is applied, the frequency or the phase should change in accordance with the instantaneous amplitude of the modulating signal. Attention will be focused on an FM generator here, but similar strategies may be employed for a PM generator. In fact, it is possible to use an FM generator to produce PM, and vice versa.

For an ideal FM modulator or VCO, the frequency deviation $\Delta f$ should be directly proportional to the magnitude of the modulating signal voltage, as shown in Figure 7–10. The slope of the curve is the *VCO sensitivity* $K_f$, which can be specified in hertz per volt (Hz/V). (On the curve, $\Delta f$ refers to any arbitrary deviation, rather than the maximum frequency deviation as earlier notation established.) If the curve deviates from a straight line,

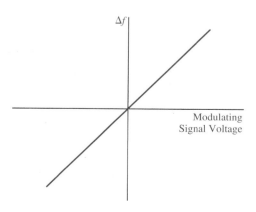

**FIGURE 7–10**

Frequency deviation versus modulating signal voltage for an ideal VCO.

the resulting FM signal will be distorted, so care must be exercised to design a circuit with as close to a straight-line characteristic as possible.

## Use of Frequency Multiplication

For wideband FM systems with large frequency deviations, it is often difficult to produce a straight-line characteristic over a wide frequency range. One widely employed strategy is to first generate the FM signal at a relatively low center frequency and deviation. By successive frequency multiplications, coupled with frequency translations, the center frequency and the deviation can both be adjusted to the final required range.

In some cases (particularly in telemetry data systems), all the modulator circuits up to the point where the FM signal is generated are directly coupled; that is, no capacitors are used as coupling components. This permits very low frequency or even dc data to be transmitted in the system, whereas it is usually more difficult to transmit such data in many AM-type systems. The ability to process dc and very low frequency data is one of the significant advantages of FM. However, in commercial FM broadcasting, it is not necessary to transmit dc, so the modulator circuits can be ac-coupled in that case.

## Variable Parameter Methods

One widely employed classical approach for FM modulators is the *variable parameter* concept. This approach has as its basis the fact that the frequency of many oscillator circuits depends on an $LC$ (or in some cases an $RC$) product. If the value of one of the critical parameters can be varied in accordance with an applied modulating signal, the instantaneous frequency can be varied. Depending on the circuit, it is possible to vary either the capacitance, the inductance, or the resistance. For the purpose of this discussion, we will assume an $LC$ oscillator and consider the capacitance as the variable parameter.

Consider the $LC$ resonant circuit shown in Figure 7–11, and assume that its resonant frequency controls the frequency of an oscillator, which is not shown. Let $L_0$ and $C_0$ represent the fixed values of inductance and capacitance, and let $\Delta C$ represent a change in capacitance. The cyclic resonant frequency $f$ of the circuit is

$$f = \frac{1}{2\pi\sqrt{L_0(C_0 + \Delta C)}} = \frac{1}{2\pi\sqrt{L_0 C_0}}\left(1 + \frac{\Delta C}{C_0}\right)^{-1/2} \tag{7-77}$$

Assume that the change in capacitance is small compared to the fixed value; that is, $\Delta C/C_0 \ll 1$. The square root in Equation 7-77 may then be approximated by the first two terms in the binomial expansion, and we have

$$f \approx \frac{1}{2\pi\sqrt{L_0 C_0}}\left(1 - \frac{\Delta C}{2C_0}\right) \tag{7-78}$$

In general, we desire that

$$f = f_0 + \Delta f = f_0\left(1 + \frac{\Delta f}{f_0}\right) \tag{7-79}$$

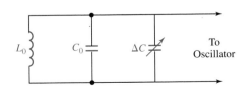

**FIGURE 7–11**

Resonant circuit used to tune oscillator in which the frequency deviation is a function of the change in capacitance.

We see then that

$$f_0 = \frac{1}{2\pi\sqrt{L_0 C_0}} \tag{7-80}$$

and

$$\frac{\Delta f}{f_0} \approx \frac{-\Delta C}{2C_0} \tag{7-81}$$

These results indicate that the change in frequency is directly proportional to the change in capacitance, provided that the change in capacitance is small compared to the fixed capacitance. The negative sign indicates that an increase in capacitance produces a decrease in frequency, as would be expected.

We have seen that varying the capacitance can be used to vary the resonant frequency of an oscillator. The next question is how to vary the capacitance as a function of the modulating voltage.

## Direct Variation of Capacitance

Direct variation of capacitance with a modulating voltage is possible with a voltage variable capacitance diode (commonly called a *varactor diode*). This device uses the property that a reverse-biased PN junction diode displays a capacitance that is a function of the applied voltage. The capacitance $C(v)$ as a function of the voltage is a nonlinear function of the form

$$C(v) = \frac{C_{jo}}{\left(1 + \frac{v}{V_{bi}}\right)^n} \tag{7-82}$$

where $C_{jo}$ is the junction capacitance with no voltage applied, $v$ is the magnitude of the reverse voltage across the diode, and $V_{bi}$ is called the "built-in potential" (typically 0.8 V for silicon and 0.4 V for germanium). The value of $n$ depends of the type of junction and ranges between about $1/2$ to $1/3$. A possible circuit is shown in Figure 7-12.

While it is not obvious from Equation 7-82, it can be shown that for small changes in $v$ relative to $V_{bi}$, the capacitance can be made to change approximately linearly with the signal voltage.

## Indirect Capacitance Variation

Indirect variation of capacitance can be achieved by connecting a capacitance across an amplifier whose gain changes as the level of the modulating signal changes. A simplified illustration of this concept is shown in Figure 7-13.

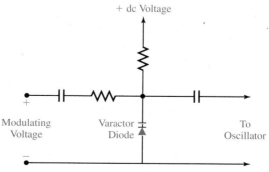

**FIGURE 7-12**

FM generator using voltage-variable capacitance diode (varactor).

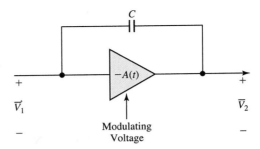

**FIGURE 7–13**
Illustration of variation of effective capacitance by varying gain.

The capacitor is connected between the input and output terminals of an inverting amplifier in which voltage gain $A(t)$ is varied in some manner as a function of the modulating signal. If the amplifier is assumed to have a very high input impedance, the total input phasor current is given by

$$\overline{I}_1 = j\omega C(\overline{V}_1 - \overline{V}_2) \tag{7-83}$$

The output voltage is related to the input voltage by

$$\overline{V}_2 = -A(t)\overline{V}_1 \tag{7-84}$$

Substitution of Equation 7-84 into Equation 7-83 yields

$$\overline{I}_1 = j\omega C[1 + A(t)]\overline{V}_1 \tag{7-85}$$

The input current is seen to lead the input voltage by 90°, so it may be considered a purely capacitive current. This result suggests that Equation 7-85 may be expressed as

$$\overline{I}_1 = j\omega C_{eq}\overline{V}_1 \tag{7-86}$$

where the equivalent capacitance $C_{eq}$ may be recognized as

$$C_{eq} = [1 + A(t)]C \tag{7-87}$$

The effective capacitance is seen to be increased by the process, and the change in capacitance is directly proportional to the gain. If the gain is made to vary linearly about some nominal gain, the resulting capacitance will also vary linearly about its nominal value.

The reader may recognize that the basic capacitance multiplier technique here is the classical *Miller effect,* in which a capacitance between the input and output of an inverting amplifier is effectively increased.

Gain variation with signal level is achieved by biasing the active device in a nonlinear portion of its characteristic curves. Care must be taken in the design to ensure that the gain variation is approximately linear with the modulating signal. This usually restricts the fractional frequency deviation that can be achieved through this concept, although frequency multiplication can be employed.

The preceding steps have been chosen to illustrate the concepts involved, rather than to provide a specific circuit. The most common circuit used through the years to obtain an indirect variation of capacitance (or inductance, for that matter) is called the *reactance modulator.*

## RC Relaxation Oscillator

At relatively low frequencies, such as encountered in subcarrier generators for multiplexing systems, *RC* relaxation oscillators may be frequency modulated by varying the trigger point in the cycle. Many of the integrated circuit timers and function generators available as VCO chips employ this concept. Since the internal operation of these chips would necessitate a more detailed explanation than desired here, the concept will be illustrated with

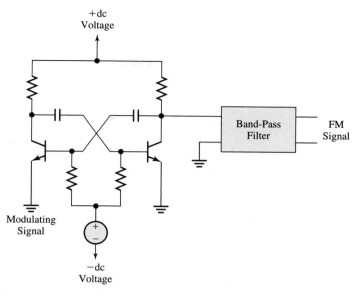

**FIGURE 7-14**
VCO implemented with an astable multivibrator.

the discrete-component astable multivibrator shown in Figure 7–14. The signal voltage either adds to or subtracts from the base return voltage. This varies the initial level at the base of one transistor as it is cut off by the negative step from the collector of the other transistor. The time between cutoff and conduction (which results when the base voltage climbs to zero) is then a function of the signal voltage. By careful design, a reasonably wide deviation can be achieved with good linearity.

The initial output is a frequency-modulated square wave, but an appropriate band-pass filter at the output converts the FM square wave to an FM sine wave. The filter must be capable of passing all the sidebands about the fundamental, but must reject all the harmonics and their associated sidebands.

## NBFM Generator

One additional circuit deserves special recognition due to its historical significance. This is the Armstrong method, named after E.H. Armstrong, a major contributor to the development of FM and other communication areas. It is referred to as a **narrow-band FM (NBFM)** generator.

Consider the basic FM equation

$$v(t) = A \cos[\omega_c t + \theta(t)] \tag{7-88}$$

as developed earlier in the chapter, with

$$\theta_s(t) = 2\pi K_f \int_0^t v_m(t)\, dt \tag{7-89}$$

The expression of Equation 7-88 can be expanded by the identity given in Equation 7-41 to yield

$$v(t) = A \cos \omega_c t \cos \theta_s(t) - A \sin \omega_c t \sin \theta_s(t) \tag{7-90}$$

Assume narrowband FM, which requires that $\theta_s(t)$ be very small. In this case, the following approximations are reasonable:

$$\cos \theta_s(t) \approx 1 \tag{7-91}$$

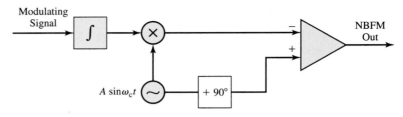

**FIGURE 7–15**
Generation of NBFM signal using the Armstrong method.

and

$$\sin \theta_s(t) \approx \theta_s(t) \tag{7-92}$$

The result of Equation 7-90 can then be closely approximated by

$$v(t) = A \cos \omega_c t - A\theta_s(t) \sin \omega_c t \tag{7-93}$$

A circuit for implementing this form is shown in Figure 7–15. The modulating signal $v_m(t)$ is integrated in accordance with Equation 7-89, and applied to a balanced modulator along with a carrier $\sin \omega_c t$. The carrier is then shifted by 90° and combined with the output of the multiplier. The result is an NBFM signal. A wideband FM signal is obtained by successive frequency multiplication of the NBFM signal. A typical system employing this concept will be explored in Example 7-17.

Since the generation of the NBFM signal involves the addition of a carrier to the output of a balanced modulator, one might naturally question why the result is not conventional AM. The answer lies in the phase relationships between the carrier and the sidebands. With AM, the phase relationship is such that the net amplitude varies with the modulating signal, but no intelligence is applied to the phase. With NBFM, essentially all the intelligence is applied to the phase angle, but there is very little variation of the amplitude.

---

**▌▌ EXAMPLE 7-16**

A VCO has an unmodulated center frequency of 100 kHz and a sensitivity of 4 kHz/V. Determine the instantaneous frequency when the input signal has an amplitude of (a) 2 V and (b) −3 V.

**SOLUTION**

(a)  The change in frequency is

$$\Delta f = K_f v_m = 4 \ \text{kHz/V} \times 2 \ \text{V} = 8 \ \text{kHz} \tag{7-94}$$

The frequency corresponding to this input voltage is

$$f = f_0 + \Delta f = 100 \ \text{kHz} + 8 \ \text{kHz} = 108 \ \text{kHz} \tag{7-95}$$

(b)  In this case, the change in frequency is negative, and is

$$\Delta f = K_f v_m = 4 \ \text{kHz/V} \times (-3 \ \text{V}) = -12 \ \text{kHz} \tag{7-96}$$

The frequency then becomes

$$f = f_0 + \Delta f = 100 \ \text{kHz} - 12 \ \text{kHz} = 88 \ \text{kHz} \tag{7-97}$$

---

**▌▌ EXAMPLE 7-17**

A commercial FM transmitter is to be designed to operate at a center frequency of 100 MHz and with a deviation of ±75 kHz. The modulating frequency range is from 50 Hz to 15 kHz. The Armstrong system is to be employed. The center frequency of the basic modulator is selected to be at 100 kHz, and tests on the modulator indicate that a maximum phase deviation of $\Delta\theta_{max} = 0.5$ rad can be tolerated in the modulator itself. Formulate a possible block diagram design layout for the transmitter, with all frequency multiplication restricted to the use of doublers, triplers, and quadruplers. Prepare a table listing the center frequency, frequency deviation, deviation ratio, and transmission bandwidth at the output of each stage.

**SOLUTION**  Like many design problems, there are a number of possible solutions, so the one presented should be interpreted as representative, rather than the only solution. It was determined after some trial and error.

First, we must determine the maximum frequency deviation at the output of the modulator. Since the modulator must produce FM at the output, the modulating signal is first integrated, and the value of $\Delta\theta$ for the modulator is interpreted as $\beta$. We are given that $\Delta\theta_{max} = 0.5$ rad, so this means that $\beta_{max} = 0.5$. Since $\beta = \Delta f/f_m$ and since $\Delta f$ is fixed for FM, then the maximum value of $\beta$ occurs at the lowest modulating frequency $f_{m,min}$ or

$$\beta_{max} = \frac{\Delta f}{f_{m,min}} \tag{7-98}$$

Solving for the maximum frequency deviation permitted at the output of the modulator, we have

$$\Delta f = \beta_{max} f_{m,min} = 0.5 \times 50 = 25 \text{ Hz} \tag{7-99}$$

The final stage of the transmitter must have a maximum frequency deviation of 75 kHz, so it is necessary to multiply the center frequency by $75{,}000/25 = 3000$ in order to achieve the desired output deviation. However, the center frequency need only be multiplied by a factor of $100 \text{ MHz}/100 \text{ kHz} = 1000$. This means that a down-conversion, which does not affect the deviation, will be required somewhere in the multiplier chain so that we do not "overshoot" the desired output center frequency. One could suggest the possibility of performing all the multiplication first and then down-converting, but this would involve a down-conversion from near 300 MHz to 100 MHz. This process could certainly be achieved, but 300 MHz is at the edge of the UHF frequency range, and implementation would be somewhat easier at a lower frequency. Consequently, the best choice is to make a down-conversion somewhere along the multiplier chain well below the final RF frequency.

We next select a combination of multipliers to achieve the required multiplication ratio. Some trial and error is involved, and it can be shown that a multiplication of exactly 3000 cannot be achieved using only doublers, triplers, and quadruplers. However, this need not be a cause for alarm, because we can always exceed the ratio slightly and then, by reducing the drive slightly to the modulator circuit, accomplish the desired result. Specifically, we find that the combination of five quadruplers and one tripler produces a multiplication of $4^5 \times 3 = 3072$. (Of course, two doublers could replace each quadrupler if desired, but we will assume quadruplers in the analysis.)

A block diagram of the proposed system is shown in Figure 7–16, and a tabular presentation of the data is presented in Table 7–3. The various letters in the left column of the

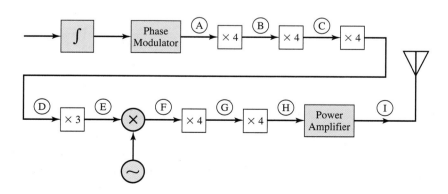

**FIGURE 7–16**
Block diagram of transmitter layout in Example 7-17.

Table 7–3   Data Concerning the Transmitter Design in Example 7-17

| Point | $f_c$ | $\Delta f$ | $D$ | $B_T$ |
|---|---|---|---|---|
| A | 100 kHz | 24.41 Hz | 0.00163 | 30 kHz |
| B | 400 kHz | 97.66 Hz | 0.00651 | 30 kHz |
| C | 1.6 MHz | 390.6 Hz | 0.026 | 30 kHz |
| D | 6.4 MHz | 1.562 kHz | 0.104 | 30 kHz |
| E | 19.2 MHz | 4.688 kHz | 0.313 | 39.38 kHz |
| F | 6.25 MHz | 4.688 kHz | 0.313 | 39.38 kHz |
| G | 25 MHz | 18.75 kHz | 1.25 | 67.5 kHz |
| H | 100 MHz | 75 kHz | 5 | 180 kHz |
| I | 100 MHz | 75 kHz | 5 | 180 kHz |

table correspond to particular points shown on the block diagram. Since the multiplication ratio is slightly greater than needed, the frequency deviation at the modulator output is reduced from 25 Hz to about 24.41 Hz. This value represents the frequency deviation that, when multiplied by 3072, will yield a maximum deviation of 75 kHz at the transmitter output. The various values in the table have been rounded for convenience.

The value of $D$ at each point in the table is the deviation ratio defined by

$$D = \frac{\Delta f}{W} = \frac{\Delta f}{15 \times 10^3} \tag{7-100}$$

This parameter, of course, is very important in determining the transmission bandwidth, which in turn is used in designing the filters that appear at the different multiplier outputs. Filters are necessary to eliminate the spurious spectral components arising from the nonlinear multiplier operations. These filters are assumed to be in the various multiplier stages and are not shown on the block diagram.

The frequency converter is placed at a point in which the center frequency is 19.2 MHz. This position is somewhat arbitrary, but reasonable. The mixing frequency is selected by first determining what the output frequency must be. Since an additional multiplication of $4 \times 4 = 16$ remains, the desired output center frequency of the mixer must be 100 MHz/16 = 6.25 MHz. The mixing oscillator must then shift the incoming signal from 19.2 MHz down to 6.25 MHz. A mixing frequency of 12.95 MHz is one possible solution. The output of the mixer will also contain an FM signal centered at 32.15 MHz, but this component is easily removed by the mixer output filter.

Note from Table 7–3 that the signal remains an NBFM signal through the first four stages, so the simpler narrowband approximation for bandwidth is used in those stages. Beginning with point E, Carson's rule is used for estimating the bandwidth. The reader is invited to verify the different values in the table.

One final note is that the FCC assigns FM channels at odd integer multiples of 100 kHz, so an actual FM transmitter would have the center frequency at either 99.9 MHz or 100.1 MHz, but the value of 100 MHz was easier to work with in the solution. The actual output center frequency could be adjusted to either of the actual frequencies mentioned by changing the mixing oscillator frequency by $\pm 100$ kHz.

## 7-7   FM Detection Circuits

In this section, we will investigate some of the concepts that are employed in circuits used for detecting or demodulating FM and PM signals. As in the last section, the discussion will concentrate primarily on FM signals. FM detector circuits are often called *FM discriminators* or *FM demodulators*.

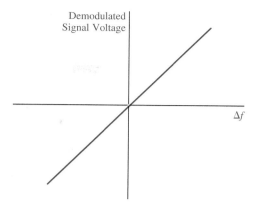

**FIGURE 7–17**
Demodulated signal voltage versus frequency deviation for an ideal FM detector.

## Ideal FM Detector

The ideal FM detector is a circuit that produces zero voltage when the frequency of the received signal is constant and equal to the carrier frequency, but in which the voltage changes linearly with changes in the input frequency. The ideal form of the output voltage versus input frequency deviation characteristic is shown in Figure 7–17. Any actual FM detector characteristic must approximate this straight-line characteristic very closely if distortion is to be minimized.

In the case of the FM modulator, we have seen that it is possible to generate the FM signal at a much lower deviation than required, and then to multiply the smaller deviation to the correct value by the use of frequency multipliers. No simple reverse process exists in the case of FM detection. Of course, the received signal may be down-converted to a lower frequency range for convenience, but the full swing of the deviation must be accommodated within the detector.

## Amplitude Limiting

An important general observation about FM signals is that the intelligence is always a function of the frequency (rate of zero crossings) only and is not a function of the amplitude. This allows various nonlinear operations to be performed on the amplitude without affecting the intelligence. The most common operation of this type is *amplitude limiting,* which is used in many FM receivers to remove most of the amplitude noise, which can create various undesirable effects in the signal processing.

The input–output characteristic of an ideal limiter is shown in Figure 7–18(a), and a simplified implementation approach using diodes is shown in Figure 7–18(b). In practice, some of the standard FM discriminator circuits accomplish a degree of limiting along with the detection, although a separate limiting stage may be used if desired.

The effect of the limiter on a noisy signal is shown in Figures 7–18(c) and (d). Observe that the output has some of the properties of a square wave, but the information, which is contained in the zero crossings, is not affected.

Although excess amplitude noise can be eliminated or reduced by this process, the FM signal is still disturbed by noise that has already affected the locations of zero crossings before reception. This noise cannot be eliminated, since its presence is embedded in the instantaneous zero crossings and cannot be distinguished from the signal itself. As will be seen later in the text, as long as the signal-to-noise ratio exceeds a certain minimum, there is a significant signal-to-noise enhancement inherent in the FM process.

## Slope Detection

A number of widely used FM detector circuits employ the concept of *slope detection.* This approach is based on the fact that certain circuits display a steep voltage-versus-frequency magnitude characteristic over a region that is equal to or can be approximated by a straight line.

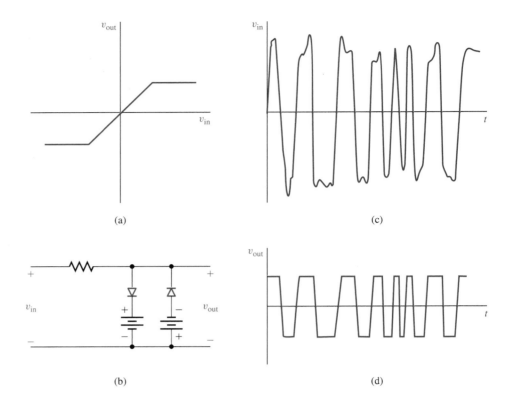

**FIGURE 7-18**
Amplitude limiter and its effect on an FM signal.

**FIGURE 7-19**
Theoretical concept of FM demodulation by differentiation and envelope detection.

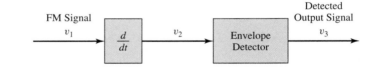

To develop an analytical basis for such a circuit, consider the block diagram shown in Figure 7–19. The block on the left is assumed to be an ideal differentiating circuit whose output $v_2$ is related to the input $v_1$ by

$$v_2 = \frac{dv_1}{dt} \qquad (7\text{-}101)$$

Assume that the input signal is of the form

$$v_1 = A \cos \left[ \omega_c t + 2\pi K_f \int_0^t v_m \, dt \right] \qquad (7\text{-}102)$$

The output of the differentiator is

$$v_2 = -A[\omega_c + 2\pi K_f v_m] \sin \left[ \omega_c t + 2\pi K_f \int_0^t v_m \, dt \right] \qquad (7\text{-}103)$$

This function contains an envelope amplitude multiplicative factor that includes a constant plus a term proportional to the intelligence $v_m(t)$.

Assume that $\omega_c + 2\pi K_f v_m > 0$, which is normally the case, and assume also that the frequency range of the sinusoidal factor is much higher than that of the modulating signal. With these assumptions, the multiplicative factor may be extracted by an envelope detector, and the output of the envelope detector contains the intelligence. This function will be denoted by $v_3$, and it will be of the form

$$v_3 = K_1 + K_2 v_m(t) \qquad (7\text{-}104)$$

The result is seen to contain the desired signal plus a dc component. If we momentarily assume that the desired signal contains no dc or very low frequency components, a simple high-pass filter could be used to eliminate the dc component, and the output would then be proportional to the modulating signal.

Although this circuit provides a sound theoretical basis for many FM detector circuits, a conventional differentiating circuit is seldom used. A conventional wideband differentiating circuit tends to accentuate high-frequency noise, due to the increasing gain of the circuit as the frequency increases. Moreover, with typical values of the operating parameters in FM systems, the dc level in Equation 7-104 turns out to be very large compared to the signal level, and since the dc level must be removed, this would eliminate the possibility of dc data transmission.

These problems can be alleviated by several possible balanced arrangements that simultaneously achieve the following objectives: (1) the dc level at the center frequency is cancelled by using a carefully balanced configuration, and (2) the differentiating characteristic (increasing voltage versus frequency) is preserved and linearized only over the range of the input signal frequency deviation. This straight-line characteristic over a band-pass region may be closely approximated by resonance curves of tuned circuits.

## Balanced Discriminator

One circuit using the preceding principles is the *balanced discriminator* shown in Figure 7–20. Consider the two halves of the secondary winding along with the capacitors as two separate parallel resonant circuits. Referring to Figure 7–21, assume that the upper

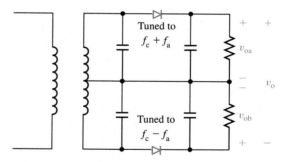

**FIGURE 7–20**
Balanced discriminator circuit.

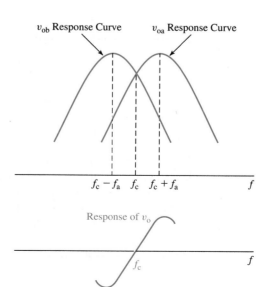

**FIGURE 7–21**
Operation of balanced discriminator circuit.

resonant circuit is tuned to $f_c + f_a$ and that the lower one is tuned to $f_c - f_a$, where $f_o$ is a carefully selected increment on each side of the carrier frequency $f_c$. The tuned circuits are assumed to be identical except for the difference in the center frequencies shown. The diodes in conjunction with the two $RC$ circuits perform an envelope detection process on the two signal components at the two respective ends of the transformer. The output voltage is the difference between these two separate voltages.

Observe that the dc levels cancel, provided that both circuits and the transformer are perfectly balanced. This eliminates one of the problems associated with the basic differentiating scheme. The high-frequency noise problem associated with a conventional differentiator is alleviated by the fact that the linear voltage variation with frequency is present only over a limited frequency range, as can be seen from Figure 7–21. The preceding qualitative discussion will now be put in quantitative terms. Assuming that the behavior is nearly linear in the operating region, the top voltage $v_{oa}$ can be expressed as

$$v_{oa} = A + Kf \tag{7-105}$$

The bottom voltage can be expressed as

$$v_{ob} = A - Kf \tag{7-106}$$

The output voltage is

$$v_o = v_{oa} - v_{ob} = 2Kf = K_d f \tag{7-107}$$

where $K_d = 2K$ is a net detector constant in volts per hertz (V/Hz).

## Foster Seely Discriminator and Ratio Detector

Two other examples of circuits that employ the slope-detection principle are the *Foster Seely discriminator* and the *ratio detector,* which are illustrated in Figure 7–22. Detailed discussions of these circuits are given in basic communications electronics circuits texts, and will not be given here.

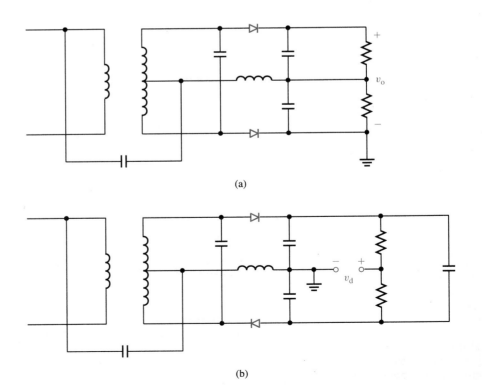

(a)

(b)

**FIGURE 7–22**
(a) Foster Seely discriminator, and (b) ratio detector.

## Pulse Averaging Discriminator

A completely different approach to FM detection is the *pulse averaging discriminator* (PAD), whose block diagram is shown in Figure 7–23. The FM signal is applied to a circuit that generates a trigger pulse for every positive-going zero crossing. (The circuit could also be implemented with negative-going zero crossings.) The triggers are used to initiate a fixed pulse from a monostable or one-shot multivibrator. The pulse width $T_m$ of the multivibrator is chosen to be one-half of the period corresponding to the center frequency of the unmodulated carrier $f_c$; that is,

$$T_m = \frac{1}{2f_c} \tag{7-108}$$

Assume that in the rest state of the multivibrator the output voltage is $-A$, and that during the duration of the pulse it is $+A$. The various waveforms associated with the operation are shown in Figures 7–24 through 7–26. The case when the input frequency is $f_c$ is shown in Figure 7–24. In this case, the voltage at the output of the multivibrator is a balanced square wave, so the net area in a cycle is zero. Thus, the output of the averaging circuit, which can be a low-pass filter, is zero.

Assume now that the instantaneous frequency changes abruptly to a value below $f_c$, as shown in Figure 7–25. The pulse width $T_m$ does not change, but the time between successive trigger pulses increases, and the negative part of the square-wave cycle is longer than the positive part. The net area is thus negative, and the output of the filter is a negative dc value.

**FIGURE 7–23**
Block diagram of pulse averaging discriminator.

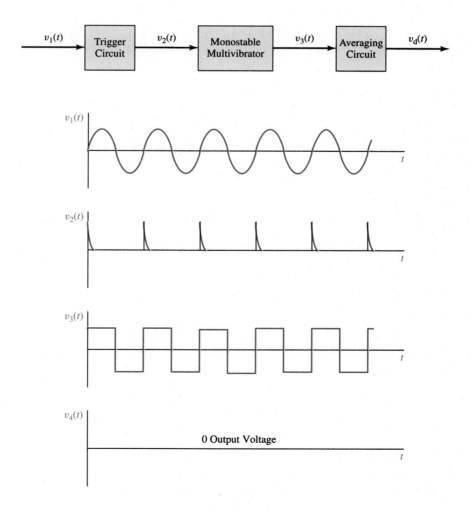

**FIGURE 7–24**
Waveforms in PAD when input frequency is at center frequency.

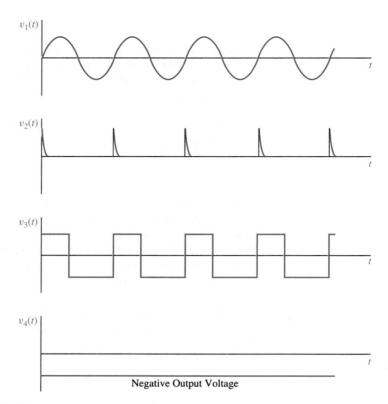

**FIGURE 7–25**
Waveforms in PAD when input frequency is below center frequency.

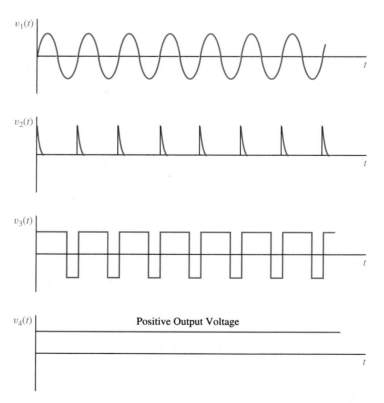

**FIGURE 7–26**
Waveforms in PAD when input frequency is above center frequency.

Finally, assume that the frequency changes abruptly to a value higher than $f_c$, as shown in Figure 7–26. In this case, the negative part of the square wave is shorter than the fixed-length positive part, and the net area is positive. Thus, the output of the filter is a positive dc value.

Operation of the circuit has been illustrated with abrupt step changes in the instantaneous input frequency. These results should demonstrate to the reader that dc data transmission is possible with FM, since the two cases of Figures 7–25 and 7–26 correspond to dc modulating signals with negative and positive values, respectively. With time-varying data, if the circuit is designed properly, the output of the filter will continually change in accordance with the instantaneous input frequency. The output filter must be chosen to have a flat passband amplitude response over the baseband modulating signal frequency range, but it must reject frequency components in the range near $f_c$ and higher.

## Phase-Locked Loop Discriminator

No modern treatment of FM detection would be complete without consideration of the *phase-locked loop (PLL) discriminator*. The PLL was introduced in Chapter 5 as an essential component of frequency synthesizers. We will now show how it can be used as an FM detector.

A block diagram of a PLL arranged for FM detection is shown in Figure 7–27. For the frequency synthesizer applications, the output was taken at the output of the VCO. For FM detection, the output is taken at the input to the VCO, which is the same as the output of the loop filter.

In qualitative terms, the feedback signal mixes with the incoming signal to force the net instantaneous signal frequency at the output of the phase comparator to be zero. This means that the instantaneous frequency of the feedback signal must be the same as that of the input signal. Since the frequency deviation at the output of the VCO is directly proportional to the input voltage, the loop action forces the input voltage to be a direct replica of the intelligence signal. In effect, the loop forces the VCO in the PLL to duplicate the action of the VCO back at the transmitter, and this means that the VCO input voltage (loop output voltage) is directly proportional to the desired intelligence.

A full treatment of the PLL requires the use of feedback control theory. There are some aspects of the system that are nonlinear, but many useful analysis and design properties of the PLL can be determined with the aid of a linearized Laplace transform model.

For the benefit of readers familiar with *s*-plane analysis using Laplace transforms, one form of this linearized model is shown in Figure 7–28. The constant $K_p$ is the phase

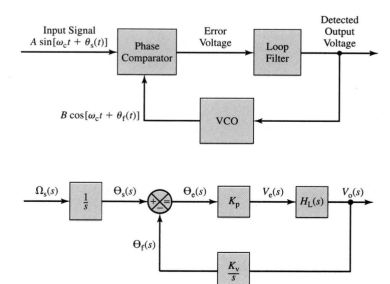

**FIGURE 7–27**
Block diagram of phase-locked loop discriminator.

**FIGURE 7–28**
Form of the linearized Laplace transform model of the PLL frequency discriminator.

comparator constant in volts per radian per second, $K_v$ is the VCO constant expressed in radians per second per volt, and the other quantities are the Laplace transform forms of the variables. If you haven't had Laplace transforms, don't worry if you don't understand the terminology here.

---

**■ EXAMPLE 7-18**

An FM detector is designed to operate at a center frequency of 100 kHz, and it has a detector constant of 2 V/kHz. Determine the output voltage corresponding to an input frequency of (a) 102.5 kHz and (b) 98.5 kHz.

**SOLUTION**    The output of the detector is characterized by

$$v_d = K_d \, \Delta f = K_d(f - f_c) = 2(f - 100) \tag{7-109}$$

where the last form requires that $f$ be expressed in kHz. (Note that the constant is expressed in V/kHz and the value subtracted from $f$ is expressed in kHz.)

(a)  The first case results in

$$v_d = 2(f - 100) = 2(102.5 - 100) = 2 \times 2.5 = 5 \text{ V} \tag{7-110}$$

(b)  For the second case, we have

$$v_d = 2(f - 100) = 2(98.5 - 100) = 2 \times (-1.5) = -3 \text{ V} \tag{7-111}$$

---

## 7-8   Frequency-Division Multiplexing

*Frequency-division muliplexing* (FDM) is a process in which a number of different baseband data signals are transmitted by one RF transmitter through the use of *subcarriers* spaced apart in frequency. A *subcarrier* is a relatively low frequency sinusoid that can be modulated with a data signal having even lower frequency components. Frequency-division multiplexing can be applied to any of the analog modulation methods, but the most common applications utilize FM or PM.

### FDM Transmitter

A block diagram illustrating the form of an FDM transmitter system is shown in Figure 7–29. Each data signal is applied to a separate modulator, and each modulator uses a

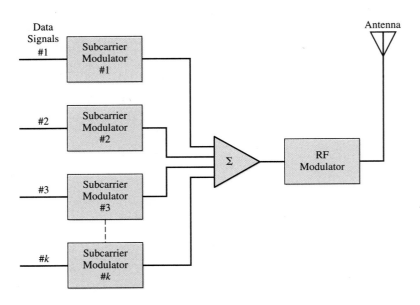

**FIGURE 7–29**
Block diagram of a frequency-division multiplexing transmitter system.

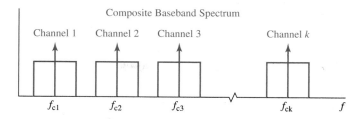

**FIGURE 7–30**
General form of composite baseband spectrum.

separate subcarrier frequency. The modulated subcarriers are all combined with a linear summing circuit, and this composite signal modulates the RF transmitter. The resulting RF signal can then be transmitted by electromagnetic means.

The general form of the composite modulating signal spectrum at the output of the summing circuit is shown in Figure 7–30. The line in the middle of each channel in this figure is used to illustrate possible carrier frequencies, and they are denoted $f_{c1}$, $f_{c2}$, and so on.

The subcarriers must be chosen so that there is no spectral overlap between the different channels. Some guard band should be used between channels to ensure that they can be separated at the receiver. Channels spaced too closely in *frequency* could result in *crosstalk,* a phenomenon in which a part of one channel leaks into a different channel.

## FDM Receiver

A block diagram illustrating the form of an FDM receiver system is shown in Figure 7–31. The front portion of the system is an RF receiver having sufficient bandwidth to accommodate the wideband FDM composite signal. The RF signal is demodulated in accordance with the form of the RF modulation, and the output of the receiver block represents the composite baseband signal, whose spectrum has the form shown in Figure 7–30. This signal is applied to a bank of band-pass filters. There is one filter for each subcarrier frequency, and the center of the passband for a given filter corresponds to the center of the corresponding subcarrier. The bandwidth of each filter must be carefully selected to pass the spectrum for the given channel while rejecting other spectra.

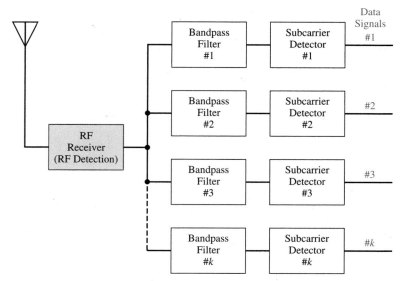

**FIGURE 7–31**
Block diagram of a frequency-division multiplexing receiver system.

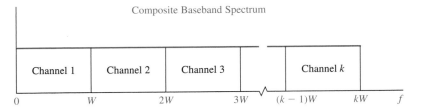

**FIGURE 7–32**
Form for hypothetical baseband composite spectrum of SSB/SSB system used in developing lower bound for transmission bandwidth.

Over the years, a variety of different modulation techniques have been employed in FDM. Notation of the form X/Y is used to designate a particular system, with X representing the subcarrier modulation and Y representing the RF carrier modulation. For example, a system designated as AM/FM implies that the subcarriers are amplitude modulated while the RF carrier is frequency modulated.

We will look carefully at only two forms: SSB/SSB and FM/FM. The first is primarily of theoretical interest, and the second has been employed in many data systems.

### SSB/SSB

This particular system utilizes the minimum amount of bandwidth for FDM systems, and is of interest for comparing with the minimum bandwidth for pulse and digital methods to be considered later. It will be assumed that there are $k$ channels, each of bandwidth $W$. Since this development is being performed for theoretical reasons, no spacing between channels will be assumed, and the composite baseband spectrum would have the form shown in Figure 7–32. Note that the first channel is not shifted and thus remains at baseband. All other channels are shifted by an integer multiple of $W$. The composite signal extends from near dc to $kW$. For SSB modulation, the RF bandwidth is the same as the bandwidth of the modulating signal. Thus,

$$B_T = kW \tag{7-112}$$

This rather intuitive result simply says that the minimum theoretical bandwidth of $k$ baseband channels with equal bandwidth is $k$ times the bandwidth of one channel. In practice, of course, guard band would be required between adjacent channels, but this result provides a theoretical lower bound for the minimum bandwidth analog process without using any form of data compression.

### FM/FM

FM has the property that it can be used for "dc" data transmission; that is, signals that vary very slowly with time. For this reason, and because FM can offer a great deal of immunity to noise, FM/FM multiplexing systems have been widely employed over the years.

Because of the widespread use of FM/FM, some standardization has been achieved in the industry. Standards have been developed by the Inter-Range Instrumentation Group (IRIG), an organization with representation from the various commercial and governmental organizations. One can purchase VCOs and discriminators operating at standard IRIG frequencies from the various telemetry firms.

One set of IRIG standards that will be used for discussion will be that of the Proportional-Bandwidth Subcarrier Channels, for which some of the pertinent data extracted from the standards is given in Table 7–4. The table shows two possible options, ±7.5% deviation and ±15% deviation. The deviation as given is the maximum percentage deviation measured from the subcarrier center frequency. The "proportional" designation

Table 7–4    Sample of Inter-Range Instrumentation Group Proportional-Bandwidth Telemetry Channels ($D = 5$)

| IRIG Channel Number | Channel Center Frequency in Hz | Deviation as ± Percentage | Frequency Deviation Limits in Hz | | Data Cutoff Frequency in Hz |
|---|---|---|---|---|---|
| 1 | 400 | 7.5 | 370 | 430 | 6 |
| 2 | 560 | 7.5 | 518 | 602 | 8 |
| 3 | 730 | 7.5 | 675 | 785 | 11 |
| 4 | 960 | 7.5 | 888 | 1032 | 14 |
| 5 | 1300 | 7.5 | 1202 | 1398 | 20 |
| 6 | 1700 | 7.5 | 1572 | 1828 | 25 |
| 7 | 2300 | 7.5 | 2127 | 2473 | 35 |
| 8 | 3000 | 7.5 | 2775 | 3225 | 45 |
| 9 | 3900 | 7.5 | 3607 | 4193 | 59 |
| 10 | 5400 | 7.5 | 4995 | 5805 | 81 |
| 11 | 7350 | 7.5 | 6799 | 7901 | 110 |
| 12 | 10,500 | 7.5 | 9712 | 11,288 | 160 |
| 13 | 14,500 | 7.5 | 13,412 | 15,588 | 220 |
| 14 | 22,000 | 7.5 | 20,350 | 23,650 | 330 |
| 15 | 30,000 | 7.5 | 27,750 | 32,250 | 450 |
| A | 22,000 | 15 | 18,700 | 25,300 | 660 |
| B | 30,000 | 15 | 25,500 | 34,500 | 900 |

refers to the fact that the bandwidth allocated to channels increases with the center frequency of the subcarrier channels. This is a natural type of allocation that tends to result in more equal implementation complexity for the different channel components. There are also Constant-Bandwidth Subcarrier Channel standards available from IRIG, in which the bandwidth allocated per channel is the same.

**EXAMPLE 7-19**

Consider an FM/FM FDM system consisting of the IRIG proportional-bandwidth channels 1 to 4 of Table 7–4, operating with a deviation ratio of $D = 5$ for each channel. (a) Develop a spectral layout of the composite baseband signal, indicating the approximate spectral limits of each channel. (b) Compute the RF transmission bandwidth if the deviation ratio of the RF transmitter is also 5.

**SOLUTION**

(a) To simplify the organization of the data developed, Table 7–5 has been prepared. The process will be illustrated with channel 1. The maximum deviation of ±7.5% corresponds to an actual subcarrier deviation $\pm\Delta f_{sc}$ given by

$$\pm\Delta f_{sc} = \pm 0.075 \times 400 = \pm 30 \text{ Hz} \qquad (7\text{-}113)$$

Table 7–5    Data for Example 7-19

| Channel | Modulating Frequency Hz | Frequency Deviation Hz | Bandwidth Hz | Lower Frequency Hz | Center Frequency Hz | Upper Frequency Hz |
|---|---|---|---|---|---|---|
| 1 | 6 | ±30 | 72 | 364 | 400 | 436 |
| 2 | 8 | ±42 | 100 | 510 | 560 | 610 |
| 3 | 11 | ±55 | 132 | 664 | 730 | 796 |
| 4 | 14 | ±72 | 172 | 874 | 960 | 1046 |

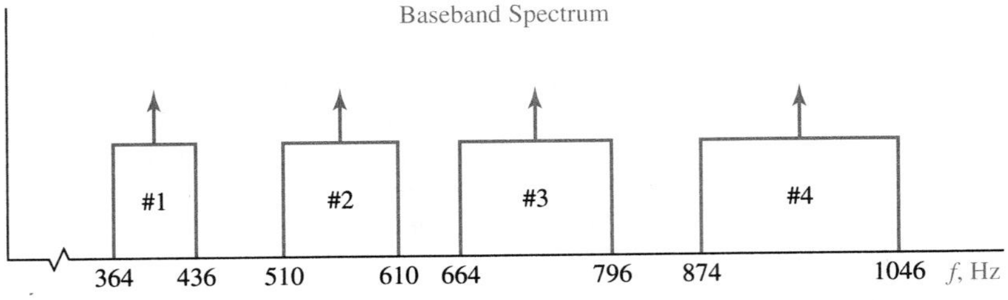

**FIGURE 7–33**
Composite baseband spectrum for FM/FM system of Example 7-19.

This means that the instantaneous frequency can vary from 370 Hz to 430 Hz, as indicated in Table 7–4. For the subcarrier deviation ratio $D_{sc} = 5$, the baseband bandwidth (highest modulating frequency) is given by

$$W = \frac{\Delta f_{sc}}{D_{sc}} = \frac{30 \text{ Hz}}{5} = 6 \text{ Hz} \tag{7-114}$$

The bandwidth $B_{sc}$ for this subcarrier channel is given by Carson's rule as follows:

$$B_{sc} = 2(\Delta f_{sc} + W) = 2(30 + 6) = 72 \text{ Hz} \tag{7-115}$$

The lower limit of the spectrum is $400 - 72/2 = 364$ Hz, and the upper limit is $400 + 72/2 = 436$ Hz. These results are summarized in the first line of Table 7–5. The reader is invited to verify the values for the other channels.

The form of the composite baseband spectrum is shown in Figure 7–33. Observe that the horizontal scale begins just below the lower limit of channel 1, so the open area from dc to that frequency is not shown. All the channel spectra are shown here in continuous "block" form for ease of illustration, with the carrier assumed to be in the middle. The actual spectra for realistic modulating signals will usually show a significant rolloff rate near the band edges.

(b) The composite modulating signal consisting of the four channels has the highest frequency at 1046 Hz. Since the RF carrier deviation ratio is $D_{rf} = 5$, the RF frequency deviation $\Delta f_{rf}$ is

$$\Delta f_{rf} = 5 \times 1046 = 5230 \text{ Hz} \tag{7-116}$$

The RF transmission bandwidth $B_{rf}$ is then determined as

$$B_{rf} = 2(5230 + 1046) = 12,552 \text{ Hz} = 12.552 \text{ kHz} \tag{7-117}$$

## 7-9  Multisim® Examples (Optional)

The Multisim examples of this section will focus on FM signals. The first example will investigate the FM signal source and the second example will deal with a Fourier analysis of its output. The third example will deal with the simulation of a quadrature FM detector.

**▮▮ MULTISIM EXAMPLE 7-1**    Investigate the frequency-modulated voltage source and display the waveform for a modulation index of 5, based on a carrier frequency of 1 kHz and a modulation frequency of 100 Hz.

**SOLUTION**    The FM waveform is obtained from the **Signal Source Family Bar** and it has the name **FM Voltage Source.** The simple circuit used to investigate the signal is

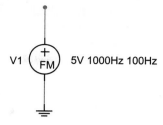

**FIGURE 7–34**
Circuit of Multisim Example 7-1.

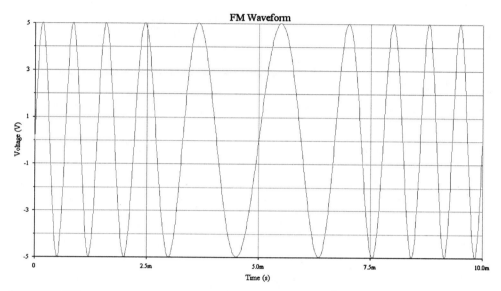

**FIGURE 7–35**
FM waveform of Multisim Example 7-1.

shown in Figure 7–34. Along with the FM signal, a **Junction** is used to observe the signal and provide a valid node number.

A double left-click on the signal opens the properties window. The parameters that were listed in the example statement are the default parameters, which are **Voltage Amplitude = 5 V, Voltage Offset = 0, Carrier Frequency = 1 kHz, Modulation Index = 5,** and **Intelligence Frequency = 100 Hz.** A **Transient Analysis** is run from $t = 0$ to $t = 10$ ms, which represents one cycle at the modulating frequency. The result is shown in Figure 7–35.

---

**⫼ MULTISIM EXAMPLE 7-2**

Perform a **Fourier Analysis** of the FM signal of Multisim Example 7-1.

**SOLUTION**  Although the spectrum is centered at **1 kHz**, the fundamental is **100 Hz**. In order to view the complete spectrum, the frequency range must extend to well above 1 kHz. Arbitrarily, the **Number of harmonics** was set to **20** and the **Stopping time for sampling** was set to **0.01 s**, which represents one cycle at the modulating frequency.

The resulting spectrum is shown in Figure 7–36. In this case there are a significant number of sidebands, as expected for an FM signal with $\beta = 5$. By Carson's rule, there should be 6 sidebands on either side of the carrier that are significant. This would suggest that the bandwidth would extend from about 400 Hz to about 1600 Hz. As can be seen, there are some sidebands that fall outside of this range, but they are relatively small.

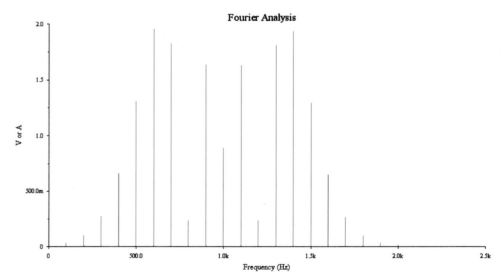

```
Fourier analysis for V(1):
DC component: 2.57195e-005

   No. Harmonics: 20, THD: 19193.9 %, Gridsize: 512, Interpolation Degree: 1

Harmonic Frequency   Magnitude   Phase          Norm. Mag    Norm. Phase
--------  ---------   ---------   -----          ---------    -----------
   1       100        0.0258321    179.91        1             0
   2       200        0.0916318    0.011505      3.5472        -179.9
   3       300        0.266619    -180           10.3212       -359.91
   4       400        0.654456    -0.00054155    25.335        -179.91
   5       500        1.30326     -180           50.4512       -359.91
   6       600        1.95092     -9.325e-005    75.5229       -179.91
   7       700        1.81749     -180           70.3576       -359.91
   8       800        0.23174      0.00331482    8.97101       -179.91
   9       900        1.62808      0.00114181    63.0254       -179.91
  10      1000        0.88125      179.994       34.1145       0.0846962
  11      1100        1.62294      179.995       62.8264       0.0849149
  12      1200        0.230627     0.0503932     8.9279        -179.86
  13      1300        1.80151      0.0050644     69.7391       -179.9
  14      1400        1.92776      0.000224089   74.6265       -179.91
  15      1500        1.2838      -0.0053408     49.6978       -179.91
  16      1600        0.64277     -0.0056207     24.8826       -179.92
  17      1700        0.261278     0.0128003     10.1145       -179.9
  18      1800        0.0898404   -0.0053151     3.47786       -179.91
  19      1900        0.0268055   -0.119         1.03768       -180.03
  20      2000        0.00718048   0.163319      0.277967      -179.75
```

**FIGURE 7–36**
Spectral data of Multisim Example 7-2.

▌▌ MULTISIM EXAMPLE 7-3

The circuit of Figure 7–37 represents a simplified model of an FM quadrature detector, set for a center frequency of 10 MHz. Set the FM input carrier frequency to 10 MHz, the modulating frequency to 10 kHz, and the modulation index to 5. Perform a **Transient Analysis,** and demonstrate that the modulating signal is recovered at the output.

SOLUTION    It should be emphasized at the outset that the circuit given is not intended to represent an optimum design, but rather is intended to demonstrate the concept involved. Quadrature FM detectors are available as off-the-shelf items, and many combine analog and digital technology to enhance the detection process.

A brief description of the circuit follows. The elements of the parallel resonant circuit consisting of L1, C1, and R1 and the capacitor C2 are carefully selected so that the voltages at nodes 1 and 2 are exactly 90° out of phase with each other at the input center frequency. The two signals are multiplied together by the multiplier. At the center frequency of 10 MHz, the average value at the output of the multiplier should be zero. The low-pass filter on the right removes the high-frequency content.

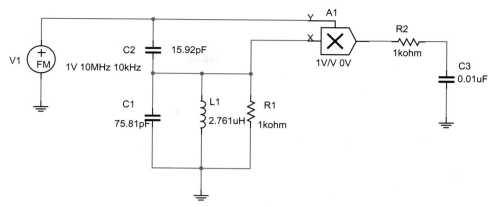

**FIGURE 7–37**
Quadrature detector of Multisim Example 7-3.

As the frequency deviates on either side of the center frequency, the phase shift between the two signals at the multiplier input changes from 90° in one direction or the other. The resulting product will then assume either a positive or a negative average value depending on the direction of the frequency swing. With proper design, the low-pass filter will follow the modulation but will reject the high-frequency content.

To set the input parameters, double left-click on the source and open the properties window. The parameters are set as follows: **Voltage Amplitude = 1 V, Voltage Offset = 0, Carrier Frequency = 10 MHz, Modulation Index = 5,** and **Intelligence Frequency = 10 kHz.**

In view of the high frequency involved for the carrier, a careful scrutiny of the transient setup is required. Quite a few trial runs were made in the adjustment process, but this is often the case in computer simulation. Successful results were obtained with the following **Transient Analysis** parameters: **Start time = 100e-6** (100 μs), **End time = 300e-6** (300 μs), and **Maximum time step = 1e-9** (1 ns). This corresponds to two cycles of the modulating signal. It would be perfectly fine to start the observation at $t = 0$, but there is an initial transient or settling interval, and by waiting for one cycle to begin observation, the output is in a steady-state condition. The resulting output waveform is shown in Figure 7–38.

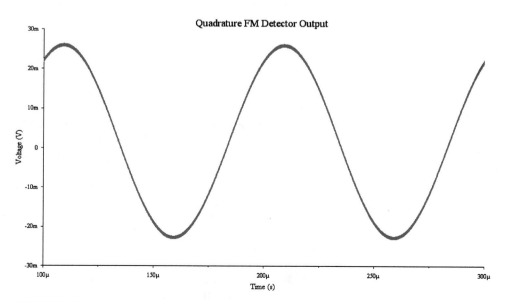

**FIGURE 7–38**
Detector output of Multisim Example 7-3.

## 7-10   MATLAB® Examples (Optional)

The first example in this section will utilize MATLAB to plot a tone-modulated FM signal. The second example will then use the FFT program to compute and plot the spectrum of the signal.

---

**III MATLAB EXAMPLE 7-1**     Use MATLAB to generate a tone-modulated FM signal with a carrier frequency of 10 kHz, a modulating frequency of 1 kHz, and a modulation index of 5. Plot the function for one cycle of the modulating signal.

**SOLUTION**    Keeping in mind the desirability to use an even number of points for the next example, a time step of 1 μs (1e-6) will be used, and we will stop one time step away from the end of one cycle. The time vector is then determined as

```
>> t = 0:1e-6:1e-3-1e-6;
```

As in the case of the AM examples, the sine function will be used for both the carrier and the sinusoidal argument. The composite FM signal **v** will be generated in one step by the command

```
>> v = sin(2*pi*1e4*t + 5*sin(2*pi*1000*t));
```

The function is then plotted by the command

```
>> plot(t,v);
```

The resulting FM function, after appropriate labeling, is shown in Figure 7–39. While we have referred to the signal as an FM signal, it could be interpreted as a PM signal as well.

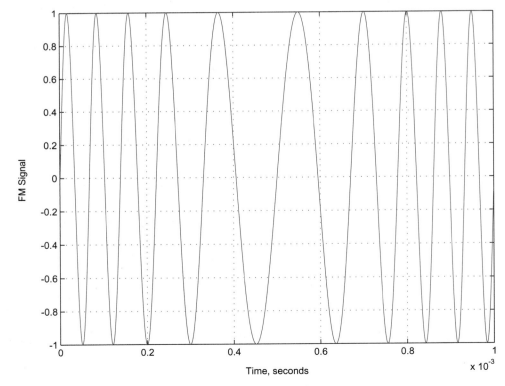

**FIGURE 7–39**
Plot of FM signal.

The difference depends on the relationship of the angle argument with respect to the modulating signal.

**■ MATLAB EXAMPLE 7-2**    Keeping the FM waveform of the previous example in memory, use the program **fourier_series_1** to determine the amplitude spectrum.

**SOLUTION**    Since the waveform of the previous example was defined as **v**, we do not need to redefine it and we may immediately initiate the spectral analysis by the command

```
>> fourier_series_1
```

The dialogue requires that we enter the time step again (1e-6). The frequency step in this case is indicated as 1000 Hz. The lowest integer value entered is 0. If the highest possible value (499) were entered, the spectrum would be extremely crowded around the lower frequency range. Instead, the value 20 is entered, which provides a spectral display from dc to 20 kHz.

The resulting spectrum is shown in Figure 7–40. By Carson's rule, the number of significant components on either side of the carrier should be $\beta + 1 = 5 + 1 = 6$, but it is obvious from the display that components do extend beyond these limits. However, remember that Carson's rule is based on the criterion that 98% of the power is contained in the carrier plus $\beta + 1$ components on either side. Thus, the actual spectrum extends farther than that predicted by Carson's rule, but the effects of the additional components should be negligible for most applications.

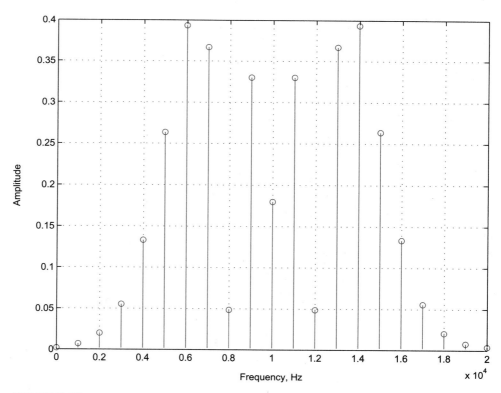

**FIGURE 7–40**

Amplitude spectrum based on one-sided form using FFT.

## SystemVue™ Closing Application (Optional)

Insert the text CD in a computer having SystemVue™ installed and activate the program. Open the CD folder entitled **SystemVue Systems** and load the file entitled 7-1.

### Review of System

The system shown on the screen was described in the chapter opening application, and you may wish to review the discussion there. Recall that the input was a 1-Hz sinusoid that frequency modulated a 10-Hz carrier.

### Change to a PM Signal

1. Double-click on the **FM** block to open the properties window.
2. Left-click on the **PM** button in order to change the waveform from FM to PM. Then left-click on **Parameters** and another window will open. Set **Mod Gain** to **1** and left-click **OK** on two successive windows to return to the system.
3. Perform a run and compare the waveforms in the same manner as in the chapter opening application. Can you detect any difference in the forms of the signals?
4. When you have completed this task, change the modulated signal back to **FM** for subsequent work. When you reach the **Parameters** window, reset **Mod Gain** to **5**. Then left-click **OK** on two successive windows to return to the system.

### Modulation with a Square Wave

1. Double-click on the source block to open the properties window.
2. Left-click on **Pulse Train** and left-click on the **Properties** window.
3. Set the **Amplitude** to **2 V**, the **Frequency** to **1 Hz**, and the **Pulse Width** to **0.5 s**. Left-click on **Square Wave** and then left-click **OK** on two successive windows to return to the system.
4. Perform a run and compare the modulating and modulated waveforms using the **Tile Horizontal** control.

### What You See

When the FM signal is modulated with a square wave, the frequency undergoes abrupt changes for each level change of the square wave. This type of modulation is very important in digital communications, and is called **frequency-shift keying** (FSK). Each discrete frequency represents a particular digital level. This type of modulation will be studied in Chapter 9.

### Additional Studies

1. Investigate the nature of the modulated signal when the modulating signal is a sawtooth waveform.
2. Investigate the effect of changing the amplitude of the modulating signal and of changing the **Mod Gain** (for "modulation gain") of the VCO.

## PROBLEMS

**7-1**  A function generator can be angle modulated with various waveforms. Based on a 300-Hz sinusoid before modulation is applied, the output signal with modulation is

$$v(t) = 20 \cos [600\pi t + 40\pi(1 - e^{-2t})]$$

Determine (a) instantaneous signal phase and (b) instantaneous signal frequency (radian and cyclic). (c) Sketch the forms of the instantaneous signal phase and frequency.

**7-2**  A function generator can be angle modulated with various waveforms. Based on a 50-Hz sinusoid before modulation is applied, the output signal with modulation is

$$v(t) = 20 \cos(100\pi t + 6\pi t)$$

Determine (a) instantaneous signal phase and (b) instantaneous signal frequency (radian and cyclic). (c) Sketch the forms of the instantaneous signal phase and frequency.

**7-3**  A single-tone angle-modulated signal is given by

$$v(t) = 20 \cos[6 \times 10^8 t + 40 \sin(1000\pi t)]$$

Assume that the function is an FM signal. Determine (a) the unmodulated carrier frequency, (b) the modulating frequency, (c) the modulation index, and (d) the maximum frequency deviation. Express all frequencies in hertz. (e) Determine the average power dissipated in a 50-Ω resistor.

**7-4**  A single-tone angle-modulated signal is given by

$$v(t) = 150 \cos[2\pi \times 110 \times 10^6 t + 60 \sin(1000t)]$$

Assume that the function is an FM signal. Determine (a) unmodulated carrier frequency, (b) modulating frequency, (c) modulation index, and (d) maximum frequency deviation. Express all frequencies in hertz. (e) Determine the average power dissipated in a 50-Ω resistor.

**7-5**  Assume that the function in Example 7-3 is a PM signal. Determine the maximum phase deviation.

**7-6**  Assume that the function in Example 7-4 is a PM signal. Determine the maximum phase deviation.

**7-7**  Assume that the modulating frequency in the function of Example 7-3 is halved, and the signal remains an FM signal with the same frequency deviation. Write an equation for the signal.

**7-8**  Assume that the modulating frequency in the function of Example 7-4 is tripled, and the signal remains an FM signal with the same frequency deviation. Write an equation for the signal.

**7-9**  Assume that the modulating frequency in the function of Example 7-5 is halved, and that the signal is PM. Write an equation for the signal.

**7-10**  Assume that the modulating frequency in the function of Example 7-6 is tripled, and that the signal is PM. Write an equation for the signal.

**7-11**  A 90-MHz carrier is to be frequency modulated by a 2-kHz tone. The maximum frequency deviation is to be adjusted to 50 kHz. If the peak amplitude of the carrier is 8 V, write an equation for the function.

**7-12**  A 105-MHz carrier is to be frequency modulated by a 3-kHz tone. The maximum frequency deviation is to be adjusted to 45 kHz. If the peak amplitude of the carrier is 40 V, write an equation for the function.

*Pg. 231*
*Example 7-6*

**7-13**  An FM tone-modulated signal source has a center frequency of 200 kHz, a modulating frequency of 3.5 kHz, and a frequency deviation of 700 Hz. Determine the transmission bandwidth.

*Pg. 235*
*Example 7-8*

**7-14**  An FM tone-modulated signal source has a center frequency of 100 MHz, a modulating frequency of 4 kHz, and a frequency deviation of 500 Hz. Determine the transmission bandwidth.

**7-15**  An FM tone-modulated signal has a frequency deviation of 12 kHz and a modulating frequency of 200 Hz. Determine the transmission bandwidth.

**7-16**  An FM tone-modulated signal has a modulating frequency of 1 kHz and a frequency deviation of 60 kHz. Determine the transmission bandwidth.

**7-17**  A baseband signal having components from near dc to 5 kHz is used to frequency modulate a high-frequency carrier, and the maximum frequency deviation is set at ±14 kHz. Determine the approximate transmission bandwidth.

**7-18**  A baseband signal having components from near dc to 8 kHz is used to frequency modulate a high-frequency carrier, and the maximum frequency deviation is set at ±12 kHz. Determine the approximate transmission bandwidth.

*Pg. 238–239*
*Example 7-10*

**7-19**  The baseband signal of Problem 7-17 is used to phase modulate a high-frequency carrier, and maximum phase deviation is set at ±2 rad. Determine the approximate transmission bandwidth.

**7-20**  The baseband signal of Problem 7-18 is used to phase modulate a high-frequency carrier, and maximum phase deviation is set at ±4 rad. Determine the approximate transmission bandwidth.

*Pg. 239*
*Example 7-11*

**7-21**  A baseband signal with frequency components from near dc to 10 kHz frequency modulates a 200-kHz carrier, and the frequency deviation is ±1.5 kHz. The signal is applied as the input to a cascade of eight frequency doublers. Determine the output (a) center frequency, (b) frequency deviation, and (c) deviation ratio.

*Pg 242*
*Example 7-14*

**7-22**  A baseband signal with frequency components from near dc to 15 kHz frequency modulates a 200-kHz carrier, and the frequency deviation is ±3 kHz. The signal is applied as the input to a cascade of two frequency triplers and one frequency doubler. Determine the output (a) center frequency, (b) frequency deviation, and (c) deviation ratio.

**7-23**  It is desired to multiply the frequency and deviation of an FM signal by a factor of 324. Using only frequency doublers and triplers, show how this can be done.

**7-24**  It is desired to multiply the frequency and deviation of an FM signal by a factor of 864. Using only frequency doublers and triplers, show how this can be done.

**7-25**  A VCO has an unmodulated carrier frequency of 5 MHz. When a control voltage of 2 V is applied to the input terminals, the frequency shifts to 4.99 MHz. Determine the VCO sensitivity in kHz/V.

**7-26**  A VCO has an unmodulated carrier frequency of 100 MHz and sensitivity of $-12$ kHz/V. Determine the frequency when the input signal is set at 5 V.

**7-27**  An FM detector has zero output voltage when the input is a carrier of 10.7 MHz. When the input signal shifts to 10.75 MHz, the output voltage is 6 V. Determine the detector sensitivity in V/kHz.

**7-28**  An FM detector is designed to operate at a center frequency of 90 MHz, and it has a detector sensitivity of 0.05 V/kHz. Determine the output voltage when the frequency changes to 90.04 MHz.

# Pulse Modulation and Time-Division Multiplexing

# 8

## OVERVIEW AND OBJECTIVES

All modulated waveforms considered thus far have been continuous-time signals—or, in more casual terminology, analog modulated signals. We will now begin the study of the classes of modulation that involve pulse and/or digital modulation techniques. Pulse and digital modulation signals are characterized by a series of *samples* of the signal rather than the complete signal. For pulse modulation techniques, the samples are encoded as the amplitude, width, or relative position of a sequence of pulses. For digital modulation techniques, the samples are encoded as digital numbers.

The resulting signals for both pulse and digital modulation are defined only at discrete values of time, and portions of the signal are obviously "missed" by this process. However, it will be shown that if the signal is bandlimited, and if a certain minimum number of samples are taken, the resulting signal can be reconstructed at the receiver to within acceptable limits.

This chapter will deal primarily with the sampling process that serves as a basis for all pulse and digital methods. Some consideration will also be given to some of the traditional pulse modulation methods. However, the major goal is to provide the background required for subsequent chapters that are devoted solely to digital communication concepts. Clearly, digital communication techniques are at the forefront for new developments, and the projected growth in this area is expected to be phenomenal.

## Objectives

After completing this chapter, the reader should be able to:

1. Discuss the concept of a *sampled-data signal* and sketch its form.
2. Display the form of the spectrum of a sampled signal.
3. State the *sampling theorem* and discuss its significance.
4. Define *aliasing* and discuss its effects.
5. Define the *Nyquist rate* and determine the *folding frequency*.
6. Discuss the form of an ideal *impulse-sampled signal* and display the form of its spectrum.
7. Discuss the concept of a *pulse-amplitude modulated* (PAM) signal.
8. Compare the properties of a *natural-sampled* PAM signal and a *flat-top* PAM signal.
9. Discuss the process of the reconstruction of a PAM signal, including the use of a *holding circuit*.
10. Discuss the concept of *time-division multiplexing* (TDM) and show how it is achieved.
11. Discuss the forms of *pulse-time modulation* including *pulse-width modulation* (PWM) and *pulse-position modulation* (PPM).

## SystemVue™ Opening Application (Optional)

Insert the text CD in a computer having SystemVue™ installed and activate the program. Open the CD folder entitled **SystemVue Systems** and load the file entitled 8-1.

### Sink Tokens

| Number | Name | Token Monitored |
|--------|------|-----------------|
| 0 | Input | 4 |
| 1 | Pulse Train | 5 |
| 2 | Sampled Signal | 6 |
| 3 | Output | 7 |

### Operational Tokens

| Number | Name | Function |
|--------|------|----------|
| 4 | Input | sinusoidal source set to 1 Hz |
| 5 | Pulse Train | pulse train set to 20 Hz with a duty cycle of 0.1 |
| 6 | Multiplier | samples input signal |
| 7 | Filter | low-pass filter |

*Note:* The filter and its output (Sinks 3 and 7) will not be discussed at this point, but will be considered at the end of the chapter.

The preset values for the run are from a starting time of 0 to a final time of 2 s, with a time step of 1 ms. Run the simulation and observe the waveforms at Tokens 0, 1, and 2.

### What You See

The 1-Hz sinusoidal source **Input** is being sampled at a rate of 20 samples per second. The **Pulse Train** multiplies the input signal by either 0 or 1, and a sample is created during each short interval in which the pulse generator has a value of 1. The **PAM Signal** is the output of the multiplier, and it consists of short samples of the input signal. The acronym PAM represents *pulse-amplitude modulation,* meaning that the signal is represented by a train of pulses, each of which is proportional to the amplitude of the modulating signal.

### How This Demonstration Relates to the Chapter Learning Objectives

In Chapter 8, you will learn how the sampling process can be used to generate discrete samples of an analog signal. This process is the basis for all pulse and digital modulation methods. You will learn that the signal can be reconstructed from the finite samples if the sampling rate meets certain minimum requirements.

# 8-1  Sampling Theorem

The starting point for all pulse and digital methods is the concept of a *sampled-data signal.* A sampled-data signal (also called a *sampled signal*) is one that consists of a regular sequence of encoded samples of a reference continuous-time (analog) signal.

Consider the arbitrary analog signal $v(t)$ shown in Figure 8–1(a). The form of a sampled data signal $v_s(t)$ representing the analog signal is shown in Figure 8–1(c). The sampled signal is obtained by observing $v(t)$ during short intervals of time of width $\tau$ seconds. The *sampling rate* is designated $f_s$, and is

$$f_s = \frac{1}{T} \tag{8-1}$$

where $T$ is the sampling period and is the time between the beginning of one sample and the beginning of the next sample. During the interval between samples, $v(t)$ is not observed at all. However, samples of other signals could be inserted in the "open space," as will be discussed later.

Note that since the sampling rate is a frequency, it is proper to express it in hertz. *One hertz* is interpreted as *one sample per second* in this context. While hertz will generally be employed for sampling rates, the term *"samples per second"* will be used when it adds clarity to a discussion.

The form of the signal shown in Figure 8–1(c) can be described as a *nonzero-width, natural-sampled signal.* The "nonzero width" description represents the fact that the sampling pulses occupy some width, however small, and this is always the case with actual samples of analog signals. However, the concept of *impulse sampling,* in which the samples are assumed to have zero width, is very useful in certain analytical developments, and will be considered in the next section. The "natural sampled" description refers to the fact that the top of each sample in Figure 8–1 follows the analog signal during the short interval $\tau$. This form is the easiest to analyze mathematically, and serves as a starting point for other forms.

It is convenient to express $v_s(t)$ as the product of $v(t)$ and the periodic pulse train $p(t)$ shown in Figure 8–1(b). Since the pulses assume only the values 0 and 1, multiplication of $v(t)$ by $p(t)$ is equivalent to the sampling operation shown; that is,

$$v_s(t) = v(t)p(t) \tag{8-2}$$

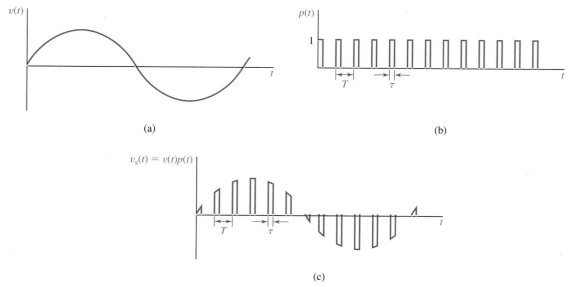

**FIGURE 8–1**

Development of sampled-data signal using nonzero-width pulses and natural sampling.

## Spectrum of Sampled Signal

It is very important to understand the form of the spectrum of the sampled-data signal. We will assume a baseband signal for the purpose of this development. An arbitrary baseband amplitude spectrum $V(f)$ is shown in Figure 8–2(a). The spectrum of the periodic baseband pulse train $p(t)$ is shown in Figure 8–2(b). This latter spectrum contains a dc component, a fundamental component at $f_s = 1/T$, and an infinite number of harmonics at integer multiples of the sampling frequency. Based on the work of Chapter 2, the exponential form of the Fourier series of the pulse train can be expressed as

$$p(t) = \sum_{-\infty}^{\infty} \overline{P}_n e^{jn\omega_s t} \tag{8-3}$$

The amplitude spectrum corresponding to $\overline{P}_n$ is given by

$$\overline{P}_n = d\frac{\sin n\pi d}{n\pi d} \tag{8-4}$$

where $d = \tau/T$ is the duty cycle.

With the Fourier series form of Equation 8-3 substituted in Equation 8-2, the sampled signal can be expressed as

$$v_s(t) = v(t)\sum_{-\infty}^{\infty} \overline{P}_n e^{jn\omega_s t} = \sum_{-\infty}^{\infty} \overline{P}_n v(t) e^{jn\omega_s t} \tag{8-5}$$

Fourier transformation of Equation 8-5 utilizing the modulation theorem yields

$$\overline{V}_s(f) = \sum_{-\infty}^{\infty} \overline{P}_n \overline{V}(f - nf_s) \tag{8-6}$$

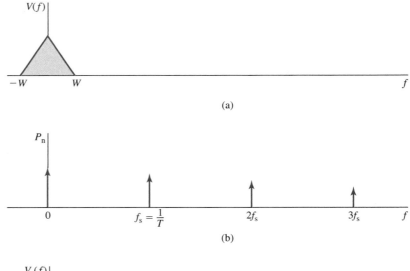

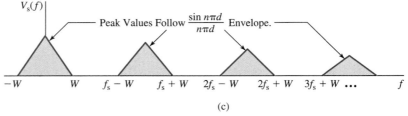

**FIGURE 8–2**
Spectra of baseband signal, pulse train, and sampled data signal. (The three parts of this figure correspond to the three parts of Figure 8–1 for an arbitrary signal.)

The form and *relative* magnitude (but not necessarily the actual magnitude levels) of $\overline{V}_s(f)$ are illustrated in Figure 8–2(c). (Only a limited portion of the negative frequency range is shown.) The spectrum of the sampled-data signal consists of the original baseband spectrum plus an infinite number of shifted or translated versions of the original spectrum. The translated components are shifted in frequency by increments equal to the sampling frequency and its harmonics. The magnitudes of the spectral components are multiplied by the $\overline{P}_n$ coefficient magnitudes, and they gradually diminish with increasing frequency. However, for a very short duty cycle ($\tau \ll T$), the relative magnitudes of the components diminish very slowly and the overall spectrum can be quite broad in form. (The *actual* magnitudes are small for a short duty cycle, but the *relative* magnitudes remain nearly the same for a wide frequency range.)

From the nature of the spectrum of the sampled signal, it is clear that the form of the original baseband spectrum is preserved in the component corresponding to $n = 0$ in Equation 8-6. This component is multiplied by $P_0$, which may be small compared with unity if $d \ll 1$, so the level may be reduced. However, since the shape is preserved, the form of the spectrum is maintained in the frequency range from dc to $W$ *provided that there is no overlap in the spectrum from any of the other components.*

## Minimum Sampling Rate

An intuitive explanation of the sampling theorem will now be developed. From Figure 8–2(c), it is observed that the original spectrum extends from dc to $W$ Hz, while the lowest portion of the first shifted component is at a frequency $f_s - W$. To be able to recover the original signal from the sampled signal, it is necessary that no portion of the first translated component overlap the original spectrum, which means that

$$f_s - W \geq W \tag{8-7}$$

or

$$f_s \geq 2W \tag{8-8}$$

Equation 8-8 is a statement of the *sampling theorem,* which serves as the foundation of all sampled-data, pulse, and digital signal and modulation systems. In words, it states that *a baseband signal must be uniformly sampled at a rate at least twice the highest frequency in a spectrum in order to be recoverable by direct low-pass filtering.* There are some special ways in which the strict requirement can be relaxed for band-pass signals, but for the more common baseband signal case, the requirement stated in Equation 8-8 will be assumed.

Although the "greater than or equal" statement has theoretical significance, in actual applications of the theorem for baseband signals, a sampling rate somewhat greater than the theoretical minimum is employed. The reason for this can be readily deduced from Figure 8–2(c) by noting that if $f_s = 2W$, there is no gap in the spectrum between the original spectrum and the first translated component, and a perfect block filter will be required to separate the components. Thus, some frequency interval, called a *guard band,* should be provided in order that the first translated component (as well as all higher-order components) can be rejected by a realistic filter. A typical example of actual sampling rates is that of some commercial voice-grade telephone sampling systems. Based on an assumed value of $W$ of about 3.4 kHz, a sampling rate $f_s = 8$ kHz is used, while the theoretical minimum would be about 6.8 kHz.

## Alternate Form

An alternate form of the sampling theorem can be stated in terms of the time interval between successive samples. Let $T = 1/f_s$ represent the time interval between successive samples. The alternate form is obtained by taking the reciprocal of both sides of Equation 8-8. Because this is an inequality, however, the sense of the inequality is reversed and we have

$$T \leq \frac{1}{2W} \tag{8-9}$$

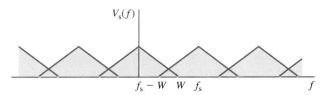

**FIGURE 8–3**
Illustration of aliasing resulting from inadequate sampling rate.

Stated in words, this form indicates that the spacing between samples can be no greater than $1/(2W) = 0.5/W$. Both forms are useful.

Although in practice it is necessary to sample at a rate somewhat greater than the theoretical minimum (or equivalently, have the time interval between samples to be less than the theoretical maximum), some of the theoretical developments that follow will employ the minimum theoretical sample rate for convenience. This theoretical minimum rate $2W$ is called the *Nyquist rate.*

## Aliasing

If the actual sampling rate is less than the theoretical minimum, some of the spectral components will overlap, as illustrated by Figure 8–3. In this case, components of the original spectrum appear at the same locations as other components and cannot be uniquely determined or separated. This process is called *aliasing,* and as the name implies, spectral components may appear to be different than what they really are.

A convenient definition in sampling theory is the *folding frequency* $f_0$. It is given by

$$f_0 = \frac{f_s}{2} = \frac{1}{2T} \tag{8-10}$$

The folding frequency is simply the highest theoretical frequency that can be processed without aliasing by a sampled-data system with a sampling rate $f_s$. Thus, *the Nyquist rate is twice the folding frequency.*

In a sampled-data system, it is important that the minimum sampling rate be satisfied for *all* spectral components in a given signal, even though some may be out of the frequency range of interest. One might erroneously believe that as long as the minimum sampling rate is met for all components of interest, that the desired portion of the signal could be reconstructed. However, if there are unwanted components having a higher frequency than the folding frequency, aliasing will occur, and portions of the unwanted spectrum will fold into the desired spectral range. (This concept will be illustrated in Example 8-4.)

Since it is not always possible to predict precisely the upper frequency limit of a given complex signal, a common practice employed in sampled-data baseband systems is to pass the analog signal through a low-pass filter prior to sampling. Such a filter is called an *anti-aliasing filter,* and should have a cutoff frequency less than, or at least no greater than, the folding frequency.

## Total Number of Samples

Let $N$ represent the total number of samples required to reproduce a signal of duration $t_p$ seconds. Since there are $f_s$ samples per second, the total number required is

$$N = f_s t_p = \frac{t_p}{T} \tag{8-11}$$

**EXAMPLE 8-1**

A baseband signal has frequency components from dc to 5 kHz. Determine (a) the theoretical minimum sampling rate and (b) the maximum time interval between successive samples.

**SOLUTION**

(a)  The minimum sampling rate is

$$f_s = 2W = 2 \times 5 \text{ kHz} = 10 \text{ kHz} \tag{8-12}$$

Thus, it is necessary to obtain no less than 10,000 samples per second.

(b)  The maximum interval between samples is

$$T = \frac{1}{f_s} = \frac{1}{10,000} = 1 \times 10^{-4} \text{ s} = 100 \text{ μs} \tag{8-13}$$

**EXAMPLE 8-2**

To provide some guard band, assume that the signal in Example 8-1 is sampled at a rate 25% above the theoretical minimum. Determine (a) the sampling rate and (b) the time interval between successive samples.

**SOLUTION**

(a)  The sampling rate will now be set at

$$f_s = 1.25 \times 2W = 1.25 \times 10,000 = 12,500 \text{ Hz} = 12.5 \text{ kHz} \tag{8-14}$$

(b)  The time interval between samples is

$$T = \frac{1}{f_s} = \frac{1}{12,500} = 80 \times 10^{-6} \text{ s} = 80 \text{ μs} \tag{8-15}$$

**EXAMPLE 8-3**

For the signal of Examples 8-1 and 8-2, and with the more practical sampling rate of Example 8-2, assume that the signal has a duration of 30 minutes. Determine the total number of samples that must be taken.

**SOLUTION**   Assuming the most basic units, the sampling rate is 12,500 samples per second. Therefore, the total duration must be expressed in seconds, and that value is 30 minutes × 60 seconds/minute = 1800 seconds. Hence,

$$N = f_s t_p = 12,500 \times 1800 = 22.5 \times 10^6 \text{ samples} \tag{8-16}$$

**EXAMPLE 8-4**

A signal $v(t)$ consists of two components $v_1(t)$ and $v_2(t)$, with $v = v_1 + v_2$. The portion of interest is $v_1$, and its amplitude spectrum $V_1(f)$ is band-limited from dc to 800 Hz. As shown in Figure 8–4(a). The amplitude spectrum $V_2(f)$ of the second component occupies a spectrum from 1200 to 1500 Hz, as shown. It is necessary to sample the signal so that it can be processed by a pulse or digital modulation system. An inexperienced young engineer believes that since the component of interest extends only to 800 Hz, a sampling rate greater than 1600 samples per second should suffice. To provide some guard band, he or she selects a sampling rate of 2000 samples per second. (a) Show by constructing a spectral diagram for the sampled signal that there is a fallacy in the reasoning, and that it will not be possible to reconstruct the signal. (b) What steps can be taken to rectify the problem?

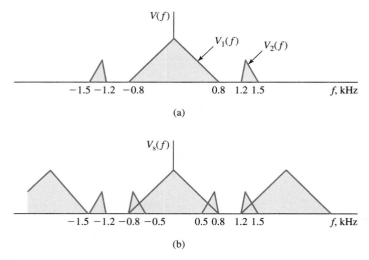

**FIGURE 8–4**
Spectral diagrams for Example 8-4.

**SOLUTION**

(a) Let $v_s(t)$ represent the sampled version of the analog signal $v(t)$. The amplitude spectrum $V_s(f)$ is obtained by sketching the form of the original spectrum plus successive shifts on the original spectrum as indicated by Equation 8-6. In most cases, enough information about the form of the composite spectrum can be deduced from one or two shifted components.

The forms of the baseband and sampled spectra over a reasonable frequency range are shown in Figure 8–4(b). (The amplitude scales of (a) and (b) are different.) Observe that the lower portion of the first translated component, which corresponds to $V_2(f)$, overlaps the upper 300-Hz range of $V_1(f)$. There is no way that these parts of the total spectrum can be separated. Thus, aliasing has occurred.

(b) There are two approaches to solving the problem. The first is to increase the sampling rate to a value greater than twice the highest frequency of the composite spectrum (i.e., greater than $2 \times 1500 = 3000$ Hz.) However, this solution would probably not be the best approach, unless there is some reason to preserve $v_2$. A better solution probably would be to pass the composite analog signal through a presampling anti-aliasing filter with a fairly sharp cutoff just above 800 Hz, and in which the attenuation is quite high in the range from 1200 to 1500 Hz. This filter would essentially eliminate $v_2$, and then 2000 samples per second should be adequate.

## 8-2   Ideal Impulse Sampling

The form of the sampled-data signal in the last section was derived with the assumption that each of the samples had a nonzero width $\tau$. We will now consider the limiting case as $\tau$ approaches zero. In this case, the samples can be conveniently represented as a sequence of *impulse* functions, for which the weight of each is the value of the signal at that point. This type of signal will be referred to as an *impulse-sampled signal.*

While we understand that no real-life analog pulses could have zero width, the concept of the impulse-sampled signal is very important in the study of *digital signal processing.* When an analog signal is converted into a digital signal and subsequently processed on a computer or microprocessor, it may be considered as a sequence of numbers, for which an assumed "width" is somewhat meaningless. A very convenient way of modeling such a signal is by means of an impulse-sampled signal.

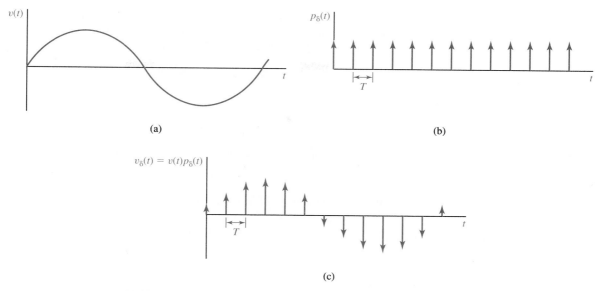

**FIGURE 8–5**
Development of sampled-data signal using ideal impulse sampling.

The form of the ideal impulse sampled signal is illustrated in Figure 8–5. An arbitrary analog signal shown in Figure 8–5(a) is sampled (or modulated) by the periodic impulse train shown in Figure 8–5(b). The resulting ideal representation of the impulse-sampled signal is shown in Figure 8–5(c).

## Spectrum of Impulse Train

As a necessary prelude to developing the form of the spectrum of the impulse sampled signal, the properties of the impulse train shown in Figure 8–5(b) will be investigated. This function can be expressed as

$$p_\delta(t) = \sum_{-\infty}^{\infty} \delta(t - nT) \tag{8-17}$$

where $\delta(t - nT)$ is the mathematical symbol for an impulse occurring at $t = nT$.

Since the impulse train is periodic in the time domain, it can be expanded in a complex exponential Fourier series of the form

$$p_\delta(t) = \sum_{-\infty}^{\infty} \overline{P}_{\delta n} e^{jn\omega_s t} \tag{8-18}$$

where $\overline{P}_{\delta n}$ represents the coefficients of the Fourier series expansion. From the work of Chapter 2, the exponential series coefficients can be determined from the following integral:

$$\overline{P}_{\delta n} = \frac{1}{T} \int_{-T/2}^{T/2} p_\delta(t) e^{-jn\omega_s t} \, dt \tag{8-19}$$

In the interval of one cycle centered at $t = 0$, $p_\delta(t)$ contains only the single impulse $\delta(t)$. Thus, from the basic definition of the impulse function given in Chapter 3, $\overline{P}_{\delta n}$ is obtained as

$$\overline{P}_{\delta n} = \frac{1}{T} \int_{-T/2}^{T/2} \delta(t) e^{-jn\omega_s t} \, dt = \frac{1}{T} = f_s \tag{8-20}$$

From this result, it is deduced that the Fourier coefficients of the impulse train all have equal weight, and there is no convergence. This is compatible with the nonperiodic single impulse function considered in Chapter 3, in which the spectrum was determined to be a constant at all frequencies. For the periodic impulse train, however, the spectrum is a constant, but exists only at integer multiples of the sampling frequency.

To summarize, the Fourier series for the periodic impulse train can be expressed as

$$p_\delta(t) = \sum_{-\infty}^{\infty} \frac{1}{T} e^{jn\omega_s t} \tag{8-21}$$

## Spectrum of Impulse-Sampled Signal

The impulse-sampled signal will be denoted $v_\delta(t)$, and it can be expressed as

$$v_\delta(t) = v(t) p_\delta(t) \tag{8-22}$$

Substitution of Equation 8-21 in Equation 8-22 results in

$$v_\delta(t) = v(t) \sum_{-\infty}^{\infty} \frac{1}{T} e^{jn\omega_s t} \tag{8-23}$$

or

$$v_\delta(t) = \frac{1}{T} \sum_{-\infty}^{\infty} v(t) e^{jn\omega_s t} \tag{8-24}$$

By Fourier transformation of both sides of Equation 8-24 using the modulation theorem, we obtain

$$\overline{V}_\delta(f) = \frac{1}{T} \sum_{-\infty}^{\infty} \overline{V}(f - nf_s) \tag{8-25}$$

The frequency-domain equivalents of the preceding operations are illustrated in Figure 8–6. An assumed arbitrary baseband amplitude spectrum $V(f)$ is shown in Figure 8–6(a), and the discrete spectrum of the periodic pulse train is shown in Figure 8–6(b). The form of the

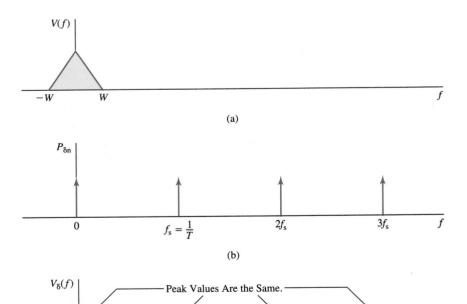

**FIGURE 8–6**
Spectra of baseband signal, impulse train, and impulse-sampled signal. (The three parts of this figure correspond to the three parts of Figure 8–5 for an arbitrary signal.)

amplitude spectrum of the ideal impulse sampled signal is shown in Figure 8–6(c). The general form is similar to the form derived from the nonzero-width natural sampling process shown in Figure 8–2, and the basic sampling requirements developed in the last section apply here. However, a comparison of the figures and the equations reveals one major difference between the spectral forms. The spectral components derived with non-zero pulse widths gradually diminish with frequency, and the magnitudes follow a $\sin x/x$ function envelope. However, components derived from impulse sampling are all of equal magnitude, and do not diminish with frequency.

An important deduction from this discussion is that the spectrum of an impulse-sampled signal is a periodic function of frequency. The "period" in the frequency domain is equal to the sampling frequency $f_s$. *Thus, ideal impulse sampling in the time domain leads to a periodic spectrum in the frequency domain.*

This development is as far as we will go with the concept of impulse sampling. However, the reader who pursues the field of *digital signal processing* will likely encounter the concept again. For the remainder of this chapter, all pulses considered will be assumed to have nonzero width.

---

**▌▌ EXAMPLE 8-5**

A pure sinusoid with a frequency of 1 kHz is sampled at intervals of 0.1 ms, and converted into digital numbers to be processed on a computer. List all positive frequencies below 45 kHz in the spectrum

**SOLUTION**    The basic sampling rate is

$$f_s = \frac{1}{T} = \frac{1}{0.1 \times 10^{-3}} = 10 \text{ kHz} \qquad (8\text{-}26)$$

The process may be considered impulse sampling. The spectrum consists of the original component at 1 kHz, and the sum and difference frequencies about the sampling frequency and all of its harmonics. The positive frequencies below 45 kHz are

  1 kHz,
  9 kHz, 11 kHz
  19 kHz, 21 kHz
  29 kHz, 31 kHz
  39 kHz, 41 kHz

▌▌

---

## 8-3 Pulse-Amplitude Modulation

The emphasis thus far in the chapter has been on establishing the theoretical basis for the sampling process. The sampling theorem applies to all forms of pulse and digital modulation, whenever an analog signal is represented as a series of samples.

The first and most basic sampling method for consideration is that of *pulse-amplitude modulation* (PAM). A PAM signal consists of samples of the analog signal, in which the amplitude of each pulse is proportional to the analog signal at that point, but in which the pulses have a fixed width. We immediately recognize that the form of the signal used in the basic development of the sampled-data concept in Section 8-1 fits this definition. Thus, the waveform shown in Figure 8–1(c) is a form of a PAM signal. This form, in which the top of the pulse follows the analog signal during the pulse width, is called a *natural-sampled PAM signal.*

In actual practice, PAM signals (along with converted digital values) are usually based on a *flat-topped PAM signal,* and this form is illustrated in Figure 8–7. The flat-top representation is a constant value that, for example, could represent the analog signal at *one* particular point in the sampling interval.

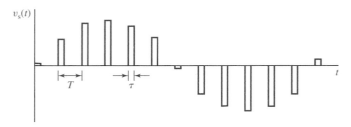

**FIGURE 8–7**
Form of flat-top sampled PAM signal.

It can be shown that the spectrum of the flat-top signal contains some distortion with respect to the ideal natural-sampled signal. However, if the widths of the pulses are small compared to the sampling interval, the resulting distortion can be ignored, or compensation may be applied during the reconstruction process.

A PAM signal may be used as a specific type of desired signal for pulse encoding and transmission. Also, other forms of pulse and/or digital modulation start with a PAM signal, and the signal is then converted to another form.

For PAM signals (as well as for other pulse-modulated methods to be considered), there is one important difference in the spectral form as compared with the analog methods considered earlier in the text. All the analog methods considered involved translating the modulating spectrum to a higher frequency, usually in the RF range, and eliminating all low-frequency components. However, the signal spectral forms shown in previous figures in this chapter show that the final PAM spectrum starts in the baseband range, and may extend all the way down to dc. Thus, PAM is not suitable in its basic form for RF transmission. In the various forms of pulse modulation, the primary objective of the modulation process is to replace the continuous signal by a sequence of samples, with the primary goal of sharing the transmission medium with other signals.

Some pulse-modulated waveforms are transmitted directly at baseband over wire links. However, when RF transmission is required, the PAM signal could be applied as a complex baseband modulating signal to a high-frequency transmitter, and the resulting spectrum would be shifted up to a higher frequency range. The signal would have undergone two levels of modulation, one of which converted the continuous signal to a sequence of discrete samples, and the other of which shifted the spectrum of the sampled signal to the RF range for transmission purposes.

## PAM Bandwidth

A fundamental question with PAM concerns the bandwidth required to transmit the encoded signal. Referring to Figure 8–2, it is observed again that not only does the original spectrum appear from dc to $W$, but shifted versions of the spectrum appear about all frequencies that are integer multiples of the sampling rate. The peak values of these shifted components slowly reduce in magnitude as a function of frequency, since these peaks are proportional to the $\sin x/x$ function.

A convenient way of looking at the PAM signal is through the concept of pulse transmission, as discussed in Chapter 4. To preserve the quality of a PAM signal, the pulse amplitudes must be preserved, and they must not be allowed to spread excessively. On the other hand, perfect reproduction of the beginning and ending of the pulses has been found to require more bandwidth than desired for most of the types of applications for which PAM is used. Specifically, a "coarse" reproduction criteria has been found to be sufficient, and the baseband bandwidth $B_T$ can be expressed as

$$B_T = \frac{K_1}{\tau} \tag{8-27}$$

where $\tau$ is the width of each pulse sample.

The constant $K_1$ depends on the spacing between adjacent pulses, the level of acceptable adjacent pulse crosstalk, the sharpness of the cutoff rate, and other factors. In certain theoretical developments, the value $K_1 = 0.5$ is used. In this case, the result of Equation 8-27 is in perfect agreement with the "coarse" pulse transmission criterion of Chapter 4.

The generation of a PAM signal may be achieved by gating on the analog signal periodically for an interval $\tau$, and turning it off for the remainder of each sampling period. A balanced modulator utilizing a pulse train for the second input could be used, or an analog switch. PAM signals are formed as a portion of a time-division multiplexing system, and further discussion will be given in Section 8-4.

## Signal Reconstruction

The reconstruction of a continuous signal from a PAM signal represents an important process. Refer to the spectral diagrams of Figure 8–2. It is seen that the portion of the spectrum of the sampled signal from dc to $W$ has exactly the same form as the original signal. (It may have a different magnitude, but the equivalence of the two spectral shapes is the important property.) Therefore, if all the components above $W$ are eliminated, the spectrum of the original unsampled signal will remain, and the form of the original signal will be preserved. Thus, *passing a PAM signal through a low-pass filter restores the form of the original analog signal.* The filter should have a flat passband from dc to $W$, and it should display a sharp cutoff between $W$ and $f_s - W$. Thus, some reasonable guard band should be provided.

## Holding Circuits

The actual level of the recovered signal may be quite small compared to the original signal, due to the loss of energy resulting from filtering. Frequently, *holding circuits* are used in conjunction with the filter as a means of maintaining a reasonable level of energy in the filtered signal, as well as for easing a portion of the filtering requirements.

While there are different orders for holding circuits, we will focus on the *zero-order* form, which is the simplest. The operation of a zero-order holding circuit is illustrated by the waveforms of Figure 8–8. The waveform of Figure 8–8(a) represents a single flat-top PAM signal for which restoration is desired. The output of a zero-order holding circuit is shown in Figure 8–8(b). Observe that a given pulse establishes in a very short time an output proportional to the pulse level. This level is retained at this value until the next pulse arrives.

The signal at the output of the zero-order holding circuit is a type of "staircase" approximation of the original analog signal. Therefore, it contains much less high-frequency content than the PAM signal, but it clearly displays some distortion due to the "steps." Additional low-pass filtering is required to eliminate the remaining high-frequency content, and this provides smoothing between the various levels.

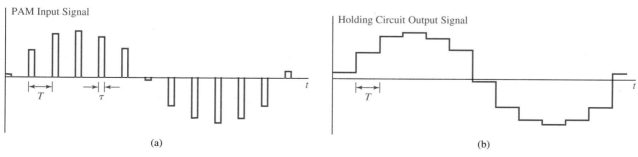

(a)

(b)

**FIGURE 8–8**
Operation of zero-order holding circuit.

**■ EXAMPLE 8-6**

An analog signal $v(t)$ whose amplitude spectrum is shown in Figure 8–9(a) is sampled at a rate $f_s = 5$ kHz. The width of each sample is $\tau = 40$ μs. (a) Sketch the form of the sampled spectrum from dc to the first zero crossing of the pulse train spectrum. (b) Compute the approximate transmission baseband bandwidth for both $K_1 = 0.5$ and $K_1 = 1$ in Equation 8-27.

### SOLUTION

(a) The basic concept to remember is that the spectrum of the sampled signal is obtained by reproducing the form of the spectrum of the unmodulated baseband signal plus translated versions at integer multiples of the sampling frequency. In this example, the relative convergence of the spectral components will be investigated.

First, we observe the spectrum of the pulse train, which may be considered to be multiplied by the continuous signal to yield the sampled signal. Since the sampling rate is $f_s = 5$ kHz, the time interval between successive pulses is $T = 1/f_s = 1/(5 \times 10^3) = 200$ μs. The duty cycle is $d = \tau/T = 40$ μs/200 μs $= 0.2$. Recalling the spectrum of a periodic pulse train from Chapter 3, the form for $d = 0.2$ is shown in Figure 8–9(b) up to the vicinity of the first zero crossing, which occurs at a frequency $f = 1/\tau = 1/40$ μs $= 25$ kHz. Observe that the magnitudes of the terms reduce in accordance with the $\sin x/x$ function as the frequency increases.

The spectrum of the sampled signal is obtained by centering the spectrum of the unsampled signal about each of the pulse train spectral frequencies and multiplying by the respective coefficient amplitudes. The amplitude scale of the pulse spectrum and the sampled signal spectrum are both amplified in order to enhance the illustration.

(b) For a choice $K_1 = 0.5$ in Equation 8-27, the bandwidth required is

$$B_T = \frac{0.5}{\tau} = \frac{0.5}{40 \times 10^{-6} \text{ s}} = 12.5 \text{ kHz} \tag{8-28}$$

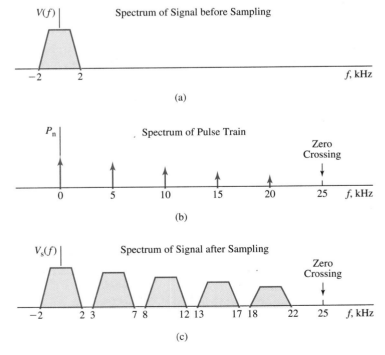

**FIGURE 8–9**
Frequency spectra of Example 8-6.

This is the minimum bandwidth used in system-level calculations, and it can be achieved only under idealized filter conditions. In many systems, a constant closer to $K_1 = 1$ is used. For this choice, the bandwidth is

$$B_T = \frac{1}{\tau} = \frac{1}{40 \times 10^{-6} \text{ s}} = 25 \text{ kHz} \qquad (8\text{-}29)$$

## 8-4  Time-Division Multiplexing

The primary motivation for PAM (as well as other pulse-modulation forms) is a process called *time-division multiplexing* (TDM), in which a number of separate signals can be sent over the same transmission link by alternately sampling the signals in a successive pattern.

The process of TDM is illustrated in Figure 8–10. A *commutator* is required at the input to the transmission medium and a *decommutator* is required at the output. In most systems, these two processes are achieved electronically, but it is easier to illustrate with mechanical components. At the input, the commutator sequentially samples each data signal in order. The decommutator at the receiving end is assumed to be synchronized with the one at the transmitter, and routes each signal to its proper destination. Thus, as far as each individual output is concerned, the signal appears as a sampled PAM signal, and it can be reconstructed by proper filtering.

Each signal must be sampled at the minimum Nyquist rate for its particular bandwidth. For the moment, we will assume that each analog signal being sampled has the same bandwidth $W$. This means that the commutators must "rotate" at a rate no less than $2W$ revolutions per second.

A typical layout of a TDM composite signal is shown in Figure 8–11. While a variety of examples could be shown, this particular example has seven separate data signals all sampled at the same rate. A space equal to a sample pulse duration is maintained between any two successive pulses, and flat-top sampling is employed. A *frame* is one complete

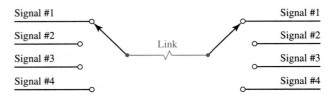

**FIGURE 8–10**

Concept of time-division multiplexing, illustrated with mechanical commutator and decommutator.

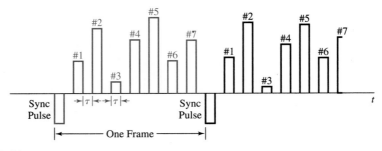

**FIGURE 8–11**

Representative PAM TDM signal format.

structured time interval in which each of the data pulses, plus any additional system information, is provided. Along with samples of the seven data signals, one synchronizing pulse is transmitted in each frame for this system. While a number of possible methods for synchronization can be used, in this system a dc bias level is added to all data channels so that they are always positive. The sync pulse is then transmitted as a negative pulse so that the receiver can readily recognize its presence. Each time a sync pulse appears, the receiver decommutator is realigned with the commutator at the transmitter by pulse-selective circuits.

For the illustration shown, there is a short "dead space" between successive samples. The unused space is often desirable in order to allow some spreading of the pulses resulting from the finite bandwidth limitations of any realistic transmission system. Without some space, there may be a noticeable amount of crosstalk between adjacent channels.

## PAM Minimum Bandwidth

A theoretical lower bound for the bandwidth of a TDM system utilizing PAM will now be developed. While unrealistic in practice, the result provides some insight into time–bandwidth interchange, and it can be compared with some analog results.

Assume that $k$ baseband signals, each with bandwidth $W$, are to be sampled and multiplexed over some transmission system. The following idealized assumptions will be made: (1) sampling set at minimum Nyquist rate, (2) no spacing between adjacent pulses, (3) no sync pulses, and (4) lower bandwidth estimate of $0.5/\tau$. The assumed signal form is shown in Figure 8–12. The value $T_f$ represents the *frame period*.

Each signal must be sampled no less than $2W$ times per second. Since $T_f$ represents the time between samples, $T_f \le 1/2W$. Based on the minimum Nyquist rate,

$$T_f = \frac{1}{2W} \tag{8-30}$$

A given frame contains $k$ samples, each having width $\tau$, which means that

$$\tau = \frac{T_f}{k} \tag{8-31}$$

The minimum transmission bandwidth is then

$$B_T = \frac{0.5}{\tau} = \frac{0.5}{1/(2kW)} = kW \tag{8-32}$$

Stated in words, *the total transmission bandwidth is the number of signals being multiplexed times the bandwidth per signal.* Thus, the more signals to be shared on a line, the greater the bandwidth required.

It should be noted that the minimum theoretical bandwidth for TDM with PAM is the same as for the SSB/SSB FDM concept considered in Chapter 7. This means that from a

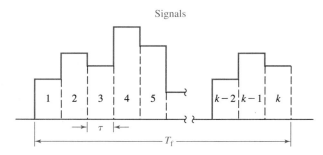

**FIGURE 8–12**
Form for hypothetical PAM signal, used in developing lower bound for transmission bandwidth.

theoretical point of view, the minimum bandwidth for time-division multiplexing is the same as for frequency-division multiplexing. Of course, there are many factors to consider in selecting FDM versus TDM, but the tradeoff between bandwidth and time sets certain basic boundary conditions that must be considered in designing any communication system.

## Multirate Sampling

Thus far in the study of multiplexing, we have considered all of the signals to have the same bandwidth. However, suppose that the different analog signals have different bandwidths. One obvious solution is to sample all signals at or above the Nyquist rate of the signal having the highest bandwidth. This solution is somewhat inefficient if there are wide differences in the bandwidths.

A more attractive solution in some cases is to devise a frame sequence that samples the higher-frequency signals more often than the lower-frequency signals. While the timing and synchronizing of the signals may be more complex, the overall benefits may be worth the extra complexity. An example of such an arrangement will be given in the next chapter.

## PAM Generation

Generation of a PAM multiplexed signal is most easily implemented with a combination of analog switches and multiplexing circuits. A circuit diagram of a system to combine four channels is shown in Figure 8–13. The switches are special $P$-channel JFETs, which are available as individual units or as a combination of several on a single IC chip. In the latter form, an extra compensating JFET (shown at the top) is also included.

A given $P$-channel JFET acts as an open circuit when the gate voltage is more positive than a minimum voltage (related to the pinch-off voltage). Conversely, when the gate voltage is nearly zero, the JFET is turned on and acts like a small series resistance (typically less than 100 $\Omega$). By connecting the gate terminals (designated A, B, C, and D) to suitable

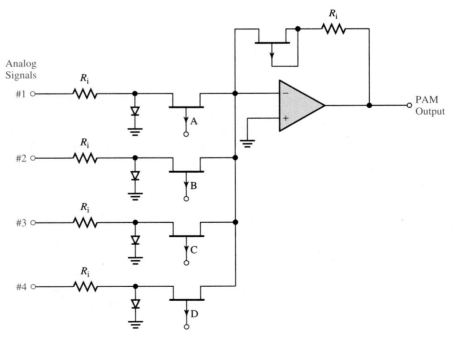

**FIGURE 8–13**
Implementation of a four-channel PAM multiplexed signal using $P$-channel JFET switches.

counter-type circuits, the gates can be turned on in sequence in accordance with a master clock reference. The gate voltage levels of many analog switch circuits such as this are compatible with basic digital logic levels such as TTL, which minimize the interfacing problems. For example, a TTL logic 1 might open the switch and a TTL logic 0 might close the switch.

The samples of the different signals are combined in the analog operational amplifier summing circuit. The compensating switch acts as a series resistance approximately equal to the series resistance of either of the series switches when they are on, and this compensates for the uncertainty in the gain level due to the switch resistance.

---

**EXAMPLE 8-7**

Consider a PAM time-division system with seven signals. Each signal has a baseband bandwidth of 1 kHz. Based on the idealized criteria discussed in this section, determine the minimum bandwidth.

**SOLUTION**   The idealized theoretical minimum bandwidth is the number of channels times the bandwidth per channel, as given by Equation 8-32. Hence,

$$B_T = kW = 7 \times 1 \text{ kHz} = 7 \text{ kHz} \tag{8-33}$$

---

**EXAMPLE 8-8**

Assume that the system of Example 8-7 is implemented in the format of the system of Figure 8–11 with the sync pulse added. Assume that the sampling rate is set to be 25% greater than the theoretical minimum Nyquist rate. (a) Based on the $0.5/\tau$ rule, determine the approximate baseband bandwidth required for the composite baseband signal. (b) If the composite PAM signal is used to amplitude modulate a high-frequency RF carrier, determine the RF bandwidth required for the high-frequency signal.

**SOLUTION**

(a) The sampling rate is to be set at 1.25 times the theoretical minimum, so the sampling rate for each signal is

$$f_s = 1.25 \times 2W = 1.25 \times 2 \times 1 \text{ kHz} = 2.5 \text{ kHz} \tag{8-34}$$

The frame time $T_f$ is

$$T_f = \frac{1}{f_s} = \frac{1}{2.5 \times 10^3} = 0.4 \text{ ms} \tag{8-35}$$

The bandwidth is determined from the minimum pulse width. From Figure 8–11, it is noted that there are 16 intervals of width $\tau$ in a given frame (7 data pulses, 1 sync pulse, and 8 open spaces). The pulse width $\tau$ is

$$\tau = \frac{T_f}{16} = \frac{0.4 \text{ ms}}{16} = 25 \text{ μs} \tag{8-36}$$

The composite baseband bandwidth is

$$B_T = \frac{0.5}{\tau} = \frac{0.5}{25 \times 10^{-6}} = 20 \text{ kHz} \tag{8-37}$$

This is considerably greater than the idealistic minimum bandwidth.

(b) If the composite baseband signal amplitude modulates a high-frequency carrier, the RF bandwidth is

$$B_T = 2 \times 20 \text{ kHz} = 40 \text{ kHz} \tag{8-38}$$

## 8-5  Pulse-Time Modulation

PAM was studied because it establishes the basic foundation for the sampling process, a requirement for all pulse and digital modulation systems. It is obvious that in today's world, and in the world of the future, digital technology will dominate the communications field. However, there are a few special applications in which other forms of pulse modulation are used, so a brief introduction to some of these concepts is worthwhile. In the pulse-time methods to be surveyed in this section, the amplitude of the pulse is maintained at a constant level, but either the width or the position of the pulse is dependent on the modulating signal level.

### Pulse-Width Modulation

*Pulse-width modulation* (PWM) is a sampled-data process in which the *width* of each pulse is varied directly in accordance with the amplitude of the modulating signal at the particular sample point. Pulse-width modulation is also called *pulse-duration modulation* (PDM). Both terms appear in the literature.

The PWM process is illustrated in Figure 8–14. The assumed analog signal shown in Figure 8–14(a) is converted by appropriate electronic circuits to the pulse train shown in Figure 8–14(b). Observe that the most positive peak corresponds to the widest pulse, and the most negative peak corresponds to the narrowest pulse. A 555 integrated circuit timer may be modulated in this fashion.

### Pulse-Position Modulation

*Pulse-position modulation* (PPM) is a sampled-data process in which the *position* of each pulse is varied directly in accordance with the amplitude of the modulating signal at the particular sample point. This process is illustrated in Figure 8–14(c) for the analog signal shown in Figure 8–14(a). In this case, the most positive peak corresponds to the maximum shift from a reference point (the beginning in this case) in each pulse interval, while the most negative peak corresponds to no shift. As in the case of PWM, it is possible to generate PPM with a 555 timer.

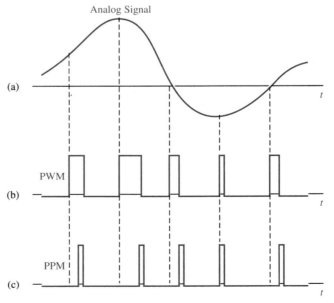

**FIGURE 8–14**
Forms for pulse-width modulated (PWM) and pulse-position modulated (PPM) signals.

## PWM and PPM Bandwidth

An exact spectral analysis of either a PWM or a PPM signal is rather complex, but for the purpose of this brief introduction, a reasonable estimation process will suffice. Since the exact width or position of the pulse is critical to the modulation process, the coarse bandwidth approximation utilized for PAM is inadequate. Instead, a "fine" approximation for the bandwidth should be used. Said differently, the allowable rise time of the pulse, which serves to locate the exact width or position of the pulse, is the criterion of reproduction. This means that for a given analog bandwidth, the required transmission bandwidth for either PWM or PPM is much greater than for PAM.

## Analog and Pulse Modulation Comparisons

Although the analogies are not perfect, there are some similarities between analog and pulse modulation processes in the following sense:

| Analog Modulation | Pulse Modulation |
|---|---|
| AM | PAM |
| FM | PWM |
| PM | PPM |

The reader is advised not to take these analogies too literally, because they are all totally different processes. However, this categorization is helpful in remembering the different methods.

---

**III EXAMPLE 8-9**

An analog signal having a baseband bandwidth of 10 kHz is to be sampled and converted to a PWM format. Specifications dictate that each pulse must be reproduced to a sufficient accuracy such that the rise time cannot exceed 1% of the sample time interval. Determine the approximate transmission bandwidth. Assume the minimum Nyquist sampling rate.

**SOLUTION** The minimum sampling rate required is

$$f_s = 2W = 2 \times 10 \text{ kHz} = 20 \text{ kHz} \tag{8-39}$$

The frame time is

$$T_f = \frac{1}{f_s} = \frac{1}{20 \text{ kHz}} = 50 \text{ μs} \tag{8-40}$$

The maximum rise time permitted is

$$t_r = 0.01 \times 50 \text{ μs} = 0.5 \text{ μs} \tag{8-41}$$

The approximate transmission bandwidth is then

$$B_T = \frac{0.5}{t_r} = \frac{0.5}{0.5 \text{ μs}} = 1 \text{ MHz} \tag{8-42}$$

---

## 8-6 Brief Introduction to Pulse-Code Modulation

Much of the next several chapters will be devoted to a detailed analysis of various forms of digital communication systems. However, within the context of sampling, it is appropriate to introduce here the idea of representing samples of an analog signal in terms of digital numbers. The basic process is called *pulse-code modulation* (*PCM*).

The most basic form of PCM utilizes the binary number system, which assumes only two states. While the actual values depend on the types of circuits involved, the levels are usually referred to as **0** and **1**. A given **bi**nary digi**t** is called a **bit**.

With binary PCM, each analog sample of the signal is represented by an $N$-bit binary word. The number of levels $M$ associated with $N$ bits is

$$M = 2^N \tag{8-43}$$

With analog signals or with pulse modulation forms, it is theoretically possible to represent an infinite number of levels. However, with PCM, the number of levels is always finite. However, this does not pose any realistic performance limitation, since increasing the number of bits in each sample will increase the number of levels to any arbitrary number.

A common example of PCM is that of commercial music compact discs (CDs). The samples of the signals recorded on CDs utilize 16 bits, which permits the possibility of $2^{16} = 65,536$ levels! Anyone owning a CD player is well aware of the high quality of the sound.

One of the great advantages of binary PCM is the fact that the receiver need only distinguish between two possible levels. Thus, as long as the noise is within reasonable limits, it is possible to determine whether a given pulse is a 1 or a 0, and the noisy and distorted pulses can be "cleaned up" and restored by creating replicas of the original pulses.

Since each sample of the signal requires $N$ bits, there must be $N$ pulses produced within a given sample length. This increases the bandwidth of binary PCM by a factor of $N$ over that required for PAM.

This brief section has been placed in this chapter simply to alert the reader to the fact that digital communications is part of the overall picture relating to pulse and sampling theory. Because of its importance in modern communications systems, the next several chapters will be devoted to a detailed development of various forms of digital communication techniques.

## 8-7 Multisim® Examples (Optional)

The Multisim examples of this section will focus on the sampling theorem and its properties. In Multisim Example 8-1, a PAM signal will be generated by sampling a sinusoidal function. The spectrum of the sample signal will be investigated in Multisim Example 8-2. Finally, the original analog signal will be reconstructed by low-pass filtering in Multisim Example 8-3.

---

**▌▌ MULTISIM EXAMPLE 8-1**

Develop a Multisim circuit for modeling the generation of a PAM signal. The input signal is to be a 1-kHz sinusoid with an amplitude of 1 V. The sampling rate is to be 10 kHz, and the width of the samples should be 10% of the period of the sampling function.

**SOLUTION**    Refer to Figure 8–15 for the discussion that follows. The sampling process can be readily implemented with a **MULTIPLIER,** which is obtained from the **CONTROL_ FUNCTIONS** option in the **Sources** toolbar. The sampling function is the **Pulse Voltage**

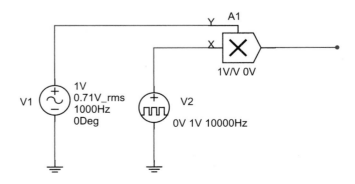

**FIGURE 8–15**
Circuit of Multisim Example 8-1.

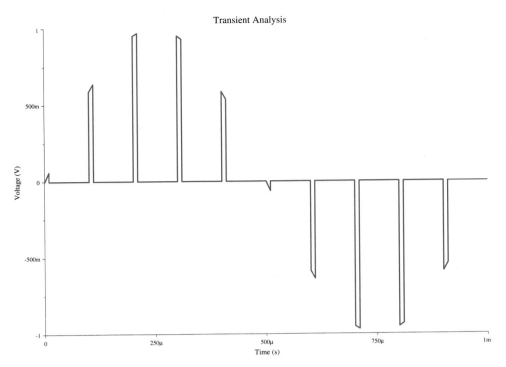

**FIGURE 8–16**
PAM signal of Multisim Example 8-1.

**Source.** The **Initial Value** is set to **0 V**, the **Pulsed Value** is set to **1 V**, and the **Delay Time** is set to **0**. The values of **Rise Time** and **Fall Time** should be small compared with the pulse width, and they have been left at their default values of **1 ns**, which should be acceptable. The **Period** is set to $1/1e4 = \textbf{0.1 ms}$. Since the sample width is to occupy 10% of the period, the **Pulse Width** is set to **0.01 ms**. Of course, the signal to be sampled is an ac sinusoidal voltage set to a frequency of 1 kHz and an amplitude of 1 V. A **Junction** is used at the output of the multiplier.

A **Transient Analysis** is performed over the time interval of 1 cycle of the input sinusoid, and the result is shown in Figure 8–16. The resulting waveform is a natural-sampled PAM signal.

---

**■|| MULTISIM EXAMPLE 8-2**      Perform a spectral analysis on the waveform of Multisim Example 8-1.

**SOLUTION**   As we know, the spectrum will be very wide, so we will choose to observe the spectral terms over the frequency range up to about 50 kHz. The input signal has a frequency of 1 kHz, which will yield spectral terms that are 1 kHz apart. The **Frequency resolution** was set to **1000 Hz** and the **Number of harmonics** was set to **49**. Finally, the **Stop time for sampling** was set to **1e-3**, which represents one cycle of the input signal.

The spectrum from dc to 50 kHz is shown in Figure 8–17. Along with the original signal at 1 kHz, we have the sum and difference frequencies between the sampling frequency and the input frequency. Thus, there are components at 9 and 11 kHz, 19 and 21 kHz, 29 and 31 kHz, and so on. Note that for higher frequencies, the components are beginning to diminish in amplitude, as expected. The first zero crossing in the spectral envelope would occur at 100 kHz.

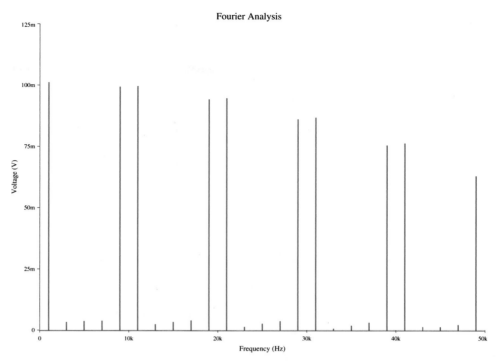

FIGURE 8–17
Spectral data of Multisim Example 8-2.

---

**▮▮ MULTISIM EXAMPLE 8-3**    The circuit of Figure 8–18 represents a combination of the PAM sampled signal of Multisim Example 8-1 followed by a Chebyshev three-pole low-pass filter with a 1-dB bandwidth of 2 kHz. Perform a transient analysis of the circuit, and show that the form of the original signal is reconstructed.

**SOLUTION**    If the composite spectrum shown in Figure 8–17 is reviewed, it can be deduced that a bandwidth of about 2 kHz should easily pass the original signal (the 1-kHz component), but it should essentially reject all the other components. A **Transient**

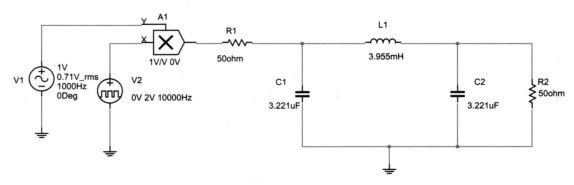

FIGURE 8–18
Circuit of Multisim Example 8-3.

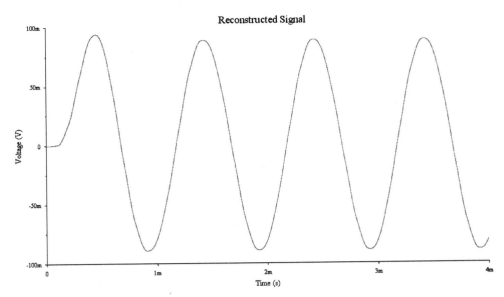

**FIGURE 8–19**
Output waveform of Multisim Example 8-3.

**Analysis** is performed over an interval from $t = 0$ to $t = 4$ ms and the result is shown in Figure 8–19.

As is the case with any filter when it is initially excited, there is a transient interval or settling time. However, the steady-state output does appear to be sinusoidal and the period can be readily measured as 1 ms, which means that the frequency is 1 kHz as expected. There is a significant loss in amplitude, since no holding circuit was employed, but the form of the input signal definitely has been recovered.

## 8-8  MATLAB® Examples (Optional)

Three MATLAB examples illustrating the sampling and reconstruction processes will be given in this section. The first example will apply the sampling process to generate a natural-sampled PAM signal; the second example will display the spectrum of the sampled signal; and the third example will demonstrate the reconstruction of the original signal by low-pass filtering.

**▌ MATLAB EXAMPLE 8-1**

Sample an input 1-kHz sinusoidal signal at a 10-kHz rate using a pulse train having a duty cycle of 0.1 for five cycles of the input signal. Plot the function over one input cycle.

**SOLUTION**  If we only desired to sample the signal and show its spectrum, one cycle would suffice. However, in MATLAB Example 8-3, we will reconstruct the original signal by filtering, and several cycles are necessary due to transient effects. We will, therefore, sample the input signal over 5 cycles, and will choose a time increment of 1 μs and stop 1 μs short of the end of 5 cycles of the input signal. The time vector **t** is

```
>> t = 0:1e-6:5e-3-1e-6;
```

The input signal voltage **vin** is generated by

```
>> vin = sin(2*pi*1000*t);
```

We now need to generate the sampling function. Since it is required to be at a frequency of 10 kHz with a duty cycle of 0.1, there is a need to generate 50 short pulses over the interval of 5 ms. The process could be done manually, but we will use a trick. Start by defining one cycle **ps1** of a periodic sampling pulse train.

```
>> ps1 = [ones(1,10) zeros(1,90)];
```

This generates one cycle of the pulse train **ps** based on a time interval of $100 \times 1e-6 = 0.1$ ms. Next define **ps** as the first cycle by the command

```
ps = ps1;
```

Now comes the trick. We will form a **for** loop by the following command:

```
>> for n = 1:49
>> ps=[ps ps1];
>> end
```

The 49 steps that follow the beginning step will add one cycle per loop, so that the final function **ps** will contain 50 identical cycles. (MATLAB is loaded with little tricks of this sort that you can learn to use.)

We now generate the sampled signal **vsampled** by the point-by-point multiplication of **vin** by **ps**.

```
>> vsampled = vin.*ps;
```

Note that the period is required after the first vector to force multiplication on a point-by-point basis. The entire time function over the five cycles of the input signal can be easily plotted, but the results tend to get crowded. Instead, we will show the plot over one cycle of the input signal by the command

```
>> plot(t(1:1000),vsampled(1:1000))
```

The result, with appropriate labeling, is shown in Figure 8–20.

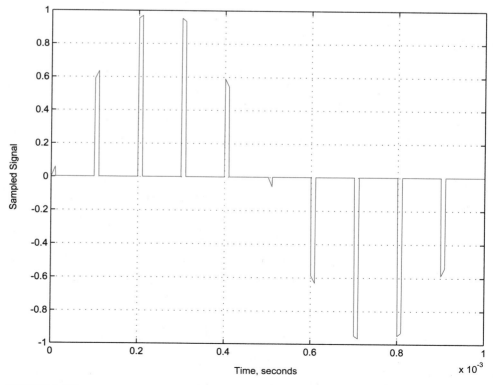

**FIGURE 8–20**

Pulse-amplitude modulated signal using natural sampling.

▌▌ MATLAB EXAMPLE 8-2

Use the program **fourier_series_1** to plot the spectrum of the sampled function of MATLAB Example 8-1.

SOLUTION    We first need to redefine the function **vsampled** as **v**, which is easily achieved by the command

```
>> v = vsampled;
```

We then initiate the special FFT program by the command

```
>> fourier_series_1
```

The first required entry is the time step, which is 1e-6 s. The frequency increment is 200 Hz, so for the integers requested, the first value is entered as 0 and the last value is entered as 175. This will result in the spectrum being displayed from dc to 35 kHz, and the result is shown in Figure 8–21. Note that the original spectrum (1 kHz) appears, as do the sum and difference frequencies about the sampling frequency. Note that the components start to get smaller as the frequency increases, which is expected.

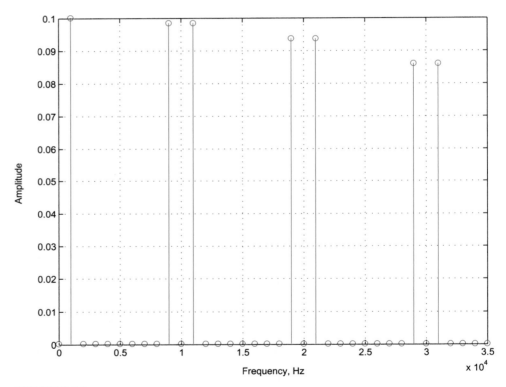

FIGURE 8–21
Amplitude spectrum based on one-sided form using FFT.

▌▌ MATLAB EXAMPLE 8-3

Apply the signal of MATLAB Example 8-1 to a low-pass filter, and show that the original sinusoidal signal is reconstructed.

SOLUTION    Based on the spectrum shown in Figure 8–21, the cutoff frequency of the low-pass filter should be somewhere above 1 kHz and well below 9 kHz. We will arbitrarily select a Butterworth five-pole low-pass filter with a cutoff frequency of 1500 Hz. Assuming that the **Signal Processing Toolbox** is installed, the numerator **n** and denominator

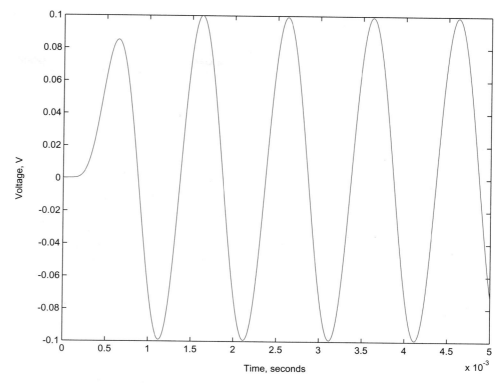

**FIGURE 8–22**
Output of low-pass filter with sampled signal as input.

**d** coefficients of the filter are generated by the command

```
>> [n d] = butter(5,2*pi*1500,'s');
```

We can then filter the signal to determine an output **vout** by the command

```
>> vout = lsim(n, d, vsampled, t);
```

The result is shown in Figure 8–22. All filters require a certain amount of settling time, based on the transient response of the filter, and this is evident in the first cycle. However, the steady-state output of the filter is a 1-kHz sinusoid, as expected.

No holding circuit was used in this example, so the output amplitude is reduced with respect to the input, as expected. However, the form of the output is proportional to the original 1-kHz sinusoid.

## SystemVue™ Closing Application (Optional)

Insert the text CD in a computer having SystemVue™ installed and activate the program. Open the CD folder entitled **SystemVue Systems** and load the file entitled **8-1**.

### Review of System

The system shown on the screen was described in the chapter opening application, and you may wish to review the discussion and system configuration provided there. Recall that the input signal is a 1-Hz sinusoid that is sampled at a rate of 20 samples per second, and each sample has a duration of 0.1 s.

## Reconstruction of the Analog Signal

*Note:* All filters have a transient or settling interval, which is of the order of the reciprocal of the bandwidth. Therefore, form any conclusions after an allowance has been made for this phenomenon. Moreover, since a holding circuit has not been used in this demonstration, the actual level of the output will be much smaller than the level of the signal prior to sampling.

1. Check the parameter values of the output low-pass filter. It should be initially set as a 5-pole low-pass Butterworth filter with a cutoff frequency of 5 Hz.

2. Perform a simulation run and observe the **Output** waveform (Token 3). After a short settling interval, what can you say about this waveform in relationship to the input waveform before sampling?

3. Experiment with different cutoff frequencies for the low-pass filter. Determine the approximate lowest cutoff frequency that does not cause severe distortion to the signal. Likewise, determine the highest cutoff frequency that does not cause severe distortion.

## Aliasing

1. Reset the cutoff frequency of the filter to a value of 5 Hz.

2. Increase the input signal frequency to a value of 19 Hz and observe the output. What actual frequency do you see on the graph, and why is it not the input frequency?

## PROBLEMS

**8-1**  A baseband signal has frequency components from dc to 8 kHz. Determine (a) the theoretical minimum sampling rate and (b) the maximum time interval between successive samples.

**8-2**  A baseband signal has frequency components from dc to 20 kHz. Determine (a) the theoretical minimum sampling rate and (b) the maximum time interval between successive samples.

**8-3**  Assume that the signal of Example 8-1 is sampled at a rate 25% above the theoretical minimum. Determine (a) the sampling rate and (b) the maximum time interval between successive samples.

**8-4**  Assume that the signal of Example 8-2 is sampled at a rate 40% above the theoretical minimum. Determine (a) the sampling rate and (b) the maximum time interval between successive samples.

**8-5**  For the signal of Problem 8-1, determine the minimum sampling rate such that the *guard band* is 6 kHz. (The *guard band* is the frequency difference between the highest baseband frequency and the lowest frequency of the first shifted component.)

**8-6**  For the signal of Problem 8-2, determine the minimum sampling rate such that the *guard band* is 10 kHz. (See the definition of *guard band* in Problem 8-5.)

**8-7**  The sampling rate for commercial CDs is 44.1 kHz. Determine the highest theoretical baseband frequency that could be reproduced.

**8-8**  A system has a sampling rate of 14 kHz. Determine the highest theoretical baseband frequency that could be reproduced.

**8-9**  An analog signal has a duration of 1 minute. The spectral content ranges from near dc to 500 Hz. The signal is to be sampled, converted to digital format, and stored in memory for subsequent processing. To assist in recovery, the sampling rate is chosen to be 25% above the theoretical minimum. (a) Determine the minimum number of samples that must be taken if reconstruction is desired. (b) Determine the time interval between successive samples.

**8-10**  An analog signal has a duration of 1 hour. The spectral content ranges from near dc to 4 kHz. The signal is to be sampled, converted to digital format, and stored in memory for subsequent processing. To assist in recovery, the sampling rate is chosen to be 50% above the theoretical minimum. (a) Determine the minimum number of samples that must be taken if reconstruction is desired. (b) Determine the time interval between successive samples.

**8-11**  A digital signal processing system operates at a sampling rate of 10,000 samples/s. Determine the theoretical highest frequency permitted in a baseband signal if a guard band of 4 kHz is to be established between the upper range of the baseband signal and the lower range of the first translated component.

**8-12**  A digital signal processing system operates at a sampling rate of 9000 samples/s. Determine the theoretical highest frequency permitted in a baseband signal if the lowest frequency of the first translated component is to be 25% higher than the highest frequency of the baseband component.

**8-13**  A sinusoid with a frequency of 2 kHz is sampled at a rate of 16 kHz and converted into digital numbers to be processed on a computer. List all positive frequencies in the spectrum below 60 kHz.

**8-14**  A sinusoid with a frequency of 500 Hz is sampled at intervals of 0.25 ms and converted into digital numbers to be processed on a computer. List all positive frequencies in the spectrum below 18 kHz.

**8-15**  A signal consists of two sinusoidal components at frequencies of 1 kHz and 2 kHz. It is sampled at intervals of 0.1 ms and converted into digital numbers to be processed on a computer. List all positive frequencies below 35 kHz.

**8-16**  A signal consists of two sinusoidal components at frequencies of 500 Hz and 1.5 kHz. It is sampled at a rate of 5 kHz and converted into digital numbers to be processed on a computer. List all positive frequencies below 18 kHz.

**8-17**  A PAM TDM system has six signals plus synchronization with a frame format defined as follows: Sync pulse occupies an interval $2\tau$, while each signal sample has a width $\tau$ and a space of width $\tau$ is placed between successive pulses. The receiver detects the wider sync pulse once per frame, and realigns the timing sequence accordingly. (a) Determine the approximate baseband bandwidth required for the composite PAM signal if each channel has a 1-kHz bandwidth. (b) If the composite PAM signal is used to amplitude modulate a high-frequency carrier, determine the RF bandwidth required for the resulting high-frequency signal.

**8-18**  For the PAM TDM system of Example 8-8, repeat all computations if the sampling rate is increased to twice the theoretical minimum rate.

**8-19**  For the PWM TDM system of Example 8-9, determine the approximate transmission bandwidth if the sampling rate is set at 25% above the theoretical minimum.

**8-20**  For the PWM TDM system of Example 8-9, determine the approximate transmission bandwidth if the rise time specification is changed so that it cannot exceed 0.2% of the time allocated to the given channel. All other specifications are the same as in Example 8-9.

# Digital Communications I: Binary Systems

# 9

## OVERVIEW AND OBJECTIVES

The emphasis in this chapter will be directed toward the development, operation, and general comparison of basic digital communication systems utilizing **binary encoding.** The binary process is the basis for all digital communications, and it was universally employed in earlier digital communication systems. In the next chapter, the concept will be extended to encoding schemes involving more levels, but it is necessary to understand the binary process as a prerequisite to the more recent higher data rate systems.

The basic concept of *pulse-code modulation,* which is the foundation for encoding analog signals in digital form, will be explored in detail in this chapter. Various data formats and conversion techniques will be discussed. Modulation methods used for encoding and transmitting digital data at higher frequencies will be presented.

## Objectives

After completing this chapter, the reader should be able to:

1. Define the basic form of a *pulse-code modulated* (PCM) signal and show how it is created.

2. Define the strategies involved in *unipolar* and *bipolar offset encoding* of A/D converters.

3. State and apply the relationships for step size and error for both unipolar and bipolar offset encoding.

4. Define *compression, expansion, companding,* and the *μ-law characteristic.*

5. Discuss the various baseband encoding forms, including NRZ-L, NRZ-M, NRZ-S, RZ, Biphase-L, Biphase-M, and Biphase-S.

6. Determine the concept of *time-division multiplexing* (TDM) as it relates to PCM, and determine the transmission bandwidth.

7. Define *amplitude-shift keying* (ASK), and discuss its generation, detection, and bandwidth requirements.

8. Define *frequency-shift keying* (FSK), and discuss its generation, detection, and bandwidth requirements.

9. Define *binary phase-shift keying* (BPSK), and discuss its generation, detection, and bandwidth requirements.

10. Define *differential phase-shift keying* (DPSK), and discuss its properties and special advantages.

## SystemVue™ Opening Application (Optional)

Insert the text CD in a computer having SystemVue™ installed and activate the program. Open the CD folder entitled **SystemVue Systems** and load the file entitled 9-1.

### Sink Tokens

| Number | Name | Token Monitored |
|--------|------|-----------------|
| 0 | Input | 2 |
| 1 | Output | 3 |

### Operational Tokens

| Number | Name | Function |
|--------|------|----------|
| 2 | Input | sawtooth waveform varying from −1 V to 1 V at a frequency of 1 Hz. |
| 3 | Quantizer | quantizer set for 2 bits (4 levels) over range from −1 V to 1 V. |

The preset values for the demonstration are from a starting time of 0 to a final time of 2 s, with a time step of 1 ms. Run the simulation and observe the waveforms at sinks 0 and 1. Next left-click on the **Sink Calculator** at the lower left end of the screen. It can be identified by the symbol $\sqrt{\alpha}$. A window with the appropriate name will then open.

Depending on the manner in which the initial graphs were displayed, some of the settings that follow may already be in place, but the sequence is provided for convenience. First, left-click on **Operators.** The upper window on the right has the title **Select.** In this case, there are only two windows. Depress the **Ctrl** button and hold it while selecting both of the windows. Left-click the **Overlay Plots** button and then left-click **OK.** A window should open with both curves on the same scale. If necessary maximize this window.

### What You See

The trapezoidal analog waveform ramps from a level of −1 V to 1 V at a rate of 1 Hz. Based on only two bits, the quantizer permits only four possible levels for the output signal. Of course, this is a rather poor representation, but it has been chosen to illustrate the process very clearly.

### How This Demonstration Relates to the Chapter Learning Objectives

In Chapter 9, you will learn how an analog-to-digital (A/D) converter can convert an analog signal to a digital signal, and how a digital-to-analog (D/A) converter can convert the digital signal back to an analog signal. A fundamental step in the process is that of quantization, in which the infinite number of possible analog levels is converted to a finite number of quantized levels. Though finite, the number of levels can be increased to a very large number for very fine representation. For example, in commercial compact disc recordings, the number of possible levels is 65,536.

At the end of the chapter you will return to this system and investigate some of its quantitative behavior, along with the effect of changing the number of levels and the type of waveform.

## 9-1    Pulse-Code Modulation

We will begin the study of digital communications by considering those forms in which an analog signal is to be approximated by a sequence of samples, each of which is converted to a binary word. The basic form of the resulting digital signal is referred to as *pulse-code modulation.*

The block diagram shown in Figure 9–1 illustrates the concept of PCM generation. (The elements in the block are identified in a form that illustrates the processes involved rather than showing an actual implementation. In practice, several of these operations may occur in the same circuit.) Assume an analog signal $v(t)$, band-limited from dc to $W$ hertz, which is to be converted to PCM form. The signal is first filtered by an *anti-aliasing* analog filter whose function is to remove any superfluous frequency components above $W$ that might appear at the input and be shifted into the data band by the aliasing effect. The signal is then sampled at a rate $f_s > 2W$. At this point in the block diagram, the signal has essentially the same characteristics as a sampled-data PAM signal as discussed in the preceding chapter.

### Quantization

A process unique to all digital representation of analog signals is now performed. Each sampled-data pulse is converted into (or replaced by) one of a finite number of possible values. This process is called *quantization,* and the quantizer block in the diagram of Figure 9–1 is assumed to accomplish this purpose. Following the quantization process, each of the standard values generated is encoded into the proper form for transmission or subsequent signal processing. Each of the composite encoded samples appearing at the output of the encoder is called a *PCM word.* Thus, the infinite number of possible amplitude levels of the sampled signal are converted to a finite number of possible PCM words.

### Binary Number System

The basic format for PCM word representation is the *binary* number system. Each word then consists of a combination of two logic levels: 0 and 1. The actual voltage levels associated with these binary values vary with the type of logic circuits utilized.

The presence of only two levels means that the transmission channel and receiver circuitry need only be capable of recognizing two possible signal conditions. This results in some significant advantages, as we will see later.

**Binary Digits**

Let $N$ represent the number of <u>binary digits</u> (bits) used in each word, and let $M$ represent the number of possible distinct words or values that can be generated. We have

$$M = 2^N \qquad (9\text{-}1)$$

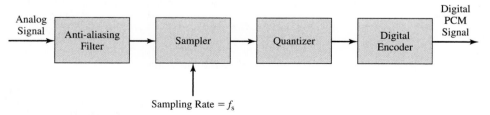

**FIGURE 9–1**
Block diagram illustrating the steps involved in generating a digital PCM signal. (In practice, several of the steps may be accomplished in the same circuit.)

(It turns out that we will also use the symbol $N$ with various subscripts later to represent noise power, but it should be obvious to the reader which variable is under consideration.)

The number of possible words is very limited for only a few bits, but as the number of bits increases, the number of possible words increases exponentially. Thus, while PCM requires all signal levels to be encoded or quantized to a finite number of standard levels, the degree of accuracy in this process can, in theory, be made as close as desired. In practice, there are limits to the ultimate accuracy attainable, but the same is true of analog systems as well.

If the number of possible words $M$ is known, the number of bits $N$ is given by

$$N = \log_2 M = 3.32 \log_{10} M \tag{9-2}$$

where the latter form permits the logarithm to the base 2 to be computed in terms of the logarithm to the base 10. (This formula will be derived in Chapter 10.)

Since the initial value of $N$ as calculated in Equation 9-2 may not be an integer, it must be rounded to an integer value. To achieve the accuracy originally desired, the value would be rounded *upward* to the next largest integer. For that reason, system designers usually specify the number of possible PCM levels directly as an integer power of 2.

The end result of the basic sampling, quantization, and encoding process in a binary PCM system then is a sequence of binary words, each having $N$ bits. The value of each binary word represents in some predetermined way the amplitude of the analog signal at the point of sampling. In effect, the basic process is an *analog-to-digital* (A/D) conversion. Depending on the system requirements and other factors, some of the readily available A/D converters could be used for the PCM transmitter encoding process.

## Decoding

Skipping for the moment all the detailed steps between the encoding process and the receiver, an inverse decoding process must be employed at the receiver to convert the received digital words into a usable analog form, if the desired output is an analog signal. This process is a form of *digital-to-analog* (D/A) conversion, and some of the available D/A converters could be used for the receiver decoding purpose.

## Signal Restoration

With analog communication processes, there are various ways in which a signal may be degraded or changed in form in the transmission medium. With binary digital or PCM systems, *as long as the signal level is sufficiently large that it is possible to tell the difference between the two values transmitted, the signal can theoretically be reconstructed to the accuracy inherent in the sampling and quantization processes.* This is one of the outstanding advantages of digital systems.

---

**▌▌ EXAMPLE 9-1**

Determine the number of possible PCM word values that can be encoded for (a) 4 bits, (b) 8 bits, and (c) 16 bits.

**SOLUTION**    The basic relationship is that of Equation 9-1.

(a)  $M = 2^4 = 16$ values $\tag{9-3}$

(b)  $M = 2^8 = 256$ values $\tag{9-4}$

(c)  $M = 2^{16} = 65,536$ values $\tag{9-5}$

As noted, the number of possible levels increases very rapidly as the number of bits increases.

---

**▌▌ EXAMPLE 9-2**

It is desired to represent an analog signal in no less than 100 fixed values. Determine the minimum number of bits used for each word.

**SOLUTION** If we employ the basic relationship of Equation 9-2, we obtain

$$N = \log_2 100 = 3.32 \log_{10} 100 = 3.32 \times 2 = 6.64 \text{ bits} \qquad (9\text{-}6)$$

However, the number of bits must be an integer number. Thus, we round upward to **7 bits,** which will actually provide **128 values.**

## 9-2 Basic PCM Encoding and Quantization

In this section, we will consider the manner in which a PCM signal is encoded in its basic form, and the corresponding quantization characteristics. By "basic" encoding, we refer to the natural process of representing numbers in the conventional or *natural binary* number system form, as well as conversions or interpretations of those numbers in normalized or fractional forms. The encoding or conversion to special data forms will be considered in later sections.

To simplify the discussion in this section, $N = 4$ bits will be used for illustration. This choice results in $M = 2^4 = 16$ words, which is sufficiently large that the general trend can be deduced and yet is sufficiently small that the results can be readily shown in graphical and tabular form.

The 16 natural binary words attainable with 4 bits and their corresponding decimal values are shown in the two left-hand columns of Table 9–1. (The two right-hand columns will be explained later.) Note that the smallest value is at the bottom of the table, and the largest value is at the top. For each binary number, the bit farthest to the left is called the *most significant bit* (*MSB*), and the bit farthest to the right is called the *least significant bit* (*LSB*).

Depending on the exact circuitry involved, the dynamic range of the signal, and other factors, the 16 possible levels could be made to correspond to particular values of the analog signal, as desired. However, some standardization has been achieved with many of the commercially available A/D and D/A converters, and the discussion here will concentrate on some of these standard forms. Typically, A/D and D/A converters are designed to operate with maximum analog voltage levels of 2.5, 5, 10, or 20 V, although means are often provided for changing to other values if desired.

**Table 9–1** Natural Binary Numbers, Decimal Values, and Unipolar and Bipolar Offset Values for 4 Bits

| Natural Binary Number | Decimal Value[1] | Unipolar Normalized Decimal Value[2] | Bipolar Offset Normalized Decimal Value[2] |
|---|---|---|---|
| 1111 | 15 | 15/16 = 0.9375 | 7/8 = 0.875 |
| 1110 | 14 | 14/16 = 0.875 | 6/8 = 0.75 |
| 1101 | 13 | 13/16 = 0.8125 | 5/8 = 0.625 |
| 1100 | 12 | 12/16 = 0.75 | 4/8 = 0.5 |
| 1011 | 11 | 11/16 = 0.6875 | 3/8 = 0.375 |
| 1010 | 10 | 10/16 = 0.625 | 2/8 = 0.25 |
| 1001 | 9 | 9/16 = 0.5625 | 1/8 = 0.125 |
| 1000 | 8 | 8/16 = 0.5 | 0 |
| 0111 | 7 | 7/16 = 0.4375 | −1/8 = −0.125 |
| 0110 | 6 | 6/16 = 0.375 | −2/8 = −0.25 |
| 0101 | 5 | 5/16 = 0.3125 | −3/8 = −0.375 |
| 0100 | 4 | 4/16 = 0.25 | −4/8 = −0.5 |
| 0011 | 3 | 3/16 = 0.1875 | −5/8 = −0.625 |
| 0010 | 2 | 2/16 = 0.125 | −6/8 = −0.75 |
| 0001 | 1 | 1/16 = 0.0625 | −7/8 = −0.875 |
| 0000 | 0 | 0 | −8/8 = −1 |

[1]Decimal value for integer representation of binary number.

[2]Decimal value with binary point understood on left-hand side of binary number.

## Normalization

Because of these different voltage levels, and the widely different decimal values of the binary number system as the number of bits is changed, it is frequently desirable to *normalize* the levels of both the analog signal and digital words so that the maximum magnitudes of both forms have (or at least approach) unity. The *normalized input analog voltage* is defined as

$$\text{normalized input analog voltage} = \frac{\text{actual input analog voltage}}{\text{full-scale voltage of A/D converter}} \qquad (9\text{-}7)$$

where the full-scale voltage of the A/D converter is typically 2.5, 5, 10, or 20 V. At the D/A converter in the receiver, the output voltage is

$$\begin{pmatrix} \text{actual output} \\ \text{analog voltage} \end{pmatrix} = \begin{pmatrix} \text{normalized value} \\ \text{of digital word} \end{pmatrix} \times \begin{pmatrix} \text{full-scale voltage} \\ \text{of D/A converter} \end{pmatrix} \qquad (9\text{-}8)$$

Most of the subsequent discussions in this section will be based on the normalized forms of both the analog and digital values. The normalized voltage will be denoted $x(t)$, with subscripts added as appropriate. The full-scale voltage will be denoted $V_{\text{fs}}$.

Normalization of the values of the binary words in Table 9–1 to a range less than 1 is achieved by adding a binary point to the left of the values given in the table. To simplify the notation, the binary point will usually be omitted, but it will be understood in all discussions in which the normalized form is assumed.

We will now consider the manner in which the normalized analog level can be related to the normalized digital level. The two most common forms employed in A/D conversion are (1) the *unipolar representation* and (2) the *bipolar offset representation*. The subscript "u" will be used for *unipolar* parameters, and the subscript "b" will be used for *bipolar* parameters.

## Unipolar Encoding

The *unipolar* representation is most appropriate when the analog signal is always of one polarity (including zero). If the signal is negative, it can be inverted before sampling, so assume that the range of the normalized analog signal $x(t)$ is $0 \leq x < 1$. Let $X_{\text{u}}$ represent the unipolar normalized quantized decimal representation of $x$ following the A/D conversion. The 16 possible values of $X_{\text{u}}$ for the case of 4 bits are shown in the third column of Table 9–1. Note that 0000 in binary corresponds to true decimal 0, and that the binary value 1000 corresponds to the exact decimal mid-range value 0.5. However, the upper value of binary 1111 does not actually reach decimal 1, but instead has the decimal value $15/16 = 0.9375$.

Let $\Delta X_{\text{u}}$ represent the normalized step size, which represents on a decimal basis the difference between successive levels. This value is also the decimal value corresponding to 1 LSB and is, in the general case,

$$\Delta X_{\text{u}} = 2^{-N} \qquad (9\text{-}9)$$

The actual step size in volts $\Delta v_{\text{u}}$ is obtained by multiplying the normalized step size by the full-scale voltage,

$$\Delta v_{\text{u}} = \Delta X_{\text{u}} V_{\text{fs}} = 2^{-N} V_{\text{fs}} \qquad (9\text{-}10)$$

for the *unipolar* system. The largest normalized quantized decimal value attainable $X_{\text{u}}(\text{max})$ differs from unity by the value of 1 LSB, and is

$$X_{\text{u}}(\text{max}) = 1 - \Delta X_{\text{u}} = 1 - 2^{-N} \qquad (9\text{-}11)$$

for the *unipolar* system. Thus, unity can be approached to an arbitrarily close value, but it can never be completely reached. The resulting distribution about the midpoint is not completely symmetrical, but it has the advantages that the absolute zero levels of both number systems are identical, and the exact midpoint of the analog voltage corresponds to the binary value whose MSB is 1 and whose other bits are all 0.

## Bipolar Offset Encoding

The *bipolar offset representation* is most appropriate when the analog signal has both polarities. Specifically, it assumes that the normalized range of the analog signal is $-1 \leq x < 1$. Let $X_b$ represent the bipolar normalized quantized decimal representation of $x$, following the A/D conversion. The 16 possible values of $X_b$ for the case of 4 bits are shown in the fourth column of Table 9–1. Note that 0000 in binary corresponds exactly to the decimal value of $-1$, while the binary value 1000 corresponds to the exact decimal value of 0. However, the upper value of binary 1111 does not actually reach the decimal value 1 but instead has the decimal value $7/8 = 0.875$.

Let $\Delta X_b$ represent the normalized step size or value corresponding to 1 LSB for the bipolar case. This value is

$$\Delta X_b = 2^{-(N-1)} = 2^{-N+1} \tag{9-12}$$

The actual step size $\Delta v_b$ is

$$\Delta v_b = \Delta X_b V_{fs} = 2^{-N+1} V_{fs} \tag{9-13}$$

for the *bipolar offset* case, which is twice as large as for the unipolar case. This value on a normalized basis appears to be twice as great as for the unipolar case, but that is because the normalized peak-to-peak value is twice as great as for the unipolar case. The largest normalized quantized decimal value attainable $X_b(\text{max})$ is

$$X_b(\text{max}) = 1 - \Delta X_b = 1 - 2^{-N+1} \tag{9-14}$$

for the *bipolar offset* case.

Observe from the fourth column of Table 9–1 that for 4 bits there are 8 binary words corresponding to negative decimal quantized levels, 7 binary words corresponding to positive decimal levels, and 1 binary word (1000) corresponding to a decimal level of 0. In general, there are $M/2 = 2^{N-1}$ binary words corresponding to negative quantized decimal signal levels, $M/2 - 1 = 2^{N-1} - 1$ binary words corresponding to positive decimal values, and one word corresponding to decimal 0. This last level is always represented by a binary word whose MSB is 1 and whose other bits are 0.

As an additional point of interest, the *two's-complement* representation of a given binary number can be obtained by replacing the MSB in the bipolar representation by its logical complement. In fact, the bipolar offset representation is sometimes referred to as a *modified two's-complement* representation.

## Quantization Curves

The quantized decimal values and their binary representations have now been defined for two possible forms. The exact manner in which the quantized decimal values represent different ranges of the analog signal is defined by means of the *quantization characteristic curve*. Both *rounding* and *truncation* strategies may be employed. In *rounding,* the sampled value of the analog signal is assigned to the *nearest* quantized level. In *truncation,* the sampled value is rounded down to the next lowest quantized level. For example, assume that two successive quantization levels correspond to 6.2 and 6.4 V, respectively. With rounding, a sample of value 6.29 V would be set at 6.2 V, and a sample of 6.31 V would be set at 6.4 V. With truncation, both samples would be set to 6.2 V. As might be expected, the average error associated with truncation is greater than for rounding. Nevertheless, there are applications in which truncation is used.

From this point on, we will assume the more common process of *rounding*. The quantization characteristic curve for an ideal 4-bit A/D converter employing *rounding* and *unipolar* encoding is shown in Figure 9–2. The horizontal scale represents the decimal analog signal level on a *normalized* basis. The vertical scale represents the decimal representation on a *normalized unipolar* basis. The straight line represents the true analog ideal characteristic in which the output would equal the input.

The corresponding quantization curve for *bipolar offset* encoding is shown in Figure 9–3. As in the previous case, rounding is assumed.

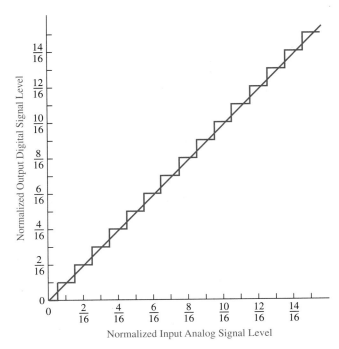

**FIGURE 9–2**
Unipolar quantization characteristic for a 4-bit A/D converter.

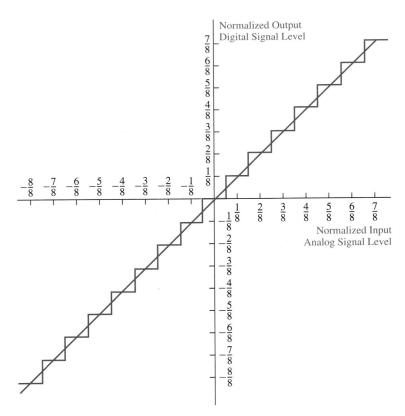

**FIGURE 9–3**
Bipolar offset quantization characteristic for a 4-bit A/D converter.

## Quantization Error

With the assumption of rounding, the theoretical peak quantization error is one-half of one LSB. Let $E_u$ represent the peak unipolar normalized error, and let $E_b$ represent the peak bipolar normalized error. These values are, respectively,

$$E_u = \frac{\Delta X_u}{2} = 2^{-(N+1)} \tag{9-15}$$

and

$$E_b = \frac{\Delta X_b}{2} = 2^{-N} \tag{9-16}$$

Let $e_u$ represent the actual peak unipolar error, and let $e_b$ represent the actual peak bipolar offset error. These values are obtained by multiplying the preceding two values by the full-scale voltage. Hence,

$$e_u = E_u V_{fs} = 2^{-(N+1)} V_{fs} \tag{9-17}$$

and

$$e_b = E_b V_{fs} = 2^{-N} V_{fs} \tag{9-18}$$

---

**▌▌ EXAMPLE 9-3**

The normalized analog waveform shown as the smooth curve in Figure 9–4 is to be converted to a PCM signal by a 4-bit A/D converter whose input–output normalized characteristic curve is given by Figure 9–3. Sampling will occur at $t = 0$ and at 1-ms intervals thereafter. Over the time interval shown, construct on the same scale the quantized form of the signal that is actually encoded and subsequently reconstructed at the receiver. In addition, list the corresponding binary words that are transmitted.

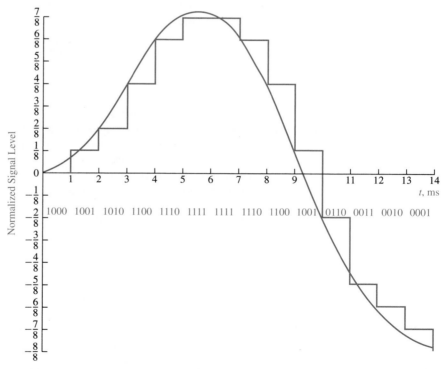

**FIGURE 9–4**

Analog and quantized signals used in Example 9-3.

Table 9–2   Data Supporting Example 9-3

| Time, ms | Closest Quantization Value | Binary Representation |
|---|---|---|
| 0 | 0 | 1000 |
| 1 | 1/8 | 1001 |
| 2 | 2/8 | 1010 |
| 3 | 4/8 | 1100 |
| 4 | 6/8 | 1110 |
| 5 | 7/8 | 1111 |
| 6 | 7/8 | 1111 |
| 7 | 6/8 | 1110 |
| 8 | 4/8 | 1100 |
| 9 | 1/8 | 1001 |
| 10 | −2/8 | 0110 |
| 11 | −5/8 | 0011 |
| 12 | −6/8 | 0010 |
| 13 | −7/8 | 0001 |
| 14 | −8/8 | 0000 |

**SOLUTION**   Since the analog signal is represented in normalized form, the values may be used directly on the quantization characteristic scales of Figure 9–3. If the actual voltage level had been given instead, it would be convenient to divide the actual voltage level by the full-scale voltage, so that the results could be used directly on the normalized scale. The quantized signal is constructed by reading the exact analog voltage at sampling points, noting which quantization level is nearest, and indicating the corresponding binary word, using Table 9–1 if necessary.

At $t = 0$, the exact analog voltage is 0, so the quantized value is also 0. The corresponding binary bipolar offset value is 1000. At $t = 1$ ms, the analog voltage is closer to 1/8 than to 0, so the quantized level is 1/8. The corresponding binary value is 1001. This process is continued for the entire duration of the signal, and the results are summarized in Table 9–2 (and partially in Figure 9–4). The reader should find it instructive to verify the results given.

**III EXAMPLE 9-4**

An 8-bit A/D converter with a full-scale voltage of 20 V is to be employed in a binary PCM system. The input analog signal is adjusted to cover the range from zero to slightly under 20 V, and the converter is connected for unipolar encoding. Rounding is employed in the quantization strategy. Determine the following quantities: (a) normalized step size, (b) actual step size in volts, (c) normalized maximum quantized analog level, (d) actual maximum quantized level in volts, (e) normalized peak error, and (f) actual peak error in volts.

**SOLUTION**   The various quantities desired may be readily calculated from the relationships developed in this section. The results, with some practical rounding, are summarized in the following.

(a)   The normalized unipolar step size is

$$\Delta X_u = 2^{-N} = 2^{-8} = 0.003906 \tag{9-19}$$

(b)   The actual step size is

$$\Delta v_u = \Delta X_u V_{fs} = 0.003906 \times 20 = 78.12 \text{ mV} \tag{9-20}$$

(c)   The normalized maximum quantized level is

$$X_u(\text{max}) = 1 - \Delta X_u = 1 - 0.003906 = 0.9961 \tag{9-21}$$

(d) The actual maximum quantized level is

$$v_u(\text{max}) = X_u(\text{max})V_{fs} = 0.9961 \times 20 = 19.92 \text{ V} \qquad (9\text{-}22)$$

Alternately, the actual step size could have been subtracted from 20 V to yield the same result.

(e) The normalized peak error is

$$E_u = \frac{\Delta X_u}{2} = \frac{0.003906}{2} = 0.001953 \qquad (9\text{-}23)$$

(f) The actual peak error is

$$e_u = E_u V_{fs} = 0.001953 \times 20 = 39.06 \text{ mV} \qquad (9\text{-}24)$$

---

**▌▌ EXAMPLE 9-5**

The 8-bit A/D converter of Example 9-4, while maintaining the same peak-to-peak range, is converted to bipolar offset form so that it can be used with an analog signal having a range from $-10$ V to just under 10 V. Repeat all the calculations of Example 9-4.

**SOLUTION**    Notice that the peak-to-peak voltage is still 20 V, as in Example 9-4. Many A/D converters utilize a unipolar internal encoding process, but accomplish the bipolar operation by adding a dc bias equal to half the peak-to-peak voltage. In this case, the bias level would be 10 V. Some manufacturers would still specify the full-scale voltage as 20 V, but with the approach given here, the full-scale voltage will be assumed as the peak positive level, that is, $V_{fs} = 10$ V.

(a) The normalized bipolar step size is

$$\Delta X_b = 2^{-N+1} = 2^{-7} = 0.007812 \qquad (9\text{-}25)$$

(b) The actual step size is

$$\Delta v_b = \Delta X_b V_{fs} = 0.007812 \times 10 = 78.12 \text{ mV} \qquad (9\text{-}26)$$

Note that while the normalized step size appears to be twice as great as for unipolar, the actual step size in volts is the same, since the peak-to-peak voltage is the same.

(c) The normalized maximum quantized level is

$$X_b(\text{max}) = 1 - \Delta X_b = 1 - 0.007812 = 0.9922 \qquad (9\text{-}27)$$

(d) The actual maximum quantized level is

$$v_b(\text{max}) = X_b(\text{max})V_{fs} = 0.9922 \times 10 = 9.922 \text{ V} \qquad (9\text{-}28)$$

Alternately, the step size could have been subtracted from 10 V to yield the same result.

(e) The normalized peak error is

$$E_b = \frac{\Delta X_b}{2} = \frac{0.007812}{2} = 0.003906 \qquad (9\text{-}29)$$

(f) The actual peak error is

$$e_b = E_b V_{fs} = 0.003906 \times 10 = 39.06 \text{ mV} \qquad (9\text{-}30)$$

which is the same as for unipolar encoding, since the peak-to-peak range is the same.

▌▌

## 9-3  Companding and μ-Law Encoding

A special problem occurs with PCM systems in which the expected dynamic range of the signal is very large. To avoid saturation and subsequent signal clipping, the level of the signal must be adjusted so that its peak value never exceeds the full-scale signal level of the

A/D converter. The resulting effect is that, during intervals in which the signal level is very low, the percent of quantization error increases significantly.

To illustrate this effect, assume that a given A/D converter has a full-scale voltage of 10 V, and assume that the peak error is about 4 mV, corresponding to a step size of 8 mV when rounding is employed. During an interval in which the signal voltage is close to 10 V, the percentage peak quantization error is in the neighborhood of (4 mV/10 V) × 100% = 0.04%. However, assume that over some interval of time, the signal levels hovers around 10 mV, corresponding to a signal dynamic range of 60 dB. In this time interval, the percentage peak quantization error is in the vicinity of (4 mV/10 mV) × 100% = 40%! The result may be intolerable for some applications.

There are two ways in which this problem can be partially alleviated. The first method is to use a variable step size that increases as the signal level increases. Thus, at very small signal levels, the step size would be smaller, permitting a lower percentage error, while at high signal levels, the step size could be made to be proportionately higher. At the receiver, the taper of the D/A converter quantization curve would have the inverse characteristic; that is, the steps would be larger at smaller signal amplitudes and smaller at larger signal amplitudes.

The second method, which accomplishes the same objective, is to compress the dynamic range of the signal at the transmitter. At the receiver, the dynamic range of the analog signal is then expanded to compensate. The nonlinear process at the transmitter is called *compression,* and the process at the receiver is called *expansion.* The combined process is referred to as signal *companding.*

The input–output form of a typical compression characteristic curve is illustrated in Figure 9–5(a). Observe that at low input signal levels, the output level is increased considerably, thus "lifting" the signal to a higher level and thereby reducing the percentage of quantization error. Of course, higher signals must be "pushed down," so to speak, to keep them in the proper range.

The input–output form of a typical expansion curve is illustrated in Figure 9–5(b). It is necessary to design the compression and expansion curves together for a particular system so that the two curves exactly offset each other. Obviously, if this condition is not met, the resulting signal will be distorted.

## μ-Compression Law

One of the most common compression laws utilized in the commercial telephone industry is the so-call μ-*compression law* curve, which is given by

$$v_o = V_{o\,max} \frac{\ln\left(1 + \mu \frac{v_i}{V_{i\,max}}\right)}{\ln(1 + \mu)} \quad \text{for } v_i \geq 0 \tag{9-31}$$

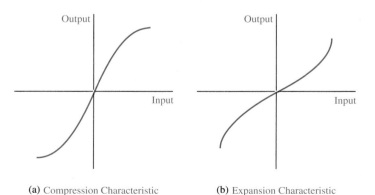

(a) Compression Characteristic     (b) Expansion Characteristic

**FIGURE 9–5**
Examples of compression and expansion curves used in PCM systems to reduce relative quantization error effects.

Compression Characteristic

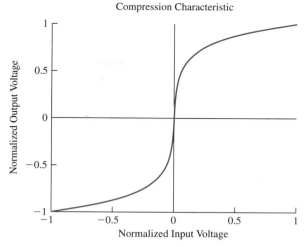

**FIGURE 9–6**
Normalized μ-compression law characteristic curve.

where

$$v_i = \text{input voltage}$$
$$v_o = \text{output voltage}$$
$$V_{i\,max} = \text{maximum value of input voltage}$$
$$V_{o\,max} = \text{maximum value of output voltage}$$
$$\mu = \text{compression parameter} = 255 \text{ in many systems}$$

For negative input voltages, the curve has the same form but is inverted. A complete curve of this compression characteristic in normalized form for both positive and negative voltages is shown in Figure 9–6.

At the receiver, the expansion curve required to compensate for this compression would have the form

$$v_{ie} = \frac{V_{i\,max}}{\mu}\left[(1 + \mu)^{\frac{v_r}{V_{o\,max}}} - 1\right] \quad \text{for } v_r \geq 0 \tag{9-32}$$

where $v_{ie}$ is the expanded version of the received signal $v_r$ after decoding. All other parameters are the same as previously defined. For negative values, the expansion curve is the reverse mirror image, as in the case of the compression curve.

## Codec

A device that performs the process of compression and expansion is often referred to as a *codec*. The term *codec* is a contraction for <u>co</u>der and <u>dec</u>oder. Much of the process can be performed through digital signal processing.

---

**EXAMPLE 9-6**

A μ-compression law encoder has the following parameters: $V_{i\,max} = 16$ V, $V_{o\,max} = 2$ V, and $\mu = 255$. Determine the output voltage for each of the following input voltage levels: (a) 2 V, (b) 4 V, (c) 8 V, (d) 16 V.

**SOLUTION**   Substitution of the various parameter values in Equation 9-31 yields

$$v_o = 2\frac{\ln\left(1 + \frac{255 v_i}{16}\right)}{\ln(1 + 255)} = 0.3607 \ln(1 + 15.94 v_i) \tag{9-33}$$

in which the latter form resulted from combining and simplifying some of the constants. The four values of input voltage may now be substituted in Equation 9-33. The results are provided in the table that follows.

|     | $v_i$, volts | $v_o$, volts |
| --- | --- | --- |
| (a) | 2 | 1.260 |
| (b) | 4 | 1.504 |
| (c) | 8 | 1.751 |
| (d) | 16 | 2.000 |

## 9-4 Baseband Encoding Forms

So far, all the binary PCM signal forms considered have employed natural binary encoding. While some systems transmit the natural binary words directly, other systems convert the natural binary stream to special formats prior to actual transmission or high-frequency modulation. Such codes are referred to as *line codes*. These codes have advantages in particular situations in terms of ease of data processing, bandwidth requirements, synchronization, and the elimination of dc components in the spectrum. Several of these baseband line codes will be discussed in this section.

For illustrations of some of the methods involved, refer to Figure 9–7. A binary pattern of 10110001101 is shown at the top, and some of the different forms in which this bit stream can be encoded are shown below. For all the waveforms given, no absolute levels are indicated, since this would depend on the particular voltage or current levels employed. It will suffice to say that there are always two levels, one of which is a binary 1 and the other of which is a binary 0. In the context of usage, *a binary* **1** *is called a* **mark,** *and a binary* **0** *is called a* **space.**

In the discussion that follows, let $\tau$ represent the width in seconds of the data bit to be encoded. The actual encoded pulse may have a width $\tau$ or some fraction of $\tau$, depending on the particular scheme employed. Several methods will now be discussed.

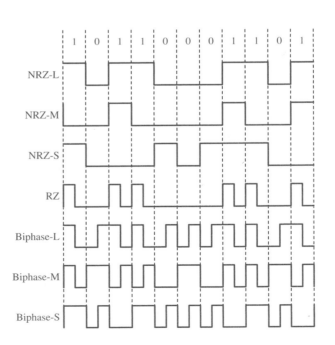

**FIGURE 9–7**

Examples of different baseband encoding formats (line codes).

## NRZ-L

This designation refers to *non-return-to-zero level*. A 1 is always represented by one fixed level of width $\tau$, and a 0 is always represented by the other level, also of width $\tau$. This form is thus equivalent to a conventional binary representation, and each word would be a natural binary value if the signal to be encoded were expressed in natural binary form. Most of the other forms to be discussed are obtained by first generating an NRZ-L signal and converting it to one of the other forms by appropriate circuitry. For the data given in Figure 9–7, the more positive level is a mark or 1 and the more negative level is a space or 0. This corresponds to *positive logic*. Some systems employ *negative logic,* in which the more positive level is a 0 and the more negative level is a 1.

## NRZ-M

This designation refers to *non-return-to-zero mark*. If a 1 is to be transmitted, there is always a change at the beginning of the bit interval. However, if a 0 is to be transmitted, there is no change. The reader can verify that this scheme is satisfied for the data shown in Figure 9–7. The signal level just prior to the first data bit would be a 1 for the case shown. It will be seen later that this initial level is usually arbitrary.

## NRZ-S

This designation refers to *non-return-to-zero space*. If a 0 is to be transmitted, there is always a change at the beginning of the bit interval. However, if a 1 is to be transmitted, there is no change. As in NRZ-M, the beginning level is arbitrary.

## RZ

This designation refers to *return-to-zero*. If a 0 is to be transmitted, the signal remains at the 0 level (lower level in this case) for the entire bit interval. However, if a 1 is to be transmitted, a pulse of width $\tau/2$ is inserted in a designated part of the bit interval.

There is another form of RZ (not shown) that essentially utilizes three levels: no transmission, a positive pulse, and a negative pulse. The positive and negative pulses represent the 1 and 0 levels, and these pulses each have a width of $\tau/2$.

## Biphase-L

This designation refers to *biphase level*. Within each data bit interval, there are always two states, each of width $\tau/2$. If the data bit is a 0, the logical sequence 01 is inserted. Conversely, if the data bit is a 1, the sequence 10 is inserted. This encoding scheme, along with the next two, are referred to as *split-phase* or *Manchester* codes.

## Biphase-M

This designation refers to *biphase mark*. A transition always occurs at the beginning of a bit interval. If the data bit is a 1, a second transition occurs at a time $\tau/2$ later. If the data bit is a 0, no further transition occurs until the beginning of the next bit interval. Referring to Figure 9–7, the level just prior to the beginning of the data stream was arbitrarily selected.

## Biphase-S

This designation refers to *biphase space*. A transition always occurs at the beginning of a bit interval. If the data bit is a 0, a second transition occurs at a time $\tau/2$ later. If the data bit is a 1, no further transition occurs until the beginning of the next bit interval. As in the case of biphase-M, the initial level was arbitrarily selected.

## Comparisons

Although it is not widely known how all of these forms evolved, a few of the significant properties of some of these formats will be discussed. Two particular comparison areas are the relative bandwidth requirements and the relative ease of synchronization.

The RZ and NRZ forms require the transmission of a dc component in the spectrum. This can be deduced by noting that for either form, there is at least one possible transmission condition in which only one level would be transmitted for a long time (e.g., long string of 1s in NRZ-S). This would eliminate the use of transformer coupling or ac coupled amplifiers. In contrast, the biphase forms always have at least one transition per bit interval, so there is no need to retain the dc level.

Another advantage of the biphase forms is that at least one orderly transition occurs per data bit interval, and this property may be used to derive a synchronized clock reference at the receiver. In fact, the biphase forms may be generated by combining a data signal in an NRZ form with a reference clock, so that the clock is actually transmitted with the signal.

While the preceding arguments would tend to favor the biphase data forms, there is, however, a distinct disadvantage in terms of the total required bandwidth. For the NRZ forms, the shortest pulse width is always equal to the data-bit interval width $\tau$. In contrast, the biphase forms exhibit pulses that have a width $\tau/2$. Since the transmission bandwidth of a pulse is inversely proportional to the pulse width, the required transmission bandwidth for the biphase forms is greater, if all other factors are equal.

The NRZ-L and NRZ-M formats are used in the encoding and detection of *differentially encoded phase-shift keying* (DPSK) data. This concept will be explored in some depth in Section 9-9.

Irrespective of the particular data format employed in the encoding process, it is convenient to classify the final PCM signal, from the viewpoint of the relative levels used in the actual pulses, as either a *unipolar* signal or a *bipolar* signal. A *unipolar* signal is one in which one of the two possible levels corresponds to an actual zero voltage or current level (i.e., nothing transmitted), and the other level is either a positive or negative value as established in the design process. A *bipolar* signal employs a positive voltage or current as one level and a negative voltage or current as the other level. These terms as employed here have nothing to do with the *unipolar* and *bipolar offset* terminology employed in A/D converters.

## 9-5   Time-Division Multiplexing of PCM Signals

As in the case of other sampled-data signals discussed in Chapter 8, PCM signals from a number of different information sources may be combined and transmitted over a common channel by means of *time-division multiplexing* (TDM). This process is illustrated for PCM in Figure 9–8. In this particular simplified system, each of the individual data channels is sampled at the same rate, and a given word representing the encoded value of the particular

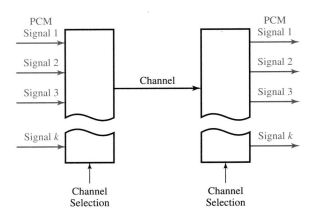

**FIGURE 9–8**

Multiplexing and demultiplexing of PCM signals.

sample is transmitted between the point of sampling of that channel and of the next channel in the sequence. As in the case of pulse modulation, there are numerous variations on this basic scheme that can be devised when data sources have different bandwidths.

To ensure that the electronic commutation at the receiver is exactly in step with that at the source, some means of synchronization must be employed. One way this can be achieved is by means of a special synchronization word or bit combination that is inserted at the beginning of a frame. The synchronized word is chosen to have a different pattern (or value) than other possible words that could be transmitted. At the receiver, all incoming words are sensed by digital circuitry, which establishes the beginning of a new frame when the synchronization word is received.

## Lower Bound for Bandwidth

Since a number of pulses will have to be used to represent one data value with PCM, the effective baseband bandwidth of a TDM PCM signal is much greater than for a TDM PAM signal. In Chapter 8, an approximate lower bound for the bandwidth of a PAM signal was developed, with idealized assumptions employed in the process. A similar development will now be performed for PCM.

A PCM TDM system in which there are $k$ signals to be multiplexed will be considered, and each signal will be assumed to have a baseband bandwidth $W$. The same several idealized limiting-case assumptions made in Chapter 8 will be made again. They are (1) no spacing between successive pulses, (2) sampling rate set at minimum Nyquist rate, and (3) no sync information. An additional assumption made in this development is that the actual bit pulses occupy the full width of the bit interval. As we saw in Section 9-4, some PCM encoded forms employ pulses shorter than the actual bit interval, and such systems would generally require greater bandwidths.

Referring to Figure 9–9, there are three time intervals of significance: $T_f$, $T_w$, and $\tau$. The *frame time* $T_f$ is the total interval in which all $k$ signals are sampled once. At the minimum Nyquist rate, $T_f$ must satisfy

$$T_f = \frac{1}{2W} \tag{9-34}$$

The *word time* $T_w$ represents the duration of one particular encoded sample, and is

$$T_w = \frac{T_f}{k} = \frac{1}{2kW} \tag{9-35}$$

Each word is assumed to be represented by $N$ bits, each of width $\tau$. Thus,

$$\tau = \frac{T_w}{N} = \frac{1}{2kNW} \tag{9-36}$$

The minimum baseband transmission bandwidth for the PCM TDM signal is then assumed to be

$$B_T = \frac{0.5}{\tau} = 0.5(2kNW) = kNW = kW \log_2 M \tag{9-37}$$

The lower bound for the transmission bandwidth in this idealized case is simply *the product of the number of signals times the bandwidth per signal times the number of bits per*

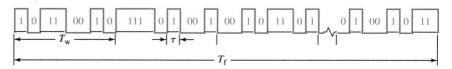

**FIGURE 9–9**
Frame format used in developing lower bound for transmission bandwidth of PCM TDM system.

*word.* Referring back to the development for PAM, it is noted that the bandwidth for PCM is $N$ times the bandwidth required for PAM with all other factors equal. The particular point could have also been deduced very simply by realizing that in PCM, $N$ pulses must be inserted in the same time interval that only one pulse is required for PAM. The preceding development, however, provides some additional insight into the framing process.

## Tradeoff between Accuracy and Bandwidth

The last expression in Equation 9-37 displays an important result that should be emphasized. The error in the PCM quantization process is inversely proportional to the number of levels $M$. This error can be viewed as a type of noise. Thus, the signal-to-noise ratio for the PCM signal increases rapidly with increasing $M$. However, the required transmission bandwidth increases as the logarithm of the number of levels, and this increase is much more moderate. This suggests that a moderate increase in transmission bandwidth can result in a significant increase in the signal-to-noise ratio. For example, in changing from an 8-bit system to a 9-bit system, the transmission bandwidth is increased by only 12.5%, but the quantization error is reduced by 50%. There are other factors that must be considered, but this example represents a special case of a rather general trade-off that appears in many communication processes. With proper encoding and/or modulation, improved signal-to-noise ratio may be achieved at the expense of higher bandwidth, but the percentage of improvement may be greater than the percentage of increase in bandwidth.

## Multirate Sampling

Many multiplexing systems require the transmission of signals with widely different bandwidths. For example, the signals transmitted from space vehicles may include such diverse data as the outputs of slowly varying instruments, voice data, and much higher frequency video data. If all the signals were multiplexed in a simple frame structure, it would be necessary to sample all signals at a rate governed by the signal with the highest bandwidth. The result would be a very inefficient utilization of bandwidth.

In situations where the bandwidths of different signals vary widely, a more complex frame structure may be devised whereby some signals are sampled more than once per frame (i.e., at a "super" rate), and other signals are sampled less than once per frame (i.e., at a "sub" rate). A representative example will be discussed to illustrate some of the general features.

Consider the three-commutator system shown in Figure 9–10. The 25-channel commutator shown on the right will be referred to as the *prime commutator,* and one complete

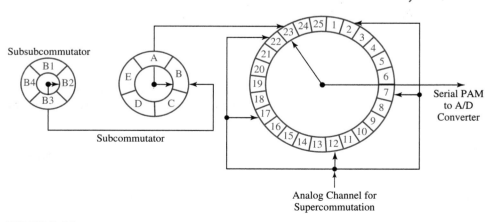

**FIGURE 9–10**
Sampling scheme employing several levels of commutation.

cycle of this unit is called a *prime frame*. The primary commutator has 25 slots. Notice, however, that every fifth slot (i.e., numbers 2, 7, 12, 17, and 22) is connected. The process applied to this signal is called *supercommutation,* and this analog signal is being sampled at five times the prime frame rate.

The commutator in the middle is called a *subcommutator*. The sampling rate of the subcommutator is one-fifth the rate of the prime commutator. Thus, each time the prime commutator samples channel 23, the subcommutator has advanced to the next of its five possible positions. Each subcommutator channel then is being sampled at a rate equal to one-fifth of the prime sampling rate.

The commutator on the left is called the *subsubcommutator*. The sampling rate of the subsubcommutator is one-fourth the rate of the subcommutator or $(1/5) \times (1/4) = 1/20$ of the rate of the prime commutator. Each time the subcommutator samples channel B, the subsubcommutator has advanced to the next of its four possible positions. Each subsubcommutator channel then is being sampled at a rate equal to one-twentieth of the prime sampling rate.

The ratio of the bandwidth permitted in the highest-frequency data channel to the bandwidth permitted in the lowest-frequency channel is the ratio of the sampling rate of the supercommutated channel to the sampling rate of the subsubcommutated channel, and is $5/(1/20) = 100$. Note that it takes 20 complete cycles of the prime frame before all the signals are sampled at least once. This complete "supercycle" is called a *data field*.

---

**▮▮▮ EXAMPLE 9-7**

Consider a PCM TDM system in which 19 signals are to be processed. Each of the signals has a baseband bandwidth $W = 5$ kHz, and 8 bits are to be used in each word. Conventional NRZ-L encoding will be used, and an additional 8-bit sync word will be placed in each frame. Determine the theoretical minimum bandwidth required.

**SOLUTION**   To determine the bandwidth, it is necessary to determine the width of the shortest possible pulse. The sampling rate is

$$f_s = 2W = 2 \times 5 \text{ kHz} = 10 \text{ kHz} \qquad (9\text{-}38)$$

The frame time is

$$T_f = \frac{1}{f_s} = \frac{1}{10 \times 10^3} = 0.1 \text{ ms} \qquad (9\text{-}39)$$

The word time is

$$T_w = \frac{T_f}{k} = \frac{0.1 \text{ ms}}{20} = \frac{100 \text{ μs}}{20} = 5 \text{ μs} \qquad (9\text{-}40)$$

where the value of 20 represents the 19 data words plus the sync word for each frame. The bit interval is

$$\tau = \frac{T_w}{N} = \frac{5 \text{ μs}}{8} = 0.625 \text{ μs} \qquad (9\text{-}41)$$

The approximate transmission bandwidth is

$$B_T = \frac{0.5}{\tau} = \frac{0.5}{0.625 \times 10^{-6}} = 800 \text{ kHz} \qquad (9\text{-}42)$$

The reader can readily verify that the result can be obtained in one step by using Equation 9-37. However, the value of this exercise was to go through the various steps to determine the minimum pulse width required, based on the frame format, and to show that that minimum pulse width determines the bandwidth. Moreover, Equation 9-37 is a minimum bound based on a theoretical minimum sampling rate and other idealistic assumptions. When practical additional factors are considered, the analysis will be somewhat more involved, and the bandwidth will usually be greater than the theoretical minimum.   ▮▮▮

## 9-6 Amplitude-Shift Keying

In this and the next several sections, we will investigate the means by which digital signals can be shifted to a higher frequency range for transmission. This could be the radio frequency range when electromagnetic wave propagation is desired, or it could be the audio frequency range for transmission over telephone lines and the internet.

The discussion in this section will focus on *amplitude-shift keying* (ASK). In a sense, ASK corresponds to AM for analog communications. Digital forms involving frequency and phase modulation will be considered in later sections.

Digital modulation systems in the RF range can be classified as either *coherent* or *noncoherent*. *Coherent* methods require a reference carrier at the receiver having the exact frequency and phase of the transmitter carrier. This process is similar to the concept of DSB synchronous detection encountered in Chapter 6. *Noncoherent* methods do not require a reference carrier, and can be detected by other means.

As a general rule, coherent systems tend to provide better performance in the presence of noise, if all other factors are the same. However, noncoherent systems generally are less complex in design and operation.

### Change to Sine Function Reference

In studying the various forms of analog modulation, it was more convenient to use the cosine function as the basis for analysis, since the sign patterns of the expansions, particularly those employing the complex exponential form, were "cleaner." We will now somewhat arbitrarily shift to the sine function for many of the modulation operations involving digital data. The reason is that the abrupt changes encountered in digital data suggest switching at zero crossings of the carrier time function whenever possible. The sine function crosses zero at $t = 0$ and is more natural for describing a digitally oriented change in amplitude, frequency, or phase.

### ASK Form

The process of amplitude-shift keying is probably the simplest and most intuitive form of higher frequency encoding. In this process, turning the carrier on for the bit duration represents a 1, and leaving the carrier off for the bit duration represents a 0. Thus, a fixed-frequency RF carrier is simply gated off and on in accordance with the bit value.

Consider the arbitrary baseband bit stream shown in Figure 9–11(a). The form of an ASK signal is illustrated in Figure 9–11(b). Bear in mind that in this figure (and in others that will be shown later), it is not feasible to show more than a few cycles. In audio frequency situations such as encountered on telephone lines, there may only be a few cycles

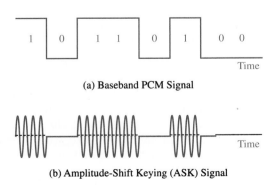

(a) Baseband PCM Signal

(b) Amplitude-Shift Keying (ASK) Signal

**FIGURE 9–11**
Baseband PCM signal and the amplitude-shift keying (ASK) modulated signal.

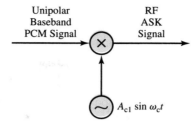

**FIGURE 9–12**

Generation of an ASK signal using a balanced modulator.

within the width of a pulse, but in the RF frequency range, there may be hundreds or thousands of cycles in a pulse width.

The ASK signal is sometimes referred to as *on-off keying* (OOK). Classical telegraphy can be thought of as a form of ASK, although that medium employs both short (*dot*) and longer (*dash*) intervals in which the carrier is turned on.

## ASK Generation

An ASK signal can be generated by essentially the same process utilized in DSB analog systems; that is, a balanced modulator or multiplier. Consider the system shown in Figure 9–12. A PCM signal having a unipolar level form is applied as the data input, and this signal is multiplied by the carrier. The carrier is thus multiplied by either a positive constant, which results in an RF burst, or by zero, which results in no transmission. Alternately, the RF oscillator (or a buffer) can be turned on and off by the digital levels of the signal.

## ASK Noncoherent Detection

An ASK signal can be detected either noncoherently or coherently. *Noncoherent* detection can be achieved by the simple process of envelope detection, as illustrated by the block diagram of Figure 9–13(a). Following the actual extraction of the envelope, the signal is smoothed by a filter. The resulting signal at this point is no longer a simple binary two-state signal, since the finite bandwidth limitations of the channel have rounded the various pulses. By employing threshold circuits such as comparators, the imperfect pulses can be converted back to ideal forms. The block entitled "binary restoration" accomplishes this

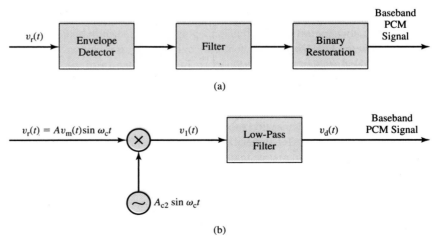

**FIGURE 9–13**

Forms of (a) noncoherent and (b) coherent detection of ASK signals.

purpose, and the output is a reconstructed PCM replica of the transmitted signal (assuming no decision errors for bits).

The reader should note that, while envelope detection of DSB signals with analog modulated signals was specifically prohibited in Chapter 6, envelope detection of ASK digital data using an equivalent modulation process is perfectly permissible! Why?

## ASK Coherent Detection

Coherent detection of ASK data is also possible, and this approach is indicated by the block diagram of Figure 9–13(b). A coherent reference must be provided for this purpose. The actual process is essentially equivalent to synchronous detection of a DSB signal as discussed in Chapter 6.

## ASK Bandwidth

An estimate of the minimum bandwidth for an ASK signal may be developed by assuming an alternating periodic pattern of a 1 followed by a 0. This type of signal is called a *dotting pattern* and is used in testing digital systems. It represents a worst-case condition, in the sense that if the system will successfully transmit this pattern, it should handle a more realistic signal.

An ASK signal modulated with the dotting pattern is shown in Figure 9–14(a). Let $\tau$ represent the width of each pulse and let $T_p$ represent the period, which is related to the pulse width by

$$T_p = 2\tau \tag{9-43}$$

Let $f_c$ represent the frequency of the gated sinusoid. The ASK signal can be represented by the product of a continuous sinusoid with frequency $f_c$ multiplied by the baseband square wave of Figure 9–14(b), which oscillates between 1 and 0. The spectral properties of the baseband square wave were developed in Chapter 2, and it can be recalled that the Fourier series consist of a dc component, a fundamental, and odd-numbered harmonics. The latter property is due to the fact that when the dc component is subtracted out, the resulting waveform will have half-wave symmetry. The fundamental frequency $f_p$ of this square wave is

$$f_p = \frac{1}{T_p} = \frac{1}{2\tau} = \frac{0.5}{\tau} \tag{9-44}$$

Let $R$ represent the *data rate* in bits per second (bps), which is given by

$$R = \frac{1}{\tau} = \frac{1}{T_p/2} = \frac{2}{T_p} = 2f_p \tag{9-45}$$

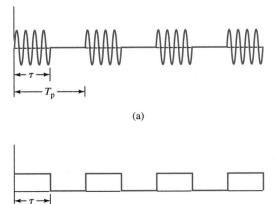

(a)

(b)

**FIGURE 9–14**

Time-domain waveforms used in developing estimate of ASK bandwidth.

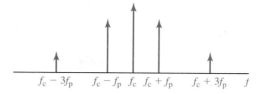

**FIGURE 9–15**
Spectral form used in developing estimate of ASK bandwidth.

The spectrum of this signal may readily be predicted from the work of Chapter 3 and is shown in Figure 9–15. It consists of a line at the frequency $f_c$, components spaced $f_p$ on either side of the center and at odd integer multiples of $f_p$ on either side.

An estimate of the bandwidth may be made by assuming that a bandwidth sufficient to transmit the center component and the most adjacent sidebands on either side will be adequate for pulse recognition at the receiver ("coarse" reproduction). This bandwidth $B_T$ is then given by

$$B_T = 2f_p = R \qquad (9\text{-}46)$$

Stated in words, *the minimum bandwidth for ASK is equal to the data rate in bits/second.*

**▌▌ EXAMPLE 9-8**

A binary NRZ-L PCM signal having a data rate of 200 kbits/s modulates an RF carrier in an ASK format. Determine the transmission bandwidth.

**SOLUTION**    From Equation 9-46, the bandwidth is simply

$$B_T = R = 200 \text{ kHz} \qquad (9\text{-}47)$$

The spectrum will be centered at the RF carrier frequency.    ▌▌

## 9-7   Frequency-Shift Keying

The process of binary *frequency-shift keying* (*FSK*) involves the transmission of one or the other of two distinct frequencies. A 0 is represented by the transmission of a frequency $f_0$ and a 1 is represented by the transmission of a frequency $f_1$. The transmitter oscillator is thus required to switch back and forth between two separate frequencies at the data rate.

For the same data pattern used in explaining ASK, the form of the corresponding FSK signal is shown in Figure 9–16.

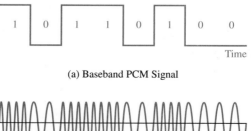

(a) Baseband PCM Signal

(b) Frequency-Shift Keying (FSK) Signal

**FIGURE 9–16**
Baseband PCM signal and the frequency-shift keying (FSK) modulated signal.

**FIGURE 9–17**
Generation of FSK signal using a voltage-controlled oscillator (VCO).

## FSK Generation

An FSK signal can be generated by using the baseband data signal as the control signal for a voltage-controlled oscillator (VCO) as illustrated in Figure 9–17. Since the baseband signal assumes only one of two values, the VCO generates only one of two possible instantaneous frequencies. The level of the control voltage is adjusted in accordance with the VCO circuitry so that the two possible frequencies are produced at the proper levels of the PCM signal. It should be understood that while the FSK signal itself assumes only two instantaneous frequencies, the spectrum is much more complex and contains many frequencies.

## FSK Noncoherent Detection

As in the case of the ASK signal, the FSK signal can be detected either noncoherently or coherently. *Noncoherent* detection can be achieved through the process illustrated in Figure 9–18(a). The signal is simultaneously applied to two band-pass filters, one tuned to $f_1$ and the other to $f_0$. When a 1 is transmitted, the output of the upper filter is maximum,

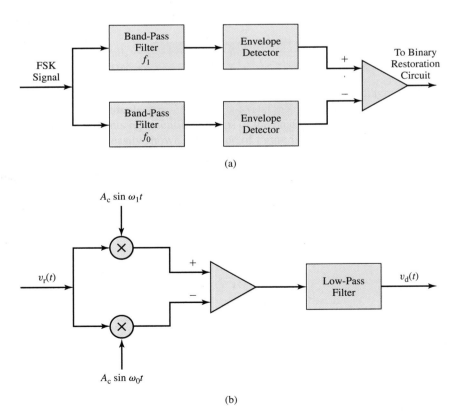

**FIGURE 9–18**
Forms of (a) noncoherent and (b) coherent detection of FSK signals.

while the output of the lower filter should be small (assuming reasonably high signal-to-noise ratio and proper spacing between channels). When a 0 is transmitted, the opposite result occurs. The envelope detector in each path converts the corresponding RF pulses to baseband pulses. Finally, pulses from both paths are combined to yield a baseband data signal. The pulses from the lower path are subtracted from those in the upper path so that a bipolar baseband pulse stream is constructed. The resulting pulses are distorted as well as corrupted by noise, so a binary restoration process should be used to generate a more ideal PCM data signal before D/A conversion.

### FSK Coherent Detection

Coherent detection of FSK signals can be achieved by the process shown in Figure 9–18(b). This system requires local generated carriers having the same frequencies and phases as those used at the transmitter. For the purpose of illustration, assume that a 1 is transmitted, which means that the received signal $v_r(t)$ is of the form

$$v_r(t) = A \sin \omega_1 t \tag{9-48}$$

Along the upper path, the output of the mixer is

$$v_1(t) = A A_c \sin^2 \omega_1 t = \frac{A A_c}{2}(1 - \cos 2\omega_1 t) \tag{9-49}$$

The last term on the right-hand side of Equation 9-49 is easily removed by the low-pass filter, so the remaining term at the output is

$$v_d(t) = \frac{A A_c}{2} = \text{positive constant} \tag{9-50}$$

This is simply a positive value indicating a 1. A similar development for the lower path would produce a negative constant in the output for the case when a 0 is transmitted.

A basic question concerns the level of *crosstalk*, the effect produced in one channel when the signal corresponding to the other channel is transmitted. Assuming again that a 1 is transmitted, the output $v_2(t)$ of the lower channel is

$$v_2(t) = A A_c \sin \omega_1 t \sin \omega_0 t = \frac{A A_c}{2} \cos(\omega_1 - \omega_0)t - \frac{A A_c}{2} \cos(\omega_1 + \omega_0)t \tag{9-51}$$

The last term in Equation 9-51 is well above the passband of the low-pass filter and is easily removed. However, the other term has a frequency $f_1 - f_0$, which may be in the range of the low-pass filter. Designing the low-pass filter to have a "notch"—or, at least, very high attenuation—at this frequency can circumvent the problem. Thus, the crosstalk level can be managed by proper design of the signal processing circuitry. The resulting detected signal $v_d(t)$ can then be applied to a circuit for proper binary restoration.

### FSK Bandwidth

An estimate of the bandwidth required for FSK may be determined by reverting back to Carson's rule as discussed for analog FM. After all, FSK is a form of FM in which the modulating signal may assume only two possible levels. Although Carson's rule was specifically determined on the basis of sinusoidal modulation, it can be used (with some caution) for FSK.

Carson's rule from Chapter 7 can be stated as

$$B_T = 2(\Delta f + f_m) \tag{9-52}$$

where $\Delta f$ is the frequency deviation on either side of the center frequency, and $f_m$ is the modulating frequency.

Strictly speaking, there is no "center frequency" for FSK in the same sense as for FM, since the instantaneous frequency swings back and forth between two extremes. However, to accommodate Equation 9-52, the "center frequency" $f_c$ may be considered as the mid-point between $f_1$ and $f_0$, which is

$$f_c = \frac{f_0 + f_1}{2} \qquad (9\text{-}53)$$

The frequency deviation can then be defined as half of the total distance between the two frequencies,

$$\Delta f = \frac{f_1 - f_0}{2} \qquad (9\text{-}54)$$

The modulating frequency may be defined as the fundamental frequency of the dotting pattern; that is,

$$f_m = f_p = 0.5R \qquad (9\text{-}55)$$

Substituting Equations 9-54 and 9-55 in Equation 9-52, we obtain the following estimate for the minimum bandwidth:

$$B_T = f_1 - f_0 + R \qquad (9\text{-}56)$$

The spectrum is centered at $f_c$, as given by Equation 9-53. Note that the minimum bandwidth for FSK is greater than for ASK.

---

▌▐ EXAMPLE 9-9

A binary NRZ-L PCM signal having a data rate of 200 kbits/s modulates an RF carrier in an FSK format. The two RF frequencies are spaced apart by 150 kHz. Determine the transmission bandwidth.

**SOLUTION**   From Equation 9-56, the bandwidth is

$$B_T = f_1 - f_0 + R = 150 + 200 = 350 \text{ kHz} \qquad (9\text{-}57)$$

The spectrum will be centered at the midpoint between the two separate frequencies.   ▌▐

---

## 9-8   Binary Phase-Shift Keying

In many texts, *binary phase-shift keying* (*BPSK*) is referred to simply as *phase-shift keying* (*PSK*). However, to distinguish it from forms utilizing more than two levels, we will use the first reference in this text.

The process of BPSK utilizes a fixed-frequency sinusoid whose relative phase shift can assume one of two possible values. The values assumed will be 0° and 180°. For the same data stream used in explaining ASK and FSK, the form of a BPSK signal is shown in Figure 9–19. Note how a change in a bit value results in a sudden change in the phase of the sinusoid.

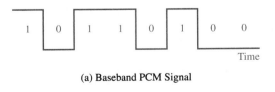

(a) Baseband PCM Signal

**FIGURE 9–19**
Baseband PCM signal and the binary phase-shift keying (BPSK) modulated signal.

(b) Binary Phase-Shift Keying (BPSK) Signal

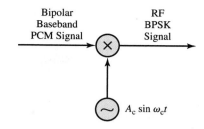

**FIGURE 9–20**
Generation of a BPSK signal using a balanced modulator.

## BPSK Generation

A BPSK signal can be generated by the process shown in Figure 9–20. Note that this is almost the same process used for generating ASK. However, one essential difference is that the PCM data stream has a *bipolar* form; that is, a 1 is represented by a voltage of one polarity and a 0 by a voltage of the opposite polarity.

In the case of BPSK, the modulating signal $v_m(t)$ assumes one of two forms:

$$v_m(t) = +A \quad \text{or} \quad -A \tag{9-58}$$

The resulting modulating signal can be expressed as

$$v_o(t) = A_c v_m(t) \sin \omega_c t \tag{9-59}$$

which is, of course, a DSB signal. Alternatively, $v_o(t)$ can be expressed as

$$v_o(t) = A A_c \sin(\omega_c t + \theta) = B \sin(\omega_c t + \theta) \tag{9-60}$$

where $\theta = 0°$ or $180°$ and $B = A A_c$.

## BPSK Detection

While ASK or FSK can be detected either noncoherently or coherently, BPSK can be detected only by a *coherent* process. One particular process used for detecting BPSK actually derives a phase-coherent reference from the composite PCM signal, and this process will now be explored. Consider the system shown in Figure 9–21, and assume that the received signal $v_r(t)$ is of the form

$$v_r(t) = K_r v_m(t) \sin \omega_c t \tag{9-61}$$

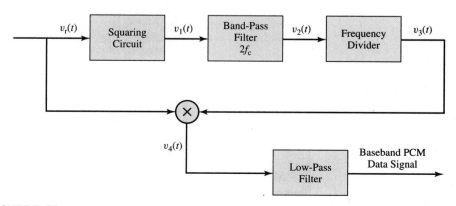

**FIGURE 9–21**
Detection of a BPSK signal by extracting a phase-coherent reference from the signal.

The signal is first squared by a nonlinear circuit, yielding

$$v_1(t) = K^2 v_m^2(t) \sin^2 \omega_c t = \frac{K^2}{2} v_m^2(t)(1 - \cos 2\omega_c t)$$

$$= \frac{K^2}{2} v_m^2(t) - \frac{K^2}{2} v_m^2(t) \cos 2\omega_c t \tag{9-62}$$

The original modulating signal $v_m(t)$ assumed both positive and negative values. However, the function $v_m^2(t) = A^2$ is positive irrespective of whether $v_m(t) = +A$ or $-A$. This means that the last term in Equation 9-62, in the ideal case, is simply $(-K^2 A^2/2) \cos 2\omega_c t$. There is some distortion, and noise is present, so a band-pass filter (or phase-locked loop) having a very narrow passband and center frequency $2f_c$ is used to separate this component. The output $v_c$ of the narrowband filter (with assumed phase inversion) is then

$$v_2(t) = K_2 \cos 2\omega_c t \tag{9-63}$$

The frequency of this signal is twice the required frequency for detection, but it has phase coherency with respect to the original carrier, so the frequency is divided by 2 and the angle is shifted in phase to yield a function given by

$$v_3(t) = K_3 \sin \omega_c t \tag{9-64}$$

The function $v_3$ is multiplied by the received signal to yield

$$v_4(t) = v_3(t)v_r(t) = K_3 K_r v_m(t) \sin^2 \omega_c t = C v_m(t)(1 - \cos 2\omega_c t) \tag{9-65}$$

where C is a constant. The last part of the last term in Equation 9-65 is a DSB signal centered at $2f_c$, which is easily removed by a low-pass filter. The detected output is then

$$v_d(t) = C v_m(t) \tag{9-66}$$

which is a constant times the modulating digital signal.

## BPSK Bandwidth

An estimate of the minimum bandwidth for a BPSK signal may be developed by again assuming a dotting pattern for the modulating signal. In this case, however, the resulting waveform will appear as in Figure 9–22(a). Using the same notation developed for ASK, the BPSK signal may be represented as the product of a continuous sine wave with frequency $f_c$ and a square wave of frequency $f_p$. However, the square wave in this case will not have a dc component, and will have the form shown in Figure 9–22(b).

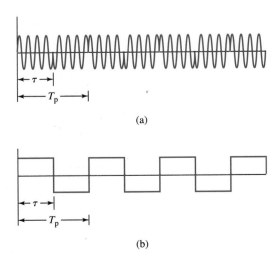

(a)

(b)

**FIGURE 9–22**
Time-domain waveforms used in developing estimate of BPSK bandwidth.

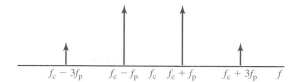

**FIGURE 9–23**
Spectral form used in developing estimate of BPSK bandwidth.

The spectrum of this signal may be predicted from Section 3-6, and is shown in Figure 9–23. In this case, there is no component at the frequency $f_c$, but the sideband pattern is the same as for ASK. An estimate of the bandwidth may be made by assuming that a bandwidth sufficient to transmit the two major sidebands will be sufficient for pulse recognition at the receiver ("coarse" reproduction). This bandwidth $B_T$ is given by

$$B_T = 2f_p = R \tag{9-67}$$

Stated in words, *the minimum bandwidth for BPSK is equal to the data rate in bits/second, and is the same as ASK for the same data rate.*

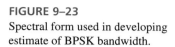 **EXAMPLE 9-10**

A binary NRZ-L PCM signal having a data rate of 200 kbits/s modulates an RF carrier in a BPSK format. Determine the transmission bandwidth.

**SOLUTION**   From Equation 9-67, the bandwidth is simply

$$B_T = R = 200 \text{ kHz} \tag{9-68}$$

Comparing Examples 9-8, 9-9, and 9-10 for the same data rate, it can be concluded that the transmission bandwidth values for ASK and BPSK are the same, but FSK requires a greater bandwidth.

## 9-9   Differentially Encoded Phase-Shift Keying

A special form of PSK offering certain advantages is *differentially encoded phase-shift keying,* hereafter referred to by the acronym *DPSK*. This signal form is obtained by first converting the basic NRZ-L form of the signal to either NRZ-M or NRZ-S before phase-modulating the RF carrier. The process will be illustrated with the NRZ-S format.

### DPSK Generation

Consider the system shown in Figure 9–24. The signal $v_1(t)$ is the input data stream, which is assumed to be encoded in the basic NRZ-L format. The circuit preceding the balanced

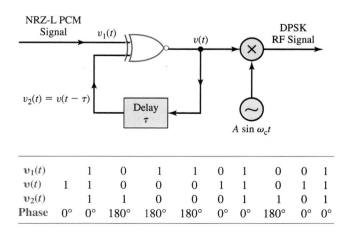

**FIGURE 9–24**
System used to generate differentially encoded phase-shift keying (DPSK) signal.

| $v_1(t)$ | | 1 | 0 | 1 | 1 | 0 | 1 | 0 | 0 | 1 |
|---|---|---|---|---|---|---|---|---|---|---|
| $v(t)$ | 1 | 1 | 0 | 0 | 0 | 1 | 1 | 0 | 1 | 1 |
| $v_2(t)$ | | 1 | 1 | 0 | 0 | 0 | 1 | 1 | 0 | 1 |
| **Phase** | 0° | 0° | 180° | 180° | 180° | 0° | 0° | 180° | 0° | 0° |

modulator consists of an Exclusive-Nor circuit and a 1-bit shift register (or delay element). Let $\tau$ represent the width of the bit interval. Observe that the output of the Exclusive-Nor circuit is delayed by 1 bit interval and applied as one of the inputs.

The logic for the Exclusive-Nor circuit is that the output is 1 when both inputs are the same and 0 when they are different. For illustration, assume the data stream $v_1(t)$ shown beneath the figure. We will arbitrarily assume a 1 for the output $v(t)$ of the Exclusive-Nor circuit one bit interval $\tau$ before the first data bit appears. (This choice does not make any difference as far as the final detected value is concerned.)

The initial 1 at the output is delayed one bit interval and appears at the input as $v(t - \tau)$ when the first data bit appears. It is compared with the first data bit, which is a 1, and the resulting output is also 1. This value is delayed by $\tau$ and compared with the next input bit, which is 0, and this produces a 0 for that output. The tabulated values appearing below the figure should permit the reader to follow the pattern.

By carefully observing the relationship between the input data stream $v_1(t - \tau)$ and the modified data stream $v(t)$, it can be deduced that $v(t)$ is an NRZ-S representation for the signal. (Refer back to Figure 9–7 if necessary.) Thus, this circuit represents one way for converting NRZ-L data to NRZ-S data.

The NRZ-S signal can now be applied to the balanced modulator, and the resulting output is a form of a PSK signal. However, since the signal was differentially encoded, the more appropriate label DPSK is used. The last row in the tabulated data of Figure 9–24 represents the relative phase of the reference sinusoidal carrier, which is assumed to be either 0° or 180°.

## DPSK Detection

The process for generating the DPSK signal is interesting, but it offers no insight as to why such a seemingly complex encoding strategy is used. The answer lies in the detection process, which will now be discussed. Consider the block diagram shown in Figure 9–25, and assume an ideal received signal $v_r(t)$ with no noise present. The received signal is delayed by the bit duration $\tau$ and applied as one input to an ideal multiplier, and the other input is the undelayed signal. The output $v_1(t)$ of the multiplier is given by

$$v_1(t) = [Av(t)\sin\omega_c t] \cdot [Av(t - \tau)\sin\omega_c(t - \tau)]$$

$$= A^2 v(t)v(t - \tau)\sin\omega_c t \sin\omega_c(t - \tau) \tag{9-69}$$

By using a standard trigonometric identity, this equation can be expanded as

$$v_1(t) = \frac{A^2}{2}v(t)v(t - \tau)\cos\omega_c\tau - \frac{A^2}{2}v(t)v(t - \tau)\cos(2\omega_c t - \omega_c\tau) \tag{9-70}$$

The second term in Equation 9-70 is a DSB signal centered at $2f_c$ and is easily eliminated with a low-pass filter—assuming, of course, that the carrier frequency range is much higher than the highest frequency in the baseband spectrum.

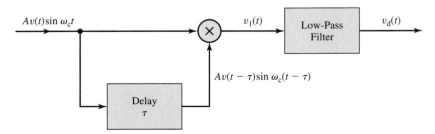

**FIGURE 9–25**
Detection system for DPSK signals.

The first term in Equation 9-70 contains the constant factor $\cos \omega_c \tau$. Since the signal output should be as large as possible, an optimum design would call for

$$\omega_c \tau = 2\pi f_c \tau = 2\pi n \tag{9-71}$$

where $n$ is an integer. This leads to

$$f_c = \frac{n}{\tau} \tag{9-72}$$

or

$$\tau = \frac{n}{f_c} = nT_c \tag{9-73}$$

where $T_c = 1/f_c$ is the carrier period. Stated in words, the bit interval $\tau$ should be selected so that it contains an *integer* number of cycles of the RF carrier (or the IF carrier if the signal has been down-converted). With this assumption, the detected output is

$$v_d(t) = \frac{A^2}{2} v(t)v(t - \tau) \tag{9-74}$$

At this point, we are still a little puzzled about the result, since Equation 9-74 does not look proper. However, let's see what this result means. Ignoring the $A^2/2$ factor, the product $v(t)v(t - \tau)$ is shown in Figure 9–26. When the result is compared with the original data stream of Figure 9–24(a), we see that the product $v(t)v(t - \tau) = v_1(t)$ is in bipolar form, a most fascinating result! We now see the primary advantage of the DPSK encoding process. The RF signal acts as its own synchronous reference for detection purposes, provided that the reference is delayed by one bit interval.

The error rate with DPSK when noise is present is higher than for conventional PSK under the most optimum conditions. It turns out that errors tend to occur in pairs of two, as will be illustrated in Figure 9–27. The waveform in Figure 9–27(a) is identical to the NRZ-S representation in Figure 9–26, except for the one bit error as shown. When the signal is delayed as shown in Figure 9–27(b), the same bit is obviously still incorrect. The resulting product shown in Figure 9–27(c) contains two bit errors. In spite of the increase in the number of errors, DPSK offers significant advantages in implementation and ease of detection.

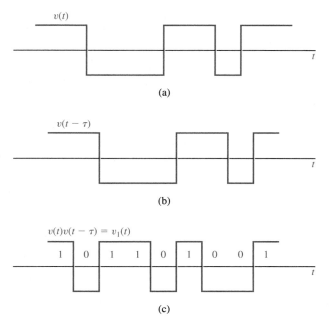

**FIGURE 9–26**
Waveforms pertaining to discussion of DPSK detection scheme.

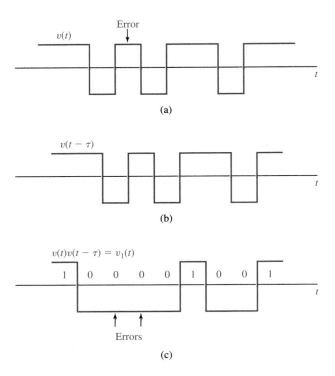

**FIGURE 9–27**
Waveforms illustrating how errors tend to occur in pairs when detecting DPSK signals.

## 9-10  Multisim® Examples (Optional)

The examples in this section will deal with the design, implementation, and analysis of circuits that generate and detect differentially encoded phase-shift keying signals. The process involves the utilization of certain digital circuits and modules, for which a complete treatment is outside of the scope of the text. The circuits will be analyzed to the extent feasible, and readers having a reasonable background in digital circuits and systems should be able to follow the development. Otherwise, the examples may be omitted without loss of continuity.

The circuit designs should not necessarily be considered as optimum but were chosen to readily illustrate the concepts with common Multisim components.

---

**⦀ MULTISIM EXAMPLE 9-1**

The circuit of Figure 9–28 represents a data-domain implementation of an NRZ-S encoder, which is the first step required in generating a DPSK signal. Generate the bit pattern of $v_1(t)$ in Figure 9–24, and show that it is converted to an NRZ-S data stream.

**SOLUTION**  The block on the left represents a **Word Generator** module, and the block on the right represents a **Logic Analyzer,** both of which are obtained from the **Instruments Toolbar.** The digital logic components are all obtained from the **TTL** parts bin under the heading of **74STD.** The actual circuit consists of a **7486N Exclusive-Or** circuit, a **7404N Inverter,** and a **7474N D Flip-Flop.** The first two circuits form an **Exclusive-Nor** operation and the latter circuit performs the delay required for the feedback signal. A **Clock Voltage Source** is obtained from the **Signal Source Family Bar** and used as the timing clock for the circuit. It was left at the default value of **1 kHz,** which will result in a data rate of 1 kbit/s.

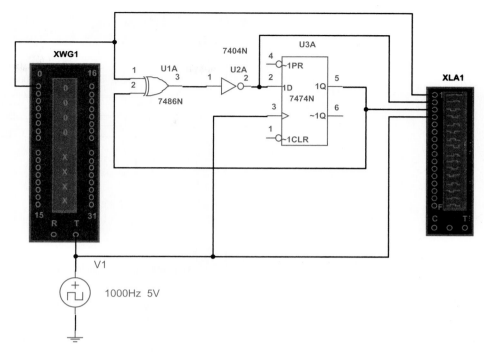

**FIGURE 9–28**
NRZ-S encoder of Multisim Example 9-1.

Refer to Figure 9–29 for the discussion that follows. A double-click on both the **Word Generator** and the **Logic Analyzer** will open the windows shown, but the latter one will not show any activity until the circuit is energized. The bit pattern is entered in the **Word Generator,** and various parameters and settings are shown. Left-click next to the beginning data point and a window will open. Then left-click on **Set Initial Postion** and the first data point will be identified. A similar procedure is followed at the last point except the command in this case is **Set Final Position.** For each run, a left-click on the first point should be made followed by a left-click on **Set Cursor.**

The switch just above the Circuit Window that is identified with a **0** and a **1** is used to activate the logic data domain analysis, although some of the types of runs could be made directly within the **Logic Analyzer.** Depressing the switch to **1** activates the circuit, and if the **Logic Analyzer** window is open, the responses of the various stages may be observed there. After the analysis is complete, the switch should be returned to the **0** position before making other changes. The logic pattern on the analyzer remains in the window, and it can be cleared before a new run is made. Before a new run is made, left-click on **Reset.**

The results are displayed in the window at the top of Figure 9–29. The input signal to the decoder is the top one and is indicated as A. The resulting output is shown as B. The first bit shown in Figure 9–24 is not present, since it is assumed to occur prior to the timing diagram of Figure 9–29. However, all other bits have been converted to the NRZ-S format shown in Figure 9–24. The feedback signal is shown in C, and the circuit clock is shown in D. The clock driving the analyzer has been set to **16** bits per division, so it appears as a solid segment near the bottom of the window.

---

**⫿⫿ MULTISIM EXAMPLE 9-2**

Continuing with a data-domain analysis, connect the output of the circuit of Multisim Circuit 9-1 to a similar circuit, and show that the original data signal is obtained.

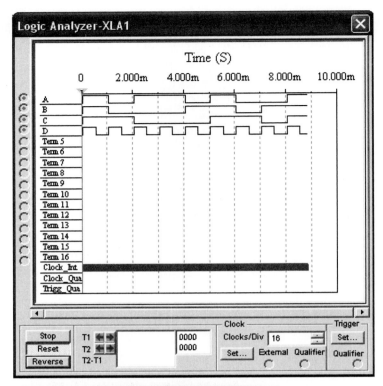

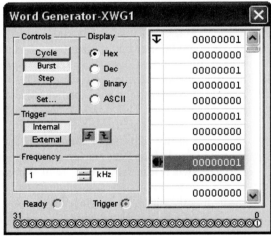

**FIGURE 9–29**
Results for Multisim Example 9-1.

**SOLUTION**    The circuit is shown in Figure 9–30. The portion to the right of the **Logic Analyzer** is essentially a replica of the circuit used to generate the NRZ-S signal.

The bit patterns are shown in Figure 9–31. The normal delay that would be present in any real system is not present here, since the output of the NRZ-S circuit is directly connected to another similar circuit. The first four waveforms are the same as in Figure 9–29.

The output of the right-hand D flip-flop is shown in E, and the reconstructed signal is shown in F. As can be seen, this bit pattern is the same as the input shown in A.

---

**▌▌ MULTISIM EXAMPLE 9-3**    The circuit of Figure 9–32 represents a complete DPSK system utilizing circuit voltages and currents (as opposed to the data-domain logic analysis). Perform a complete **Transient Analysis** on the system based on the input bit pattern used in Figure 9–24.

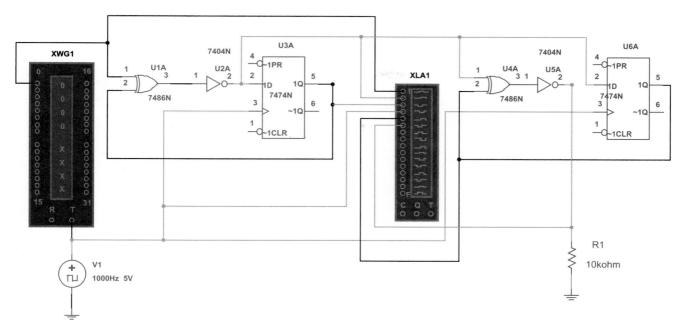

**FIGURE 9–30**
NRZ-S encoder and decoder of Multisim Example 9-2.

**SOLUTION**  When circuits begin to reach a size that would prohibit showing all the individual components in the **Circuit Window,** the use of **Subcircuits** becomes a feasible option. A **Subcircuit** is a portion of some larger circuit that has been replaced by a block having input and output terminals. To create a subcircuit, all the components are wired in the fashion required, and it should be tested to ensure that it is working correctly. Then left-click on **Place** on the top row. Input and output terminals are established by **HB/SB** (hierarchical blocks/subcircuit blocks) obtained in the same manner as connectors. Choose a corner of the circuit at which to begin. Select the **Enter Circuit** by holding down the left button, and move the resulting dashed line around the entire circuit. Next, left-click again on **Place** on the **Menu Row** and left-click on **Replace by Subcircuit.** A window will then open in which you give a name to the subcircuit. This procedure is repeated as many times as desired in order to form a complete system. The subcircuits may be saved in the same manner as complete circuits, and connected in the same manner as components.

A double-click on any one of the modules will allow its contents to be inspected, and edited if necessary by the **Edit** command. The NRZ-S encoder is shown in Figure 9–33, and it follows essentially the same pattern as in Multisim Examples 9-1 and 9-2, except that the signal has been generated with the PWL source. The DPSK modulator is shown in Figure 9–34, and it consists of a multiplier and a sinusoidal "carrier." A bias of –2.5 V is added as an offset in the **Y** path within the modulator, to give the signal a bipolar nature. The signal is then applied to the DPSK demodulator (Figure 9–35), representing the first step in the detection process. Since the signal at this point is analog in nature, the required delay in the detection process is achieved with a lossless section of transmission line having a delay equal to one bit width; that is, by 1 ms. (The properties of the Multisim transmission-line model will be developed in Section 14-9.)

The schematic of the active filter is shown in Figure 9–36. It is a two-pole Butterworth filter with a 3-dB cutoff frequency of 1.5 kHz. This filter attenuates the components centered at 2 kHz. The resulting analog signal is then applied to the digital reconstruction circuit shown in Figure 9–37. This circuit consists of a comparator, which converts the signal back to its digital form. Finally, the digital output is shown in Figure 9–38, based on a necessary shift of one bit width in order to establish the synchronous detection process. There is also a slight additional delay due to the filtering process, but the restored signal is seen to be equivalent to the original data stream.

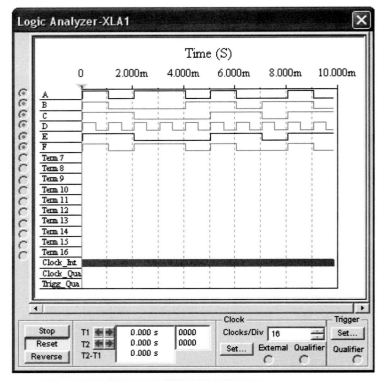

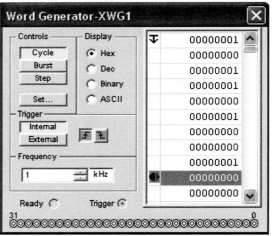

**FIGURE 9–31**
Results for Multisim Example 9-2.

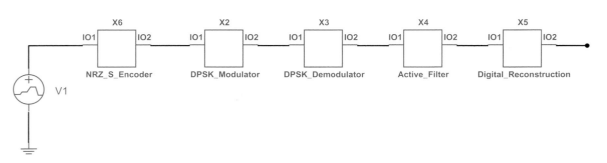

**FIGURE 9–32**
DPSK system of Multisim Example 9-3.

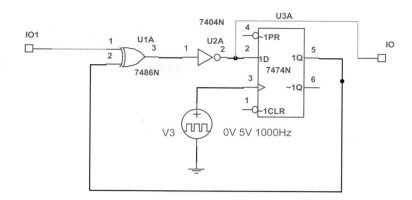

**FIGURE 9–33**
DPSK generator circuit.

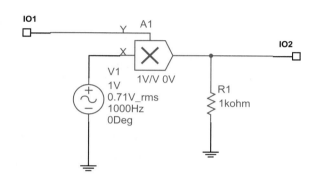

**FIGURE 9–34**
DPSK modulator circuit.

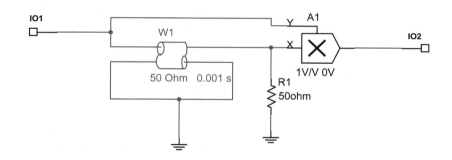

**FIGURE 9–35**
DPSK demodulator circuit.

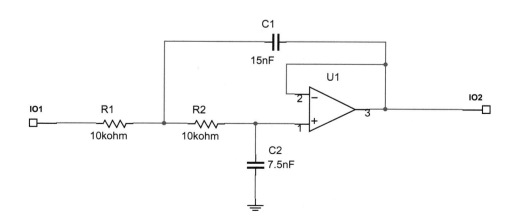

**FIGURE 9–36**
Active filter circuit.

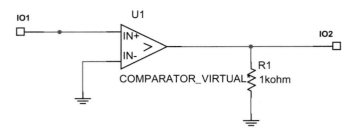

**FIGURE 9–37**
Digital reconstruction circuit.

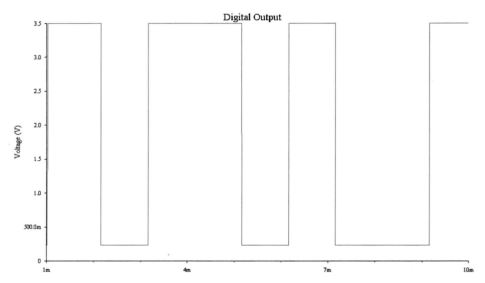

**FIGURE 9–38**
Output of DPSK reconstruction process.

## 9-11    MATLAB® Example (Optional)

An example illustrating the simulation of a digital binary FSK system will be developed in this section. Due to the length of this example, we will utilize an **M-file** instead of the **Command Window.** An M-file is basically a computer program written with MATLAB code. For longer analysis purposes, it is much easier to troubleshoot an M-file than to determine possible errors in the Command Window. Further information concerning the development of M-files can be found in Appendix C.

**MATLAB EXAMPLE 9-1**    Consider an FSK system with the following parameters:

Space or 0:    1 kHz

Mark or 1:    2 kHz

Data Rate:    1 kb/s

Write an M-file program that will generate a dotting pattern for 10 bit widths, and apply the signal to two bandpass filters, each of which provides the maximum response for the two possible signal values.

**SOLUTION**    The M-file program is shown in Figure 9–39, and reference to it will be made in the discussion that follows. First, the time step **delt** is somewhat arbitrarily

```
%Program fsk.m
delt=1e-6;
t=0:delt:.012-delt;
tbit=0:delt:1e-3-delt;
v0=sin(2*pi*1000*tbit);
v1=sin(2*pi*2000*tbit);
v=[v0 v1];
for n=2:5
    v=[v v0 v1];
end
v=[v zeros (1,2000)];
[n0 d0]=butter(2,2*pi*[500 1500], 's');
[n1 d1]=butter(2,2*pi*[1500 2500], 's');
y0=1sim(n0,d0,v,t);
y1=1sim(n1,d1,v,t);
figure(1)
plot(t,v)
xlabel('Time, seconds')
ylabel('Voltage, volts')
title('FSK Signal Based on Dotting Pattern')
grid
pause
figure(2)
plot(t(1:2000),v(1:2000))
xlabel('Time, seconds')
ylabel('Voltage, volts')
title('FSK Signal For First Two Bits')
grid
pause
figure(3)
plot(t,y0)
xlabel('Time, seconds')
ylabel('Voltage, volts')
title('Output of Space Filter')
grid
pause
figure(4)
plot(t,y1)
xlabel('Time, seconds')
ylabel('Voltage, volts')
title('Output of Mark Filter')
grid
pause
```

**FIGURE 9–39**
M-file program of MATLAB Example 9-1.

selected as 1 µs by the command

```
>> delt = 1e-6;
```

The bit width **tbit** is the reciprocal of the data rate and is $1/1000 = 1$ ms. It is generated by 1000 steps of **delt** by the command

```
>> tbit = 0:delt:1e-3-delt;
```

Note that the last point is one time step below 1 ms, which results in 1000 simulation points per bit.

While the length of 10 bits is 10 ms, there will be some time delays in the output filters, and in order to best observe the complete pattern, the total time duration is chosen to be 12 ms. The time vector **t** is then generated by the command

```
>> t = 0:delt:0.012e-3-delt;
```

This results in 12,000 total points for the signal.

A space (or 0) for one bit width is denoted **v0**, and is generated by the command

```
>> v0 = sin(2*pi*1000*tbit);
```

A mark (or 1) for one bit width is denoted **v1**, and is generated by

```
>> v1 = sin(2*pi*2000*tbit);
```

The combined signal is denoted **v**, and the first two bits are

```
>> v = [v0 v1];
```

A **for loop** is then used to generate the next 8 bits.

```
>> for n = 2:5
>> v =[v v0 v1];
>> end
```

Finally, zeros are added for the last two bit widths.

```
>> v = [v zeros(1,2000)];
```

A computation of the approximate bandwidth results in a value of 2 kHz. If the hypothetical center frequency is then assumed to be 1.5 kHz, this would suggest that the total bandwidth would extend from about 500 Hz to about 2.5 kHz. The space output filter should have its center frequency at 1 kHz, and the mark output filter should have its center frequency at 2 kHz. Assuming a bandwidth of about 1 kHz each for reasonable separation, this would suggest that the space fitter band-edge frequencies should be set at about 500 Hz and 1500 Hz. The mark filter band-edge frequencies should then be set at about 1500 and 2500 Hz.

Based on the preceding ideas, two separate two pole-pair Butterworth filters are implemented with MATLAB. For the space filter, the numerator and denominator polynomials are denoted **n0** and **d0**, respectively. For the mark filter, the corresponding polynomials are denoted **n1** and **d1**. The commands to generate these filters are

```
>> [n0 d0] = butter(2, 2*pi*[500 1500], 's')
>> [n1 d1] = butter(2, 2*pi*[1500 2500],'s')
```

The remaining part of the program consists of plots of the dotting pattern input and outputs of the two filters. First, the FSK input signal is shown for a 10-ms interval in Figure 9–40. An expanded view of the first two bits is shown in Figure 9–41.

The output of the space filter is shown in Figure 9–42, and the output of the mark filter is shown in Figure 9–43. It is necessary to understand the action of the filters in order to interpret these figures. In each case, there is a delay of the order of 1 ms before the output reaches a peak value. Thus, the first peak of the space filter occurs at about 1 ms, and other maxima occur at 2 ms intervals thereafter. Likewise, the first peak of the mark filter occurs at about 2 ms, and other maxima occur at 2 ms intervals thereafter.

We have only dealt with the first step of the detection process in this example. Subsequent processing by a peak detector or some similar operation would be required, along with reconstruction of the digital signal. The delay process is always a reality, since it takes time for a signal to reach a receiver, and filtering introduces additional delays. As will be discussed later, either a separate synchronizing signal or pattern must be used, or a self-synchronizing (asynchronous) approach is employed.

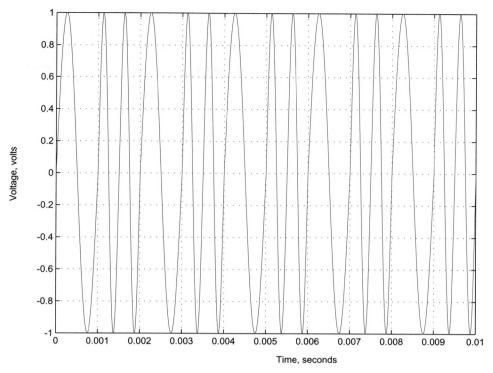

**FIGURE 9–40**
FSK signal based on dotting pattern.

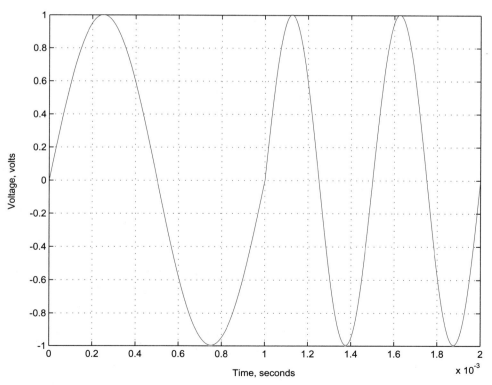

**FIGURE 9–41**
FSK signal for first two bits.

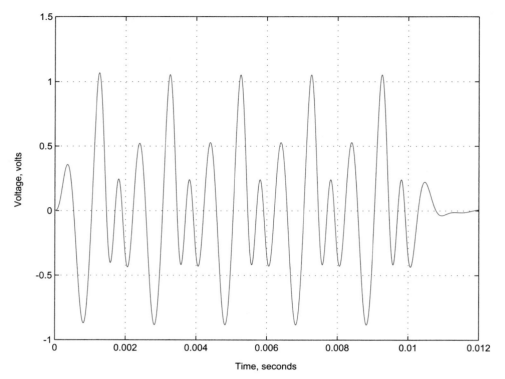

**FIGURE 9–42**
Output of space filter.

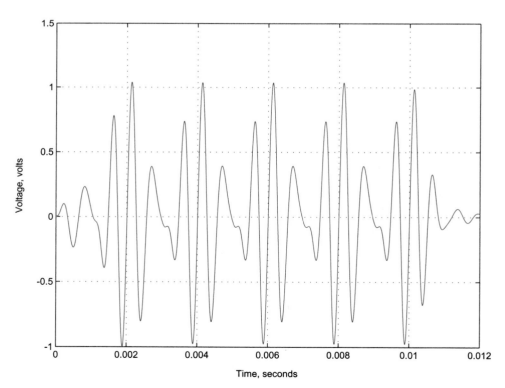

**FIGURE 9–43**
Output of mark filter.

## SystemVue™ Closing Application (Optional)

Insert the text CD in a computer having SystemVue™ installed and activate the program. Open the CD folder entitled **SystemVue Systems** and load the file entitled 9-1.

### Review of System

The system shown on the screen was described in the chapter opening application, and you may wish to review the discussion and system configuration provided there. Perform a simulation based on the initial parameters, and overlay the two plots according to the procedure given in the opening application.

### Quantization as a Function of the Number of Bits

1. Based on the material covered in the chapter, identify some of the parameters here. Is this unipolar or bipolar encoding? Is the quantization based on truncation or rounding? What is the step size? Except for the larger step at the top, which is expected, what is the maximum error?

2. Change the quantizer to 4 bits and perform a simulation. Except for the larger step at the top, what is the maximum error?

3. Repeat the preceding step with 8 bits.

4. Change the input to a sinusoidal function having a peak value of 1 V (peak-to-peak value of 2 V), and observe the input and output waveforms.

## PROBLEMS

**9-1**   Determine the number of possible PCM values that can be encoded for (a) 6 bits and (b) 10 bits.

**9-2**   Determine the number of possible PCM values that can be encoded for (a) 12 bits and (b) 14 bits.

**9-3**   It is desired to represent an analog signal in no less than 500 values. Determine the minimum number of bits required for each word.

**9-4**   It is desired to represent an analog signal in no less than 5000 values. Determine the minimum number of bits required for each word.

**9-5**   A unipolar analog signal is sampled at intervals of 1 ms by a 4-bit A/D converter whose normalized input–output characteristic is given by Figure 9–2. The voltage values at the sampling points are provided in the table that follows, and the full-scale voltage is 10 V. Determine the 4-bit digital word that would be generated at each point.

| time, ms | 0 | 1 | 2 | 3 | 4 | 5 | 6 |
|---|---|---|---|---|---|---|---|
| value, V | 0 | 1.5 | 2 | 3.5 | 5.5 | 7.7 | 9.4 |
| digital word | | | | | | | |

**9-6**   A unipolar analog signal is sampled at intervals of 1 ms by a 4-bit A/D converter whose normalized input–output characteristic is given by Figure 9–2. The voltage values

| time, ms | 0 | 1 | 2 | 3 | 4 | 5 | 6 |
|---|---|---|---|---|---|---|---|
| value, V | 0 | 1.5 | 2.4 | 3.5 | 4.5 | 4 | 0.5 |
| digital word | | | | | | | |

at the sampling points are provided in the table above, and the full-scale voltage is 5 V. Determine the 4-bit digital word that would be generated at each point.

**9-7**   A bipolar analog signal is sampled at intervals of 1 ms by a 4-bit A/D converter whose input–output characteristic is given by Figure 9–3. The voltage values at the sampling points are provided in the table that follows, and the full-scale voltage is 5 V (based on a peak-to-peak range of 10 V) Determine the 4-bit digital word that would be generated at each point.

| time, ms | 0 | 1 | 2 | 3 | 4 | 5 | 6 |
|---|---|---|---|---|---|---|---|
| value, V | 3.8 | 2.7 | 2.3 | 0.25 | −2 | −3.5 | −4.8 |
| digital word | | | | | | | |

**9-8**   A bipolar analog signal is sampled at intervals of 1 ms by a 4-bit A/D converter whose input–output characteristic is given by Figure 9–3. The voltage values at the sampling points are provided in the table that follows, and the full-scale voltage is 10 V (based on a peak-to-peak range of

| time, ms | 0 | 1 | 2 | 3 | 4 | 5 | 6 |
|---|---|---|---|---|---|---|---|
| value, V | −9 | −6 | −1 | 0 | 4 | 6 | 8 |
| digital word | | | | | | | |

20 V) Determine the 4-bit digital word that would be generated at each point.

**9-9** A 16-bit A/D converter with a full-scale voltage of 20 V is to be employed in a binary PCM system. The input analog signal is adjusted to cover the range from zero to slightly under 20 V, and the converter is connected for unipolar encoding. Rounding is employed in the quantization strategy. Determine the following quantities: (a) normalized step size, (b) actual step size in volts, (c) normalized maximum quantized analog level, (d) actual maximum quantized level in volts, (e) normalized peak error, and (f) actual peak error in volts.

**9-10** A 6-bit A/D converter with a full-scale voltage of 20 V is to be employed in a binary PCM system. The input analog signal is adjusted to cover the range from zero to slightly under 20 V, and the converter is connected for unipolar encoding. Rounding is employed in the quantization strategy. Determine the following quantities: (a) normalized step size, (b) actual step size in volts, (c) normalized maximum quantized analog level, (d) actual maximum quantized level in volts, (e) normalized peak error, and (f) actual peak error in volts.

**9-11** The 16-bit A/D converter of Problem 9-9, while maintaining the same peak-to-peak range, is converted to bipolar offset form so that it can be used with an analog signal having a range from −10 V to just under 10 V. Repeat all the calculations of Problem 9-9.

**9-12** The 6-bit A/D converter of Problem 9-10, while maintaining the same peak-to-peak range, is converted to bipolar offset form so that it can be used with an analog signal having a range from −10 V to just under 10 V. Repeat all the calculations of Problem 9-10.

**9-13** A $\mu$-compression law encoder has the following parameters: $V_{i\,max} = 8$ V, $V_{o\,max} = 5$ V, $\mu = 255$. Determine the output voltage for each of the following input voltage levels: (a) 2 V, (b) 4 V, (c) 8 V.

**9-14** From the compressed values in the encoder of Example 9-6, use the expansion function of Equation 9-32 to determine the original values for each of the compressed values. (*Note:* Expect more than the average amount of roundoff due to the exponentiation process.)

**9-15** Consider the bit stream 01001110. Sketch the forms for each of the following data formats: (a) NRZ-L, (b) NRZ-M, (c) NRZ-S, (d) RZ, (e) biphase-L, (f) biphase-M, (g) biphase-S. For encoding schemes in which the initial level is aribitrary, assume a zero level just before the first bit is received.

**9-16** Repeat the work of Problem 9-15 for the bit stream 101101001.

**9-17** Design a circuit using one or more basic logic gates to convert an NRZ-L signal to an RZ signal. Assume standard TTL levels, and assume that a synchronous square

wave having twice the data bit rate frequency is available. Show waveforms to verify the result.

**9-18** Design a circuit using one or more basic logic gates to convert an NRZ-L signal to a biphase-L signal. Assume standard TTL levels, and assume that a synchronous square wave having twice the data bit-rate frequency is available. Show waveforms to verify the result.

**9-19** Consider a PCM TDM system in which 24 signals are to be processed. Each signal has a baseband bandwidth $W = 3$ kHz. The sampling rate is to be 33.3% higher than the theoretical minimum, and 8 bits are to be used in each word. Conventional NRZ-L encoding will be used, and an extra bit is added to each frame for sync. (a) Determine the bit rate. (b) Determine the approximate minimum baseband transmission bandwidth. (*Note:* This is the format used in the commercial T1 telephone system.)

**9-20** A 12-bit PCM TDM system must be designed to process six data channels. Channels 1 to 4 each have a baseband bandwidth $W = 1$ kHz, while channels 5 and 6 each have $W = 2$ kHz. To provide guard band, a sampling rate 25% above the theoretical minimum is to be employed. (a) Using only one commutator, devise a scheme whereby each of the channels is uniformly sampled at the proper rate. (b) Determine the approximate minimum baseband transmission bandwidth.

**9-21** Consider the complex sampling scheme illustrated in Figure 9–10. Assume that the prime frame sampling rate is 2 kHz; that is, the prime commutator makes one "rotation" in 0.5 ms. (a) How many separate signals can be processed with the system as it is connected? (b) Assuming sampling at the Nyquist rate with no guard band provided, determine the maximum baseband data frequency for each of the available channels.

**9-22** In this problem, you are to make a brief study of PCM peak quantization error as a function of the required transmission bandwidth. The result should vividly display the exponential trade-off between improved accuracy (lower noise) and transmission bandwidth. Consider a PCM system with $N$ bits and assume (1) minimum Nyquist rate, (2) no spacing between pulses, and (3) $0.5/\tau$ bandwidth rule. The peak quantization error will be defined as the magnitude of the percentage error for this purpose, and unipolar encoding will be assumed. Compute enough data to plot a curve of the percentage of *peak quantization error* as a function of the ratio $B_T/W$, where $B_T$ is the approximate baseband bandwidth. Use *semilog* paper with error on the log scale. While your result will be computed for integer values of $N$, extrapolate between points so that the nature of the relationship can be clearly seen.

**9-23** A single-channel binary NRZ-L PCM signal having a sampling rate of 4000 samples per second utilizes 8-bit words and modulates an RF carrier in an ASK format. Determine the RF transmission bandwidth.

**9-24** A single-channel binary NRZ-L PCM signal having a sampling rate of 20,000 samples per second utilizes 16-bit words and modulates an RF carrier in an ASK format. Determine the RF transmission bandwidth.

**9-25** Repeat the analysis of Problem 9-23 if FSK is utilized and the two frequencies are separated by 25 kHz.

**9-26** Repeat the analysis of Problem 9-24 if FSK is utilized and the two frequencies are separated by 250 kHz.

**9-27** Repeat the analysis of Problem 9-23 if BPSK is utilized.

**9-28** Repeat the analysis of Problem 9-24 if BPSK is utilized.

**9-29** Consider the bit stream used in explaining the DPSK NRZ-S PCM in Figures 9–24 and 9–26. (a) Develop a new table of entries for the case in which the initial bit for $v(t)$ is assumed to be a 0. (b) By repeating the development of Figure 9–26, verify that the original bit stream is recovered.

**9-30** Consider the bit stream used in the NRZ-S system of Figures 9–24 and 9–26. Replace the Exclusive-Nor circuit by an Exclusive-Or circuit, and show that the resulting signal is an NRZ-M signal. Assume an initial 0 for $v(t)$.

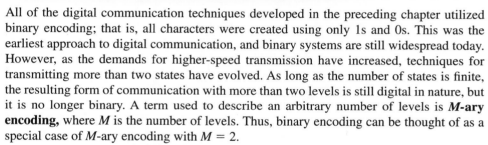

# Digital Communications II: M-ary Systems

## 10

### OVERVIEW AND OBJECTIVES

All of the digital communication techniques developed in the preceding chapter utilized binary encoding; that is, all characters were created using only 1s and 0s. This was the earliest approach to digital communication, and binary systems are still widespread today. However, as the demands for higher-speed transmission have increased, techniques for transmitting more than two states have evolved. As long as the number of states is finite, the resulting form of communication with more than two levels is still digital in nature, but it is no longer binary. A term used to describe an arbitrary number of levels is **M-ary encoding,** where $M$ is the number of levels. Thus, binary encoding can be thought of as a special case of $M$-ary encoding with $M = 2$.

The process of creating an arbitrary number of transmission levels is a rapidly changing technological area. Much of this evolution has been driven by the internet, in which the growth continues to stretch the limits of transmission rates. It is virtually impossible to produce a textbook that could relate to all of the most recent developments in this area, since many are proprietary in nature. However, by focusing on the principles of higher data rate encoding, the reader will develop a facility for understanding and dealing with the newer developments as they arise.

## Objectives

After completing this chapter, the reader should be able to:

1. Define *M-ary encoding* and discuss its advantages and disadvantages.
2. Define *baud* or *symbol* rate.
3. Define *information rate* or *bit rate* and its relationship to *baud rate* or *symbol rate.*
4. State the Shannon-Hartley theorem and discuss its implications.
5. Apply the Shannon-Hartley theorem to determine *channel capacity.*
6. State the Shannon limit and discuss its implications.
7. Apply the Shannon limit to determine *channel capacity.*
8. Discuss *quadriphase shift keying* (QPSK) and its generation and detection.
9. Discuss the concept of *quadrature amplitude modulation* (QAM) and list some of its forms.
10. Draw constellation diagrams for different forms of *M-ary* encoding.

## SystemVue™ Opening Application (Optional)

Insert the text CD in a computer having SystemVue™ installed and activate the program. Open the CD folder entitled **SystemVue Systems** and load the file entitled 10-1.

### Sink Tokens

| Number | Name | Signal Monitored |
|--------|------|------------------|
| 0 | Bit Train | 2 at output 0 |
| 1 | BPSK Signal | 2 at output 1 |

### Operational Tokens

| Number | Name | Function |
|--------|------|----------|
| 2 | BPSK Signal | bit train (1 Hz) and carrier (2 Hz) |

The preset values for the run are from a starting time of 0 to a final time of 10 s, with a time step of 1 ms. Run the simulation and observe both the **Bit Train** and the **BPSK Signal.** Align the two curves using the **Cascade Horizontal** control.

### What You See

Although the bit train is ultimately periodic, its period is very long and it appears to be random in nature. This type of waveform is referred to as a *pseudorandom* signal. You may wish to initiate the simulation a number of times in order to see various patterns that appear to be random.

The input digital signal has two possible levels. It modulates the carrier with two possible phase shifts. Observe the various segments of the modulated signal to see how each of the two possible levels creates a different phase relationship for the modulated signal. Note that since there are an integer number of cycles of the carrier in a bit interval, the phase reversals occur at zero crossings of the waveform.

### How This Demonstration Relates to Chapters 9 and 10 Learning Objectives

You have already studied *binary phase-shift keying* (BPSK), so this demonstration should help to solidify some of the concepts of Chapter 9. In Chapter 10, you will learn how utilizing more than the two levels encountered with binary signals can increase the data rate for digital transmission. At the end of the chapter, you will return to this system and modify it to produce more than two levels.

## 10-1 Information Measures

The process of binary PCM encoding was the basis for all the work in the preceding chapter. For basic binary encoding, each transmitted symbol represents one bit; that is, one of only two possible states. Suppose, however, that it is possible to transmit and detect more than two possible states. This could effectively increase the transmission rate for a given bandwidth. For example, suppose that it were possible to transmit and detect eight possible states at the receiver. For binary transmission, encoding in eight possible states requires three bits. In effect, the data transmission rate for a given bandwidth would be tripled with eight levels.

### M-ary Encoding

Much research and development in recent years have concentrated on *M*-ary systems, in which the number of distinguishable levels can be increased over that of basic binary transmission. It is this type of technology that permits high-speed data transmission over telephone lines that are limited in bandwidth.

Assume that a signaling scheme permits *M* levels to be detected at the receiver. The number of bits *N* associated with this number of levels is

$$N = \log_2 M \tag{10-1}$$

Conversely,

$$M = 2^N \tag{10-2}$$

Unless *M* is an integer power of 2, the number of bits *N* will not be an integer, but the definition is still used as given. In most practical systems, *M* is chosen as an integer power of 2, in which case both *M* and *N* are integers.

### Computing Base-2 Logarithms

Most calculators do not have base-2 logarithmic functions. However, most scientific calculators do have base-10 functions. The following relationship allows the determination of a base-2 logarithm in terms of a base-10 logarithm:

$$N = \frac{\log_{10} M}{\log_{10} 2} = \frac{\log_{10} M}{0.3010} = 3.32 \log_{10} M \tag{10-3}$$

### Baud Rate

A term that has been widely used in digital data communications is the *baud rate*. This term is also referred to both as the *symbol rate* and the *signaling rate*. The *baud* or *symbol rate* is defined as the number of distinct symbols transmitted per second, irrespective of the form of encoding. Thus, the baud rate is independent of the number of possible levels that the signal can assume, and depends only on the number of transitions per unit time.

Recall from earlier chapters that the basic limitation on the minimum width of a pulse is the bandwidth. Specifically, for baseband transmission, the minimum width of a pulse for "coarse" reproduction has been assumed as $0.5/B$, where $B$ is the baseband bandwidth. Accepting this limitation, the maximum number of pulses that can be transmitted per second is the reciprocal of the minimum pulse width, and this serves as the basis for the maximum baud rate. Hence, for **baseband** transmission

$$\text{maximum baud rate} = 2B \tag{10-4}$$

Note that this equation says nothing about the number of possible levels. The assumption is that if the bandwidth is sufficient to permit a single pulse, it is capable of discerning different levels of pulses.

### Information Rate and Channel Capacity

Assume now that each pulse can assume *M* possible levels that can be detected at the receiver. This means that the system is equivalent to a binary system in which there are more bits than the number of pulses. Let *R* represent the *information* or *data rate*, which is measured in bits/second (bits/s). The information rate in general is expressed in terms of the baud rate as

$$R = \text{baud rate} \times \text{bits per baud}$$
$$= \text{baud rate} \times N$$
$$= \text{baud rate} \times \log_2 M \tag{10-5}$$

Note that when **basic binary encoding** is used,

$$R = \text{baud rate} \tag{10-6}$$

Thus, *for basic binary encoding, the baud rate and the information rate in bits per second are the same.* In more complex multilevel systems, however, the information rate in bits per second is greater than the baud rate. This situation frequently results in confusion in casual usage, when the defining terms and units are not used properly.

## Shannon-Hartley Theorem

The Shannon-Hartley theorem ties together the preceding concepts into an equation providing the maximum data rate limitation for a channel based on *baseband* bandwidth limitations. Let $C$ represent the *maximum channel capacity* in bits/s. Utilizing the maximum baud rate given by Equation 10-4 along with the result of Equation 10-5, the maximum channel capacity is given by

$$C = 2B \log_2 M \tag{10-7}$$

Strictly speaking, this equation ignores any effects of noise, and assumes that all $M$ levels are discernible at the receiver. In practice, the presence of noise may be such that the number of levels must be reduced, in which case the channel capacity predicted by Equation 10-7 may be unrealizable. (The next section will provide further information on this subject.)

Note that for basic binary transmission,

$$C = 2B \tag{10-8}$$

Thus, for basic binary baseband transmission, the channel capacity in bits per second is twice the bandwidth.

---

**▮▮ EXAMPLE 10-1**

Assume that a given baseband transmission channel has a bandwidth of 4 kHz. Neglecting the effects of noise, determine the channel capacity for (a) 2-level binary encoding, (b) 4-level encoding, and (c) 128-level encoding.

**SOLUTION**

(a)  For basic binary encoding, we can use the simple result of Equation 10-8.

$$C = 2B = 2 \times 4000 = 8 \text{ kbits/s} \tag{10-9}$$

(b)  For 4-level encoding, we use the result of Equation 10-7.

$$C = 2B \log_2 4 = 2 \times 4000 \times 2 = 16 \text{ kbits/s} \tag{10-10}$$

(c)  For 128-level encoding, we have

$$C = 2 \times 4000 \times \log_2 128 = 8000 \times 7 = 56 \text{ kbits/s} \tag{10-11}$$

▮▮

---

## 10-2  Effects of Noise on Data Rate

The basic models for analyzing noise will be developed in full detail in Chapter 12. However, as early as Chapter 1, the concept of signal-to-noise ratio was introduced, primarily to show at that point how the decibel basis of measurement was utilized. As we develop the concept of information measure, it is desirable to again utilize the concept of noise as it affects data rate.

The *absolute signal-to-noise power ratio* will be denoted $(S/N)$, which may be defined as

$$(S/N) = \frac{\text{signal power}}{\text{noise power}} \tag{10-12}$$

with both signal power and noise power expressed in the same units.

The corresponding decibel measure may be defined as

$$(S/N)_{\text{dB}} = 10 \log(S/N) \tag{10-13}$$

A given transmission system operating with a fixed signal power level and a relatively stable noise power average may be characterized as operating at a specified signal-to-noise ratio level.

## Shannon Limit

When random noise exists on a communications channel, errors with digital transmission may occur in two ways: (1) a positive spike of noise may add to the signal and cause the transmitted level to be interpreted as a higher level, or (2) a negative spike may subtract from the signal and cause the transmitted level to be interpreted as a lower level. The maximum data rate that can theoretically be transmitted without error is based on the signal-to-noise ratio and the bandwidth, and is called the *Shannon limit*. Expressed as a channel capacity $C$ in bits/s, it is given by

$$C = B \log_2[1 + (S/N)] \tag{10-14}$$

Note that it is the *absolute signal-to-noise ratio* used in Equation 10-14 and *not the decibel form*.

Between the Shannon-Hartley law of Equation 10-7 and the Shannon limit of Equation 10-14, there are two ways of characterizing the baseband channel capacity. The Shannon-Hartley law provides a bound based on bandwidth and the number of levels, which would be the limit in a noise-free environment. However, the Shannon limit places an upper bound based on the signal-to-noise ratio. In practice, *the result providing the smallest value of channel capacity is the one that establishes the limit.*

The Shannon-Hartley law and the Shannon limit should be interpreted more as theoretical or even "philosophical" concepts than practical design equations. They have been widely used in establishing operating bounds for information transmission, but they are guidelines rather than exact operational formulas.

---

▌▌ EXAMPLE 10-2

For a communication channel with a bandwidth of 4 kHz, determine the maximum channel capacity for each of the following signal-to-noise ratios: (a) 20 dB, (b) 30 dB, and (c) 40 dB.

SOLUTION  Each of the $S/N$ ratios in decibels must be converted to absolute values. The respective values for (a), (b), and (c) are 100, 1000, and 10,000. In each case, the Shannon limit of Equation 10-14 is used to compute the channel capacity.

(a) For 20 dB, we have

$$C = B \log_2[1 + (S/N)] = 4 \times 10^3 \times \log_2(1 + 100)$$
$$= 4 \times 10^3 \times 3.32 \times \log_{10}(101) = 4 \times 10^3 \times 3.32 \times 2.004$$
$$= 26.6 \text{ kbits/s} \tag{10-15}$$

(b) For 30 dB,

$$C = B \log_2[1 + (S/N)] = 4 \times 10^3 \times \log_2(1 + 1000)$$
$$= 4 \times 10^3 \times 3.32 \times 3.000$$
$$= 39.8 \text{ kbits/s} \tag{10-16}$$

(c)  For 40 dB,

$$C = B \log_2 [1 + (S/N)] = 4 \times 10^3 \times \log_2(1 + 10^4)$$

$$= 4 \times 10^3 \times 3.32 \times 4.000$$

$$= 53.1 \text{ kbits/s} \tag{10-17}$$

---

**▍▍ EXAMPLE 10-3**

Examples 10-1 and 10-2 both considered a transmission system having a bandwidth of 4 kHz. In Example 10-1, the data rates for different encoding levels assuming a noise-free channel were computed. In Example 10-2, the maximum data rates for different signal-to-noise ratios were computed, without regard to the number of different encoding levels. In this example and the next one, both effects will be considered together to see how the limiting effects arise.

Assume a 4-kHz transmission channel, 4-level encoding, and a signal-to-noise ratio of 30 dB. Determine the maximum allowable information rate.

**SOLUTION**    It is necessary to check the channel capacity using both the Shannon-Hartley Law and the Shannon limit. The one having the smallest value will then dictate the limit.

From Example 10-1, the channel capacity based on 4 levels of encoding was determined to be 16 kbits/s. From Example 10-2, the channel capacity based on a signal-to-noise ratio of 30 dB was determined to be 39.8 kbits. Thus, the smaller value dictates the maximum information rate for this channel, which is

$$R = 16 \text{ kbits/s} \tag{10-18}$$

The data rate is thus limited by the number of levels employed rather than by the noise. Said differently, it is theoretically possible to transmit information at a higher data rate over this channel if more levels were employed.

---

**▍▍ EXAMPLE 10-4**

For the system of Example 10-3, assume that a more complex encoding scheme permitting 128 levels is employed. All other parameters are the same as in Example 10-3. Determine the maximum information rate.

**SOLUTION**    From the results of Example 10-1, the channel capacity based on 128 levels was determined as 56 kbits/s. However, from Example 10-2, the channel capacity based on a signal-to-noise ratio of 30 dB was determined as 39.8 kbits/s. The latter value is the smallest one in this case. Thus, the maximum information rate in this case is

$$R = 39.8 \text{ kbits/s} \tag{10-19}$$

The information rate in this case is limited by the noise on the channel. Said differently, it would be impossible to utilize the full bandwidth and level capabilities of this channel without increasing the signal-to-noise ratio.

---

**▍▍ EXAMPLE 10-5**

A digital communications system is being designed with the goal of transmitting 160 kb/s over a baseband channel having a bandwidth of 20 kHz. Determine (a) the number of encoding levels required and (b) the minimum signal-to-noise ratio required in the channel.

**SOLUTION**

(a)  First, we use the Shannon-Hartley law to determine the minimum number of levels required. We have

$$C = 2B \log_2 M \tag{10-20}$$

or

$$160 \times 10^3 = 2 \times 20 \times 10^3 \log_2 M \qquad (10\text{-}21)$$

Solving for $M$,

$$M = 2^4 = 16 \qquad (10\text{-}22)$$

Thus 16 levels are required. This corresponds to $\log_2 16 = 4$ bits/baud. If this number had not turned out to be an integer, it would be necessary to round it up to the next highest integer.

(b) To determine the signal-to-noise ratio required, we begin with

$$C = B \log_2 [1 + (S/N)] \qquad (10\text{-}23)$$

or

$$160 \times 10^3 = 20 \times 10^3 \log_2 [1 + (S/N)] \qquad (10\text{-}24)$$

Solving for the $S/N$ ratio, we have

$$(S/N) = 2^8 - 1 = 256 - 1 = 255 \qquad (10\text{-}25)$$

This corresponds to a decibel value of

$$(S/N)_{dB} = 10 \log_{10}(255) = 24.07 \text{ dB} \qquad (10\text{-}26)$$

Thus, to transmit digital information at a rate of 160 kb/s over a baseband channel having a bandwidth of 20 kHz, it would be necessary to employ an encoding scheme having 4 bits per symbol (16 possible levels) and to maintain a signal-to-noise ratio exceeding 24 dB. Accept these results as theoretical bounds based on idealized assumptions rather than exact design specifications.

## 10-3 Quadriphase Shift Keying

Other than binary, the earliest form of $M$-ary encoding, and one still widely employed, is that of 4-level PCM, which is called *quaternary* encoding. The most common way in which quaternary encoding is converted to RF (or audio) for transmission is with phase-shift keying, and the resulting process is called *quadriphase shift keying* (QPSK). (Recall from Chapter 9 that binary phase-shift keying is referred to as BPSK.)

A QPSK signal is required to have four distinct states. This can be achieved by utilizing four separate reference phase shifts for the transmitted signal. Refer to the relative phase diagram of Figure 10–1 for one possible format. The four phasor forms and the corresponding

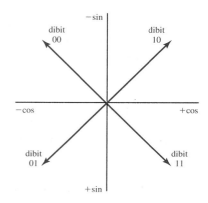

**FIGURE 10–1**
Relative phase sequence of four states (dibits) used in QPSK.

cosine and sine references are shown on the diagram. The individual phase shifts differ by 90°; that is, they are all either in phase quadrature or 180° out of phase with each other.

Each of the four possible states is referred to as a *dibit*. Since 2 bits would be required to represent four possible levels, each dibit represents the equivalent of 2 bits. Observe the 2-bit binary representation for the 4 dibits. These representations are meaningful in establishing a relationship between a binary representation and the corresponding dibit representation.

## QPSK Generation

A means of generating the RF signal is shown in Figure 10–2. The input signal is a binary PCM representation of the data signal (e.g., output of an A/D converter). The signal is first applied to a 2-bit serial-to-parallel converter. Let $\tau$ represent the bit duration of the input binary signal. After parallel conversion of the first 2 bits, the first bit is applied to the upper path, and the second bit is applied to the lower path. The duration of these two bits can be $2\tau$, since there is no need to change until two more binary bits have been received.

The process of dividing $v(t)$ into two more slowly changing signals $v_1(t)$ and $v_2(t)$ is illustrated in Figure 10–3. The *delay time* required to perform the serial-to-parallel

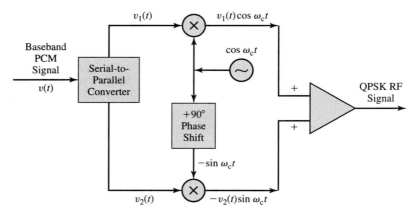

**FIGURE 10–2**
System used to generate QPSK signal.

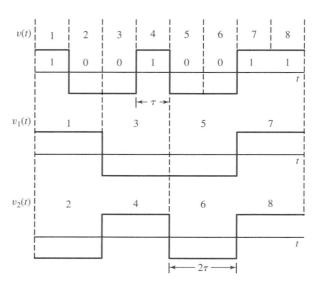

**FIGURE 10–3**
Waveforms illustrating channeling of successive bits in QPSK.

conversion is not shown on the figure, so $v_1(t)$ and $v_2(t)$ would, in general, be delayed with respect to $v(t)$. The decimal numbers shown on the figure are used to label the bits for clarity. Observe that $v_1(t)$ assumes the pattern of the odd bits while $v_2(t)$ assumes the even bits. However, $v_1(t)$ and $v_2(t)$ each have bit periods of $2\tau$, so the transmission bandwidth of these signals should be half the bandwidth that would have been required if $v(t)$ had been transmitted directly.

Refer now to Figure 10–2. The signal $v_1(t)$ is multiplied by $\cos \omega_c t$, while the signal $v_2(t)$ is multiplied by $-\sin \omega_c t$. The results will be denoted $v_{o1}(t)$ and $v_{o2}(t)$, and they are added to form the composite transmitted signal $v_o(t)$. Thus,

$$v_o(t) = v_{o1}(t) + v_{o2}(t)$$

$$= v_1(t) \cos \omega_c t - v_2(t) \sin \omega_c t \tag{10-27}$$

Both $v_{o1}(t)$ and $v_{o2}(t)$ are DSB signals centered at $f_c$, so their spectra overlap. Why then do they not interfere with each other? Stated differently, how can $v_{o1}(t)$ and $v_{o2}(t)$ be separated at the receiver when the frequency components occupy the same range? The answer lies in the fact that the two signals are in *phase quadrature;* that is, all spectral components of $v_{o2}(t)$ are 90° out of phase with those of $v_{o1}(t)$, and they can be separated by coherent detection, as we will see shortly.

## QPSK Detection

The detection process for QPSK is shown in Figure 10–4. The signal is simultaneously applied to two channels. Assume that the form of the received signal $v_r(t)$ is

$$v_r(t) = A v_1(t) \cos \omega_c t - A v_2(t) \sin \omega_c t \tag{10-28}$$

Phase-coherent references $\cos \omega_c t$ and $-\sin \omega_c t$ are used for synchronous detection in the two channels. Let $v_{r1}(t)$ and $v_{r2}(t)$ represent the outputs of the upper and lower synchronous detectors, respectively. We have

$$v_{r1}(t) = A v_1(t) \cos^2 \omega_c t - A v_2(t) \sin \omega_c t \cos \omega_c t$$

$$= \frac{A}{2} v_1(t) + \frac{A}{2} v_1(t) \cos 2\omega_c t - \frac{A}{2} v_2(t) \sin 2\omega_c t \tag{10-29}$$

and

$$v_{r2}(t) = -A v_1(t) \cos \omega_c t \sin \omega_c t + A v_2(t) \sin^2 \omega_c t$$

$$= -\frac{A}{2} v_1(t) \sin 2\omega_c t + \frac{A}{2} v_2(t) - \frac{A}{2} v_2(t) \cos 2\omega_c t \tag{10-30}$$

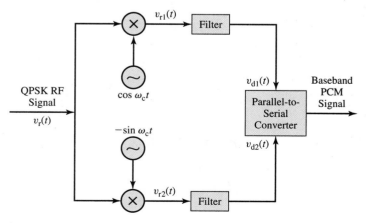

**FIGURE 10–4**
Detection system for QPSK.

Each of the preceding results contains two terms centered at $2f_c$ and one low-frequency data term. After appropriate low-pass filtering, we have the following for the two data outputs:

$$v_{d1}(t) = \frac{A}{2}v_1(t) \tag{10-31}$$

and

$$v_{d2}(t) = \frac{A}{2}v_2(t) \tag{10-32}$$

Thus, the two separate data streams are obtained. By means of a parallel-to-serial conversion, the original data stream is reconstructed.

### QPSK and DPSK

It is possible to combine the processes of differentially encoded phase shift keying (discussed in Chapter 9) and quadrature phase-shift keying as presented in this chapter. The advantages of both differential detection and the higher data rata of quadrature phase-shift keying can be realized together. The acronym QDPSK is sometimes used to describe this process utilizing the two separate concepts.

### QPSK Bandwidth

Since the pulse width can be twice as great as for BPSK, the required transmission bandwidth for a given data rate is half as great as for BPSK. Extending the results developed in Chapter 9 for BPSK, we can state that the bandwidth $B_T$ is related to the information or data rate $R$ by

$$B_T = \frac{R}{2} \tag{10-33}$$

### Constellation Diagram

Figure 10–1 displayed the four states of QPSK in the form of a phasor diagram. As systems become increasingly complex, it is convenient to change the phasor diagram into a simpler figure showing the points representing the tips of the phasors. In that manner, variable amplitudes may be more easily shown. This type of figure is called a *constellation diagram*. The constellation diagram for QPSK is shown in Figure 10–5.

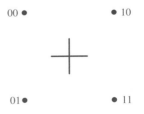

**FIGURE 10–5**
Constellation diagram for QPSK.

---

▌▌ **EXAMPLE 10-6**

A QPSK system is to be designed for a data rate of 1 Mb/s. Determine the minimum required bandwidth of the transmission link.

**SOLUTION**   From Equation 10-33, we have

$$B_T = \frac{R}{2} = \frac{10^6}{2} = 500 \text{ kHz} \tag{10-34}$$

▌▌

---

## 10-4 Quadrature Amplitude Modulation

Consider now the situation in which both the amplitude and the phase can assume two or more states. Let $K_A$ represent the number of possible amplitude states and let $K_P$ represent the number of possible phase states. If each amplitude state can assume each of the phase states, the total number of possible states $M$ is then given by

$$M = K_A K_P \tag{10-35}$$

The equivalent number of binary bits is then given by

$$N = \log_2 M = \log_2 K_A K_P \tag{10-36}$$

### 8-Level QAM

One form for 8-level QAM employs 2 amplitude states and 4 phase states; the constellation diagram is shown in Figure 10–6. Since each amplitude state can assume all possible phase states, the total number of states is $2 \times 4 = 8$, and the number of bits per baud is 3.

### 16-Level QAM

One form for 16-level QAM employs a rectangular type of constellation diagram as shown in Figure 10–7. In this case, some of the amplitude states do not assume all the phase states, so it is necessary to inspect the diagram to determine the number of states. The total number of states is 16, and the number of bits per baud is 4.

An alternate 16-level QAM form with 2 amplitude states and 8 phase states, in which each amplitude state can assume each phase state, is shown in Figure 10–8.

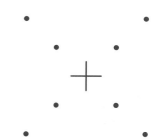

**FIGURE 10–6**
Constellation diagram for 8-level QAM.

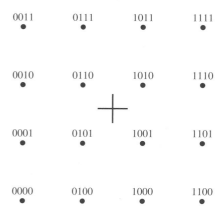

**FIGURE 10–7**
Constellation diagram for 16-level QAM.

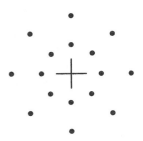

**FIGURE 10-8**
Alternate constellation diagram for 16-level QAM.

## 10-5   Multisim® Examples (Optional)

The examples in this section will deal with the design, implementation, and analysis of circuits that generate and detect quadrature phase-shift keying signals. As was the case in Chapter 9, the process involves the utilization of various digital circuits and modules that fall outside of the scope of the text. Again, the section may be bypassed without loss of continuity.

---

**∎∎ MULTISIM EXAMPLE 10-1**   The circuit of Figure 10–9 represents a data-domain implementation of an encoder that converts a serial train of data to a parallel train of alternate bits as the first step in generating a QPSK signal. Consider the bit pattern of $v(t)$ in Figure 10–3, and show that it is converted to the parallel bit trains shown for the other two waveforms.

**SOLUTION**   Along with the **Word Generator** and **Logic Analyzer,** the circuit involved utilizes three **7474N D Flip-Flops.** Since it is necessary to sort each pair of serial bits into

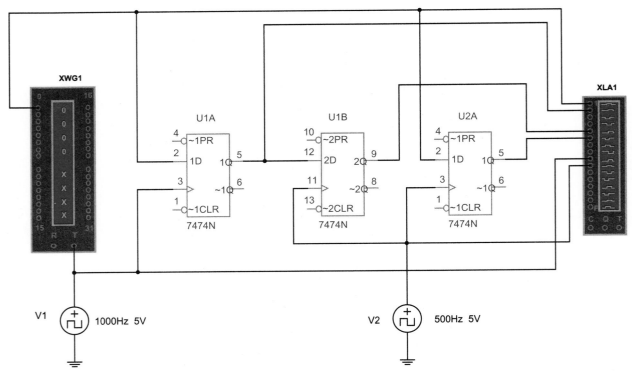

**FIGURE 10-9**
Serial-to-parallel converter of Multisim Example 10-1.

two parallel bits, some shifting or delay is required in the bit pattern. This delay was not shown in Figure 10–3.

As in the DPSK simulation of Chapter 9, the input data bits are 1 ms in width, thus yielding a data rate of 1 kbit/s. However, the parallel conversion results in bits of width 2 ms, and this means that the baud rate will be 500 Hz. Therefore, the frequency for the right-hand clock is 500 Hz.

The input bit pattern, which is of the form of Figure 10–3, is entered in the **Word Generator,** and the form is shown as the top stream (**1**) in Figure 10–10. A delayed version is shown as **2**. The resulting parallel outputs are shown in **3** and **4**. Ignore the first two bits for each waveform. The odd-numbered input bits begin at the third bit position in **3** and the even-number input bits begin at the third bit position in **4**.

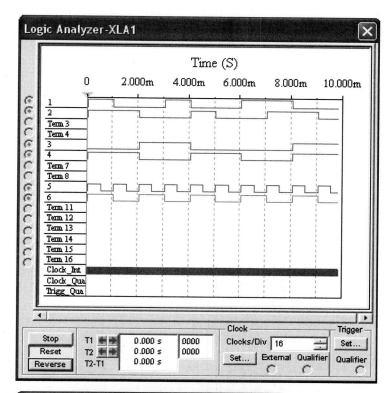

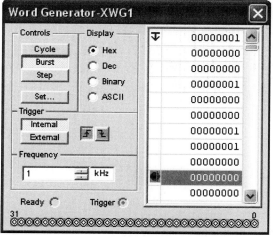

**FIGURE 10–10**

Results for Multisim Example 10-1.

▌▌ **MULTISIM EXAMPLE 10-2**    Continuing with a data-domain analysis, connect the output of the circuit of Multisim Example 10-1 to a parallel-to-serial converter, and show that the original data signal is obtained.

**SOLUTION**    The circuit is shown in Figure 10–11. The portion to the right of the **Logic Analyzer** utilizes a **74157N** circuit to reconstruct the serial bit pattern.

The various waveforms in the **Logic Analyzer** are shown in Figure 10–12. The output signal has a net delay of 2 ms without considering any actual transmission delay. The output is shown as **4** and it is seen to be a delayed version of the original bit train after the 2-ms delay is taken into consideration. The clock signals are shown below the output.

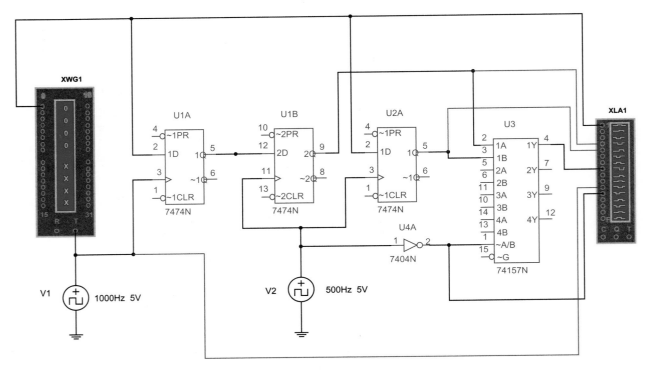

**FIGURE 10–11**
Serial-to-parallel and parallel-to-serial converters of Multisim Example 10-2.

▌▌ **MULTISIM EXAMPLE 10-3**    The circuit of Figure 10–13 represents a complete QPSK system utilizing circuit voltages and currents (as opposed to the data-domain logic analysis). Perform a complete analysis on the system based on the input bit pattern used in Figure 10–3.

**SOLUTION**    As in the case of the DPSK system of Chapter 9, the block diagram consists of modules that were implemented in the normal fashion and then converted to **Subcircuits.** The procedure may be reviewed in the discussion of Multisim Example 9-3 if necessary.

The serial-to-parallel converter is shown in Figure 10–14, and it is designed around the same circuit components as in Multisim Examples 10-1 and 10-2. The QPSK modulator shown in Figure 10–15 utilizes two multipliers, in which the local oscillators are 90° out of phase with each other (as required). A bias level of −2.5 V is added within each modulator to provide a bipolar output. The multiplier output signals are then added together.

The detection process begins with the QPSK demodulator, whose schematic diagram is shown in Figure 10–16. As in the modulator, the two multiplying carriers are 90° out of phase with each other. Each signal is then filtered by an active low-pass filter whose schematic diagram is shown in Figure 10–17. This is the same filter circuit used in the DPSK circuit of the preceding chapter, and it has a cutoff frequency of 1.5 kHz.

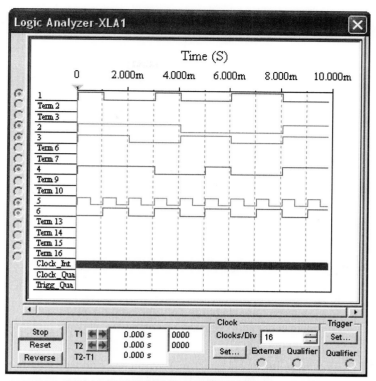

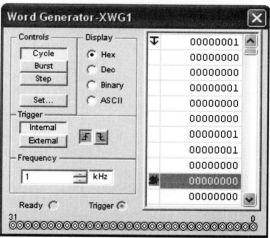

**FIGURE 10–12**
Results for Multisim Example 10-2.

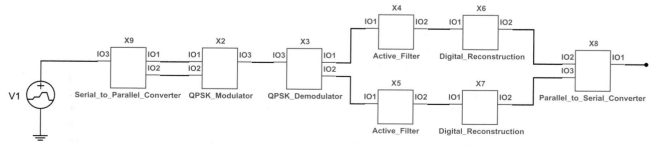

**FIGURE 10–13**
QPSK system of Multisim Example 10-3.

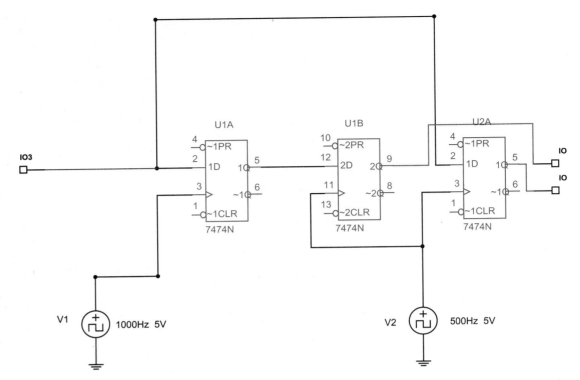

**FIGURE 10–14**
Serial-to-parallel converter.

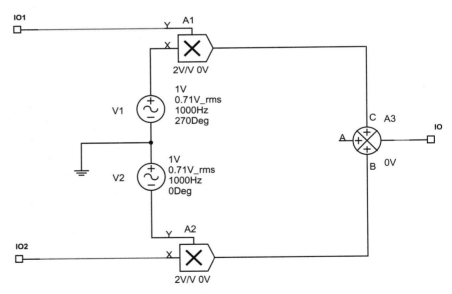

**FIGURE 10–15**
QPSK modulator.

Along each path there is a digital restoration circuit consisting of a comparator; the schematic is shown in Figure 10–18. The two bit trains are then combined in the parallel-to-serial converter of Figure 10–19. The final digital signal is shown in Figure 10–20. Since there is an inherent delay of 2 ms, this **Transient Response** is shown with a beginning time of 2 ms. Note the small additional delay resulting primarily from the filter characteristics.

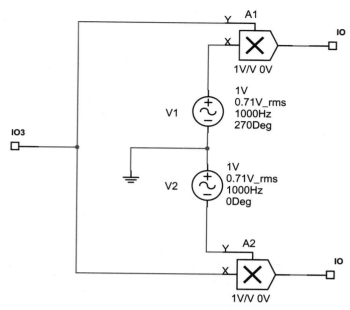

**FIGURE 10–16**
QPSK demodulator.

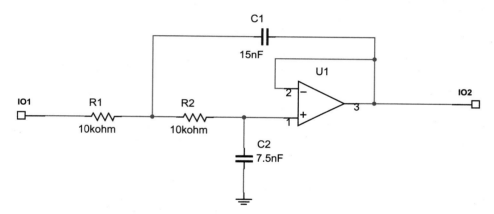

**FIGURE 10–17**
Active filter.

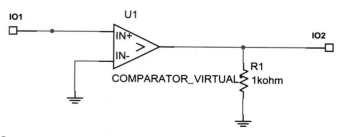

**FIGURE 10–18**
Digital reconstruction circuit.

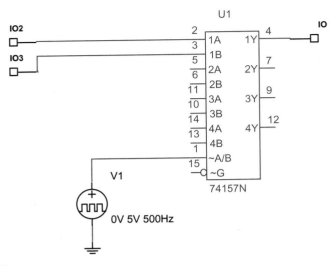

**FIGURE 10–19**
Parallel-to-serial converter.

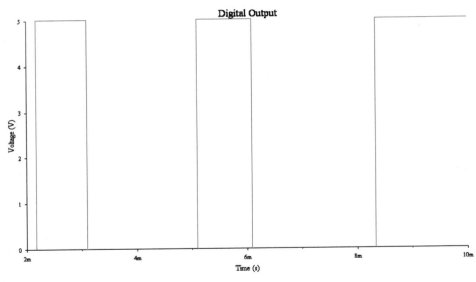

**FIGURE 10–20**
Output of QPSK reconstruction process.

## 10-6 MATLAB® Example (Optional)

To aid in the visualization of an *M*-ary signal, we will generate a QPSK signal in this section. The numbers selected will be close to those employed in the Bell 212A system, but to make the process slightly easier to visualize, the values will be changed somewhat. The actual Bell 212A modem employs a baud rate of 600 and a data rate of 1200 b/s. For our purpose, we will use a baud rate of 500 and a data rate of 1000 bits (b/s).

**▌▌ MATLAB EXAMPLE 10-1**     Consider a QPSK system with the following parameters:

Baud rate: 500 baud

Data rate: 1000 b/s

```
%Program qpsk.m
%Program to observe behavior of 1000 Hz signal in 4 states.
tbaud=1/500;
delt=tbaud/1000;
t=0:delt:1/500-delt;
arg=2*pi*1000*t;
v00=cos(arg+.75*pi);
v01=cos(arg-.75*pi);
v10=cos(arg+.25*pi);
v11=cos(arg-.25*pi);
subplot(2,2,1)
plot(t,v00)
xlabel('Time, seconds')
ylabel('v, volts')
gtext('00')
subplot(2,2,2)
plot(t,v01);
xlabel('Time, seconds')
ylabel('v, volts')
gtext('01')
subplot(2,2,3)
plot(t,v10)
xlabel('Time, seconds')
ylabel('v, volts')
gtext('10')
subplot(2,2,4)
plot(t,v11)
xlabel('Time, seconds')
ylabel('v, volts')
gtext('11')
```

**FIGURE 10–21**
M-file program of MATLAB Example 10-1.

Originate mode frequency: 1 kHz

Answer mode frequency: 2 kHz

Write an M-file program using subplots that will display the four possible values over one baud interval for each of the four possible dibits: 00, 01, 10, and 11 in the originate mode.

**SOLUTION**  The M-file program for generating the four possible values is shown in Figure 10–21. Since the baud rate is 500 Hz, the baud interval **tbaud** is

```
>> tbaud = 1/500;
```

The time step **delt** is arbitrarily selected as 1/1000 of the baud width:

```
>> delt = tbaud/1000;
```

The time vector **t** is then generated over a baud width as

```
>> t = 0:delt:tbaud-delt;
```

The basic argument **arg** without a phase angle is then generated as

```
>> arg = 2*pi*1000*t;
```

For each of the four dibits, a peak value of unity is assumed. The resulting sinusoid functions, after combining the sine and cosine with appropriate phase angles, are generated by the following four commands:

```
>> v00 = cos(arg+0.75*pi);
>> v01 = cos(arg-0.75*pi);
>> v10 = cos(arg+0.25*pi);
>> v11 = cos(arg-0.25*pi);
```

The remainder of the work consists of using four subplots to plot the 4 dibits. The results are shown in Figure 10–22.

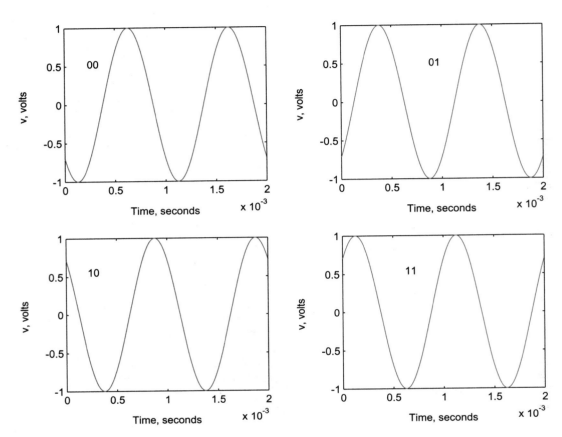

**FIGURE 10–22**
The four dibits of MATLAB Example 10-1.

## SystemVue™ Closing Application (Optional)

Insert the text CD in a computer having SystemVue™ installed and activate the program. Open the CD folder entitled **SystemVue Systems** and load the file entitled 10-1.

### Review of System

The system shown on the screen was described in the chapter opening application, and you may wish to review the discussion and system configuration provided there. Recall that the signal is a *binary phase-shift keying* (BPSK) signal. You may wish to make a few additional runs to ensure that you understand the nature of the two signals being monitored.

## QPSK Signal

1. Prior to making any changes, what is the initial data rate in bits/s for the BPSK signal?

2. Modify the PSK source by increasing **No. Symbols** (number of symbols) from **2** to **4.** The result for the modulated waveform is that of a *quadriphase shift keying* (QPSK) signal.

3. Perform a simulation run and observe the input and output waveforms. Occasionally there may be a run that does not show all four levels of the bit train. Should that happen, make additional runs until you can actually see the four possible levels. What is the data rate in this case?

3. Observe how each of the four levels results in a different pattern for the phase of the modulated signal.

4. Increase the number of symbols from 4 to 8. The result in this case is an *8-ary phase-shift keying* signal.

5. Observe the waveforms for this latter case. What is the data rate in this case?

## PROBLEMS

(*Note:* In Problems 10-1 through 10-10, interpret the Shannon-Hartley theorem and the Shannon limit as exact limitations based on baseband transmission.)

**10-1** A transmission channel has a bandwidth of 1 MHz. Neglecting the effects of noise, determine the channel capacity for (a) 2-level binary encoding, (b) 8-level encoding, and (c) 64-level encoding.

**10-2** A transmission channel has a bandwidth of 5 kHz. Neglecting the effects of noise, determine the channel capacity for (a) 2-level binary encoding, (b) 4-level encoding, and (c) 32-level encoding.

**10-3** For the transmission channel of Problem 10-1, determine the channel capacity for each of the following signal-to-noise ratios: (a) 15 dB, (b) 25 dB, and (c) 35 dB.

**10-4** For the transmission channel of Problem 10-2, determine the channel capacity for each of the following signal-to-noise ratios: (a) 10 dB, (b) 20 dB, and (c) 30 dB.

**10-5** Problems 10-1 and 10-3 both considered a transmission system having a bandwidth of 1 MHz. For this channel with 64-level encoding, and a signal-to-noise ratio of 15 dB, determine the maximum allowable information rate.

**10-6** Problems 10-2 and 10-4 both considered a transmission system having a bandwidth of 5 kHz. For this channel with 4-level encoding, and a signal-to-noise ratio of 30 dB, determine the maximum allowable information rate.

**10-7** For the system of Problem 10-5, assume that the signal-to-noise ratio is increased to 35 dB and the encoding

scheme is changed to 8 levels. Determine the maximum allowable information rate.

**10-8** For the system of Problem 10-6, assume that the signal-to-noise ratio remains at 30 dB, but binary transmission is used. Determine the maximum allowable information rate.

**10-9** A digital communications system is being designed with the goal of transmitting 36 kbits/s over a channel having a bandwidth of 6 kHz. Determine (a) the number of encoding levels required and (b) the minimum signal-to-noise ratio required in the channel.

**10-10** A digital communications system is being designed with the goal of transmitting 50 kbits/s over a channel having a bandwidth of 5 kHz. Determine (a) the number of encoding levels required and (b) the minimum signal-to-noise ratio required in the channel.

**10-11** A QPSK system is to be designed for a data rate of 200 kHz. Determine the minimum bandwidth required.

**10-12** For an RF bandwidth of 1 MHz, determine the maximum information rate that can be transmitted in QPSK.

**10-13** The constellation diagram for a particular QAM system has 4 different amplitudes and each amplitude has 8 different possible phase shifts. Determine the number of bits per baud.

**10-14** The constellation diagram for a particular QAM system has 16 different amplitudes and each amplitude has 8 different possible phase shifts. Determine the number of bits per baud.

# Computer Data Communications 11

## OVERVIEW AND OBJECTIVES

The emphasis thus far in digital communications has been directed toward systems in which analog signals were first sampled, and then converted into binary words whose values represented the levels of the corresponding analog samples. There is still another broad area of digital communications in which the individual words represent *alphanumeric* characters. This includes data transmission between two computers, between a computer and a remote terminal, between a business terminal and a central accounting system, as well as most of the communication on the Internet. The key factor here is that the input and output data employ digital codes corresponding to characters, rather than samples of an analog signal. This broad and rapidly growing area of communications can be referred to as *data communications.*

Historically, some forms of data communications were in use even before the advent of radio. The Morse code, in which letters and numbers are encoded by combinations of short pulses (dots) and long pulses (dashes) dates back to the nineteenth century. Teletype transmission has also been in existence for many years.

Nothing has impacted the area of data communications more than the Internet and the personal computer. The rapid growth of this area has brought data communications into millions of households, and the future looks very bright for many new innovations and applications of data communications.

A basic consideration in all digital data communications systems is that each of the possible characters must have a unique code word. A means must be provided at the data source for converting each possible character to its associated code word, and a similar means must be provided at the destination for converting the code word back to its proper character.

In this chapter we will consider the codes used in data communications, and the way that data is stored and transferred within computers and between computers or other local devices. In Chapter 18 we will consider data communications on wired networks including the Internet, and in Chapter 19 we will consider data communications in wireless networks.

## Objectives

After completing this chapter, the reader should be able to:

1. Compare *synchronous, asynchronous,* and *isochronous* transmission, *parallel* versus *serial* transmission, and *text* versus *binary* data.

2. Define *simplex, half-duplex,* and *full-duplex.*

3. Discuss the *ASCII, EBCDIC,* and *Unicode* codes and their general formats.

4. Show the normal pattern for asynchronous serial data transmission including start and stop bits.

5. Describe serial data communications interfaces including the EIA/RS-232C and Universal Serial Bus (USB) standards.

6. Discuss parallel data communications and its applications.

## SystemVue™ Opening Application (Optional)

Insert the text CD in a computer having SystemVue™ installed and activate the program. Open the CD folder entitled **SystemVue Systems** and open the file entitled 11-1.

### Sink Tokens

| Number | Name | Token Monitored |
|---|---|---|
| 0 | Input | 1 |

### Operational Tokens

| Number | Name | Function |
|---|---|---|
| 1 | Input | generates ASCII code for WOW |

The system shown on the screen consists of just a *source* token (Token 1) and a *sink* token (Token 0). Token 1 reads the contents of a text file (ASCII_WOW.txt) containing the pulse train showing what would be transmitted if WOW were sent using asynchronous serial transmission. The first character, after some positive pulses, consists of a start pulse (negative) to signal the beginning of the character, the ASCII character "W" (1010111), least significant bit first, a parity bit for odd parity (negative), and a stop bit (positive). The second character consists of a start pulse (negative) to signal the beginning of the character, the ASCII character "O" (1001111), least significant bit first, a parity bit for odd parity (negative), and a stop bit (positive). The third character, like the first, consists of a start pulse (negative) to signal the beginning of the character, the ASCII character "W" (1010111), least significant bit first, a parity bit for odd parity (negative), and a stop bit (positive). This is followed by some more positive pulses that indicate that the channel is idle. This pulse train is then monitored by Token 0, which displays the pulse train on the screen and also in the Analysis screen.

### Checking the Settings

You will not need to modify any settings in this exercise, but you can observe the values of the parameters by placing the cursor over a token. A window will appear containing the parameters of that token.

### Activating the Simulation

Left-click the **Run System** button or press F5 to start the simulation. A moving blue line at the bottom of the screen indicates that the simulation is running. Note that when the line has reached the right end of its excursion the waveforms appear in the box on the screen to the right of Token 0. Left-click the **Analysis Window** button to open the Analysis window.

As you examine the waveform displayed, note that the waveform starts at a positive level. When an asynchronous serial channel is idle, a series of positive pulses or **MARK**s are sent. The first negative pulse is the start pulse indicating that a character is being sent. The next seven pulse periods represent the character, 1110101, which is actually 1010111, because in asynchronous serial communication the least significant bit is sent first. But 1010111 is 57 hex or the upper case W. Following the most significant bit of the character is the parity bit. There are five 1s in the character and we are using odd parity, so the parity bit is a negative pulse or **SPACE**. Finally, there is a stop bit, another MARK. The SPACE following that is the start bit for the second character, which is 1111001. Reversing the order, we see that the ASCII code is 1001111 or 4F hex, the ASCII code for the upper case O. This also has five 1s, so the parity bit that follows the character is a SPACE. After another MARK for a stop bit, a start bit followed by the third character, another W, is sent. After a short idle period, the sequence starts over.

The file entitled **ASCII_WOW.txt** contains the input for Token 1. This is simply a table of values of the output of that token at each sample time. At the time of the first sample period, the program reads the first value, 1, interprets it as 1 V, and outputs it. At the time of the second sample period, the program reads the second value, 1, interprets it as 1 V, and outputs it. It does this at each sample time. The Word file entitled **11-1 initial settings.doc** contains views of the system and of the time settings. The sample frequency is 100 Hz and the file contains 10 entries for each pulse, so this represents data sent at a rate of 10 bps. After you have completed the chapter you will see the effect of changing the sample frequency, as well as changing the data.

## 11-1 Data Transmission Terminology

Prior to developing the concepts of data transmission, some terminology will be established.

### Bits and Bytes

The fundamental element of computer data is the *bit,* which has only two possible states, usually designated 0 and 1. In various contexts, the states of the bit can also be considered *on* and *off* or *true* and *false*. For purposes of storage and transmission, bits are usually combined in groups of eight, called *bytes* (or sometimes *octets*).

### Synchronous versus Asynchronous

Data transmission may be characterized as either *synchronous* or *asynchronous*. *Synchronous* transmission is achieved by means of a master clock or timed reference, which is either transmitted with the signal or available at both ends through some other means. *Asynchronous* (also called *start-stop*) data transmission systems have the capability of initiating or terminating transmissions in a much more flexible manner. This is achieved by provided start and stop information with each transmitted data unit, usually a byte, so it is essentially self-clocking. The advantages of asynchronous transmission are obvious for such applications as computer terminals and Internet transmission, in which the beginnings and endings of data exchange are random.

## Isochronous

Both synchronous serial communication and asynchronous serial communication generally involve error checking mechanisms that request the retransmission of erroneous data and can slow down the flow of data. Some applications, such as video conferencing, require a steady stream of data and can tolerate an occasional error, but cannot tolerate delays or pauses in the transmission of data. *Isochronous* serial communication provides a fast, steady, uninterrupted flow of data in which the clocking information is included in the data stream.

## Parallel versus Serial

Data transmission may also be characterized as either *parallel* or *serial*. In parallel transmission, each bit of the transmitted data unit has a separate transmission path, and all bits are sent simultaneously. An example of parallel transmission is the connection between the parallel port of a computer and a printer or the transmission of data between a computer's CPU and memory.

In serial transmission, only one channel is used and the bits are transmitted in sequence. In general, parallel transmission is faster but is usually limited to short distances due to the requirement for multiple channels.

## Text versus Binary Data

The transmission of alphanumeric characters is usually referred to as *text data*. However, digitized audio and video are often treated in much the same manner in computer and Internet communications. These data are referred to as *binary data* in this particular context.

## Simplex or Duplex

The transmission channel and connection between two data devices such as computers or terminals may be characterized in three different ways: (1) *simplex,* (2) *half-duplex,* and (3) *full-duplex.*

(1) A *simplex* connection is one in which transmission is always in one direction only and that direction is fixed.

(2) A *half-duplex* connection is one in which transmission can take place in either direction, but not simultaneously. Thus, the sending device and receiving device must take turns.

(3) A *full-duplex* connection is one in which transmission can take place in both directions simultaneously.

## UART

The term *UART* stands for *universal asynchronous receiver-transmitter.* It is an integrated circuit that converts parallel data to serial data for transmission over a single line at the sending end, and converts the serial data at the receiving end back to parallel data.

## Data Terminal Equipment (DTE) versus Data Circuit-Terminating Equipment (DCE)

DTE is that part of a data station that converts user data into signals for transmission or that converts received signals into user data. DCE is that part of a data station that performs the functions, such as signal conversion and coding, necessary to interface the DTE to the communications channel.

### Interface versus Protocol

For data communications purposes, an interface can be considered to be the hardware and software link that enables communications to occur between two entities, such as between two computers, or between a computer and its peripherals. A protocol is the set of rules that the end points of a communication connection must follow in order to be able to communicate effectively.

## 11-2    Bits, Bytes, and Beyond

The basic unit of data in a digital circuit is the bit. Physically, a bit may be a memory location, or part of a register that has the ability to take on or hold one of two different values. It can be a switch that is either on or off; it can be a magnetic field oriented in one direction or the opposite direction; it can be a capacitor that is charged or discharged; it can be a flip-flop that is SET or CLEAR. From the viewpoint of data communications, however, a bit is a single element of data that can be either 1 or 0. Data communications, however, seldom involves transferring just one bit. For data communications purposes, bits are grouped in bytes, nibbles, words, double-words, blocks, and files.

### Bits, Bytes, and Blocks

The basic unit of data communications is the *byte,* which consists of eight bits. A byte may represent a binary number, or it may represent an encoded symbol such as a letter of the alphabet, a number, or some other character. Each bit in a byte has a specific meaning and the order of the bits is important. Usually we represent a byte with the most significant bit (MSb) to the left, as illustrated in Figure 11–1. Note the use of the lowercase b in MSb. We will use the uppercase B to designate byte and the lowercase b to designate bit. Therefore, MSB is most significant byte, LSB is least significant byte, MSb is most significant bit, and LSb is least significant bit. If the byte represents a binary number, each bit has a weight starting at $2^0$ at the right and proceeding to $2^1$, $2^2$, and so on, up to $2^7$ for the MSb. Therefore, we will designate the LSb as bit 0, the bit to the left of the LSb as bit 1, up to bit 7 for the MSb when it is necessary to refer to specific bits in a byte. Coding of symbols such as letters, numbers, or other characters is generally accomplished with one symbol per byte or one symbol in two bytes (see Section 11-3). Occasionally symbols are encoded with just four bits or two symbols per byte (see Binary Coded Decimal in Section 11-3). For convenience, four bits may be referred to as a *nibble* (sometimes spelled *nybble*). Two related bytes taken together may be called a *word* and four related bytes taken together are sometimes called a *double-word.* When storing and transferring data, multiple bytes are usually handled as a unit and referred to as a *block.* A block normally consists of $2^n$ bytes, such as 64, 128, 256, 512, and so on. Typically, data is transferred either a byte at a time or a block at a time, with some form of error checking performed when the byte or the block is received.

Accuracy is an important consideration when transferring data between components within a computer, or between a computer and other computers or peripherals (such as printers). Therefore, error checking is generally, although not always, an integral part of any data communications scheme. One of the simplest forms of error checking is performed at the byte level. Originally, letters, numbers, and punctuation were encoded using only seven bits (see ASCII in Section 11-3). The eighth bit in each byte can then be used in a process

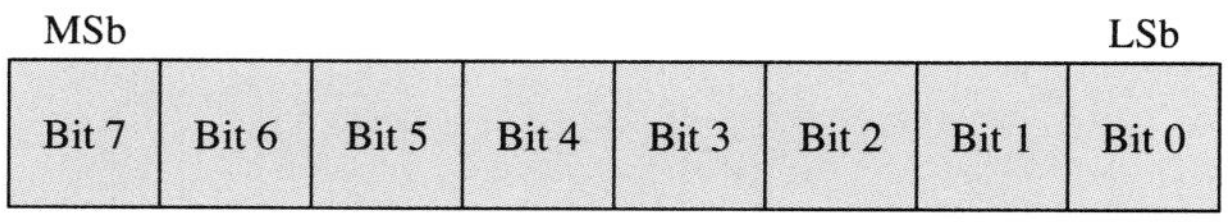

| MSb | | | | | | | LSb |
|-------|-------|-------|-------|-------|-------|-------|-------|
| Bit 7 | Bit 6 | Bit 5 | Bit 4 | Bit 3 | Bit 2 | Bit 1 | Bit 0 |

**FIGURE 11–1**
Ordering of bits in a byte.

| Parity Bit | Bit 6 | Bit 5 | Bit 4 | Bit 3 | Bit 2 | Bit 1 | Bit 0 |
|------------|-------|-------|-------|-------|-------|-------|-------|
| | | | | Data | | | |

**FIGURE 11–2**
Insertion of parity bit.

called parity checking to detect errors. The process involves counting the number of ones in the seven-bit encoded symbol. If *even parity* is being used, a zero is placed in the eighth bit if the number of ones is even and a one is placed in the eighth bit if the number of ones is odd. The intent is to make the number of ones in the byte even. When the byte is received after transmission, an odd number of ones in the byte indicates an error in transmission. If *odd parity* is being used, the number of ones in the byte is adjusted to be odd and an error has occurred if the received byte contains an even number of ones. Figure 11–2 shows a parity bit added to a byte.

Data communications has become much more reliable than it once was, and very little data is transferred byte by byte. More sophisticated error detection methods have been developed to detect errors in blocks of transmitted data. Therefore, parity checking is seldom used, and there is no need to reserve a bit to be used as a parity bit. This makes it possible to use all eight bits in a byte to encode symbols. Error checking is still an important consideration, but is now generally performed at the block or file level, rather than at the byte level.

## Files and Error Checking

Individual letters, numbers, punctuation, and other symbols may be represented by binary data, generally one symbol per byte. These symbols can then be joined together to form a complete document such as a letter, a paper, or a book. If the symbols involved are textual, the result is a text file. It is also possible to encode pictures, audio, and video, and represent them by binary data. When this is done, the result is called a binary file. The programs necessary to operate a computer are also stored in binary files.

Blocks of data are usually transferred between computers. A form of error checking can be accomplished on files by considering the bytes that make up a block as binary numbers, and performing arithmetic operations on those numbers. The result of the operation is sent along with the file to the receiver. The same operations are performed on the received file, and the results compared to the transmitted results. If they are different, an error in transmission must have occurred. A simple arithmetic operation would be to add the binary number consisting of the first sixteen bytes of the file to the binary number consisting of the second sixteen bytes. Each succeeding group of sixteen bytes is added to the previous sum until the end of the file is reached. When the data block is transmitted, additional information is sent with it in the form of a header field and the calculated *checksum*. The result is called a frame, and is illustrated in Figure 11–3. (Headers will be described in more detail when specific protocols are discussed.) The transmitted *checksum* can be compared to a checksum calculated when the frame is received, to determine whether an error in transmission occurred. If so, a request to retransmit the frame is made. When transmitting a file from one computer to another, the file is usually divided into segments, and each segment sent in a separate frame. The file is reassembled at the remote computer. Then, if a

| Header | Data | Checksum |
|--------|------|----------|

**FIGURE 11–3**
Example frame with header, data, and checksum.

Table 11–1   Binary Coded Decimal (BCD) Numbers

| Decimal Number | BCD Code | XS3 Code |
|---|---|---|
| 0 | 0000 | 0011 |
| 1 | 0001 | 0100 |
| 2 | 0010 | 0101 |
| 3 | 0011 | 0110 |
| 4 | 0100 | 0111 |
| 5 | 0101 | 1000 |
| 6 | 0110 | 1001 |
| 7 | 0111 | 1010 |
| 8 | 1000 | 1011 |
| 9 | 1001 | 1100 |

frame is lost or corrupted, it is only necessary to retransmit a relatively short frame, not the entire file. How large the segments should be depends largely upon the reliability of the transmission method.

### Files and Compression

Two factors affect the length of time required to transfer data: the amount of data to be transferred, and the speed of transfer. Regardless of the speed of transfer, if the same amount of information can be included in a smaller file, transmission time can be reduced. For this reason, data compression is often used when transferring files, and improved methods of data compression are constantly being sought. Most data compression is based on the fact that certain patterns of bits tend to occur repeatedly in files. If a summary of a file containing a list of the recurring patterns and their locations in the file can be created, it may be smaller than the file itself. That summary can then be transmitted and the file rebuilt from the summary at the receiver. Although the effectiveness of compression depends upon the type of file and the actual contents of the file, many files can be compressed significantly. The time required to transmit such compressed files is thus significantly reduced.

## 11-3   Data Encoding

### Binary Coded Decimal (BCD)

If we convert the decimal numbers 0 through 9, to binary and represent each of them as a four-bit binary number, the representations in Table 11–1 are obtained. Each digit in a multidigit decimal number is separately encoded. Thus, the BCD representation of the decimal number 637 would be 0110 0011 0111. BCD may be used to store decimal numbers in computers and to perform arithmetic operations upon those numbers, although this is not as common as it once was. A variation on BCD is to represent the decimal number with the binary number that is three higher than the decimal number. Thus, 6 would be represented as 1001 rather than 0110, 3 would be represented as 0110, and 7 would be represented as 1010. Decimal 637 then would become 1001 0110 1010. This method of representing decimal numbers, called XS3, has advantages when arithmetic operations are to be performed on the numbers.

**⫼ EXAMPLE 11-1**

Code the decimal number 298 as BCD.

**SOLUTION**   Refer to Table 11–1. Decimal 2 is coded as BCD 0010, Decimal 9 is coded as BCD 1001, and decimal 8 is coded as BCD 1000. Therefore, decimal 298 is coded as BCD 0010 1001 1000.

**⫼ EXAMPLE 11-2**

Code the decimal number 298 as XS3.

**SOLUTION**   Refer to Table 11–1. Decimal 2 is coded as XS3 0101, Decimal 9 is coded as XS3 1100, and decimal 8 is coded as XS3 1011. Therefore, decimal 298 is coded as XS3 0101 1100 1011.                                                ⫼

### American Standard Code for Information Interchange (ASCII)

The most widely used encoding process in modern digital data systems is the *American Standard Code for Information Interchange,* hereafter referred to as the *ASCII* code. Originally, the ASCII code was a 7-bit code. With 7 bits, the number of possible distinct codes is

$2^7 = 128$. This value is sufficient to encode all required letters (uppercase and lowercase), numerals, punctuation marks, and a number of special characters used on keyboards and in data transmission. Many applications use the eighth bit in a byte as a parity check bit. If parity checking is not necessary, the eighth bit in a byte can be used to encode an additional 128 symbols. This extended ASCII is commonly used in modern computers.

A complete summary of the basic 7-bit ASCII code is provided in Table 11–2. The 128 possible binary codes, the corresponding characters they represent, and the definitions of special characters are listed there. Many of the special characters employ terminology peculiar to data processing procedures, and will not be discussed here.

**Table 11–2**   American Standard Code for Information Interchange (ASCII)

| Bits 6 | 0 | 0 | 0 | 0 | 1 | 1 | 1 | 1 |
|---|---|---|---|---|---|---|---|---|
| 5 | 0 | 0 | 1 | 1 | 0 | 0 | 1 | 1 |
| 4 | 0 | 1 | 0 | 1 | 0 | 1 | 0 | 1 |
| **Bits 3210** | | | | | | | | |
| 0000 | NUL | DLE | SP | 0 | @ | P | ` | p |
| 0001 | SOH | DC1 | ! | 1 | A | Q | a | q |
| 0010 | STX | DC2 | " | 2 | B | R | b | r |
| 0011 | ETX | DC3 | # | 3 | C | S | c | s |
| 0100 | EOT | DC4 | $ | 4 | D | T | d | t |
| 0101 | ENQ | NAK | % | 5 | E | U | e | u |
| 0110 | ACK | SYN | & | 6 | F | V | f | v |
| 0111 | BEL | ETB | ' | 7 | G | W | g | w |
| 1000 | BS | CAN | ( | 8 | H | X | h | x |
| 1001 | HT | EM | ) | 9 | I | Y | i | y |
| 1010 | LF | SUB | * | : | J | Z | j | z |
| 1011 | VT | ESC | + | ; | K | [ | k | { |
| 1100 | FF | FS | , | < | L | \ | l | \| |
| 1101 | CR | GS | - | = | M | ] | m | } |
| 1110 | SO | RS | . | > | N | ^ | n | ~ |
| 1111 | SI | US | / | ? | O | _ | o | DEL |

| | | | | | |
|---|---|---|---|---|---|
| NUL | Null | DLE | Data Link Escape | DEL | Delete |
| SOH | Start of Heading | DC1 | Device Control 1 | SP | Space |
| STX | Start of Text | DC2 | Device Control 2 | | |
| ETX | End of Text | DC3 | Device Control 3 | | |
| EOT | End of Transmission | DC4 | Device Control 4 | | |
| ENQ | Enquiry | NAK | Negative Acknowledge | | |
| ACK | Acknowledge | SYN | Synchronous Idle | | |
| BEL | Bell | ETB | End of Transmission Block | | |
| BS | Backspace | CAN | Cancel | | |
| HT | Horizontal Tabulation | EM | End of Medium | | |
| LF | Line Feed | SUB | Substitute | | |
| VT | Vertical Tabulation | ESC | Escape | | |
| FF | Form Feed | FS | File Separator | | |
| CR | Carriage Return | GS | Group Separator | | |
| SO | Shift Out | RS | Record Separator | | |
| SI | Shift In | US | Unit Separator | | |

Bit 6 is considered the *most significant bit* (*MSb*) and bit 0 is considered the *least significant bit* (*LSb*). As we will see shortly, however, the usual manner of serial transmission is with the *LSb transmitted first and the MSb transmitted last.* One could make an argument that since ASCII words represent characters rather than values, it is immaterial which bits are considered MSb and LSb as long as the order of transmission is understood. However, there are certain properties that have a common pattern with respect to basic binary encoding, so it is desirable to have a defined order based on the MSb to the LSb.

**▌▌ EXAMPLE 11-3**

Write the binary codes for uppercase *W* and lowercase *w* and compare them. What can you conclude?

**SOLUTION**  The two characters, beginning with the MSb (bit 6), are obtained from Table 11–2 and compared as follows:

W    1010111

w    1110111

The two codes differ only in bit 5 (the one following the MSb). This pattern is true for all of the letters of the alphabet. An uppercase letter may be converted to a lowercase letter by complementing bit 5.

**▌▌ EXAMPLE 11-4**

Write down the ASCII code for the number 9 and indicate any pattern that is observed.

**SOLUTION**  From Table 11–2, the ASCII code for the number 9 is 0111001.

The last four digits of the binary form represent the corresponding 4-bit binary representation of the number 9. This pattern is true for all the numbers from 0 through 9.   **▌▌**

## EBCDIC

IBM independently developed an encoding scheme for letters, numbers, and punctuation, which it called *Extended Binary Coded Decimal Interchange Code,* EBCDIC. EBCDIC is an 8-bit extension of BCD, the 4-bit encoding of the digits 0 through 9. A partial listing of the EBCDIC codes is shown in Table 11–3. Note that all of the codes for letters and numbers are from the upper 128 codes in the range of 256 codes possible with eight bits. EBCDIC also includes codes for punctuation and special characters like those included in ASCII, but those are encoded in the lower 128 codes.

**Table 11–3**  Partial Table of EBCDIC Code

|      | 1000 | 1001 | 1010 | 1011 | 1100 | 1101 | 1110 | 1111 |
|------|------|------|------|------|------|------|------|------|
| 0000 |      |      |      |      |      |      |      | 0 |
| 0001 | a | j |   |   | A | J |   | 1 |
| 0010 | b | k | s |   | B | K | S | 2 |
| 0011 | c | l | t |   | C | L | T | 3 |
| 0100 | d | m | u |   | D | M | U | 4 |
| 0101 | e | n | v |   | E | N | V | 5 |
| 0110 | f | o | w |   | F | O | W | 6 |
| 0111 | g | p | x |   | G | P | X | 7 |
| 1000 | h | q | y |   | H | Q | Y | 8 |
| 1001 | i | r | z |   | I | R | Z | 9 |

**▌▌ EXAMPLE 11-5**

Write the binary codes for uppercase *W* and lowercase *w* from Table 11–3 and compare them. What can you conclude?

**SOLUTION**  The two characters, beginning with the MSb (bit 8), are obtained from Table 11–3 and compared as follows:

W    11100110

w    10100110

The two codes differ only in bit 6 (the one following the MSb). This pattern is true for all of the letters. An uppercase letter may be converted to a lowercase letter by complementing bit 6.

---

**▐▌ EXAMPLE 11-6**    Write the EBCDIC code for the number 9 and indicate any pattern that is observed.

**SOLUTION**    From Table 11–3, the EBCDIC code for the number 9 is 11111001.

The last four digits of the binary form represent the corresponding 4-bit binary representation of the number 9. This pattern is true for all the numbers from 0 through 9.    ▐▌

---

## Unicode

The 128 characters that can be encoded with ASCII are sufficient for documents written using the Latin alphabet, but the computer is now an international machine. A new code was needed to be able to encode all of the characters used in various languages. Using a 16-bit code, it is possible to encode $2^{16} = 65,536$ symbols. For compatibility with the existing ASCII, the ASCII code has been incorporated into Unicode using 00000000 as the first byte and the existing ASCII coding for the second byte. Arabic-language characters are encoded using 00000110 for the first byte and the second byte for the individual characters. This is the standard pattern. A different set of characters is designated by the first byte and the individual characters within the set are encoded by the numbers of the second byte. Of course, a program must include the proper instructions in order to be able to use Unicode.

## 11-4  Parallel Data Communications

Parallel data communication must be considered in two contexts. Data communication between components within a computer is accomplished over buses that transfer the data in parallel. In addition, parallel data communication is used as a relatively high-speed method of transferring data from a computer to external peripherals. Although the use of parallel paths internally is as important as ever, the use of parallel ports to communicate with external devices is being replaced by high-speed serial transmission. Figure 11–4 shows two devices joined by an eight-bit parallel bus.

### Front Side Bus

Two parallel bus characteristics determine how fast data can be transferred on a bus: bus speed and bus width. Computers commonly have several different buses, which may operate at different speeds and have different bus widths. The product of bus width, in bits, and bus speed, in MHz, is the bus bandwidth, in Mb/s (megabits per second). If the bus bandwidth in Mb/s is divided by eight, the result is the bus bandwidth in MB/s (megabytes per second). Modern PC processors connect to the support chips using a bus called the front side bus (FSB). FSB speed defines the speed at which the processor communicates with its support chips, and ranges from 66 MHz to as high as 800 MHz. Modern processors commonly use 64-bit bus widths.

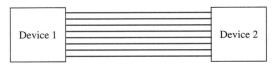

**FIGURE 11–4**
Parallel connection of devices.

**▮▮ EXAMPLE 11-7**

A Pentium 4 processor has a 64-bit bus width and 800 MHz FSB speed. What is the processor bandwidth?

**SOLUTION** The processor bandwidth is the product of bus speed and bus width.

$$64 \times 800 \text{ MHz} = 51200 \text{ Mb/s or } 6400 \text{ MB/s}$$

## Memory Bus

Data transferred from memory to the processor usually does not go directly from memory to the processor. It is transferred from memory to the memory controller on the memory bus, and then transferred from the memory controller to the processor on the FSB. Memory bus speeds are typically much slower than FSB speeds, and depend upon the speed of the memory chips used. Figure 11–5 shows a typical FSB and memory bus.

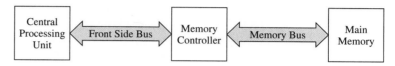

**FIGURE 11–5**
FSB and memory bus.

**▮▮ EXAMPLE 11-8**

A DDR-SDRAM chipset uses a 64-bit bus width and memory bus speed of 400 MHz. What is the memory bus bandwidth?

**SOLUTION** The memory bus bandwidth is the product of the bus speed and the bus width.

$$64 \times 400 \text{ MHz} = 25600 \text{ Mb/s or } 3200 \text{ MB/s}$$

Comparing the results of Examples 11-7 and 11-8, we see that the memory bus is a bottleneck because it cannot provide data to the processor as fast as the processor can process the data. To solve this problem, some chipsets support a dual-channel memory bus that doubles the width of the memory bus, as shown in Figure 11–6.

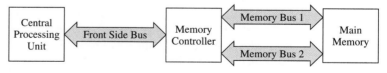

**FIGURE 11–6**
FSB and dual-channel memory bus.

**▮▮ EXAMPLE 11-9**

A DDR-SDRAM chipset supports a dual-channel memory bus operating at 400 MHz. What is the effective memory bandwidth?

**SOLUTION** The memory bus bandwidth is the product of the bus speed and the bus width.

$$128 \times 400 \text{ MHz} = 6400 \text{ MB/s}$$

Thus, this memory throughput matches that of the Pentium 4 of Example 11-7, which is an ideal situation.

## I/O Busses

Although modern PC motherboards often include functions such as video, sound, and network interfaces that formerly were provided by expansion boards, expansion slots are still included on motherboards. Input/output (I/O) busses connect those expansion slots to the chipset. Currently, the only I/O busses commonly used are the peripheral component interconnect (PCI) bus and the accelerated graphics port (AGP) bus. Currently, PCI busses are 64 bits wide and operate at speeds up to 133 MHz, providing data transfer rates of up to 1066 MB/s. The AGP bus is used only for video data. It has a 32-bit width and operates at a speed of 66 MHz. However, 4X AGP transfers data 4 times per clock cycle to provide an effective bandwidth of 1066 MB/s, and 8X AGP transfers data 8 times per clock cycle to provide an effective bandwidth of 2132 MB/s.

## The Parallel Port

Although USB is a popular data communication method for new peripherals such as printers and scanners, there are still many peripherals designed to be accessed using the parallel port. As originally designed by Centronics in the 1960s and updated by IBM in 1981, the parallel port was intended for use only from the computer to a printer and for cable lengths up to only about six feet. The IEEE 1284 standard, approved in 1994, defines three parallel port standards, Standard Parallel Port (SPP), Enhanced Parallel Port (EPP), and Extended Capabilities Port (ECP), based upon improvement efforts made by different vendors. SPP is a standardization of the Centronics unidirectional parallel port. The eight data lines in SPP can transfer data in one direction only, from the computer to the peripheral, and the maximum transfer rate is limited to about 150 KB/s. The EPP standard incorporates additional registers and defines control lines for bidirectional transfer of data. Data lines are bidirectional and also serve as address lines. Data rates up to about 10 MB/s can be achieved. ECP utilizes a different combination of registers and control signals to accomplish bidirectional transfer of data. Data/address lines are bidirectional and data rates up to about 2.4 MB/s can be achieved.

## 11-5 Serial Data Transmission

Instead of the multiple bit paths provided for data by parallel data transmission, serial data transmission offers a single bit path for data as shown in Figure 11–7. Parallel data in the computing machine is converted to serial data by the universal asynchronous receiver-transmitter (UART). The serial data is then transferred to the other computing machine where it is converted back to parallel data by another UART. As previously noted, serial data transmission may be either asynchronous or synchronous, depending upon whether the clocks at transmitter and receiver must be synchronized or not. Asynchronous serial data transmission does not require that provision be made to synchronize the transmitter and receiver clocks, although the clock speeds must be approximately the same. In synchronous serial data transmission, either a separate clock signal is provided, or the means of synchronizing the clock is incorporated into the data. The UART is used for asynchronous data transmission.

**FIGURE 11–7**
UARTs used in serial data communication.

## Asynchronous Serial Data Transmission Format

One manner in which a given character is encoded for asynchronous serial transmission will now be illustrated. Refer to Figure 11–8, in which the uppercase ASCII letter *W* is used as the example. From Table 11–2 or from Example 11-3, the form of the binary code with the MSb first is 1010111. For data transmission, the normal order of transmission will be *with the LSb first;* that is, as 1110101. It is also necessary to add additional start and stop bits, as discussed below. A start bit is the same length as each of the data bits. There is no limit to the maximum length of a stop bit, but the minimum length depends on the data rate and typically ranges from 1 to 2 bit intervals.

For asynchronous transmission, the normal level of the signal prior to the beginning of the word is the *mark* or "1" level. (There is an old expression "marking time" that can help jog the memory.) The actual level associated with the mark level can vary with the type of system and may even be a negative voltage (as we will see later). However, Figure 11–8 shows the mark level as the upper level, in accordance with the simplest type of interpretation.

The beginning of a new character is indicated by a transition between the *mark* level and the *space* level, as shown. This transition starts the timing process at the receiver for the given word. Assuming normal tolerances of components, the synchronization should certainly last through the remainder of the word transmission. The clock speed of both transmitter and receiver must be approximately the same but the clocks need not be synchronized.

Following the start bit, the 7 bits representing the character *W* are now transmitted, with the LSb transmitted first and MSb last. From Table 11–2, the 7 bits arranged in the order of transmission from left to right are 1110101.

After the completion of the 7 data bits, an odd parity check bit is inserted. In this particular example a 0 is inserted since there are already an odd number (5) of 1s. A *stop* signal is now transmitted. The stop signal is always at the mark level. As previously mentioned, it must have a minimum length, which in this example is assumed to be 1 bit interval. The signal then remains at the mark level until the next start bit is received. This could represent any arbitrary length of time, provided that the minimum stop element length has been satisfied.

## The EIA-232 Standard

A widely employed standard in the serial transmission of data is the EIA-232 (formerly the RS-232) standard, which has evolved through several versions since its inception. The EIA-232C seems to be the version most widely used. This standard describes the physical interface and protocol for relatively low-speed serial data communication between computers and related devices. The *serial port* on most computers follows this standard. Prior to the development of the universal serial bus (USB), the EIA-232C was the interface that most

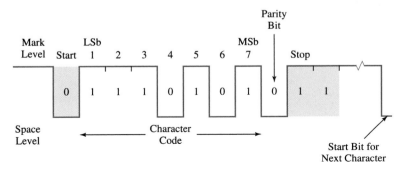

**FIGURE 11–8**
Asynchronous serial data communication.

computers used to talk to and exchange data with modems and other serial devices. Within the computer, a UART chip on the motherboard takes the data from the computer and transmits it to an internal or external modem. Since data in the computer flows along parallel circuits, and serial devices can handle only one bit at a time, the UART converts the groups of bits in parallel to a serial stream of bits. In this implementation, the computer is the *data terminal equipment* (DTE), and the modem is the *data communications equipment* (DCE).

The signal voltage levels on the EIA-232C utilize "negative logic," and are defined as follows:

**Transmitted Signal Voltage Levels:**

Binary 0 or Space: $+5$ to $+15$ V

Binary 1 or Mark: $-5$ to $-15$ V

**Received Signal Voltage Levels:**

Binary 0 or Space: $+3$ to $+13$ V

Binary 1 or Mark: $-3$ to $-13$ V

This difference between the transmitted and received signal voltage levels provides a tolerance for loss along the line between transmitter and receiver. For example, if $+5$ volts is the signal voltage level of a binary 0 from the transmitter, 2 volts can be dropped in the transmission line and still be recognized by the receiver as a binary 0.

In contrast to the data lines, the control lines use what could be described as "positive logic"; that is, a line is *asserted* (active) when it carries a positive voltage, which is actually logical zero in the 232 standard. This situation necessarily results in a great deal of confusion. Spot checks of several references by one of the authors showed a great deal of inconsistency and errors in the literature.

In connecting a DTE to a DCE, each device has a Transmitted Data (TxD) connection and a Received Data (RxD) connection. The cable between the two devices connects the DTE TxD to the DCE RxD and the DTE RxD to the DCE TxD. Unless data is being sent, each line is maintained at the mark level (logic 1 or negative voltage). In addition to the data lines, each device has a Request to Send (RTS), Clear to Send (CTS), Data Terminal Ready (DTR), Data Set Ready (DSR), and Data Carrier Detect (DCD) connection. The cable between the devices connects each RTS to the opposite CTS and each DTR to the opposite DSR. The two DCDs are connected together.

Hardware flow control is accomplished mainly with the RTS and CTS lines. When a device has something to send it will assert (apply a positive voltage to) its RTS line. This asserts the other device's CTS line. If the other device is ready to receive data it will assert its RTS line which asserts the first device's CTS line. The first device will then start sending data. When the receiving device is no longer ready to receive data (its buffer is full) it removes the positive voltage from its RTS line and thus from the transmitting device's CTS connection. The transmitting device then stops transmitting until its CTS is asserted again.

## 11-6    High-Level Data Link Control

The high-level data link control (HDLC) protocol is a set of rules for establishing serial communications links between two *stations* in order to transfer information from one to the other. HDLC, which is documented in ISO 3309 and ISO 4335, and protocols derived from it, are the most common serial data communications protocols in use. Under HDLC there are three types of stations: *primary, secondary,* and *combined.* A primary station controls operation of a communication link and issues commands to secondary stations. A secondary station operates under the control of a primary station and sends responses to that primary station. A combined station combines the functions of primary and secondary stations and can issue both commands and responses. The communication links can be either

| Flag 8 Bits | Address 8 Bits | Control 8 or 16 Bits | Information (Optional) 8*n Bits | CRC 16 or 32 Bits | Flag 8 Bits |

**FIGURE 11–9**
HDLC frame.

*balanced* or *unbalanced* and either *full-duplex* or *half-duplex*. The balanced configuration consists of two combined stations and the unbalanced configuration consists of a primary station and one or more secondary stations. Both configurations support either full-duplex or half-duplex.

Primary stations issue *commands* and secondary stations issue *responses*. Combined stations issue both commands and responses. Through the exchange of commands and responses, stations set up serial data communications links, control the flow of information over the links, and then disconnect the links. The commands and responses are contained in frames, and the basic HDLC frame structure is shown in Figure 11–9. Each of the fields of the HDLC frame is discussed in detail below.

## HDLC Operation

HDLC operation consists of three phases: initialization, data transfer, and disconnection. Initialization establishes a communication link between two stations and defines the rules governing the flow of information between the stations. During the data transfer phase, control information to regulate data flow and error handling is exchanged along with the data being transferred. Finally, when the link has served its purpose, it is disconnected. Either station can request initialization and initiate a disconnect.

There are three types of frames that are identified by a code in the control field: information (I), supervisory (S), and unnumbered (U). Information frames, called I frames, have a binary 0 in the first bit of the control field and are used to transfer user data between stations. I frames can also contain flow and error control information in their control fields. Supervisory frames, called S frames, are identified by a binary 10 in the first two bits of the control field and contain flow and error control information. U frames are identified by a binary 11 in the first two bits of the control field and are used primarily in the initialization and disconnect phases of HDLC operation. The control functions are exchanged by codes included in the control field of the frames. Table 11–4 lists the flow and error control functions that are conveyed in bits 3 and 4 of the control field of S frames. Table 11–5 lists the functions that are conveyed in bits 3, 4, 6, 7, and 8 of U frames.

HDLC defines three data transfer modes: normal response mode (NRM), asynchronous balanced mode (ABM), and asynchronous response mode (ARM). NRM and ARM are used with unbalanced configurations (primary and secondary stations). In NRM, a secondary can only transmit data in response to a command from the primary. In ARM, a secondary can transmit data without permission from the primary, but the primary still controls initialization, error control, and disconnection. In ABM, either combined station can transmit data without permission from the other station.

Table 11–4   Control Field Bits 3 and 4 of HDLC S Frames

| Bits 3 and 4 | Function | Abbreviation |
| --- | --- | --- |
| 00 | Receive Ready | RR |
| 01 | Receive Not Ready | RNR |
| 10 | Reject | REJ |
| 11 | Selective Reject | SREJ |

Table 11–5    Control Field Bits 3, 4, and 6 through 8 of HDLC U Frames

| Bits 34x678 | Function | Abbreviation |
|---|---|---|
| 00x001 | Set Normal Response Mode | SNRM |
| 11x000 | Set Asynchronous Response Mode | SARM |
| 11x100 | Set Asynchronous Balanced Mode | SABM |
| 00x010 | Disconnect | DISC |
| 11x011 | Set Normal Response Mode Extended | SNRME |
| 11x010 | Set Asynchronous Response Mode Extended | SARME |
| 11x110 | Set Asynchronous Balanced Mode Extended | SABME |
| 10x010 | Set Initialization Mode | SIM |
| 00x100 | Unnumbered Poll | UP |
| 00x000 | Unnumbered Information | UI |
| 11x101 | Exchange Identification | XID |
| 11x001 | Reset | RESET |
| 00x111 | Test | TEST |
| 11x000 | Disconnected Mode | DM |
| 00x110 | Unnumbered Acknowledgment | UA |
| 10x001 | Frame Reject | FRMR |
| 00x010 | Request Disconnect | RD |
| 10x000 | Request Initialization Mode | RIM |

Three of the functions listed in Table 11–5 are Set Normal Response Mode (SNRM), Set Asynchronous Balanced Mode (SABM), and Set Asynchronous Response Mode (SARM). A station issuing a command containing one of these codes in the control field of a U frame is requesting that a link be established using that mode of operation and 3-bit sequence numbers. Sequence numbers are used in I frames and S frames to identify specific I frames so that the receipt of the frame can be acknowledged. Bits 2 through 4 of the control frame of an I frame contain the sequence number of that frame when 3-bit sequence numbers are being used. It is also possible to use 7-bit sequence numbers. This is requested by issuing a command frame with Set Normal Response Mode Extended (SNRME), Set Asynchronous Balanced Mode Extended (SABME), or Set Asynchronous Response Mode Extended (SARME). A communication link is established by one station issuing a command (U frame) with one of the above six functions in its control field and the other station responding by issuing a U frame with the unnumbered acknowledgment (UA) function in its control field.

Once the communication link is established, the stations can exchange I frames. I frames contain two types of sequence numbers, send sequence numbers, N(S), and receive sequence numbers, N(R). N(S) in bits 2 through 4 of the control field is the sequence number of the current I frame. N(R) in bits 6 through 8 of the control field is the sequence number of the next I frame the station expects to receive. Thus, when a station has received a frame with N(S) = 3 in its response frame it will send N(R) = 4, which acknowledges receipt of frame 3 and indicates that it is expecting frame 4 next. In addition to I frames, stations may exchange S frames to control flow. If a station cannot process I frames as fast as they are received, its buffer gets full. It will issue a Receive Not Ready (RNR) S frame (00 in bits 3 and 4 of the control field). The RNR will contain the N(R) sequence number of the next I frame it can handle as an acknowledgment of all previously sent I frames. For example, if a station sends RNR, 3, this acknowledges receipt of I frames with sequence number 2 and below. Even if the other station already sent frame 3, that is the frame it must send next. When the station issuing the RNR frame is ready to receive frames again it will issue a Receive Ready (RR) frame that also contains the N(R) sequence number of the next frame it expects to receive.

During the period after the RNR frame was sent, the station receiving that frame may poll the other station by sending RR frames containing a 1 in bit 5 of the control field. Bit 5 is the poll/final (P/F) bit. When the P/F bit is a 1 the station receiving the frame is required to respond with either a RR or a RNR frame (with a 0 in the P/F bit).

**■ EXAMPLE 11-10**

Station A has transmitted three consecutive I frames with N(S) of 0, 1, and 2. Station B has received all three frames and verified that the data is valid. What response does Station B send?

**SOLUTION**   Station B could issue an I frame with N(R) = 3 to acknowledge the receipt of frames 0, 1, and 2 from Station A and to send data to Station A at the same time. It could issue an RR U frame with N(R) = 3 if it was ready to receive more data and had no data to send to Station A. It could also issue an RNR U frame with N(R) = 3 if it was not ready to receive more frames but is acknowledging the receipt of frames 0, 1, and 2.

As stated above, in NRM a secondary station can only transfer data in response to a command from the primary station. The primary station polls the secondary station by periodically issuing a U frame with the P/F bit set to 1. If the secondary station has data to transmit, it transmits the data, usually as a sequence of I frames. If the secondary station does not have data to transmit, it issues a RNR frame with 1 in the P/F bit.

It is possible for a frame to be lost in transit. One station might send frames 0, 1, and 2. It is usually not necessary to wait for an acknowledgment for each frame before sending the next. If the other station received frames 0 and 2, but did not receive frame 1, it would discard frame 2 when it was received because it is out of order. It would send an S frame with binary 10 in bits 3 and 4 of the control field (REJ) and binary 001 (assuming 3-bit sequence numbers) in bits 6 through 8. This would tell the first station that it must resend frame 1 and all following frames. When the communication link is no longer needed, one station will send a U frame with the disconnect code in its control field (a DISC frame), and the other station will respond with a UA frame.

**■ EXAMPLE 11-11**

Station A has transmitted three consecutive I frames numbered 0, 1, and 2. Station B received frames 0 and 2, but not frame 1. What response does Station B send?

**SOLUTION**   Station B will discard frame 2 because it is out of sequence. If it ready to receive more frames it will send either an I frame or an RR frame with N(R) = 1 to indicate that all frames after 0 must be retransmitted. Of course, if it is not ready to receive more frames it would send an RNR frame with N(R) = 1.                                               ■

## Other Fields in the HDLC Frame

Flow and error control are the functions of the control field. The commands and responses that are included in that field enable the stations to regulate the flow of data between them and retransmit lost or damaged frames. Each of the other fields in an HDLC frame—flag, address, information, and frame check sequence (or cyclic redundancy check)—has a purpose as well.

The flag field defines the beginning and end of the frame with a specific code, 01111110. A sequence of six consecutive 1s is not allowed to exist anywhere else in the frame, including the information field. This is prevented by using a technique called bit stuffing. Bit stuffing involves inserting a 0 after each five consecutive 1s anywhere in the frame except in the flag field. A station receiving a frame with five consecutive 1s examines the next bit and discards it if it is a 0. If the next bit is a 1 and it is followed by a 0 it is recognized as the flag character. If, instead, the next bit is a 1 followed by another 1, this is recognized as the abort signal and the frame is discarded. Between frames, when the link is idle, synchronous links usually transmit a constant stream of data. This data can be all 1s (called *mark idle*) or a series of flag characters (called *flag idle*).

The address field contains the address of either the station to which the frame is being sent (command frame) or the station sending the frame (response frame). The order of the bits in the address field is LSb to the left and MSb to the right. A 0 in the MSb of the address field indicates that the frame is intended for one station; a 1 in that bit indicates it is intended for multiple stations. Although Figure 11–9 shows the address field as 8 bits there may be additional bytes in that field. A 1 in the LSb of that first byte indicates that it is the only byte in the address field. A 0 in the LSb indicates that the address field has an additional byte. An address field that is all 1s (11111111) is a broadcast address.

The CRC or Frame Check Sequence (FCS) field is used to detect errors in transmission. A cyclic redundancy check involves an arithmetic operation performed on the data being sent, which can be duplicated at the receiver to verify the accuracy of the received data. A detailed discussion of the CRC process is given in the next section.

## 11-7  Cyclic Redundancy Check

The basic concept of the cyclic redundancy check (CRC) is very simple. The data being sent is treated as a number, and is divided by another predetermined number. The remainder of this division is sent along with the data. The receiver divides the data by the same predetermined number and if the same remainder is obtained, the data is assumed to be valid. For a very simplified example of this, assume that the number being transmitted is 167. If the predetermined divisor is 11, the remainder is 2. If a transmission error caused the number to change to 163 during transmission, the remainder, after the division by 11 in the receiver, would be 9, indicating that an error in transmission had occurred. Alternately, we could subtract the remainder from the data and perform the division on the number minus the remainder. Then if the remainder from the division in the receiver is zero, this indicates that the data is valid.

Of course, the numbers used in these calculations are binary, not decimal. Modulus-2 arithmetic is used in the calculations, in which both addition and subtraction involve the exclusive-OR (XOR) operation. In modulus-2 arithmetic, $0 + 1 = 1$ and $0 - 1 = 1$, $1 + 0 = 1$ and $1 - 0 = 1$, $0 + 0 = 0$ and $0 - 0 = 0$, and $1 + 1 = 0$ and $1 - 1 = 0$. The division process is somewhat different from ordinary long division. Assume that the data to be transmitted consists of $n$ bits. Assume that we are going to use a $k + 1$ length predetermined number as the divisor. We append $k$ binary 0s at the end of the data and perform the division. An example will illustrate this.

---

**■ EXAMPLE 11-12**

Assume that we want to transmit 10011 and ensure that the data is valid when it is received. We will use 1001 as the predetermined divisor. What is the remainder?

**SOLUTION**  Figure 11–10 shows the process: We first append 000 to the data ($k$ 0s), which is equivalent to shifting the data three places to the left. We then divide by 1001. There will be a three-bit remainder, which is 001 in this case.

---

Data to be transmitted = 1 0 0 1 1

CRC generating function = 1 0 0 1

　　Shifted left three bits = 1 0 0 1 1 0 0 0

Divide shifted data by generating
function using modulus-2 arithmetic:

```
                 1 0 0 0 1
     1 0 0 1)1 0 0 1 1 0 0 0
             1 0 0 1
             0 0 0 0 1 0 0 0
                     1 0 0 1
                     0 0 0 1
```

　　　　　Remainder = 001
　　　　　This is the CRC.

---

**FIGURE 11–10**
CRC calculation example.

To verify that the transferred data is valid we could repeat the division and compare the result to the value in the CRC field. However, if we subtract the remainder from the original data and divide by the divisor, the remainder of that operation should be zero if there was no error in transmission. Using modulus-2 arithmetic, subtracting the remainder can be accomplished just by appending the remainder to the data instead of shifting the data to the left. Thus, if we divide the combined data and CRC fields by the divisor, the result should be zero.

**EXAMPLE 11-13**

Assume that the transmitted data field is 10011 and the CRC field is 001. Use 1001 as the predetermined divisor. Verify that the transmitted value is valid.

**SOLUTION**   Figure 11–11 shows the process: 001 is appended to 10011 to form 10011001. We then divide by 1001. The remainder of zero indicates that the data is valid.

If it was necessary to do long division like this to verify the data, the process would be burdensome. However, because subtracting in modulus-2 arithmetic is accomplished by using the XOR process, the division of Figures 11–10 and 11–11 can be accomplished easily with logic circuitry as the data is received. In Example 11-13, when the first four bits of the data are received (1001), they are XORed with 1001 and the result is 0000. We append the fifth and later received bits (including the CRC) to the results of the XOR operation and then left-shift until the MSb is a binary 1. That MSb and the three bits following it are then XORed with 1001. In this case 1001 (the 1 of the data and the 001 of the CRC) is XORed with the 1001 divisor. The result is also 0000. Since all of the data and the CRC fields have been processed, we know that the transmitted data is valid.

---

Transmitted data = 1 0 0 1 1 0 0 1

CRC generating function = 1 0 0 1

Divide transmitted data by generating
function using modulus-2 arithmetic:

```
              1 0 0 0 1
    1 0 0 1 ) 1 0 0 1 1 0 0 1
              1 0 0 1
              0 0 0 0 1 0 0 1
                      1 0 0 1
                      0 0 0 0
```

Remainder = 000
Zero remainder indicates that
data was transmitted without error.

---

**FIGURE 11–11**
CRC verification example.

## 11-8   The Universal Serial Bus

Data communications between a computer and its peripherals using the RS-232 standard are too slow for modern computer applications, and a separate serial port is required for each device that communicates with a computer using the standard. The *universal serial bus* (USB) was developed to overcome both of these limitations. The USB 2.0 standard defines three speeds for different types of applications. Low-speed, up to 1.5 Mb/s, is intended for applications such as the keyboard, mouse, and game peripherals. Full-speed, up to

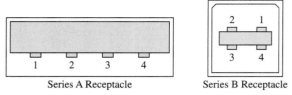

FIGURE 11–12
USB Series A and B connectors.

12 Mb/s, is intended for applications such as modems, broadband, and audio. High-speed, up to 480 Mb/s, is intended for applications such as video, storage, imaging, and broadband.

USB uses a much simpler connector and cable than RS-232. Figure 11–12 shows the USB Series A and Series B ports and Figure 11–13 shows a USB cable. In the USB cable, D+ and D– are the differential data lines and VBUS and GND are the power lines. In addition to increased data transfer rates, USB provides the ability to connect up to 127 devices to a USB port through a series of hubs, as shown in Figure 11–14, and provides a *plug-and-play* capability. With plug-and-play it is not necessary to configure a computer for a newly installed device, nor is it necessary to reboot the computer after the device is installed. A plug-and-play compatible computer will identify the newly installed device and configure itself to use the device. When the device is to be removed, the computer is so notified with a couple of mouse clicks and the device can be unplugged without shutting down the computer first. In addition to providing high-speed serial data communications with peripheral devices, USB can provide power to those devices as well, as long as their power requirements are low.

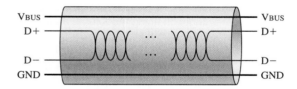

FIGURE 11–13
USB cable.

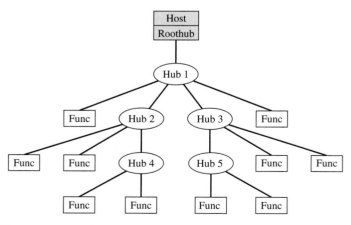

FIGURE 11–14
Connection of USB hubs and functions.

## USB Structure

A USB system consists of a host, a host controller, devices, client software, system software, and the physical topology. The host is the computer that uses USB to communicate with its peripherals. The host controller consists of the hardware and software in the host that enable USB devices to be connected to the host. The physical devices are the peripherals with which the host will communicate. The system software is that part of the operating system that enables it to support USB. The physical topology consists of the hubs and cables that connect the devices to the host.

The USB host is the nerve center for the USB system. The host is responsible for control of the system, including detecting when USB devices are connected or disconnected from the bus, managing data and control flow between the host and devices, collecting statistics on status and activity, and providing power to USB devices. There can be only one host in any USB system. The interface between the USB bus and the host computer system is a combination of hardware, software, and firmware called the host controller. The host can also supply a limited amount of power to devices that are directly connected to it (bus-powered devices) or devices may have their own power sources (self-powered devices).

USB devices consist of hubs, which provide attachment points to the USB bus, and functions, which add functionality to the system. Examples of functions include keyboards, mice, monitors, printers, scanners, external hard drives, digital cameras, and other peripherals. A device can be both a hub and a function. For example, a keyboard can contain additional attachment points to the bus, called ports. The Series A port of Figure 11–12 is called a *downstream connection* or *downstream port* and the Series B port is called an *upstream connection* or *upstream port*. All USB devices have an upstream connection but only hubs have downstream connections. Hubs are self-powered devices that can also supply power to bus-powered devices connected to them.

Figure 11–13 shows the configuration of a USB cable. The D+ and D− lines (Data+ and Data− lines) are connected to transceivers at each end of the line. Signaling involves three states, designated as the *J state* (or J), the *K state* (or K), and the *idle state*. To signal a J, a transceiver directs current into the D+ line. To signal a K, a transceiver directs current into the D− line. The idle state exists when neither transceiver drives current into the D+ and D− lines. Non return to zero inverted (NRZI) data encoding is used. A binary 0 is represented by a transition between J and K or between K and J. A binary 1 is represented by no transition. Thus, a series of binary 0s consists of a JKJKJKJKJK sequence, while a series of binary 1s would consist of a JJJJJJJJJJ sequence or a KKKKKKKKKK sequence. JK transitions are required to synchronize the local clock to the received signal, so a long series of binary 1s creates a problem. To prevent this, bit stuffing is used in the transmitter. A binary 0 is inserted after every sequence of six binary 1s. These stuffed bits are removed as part of the decoding process at the receiver.

Figure 11–14 shows how hubs and functions are connected to form the bus. Each USB hub or function is assigned a unique address by the USB system software when the device is connected to the bus. When a device is initially powered up, its address defaults to 0000000, a 7-bit address. Then, when it is recognized by the host as being connected, the host assigns a 7-bit, nonzero address. With seven bits available for the address, there are 128 unique addresses possible, but since 0000000 is reserved for the default address, only 127 devices can be connected to a USB bus. This process of identifying and assigning unique addresses to the devices connected to the bus is called *bus enumeration*. Bus enumeration also includes detecting and processing removals. The USB system software also creates one or more associations between itself and each device on the bus used to communicate with the device. These associations are called *pipes,* and may further be designated *message pipes,* and *stream pipes.* A pipe to a device is designated by a 4-bit *endpoint*. Once a device is powered up, a message pipe called the default control pipe at endpoint 0000 always exists so that the host can have access to the device's configuration, status, and control information. Full- and low-speed devices may have up to two additional pipes. High-speed devices may have up to fifteen additional pipes.

## USB Operation

Unlike the EIA-232C standard, which controls data flow using voltage levels on separate wires to control communication between devices, USB uses control signals on the data wires to accomplish flow control. This reduces cable size and increases flexibility. USB is a polled system, meaning that devices cannot communicate on the bus without being invited to do so by the host. There are a total of fifteen types of packets used in USB operation, with the type of packet designated by the packet identifier field (PID). These fifteen types are divided into four groups: token packets, data packets, handshake packets, and special packets. Each group has a different packet structure, but there are just a few field types used in the packets.

Every packet begins with a synchronization (SYNC) field. For high-speed the SYNC field consists of 32 bits, fifteen KJ transitions followed by KK. For full- and low-speed, the SYNC field consists of eight bits, three KJ transitions followed by KK.

The field following the SYNC field is the PID field, which consists of eight bits. The first four bits identify the type of packet or PID and the last four bits are the complement of the PID. The PID for an OUT token packet is 0001B (B indicates binary), where the first 0 is the MSb (most significant bit) and the 1 is the LSb (least significant bit). The MSb is designated $PID_3$ and the LSb is designated $PID_0$. In this sequence, $PID_1$ and $PID_0$, 01, designate the packet as a token packet and $PID_3$ and $PID_2$, 00, designate it as an OUT packet. Table 11–6 shows the PIDs used in USB packets. Note that in each case, the group is designated by $PID_1$ and $PID_0$.

The address field, when used, is seven bits long, with the LSb designated $Addr_0$ and the MSb designated $Addr_6$. When transmitted, the LSb is always transmitted first and the MSb transmitted last. Associated with the address field there is an endpoint field that is four bits long, with the LSb designated as $Endp_0$ and the MSb designated as $Endp_3$.

**Table 11–6**   PID Field of USB Packets

| PID Type | PID Name | PID | Description |
|---|---|---|---|
| Token | OUT | 0001 | Addr & Endp in host $\Rightarrow$ function transaction |
| | IN | 1001 | Addr & Endp in function $\Rightarrow$ host transaction |
| | SOF | 0101 | Start-of-frame frame number |
| | SETUP | 1101 | Addr & Endp in setup transaction |
| Data | DATA0 | 0011 | Even PID data packet |
| | DATA1 | 1011 | Odd PID data packet |
| | DATA2 | 0111 | Data packet for high-speed, high-bandwidth isochronous transactions |
| | MDATA | 1111 | Data packet for high-speed, high-bandwidth isochronous transactions and high-speed split transactions |
| Handshake | ACK | 0010 | Error-free packet received |
| | NAK | 1010 | Receiver can't accept data or transmitter can't send data |
| | STALL | 0110 | Endpoint halted or control pipe doesn't support transaction type |
| | NYET | 1110 | No response yet from receiver |
| Special | PRE | 1100 | Token to enable downstream bus traffic to low-speed functions |
| | ERR | 1100 | Split transaction error handshake |
| | SPLIT | 1000 | High-speed split transaction token |
| | PING | 0100 | High-speed flow control probe for bulk/control endpoint |
| | Reserved | 0000 | Reserved |

The frame number field, when used, is eleven bits long. The host increments the frame number on a per-frame basis until the maximum frame number, 7FFH (H for hexidecimal) is reached. Then the frame number rolls over to 000H. The frame number field is used only in SOF tokens.

The data field has a variable length, from 0 to 1,024 bytes (*note:* bytes not bits). Within each byte, the bits are transmitted LSb first and MSb last.

All fields in a packet are protected by cyclic redundancy checks except the PID, which is protected by including the complement of the PID in the PID field. CRCs are generated in a transmitter before bit stuffing is performed, and are decoded in a receiver after stuffed bits are removed. Tokens use a 5-bit CRC. A 16-bit CRC is used for data.

IN, OUT, and SETUP token packets, and PING special token packets, consist of three bytes or 24 bits in four fields: PID, address, endpoint, and CRC. This is eight bits for the PID, seven bits for the address, four bits for the endpoint, and five bits for the CRC. The PID, of course, designates the packet as IN, OUT, or SETUP. For IN and SETUP packets, the address and endpoint designate the device and endpoint that will receive a subsequent data packet. For OUT packets, the address and endpoint designate the device and endpoint that will transmit a subsequent data packet. For PING packets, the address and endpoint designate the device and endpoint that will respond with a handshake packet. Only the host can issue token packets.

The USB 2.0 standard anticipates that there will be full- and low-speed devices connected to hubs on the USB bus. In such a situation, operation is optimized if communication between the host and the hub is at high speed. Therefore a split operation is defined in which the hub communicates with the device using full- or low-speed as appropriate, and once that communication is complete, sends any data or status received from the device to the host at high-speed. This SPLIT operation utilizes two special 4-byte tokens, start split (SSPLIT) and complete split (CSPLIT). These packets contain the PID (eight bits), the address of the hub (seven bits), a one bit start/complete field, a 7-bit port field to designate the port to which the full- or low-speed device is connected, a 1-bit speed field, a 1-bit E or U field, a 2-bit endpoint type field, and a 5-bit CRC field.

A data packet is of variable length and contains three fields: the PID field, the data field, and the CRC field. The PID field is one byte long, the data field can be zero to 1,024 bytes long, and the CRC field is two bytes long. Although the data field has a variable length it must be an integral number of bytes long. As shown in Table 11–6, there are four types of data packet, DATA0, DATA1, DATA2, and MDATA. All four data packets may be used for high-speed isochronous communications, while DATA0, DATA1, and MDATA are used for split transactions.

Handshake packets consist of only the PID and are returned in the handshake phase of a transaction, and may be returned in the data phase. An ACK packet is used when data has been received, and indicates that the data packet was received without error. The host would return an ACK packet for an IN transaction and a function would return an ACK packet for an OUT, SETUP, or PING transaction. NAK packets are always issued by functions and indicate that the function could not receive data in an OUT transaction, or that the function has no data to send in the data phase of an IN transaction or the handshake phase of an OUT or PING transaction. The return of a STALL packet in response to an IN packet or a PING packet, or after the data phase of an OUT transaction, indicates that the function is unable to receive or transmit data. It can also indicate that a control pipe request is not supported. NYET is used during split transactions to indicate that the full- or low-speed transaction is not yet completed or that the hub can not handle the split transaction. The ERR packet is used to report an error on a full- or low-speed bus.

Transactions are classified based upon endpoint types. There are four endpoint types: bulk, control, interrupt, and isochronous. Each has different requirements and is used for different purposes, as will be discussed below. In most cases, each transaction consists of three phases: token, data, and handshake.

Bulk transactions guarantee error-free transmission of data between a host and a function. For example, if the function is a printer, the host will use OUT tokens to send the data to the printer. The transaction is initiated by the host issuing an OUT token that identifies the function (by address) and pipe (by address and endpoint), followed by a data packet. If the data is received without error, the function will return an ACK handshake. If there is an error in the received data packet, the function will discard it and not respond to the host. The host waits a specified time for a response, and if none is received, reissues the previous data packet. If the data packet was received without error but the function could not process it (for example, its buffer was full), it responds with a NAK handshake to indicate that the host should reissue the previous data packet. If the endpoint was halted, the function returns a STALL handshake indicating that the host should not try to reissue the data packet and that intervention by the USB system software is required.

If, on the other hand, the host is ready to receive data in a bulk transaction, the host issues an IN token specifying the function and endpoint that should transmit the data. The function addressed responds by issuing either a data packet or a NAK or STALL handshake. The NAK indicates that the function is temporarily unable to send data, perhaps because it has none to send. The STALL indicates a halt condition that requires USB system software intervention. If data is sent and received by the host without error, the host issues an ACK handshake. If the data was corrupted, the host discards it and does nothing. If the function does not receive an ACK in the requisite time, it will reissue the data packet.

---

**■ EXAMPLE 11-14**

A printer is connected to a computer through the USB bus. What tokens, packets, or handshakes would the computer send to the printer in order to print a document, and how would the printer respond?

**SOLUTION** The computer would send the OUT token followed by a data packet containing data to be printed. The printer would respond to the data packet by sending an ACK handshake if the data was received and could be processed. If the data was received but could not be processed because the printer's buffer was full, the printer would send a NAK handshake. If the data received was corrupt (PID or CRC bad), the printer would not respond and the computer would resend the data after a delay.

In a control transaction, there will be a minimum of two stages (stages, not phases): the setup stage and the status stage. During the setup stage, a SETUP token is sent, followed by a DATA0 packet containing the setup information. The function responds with an ACK if the data is received without error, or does not respond if there is an error in transmission. The status stage would consist of the host issuing an IN token, the function issuing a DATA1 packet containing the status information, and the host issuing an ACK if the data was received without error.

There can also be a data stage as part of a control transaction, which would follow the setup stage previously described. This data stage can be either a control write in which the host issues OUT tokens followed by data packets that must be acknowledged by the function, or a control read in which the host issues IN tokens to which the function responds by issuing data packets that the host must acknowledge. If there is a data stage, the direction of the status stage is opposite that of the data stage. IN tokens are issued by the host during the status stage of control writes and OUT tokens are issued by the host during the status stage of control reads.

Interrupt transactions are used to process interrupts from functions. A host periodically polls functions that may have pending interrupts, by sending an IN token. If the function has a pending interrupt it returns the interrupt information as a data packet. The host acknowledges appropriately. If the function doesn't have a pending interrupt it returns a NAK handshake. An interrupt transaction could also involve the host issuing an OUT token followed by a data packet to which the function would respond appropriately.

Isochronous transactions are used for the transfer of data such as voice or video, for which delays in transmitting the data cannot be tolerated, but an occasional error can be tolerated. Isochronous transactions consist of a token phase and a data phase, but no handshake phase and no reissuing of data packets. ■

## SystemVue™ Closing Application (Optional)

Insert the text CD in a computer having SystemVue™ installed and activate the program. Open the CD folder entitled **SystemVue Systems** and load the file entitled 11-1.

### Description of System

The system shown on the screen is the same one that was used in the chapter opening application. In this exercise, however, you will change the sample time to see what effect that has on the waveforms. You will also modify the data in the file entitled **ASCII_WOW.txt** that is available on the CD, and change the file name that is used by Token 1, the input token, and observe the effect.

First, left-click the **System Time** button. This is the button near the right end of the toolbar that looks like a clock. The **System Time Specification** window opens. The value in the **Sample Rate (Hz)** text box is 100. Change it to 1000 and left-click **OK.** Press F5 to run the simulation. You should see either in the box to the right of Sink 0 or in the Analysis window that the waveform has not changed. What has changed is how long it takes to run the simulation. Open the **System Time Specification** window again. When the sample rate is 100 Hz it takes 5.11 seconds to run the simulation because we are displaying 512 samples, one sample every 0.01 second. When the sample rate is 1000 Hz it takes only 0.511 seconds to run the simulation because we are displaying 512 samples, one every millisecond. This can be determined by looking at the Start Time and Stop Time text boxes in the System Time Specification window.

Now, open **Notepad** and create a new text file. Type a 1 and then press Enter. Repeat this nine more times, so that you have ten lines in the file, each containing a 1. You have entered a MARK at a data rate that is 1/10 that of the sample rate. Then type in −1 and press enter. Repeat this nine more times so that you have another ten lines, each containing a −1. You have entered a SPACE at a data rate that is 1/10 that of the sample rate. You can cut and paste MARKs and SPACEs until you have entered 1010100010111, including the first MARK and SPACE. This is the asynchronous serial coding for the uppercase *E*. The first 0 is the start bit. 1010001 is the hexidecimal 45 backwards, which is the ASCII code for *E*. The sixth 0 is the parity bit, and the 1 following it is the stop bit. The other two 1s are just marks in the idle channel. Save the file as **ASCII_E.txt.**

Now, right-click Token 1 and select **Edit Parameters** from the menu. Note that **Text** is selected in the **Open External Source File** window that appears. Left-click the **Select File** button. Find and select the **ASCII_E.txt** file and left-click **Open.** Ensure that **Repeat if EOF** is selected in the **Open External Source File** window and left-click **OK. Repeat if EOF** will ensure that the data you entered, which is less than the 512 samples the program needs, will provide the 512 samples. If you do not select **Repeat if EOF** you will get an error message telling you that you do not have enough samples and **Repeat if EOF** is not selected. You will be given the choice of padding the simulation with 0s, which means that you will see only one *E* in the waveform, or not running the simulation until you have selected **Repeat if EOF.**

Now, run the simulation by pressing F5 or left-clicking the **Run System** button.

## PROBLEMS

**11-1**    Code the decimal number 356 as BCD.

**11-2**    Code the decimal number 184 as BCD.

**11-3**    Code the decimal number 356 as XS3.

**11-4**    Code the decimal number 184 as XS3.

**11-5**    Write the ASCII code for the uppercase letter *D* with the most significant bit first.

**11-6**    Write the ASCII code for the uppercase letter *Q* with the most significant bit first.

**11-7**   Write the EBCDIC code for the uppercase letter *D* with the most significant bit first.

**11-8**   Write the EBCDIC code for the uppercase letter *Q* with the most significant bit first.

**11-9**   A computer has a 32-bit-wide data bus and a 66-MHz bus speed. How many bits per second can be transferred across this bus (i.e., what is the bandwidth of this bus)?

**11-10**  A computer has a 32-bit-wide data bus and a 100-MHz bus speed. How many bits per second can be transferred across this bus (i.e., what is the bandwidth of this bus)?

**11-11**  A microcontroller has an 8-bit-wide data bus and a 4-MHz bus speed. How many bits per second can be transferred across this bus (i.e., what is the bandwidth of this bus)?

**11-12**  A microcontroller has a 16-bit-wide data bus and a 4-MHz bus speed. How many bits per second can be transferred across this bus (i.e., what is the bandwidth of this bus)?

**11-13**  Based on the normal serial transmission format, start with a mark level, and draw a sketch similar to Figure 11–8 of the transmission of the uppercase letter *D*, including an odd parity bit and a stop bit.

**11-14**  Based on the normal serial transmission format, start with a mark level, and draw a sketch similar to Figure 11–8 of the transmission of the upper-case letter Q, including an odd parity bit and a stop bit.

*Note:* In Problems 11-15 through 11-18, assume that a 1 represents a mark and a 0 represents a space, and that *asynchronous* transmission is used. All characters are based on a start bit, the 7-bit ASCII code, an *odd* parity bit, and a stop bit. Remember that there may be long pauses between characters. Refer back to Figure 11–8 for the general format.

**11-15**  The asynchronous serial input of a computer has the following bit train:

````
----111101001101111111----
````

Identify the one character transmitted.

**11-16**  The asynchronous serial input of a computer has the following bit train:

````
----111101100101111111----
````

Identify the one character transmitted.

**11-17**  Decode the following asynchronous message:

````
--11010010010111--11000110010111--
11010101011111--11001101011111--
11010101011111--
````

**11-18**  Decode the following asynchronous message:

````
--11000010011111--11010100010111--
11000110010111--11000110010111--
11011110010111--
````

**11-19**  Using HDLC, station A is a primary station and station B is a secondary station. What command frame should station A issue to initialize an NRM communication link with station B? What should be station B's response?

**11-20**  Using HDLC, station A and station B are both combined stations. What frame should station A issue to initialize an ABM communication link with station B? What should be station B's response?

**11-21**  Using HDLC, station A has issued 3 consecutive I frames numbered 2, 3, and 4. Station B has no data to send to station A and is ready to receive further frames. What response should station B send?

**11-22**  Using HDLC, station A has issued 3 consecutive I frames numbered 1, 2, and 3. Station B has no data to send to station A, but its buffer is full and it is not ready to receive further frames yet. What response should station B send?

**11-23**  Using HDLC, station A is sending the following data to station B: 1001011010110. The CRC divisor is 1011. What will be in the FCS frame of the frame containing the data?

**11-24**  Using HDLC, station A is sending the following data to station B: 1001001010101. The CRC divisor is 1011. What will be in the FCS frame of the frame containing the data?

**11-25**  A scanner is connected to a printer using the USB bus. What tokens, packets, or handshakes would the computer send to the scanner to retrieve a scanned document from it, and how would the scanner respond?

**11-26**  A video camera is connected to a computer using the USB bus. What tokens, packets, or handshakes would the computer send to the video camera to input the video from the camera, and how would the video camera respond?

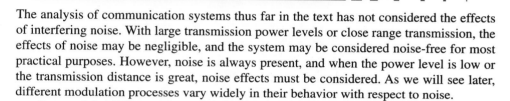

# Noise in Communication Systems 12

## OVERVIEW AND OBJECTIVES

The analysis of communication systems thus far in the text has not considered the effects of interfering noise. With large transmission power levels or close range transmission, the effects of noise may be negligible, and the system may be considered noise-free for most practical purposes. However, noise is always present, and when the power level is low or the transmission distance is great, noise effects must be considered. As we will see later, different modulation processes vary widely in their behavior with respect to noise.

Some of the different types of electrical noise will be surveyed early in this chapter. The techniques for analyzing noise effects will be developed through the use of thermal noise models. The concepts of effective noise temperature and noise figure will be introduced as a means of characterizing the overall noise present at the output of a receiver.

## Objectives

After completing this chapter, the reader should be able to:

1. List the different forms of *external* and *internal noise* and their properties.
2. Discuss *thermal noise* and its behavior.
3. State and apply the relationships for thermal noise voltage or current produced by a resistance, and draw the corresponding Thevenin and Norton equivalent circuit models.
4. Determine the noise voltage produced by an arbitrary combination of resistances.
5. Determine the *available noise power* produced by a resistance.
6. Determine the *noise power spectral density* produced by a resistance.
7. Utilize system gain to predict output noise power, voltage, and current.
8. Define *effective noise temperature* and apply it to determine the output noise produced by a system.
9. Define *noise figure,* both in absolute and decibel forms, and apply it to determine the output noise produced by a system.
10. Convert between *noise temperature* and *noise figure.*
11. Determine the noise temperature or noise figure for an attenuator or lossy component.
12. Determine the effective noise temperature or noise figure of a cascade of amplifier or attenuator stages.

## SystemVue™ Opening Application (Optional)

Insert the text CD in a computer having SystemVue™ installed and activate the program. Open the CD folder entitled **SystemVue Systems** and load the file entitled 12-1.

### Sink Token

| Number | Name | Token Monitored |
|--------|------|-----------------|
| 0 | Noise | 1 |

### Operational Token

| Number | Name | Function |
|--------|------|----------|
| 1 | AWGN | noise source |

The preset values for the run are from a starting time of 0 to a final time of 1 s with a time step of 1 ms. Run the simulation and observe the waveform at Sink 0.

### What You See

You are observing random noise having a gaussian statistical distribution. This type of noise arises in all communication systems. It is of special interest because of the very small signal levels encountered in communication systems, and because different types of modulation systems exhibit different types of behavior in the presence of noise. A term widely employed is *additive white gaussian noise,* abbreviated AWGN. As in the case of some earlier digital signals, the function on the screen is a segment of a pseudo-random signal. Different runs will produce different patterns but the statistical properties should remain the same.

The particular segment on the screen consists of 1001 points. The theoretical mean value is 0 and the theoretical rms value is 1.

### How This Demonstration Relates to Chapter 12 Learning Objectives

In this chapter you will learn the major properties of additive white gaussian noise as far as communication systems are concerned. You will learn the physical significance of the mean and rms values of noise. This material will set the stage for determining the behavior of different types of analog and digital communication systems in the presence of noise.

## 12-1  Noise Classifications

On the broadest scale, noise can be classified as either **external** or **internal.** Each category consists of several different types.

### External Noise

*External noise* represents all the different types that arise *outside* of the communication system components. It includes *atmospheric noise, galactic noise, human-made* noise, and *interference* from other communication sources.

## Internal Noise

*Internal noise* represents all the different types that arise *inside* of the communication system components. It includes *thermal noise, shot noise,* and *flicker noise.* Although the components of both the transmitter and receiver are included in the definition, the region of primary concern is from the receiving antenna through the first several stages of the receiver. It is in this region of small signal amplitudes that internal noise is most troublesome.

Some of the most common types of external and internal noise will be described in the remainder of this section.

## Atmospheric Noise

*Atmospheric noise* is produced mostly by lightning discharges in thunderstorms. It is usually the dominating external noise source in relatively quiet locations at frequencies below about 20 MHz or so. However, the power spectrum of atmospheric noise decreases rapidly as the frequency increases, and the effect becomes relatively insignificant at frequencies well above this value. The level of atmospheric noise also decreases with increasing latitude on the surface of the globe, and it is particularly severe during the rainy season in regions near the equator.

## Galactic Noise

*Galactic noise* is caused by disturbances originating outside the earth's atmosphere. The primary sources of galactic noise are the sun, background radiation along the galactic plane, and the many cosmic sources distributed along the galactic plane. The primary frequency range in which galactic noise is significant is from about 15 MHz to perhaps 500 MHz, and its power spectrum decreases with increasing frequency.

## Human-Made Noise

*Human-made noise* is somewhat obvious from its title, and consists of any source of electrical noise resulting from a human-made device or system. Among the chief offenders in this category are electric motors, automobile ignition systems, neon signs, and power lines. As one would likely suspect, the average level of human-made noise is significantly higher in urban areas than in rural areas. This fact has led to the selection of certain remote rural areas for the locations of many satellite tracking stations and radio astronomy observatories. The power spectrum of human-made noise decreases as the frequency increases, but the exact frequency range at which it becomes negligible is a function of its relative level. For example, in quite remote locations, the noise level from human-made sources will usually be below galactic noise in the frequency range from about 10 MHz or so.

## Interference

One can debate as to whether or not *interference from other communication sources* should be classified as "noise." However, it produces many of the same interfering effects, and can thus be classified as noise as far as the desired signal is concerned.

## Thermal Noise

*Thermal noise* is the result of the random motion of charged particles (usually electrons) in a conducting medium such as a resistor. Since all circuits necessarily contain resistive devices, thermal noise sources appear throughout all electrical circuits. The power spectrum of thermal noise is quite wide, and is essentially uniform over the RF spectrum of interest for most communication applications. Mathematical models for analyzing thermal noise are used either directly or indirectly for dealing with a variety of different types of noise.

Consequently, much of this chapter is devoted to dealing with the development and application of thermal noise models, so we will postpone further consideration until the next section.

### Shot Noise

*Shot noise* arises from the discrete nature of current flow in electronic devices such as transistors and tubes. For example, the electrons or holes crossing a semiconductor junction display a random variation of the time corresponding to the crossing, which in turn produces a random fluctuation of the current. The associated random variation in the current appears as a disturbance to the signal being processed by the device, and so the result is a form of noise. The power spectrum of shot noise is similar to that of thermal noise, and the two effects are usually lumped together for system analysis.

### Flicker Noise

*Flicker noise* (also called $1/f$ noise) is a somewhat vaguely understood form of noise occurring in active devices such as transistors at very low frequencies. It is most significant near dc and at a few hertz, and is usually negligible above about 1 kHz or so. Flicker noise is often a limiting factor for the minimum signal level that can be processed by a direct-coupled (dc) amplifier.

There are other forms of noise that are peculiar to certain types of modulation systems, which will be discussed as the need arises. For example, digital PCM systems exhibit an inherent noiselike uncertainty called *quantization noise.*

### Noise Pollution

There are many sources of pollution of which most people are aware—for example, air pollution and water pollution. However, another form of pollution of major interest to the communications engineer or technologist, of which many people are not aware, is that of *spectral pollution.* Because of the large number of electromagnetic transmission sources, a large amount of spectral background radiation exists. The available frequency spectrum is another natural resource that is rapidly being depleted by the increasing utilization of so many different types of communications systems.

Two widely employed acronyms of interest in discussing noise effects are **RFI** (radio frequency interference) and **EMC** (electromagnetic compatibility). The latter term relates to the process of ensuring that different electronic equipment can coexist in the same environment without unwanted interference between units. It is sufficiently important that some engineers are classified as *"EMC engineers."*

## 12-2 Thermal Noise

In basic circuit theory, a resistance is considered a passive device containing no energy. In reality, however, all resistive devices generate a small level of thermal noise as a result of the random motions of the electrons within the device. This effect is usually insignificant in applications in which the signal levels are moderate to large. However, in communication systems, signal levels at the antenna and within the first few stages of a receiver are often of the order of microvolts. Thermal noise may completely overshadow a small signal and render it completely unintelligible.

### Noise Properties

Consider the simple resistance $R$ shown in Figure 12–1(a). The waveform of part (b) illustrates the random behavior of the small thermal voltage existing across the terminals of the

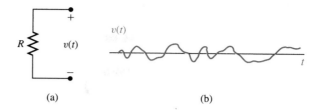

**FIGURE 12–1**

(a) Simple resistance, and (b) small random thermal noise voltage appearing across it.

resistance. The properties of the random thermal noise voltage have been studied extensively, both theoretically and practically, and the following properties have been deduced:

1. The *dc value* (*mean* or *average value*) of the voltage is zero; that is,

$$v_{dc} = 0 \qquad (12\text{-}1)$$

2. For a given set of operating conditions, the noise voltage may be described by an *rms* or *effective value* that can be used in describing or characterizing the voltage. This topic will be developed shortly.

3. The power spectrum of the noise before filtering is essentially constant over a wide frequency range, encompassing virtually all the range used in conventional RF communications.

4. The voltage is random in nature and its instantaneous behavior can only be described on a statistical basis. We will sidestep the use of statistics within the text, but the interested reader may refer to some of the more advanced references on communication theory or random variables for further details.

Because of the presence of a broad spectrum and the corresponding analogy with white light, thermal noise is frequently referred to as *white noise*.

## Mean-Square or RMS Noise

It turns out that the noise voltage generated by a resistance is a function of the resistance, the absolute temperature, and the bandwidth over which the noise is transferred to an external circuit. The bandwidth concept needs some further clarification, since the simple model shown in Figure 12–1 shows no bandwidth limiting parameters. However, in the actual transfer of noise power, there will be a finite bandwidth imposed by the receiver, amplifier, or instrument responding to the noise. For the moment, accept this notion, and further clarification of the exact meaning will be provided later.

Let $v_{rms}$ represent the *rms* value of the noise voltage and let $v_{rms}^2$ represent the square of the *rms* value. This latter quantity is known in statistics as the *mean-squared value*. The *mean-squared value* of the noise voltage is given by

$$v_{rms}^2 = 4RkTB \qquad (12\text{-}2)$$

The corresponding *rms* value is then

$$v_{rms} = \sqrt{v_{rms}^2} = \sqrt{4RkTB} \qquad (12\text{-}3)$$

The parameters in these equations are as follows:

$R$ = resistance in ohms

$k$ = Boltzmann's constant = $1.38 \times 10^{-23}$ joules per kelvin (J/K)

$T$ = absolute temperature in kelvins

$B$ = bandwidth in hertz

The absolute temperature $T$ in kelvins is determined from the more familiar Celsius temperature $T_C$ by the addition of 273 K; that is,

$$T = T_C + 273 \tag{12-4}$$

It should be stressed that this voltage is the value that would be measured under *open-circuit* conditions in a bandwidth $B$; in other words, there is no loading by the external circuit used to measure the voltage. Loading effects will be considered later.

## "Standard" Reference Temperature

While the preceding equations hold for any arbitrary temperature $T$, many noise calculations and measurements utilize a "standard" reference temperature of $T_0 = 290$ K. This value corresponds to 17°C or 62.6°F, and is widely used in system noise analysis and measurements. For the remainder of the text, the term "standard temperature" will refer to this value, and the symbol $T_0$ will be used.

For $T = T_0 = 290$ K, the product $kT_0 = 1.38 \times 10^{-23}$ J/K $\times$ 290 K $= 4 \times 10^{-21}$ J. With the additional factor of 4 in Equation 12-1, the expression for the mean-square noise voltage at $T = T_0$ reduces to

$$v_{rms}^2 = 16 \times 10^{-21} \, RB \tag{12-5}$$

---

**▮▮ EXAMPLE 12-1**

The purpose of this example is to illustrate typical values of the rms noise voltage for different values of resistance at a fixed bandwidth. Assume the bandwidth over which the noise is measured is 1 MHz and that the temperature is $T_0$. Determine the mean-square and rms values of voltage for the following values of $R$: (a) 1 k$\Omega$, (b) 100 k$\Omega$, and (c) 10 M$\Omega$.

**SOLUTION**   Since $T = T_0 = 290$ K, the simplified form of Equation 12-5 will be used.

(a) $R = 1 \, k\Omega$

$$v_{rms}^2 = 16 \times 10^{-21} \times 10^3 \times 10^6 = 16 \times 10^{-12} \text{ V}^2 \tag{12-6}$$

$$v_{rms} = \sqrt{16 \times 10^{-12}} = 4 \times 10^{-6} \text{ V} = 4 \ \mu\text{V} \tag{12-7}$$

(b) $R = 100 \, k\Omega$

$$v_{rms}^2 = 16 \times 10^{-21} \times 10^5 \times 10^6 = 16 \times 10^{-10} \text{ V}^2 \tag{12-8}$$

$$v_{rms} = \sqrt{16 \times 10^{-10}} = 4 \times 10^{-5} \text{ V} = 40 \ \mu\text{V} \tag{12-9}$$

(c) $R = 10 \, M\Omega$

$$v_{rms}^2 = 16 \times 10^{-21} \times 10^7 \times 10^6 = 16 \times 10^{-8} \text{ V}^2 \tag{12-10}$$

$$v_{rms} = \sqrt{16 \times 10^{-8}} = 4 \times 10^{-4} \text{ V} = 400 \ \mu\text{V} \tag{12-11}$$

Before the reader decides to visit an electronics laboratory to attempt to verify some of the preceding results, it should be noted that the measurement of a noise voltage is not a routine process. First, the typical values are quite small and well below the range of most voltmeters, except when resistance values and the bandwidth are very large. Second, if one attempts to amplify the noise to enhance the measurement, additional noise and stray pickup are introduced by the amplifier, which makes it difficult to separate the noise produced by the resistor. Third, the appropriate bandwidth parameter $B$ for the system must be known. Finally, a true rms instrument with a wide bandwidth is required. Special noise-measuring instrumentation systems are available to accomplish this task.

**▮▮ EXAMPLE 12-2**

Consider the 10-M$\Omega$ resistor of Example 12-1(c), and assume that it is connected across the input of an ideal noise-free amplifier with a voltage gain of 5000 and a bandwidth of 1 MHz. Assume that the input impedance of the amplifier is infinite so that there is no loading effect. Determine the rms value of the output noise voltage.

**SOLUTION**  Let $v_{\text{orms}}$ represent the rms value of the amplifier output voltage, which is determined by multiplying the input rms voltage by the voltage gain of 5000,

$$v_{\text{orms}} = 400 \times 10^{-6} \times 5000 = 2 \text{ V} \tag{12-12}$$

where the input voltage is 400 $\mu$V as given by Equation 12-11. This voltage is relatively large in comparison with many signal levels, and illustrates the potential difficulty with large resistances and wide bandwidths.

This example assumed an infinite input impedance for the amplifier in order to make a point about the noise voltage level. As we will see later, most communication circuits utilize matched impedances at input and output, and it is better to work with power transfer in those cases.

▮▮

## 12-3  Noise Produced by a Combination of Resistances

In the last section, the *open-circuit* noise voltage across a resistance was the quantity of interest. Suppose, however, that there is a loading effect produced by another resistance. This leads us to some general models for representing the random noise based on standard circuit theorems.

### Thevenin Noise Model

Both Thevenin's and Norton's theorems may be used to develop noise models for a resistance, although it is easier to apply Thevenin's theorem first, due to the fact that the noise has been stated so far in terms of voltage. In this case the mean-square open circuit voltage is computed over a hypothetical rectangular bandwidth $B$, as given by Equation 12-2. With this source deenergized, the equivalent impedance is readily determined to be simply the resistance $R$. Thus, the Thevenin equivalent circuit of the resistance plus thermal noise is shown in Figure 12–2(a). For noise analysis, it is best to work with mean-square voltage (rms value squared), so the usual plus and minus signs are irrelevant.

### Norton Noise Model

Norton's theorem may now be applied directly to the Thevenin model by measuring the short-circuit current. However, it is best with noise analysis to work with mean-square values, so let $i_{\text{rms}}$ represent the rms noise current, and let $i_{\text{rms}}^2$ represent the mean-square value of the current. Since we are working with squared quantities, we have

$$i_{\text{rms}}^2 = \frac{v_{\text{rms}}^2}{R^2} = \frac{4RkTB}{R^2} = \frac{4kTB}{R} = 4GTB \tag{12-13}$$

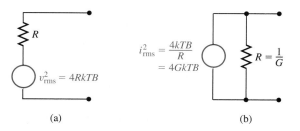

(a)                                                           (b)

**FIGURE 12–2**

(a) Thevenin and (b) Norton equivalent circuits for thermal noise sources associated with resistance.

where $G = 1/R$ is the conductance value in siemens associated with the resistance. The corresponding Norton equivalent circuit is shown in Figure 12–2(b). (Don't confuse the $G$ symbol for conductance with the fact that $G$ is also used to represent power gain. It should be clear in a given case which variable is under consideration.)

## Addition of Noise Voltages

The two equivalent circuits developed in this section can be used to simplify noise computations in circuits containing several resistors. One important rule should be remembered when combining the effects of sources contained in more than one resistor: *The net mean-square (or power) effect produced by more than one independent noise source is obtained by adding individual mean-square (or power) effects.* This concept is based on the fact that the random voltages produced by the individual resistances have no dependency between them. (In statistical terms, they are *statistically independent.*)

For example, suppose that the rms noise voltages produced by two resistances are 3 V and 4 V, respectively. If the resistances are connected in series, one might be led to believe that the net voltage would be 7 V, but that is *not* the case. Instead, the squared values are added; that is, $(3)^2 = 9\ V^2$ is added to $(4)^2 = 16\ V^2$ to yield 25 $V^2$. Taking the square root, the net rms voltage is 5 V.

The process just described is the same as employed in circuit theory for determining the net rms value of two sine waves at different frequencies. This is the main reason why it is easier to work with mean-square or power values in noise analysis, since they add directly.

## Combining Resistances

To illustrate some of the preceding concepts, consider the simple series connection of two resistors, both at the same temperature $T$, as shown in Figure 12–3(a). The two resistors can be represented by their Thevenin models as shown in Figure 12–3(b). In view of the simple series connection, both the respective resistances and the mean-square voltages can be added. Thus, the net resistance $R$ is given by

$$R = R_1 + R_2 \tag{12-14}$$

The net mean-square voltage $v_{rms}^2$ is expressed in terms of the separate mean-square voltages as

$$v_{rms}^2 = v_{1rms}^2 + v_{2rms}^2 = 4R_1kTB + 4R_2kTB = 4(R_1 + R_2)kTB = 4RkTB \tag{12-15}$$

A resulting equivalent circuit is shown in Figure 12–3(c).

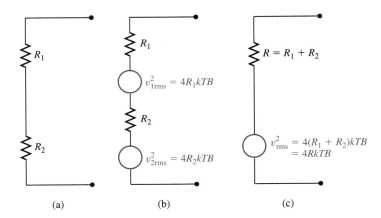

(a)          (b)          (c)

**FIGURE 12–3**
Series combination of two resistors, and equivalent models that represent the effect of thermal noise.

Several comments about this development will now be made. First, note in Equation 12-15 that the combined effect of the sources was obtained by adding *mean-square* values as previously discussed. Polarity is not important for this purpose since the mean-square values are all positive. Next, note that the net effective resistance is simply the sum of the individual resistances as noted in Equation 12-15, which is the equivalent resistance of two resistors in series. While this circuit represents a single simple example, it turns out that the concept is more general and can be stated as follows: *The net thermal noise effect of any arbitrary combination of simple resistances, all at the same temperature, is the same as that of a single resistance whose value is the equivalent resistance of the combination at the reference terminals of interest.* This concept reduces considerably the amount of effort involved with analyzing thermal noise in circuits containing several resistors. If the temperatures are different, it is necessary to utilize the Thevenin or Norton models and perform a more complex analysis.

---

**▌▌ EXAMPLE 12-3**

Consider the resistance combination shown in Figure 12–4 and assume that all resistors have a temperature $T = T_0$. Determine the net rms voltage across the terminals $A–A'$ over a bandwidth of 2 MHz.

**SOLUTION**   The first step in the problem is an exercise in simple dc circuit analysis, consisting of determining the equivalent resistance at the terminals. First, the parallel combination of 4 MΩ and 12 MΩ is determined as 3 MΩ. Next this value is series combined with 12 MΩ to yield 15 MΩ. This value is parallel combined with 10 MΩ to yield 6 MΩ. The net mean-square noise is then

$$v_{rms}^2 = 16 \times 10^{-21} R_{eq} B = 16 \times 10^{-21} \times 6 \times 10^6 \times 2 \times 10^6 = 192 \times 10^{-9} \text{ V}^2 \qquad (12\text{-}16)$$

The rms voltage is then

$$v_{rms} = \sqrt{192 \times 10^{-9}} = 438.2 \ \mu\text{V} \qquad (12\text{-}17)$$

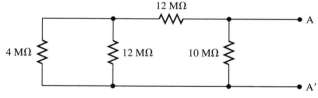

**FIGURE 12–4**
Circuit of Example 12-3.

---

## 12-4 Noise Power

The emphasis on noise analysis up to this point has been on noise voltage and current, since these are the variables of common familiarity. However, most communication system components and subsystems are designed around the concept of maximum power transfer. Not only does this usually ensure maximum signal strength, but it also serves to minimize reflections produced by interconnecting sections of transmission lines. Beginning in this section and continuing throughout the chapter, the focus will be primarily aimed at the process of signal and noise power transfer, as opposed to voltage and current considerations.

### Available Power

All practical signal sources have an internal impedance that will limit the amount of power that can be extracted from the source. Consider the source of Figure 12–5, in which the

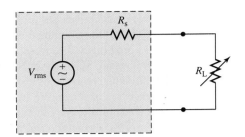

**FIGURE 12–5**

Concept of available power from source.

Thevenin equivalent circuit is a voltage source with rms value $V_{rms}$ in series with a resistive impedance $R_S$. It is important to realize in this model that $R_S$ is *internal* and cannot be changed or eliminated. Under this constraint, an *external* load resistance $R_L$ is connected to the terminals and adjusted for maximum power transfer to the load.

From the maximum power transfer theorem, maximum power is delivered to $R_L$ when its value is equal to the internal source resistance; that is, when $R_L = R_S$. For this condition, the current around the loop is

$$I = \frac{V_{rms}}{R_S + R_L} = \frac{V_{rms}}{R_S + R_S} = \frac{V_{rms}}{2R_S} \tag{12-18}$$

The power $P_L$ in the load resistance $R_L$ is

$$P_L = I^2 R_L = \left(\frac{V_{rms}}{2R_S}\right)^2 R_L = \left(\frac{V_{rms}}{2R_S}\right)^2 R_S = \frac{V_{rms}^2}{4R_S} \tag{12-19}$$

This value of power is the maximum amount that can extracted from the source into an external load, and is referred to as the *maximum available power* $P_{av}$. Hence,

$$P_{av} = \frac{V_{rms}^2}{4R_S} \tag{12-20}$$

This value is the *maximum* average power that can be extracted from the source into an external load, and that power will be delivered to the load only when the load resistance is matched to the internal resistance of the source. It should be noted that there is additional power dissipated in the internal resistance, which, at the point of maximum power transfer, is equal to the power delivered to the external load.

## Available Noise Power

Consider now a resistance $R$ at temperature $T$ and its equivalent noise circuit, as shown in Figure 12–6. Assume that an external load $R_L$ is connected to $R$ as shown. When $R_L$ is connected to $R$, some of the available noise power from $R$ is transferred to $R_L$. Maximum power will be transferred when $R_L = R$. Let $N_{av}$ represent the *available noise power,* which is

$$N_{av} = \frac{v_{rms}^2}{4R} = \frac{4RkTB}{4R} = kTB \tag{12-21}$$

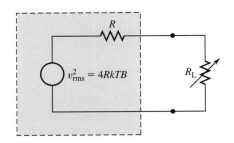

**FIGURE 12–6**

Circuit used to establish available noise power from resistance.

This result is very interesting in that *the available noise power is completely independent of the value of the resistance!* Thus, while the noise voltage and noise current are both dependent on the resistance, the available noise power is independent of the resistance. Moreover, this available power will be delivered to the load resistance if the load resistance is equal to *R*.

It should be noted that for stable thermal equillibrium, the load resistance transfers an equal amount of power back to the internal resistance. However, our concern here is with the power transferred to the load.

For $T = T_0 = 290$ K, Equation 12-21 may be expressed as

$$N_{av} = kT_0B = 4 \times 10^{-21}B \tag{12-22}$$

A result of the preceding development is that the analysis of noise transfer through a system is greatly simplified when all junctions in the system are matched. Instead of having to work with the somewhat clumsy forms of mean-square voltage and current, one can simply compute the available power *kTB* and assume that value as the basis for noise power transfer.

In many of the subsequent developments, matched conditions will be assumed and stated. Such conditions are widely assumed by communication engineers and technologists, since this is part of the design strategy.

## Power Gain

The power gain *G* of a system can be defined as

$$G = \frac{\text{output power}}{\text{input power}} \tag{12-23}$$

There are several variations on power gain definitions, depending primarily on whether the input and output ports are terminated in matched impedances. To simplify the development in this text, *we will assume, unless otherwise stated, that both input and output ports are terminated correctly for maximum power transfer.* This means that the available power from a source will be delivered to the input of the system and the available output power from the system is delivered to the load. As we will see, this assumption greatly simplifies the analysis of the system, and it is usually a condition around which much of the design is aimed. We will not need to use the conductance parameter in subsequent developments, so unless otherwise stated, the symbol *G* with appropriate subscripts will denote a power gain based on both input and output ports matched for maximum power transfer.

Let $P_i$ represent the input power and let $P_o$ represent the output power. The output power is

$$P_o = GP_i \tag{12-24}$$

where both power values represent maximum values based on the assumed conditions. When the input is a noise source with temperature *T*, the output noise power $N_o$ is then

$$N_o = GN_{av} = GkTB \tag{12-25}$$

where *B* is the bandwidth of the system.

---

**▌▌ EXAMPLE 12-4**

Determine the available noise power contained in a simple 50-$\Omega$ resistance at the standard reference temperature in a bandwidth of 2 MHz.

**SOLUTION**   Since the temperature is at the standard reference level, the form of Equation 12-22 will be used.

$$N_{av} = kT_0B = 4 \times 10^{-21} \times 2 \times 10^6 = 8 \times 10^{-15} \text{ W} = 8 \text{ fW} \tag{12-26}$$

Note that it was not necessary to specify the resistance value for the determination of available power, but the resistance value will be important in a later example.

The resistance of Example 12-4 is connected across the input of an ideal noise-free amplifier whose input and output impedances have resistive values of 50 $\Omega$. The amplifier has a matched gain of 60 dB and a bandwidth of 2 MHz. Determine the output noise power in a 50-$\Omega$ load resistance.

**SOLUTION**   The amplifier is obviously matched at both input and output. The decibel gain of 60 dB corresponds to an absolute power gain of $10^6$. The output noise power $N_0$ is

$$N_o = GN_{av} = GkT_0B = 10^6 \times 8 \times 10^{-15} = 8 \times 10^{-9} \text{ W} = 8 \text{ nW} \qquad (12\text{-}27)$$

where the result of Equation 12-26 was used as the available input power

Determine the rms noise voltage across the load resistance in the amplifier of Example 12-5.

**SOLUTION**   One might be initially tempted to return to the basic formula of Equation 12-3, but note that what is desired is the voltage *across* the load resistance resulting from the input source. Moreover, Equation 12-3 is based on an *open-circuit voltage,* while the present situation involves a terminated resistance.

The desired result can be achieved by reverting back to basic circuit analysis, and expressing the power in the load in terms of the rms voltage across the load and the load resistance. We have

$$N_o = 8 \times 10^{-9} = \frac{v_{rms}^2}{50} \qquad (12\text{-}28)$$

Solving for $v_{rms}$, we obtain

$$v_{rms} = 632.5 \ \mu\text{V} \qquad (12\text{-}29)$$

## 12-5 Power Spectrum Concepts

In all computations made thus far, a bandwidth $B$ has been assumed. In practice, the noise spectrum will be altered by the frequency response of the system, and the value of the bandwidth used for noise purposes may not be immediately evident. In this section, this phenomenon will be investigated, and the process by which an equivalent noise bandwidth is established will be described.

### Power Spectral Density

The power spectral density is based on a spectral representation of the power and is defined at any frequency as the power per unit bandwidth measured in watts per hertz (W/Hz). It is possible to use either a one-sided or a two-sided spectral form. However, the treatment here will utilize the *one-sided* form, since it relates more to the practical interpretation of the concept for our purposes.

In general, the power spectral density is a frequency-dependent function and will be denoted $S(f)$, with appropriate subscripts added as needed. For the case of a simple resistance with available noise power *kTB,* the power spectral density is easily obtained by dividing by $B$, and the resulting value will be denoted $\eta$. Hence, for a simple resistance

$$S(f) = \eta = kT \quad \text{W/Hz} \qquad (12\text{-}30)$$

as shown in Figure 12–7. At the reference temperature $T = T_0 = 290$ K, this value is

$$S(f) = \eta_0 = kT_0 = 4 \times 10^{-21} \text{ W/Hz} \qquad (12\text{-}31)$$

where additional subscripts have been added for clarity.

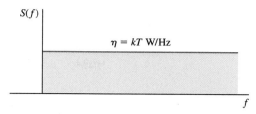

**FIGURE 12–7**
Power spectral density of unfiltered white noise on a one-sided basis.

## Common Alternate Notation

A large number of references use either $N_0$ or $N_o$ to represent a constant value of noise spectral density, and the author was very tempted to do the same. The problem is that noise spectral density is measured in watts/hertz, and many of the references also use $N$ with various subscripts to represent total noise power in watts. To avoid the potential confusion, we will use $\eta$ for the density value in watts/hertz, and $N$ with various subscripts for total noise power in watts. Just be aware in looking at other references that you may encounter $N_0$ or $N_o$ as noise density forms.

## Input and Output Spectral Densities

In some cases, it is desirable to treat power spectral density functions as the input and output variables for an amplifier or other linear system. Let $S_i(f)$ represent the input power spectral density, and let $S_o(f)$ represent the corresponding output density. For a constant power gain $G$, these functions are related in exactly the same fashion as the power values; that is,

$$S_o(f) = GS_i(f) \qquad (12\text{-}32)$$

One situation in which spectral density functions are appropriate is when the bandwidths of different stages have different values. In such situations, not all of the available power at the input is amplified and delivered to the load. However, the spectral density functions may still serve as a meaningful way to relate output to input.

## Equivalent Noise Bandwidth

For nonideal filters, a term called the *equivalent noise bandwidth* is used to predict the output noise power produced by a flat input spectrum. Assume a nonideal amplitude response $A(f) = |H(f)|$, and let $A^2(f) = |H(f)|^2$, as shown in Figure 12–8. Let $A_0^2$ represent the maximum level of the amplitude-squared response, which for the case shown is at dc.

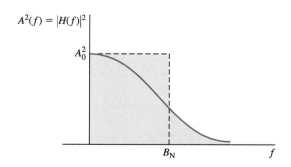

**FIGURE 12–8**
Illustration of equivalent noise bandwidth.

Table 12–1    Equivalent Noise Bandwidths of Butterworth Filters

| Poles | 1 | 2 | 3 | 4 | 5 | 6 | 7 | 8 | 9 | 10 |
|---|---|---|---|---|---|---|---|---|---|---|
| $B_N/f_1$ | 1.571 | 1.111 | 1.047 | 1.026 | 1.017 | 1.012 | 1.008 | 1.006 | 1.005 | 1.004 |

A fictitious ideal block characteristic having the same maximum level as the actual filter amplitude response is sketched on the same scale as the actual filter response. The *equivalent noise bandwidth* $B_N$ is the bandwidth of the ideal block response that would produce the same output noise power as the real filter. It can be shown that this value is

$$B_N = \frac{1}{A_0^2} \int_0^\infty A^2(f)\, df \qquad (12\text{-}33)$$

This is the quantity that has been denoted simply as $B$ in all computations involving noise power up to this point. To keep the notation as simple as possible, we will continue to use $B$, but it should be understood that it is the noise bandwidth that is being used for noise computations.

An obvious question is how $B_N$ compares with the usual bandwidth parameter for a given amplifier or other linear system. For circuits with a low to moderate rate of cutoff in the amplitude response, the equivalent noise bandwidth may be much greater than, say, the 3-dB bandwidth. As the order of the filter increases, the equivalent noise bandwidth becomes much closer to the usual bandwidth value. In the theoretical limit of an ideal block amplitude response, the two values are equal.

An abbreviated set of noise bandwidth values for low-pass Butterworth filters up to 10 poles is provided in Table 12–1. In each case, $f_c$ is the 3-dB bandwidth.

## Noise Temperature

We have seen that the available noise power density for a resistive source can be expressed simply as $kT$. In the context of a resistance $R$, the temperature $T$ is the physical temperature. The fact is that there are many noise sources that are not simple resistances. An antenna, for example, may absorb different forms of radiation within its field pattern and produce an output noise much greater than the value predicted by the $kT$ factor based on physical temperature alone. Likewise, an antenna pointed out into deep space may absorb an equivalent noise near zero.

It has been customary in the communications field to represent noise sources by an *equivalent noise temperature* that may or may not represent a true physical temperature. For any noise source that produces a flat noise power spectrum $\eta_s$, an equivalent noise source temperature $T_s$ may be defined as

$$T_s = \frac{\eta_s}{k} \qquad (12\text{-}34)$$

Once $T_s$ is known, the available noise power density may always be computed simply as

$$\eta_s = kT_s \qquad (12\text{-}35)$$

---

**▌▌ EXAMPLE 12-7**

A simple 50-$\Omega$ resistance at a temperature of 290 K is connected across the 50-$\Omega$ input of a noise-free amplifier having a matched gain of 80 dB. Determine the power spectral density at the output.

**SOLUTION**   A decibel gain of 80 dB corresponds to an absolute power gain of $10^8$. From Equation 12-31, the input power spectral density is simply $4 \times 10^{-21}$ W/Hz. The output spectral density is then

$$S_o(f) = GkT_0 = 10^8 \times 4 \times 10^{-21} = 4 \times 10^{-13} \text{ W/Hz} = 400 \text{ fW/Hz} \qquad (12\text{-}36)$$

**EXAMPLE 12-8**

A noise generator produces white noise having a power spectral density of $6 \times 10^{-18}$ W/Hz. (a) Determine the equivalent source noise temperature. (b) If the generator is connected as a matched input to a noise-free amplifier having a flat gain of 43 dB over a wide band, determine the output power spectral density over the flat region. (c) If the noise bandwidth of the amplifier is 12 MHz, determine the output noise power due to the source.

**SOLUTION**

(a)  The equivalent source temperature is determined from Equation 12-34 as

$$T_s = \frac{\eta_s}{k} = \frac{6 \times 10^{-18}}{1.38 \times 10^{-23}} = 434.8 \times 10^3 \text{ K} = 434{,}800 \text{ K} \qquad (12\text{-}37)$$

It is obvious that this is not a real physical temperature!

(b)  The absolute power gain of the amplifier is $20 \times 10^3$, and the output power spectral density over the flat region is

$$S_o(f) = \eta_o = G S_i(f) = 20 \times 10^3 \times 6 \times 10^{-18}$$

$$= 120 \times 10^{-15} \text{ W/Hz} = 120 \text{ fW/Hz} \qquad (12\text{-}38)$$

(c)  The total output noise power $N_o$ in a bandwidth of 12 MHz is

$$N_o = \eta_o B = 120 \times 10^{-15} \times 12 \times 10^6 = 1.44 \times 10^{-6} \text{ W} = 1.44 \text{ } \mu\text{W} \qquad (12\text{-}39)$$

**EXAMPLE 12-9**

Consider the system shown in Figure 12–9, and assume that impedances at all junctions are matched. The input is a thermal noise source having a one-sided power spectral density of 1 pW/Hz. Assume that any internal noise is negligible in comparison to the noise produced by the source. The bandwidths are all low-pass in nature and have different values for the three stages. Determine the output noise power.

**SOLUTION**    The net decibel gain $G_{dB}$ is readily determined as

$$G_{dB} = 10 + 15 + 25 = 50 \text{ dB} \qquad (12\text{-}40)$$

This corresponds to an absolute power gain $G = 10^5$. The bandwidths of the three stages are different but are all low-pass in nature. Therefore, the *smallest* bandwidth, 10 kHz, determines the effective bandwidth for the noise delivered to the output. Thus,

$$N_o = \eta_i G B = 10^5 \times 10^{-12} \times 10^4 = 1 \text{ mW} \qquad (12\text{-}41)$$

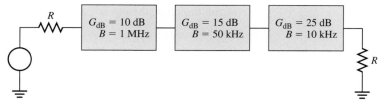

**FIGURE 12–9**
System of Example 12-9.

## 12-6  Models for Internally Generated Noise

All of the noise sources within an amplifier or other subsystem can be collectively specified as a single noise parameter, which can be used to predict the noise performance of the system. The two most widely used approaches for specifying the total noise are (1) the

**noise temperature method** and (2) the **noise figure method.** Often, specifications are given that utilize both parameters, so it is necessary to investigate both methods. The author has a preference for the noise temperature approach, and it will be considered first.

## Effective Noise Temperature Method

Consider the block diagram of an amplifier as shown in Figure 12–10, having an equivalent noise bandwidth $B$. Assume that both input and output ports are matched for maximum power transfer, and that the power gain is $G$. Assume that the resistance across the input port has a temperature $T_i$.

The total output noise power $N_o$ can be considered as the sum of two terms:

$$N_o = N_{o1} + N_{o2} \qquad (12\text{-}42)$$

The first term $N_{o1}$ represents the noise resulting from the input source, and the second term $N_{o2}$ represents the noise arising *within* the amplifier. A technique that is widely used to specify this second noise term is to define a fictitious *effective temperature* $T_e$ at the *input* to the amplifier that would produce the noise power arising within the amplifier. Such a temperature would then satisfy the equation

$$N_{o2} = GkT_e B \qquad (12\text{-}43)$$

The total output noise power is then

$$N_o = GkT_i B + GkT_e B \qquad (12\text{-}44a)$$
$$= Gk(T_i + T_e)B \qquad (12\text{-}44b)$$
$$= GkT_{sys}B \qquad (12\text{-}44c)$$

where

$$T_{sys} = T_i + T_e \qquad (12\text{-}45)$$

is called the *system temperature*. (It is also called the *operating temperature* in some references, in which case the symbol $T_{op}$ may be used.)

The process of determining the output noise power for an amplifier can be summarized as follows:

1. The system temperature is determined by adding the input source noise temperature to the amplifier effective noise temperature referred to the input.

2. The system temperature is treated in the same manner as the physical temperature of a resistance at the input (although it will often be drastically different than any real physical temperature). Thus, the equivalent available noise power at the input $kT_{sys}B$ is multiplied by the power gain $G$ of the amplifier to determine the output noise power.

It should be stressed again that the effective noise temperature of the amplifier is *not* the physical temperature of the unit. It is simply a number that is treated like a physical temperature for the purpose of computing noise power. In fact, even the source temperature $T_i$ need not correspond to a physical temperature. Source and amplifier noise temperatures vary from a few kelvins to thousands of kelvins.

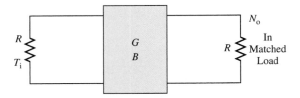

**FIGURE 12–10**
Block diagram used in defining noise temperature.

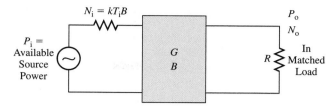

**FIGURE 12–11**
Block diagram used in defining noise figure.

## Noise Figure Method

Next, we consider the concept of the *noise figure F* (also called the *noise factor*). There are a number of variations on the definition of noise figure in the literature, depending on whether it is defined on a frequency dependent (or "spot") basis or on an overall average basis. In accordance with most manufacturers' specifications, we will consider only the average basis representation.

Refer to Figure 12–11, which is very similar to Figure 12–10 except that a few additional quantities are labeled on this figure. Along with the noise source input, a signal input source is also assumed. The available power corresponding to the signal input source is denoted $P_i$, and the available signal output power is denoted $P_o$. The output noise power $N_o$ is the same as in the noise temperature development, and $N_i = kT_iB$ is the input available noise source power defined over the equivalent noise bandwidth.

The following signal-to-noise ratios are defined:

$$(S/N)_{\text{input}} = \frac{P_i}{N_i} = \text{signal-to-noise power ratio of input source} \qquad (12\text{-}46)$$

$$(S/N)_{\text{output}} = \frac{P_o}{N_o} = \text{signal-to-noise power ratio of output} \qquad (12\text{-}47)$$

The *noise figure F* is then defined as

$$F = \frac{(S/N)_{\text{input}}}{(S/N)_{\text{output}}}\bigg]_{T_i=T_0} \qquad (12\text{-}48)$$

Stated in words, *the noise figure is the input source signal-to-noise ratio divided by the output signal-to-noise ratio determined with the input at the standard reference temperature of $T_0 = 290$ K.*

The noise figure is often stated in decibels, and the value $F_{\text{dB}}$ is given by

$$F_{\text{dB}} = 10 \log F \qquad (12\text{-}49)$$

Since both input and output power levels are measured over the same reference bandwidth, the smallest possible value of the noise figure is $F = 1$ or $F_{\text{dB}} = 0$ dB, which would correspond to a noise-free device. In general, the larger the value of the noise figure (either absolute or in dB), the more the degradation imposed by the unit.

## Use of Noise Figure at Reference Temperature

Assume for the moment that the input signal source has the reference temperature $T_i = T_0$. In this case, Equation 12-48 may be rearranged as

$$(S/N)_{\text{output}} = \frac{(S/N)_{\text{input}}}{F} \qquad (12\text{-}50)$$

The decibel form of Equation 12-50 is obtained by taking the logarithms of both sides and multiplying by 10. This leads to

$$(S/N)_{\text{output,dB}} = (S/N)_{\text{input,dB}} - F_{\text{dB}} \qquad (12\text{-}51)$$

## Misuse of Noise Figure in Literature

The preceding two equations are easy to apply and they clearly illustrate the degradation imposed by the noise figure. Unfortunately, however, they are often misused. Since the noise figure is defined at the standard reference temperature, these equations are correct only when the input source temperature is equal to the standard reference temperature. Throughout the literature, there are many examples in which these equations are used at different source temperatures, and this has led to a lot of confusion.

Provided that the temperature is not too different from the reference temperature, Equations 12-50 and 12-51 may be used as reasonable approximations. However, the potential error that can arise when the source temperature is drastically different from the standard reference temperature is the primary reason that the author prefers the noise temperature approach.

## Relationship between Noise Temperature and Noise Figure

We will now develop a relationship between the noise temperature and the noise figure, such that if one is known, the other may be determined. Start with Equation 12-48, insert the definitions of Equations 12-46 and 12-47, and rearrange as follows:

$$F = \left. \frac{P_i}{P_o} \frac{N_o}{N_i} \right]_{T_i = T_0} = \frac{N_o}{GkT_0 B} \tag{12-52}$$

where the ratio $P_o / P_i = G$ and the value $T_i = T_0$ have been substituted. Next, substitute the expression of Equation 12-44b for $N_o$ in Equation 12-52. This results in

$$F = \frac{GkT_0 B + GkT_e B}{GkT_0 B} \tag{12-53}$$

After cancellation of the surplus common factors, we obtain

$$F = \frac{T_0 + T_e}{T_0} = 1 + \frac{T_e}{T_0} = 1 + \frac{T_e}{290} \tag{12-54}$$

The inverse relationship is determined by solving for $T_e$ in terms of $F$.

$$T_e = (F - 1)T_0 = (F - 1) \times 290 \tag{12-55}$$

The preceding two relationships permit conversion from noise temperature to noise figure, and vice versa.

---

**▌▌ EXAMPLE 12-10**

A low-noise amplifier has an effective noise temperature of 50 K. Determine (a) the absolute noise figure and (b) the decibel noise figure.

**SOLUTION**

(a)  To convert noise temperature to noise figure, we use Equation 12-54:

$$F = \frac{T_0 + T_e}{T_0} = 1 + \frac{T_e}{T_0} = 1 + \frac{T_e}{290} = 1 + \frac{50}{290} = 1.172 \tag{12-56}$$

(b)  The decibel value is

$$F_{dB} = 10 \log F = 10 \log 1.172 = 0.689 \text{ dB} \tag{12-57}$$

---

**▌▌ EXAMPLE 12-11**

An amplifier has a specified noise figure of 5 dB. Determine the effective noise temperature referred to the input.

**SOLUTION**    We must first convert the decibel noise figure to the absolute value. We have

$$F = 10^{F_{dB}/10} = 10^{5/10} = 10^{0.5} = 3.162 \tag{12-58}$$

To convert from absolute noise figure to noise temperature, we use Equation 12-55:

$$T_e = (F - 1)T_o = (F - 1) \times 290 = (3.162 - 1) \times 290 = 627 \text{ K} \qquad (12\text{-}59)$$

**▌▌ EXAMPLE 12-12**

An RF amplifier has a matched gain of 50 dB, a noise figure of 9 dB, and an equivalent noise bandwidth of 2 MHz. The input signal level is 8 pW and the input source effective noise temperature is $T_i = T_0 = 290$ K. Using the noise temperature approach, determine (a) input source noise power, (b) input signal-to-noise ratio, (c) output signal power, (d) output noise power, and (e) output signal-to-noise ratio.

**SOLUTION**   As is often the case with practical problems, the values are given in decibels, and they must be converted to absolute quantities before proceeding. The given decibel and absolute values are determined as in the following table.

|  | Decibel Value | Absolute Value |
|---|---|---|
| Amplifier Gain | 50 dB | $10^5$ |
| Amplifier Noise Figure | 9 dB | 8 |

To use the noise temperature approach, we first determine the effective noise temperature of the amplifier:

$$T_e = (F - 1) \times 290 = (8 - 1) \times 290 = 2030 \text{ K} \qquad (12\text{-}60)$$

(a)  The input signal source has a temperature equal to the standard reference temperature of 290 K, in which case the available noise power is

$$N_i = kT_i B = kT_0 B = 4 \times 10^{-21} \times 2 \times 10^6 = 8 \times 10^{-15} \text{ W} = 8 \text{ fW} \qquad (12\text{-}61)$$

(b)  The input source signal-to-noise ratio is

$$(S/N)_{input} = \frac{P_i}{N_i} = \frac{8 \times 10^{-12}}{8 \times 10^{-15}} = 1000 \qquad (12\text{-}62)$$

The corresponding decibel value is

$$(S/N)_{input,dB} = 10 \log(1000) = 30 \text{ dB} \qquad (12\text{-}63)$$

(c)  The output signal power is given by

$$P_o = G P_i = 10^5 \times 8 \times 10^{-12} = 800 \text{ nW} \qquad (12\text{-}64)$$

(d)  The output noise power is given by

$$N_o = G k T_{sys} B$$
$$= Gk(T_i + T_e) = 10^5 \times 1.38 \times 10^{-23} \times (290 + 2030) \times 2 \times 10^6$$
$$= 6.4 \text{ nW} \qquad (12\text{-}65)$$

(e)  The output signal-to-noise ratio is then

$$(S/N)_{output} = \frac{800 \text{ nW}}{6.4 \text{ nW}} = 125 \qquad (12\text{-}66)$$

The corresponding dB value is

$$(S/N)_{output,dB} = 10 \log(125) = 21 \text{ dB} \qquad (12\text{-}67)$$

The signal-to-noise ratio is thus degraded by 9 dB.

▌▌ EXAMPLE 12-13

For the system of Example 12-12, use the noise figure approach to predict the output signal-to-noise ratio.

SOLUTION   Many of the results obtained in Example 12-12 will be used in this analysis. The input source signal-to-noise ratio was determined as 1000 on an absolute scale or 30 dB on a decibel scale. The output signal-to-noise ratio can then be determined most easily by Equation 12-51. This relationship yields

$$(S/N)_{\text{output,dB}} = (S/N)_{\text{input,dB}} - F_{\text{dB}} = 30 - 9 = 21 \text{ dB} \tag{12-68}$$

This result is exactly the same as obtained using the noise temperature approach, and this is a result of the fact that the source temperature is exactly the same as used in the noise figure definition. If the input source noise temperature is different than 290 K, the only truly correct approach is the noise temperature method. In practice, however, the method illustrated in this example is used casually because of its simplicity, and it yields approximate results whenever the source temperature is reasonably close to 290 K.   ▌▌

## 12-7   Noise of Cascaded Systems

In the previous section, the concepts of effective noise temperature and noise figure were established as a basis for representing the noise within a single amplifier. In this section, the concept will be extended to include any number of individual amplifiers (or possibly attenuators) connected in cascade, each of which has an individual noise temperature or figure. The ultimate goal is to determine a net noise temperature or figure that applies to the whole cascaded system, so that it may be treated in the same manner as a single unit.

### Cascade of Three Units

The concept will be developed for the case of a system with three cascaded units. This is sufficiently large to permit the concept to be generalized, and yet it is small enough to permit a straightforward solution. Consider then the system shown in Figure 12–12, containing three amplifiers and an input noise source with temperature $T_i$. It is assumed that the three effective noise temperatures of the individual stages are $T_{e1}$, $T_{e2}$, and $T_{e3}$, respectively. The corresponding matched power gains are $G_1$, $G_2$, and $G_3$. The net gain from input to output is

$$G = G_1 G_2 G_3 \tag{12-69}$$

The actual noise output $N_o$ can be represented as the sum of several separate noise effects as follows: (1) the input noise source amplified by the total gain $G$; (2) the effective noise produced by the first amplifier referred back to its input, amplified by the total gain $G$; (3) the effective noise produced by the second stage referred back to its input, multiplied by the gain of the last two stages $G_2 G_3$; and (4) the effective noise of the last stage referred back to its input, multiplied by the gain of the last stage $G_3$. In the same order as just listed, the output noise can be expressed as

$$N_o = G_1 G_2 G_3 k T_i B + G_1 G_2 G_3 k T_{e1} B + G_2 G_3 k T_{e2} B + G_3 k T_{e3} B \tag{12-70}$$

where the net gain defined in Equation 12-69 is used in the first two terms on the right.

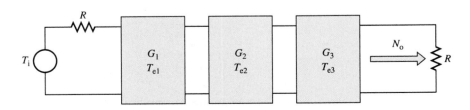

FIGURE 12–12
Three amplifiers in cascade, whose total noise effects are to be determined.

### Effective Input Noise Temperature

We now define an effective noise temperature $T_e$ for the cascade that will produce the noise effect for the system with gain $G$. This effective noise temperature must be the value that, when added to the source temperature at the input, produces the effect at the output as given by Equation 12-70. The equation will read

$$N_o = GkT_{sys}B = G_1G_2G_3k(T_i + T_e)B \qquad (12\text{-}71)$$

Equating the previous two equations and canceling common factors, we obtain

$$T_e = T_{e1} + \frac{T_{e2}}{G_1} + \frac{T_{e3}}{G_1G_2} \qquad (12\text{-}72)$$

This result is a single effective noise temperature *referred to the input of the first stage* that, in combination with the input source noise temperature, can be considered the net system temperature for predicting the output noise.

The form of Equation 12-72 can be readily generalized to the case of an arbitrary number of stages in cascade. The first term is the noise temperature of the input stage. The second term is the noise temperature of the second stage divided by the gain up to that input, which is the gain of the first stage. The third term is the noise temperature of the third stage divided by the gain up to that input, which is the product of the gains of the first two stages. In general, the additive term for any stage is the noise temperature of that stage divided by the product of all preceding gains up to, but not including, that particular stage.

### Most Critical Stages

For the moment, assume that all values of gain are greater than unity. From Equation 12-72, it is apparent that the effects of terms farther to the right are less significant, since the temperature values are being divided by successively increasing gain factors. This means that *in a cascade of amplifier stages, the input stage is usually the most important one in establishing the net noise contribution for the system.* In fact, if all the gain factors (especially $G_1$) are very large, the total noise temperature may be only slightly higher than that of the first stage. Therefore, *the first stage of a receiver should be selected to have the combination of a low noise temperature (or noise figure) and a high gain, whenever feasible.*

It should be noted that any unit that provides a loss, such as a transmission line, attenuator, or passive mixer, is treated as a gain less than 1, and this tends to increase the noise temperature. This effect will be considered shortly.

### Combined Noise Figure

Since a combined effective noise temperature can be obtained from the cascaded system, it is also possible to obtain a combined noise figure from the individual noise figures. The most direct way to accomplish this is to take Equation 12-72 and substitute for each noise temperature the corresponding expression for the noise figure. Let $F_1$, $F_2$, and $F_3$ represent the three noise figures corresponding to $T_{e1}$, $T_{e2}$, and $T_{e3}$, respectively. Using Equation 12-55 as the basis in each case, we have

$$(F - 1)T_0 = (F_1 - 1)T_0 + \frac{(F_2 - 1)T_0}{G_1} + \frac{(F_3 - 1)T_0}{G_1G_2} \qquad (12\text{-}73)$$

After cancellation and some rearrangement, we obtain

$$F = F_1 + \frac{F_2 - 1}{G_1} + \frac{F_3 - 1}{G_1G_2} \qquad (12\text{-}74)$$

A form for noise figure is obtained, similar to Equation 12-72 for the noise temperature. However, it is noted that all $F$ terms on the right except the $F_1$ term have unity subtracted from the value. There is an interesting reason for this, and it is based on the fact that the noise figure definition includes both the source noise and the amplifier noise in its definition. Since the source is applied only to the first stage, all noise figures except $F_1$ must be reduced to compensate.

## Noise Temperature of Matched Attenuator

An important topic of consideration is the effect of an attenuation element on the overall noise. For the purpose of this development, the term *matched attenuation network* will include any network having an input and an output that satisfies the following requirements: (1) impedances are matched both at the input and the output, (2) the network is *passive* in the conventional sense that the only source of energy within the network is that produced by thermal noise effects, and (3) some of the power delivered to the input is dissipated within the network, so that the power delivered to the load from the source is less than the power delivered to the input.

A block diagram of an attenuator circuit that can satisfy this condition is shown in Figure 12–13. The most common element that produces this type of effect in a receiving system is a *transmission line* (or *waveguide* at microwave frequencies). A passive transmission line is required to couple between an antenna and a receiver, and some distance physically separates the units, which introduces losses into the system. Within a receiver, many mixers also introduce losses and must be treated in the same manner as attenuators.

The most convenient way to represent the losses in a matched attenuator is through the use of an *insertion loss factor L*. Let $P_i$ represent the power delivered to the input, and let $P_o$ represent the power delivered to the output. The relationship between these two values is

$$P_o = \frac{P_i}{L} \tag{12-75}$$

The factor $L$ for a true attenuator must satisfy $L \geq 1$, with the lower bound corresponding to no attenuation at all, and the upper bound (infinity) corresponding to complete absorption of the signal in the network. The factor $L$ is often given as a decibel value $L_{dB}$, where

$$L_{dB} = 10 \log L \tag{12-76}$$

The factor $1/L$ for a matched attenuator is treated in much the same way as available gain $G$ for an amplifier. This association will help in some of the results that occur later.

Now consider the situation depicted in Figure 12–14, with the following conditions imposed: (1) the attenuator has a constant physical temperature $T_P$, which means that all lossy elements within the network possess that temperature; and (2) a noise source with an

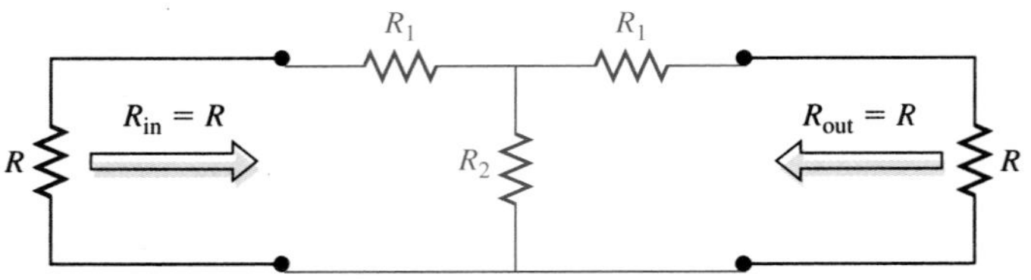

**FIGURE 12–13**
Illustration of matched attenuator system.

**FIGURE 12–14**
Matched attenuator at a physical temperature excited by a noise input source with effective temperature.

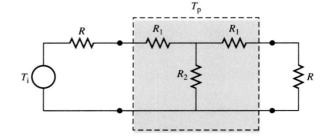

effective source temperature $T_i$ is connected to the input. Assume for the purpose of this development that $T_i \geq T_p$, although the results apply in general.

If the input noise source were at the same temperature as the network, the effective output noise would simply be $kT_pB$. However, since the input noise source is assumed to have a greater temperature, there is an *excess noise temperature* $T_i - T_p$ supplied to the input, and it is reduced by the factor $1/L$. Hence, the net output noise is given by

$$N_o = kT_pB + \frac{k(T_i - T_p)B}{L} \tag{12-77}$$

This equation may also be expressed as

$$N_o = \frac{kT_iB}{L} + \left(1 - \frac{1}{L}\right)kT_pB \tag{12-78}$$

This latter form displays the relative effects of the input temperature and the physical temperature as a function of the loss. For convenience in the development, the assumption was made that the source temperature was greater than the physical temperature, but the result applies for the opposite inequality as well.

An *effective temperature* $T_e$ referred to the input may be defined such that

$$N_o = \frac{1}{L}k(T_i + T_e)B \tag{12-79}$$

which is the standard form for evaluating the output noise power in terms of the effective temperature with the "gain" set as $1/L$. Equating Equations 12-79 and 12-78, we obtain

$$T_e = (L - 1)T_p \tag{12-80}$$

These results indicate that a matched attenuator can be handled in the same way as an amplifier by defining an effective noise temperature at the input. Note that $T_e$ increases linearly with changes in $L$. This means that *the greater the attenuation, the greater the effective noise temperature resulting from the attenuation.*

## Noise Figure of Attenuator

The noise figure $F$ corresponding to Equation 12-80 is obtained from the application of Equation 12-54. The result is

$$F = 1 + (L - 1)\frac{T_p}{T_0} = 1 + (L - 1)\frac{T_p}{290} \tag{12-81}$$

An interesting special case occurs when $T_p = T_0$. In this case, Equation 12-81 reduces to

$$F = L \tag{12-82}$$

*For the special case when the attenuator is at the standard temperature of 290 K, the noise figure is equal to the insertion power loss factor.* Because of the simplicity of this relationship, and the fact that in many systems the actual physical temperatures of lossy elements may be reasonably close to the standard reference temperature, this result is often used as an estimate of the noise figure for a variety of cases.

The noise temperature associated with a loss increases as the loss factor increases. Thus, it is very desirable in any receiving system to have a high-gain, low-noise amplifier as close to the front end as possible.

## Choice of Reference Points

Assume that there is a loss associated with the input to a receiver, such as might be encountered in the transmission line connecting the antenna to the receiver. There are two approaches to dealing with the problem: (1) The effective input temperature may be referred all the way back to the transmission line input as the reference point, using the effective noise temperature of the transmission line and treating the line with loss $L$ as if it were a gain of value $1/L$. (2) The receiver input may be used as the reference point, in which case the effective noise temperature is that of the receiver alone, but the output noise of the transmission line is considered as a linear combination of the antenna source temperature with the ohmic losses of the transmission line (as indicated by Equation 12-78). Both approaches are equally valid.

**▐▌ EXAMPLE 12-14**

For the system of Figure 12–15, determine the equivalent noise temperature referred to the input.

**SOLUTION**    Note that all the gains are given in absolute form, so we need not convert from decibel values in this example. Since there are three stages, the form of Equation 12-72 may be directly applied.

$$T_e = T_{e1} + \frac{T_{e2}}{G_1} + \frac{T_{e3}}{G_1 G_2} = 100 + \frac{200}{20} + \frac{300}{20 \times 15}$$

$$= 100 + 10 + 1 = 111 \text{ K} \tag{12-83}$$

This problem was obviously "rigged" to produce simple values, but the pattern should be of educational value. These amplifier gains are relatively low. Even so, it is clear that the most dominating noise temperature is that of the first stage. Moreover, the effects of the noise temperatures diminish as we move from the input stage to the output stage.

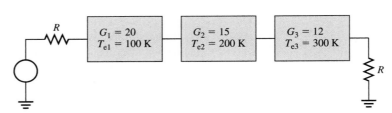

**FIGURE 12–15**
Circuit of Example 12-14.

**▐▌ EXAMPLE 12-15**

Rework Example 12-14, by first determining the noise figures for the individual stages and then determining the net noise figure.

**SOLUTION**    The noise figure for each stage may be determined from the relationship

$$F = 1 + \frac{T_e}{290} \tag{12-84}$$

The values obtained are readily determined as $F_1 = 1.345$, $F_2 = 1.690$, and $F_3 = 2.034$.

$$F = F_1 + \frac{F_2 - 1}{G_1} + \frac{F_3 - 1}{G_1 G_2} = 1.345 + \frac{1.690 - 1}{20} + \frac{2.034 - 1}{20 \times 15}$$

$$= 1.3448 + 0.0345 + 0.0034 = 1.3827 \tag{12-85}$$

As a check, the noise temperature may be determined from this value as

$$T_e = (F - 1) \times 290 = (1.3827 - 1) \times 290 = 111 \text{ K} \tag{12-86}$$

Understand, of course, that the total noise figure could have been determined from the total noise temperature in one step using Equation 12-84 and the result of Example 12-14. This exercise was provided for its educational value.

**EXAMPLE 12-16**

The circuit of Figure 12–16 represents a **bad** design, which has been somewhat exaggerated to make a point. The transmission line between the antenna and the preamplifier has a loss of 6.02 dB. The complete receiving system contains both a preamplifier and a receiver. Determine (a) the effective system noise temperature referred to the antenna output, and (b) the noise figure (absolute and in dB).

**SOLUTION** The absolute value of the loss factor is

$$L = 10^{L_{dB}/10} = 10^{6.02/10} = 4 \qquad (12\text{-}87)$$

The absolute gain of the preamplifier stage is readily determined as 100, and the absolute gain of the receiver is determined as $10^6$, although the latter value is not required at this time.

(a) The effective noise temperature $T_{eL}$ of the lossy transmission line referred to the input is determined from Equation 12-80.

$$T_{e,L} = (L - 1)T_p = (4 - 1) \times 290 = 870 \text{ K} \qquad (12\text{-}88)$$

The effective noise temperature referred to the input of the transmission line is determined from Equation 12-72. However, the "gain" of the first stage is the factor $1/L$ for the transmission line. The factor $1/L$ in the denominator brings $L$ back to the numerator, and the form of Equation 12-72 adapted to this case is

$$T_e = T_{e,L} + LT_{e,pre} + \frac{LT_{e,rec}}{G_{pre}} \qquad (12\text{-}89)$$

where $T_{e,pre}$ is the noise temperature of the preamplifier, $T_{e,rec}$ is the noise temperature of the receiver, and $G_{pre}$ is the gain of the preamplifier. Substituting these values, we have

$$T_e = 870 + 4 \times 50 + \frac{4 \times 200}{100} = 870 + 200 + 8 = 1078 \text{ K} \qquad (12\text{-}90)$$

It is obvious that the lossy transmission line has caused a drastic increase in the effective noise temperature.

(b) The corresponding noise figure is

$$F = 1 + \frac{T_e}{290} = 1 + \frac{1078}{290} = 4.717 \qquad (12\text{-}91)$$

The decibel value is

$$F_{dB} = 10 \log 4.717 = 6.74 \text{ dB} \qquad (12\text{-}92)$$

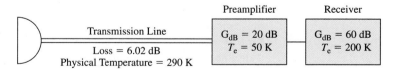

**FIGURE 12–16**
Circuit of Example 12-16.

**EXAMPLE 12-17**

A young electronics specialist who studied from this book decided to switch the arrangement of the previous circuit to the form shown in Figure 12–17. (It is assumed that the preamplifier is weather protected, and capable of being placed at the output of the antenna.) Determine (a) the effective system noise temperature referred to the antenna output, and (b) the noise figure (absolute and in dB)

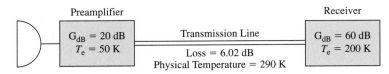

**FIGURE 12–17**
Circuit of Example 12-17.

**SOLUTION** It should be clear that the net gain between antenna output and receiver output will not be changed by reversing the order of the transmission line and the amplifier. However, let's see what happens to the noise temperature and the noise figure.

(a) Using the form of Equation 12-72 adapted to this case, the effective noise temperature is now given by

$$T_e = T_{e,\text{pre}} + \frac{T_{e,L}}{G_{\text{pre}}} + \frac{LT_{e,\text{rec}}}{G_{\text{pre}}} \tag{12-93}$$

Substituting values, we have

$$T_e = 50 + \frac{870}{100} + \frac{4 \times 200}{100} = 50 + 8.7 + 8 = 66.7 \text{ K} \tag{12-94}$$

This is a significant reduction in noise temperature as compared with the preceding system form. Yet the output signal should have the same level.

(b) The noise figure is

$$F = 1 + \frac{T_e}{290} = 1 + \frac{66.7}{290} = 1.23 \tag{12-95}$$

The decibel value is

$$F_{\text{dB}} = 10 \log 1.23 = 0.90 \text{ dB} \tag{12-96}$$

The major point illustrated by this example is that it is very undesirable to have a lossy circuit at the front end of any receiving system dealing with very low signal levels. When practical, a preamplifier can be placed right at the antenna output ahead of the lossy circuit, and this is done in many systems. For example, some satellite systems have a unit called a *low-noise block* (LNB), which provides both amplification and down-conversion for the signal at the antenna before the signal is connected to the transmission line. ▐█

## 12-8 Multisim® Examples (Optional)

The Mutisim examples of this chapter will focus on the **Noise Analysis** capability of Multisim. This operation utilizes individual thermal noise contributions of resistors, along with noise models of semiconductors, to provide estimates of the mean-square noise parameters for a given circuit. The data obtained can be a little tricky to use and interpret, so we will start with the simplest possible circuit forms and then build the approach up to more complex circuits.

First, it should be noted that the "standard" reference temperature for Multisim is slightly different than the communication "standard." This is not the result of some arbitrary decision on the part of the company; rather, most of the semiconductor industry utilizes a different reference. As we have seen, most communications system standards utilize 17°C or 290 K. However, the semiconductor industry typically uses 27°C or 300 K as the standard. This difference is reasonable because semiconductors within electronic equipment tend to operate at higher physical temperatures than the outside environment. For noise analysis purposes, this difference does not cause any significant discrepancies, and is probably within the "window of uncertainty" associated with noise estimates in any case.

Just for reference purposes, modified forms of certain major equations converted to a reference of 300 K follow. The mean-square noise voltage of Equation 12-5 becomes

$$v_{\text{rms}}^2 = 16.6 \times 10^{-21} RB \qquad (12\text{-}97)$$

The available noise power of Equation 12-22 becomes

$$N_{\text{av}} = 4.15 \times 10^{-21} B \qquad (12\text{-}98)$$

## Noise Analysis

Any circuit for which noise analysis is desired must have two specific parameters identified: (1) input noise reference source and (2) at least one output node. A bandwidth must be provided over which the noise is to be estimated.

The output data are measured in $V^2$ according to the convention of Equation 12-97, and are generally given in four different forms: (1) total noise output, (2) total noise referred back to the input based on the gain (or loss in the case of an atttenuator), (3) individual noise contributions at the points where they occur in the circuit, and (4) individual noise contributions referred back to the input.

---

**❙❙ MULTISIM EXAMPLE 12-1**

Consider the simple circuit shown in Figure 12–18, consisting of a sinusoidal source and a resistance. Perform a noise analysis of the circuit over a bandwidth from 1 Hz to 10 MHz.

**SOLUTION**  This circuit represents about the simplest form for which a noise analysis can be performed. There must be a component that produces noise and it is the resistor in this case. There must also be an ac sinsusoidal source, which establishes the input reference point for the input noise, even though the source itself contributes no noise. The actual parameters of the source are meaningless for a noise analysis, and they can be left at their default values. For that reason, we have chosen to leave off the source parameters, which was the practice followed for frequency response analysis in earlier chapters. Notice that a junction has been added to the output, and that point becomes node 2.

To perform the analysis, left-click on **Analyses** and then left-click on **Noise Analysis.** A window with the title **Noise Analysis** will then open. The default tab will probably be **Analysis Parameters,** but if not, left click on that tab. The slot entitled **Input noise reference source** represents the reference source for which the noise may be referred to on an input basis. Only one option is provided in this circuit and that is **vv1.** If the circuit contained more than one ac source, we could select a choice. The **Output node** is set to the value **2.** The **Reference node** is left at the default value of **0.**

Next, left-click on the **Frequency Parameters** tab and a window with that title will open. Because of $1/f$ noise at low frequencies in active devices, the **Start frequency** should *not* be set to 0. Instead, a **Start frequency** of **1 Hz** should be fine for the wideband

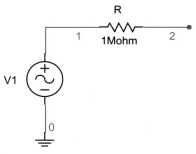

**FIGURE 12–18**
Circuit of Multisim Example 12-1.

thermal noise in this example. Set the **Stop frequency** to **10 MHz,** and for all practical purposes, we can assume a bandwidth $B = 10$ MHz for thermal noise analysis. The **Sweep type** should be left at the default setting of **Decade.** Due to the constant spectrum of thermal noise, the **Number of points per decade** can be left at the default value of **10.** The **Vertical** scale can be left at the default setting of **Logarithmic.**

For the next setting, left-click on the **Output variables** tab. In this circuit, there are only four possible variables, so for illustrative purposes, transfer all of them to the **Selected variables for analysis** block using the **Plot during simulation** operation, which has been considered numerous times in the text for different forms of analysis. An explanation of the four quantities follows:

**onoise_total** $=$ total output noise in $V^2$

**inoise_total** $=$ total noise referred back to the input in $V^2$

**onoise_r1** $=$ output noise due to the resistance R1 in $V^2$

**inoise_r1** $=$ noise due to R1 referred back to the input in $V^2$

If all of this seems confusing even for this simple circuit, the first two are arguably the most important since they provide a measure of the total output noise due to all sources, and that value referred back to the input. The second value is simply the output divided by the square of the voltage gain. All other variables, which increase in number as the circuit becomes more complex, allow one to look at the effects of different circuit parameters.

When **Simulate** is depressed, the results are shown, as in Figure 12–19. The values are measured in $V^2$. With very slight but insignificant differences, the four values in this simple circuit can be considered equal. The reason is that there is only one component that contributes noise and the voltage gain between the source and the output is 1. For most practical purposes, the noise voltage squared is about $166 \times 10^{-9}$ $V^2$. A quick check with Equation 12-97 yields the same result. The corresponding rms noise voltage is then the square root of the preceding value, and is about 407 $\mu$V.

**Integrated Noise - V^2 or A^2**

| Noise Analysis | |
|---|---|
| onoise_total_rr | 165.75720n |
| inoise_total_rr | 165.75753n |
| onoise_total | 165.75720n |
| inoise_total | 165.75753n |

**FIGURE 12–19**
Noise data of Multisim Example 12-1.

---

**▐▌ MULTISIM EXAMPLE 12-2**   Consider the circuit shown in Figure 12–20, consisting of a sinusoidal source and a voltage divider consisting of two equal resistances. Perform a noise analysis of the circuit over a bandwidth from 1 Hz to 10 MHz.

**SOLUTION**   This circuit is only slightly more complex than that of the preceding example, but as we will see shortly, the data obtained can be a bit more perplexing. Each resistor

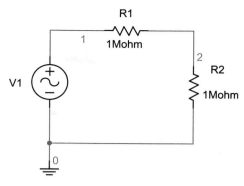

**FIGURE 12–20**
Circuit of Multisim Example 12-2.

has the same value as in the preceding example, and the bandwidth is the same, so we should be expecting something that can be related in a simple fashion to the value of $166 \times 10^{-9}$ V$^2$ appearing in that example.

The results of the noise analysis are shown in Figure 12–21. First, consider the resistance **R1**. The voltage-squared contributed by this resistance is the same as the value contributed by the 1-M$\Omega$ resistance in the previous example, and since it is effectively in series with the source, its input contribution is about $166 \times 10^{-9}$ V$^2$. However, its effect on the output is $(1/2)^2 = 0.25$ times this value or about $41 \times 10^{-9}$ V$^2$. If the ac source is temporarily de-energized by replacing it with a short circuit, it is noted that the effect of **R2** on either the output or the value referred back to the input is the same as with **R1**.

In most cases the total output noise and its value referred back to the input are the most important, since they provide measures of the overall effects. Since the input source does not contribute any noise in this circuit, and all of the noise is due to resistors at the same temperature, there is a very simple way of relating the voltage-squared at the output. Looking back from the output with the source de-energized, the net resistance is simply the equivalent resistance of two 1-M$\Omega$ resistances in parallel, which is 500 k$\Omega$. Therefore, the net voltage-squared can be estimated by Equation 12-97 to be about $83 \times 10^{-9}$ V$^2$, which agrees quite well with the value obtained at the output from the analysis. The corresponding

**Integrated Noise - V^2 or A^2**

| Noise Analysis | |
| --- | --- |
| onoise_total | 82.87868n |
| inoise_total | 331.51507n |
| onoise_total_rr1 | 41.43934n |
| inoise_total_rr1 | 165.75753n |
| onoise_total_rr2 | 41.43934n |
| inoise_total_rr2 | 165.75753n |

**FIGURE 12–21**
Noise data of Multisim Example 12-2.

value referred to the input would be $(2)^2 = 4$ times this value or about $332 \times 10^{-9}$ V$^2$. Remember the voltage "gain" of this circuit from input to output is 0.5, meaning that the power gain or voltage-squared gain is $(0.5)^2 = 0.25$. Therefore, this noise referred back to the input has to be 4 times as great as the output value. Working with voltage gain squared or with power ratios can be a little tricky when resistive divider ratios are required, since the values must be squared.

Understand that if the source actually contributed more than just the thermal noise associated with a resistance at a real physical temperature, the simplified approach dealing with the equivalent output resistance taken in the last paragraph would not be valid. Instead, the results of the noise analysis should be used to estimate the total noise.

---

**▍▍ MULTISIM EXAMPLE 12-3**

Consider the circuit of Figure 12–22, consisting of a 3.01-dB matched attenuator designed for a 600-Ω source and a 600-Ω load. The source is noise-free, and all noise is thermal in nature and is a result of circuit resistances. Determine the various noise parameters for the circuit in V$^2$ over a bandwidth from 1 Hz to 10 MHz.

**SOLUTION**   The setup follows the procedure of the preceding two examples, except that the output node in this case is **3**. To illustrate how unwieldy the various details can be, all of the parameters in the circuit were listed in the output, and the values are shown in Figure 12–23. We won't go through all of the details, but note that each resistive component

**FIGURE 12–22**
Attenuator with matched source and load.

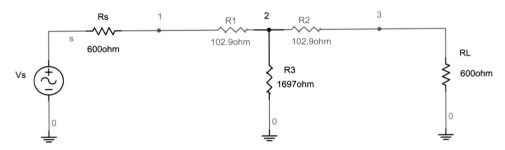

Integrated Noise - V^2 or A^2

| Noise Analysis | |
|---|---|
| onoise_total | 49.72434p |
| inoise_total | 397.74128p |
| onoise_total_rr1 | 2.13234p |
| inoise_total_rr1 | 17.05645p |
| onoise_total_rr2 | 4.26461p |
| inoise_total_rr2 | 34.11232p |
| onoise_total_rr3 | 6.03319p |
| inoise_total_rr3 | 48.25902p |
| onoise_total_rrs | 12.43349p |
| inoise_total_rrs | 99.45452p |
| onoise_total_rrl | 24.86071p |
| inoise_total_rrl | 198.85897p |

**FIGURE 12–23**
Noise data of Multisim Example 12-3.

in the circuit makes a contribution to the output noise, and each has its value referred back to the input. In each case, the corresponding input value is obtained from the output value by multiplying the latter by 8, so the bewildered reader may want to know how this value is obtained.

Since the circuit has an attenuation of 3.01 dB, the power or voltage-squared "gain" is 0.5 (corresponding to a voltage output/input ratio of 0.707). However, there is an additional power or voltage-squared transfer ratio between the source and the input to the attenuator of $(0.5)^2 = 0.25$. Therefore, the net transfer ratio or "gain" from input to output is $(0.5)(0.25) = 0.125$. Hence, the input power or voltage-squared is $1/(0.125) = 8$ times the output power or voltage-squared.

The "bottom-line" for this circuit is that the net output noise voltage-squared is about $50 \times 10^{-12}$ V$^2$.

---

**▋▌ MULTISIM EXAMPLE 12-4**

The circuit of Figure 12–24 is a Multisim schematic diagram of a noninverting operational amplifier circuit employing the popular 741, with a noninverting voltage gain of 10. Determine the various noise parameters based on a bandwidth from 1 Hz to 10 kHz.

**SOLUTION**   The actual noise in this particular circuit consists of that generated by the internal mechanisms of the op-amp plus that of the external resistances. Therefore, this is the first example in which an active device provides some contribution to the total noise.

It turns out that the 3-dB bandwidth of this particular circuit with the 741 is about 100 kHz. However, the response is essentially flat from dc to 10 kHz, the upper frequency for which the noise analysis is desired.

Note that the various nodes in this circuit have been given names that relate to their roles in the circuit. It should also be noted that the op-amp circuit diagram in Multisim has

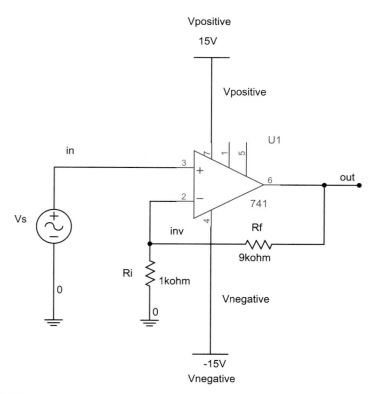

**FIGURE 12–24**
Circuit of Multisim Example 12-4.

Integrated Noise - V^2 or A^2

| Noise Analysis | |
|---|---|
| onoise_total_rri | 13.37870 p |
| inoise_total_rri | 0.13427 p |
| onoise_total | 313.86755 p |
| inoise_total | 3.14997 p |
| onoise_total_rrf | 1.48652 p |
| inoise_total_rrf | 0.01492 p |

**FIGURE 12–25**
Noise data of Multisim Example 12-4.

the noninverting input terminal on top, while much of the literature has the inverting input on top. We could "flip" the op-amp, but we have chosen to leave it in its default form.

The **Input noise reference source** for this analysis has been set at **vvs,** and the **Output node** has been set at **out.** The **Start frequency** has been set at **1 Hz** and the **Stop frequency** at **10 kHz.** In this example, the **Number of points per decade** has been set to **100.** All variables available are selected for measurement.

The resulting output is shown in Figure 12–25. Both resistances contribute to the total noise, but the largest component is that produced by the op-amp itself. The reason is that the resistance levels in this particular circuit are quite moderate and within the desirable range of operation. If the resistances were made to be very large—for example, in the order of megohms—there would be a significant increase in noise due to the resistances. On the other hand, if the resistances are made too small—in the order of hundreds of ohms—there could be excessive loading effects on the op-amp. Therefore, there is always a compromise in a circuit of this type.

A final comment is that the data sheets for a 741 were scrutinized for noise parameters, and the values given compare quite closely with those measured from the Multisim model.

## 12-9  MATLAB® Examples (Optional)

The three examples that follow will illustrate how random noise can be generated by MATLAB, and used in the simulation of communication systems. At this point, the emphasis will be on the study of the noise generation and its basic properties.

The first example will be performed in the Command Window and will consist of validating the statistical properties of a very long random noise record. Specifically, it will involve $10^6$ values (yes, one million values). Since it would be difficult to show plots with that number of values, we will then consider two examples in which we begin with 2000 points and eventually show the behavior of the last 1000 points. Example 12-2 will deal with the result of passing the noise through a low-pass filter, and Example 12-3 will deal with a bandpass filter. In these examples, M-files will be used, due to the number of commands.

**▌▌ MATLAB EXAMPLE 12-1**

The MATLAB gaussian random number generator has a **mean value of 0** and a **standard deviation of 1.** Run a simple test with one million points to check the validity of these values.

**SOLUTION**   We obviously want to suppress the printing of the output of this test! Random numbers having a gaussian or normal distribution are generated by the command **randn(m,n),** where **m** is the number of rows and **n** is the number of columns. The mean or dc value of the process is 0, and the standard deviation or rms value is 1. If a mean value other than 0 is desired, all that is required is to add a constant value to the random signal. If an rms value other than 1 is desired, the random function should be multiplied by the desired value. The latter process will be illustrated in the next example.

It should be noted that the signal is really a *pseudorandom* signal, in that it actually does have a period and eventually repeats itself. However, as far as practical applications are concerned, it can be considered random. Moreover, if the reader repeats the experiment to follow, the results may be slightly different depending on whether the function has been turned on more than once. In the next example, we will show how to maintain a fixed pattern, but our purpose here is simply to run a test.

We will choose to generate a row vector with one million points. Let **vnoise** represent the variable, which is generated by the command

```
>> vnoise = randn(1,1e6);
```

By all means, be sure the semicolon is used here!

The mean or dc value of this function is obtained from the command **mean,** and it will be denoted **vmean**.

```
>> vmean = mean(vnoise)
```

The value obtained was 0.0018.

The standard deviation or rms value is obtained from the command **std,** and it will be denoted **vrms**.

```
>> vrms = std(vnoise)
```

The value obtained was 1.0001.

While the theoretical value of the dc value should be 0, remember that this is a random process and it is very unlikely that one would obtain an exact value of 0. Indeed, the value of 0.0018 is less than 0.2% of the standard deviation.

As far as the rms value is concerned, it couldn't get much better than the value of 1.0001, since the theoretical value is 1.

Out of curiosity, let's look at the first seven values.

```
>> vnoise(1:7)

ans =

-0.4326 -1.6656 0.1253 0.2877 -1.1465 1.1909 1.1892
```

Obviously the numbers are quite random in nature.

---

**MATLAB EXAMPLE 12-2**

Write an M-file program that will perform the analysis that follows. Using a time step of 1 μs, generate a 2000-point random noise voltage with a dc or mean value of 0 and an rms value of 10 V. Then apply the signal to a Butterworth 2-pole low-pass filter with a 3-dB cutoff frequency of 2 kHz. Plot the input and output signals for the second 1000 points and measure their rms values.

**SOLUTION**   Refer to the M-file in Figure 12–26 for the discussion that follows. The program has the title **noise1.m.** The time step is established as 1 μs, which means that 2000 points will represent a time interval of 2 ms. When any filter is excited by a signal there is an initial transient response, even when the input is random noise. The settling time is of the order of the reciprocal of the bandwidth. The desired filter bandwidth is 2 kHz, and this would suggest a settling time interval of about 1/2000 = 0.5 ms. However, to maintain simpler values, we will skip over an interval of 1 ms. Actually, we don't need to do that for the input noise source, but for consistency in this example and the next one, we will use the same time interval for observation. This time interval will be from 1 ms to one time step less than 2 ms.

```
%Program noise1.m
delt=1e-6;
t_long=0:delt:2e-3-delt;
randn('seed',1)
vnoise_long=10*randn(1,2000);
[n d]=butter(2,2*pi*2000,'s');
vfil_long=lsim(n,d,vnoise_long,t_long);
t=t_long(1001:2000);
vnoise=vnoise_long(1001:2000);
vfil=vfil_long(1001:2000);
figure(1)
plot(t,vnoise)
xlabel('Time, seconds')
ylabel('Voltage, volts')
title('Unfiltered Input Noise Voltage')
grid
pause
figure(2)
plot(t,vfil)
xlabel('Time, seconds')
ylabel('Voltage, volts')
title('Filtered Output Low-Pass Noise Voltage')
grid
vrms_in=std(vnoise)
vrms_out=std(vfil)
```

**FIGURE 12–26**

M-file program of MATLAB Example 12-2.

In the program, all variables that are 2000 points long are followed by the designation **long.** Thus, the "longer" versions of time, input noise voltage, and output filtered voltage are **t_long, vnoise_long,** and **vfil_long.** The last 1000 points of each are denoted, respectively, **t, vnoise,** and **vfil.**

Although the output of the noise generator is random, it is possible to set it up so that the same pattern can be observed each time it is run. This is accomplished prior to defining the noise voltage by the following command:

```
randn('seed',1)
```

The number 1 in the argument is a value that establishes a "seed number." As long as it remains at that value, the same "random" pattern will be generated each time the program is run. A totally different random pattern can be established by changing the value, but the statistics are theoretically the same.

In order to establish an rms value of 10 V, the number generator should be multiplied by 10. Thus, the noise voltage is generated by the command

```
>> vnoise_long = 10*randn(1, 2000);
```

Assuming that the MATLAB **Signal Processing Toolbox** is installed, the low-pass filter numerator and denominator polynomials are generated by the command

```
>> [n d] = butter(2, 2*pi*2000, 's');
```

The output is then generated by the command

```
>> vfil_long = lsim(n, d, vnoise_long, t_long);
```

We then form **t, vnoise,** and **vfil** from the last 1000 points. For example, **t** is obtained by

```
t = t_long(1001:2000);
```

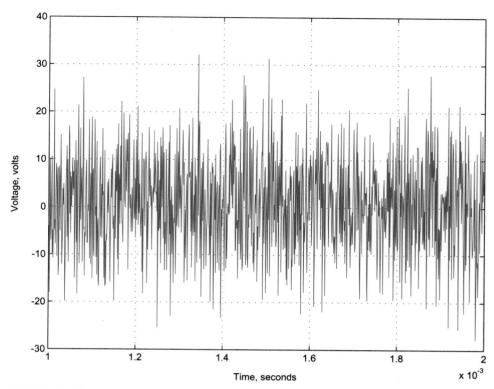

**FIGURE 12–27**
Unfiltered input noise voltage.

Similar operations are performed on the other variables, and the results are plotted and labeled. The unfiltered input noise voltage is shown in Figure 12–27, and the filtered output voltage is shown in Figure 12–28.

The input and output rms voltages are also calculated. The values obtained are

```
vrms_in = 9.8907 V
```

This value differs from the theoretical value by about 1%, which is a reasonable discrepancy based on 1000 points. As we will see shortly, the filter output for 1000 points may show a greater discrepancy.

First, the theoretical value of the output rms value will be computed. The folding frequency represents the one-sided bandwidth of the noise, and it is half of the sampling frequency or 500 kHz. From Table 12–1, the noise bandwidth of a 2-pole Butterworth filter is 1.111 times the 3-dB frequency, or $1.111 \times 2000 = 2222$ Hz. Since power varies directly with bandwidth, voltage should vary directly with the square root of the bandwidth. Assuming the theoretical input rms voltage, the output rms voltage should be about $(\sqrt{2222/5 \times 10^5}) \times 10 \approx 0.67$ V. The actual measured rms output voltage is

```
vrms_out=0.4489 V
```

While the numbers are in the same "ballpark," some readers may be a little concerned by the difference. It can be explained as follows: When the signal is filtered, the effective number of independent samples is reduced considerably. While there are 1000 independent input samples used in the estimation process, the low-pass output signal has been smoothed to the point where a much longer signal would be required to achieve an accurate estimate of the output statistics. In fact, it would be possible to change the seed values, make new runs with each new value of seed, and perform an averaging of the rms values obtained. Eventually the average of those values should converge to the theoretical value.

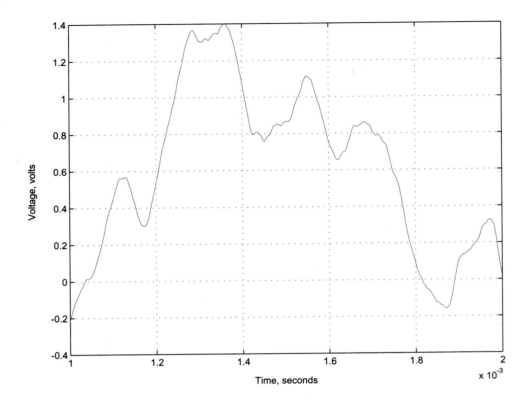

**FIGURE 12–28**

Filtered output low-pass noise voltage.

---

**MATLAB EXAMPLE 12-3**

Create a new M-file by modifying the M-file of the preceding example so that the two-pole low-pass filter with a 3-dB bandwidth of 2 kHz is replaced by a two pole-pair Butterworth band-pass filter with a center frequency of approximately 10 kHz and a bandwidth of 2 kHz. Apply the same analysis as in the past example.

**SOLUTION** It was possible to create a new program to achieve the desired objective by changing only two lines of code in the program **noise1,** and it was saved as **noise2.** The program is shown in Figure 12–29.

The most significant change is that a band-pass filter replaces the low-pass filter. Since the desired bandwidth is 2 kHz, the two 3-dB frequencies will be set at 9 kHz and 11 kHz. Strictly speaking, the "center frequency" will be at the geometric mean of these two values, which is $\sqrt{9 \times 11} \approx 9.95$ kHz, but this is close enough for our purposes. Assuming that the **Signal Processing Toolbox** is available, the code for generating the numerator and denominator polynomials in this case is

```
>> [n d] = butter(2, 2*pi*[9000 11000], 's');
```

The value of 2 in the argument for the band-pass filter represents the number of pole-pairs. The other change is to the line providing the title to the second figure, in which "low-pass" is changed to "band-pass".

The input noise voltage is the same as in the preceding example, and was shown in Figure 12–27.

The filtered output noise is shown in Figure 12–30. Note how the signal resembles somewhat a sinusoidal pattern with variations in amplitude. The reason is that the center of the filter is about 10 kHz, and since the noise contains components at all frequencies up to 500 kHz, those components in the vicinity of 10 kHz will be the strongest. To substantiate this, measure the approximate time between successive zero crossings in the same direction, and it will be noted that this time interval is about $0.1 \times 10^{-3}$ s. The corresponding frequency is $1/(0.1 \times 10^{-3}) = 10$ kHz.

```
%Program noise2.m
delt=1e-6;
t_long=0:delt:2e-3-delt;
randn('seed',1)
vnoise_long=10*randn(1,2000);
[n d]=butter(2,2*pi*[9000 11000],'s');
vfil_long=lsim(n,d,vnoise_long,t_long);
t=t_long(1001:2000);
vnoise=vnoise_long(1001:2000);
vfil=vfil_long(1001:2000);
figure(1)
plot(t,vnoise)
xlabel('Time, seconds')
ylabel('Voltage, volts')
title('Unfiltered Input Noise Voltage')
grid
pause
figure(2)
plot(t,vfil)
xlabel('Time, seconds')
ylabel('Voltage, volts')
title('Filtered Output Band-Pass Noise Voltage')
grid
vrms_in=std(vnoise)
vrms_out=std(vfil)
```

**FIGURE 12–29**
M-file program of MATLAB
Example 12-3.

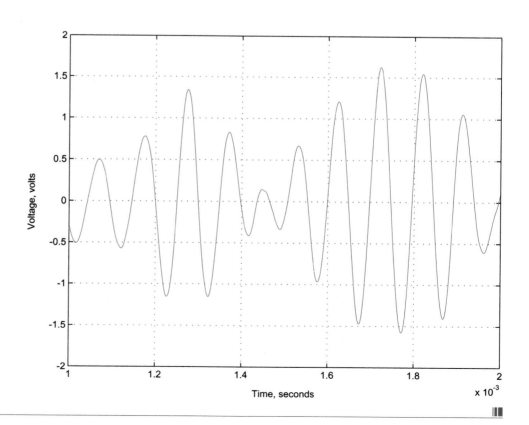

**FIGURE 12–30**
Filtered output band-pass noise
voltage.

## SystemVue™ Closing Application (Optional)

Insert the text CD in a computer having SystemVue™ installed and activate the program. Open the CD folder entitled **SystemVue Systems** and load the file entitled 12-1.

### Review of System

The system shown on the screen was described in the chapter opening application, and you may wish to review the discussion and system configuration provided there.

### Measuring the Properties of the Noise

Because of the random nature of additive white gaussian noise, statistical properties may vary from the ideal values for a finite number of points. In general, the larger the number of independent samples, the more closely the statistical parameters should approach their ideal values.

1. Double-click on the one sink token (Token 0). Left-click in the order that follows on **Numeric, Statistics,** and **OK.**

2. Make a simulation run and note the values appearing in the small window next to the token. Repeat several times to see how much variation there is in the values.

3. Left-click on **Define System Time** and set the **Stop Time** to **10 s**. Make sure that the **Number of Samples** is **10001**.

4. Make a new simulation run and note the statistical measurement values. You may want to repeat this process several times. Are these values closer to the ideal values than in step 2?

5. Double-click on the noise source, and when the window opens, left-click on **Thermal.** Open the **Parameters** window and set the **Temperature** to **290 K** and the **Resistance** to **50 Ω**.

6. Make a simulation run and measure the statistical properties. Calculate the theoretical value of the rms voltage and compare it with the value measured. The rms value is determined from the relationship that the power delivered to the resistance is $kTB = v_{rms}^2/R$. The value of the bandwidth for a sampling rate of 1000 samples/s is 500 Hz.

## PROBLEMS

**12-1**  (a) Determine the rms noise voltage produced by a 47-kΩ resistance in a 60-kHz bandwidth at the standard temperature $T_0 = 290$ K.

(b) Determine the rms noise voltage if the resistance is changed to 470 kΩ.

(c) Determine the rms noise voltage if the resistance is changed to 4.7 MΩ.

(d) With $R = 4.7$ MΩ, determine the rms noise voltage if the bandwidth is changed to 600 kHz.

(e) With $R = 4.7$ MΩ, determine the rms noise voltage if the bandwidth is changed to 6 MHz.

(f) With the values of $R$ and $B$ of part (e), determine the rms noise voltage if the temperature is increased to 310 K.

**12-2**  (a) Determine the rms noise voltage produced by a 120-kΩ resistance in an 80-kHz bandwidth at the standard temperature $T_0 = 290$ K.

(b) Determine the rms noise voltage if the resistance is changed to 1.2 MΩ.

(c) Determine the rms noise voltage if the resistance is changed to 12 MΩ.

(d) With $R = 12$ MΩ, determine the rms noise voltage if the bandwidth is changed to 800 kHz.

(e) With $R = 12$ MΩ, determine the rms noise voltage if the bandwidth is changed to 8 MHz.

(f) With the values of $R$ and $B$ of part (e), determine the rms noise voltage if the temperature is decreased to 270 K.

**12-3** The 4.7-M$\Omega$ resistance of Problem 12-1(e) is connected to the input of an ideal noise-free amplifier with a voltage gain of $10^4$, a bandwidth of 6 MHz, and infinite input impedance. Determine the output rms noise voltage.

**12-4** The 12-M$\Omega$ resistance of Problem 12-2(e) is connected to the input of an ideal noise-free amplifier with a voltage gain 2000, a bandwidth of 8 MHz, and infinite input impedance. Determine the output rms noise voltage.

**12-5** Determine the net rms voltage in a 50-kHz bandwidth appearing across the *series* combination of two 10-k$\Omega$ resistances.

**12-6** Determine the net rms voltage in a 50-kHz bandwidth appearing across the *parallel* combination of two 10-k$\Omega$ resistances.

**12-7** Determine the available noise power contained in a 600-$\Omega$ resistance at the standard reference temperature in a bandwidth of 5 MHz.

**12-8** Determine the available noise power contained in a 75-$\Omega$ resistance at the standard reference temperature in a bandwidth of 12 MHz.

**12-9** The resistance of Problem 12-7 is connected across the input of an ideal noise-free amplifier whose input and output impedances are resistive values of 600 $\Omega$. The amplifier has a matched gain of 50 dB and a bandwidth of 5 MHz. Determine the output noise power in a 600-$\Omega$ resistance.

**12-10** The resistance of Problem 12-8 is connected across the input of an ideal noise-free amplifier whose input and output impedances are resistive values of 75 $\Omega$. The amplifier has a matched gain of 46 dB and a bandwidth of 12 MHz. Determine the output noise power in a 75-$\Omega$ resistance.

**12-11** Determine the rms noise voltage across the load resistance in the amplifier of Problem 12-9.

**12-12** Determine the rms noise current in the load resistance in the amplifier of Problem 12-10.

**12-13** For the system of Problems 12-7 and 12-9, determine the power spectral densities at (a) input and (b) output.

**12-14** For the system of Problems 12-8 and 12-10, determine the power spectral densities at (a) input and (b) output.

**12-15** A source produces white noise having a power spectral density of 0.05 fW/Hz. Determine the equivalent source noise temperature.

**12-16** A white noise source has an equivalent noise temperature of $10^5$ K. Determine the power spectral density.

**12-17** Consider the system shown below, and assume that impedances at all junctions are matched. The input is a thermal noise source having an equivalent noise

temperature of 50,000 K. Assume that any internal noise is negligible in comparison to the noise produced by the source. The bandwidths are defined on the blocks. Determine the output noise power.

**12-18** Consider the system shown below, and assume that impedances at all junctions are matched. The input is a thermal noise source having a power spectral density of 5 fW/Hz. Assume that any internal noise is negligible in comparison to the noise produced by the source. The bandwidths are defined on the blocks. Determine the output noise power.

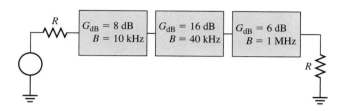

**12-19** At the standard reference temperature, the signal-to-noise ratio at the input to an amplifier is 20 dB, and the output signal-to-noise ratio is 15 dB. Determine (a) decibel noise figure and (b) absolute noise figure.

**12-20** At the standard reference temperature, the signal-to-noise ratio at the input to an amplifier is 600, and the output signal-to-noise ratio is 80. Determine (a) absolute noise figure and (b) decibel noise figure.

**12-21** Determine the effective noise temperature referred to the input of the amplifier of Problem 12-19.

**12-22** Determine the effective noise temperature referred to the input of the amplifier of Problem 12-20.

**12-23** Calculate the effective noise temperature for each of the following noise figures: (a) $F = 1$, (b) $F = 2$, and (c) $F_{dB} = 8$ dB.

**12-24** Calculate the effective noise temperature for each of the following noise figures: (a) $F = 3$, (b) $F = 4$, and (c) $F_{dB} = 7$ dB.

**12-25** Calculate the absolute noise figure $F$ and the the decibel value $F_{dB}$ for each of the following effective noise temperatures referred to the input: (a) 0 K, (b) 145 K, and (c) 290 K.

**12-26** Calculate the absolute noise figure $F$ and the the decibel value $F_{dB}$ for each of the following effective noise temperatures referred to the input: (a) 580 K, (b) 1000 K, and (c) 2900 K.

**12-27** (a) For the system shown below, determine the effective noise temperature referred to the input. (b) From the result of (a), determine the noise figure.

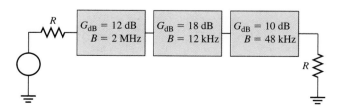

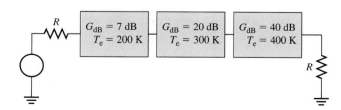

**12-28**  (a) For the system shown below, determine the effective noise temperature referred to the input. (b) From the result of (a), determine the noise figure.

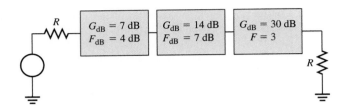

**12-29**  For the system of Problem 12-27, determine the noise figure by first determining the noise figure for each stage and then combining their effects.

**12-30**  For the system of Problem 12-28, determine the noise figure by first determining the noise figure for each stage and then combining their effects.

**12-31**  A passive mixer, matched at both input and output, has a loss of 6.02 dB and physical temperature of 290 K. Determine (a) the effective noise temperature referred to the input, and (b) the noise figure in dB.

**12-32**  A matched attenuator has a loss of 4 dB and physical temperature of 290 K. Determine (a) the effective noise temperature referred to the input, and (b) the noise figure in dB.

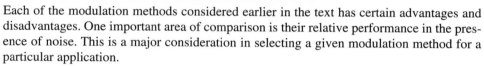

# Performance of Modulation Systems with Noise

# 13

## OVERVIEW AND OBJECTIVES

Each of the modulation methods considered earlier in the text has certain advantages and disadvantages. One important area of comparison is their relative performance in the presence of noise. This is a major consideration in selecting a given modulation method for a particular application.

The analysis of how noise affects the behavior of each of the modulation processes is quite involved mathematically, and much of the material is beyond the intended scope of this text. However, the application of the results to practical communication system analysis and design can be achieved without covering the derivations, and that approach will be emphasized in this and later chapters.

It should be stated at the outset that all comparisons are based on certain assumptions and, therefore, they should be interpreted as reasonable estimates rather than exact results. In particular, the assumed form of the modulating signal in most cases will be that of a single-frequency sinusoidal function. However, sufficient bandwidth will be assumed for a more complex modulating signal. Thus, while these assumptions are necessarily idealized, they are very useful in approximating the actual behavior; in particular, they provide a good basis for comparison. Such results are widely employed in planning and designing real-world communication systems.

## Objectives

After completing this chapter, the reader should be able to:

1. Determine the signal-to-noise ratio referred to the input of a receiver.
2. Determine the *baseband comparison gain* for a particular type of analog modulation system.
3. Determine the *receiver processing gain* for a particular type of analog modulation system.
4. Define the *threshold effect* and explain its significance.
5. Estimate the detector output signal-to-noise ratio for a particular type of analog modulation system.
6. Determine the bit energy to one-sided noise power density ratio for a digital modulation system.
7. For a PCM system, explain the *thermal noise limited region* and the *quantization noise limited region*.

8. Determine the probability of error for a digital communication system and explain its significance.

9. Estimate the detector output signal-to-noise ratio for a PCM system.

10. Compare the overall performances of different types of analog systems, and compare them with those of digital systems.

## SystemVue™ Opening Application (Optional)

Insert the text CD in a computer having SystemVue™ installed and activate the program. Open the CD folder entitled **SystemVue Systems** and load the file entitled 13-1.

### Sink Tokens

| Number | Name | Token Monitored |
|--------|------|-----------------|
| 0 | Modulation | 4 |
| 1 | Noise | 6 |
| 2 | Signal + Noise | 7 |
| 3 | Output | 10 |

### Operational Tokens

| Number | Name | Function |
|--------|------|----------|
| 4 | Modulation | sinusoidal modulating signal set to 5 Hz |
| 5 | Modulator | FM modulator with center frequency of 1 kHz and $\beta = 10$ |
| 6 | AWGN | noise source |
| 7 | Adder | sum of FM signal plus noise |
| 8 | Band-Pass Filter | band-pass filter with bandwidth of 110 Hz centered at 1 kHz |
| 9 | PLL Detector | phase-locked loop FM detector |
| 10 | Low-Pass Filter | post-detection low-pass filter with bandwidth of 6 Hz |

*Note:* The noise source is initially set to a value of 0 for this demonstration. It will be activated in the chapter closing application.

The preset values for the run are from a starting time of 0 to a final time of 1 s with a time step of 10 μs. Run the simulation and observe the waveforms at Tokens 0, 1, 2, and 3.

### What You See

The 5-Hz sinusoidal source **Modulation** seen with Token 0 is frequency modulating the **Modulator** with a frequency deviation of ±50 Hz, which corresponds to a modulation index of 10. In this opening chapter demonstration, the noise source is set to 0, so only a baseline is seen on Token 1, which monitors the output of the adder (Token 7). Token 2 then displays the FM signal, which is noise-free in this case. The band-pass filter of Token 8 is centered at 1 kHz and has a bandwidth of about 110 Hz, as predicted by

Carson's rule. Token 9 is a SystemWorks module that simulates a phase-locked loop and is connected as an FM detector. Finally, Token 10 is a low-pass filter that provides post-detection smoothing of the detected signal. It has been set with a cutoff frequency of 6 Hz, which is just above the modulating frequency. As with all filters, there is a transient interval and the peak level is a function of certain PLL constants. However, following the settling interval, the *form* of the detected output should be the same as the modulating signal.

### How This Demonstration Relates to the Chapter Learning Objectives

You have previously studied FM based on a noise-free environment and that is essentially what has been observed in this demonstration. You have also studied the properties of additive white gaussian noise. In Chapter 13, you will study the performance of different analog and digital modulation systems in the presence of noise. You will learn that among analog systems, FM is somewhat of a "champion." When you return to the system at the end of the chapter, you will add noise and study the behavior in that situation.

## 13-1 Analog System Comparisons

The block diagram shown in Figure 13–1 will be used as a basis for the development that follows. It represents a model of the input of a receiver in which both a signal and noise are present. The fictitious summing block represents the assumption that the combined effect at the input to the receiver is the signal plus additive noise. This type of noise is referred to in many references as *additive white gaussian noise* (AWGN).

The quantity $P_r$ represents the average signal power at the input, and $N_i$ represents the external average noise power at the input. This noise is basically the same as the source noise defined in Chapter 12. The effective noise power from the receiver referred to the input is $N_e$; this noise power is added to the antenna noise power, and the result is the total equivalent system input noise power.

### Antenna Noise

If the input to the receiver is the output of an antenna, the external noise source is the noise delivered by the antenna (and possibly a transmission line). It should be stressed that the effective antenna noise temperature represents a complex combination of emissions received within the radiation pattern of the antenna, and is not due to the physical temperature of the antenna. Its value varies widely with the orientation and pattern of the antenna, and may range from a few kelvins to thousands of kelvins.

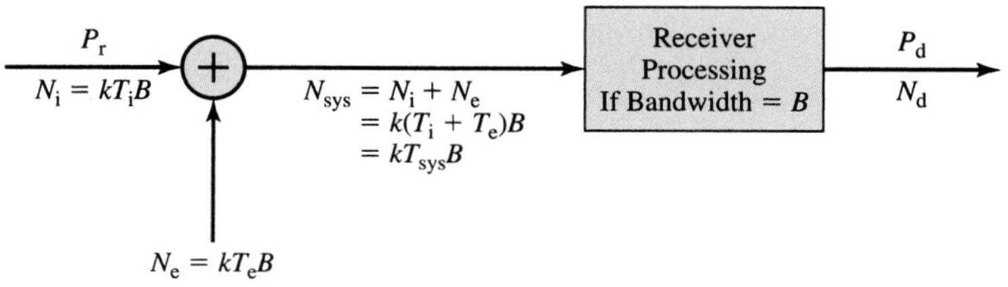

**FIGURE 13–1**
Block diagram illustrating parameters used in analog receiving system analysis.

## System Operating Temperature

The IF bandwidth $B$ will be assumed to be adjusted to a value just sufficient to pass a signal whose baseband bandwidth is $W$, and the selectivity will be assumed to be sufficiently high that the equivalent noise bandwidth and conventional bandwidth values are the same. Note that the actual received signal bandwidth $B$ will be greater than $W$ for all modulation methods except SSB.

If impedances are matched at the input, the total effective input system operating noise power $N_{\text{sys}}$ referred to the receiver input and the IF bandwidth $B$ can be expressed as

$$N_{\text{sys}} = N_{\text{i}} + N_{\text{e}} = kT_{\text{i}}B + kT_{\text{e}}B = kT_{\text{sys}}B = \eta_{\text{sys}}B \tag{13-1}$$

where

$$T_{\text{sys}} = T_{\text{i}} + T_{\text{e}} \tag{13-2}$$

is called the *system temperature* (also called the *operating temperature*), and

$$\eta_{\text{sys}} = kT_{\text{sys}} = k(T_{\text{i}} + T_{\text{e}}) \tag{13-3}$$

is the *system noise power density* referred to the same point.

## Signal-to-Noise Ratio at Receiver Input

The *net* effective signal-to-noise ratio at the receiver input will be designated $(S/N)_{\text{sys}}$, and is

$$(S/N)_{\text{sys}} = \frac{P_{\text{r}}}{N_{\text{sys}}} = \frac{P_{\text{r}}}{\eta_{\text{sys}}B} = \frac{P_{\text{r}}}{kT_{\text{sys}}B} = \frac{P_{\text{r}}}{k(T_{\text{i}} + T_{\text{e}})B} \tag{13-4}$$

The actual equivalent noise power entering the receiver RF stage may be greater than $N_{\text{sys}}$, since the RF stage bandwidth is normally greater than the minimum IF bandwidth. However, after suitable conversion and filtering, the bandwidth is reduced to $B$, so the result of Equation 13-4 is the final equivalent signal-to-noise ratio appearing at the detector input. We are assuming that the combined signal and extra noise at any point preceding the detector is not strong enough to produce an overload or to saturate any stage.

Because of its importance to this study, a point made in the previous paragraph will be repeated to ensure that its meaning is understood. Assuming that all stages between the receiver input and the detector are linear, *the signal-to-noise ratio at the receiver input is equal to the signal-to-noise ratio at the detector input.* The reason is that all predetection receiver noise has been referred back to the input, and both signal and noise will be multiplied by the same overall gain factor between the receiver input and the detector input. This property makes it easier to determine the performance without having to consider all the various stage gain factors preceding the detector input.

## Signal-to-Noise Ratio at Detector Output

The detector portion of the receiver extracts the intelligence from the input signal, and produces a useful baseband output signal representing the desired information. The resulting detected signal power will be denoted $P_{\text{d}}$, and the detected noise power will be denoted $N_{\text{d}}$. The net signal-to-noise ratio at the detector output will be designated $(S/N)_{\text{output}}$, and is

$$(S/N)_{\text{output}} = \frac{P_{\text{d}}}{N_{\text{d}}} \tag{13-5}$$

For convenience in comparison, the modulating signal in each case is assumed to be a single-frequency sinusoid. Of course, the bandwidth assumed is associated with a baseband signal of bandwidth $W$.

We will now introduce two separate measures for comparing the signal-to-noise ratios within a communication system. The first one will be called the **baseband reference**

**comparison gain** and will be denoted $G_B$. The second one will be called the **receiver processing gain** and will be denoted $G_R$. The first provides a common basis for comparing all systems without regard to the differences in transmission bandwidths, while the second provides a basis for analyzing an actual receiver with the transmission bandwidth considered as part of the analysis.

## Baseband Reference Comparison Gain

In defining the baseband comparison gain, assume that the signal could be transmitted directly as a baseband signal using only the minimum bandwidth $W$ required; for example, on a telephone line. The signal-to-noise ratio $(S/N)_{baseband}$ at the receiver would then be

$$(S/N)_{baseband} = \frac{P_r}{\eta_{sys} W} = \frac{P_r}{k T_{sys} W} \tag{13-6}$$

where it is assumed that the received power is $P_r$, and the net noise referred to the receiver input is $k T_{sys} W$.

The **baseband reference comparison gain** $G_B$ is then defined as

$$G_B = \frac{(S/N)_{output}}{(S/N)_{baseband}} \tag{13-7}$$

If $G_B > 1$, the system is superior to baseband as far as the signal-to-noise ratio is concerned, and if $G_B < 1$, the system is inferior to baseband from the same perspective. As we will see shortly, all AM systems are characterized by $G_B \leq 1$; that is, the maximum value of the baseband comparison gain is unity. In contrast, it is possible to achieve large values of $G_B$ with angle modulation systems.

## Receiver Processing Gain

Depending on the type of modulation, the type of detector, and other factors, the signal-to-noise ratio at the detector output may be less than, equal to, or greater than the signal-to-noise ratio at the receiver input. The **receiver processing gain** is defined as

$$G_R = \frac{(S/N)_{output}}{(S/N)_{sys}} \tag{13-8}$$

If $G_R > 1$, the detection process actually enhances the signal-to-noise ratio; while if $G_R < 1$, the detection process degrades the signal-to-noise ratio.

In a sense, the definition of receiver processing gain in Equation 13-8 looks similar to the reciprocal of the definition of noise figure as given in Chapter 12, but the meanings are quite different. The noise figure is based on the same bandwidth at input and output, with an assumed signal retaining its basic spectral shape throughout; that is, a linear process. However, receiver processing gain involves the extraction of the intelligence from a higher-frequency carrier, and both the signal form and the bandwidth at the output may be quite different than at the input.

## Why Both Gain Factors Are Useful

The receiver processing gain is useful in analyzing the actual performance of a given receiver as part of an operating communications system. On the other hand, the baseband comparison gain is more useful in making general comparisons between the different types of modulation systems, since all parameters are referred to the same standard.

It should be emphasized that some of the modulation processes exhibit a so-called *threshold effect,* which will be discussed in detail in Section 13-2. For any process that exhibits that effect, the results to be presented in this section apply only if the signal level at the input is above the threshold level.

Table 13–1    Signal-to-Noise Parameters for Different Analog Modulation Systems

| Modulation Method | $B$ | $G_B = \dfrac{(S/N)_{out}}{(S/N)_{baseband}}$ | $G_R = \dfrac{(S/N)_{out}}{(S/N)_{sys}}$ |
|---|---|---|---|
| SSB | $W$ | 1 | 1 |
| DSB | $2W$ | 1 | 2 |
| AM | $2W$ | $\dfrac{m^2}{2+m^2}$ | $\dfrac{2m^2}{2+m^2}$ |
| AM (100% modulation) | $2W$ | $\dfrac{1}{3}$ | $\dfrac{2}{3}$ |
| PM | $2(1+\Delta\phi)W$ | $\dfrac{(\Delta\phi)^2}{2}$ | $(1+\Delta\phi)(\Delta\phi)^2$ |
| FM | $2(1+D)W$ | $\dfrac{3}{2}D^2$ | $3(1+D)D^2$ |
| FM (preemphasis) | $2(1+D)W$ | $\dfrac{3}{2}D^2\left(\dfrac{\pi}{6}\dfrac{W}{f_1}\right)$ | $3(1+D)D^2\left(\dfrac{\pi}{6}\dfrac{W}{f_1}\right)$ |

*Note:* $(S/N)_{baseband} = \dfrac{P_r}{\eta_{sys}W}$ and $(S/N)_{sys} = \dfrac{P_r}{\eta_{sys}B}$.

## Tabulated Comparisons

A comparison of various analog modulation systems utilizing the definitions provided earlier, and assuming no threshold effects, is shown in Table 13–1. Note the definitions of the baseband signal-to-noise ratio and the receiver input signal-to-noise ratio above the table. Some of the results will now be discussed.

### SSB

For SSB, $G_B = G_R = 1$. This is the simplest of all possible cases from the viewpoint of comparison. It results from the fact that both the bandwidth and the signal processing operations essentially produce the same results at RF as at baseband.

### DSB

This case will illustrate how the two comparisons will, in general, be different since they are viewed from different perspectives. It is easier to explain by looking at the value of the receiver processing gain first, which is $G_R = 2$. This means that the detector output signal-to-noise ratio is twice as great as the receiver input signal-to-noise ratio. This results from the fact that the two signal sidebands add *coherently,* meaning that the voltages add, but in the case of the noise, the power values add. (This is essentially the same as the principle developed in Chapter 2: for voltages at different frequencies, the powers, rather than the voltages, add.) Thus, there appears to be a gain of 2, and this is true from the perspective of receiver input to detected output. However, the transmission bandwidth is twice as great as for SSB, and for a given signal power, the input signal-to-noise ratio is half as great as for SSB. Thus, the relative gain value of 2 exactly compensates for the relative input signal-to-noise ratio of 1/2, and from the baseband comparison point of view, $G_B = 1$.

Hopefully, this case will illustrate the difference between the two measures. The value of $G_B$ provides an overall comparison of all of the methods based on a common reference, while the value of $G_R$ indicates the actual performance of a receiver in a given operational condition.

### AM

For AM, the values of both $G_B$ and $G_R$ are heavily dependent on the modulation factor $m$, but are always less than unity. The maximum values of each occur for 100% modulation ($m = 1$), and these values are $G_B = \frac{1}{3}$ and $G_R = \frac{2}{3}$.

### PM

For PM, the value of $G_B$ is $(\Delta\phi)^2/2$, which allows the possibility of a baseband comparison gain exceeding unity. The expression for $G_R$ is $(1 + \Delta\phi)(\Delta\phi)^2$, which approaches $(\Delta\phi)^3$ for large values of $\Delta\phi$. Of course, the extra exponent power is a result of the larger transmission bandwidth, which makes the input signal-to-noise ratio lower than the reference baseband signal-to-noise ratio for given signal and noise power levels.

### FM

FM follows a pattern similar to PM, except that there is an additional multiplier of 3 and $\Delta\phi$ is replaced by $D$.

Plots of the baseband comparison gains for PM as a function of maximum phase deviation and for FM as a function of the deviation ratio are shown in Figure 13–2. Similar curves displaying the receiver processing gains are shown in Figure 13–3.

### FM with Preemphasis

Some FM systems employ *preemphasis*. This process is achieved by passing the modulating signal through a circuit in which higher frequencies are given a boost prior to modulation. Of course, a *deemphasis* circuit must be provided at the receiver to compensate. FM noise is more severe at higher modulating frequencies, and this process provides a further increase

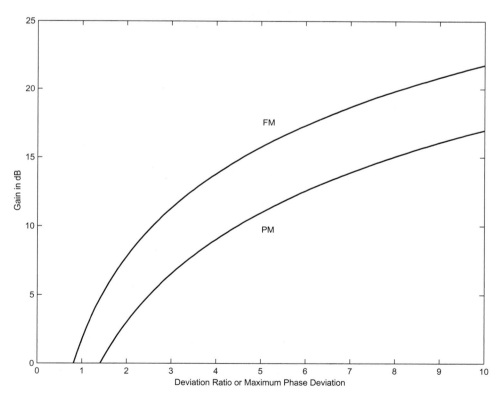

**FIGURE 13–2**

Baseband comparison gains for FM and PM.

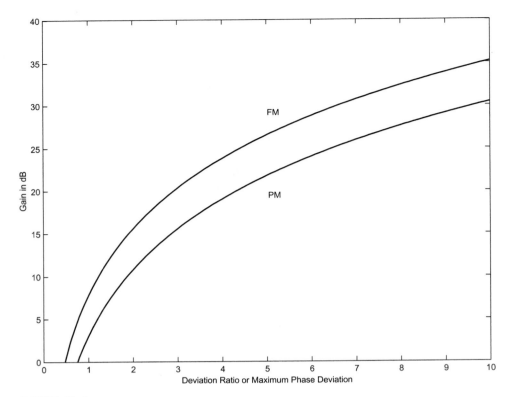

**FIGURE 13–3**
Receiver processing gains for FM and PM.

in the detected signal-to-noise ratio. The most common preemphasis circuit provides a 6 dB/octave or 20 dB/decade boost in the modulating signal above a "break frequency" $f_1$, corresponding to the 3-dB boost frequency. The improvement factor provided is a reasonable approximation and is given by $(\pi W/6f_1)$.

---

■ EXAMPLE 13-1

Calculate the values of baseband comparison gain (both absolute and decibel) for each of the following modulation methods and parameters: (a) SSB, (b) DSB, (c) AM with $m = 0.5$, (d) AM with $m = 1$, (e) PM with $\Delta\phi = 5$, (f) FM with $D = 5$, and (g) FM with $D = 5$ and preemphasis based on $W/f_1 = 7.07$ (which is the value used in commercial monaural FM).

**SOLUTION** The solution reduces to using the third column of Table 11–1 and inserting the proper values.

(a) SSB: The result is simply

$$G_B = 1 \tag{13-9}$$

and

$$G_{B,dB} = 10\log(1) = 0 \text{ dB} \tag{13-10}$$

(b) DSB: As in part (a),

$$G_B = 1 \tag{13-11}$$

and

$$G_{B,dB} = 10\log(1) = 0 \text{ dB} \tag{13-12}$$

(c)  AM ($m = 0.5$):

$$G_B = \frac{m^2}{2 + m^2} = \frac{(0.5)^2}{2 + (0.5)^2} = 0.1111 \qquad (13\text{-}13)$$

and

$$G_{B,dB} = 10\log(0.1111) = -9.54 \text{ dB} \qquad (13\text{-}14)$$

(d)  AM ($m = 1$):

$$G_B = \frac{1}{3} \qquad (13\text{-}15)$$

and

$$G_{B,dB} = 10\log\left(\frac{1}{3}\right) = -4.77 \text{ dB} \qquad (13\text{-}16)$$

(e)  PM ($\Delta\phi = 5$):

$$G_B = \frac{(\Delta\phi)^2}{2} = \frac{(5)^2}{2} = 12.5 \qquad (13\text{-}17)$$

and

$$G_{B,dB} = 10\log(12.5) = 10.97 \text{ dB} \qquad (13\text{-}18)$$

(f)  FM ($D = 5$):

$$G_B = \frac{3}{2}D^2 = \frac{3}{2}(5)^2 = 37.5 \qquad (13\text{-}19)$$

and

$$G_{B,dB} = 10\log(37.5) = 15.74 \text{ dB} \qquad (13\text{-}20)$$

(g)  FM ($D = 5$ and preemphasis with $W/f_1 = 7.07$):

$$G_B = \frac{3}{2}D^2 \left(\frac{\pi}{6}\frac{W}{f_1}\right) = 37.5 \times \left(\frac{\pi}{6} \times 7.07\right) = 138.8 \qquad (13\text{-}21)$$

and

$$G_{B,dB} = 10\log(138.8) = 21.42 \text{ dB} \qquad (13\text{-}22)$$

An alternate approach is to calculate the dB boost given by the additional factor of $(\pi W/6 f_1)$, which is $10\log(3.702) = 5.68$ dB. This value is then added to the result of part (f), which yields $15.74 + 5.68 = 21.42$ dB.

The results of this problem indicate that there is a wide variation in the performance of the different analog modulation methods. The angle modulation methods clearly show superior results to the amplitude modulation methods. However, the price that must be paid is additional bandwidth.

---

**▍▍ EXAMPLE 13-2**

For all of the systems and parameters of Example 13-1, calculate the receiver processing gain and compare the results.

**SOLUTION**  In this case, the solution reduces to using the fourth column of Table 11–1 and inserting the proper values.

(a)  SSB: The result is simply

$$G_R = 1 \qquad (13\text{-}23)$$

and

$$G_{R,dB} = 10\log(1) = 0 \text{ dB} \tag{13-24}$$

Only in this case is the receiver processing gain equal to the baseband comparison gain.

(b) DSB:

$$G_R = 2 \tag{13-25}$$

and

$$G_{R,dB} = 10\log(2) = 3.01 \text{ dB} \tag{13-26}$$

To reiterate a point made earlier, the processing gain is about 3 dB, but the input signal-to-noise ratio is about 3 dB below that of SSB for the same power and noise levels, so the effects cancel out in the baseband comparison measure.

(c) AM ($m = 0.5$)

$$G_R = \frac{2m^2}{2 + m^2} = \frac{2(0.5)^2}{2 + (0.5)^2} = 0.2222 \tag{13-27}$$

and

$$G_{R,dB} = 10\log(0.2222) = -6.53 \text{ dB} \tag{13-28}$$

(d) AM ($m = 1$):

$$G_B = \frac{2}{3} \tag{13-29}$$

and

$$G_{R,dB} = 10\log\left(\frac{2}{3}\right) = -1.76 \text{ dB} \tag{13-30}$$

(e) PM ($\Delta\phi = 5$):

$$G_R = (1 + \Delta\phi)(\Delta\phi)^2 = (1 + 5)(5)^2 = 150 \tag{13-31}$$

and

$$G_{R,dB} = 10\log(150) = 21.76 \text{ dB} \tag{13-32}$$

(f) FM ($D = 5$):

$$G_R = 3(1 + D)D^2 = 3(1 + 5)(5)^2 = 450 \tag{13-33}$$

and

$$G_{R,dB} = 10\log(450) = 26.53 \text{ dB} \tag{13-34}$$

(g) FM ($D = 5$ and preemphasis with $W/f_1 = 7.07$):

$$G_R = 3(1 + D)D^2\left(\frac{\pi}{6}\frac{W}{f_1}\right) = 3(1 + 5)(5)^2 \times \left(\frac{\pi}{6} \times 7.07\right) = 1.666 \times 10^3 \tag{13-35}$$

and

$$G_{R,dB} = 10\log(1.666 \times 10^3) = 32.22 \text{ dB} \tag{13-36}$$

An alternate approach would be to calculate the dB boost given by the additional factor of $(\pi W/6f_1)$ and add this value to the result of part (f).

A comparison of the receiver processing gains from this example and the baseband comparison gains from Example 13-1 is provided in Table 13–2. With the exception of SSB, the values of receiver processing gain are always greater than the baseband comparison

Table 13–2 Comparison of the Baseband Comparison Gains from Example 13-1 with the Receiver Processing Gains of Example 13-2

| Modulation Method | Baseband Comparison Gain $G_{B, dB}$ | Receiver Processing Gain $G_{R, dB}$ |
|---|---|---|
| SSB | 0 dB | 0 dB |
| DSB | 0 dB | 3.01 dB |
| AM (50% Modulation) | −9.54 dB | −6.53 dB |
| AM (100% Modulation) | −4.77 dB | −1.76 dB |
| PM ($\Delta\phi = 5$) | 10.97 dB | 21.76 dB |
| FM ($D = 5$) | 15.74 dB | 26.53 dB |
| FM ($D = 5$), Preemphasis, $W = 15$ kHz, $f_1 = 2.2$ kHz | 21.42 dB | 32.22 dB |

gains. This is a result of the fact that (except for SSB) the input signal-to-noise ratio at the input to a receiver is less than that of the baseband comparison signal-to-noise ratio, since the bandwidth and noise power are greater.

## 13-2 Threshold Effects in Analog Systems

Before considering the performance of any actual analog systems, it is necessary to consider a phenomenon called the *threshold effect*. A number of different types of communication processes exhibit this phenomenon, and it is a function of the type of detector in some cases. Qualitatively, the threshold effect arises if the signal-to-noise ratio at the detector input decreases below some critical level, and the result is that the detected signal becomes severely distorted or mutilated by the noise. The detected signal-to-noise ratio decreases much more rapidly than the signal level, and the result will likely be unintelligible.

All angle modulation systems exhibit the threshold effect. Consider the two curves shown in Figure 13–4. The abscissa represents the input signal-to-noise ratio, and the ordinate represents the detected signal-to-noise ratio. Thus, the slope of the curve represents the receiver processing gain.

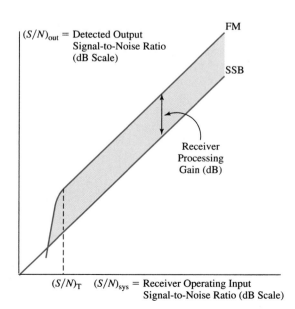

**FIGURE 13–4**
Threshold effect in the receiver processing gain for an FM detector.

The straight line has a slope of unity, meaning that the detected signal-to-noise ratio is the same as the input signal-to-noise ratio. This corresponds to SSB with synchronous product detection and there is no threshold effect.

The more complex curve represents the form of receiver processing gain for either wideband PM or FM. (The levels would be different for the different forms, but the concept is the same.) Let $(S/N)_T$ represent the input signal-to-noise ratio corresponding to threshold. As long as $(S/N)_{sys} > (S/N)_T$, this curve is a straight line when the scales are in decibels, and it is above the curve for SSB. This means, of course, that the detected signal-to-noise ratio is greater than the input signal-to-noise ratio, in accordance with Table 13–1. The difference between the two curves is the processing gain in decibels.

Observe, however, the phenomenon that arises when $(S/N)_{sys}$ drops below $(S/N)_T$. The detected signal-to-noise ratio drops very quickly and is soon below the unity signal processing gain line. In addition, the signal will likely be mutilated and unintelligible.

## Angle Modulation Threshold Level Assumption

Various mathematical developments to predict the exact behavior of the threshold phenomenon in angle modulation have been made, but the results are too involved to cover here. Instead, a simplified practical guideline will be given. It has been found experimentally that a threshold level for the signal input signal-to-noise ratio of approximately 10 dB is a reasonable rule of thumb. Since 10 dB corresponds to an absolute power ratio of 10, we will subsequently assume for angle modulation an approximate threshold level for the receiver input signal-to-noise ratio as follows:

$$(S/N)_T = 10 \text{ for angle modulation} \tag{13-37}$$

This assumption means that the receiver processing gain results of Table 13–1 are considered to be valid for PM and FM only if Equation 13-37 is satisfied. Since operation above threshold is always desirable, it is necessary to specify sufficient signal-to-noise ratio at the input to ensure operation above threshold.

The receiver input signal-to-noise ratio is smaller than the reference baseband value by the factor of $1 + D$ or $1 + \Delta\phi$. Therefore, to ensure that Equation 13-37 is satisfied, the baseband reference signal-to-noise ratio must be somewhat greater. Although it varies with the deviation ratio or phase deviation, we will somewhat arbitrarily assume a value of 20 dB for the baseband reference signal-to-noise ratio, corresponding to an absolute ratio of 100. This value is reasonable and should be adequate for subsequent analysis.

Because of the processing gain associated with angle modulation, an unusual phenomenon occurs as the input signal-to-noise ratio drops below the critical threshold level. Suppose, for example, that the processing gain is 30 dB and the input signal-to-noise ratio is slightly more than 10 dB. The detected output signal-to-noise ratio is above 40 dB, so there will be very little noise observed. However, if the signal strength drops slightly, a very pronounced drop in the detected signal-to-noise ratio will result and the signal will become obliterated. The reader may have observed this phenomenon with an FM radio in an automobile while traveling. A signal may be very strong and appear to be virtually noise free. As an increasing distance or some other phenomenon causes the input signal-to-noise ratio to drop below threshold, the noise level suddenly increases markedly and makes intelligible reception impossible.

## Amplitude Modulation Threshold Considerations

Neither of the AM processes exhibits a threshold effect when *product detection* is used. Since product detection is normally used with both SSB and DSB, the threshold effect need not be considered for such systems. However, envelope detection is used most often for conventional AM, and this form of detection does exhibit the threshold effect. However, its *relative effect* is not nearly as important as in angle modulation, for reasons that will be discussed in the next paragraph.

For a detected signal to be virtually noise free and to exhibit negligible degradation as a result of noise, the detected signal-to-noise ratio should be 30 to 40 dB or more. The signal-to-noise input threshold level for an AM envelope detector is in the neighborhood of about 10 dB. Since the processing gain for AM is less than unity, the input signal-to-noise ratio would have to be somewhat greater than 30 dB at a minimum in order to produce a good detected signal-to-noise ratio. This condition automatically guarantees operation above threshold. If the input signal-to-noise were gradually decreased down to the threshold level, the output signal-to-noise ratio would already be degraded significantly by the additive noise before threshold is observed. Thus, the threshold effect in an AM envelope detector is much more gradual than for angle modulation. To reiterate a point, the normal range for quality detected signal-to-noise ratio in an AM linear envelope detector demands an input signal-to-noise ratio that would exceed the threshold level.

Some digital modulation systems make use of AM-type signals under low signal-to-noise ratio conditions. In such cases, an exact replica of the signal is not necessary, but only the recognition of "ones" and "zeros" is important. In such cases, the use of synchronous product detection of AM can be justified, since the threshold effect could then be avoided.

---

**▌▌ EXAMPLE 13-3**

The average signal power at the input terminals of a receiver is 50 pW. The equivalent antenna input noise temperature is 150 K and the receiver noise temperature referred to the input is 325 K. For a modulating signal bandwidth of 15 kHz, determine the detected output signal-to-noise ratio for FM with $D = 5$. Assume that the IF bandwidth is just sufficient to process the signal.

SOLUTION    The first step is to compute the signal-to-noise ratio at the receiver input and ensure that we are operating above threshold. The net equivalent system operating noise temperature at the input is

$$T_{sys} = T_i + T_e = 150 + 325 = 475 \text{ K} \tag{13-38}$$

The IF bandwidth will be the same as the transmission bandwidth, and is determined from Carson's rule as

$$B = 2(1 + D)W = 2(1 + 5) \times 15 \times 10^3 = 180 \text{ kHz} \tag{13-39}$$

The net input operating noise measured over the transmission bandwidth is

$$N_{sys} = kT_{sys}B = 1.38 \times 10^{-23} \times 475 \times 180 \times 10^3$$
$$= 1.180 \times 10^{-15} \text{ W} = 1.180 \text{ fW} \tag{13-40}$$

The system operating input signal-to-noise ratio referred to the IF bandwidth is

$$(S/N)_{sys} = \frac{P_R}{N_{sys}} = \frac{50 \times 10^{-12} \text{ W}}{1.180 \times 10^{-15} \text{ W}} = 42.37 \times 10^3 \tag{13-41}$$

This value is well above the required threshold value of 10, so the receiver processing gain of Table 13–1 can be used. This value is

$$G_R = 3(1 + D)D^2 = 3(1 + 5)(5)^2 = 450 \tag{13-42}$$

The output signal-to-noise ratio is given by

$$(S/N)_{output} = G_R \times (S/N)_{sys} = 450 \times 42.37 \times 10^3 = 19.07 \times 10^6 \tag{13-43}$$

The corresponding decibel value is

$$(S/N)_{output,dB} = 10 \log(19.07 \times 10^6) = 72.80 \text{ dB} \tag{13-44}$$

An alternate approach to determine the decibel value of the output signal-to-noise ratio is to first determine the input signal-to-noise ratio, which is

$$(S/N)_{sys,dB} = 10 \log(42.37 \times 10^3) = 46.27 \text{ dB} \tag{13-45}$$

The receiver processing gain in decibels is

$$G_{R,dB} = 10 \log(450) = 26.53 \text{ dB} \qquad (13\text{-}46)$$

The detected output signal-to-noise ratio in decibels is then

$$(S/N)_{output,dB} = (S/N)_{sys,dB} + G_{R,dB} = 46.27 + 26.53 = 72.80 \text{ dB} \qquad (13\text{-}47)$$

## 13-3  Digital System Comparisons

In this section, the performance of binary digital PCM systems in the presence of noise will be discussed. Meaningful results for predicting the detected signal-to-noise ratio will be summarized.

### PCM Detected Signal-to-Noise Ratio

Assuming that the sampling rate is sufficiently high to eliminate aliasing, errors in the reconstruction of a PCM signal are based on two factors: (1) quantization noise, and (2) detection errors in the bits. The first case is built into the process and can only be reduced by increasing the number of levels. The second is a result of the noise adding to or subtracting from the signal such that, at the point where a decision on the level is made, the outcome is incorrect. Obviously, this depends greatly on the signal-to-noise ratio of the system. Errors resulting from the quantization process will be called *quantization noise,* and errors resulting from incorrect decisions at the receiver will be called *decision noise.*

Based on a single-tone modulating signal and certain other idealized assumptions, the output detected signal-to-noise ratio of a PCM binary system is given by

$$(S/N)_{output} = \frac{1.5M^2}{1 + 4M^2 P_e} \qquad (13\text{-}48)$$

The quantity $M$ is the number of levels, and is related to the number of bits $N$ in each word by the familiar relationship

$$M = 2^N \qquad (13\text{-}49)$$

The quantity $P_e$ is the probability of error in reading a bit, and requires further explanation.

### Probability of Error

Readers having a background in probability and statistics will be very familiar with the term *probability.* It is a number bounded between 0 and 1 that indicates the relative likelihood that something will occur. If the probability is 0, the event can never occur, and if the probability is 1, the event will definitely occur. In between, the greater the value, the more likely the event will occur.

In the presence of noise, there is always a probability that the noise level can overshadow the signal level. This can make a transmitted 0 appear to the receiver as a transmitted 1, or vice versa. The result will be an error in the particular bit and an incorrect decision concerning the output value.

The probability of error depends on the received signal-to-noise ratio, the particular type of digital modulation, the use of error detection and correction schemes, and many other factors. Indeed, much research has been performed in this modern space age to determine optimum means for reducing the probability of error, to permit deep space communication with a minimum power level. We will study some of the results in sections that follow.

The general shape of the detected signal-to-noise ratio for a binary PCM system is shown in Figure 13–5, without any specific values at this point. The two regions of operation will be discussed.

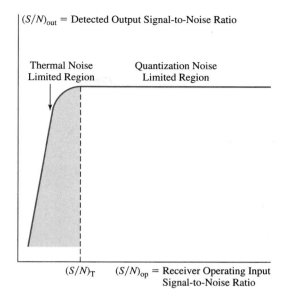

**FIGURE 13–5**
Threshold effect in a digital PCM system.

### Thermal Noise Limiting

At very low values of input signal-to-noise ratio, the value of $P_e$ may be quite large, and the detected signal-to-noise ratio is a critical function of the signal and noise levels. This region can be referred to as the *thermal noise limited region,* referring to the fact that many decision errors result from thermal noise in this region. For a digital system, this region corresponds to operation below threshold, and would be quite undesirable for system operation.

### Quantization Noise Limiting

As the input signal-to-noise ratio increases, $P_e$ eventually reduces to a negligible value, and the detected signal-to-noise ratio approaches a constant limiting value based on the number of bits. This region can be referred to as the *quantization noise limited region,* and operation in this region is desired.

### Digital System Threshold

The threshold value is a function of the probability of error $P_e$ and the number of levels $M$ used in the digital encoding process. For techniques possessing relatively high values for the probability of error, the input signal-to-noise ratio must be correspondingly higher to overcome the effects of decision noise. Likewise, as the number of levels increases, the relative effect of decoding errors is greater, so the input signal-to-noise ratio must be increased accordingly. For these reasons, it is more difficult to precisely state a general value for the threshold signal-to-noise ratio. One common criterion for the threshold value is the input signal-to-noise ratio that results in a 1-dB degradation in the detected signal-to-noise ratio from the quantization noise-limited region. Another common criterion is to specify operation at a specific probability of error, and then assume that the relative effects of decision noise will be negligible in that region. A common value is $P_e = 10^{-5}$. This means that, on average, a bit error will occur once for every 100,000 bits.

### Source Encoding

If source encoding for the purpose of detecting or correcting errors is used, the threshold may be reduced to a relatively low level. Source encoding methods employ redundancy in the form of extra bits to assist in the detection or correction of errors. At one extreme, a parity bit represents a simple form of encoding. More sophisticated methods employ several bits, which represent various combinations of the data pattern. Some of these methods require digital signal processing at the receiver to detect or correct any errors, and to sort the desired signal from the complex encoded signal.

## Detected Signal-to-Noise Ratio in Quantization Limited Region

Once the input signal-to-noise ratio exceeds the threshold level and the curve flattens out, the detected signal-to-noise ratio increases no further as the input signal-to-noise ratio continues to increase. This is in sharp contrast to analog systems, where the detected signal-to-noise ratio continues to increase. The detected signal-to-noise ratio in this region is a function only of the number of bits per word, and can be increased only by adding more bits.

The detected signal-to-noise ratio in the quantization noise limited region reduces to

$$(S/N)_{\text{output}} = 1.5M^2 = (1.5)(2)^{2N} \tag{13-50}$$

The corresponding value in decibels is

$$(S/N)_{\text{output,dB}} = 1.76 + 6.02N \text{ dB} \tag{13-51}$$

The last result is interesting in that it provides a simple and easily remembered rule of thumb. In the region above threshold, PCM provides a detected signal-to-noise ratio of *about 6 dB per bit*.

Since the signal-to-noise ratio in PCM reaches a limiting value above threshold, it is desirable to operate as close to threshold as possible in situations where the transmitted power is at a premium (e.g., space applications). The threshold level is very dependent on the probability of error $P_e$, so system designers utilize probability of error results extensively in determining appropriate power levels.

---

**III EXAMPLE 13-4**

Commercial recorded compact discs (CDs) use 16-bit words for encoding the samples. Determine the theoretical signal-to-noise ratio in decibels. Assume the presence of error-correcting codes, so that no decision errors can occur.

**SOLUTION**   Based on the assumption of no decision errors, the signal-to-noise ratio is given by

$$(S/N)_{\text{output,dB}} = 1.76 + 6.02N = 1.76 + 6.02 \times 16 = 98.08 \text{ dB} \tag{13-52}$$

This is an extremely high signal-to-noise ratio and it represents the high quality of modern recording technology.

---

**III EXAMPLE 13-5**

An instrumentation signal is to be encoded and recorded in digital format. The required signal-to-noise ratio must be at least 53 dB. Determine the minimum number of bits, if it is assumed that there are no decision errors.

**SOLUTION**   From the equation for signal-to-noise ratio in the quantization noise limited region, we have

$$(S/N)_{\text{output,dB}} = 1.76 + 6.02N = 53 \tag{13-53}$$

Solution of this equation yields $N = 8.51$ bits, which is not an integer. We need to round upward, and set

$$N = 9 \text{ bits} \tag{13-54}$$

This number of bits actually provides a theoretical signal-to-noise ratio of 55.94 dB. This margin is probably desirable, since the theoretical development is based on the assumption of a single-tone modulating signal, which tends to exaggerate the results somewhat.    ▮

## 13-4    Probability of Error Analysis

We have seen in the last section that the detected signal-to-noise ratio for PCM is a function of the probability of error and the quantization step. In the transmission of data codes (e.g., ASCII codes), quantization error is meaningless. However, decision errors on bits may occur, so the probability of error plays an important role in all digital communication methods. The determination of the probability of error for a given digital encoding system is a complex process, and many books and journal articles have been devoted to the analysis involved.

### Matched Filter

A concept that plays an important role in comparing binary PCM systems is that of the so-called *matched filter* detector. A matched filter detector is one that is optimized with respect to the shape of the transmitted digital signal. The *integrate-and-dump* circuit is an optimum matched filter for a baseband digital pulse train (e.g., an NRZ-L signal).

### Bit Energy to Noise Density Ratio

In the analysis of digital systems, it is more convenient to alter the definition of signal-to-noise ratio from the power levels used in analog systems to a different form. The signal is measured in terms of the *bit energy in joules* (J), which will be denoted $E_b$. The noise is measured in terms of the *one-sided noise density in watts/hertz* (W/Hz). The ratio of the two quantities will be called the *bit energy to one-sided noise density ratio,* which will be denoted $(E_b/\eta)$. While the units for this ratio are stated as being different, they are actually the same dimensionally. The units of *watts/hertz* is equivalent to the units of *watts × seconds,* and this is dimensionally equivalent to *joules*.

### Relationship to Power Ratios

We will show here how the bit energy to noise density ratio can be related to power ratios. For this purpose, let us define an operating power ratio $(C/N)_{sys}$ at the input to a receiver as follows:

$$(C/N)_{sys} = \frac{\text{average carrier power in watts}}{\text{average noise power in watts}} = \frac{P_r}{N_{sys}} \qquad (13\text{-}55)$$

For convenience in this development, we have redefined $P_r$ as the carrier power at the receiver input. Recall that for angle modulated systems, the net power is the same as the unmodulated carrier power, but the definition would be altered for AM systems. Thus, the $C/N$ ratio for digital systems is essentially the same as the $S/N$ ratio for analog systems, except that the numerator is the average carrier power rather than the total average power.

For binary transmission, the bit energy $E_b$ is determined from the average power $P_r$ and the bit width or duration $T_b$:

$$E_b = P_r T_b = \frac{P_r}{R} \qquad (13\text{-}56)$$

where $R$ is the data rate in bits/second for binary transmission, as given by

$$R = \frac{1}{T_b} \qquad (13\text{-}57)$$

The system noise power $N_{sys}$ is related to the one-sided noise density $\eta_{sys}$ and the bandwidth $B$ by

$$N_{sys} = \eta_{sys} B \qquad (13\text{-}58)$$

Solving for $P_r$ in Equation 13-56 and substituting that result and Equation 13-58 into Equation 13-55, we obtain

$$(C/N)_{\text{sys}} = \frac{E_b}{\eta_{\text{sys}}} \left( \frac{R}{B} \right) \tag{13-59}$$

Let us define the *bit energy to one-sided noise density ratio* as $\rho$. Using this definition and rearranging Equation 13-59, we obtain

$$\rho = \frac{E_b}{\eta_{\text{sys}}} = (C/N)_{\text{sys}} \times \left( \frac{B}{R} \right) \tag{13-60}$$

Thus, the bit energy to one-sided noise density ratio is directly proportional to the carrier power to noise power ratio, and differs only by the factor $(B/R)$ for binary systems. For some PCM systems, the data rate is exactly equal to the bandwidth, and in such cases the values of $\rho$ and $(C/N)_{\text{sys}}$ are identical. In other cases, they differ by a multiplicative constant.

The value of the bit energy to one-sided noise density ratio in decibels will be denoted $\rho_{\text{dB}}$, and is defined as

$$\rho_{\text{dB}} = 10 \log(\rho) = 10 \log \left( \frac{E_b}{\eta_{\text{sys}}} \right) = (C/N)_{\text{sys,dB}} + 10 \log \left( \frac{B}{R} \right) \tag{13-61}$$

The latter form of the preceding equation is convenient when the carrier-to-noise ratio in dB is given. We simply add the decibel form of the bandwidth to data rate ratio to the carrier-to-noise ratio to obtain the bit energy per one-sided noise density ratio in dB.

We conclude this section by reminding the reader of a point made in Chapter 12 about notation. Many references use $N_0$ or $N_o$ to represent noise density in watts/hertz, in which case the bit energy to noise density ratio is then represented as $E_b/N_0$ or $E_b/N_o$. However, we have chosen in this text to use $\eta$ for noise density, so that it will not be confused with total noise power.

## 13-5 Probability of Error Results

Some of the representative probability of error functions for binary systems will be presented in this section. While the results are not exhaustive, they will illustrate typical values that are encountered in practice.

### Binary Probability of Error Curves

Refer to the curves in Figure 13–6 for the discussion that follows.

**Curve A** corresponds to the **matched filter binary phase-shift keying (BPSK) detection** (also realizable as coherent PSK detection). This curve represents an optimum situation, and is the minimum probability of error possible without special error detection and correction. This curve also applies to **quadrature phase-shift keying (QPSK),** provided that the abscissa is interpreted as the bit energy to noise density ratio for each individual carrier. (Remember that there are two carriers operating at 90° phase shifts apart.)

**Curve B** represents the form of **differentially encoded phase-shift keying (DPSK)** developed in Chapter 12, which utilizes self-synchronized differential detection. This form has a higher probability of error than optimum PSK, which means that a higher power level would be required for DPSK to produce the same detected signal-to-noise ratio when compared with matched filter PSK. There are several other variations for DPSK detection, but the form provided is the most common type due to its ease of detection.

**Curve C** represents the probability of error for **matched filter amplitude-shift keying (ASK)** and for **matched filter frequency-shift keying (FSK).** This corresponds to coherent detection, and is superior to envelope detection from a signal-to-noise point of view. However, there is a 3-dB degradation as compared with matched filter PSK. Moreover, on average, the carrier is off for about half of the time, so the average power is half that of

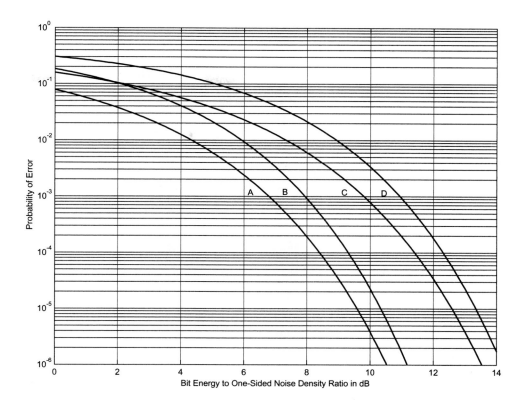

**FIGURE 13–6**
Probability of error curves for several binary modulation systems.

PSK for a given peak voltage level. The reader may readily verify the 3-dB degradation by noting that curves A and C are shifted apart horizontally by 3 dB.

**Curve D** represents the results of **envelope detection of frequency-shift keying (FSK).** As a general rule, noncoherent envelope detection processes tend to be inferior to coherent techniques, but they are often easier to implement.

These four cases provide a survey of representative probability of error results for binary PCM systems. There are other possibilities, but these are some of the common ones employed in binary PCM.

---

**⫼ EXAMPLE 13-6**

In a binary digital communication system, the average signal carrier power at the antenna terminals of the receiver is 200 fW. The equivalent antenna noise temperature is 300 K, and the receiver effective noise temperature referred to the input is 425 K. Determine the detected output signal-to-noise ratio, for binary PCM transmission with 6-bit words at a data rate of 2 Mbits/s, for (a) PSK with matched filter, (b) DPSK, (c) ASK with matched filter, and (d) envelope detected FSK.

**SOLUTION** The number of distinct binary words is given by

$$M = 2^N = 2^6 = 64 \tag{13-62}$$

The bit energy is given by

$$E_b = P_r T_b = \frac{P_r}{R} = \frac{200 \times 10^{-15}\ \text{W}}{2 \times 10^6\ \text{bits/s}} = 100 \times 10^{-21}\ \text{J} \tag{13-63}$$

The units in the preceding equation may seem strange, but division by bits/seconds is equivalent to multiplication by the width of one bit in seconds, so it works out.

The one-sided input noise power spectral density at the receiver input is given by

$$\eta_{sys} = kT_{sys} = k(T_i + T_e) = 1.38 \times 10^{-23} \times (300 + 425)$$
$$= 1 \times 10^{-20} \text{ W/Hz} \tag{13-64}$$

The bit energy to one-sided noise density ratio is

$$\rho = \frac{100 \times 10^{-21} \text{ J}}{1 \times 10^{-20} \text{ W/Hz}} = 10 \tag{13-65}$$

Once again the numerator and denominator dimensions seem strange, but 1 J is dimensionally equivalent to 1 W/Hz. The corresponding value in decibels is

$$\rho_{dB} = 10 \log \rho = 10 \log 10 = 10 \text{ dB} \tag{13-66}$$

A power ratio of 10 corresponds to a decibel ratio of 10 dB, and this is the only point where the values are numerically the same.

The procedure in each part that follows will be to first determine the probability of error for a given modulation method using Figure 13–6. The detected signal-to-noise ratio may then be determined from Equation 13-48.

(a)  From curve A of Figure 13–6, the probability of error for a bit energy to one-sided noise density ratio of 10 dB is about $P_e \approx 4 \times 10^{-6}$. Substitution of this value and $M = 64$ in Equation 13-48 yields

$$(S/N)_{output} = \frac{1.5M^2}{1 + 4M^2 P_e} = \frac{1.5 \times (64)^2}{1 + 4 \times (64)^2 \times 4 \times 10^{-6}}$$
$$= 5766 \text{ (or 37.61 dB)} \tag{13-67}$$

Operation in this case is very close to the threshold level, in which the quantization limiting signal-to-noise ratio would be $1.5 \times (64)^2 = 6144$ or 37.88 dB.

(b)  From curve B of Figure 13–6, $P_e \approx 2.3 \times 10^{-5}$. The detected signal-to-noise ratio is

$$(S/N)_{output} = \frac{1.5M^2}{1 + 4M^2 P_e} = \frac{1.5 \times (64)^2}{1 + 4 \times (64)^2 \times 2.3 \times 10^{-5}}$$
$$= 4462 \text{ (or 36.50 dB)} \tag{13-68}$$

As in part (a), operation is still reasonably near threshold, but with a slightly reduced output signal-to-noise ratio due to the higher probability of error with DPSK.

(c)  From curve C of Figure 13–6, $P_e \approx 8 \times 10^{-4}$. The detected signal-to-noise ratio is

$$(S/N)_{output} = \frac{1.5M^2}{1 + 4M^2 P_e} = \frac{1.5 \times (64)^2}{1 + 4 \times (64)^2 \times 8 \times 10^{-4}}$$
$$= 435.5 \text{ (or 26.39 dB)} \tag{13-69}$$

Operation has dropped well below threshold in this part, and the signal-to-noise ratio has been degraded significantly.

(d)  From curve D of Figure 13–6, $P_e \approx 3.5 \times 10^{-3}$. The detected signal-to-noise ratio is

$$(S/N)_{output} = \frac{1.5M^2}{1 + 4M^2 P_e} = \frac{1.5 \times (64)^2}{1 + 4 \times (64)^2 \times 3.5 \times 10^{-3}}$$
$$= 105.3 \text{ (or 20.22 dB)} \tag{13-70}$$

The degradation is even worse in part (d) than in part (c). Clearly parts (c) and (d) represent undesirable situations.

## ■▮ EXAMPLE 13-7

For the system of Example 13-6, assume that the data rate is increased to 8 Mbits/s while all other parameters are unchanged. For PSK with matched filter, determine the detected signal-to-noise ratio.

SOLUTION    The effect of an increase in the data rate is that the bit energy is reduced. Specifically, if the data rate is increased from 2 Mbits/s to 8 Mbits/s, the bit energy changes to $\frac{1}{4}$ of its original value and the value of $\rho$ is changed from 10 to $\frac{10}{4} = 2.5$. (Note that the bandwidth must increase by a factor of 4 to accomodate the higher data rate, but the noise density is assumed to remain the same.)

The new value of $\rho_{dB}$ is $10 \log(2.5) = 3.98$ dB. From curve A of Figure 13–6, the probability of error is estimated as $P_e \approx 1.3 \times 10^{-2}$. The detected signal-to-noise ratio is then given by

$$(S/N)_{\text{out}} = \frac{1.5M^2}{1 + 4M^2 P_e} = \frac{1.5 \times (64)^2}{1 + 4 \times (64)^2 \times 1.3 \times 10^{-2}}$$
$$= 28.71 \text{ (or 14.58 dB)} \tag{13-71}$$

The system is now operating well below threshold, so the signal-to-noise ratio is very sensitive to changes in the power or noise level. Comparing the result of this example with part (a) of Example 13-6, it is noted how the data rate can have a significant effect on the overall performance if operation is near threshold.

## ■▮ EXAMPLE 13-8

The design specifications in a binary data computer communication system call for the probability of error not to exceed $P_e = 10^{-5}$ at a data rate of 1 Mbit/s. The effective antenna temperature is 250 K and the effective receiver temperature referred to the input is 475 K. Determine the required receiver input average carrier power for each of the following modulation methods and conditions: (a) matched filter PSK, (b) DPSK, (c) matched filter ASK, and (d) envelope detected FSK.

SOLUTION    It should be noted that since this system transmits digital data characters (e.g., ASCII characters), the concept of quantization error is not relevant. Instead systems such as this frequently use the probability of error as a basis for design.

In each case, Figure 13–6 will be used to determine the value of $\rho_{dB}$ at which $P_e = 10^{-5}$. The corresponding value of $\rho$ will then be calculated as

$$\rho = 10^{\rho_{dB}/10} \tag{13-72}$$

in accordance with standard decibel computations. The input noise density is

$$\eta_{\text{sys}} = 1.38 \times 10^{-23} \times (250 + 475) = 1 \times 10^{-20} \text{ W/Hz} \tag{13-73}$$

Since $\rho = E_b/\eta_{\text{sys}}$, the bit energy $E_b$ is determined as

$$E_b = \eta_{\text{sys}}\rho = 1 \times 10^{-20}\rho \tag{13-74}$$

Since $E_b = P_r/R$, the carrier signal power $P_r$ is determined as

$$P_r = RE_b \tag{13-75}$$

The various calculations are tabulated in each part that follows.

(a)  From curve A of Figure 13–6, $\rho_{dB} \approx 9.6$ dB for $P_e = 10^{-5}$. The value of $\rho$ is determined from Equation 13-72.

$$\rho = 10^{9.6/10} = 9.12 \tag{13-76}$$

The bit energy is determined from Equation 13-74.

$$E_b = 1 \times 10^{-20} \times 9.12 = 91.2 \times 10^{-21} \text{ J} \tag{13-77}$$

The required received carrier power is then determined from Equation 13-75.

$$P_r = 1 \times 10^6 \times 91.2 \times 10^{-21} = 91.2 \text{ fW} \tag{13-78}$$

(b)  From curve B of Figure 13–6, $\rho_{dB} \approx 10.3$ dB for $P_e = 10^{-5}$. The remaining calculations follow.

$$\rho = 10^{10.3/10} = 10.72 \tag{13-79}$$

$$E_b = 1 \times 10^{-20} \times 10.72 = 107.2 \times 10^{-21} \text{ J} \tag{13-80}$$

$$P_r = 1 \times 10^6 \times 107.2 \times 10^{-21} = 107.2 \text{ fW} \tag{13-81}$$

(c)  From curve C of Figure 13–6, $\rho_{dB} \approx 12.6$ dB for $P_e = 10^{-5}$. The remaining calculations follow.

$$\rho = 10^{12.6/10} = 18.20 \tag{13-82}$$

$$E_b = 1 \times 10^{-20} \times 18.20 = 182.0 \times 10^{-21} \text{ J} \tag{13-83}$$

$$P_r = 1 \times 10^6 \times 182.0 \times 10^{-21} = 182.0 \text{ fW} \tag{13-84}$$

(d)  From curve D of Figure 13–6, $\rho_{dB} \approx 13.4$ dB for $P_e = 10^{-5}$. The remaining calculations follow.

$$\rho = 10^{13.4/10} = 21.88 \tag{13-85}$$

$$E_b = 1 \times 10^{-20} \times 21.88 = 218.8 \times 10^{-21} \text{ J} \tag{13-86}$$

$$P_r = 1 \times 10^6 \times 218.8 \times 10^{-21} = 218.8 \text{ fW} \tag{13-87}$$

The required power at the receiver antenna varies from less than 100 fW to more than 200 fW for the same probability of error with the methods considered.

---

**▌▊ EXAMPLE 13-9**

For the system of Example 13-8, assume that the data rate is increased to 2 Mbits/s while all other parameters are unchanged. For PSK with matched filter, determine the required receiver carrier power.

**SOLUTION**  For the same noise density, the bit energy must remain the same in order to maintain the same probability of error. Therefore, the only variable that must change is the received carrier power—which, as seen by Equation 13-75, is directly proportional to the data rate for a fixed bit energy. From part (a) of Example 13-8, the required bit energy for PSK with matched filter is 91.2 fW. Hence,

$$P_r = RE_b = 2 \times 10^6 \times 91.2 \times 10^{-21} = 182.4 \text{ fW} \tag{13-88}$$

The power level must be doubled to maintain the same probability of error if the data rate is doubled.                                                                          ▐▊

---

## 13-6  General Modulation System Comparisons

All considerations up to this point in the chapter have been directed toward the performance analysis of different modulation systems as a function of the power levels, signal-to-noise ratios, and other operational parameters of the system. In determining the response of receivers, use was made of the receiver processing gain values from Table 13–1 for analog systems, and the probability of error curves for digital systems.

In this section, some general comparisons between different types of modulation systems will be made, including a comparison between analog and digital systems. The best parameter to use for that purpose is the baseband comparison gain parameter $R_B$.

### PM

Several curves comparing the detected signal-to-noise ratio to the baseband reference signal-to-noise ratio for PM above threshold, both measured in dB, are shown in Figure 13–7. The parameter for the different curves is the maximum phase deviation. To simplify the presentation,

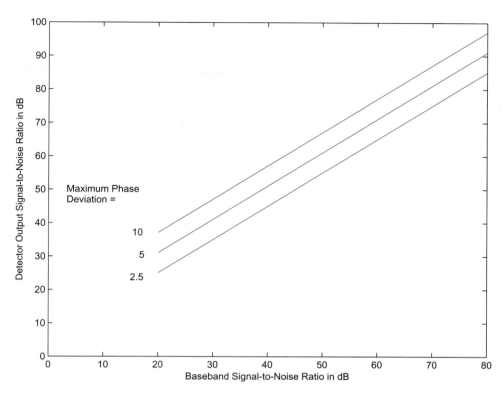

**FIGURE 13–7**
Performance of PM for three different maximum phase deviation values.

the same value for the threshold level has been assumed in all cases, since the actual value varies somewhat with the type of detection process. In the region above threshold, the variation in detected signal-to-noise ratio in dB is a linear function of the baseband reference signal-to-noise ratio in dB, with a slope of 1. However, the curves shift upward as the maximum phase deviation increases.

## FM

Several curves comparing the detected signal-to-noise ratio to the baseband reference signal-to-noise ratio for FM above threshold, both measured in dB, are shown in Figure 13–8. The parameter for the different curves in this case is the deviation ratio. As in the case of PM, the same value for the threshold level has been assumed in all cases. A pattern similar to that of PM is that the variation in detected signal-to-noise ratio in dB is a linear function of the baseband reference signal-to-noise ratio in dB, with a slope of 1. The curves shift upward as the deviation ratio increases. For an FM deviation ratio equal to a PM maximum phase deviation, the performance of FM is about 4.77 dB greater than PM. No preemphasis has been assumed for the FM cases shown here.

## Comparison of Different Analog Systems

A comparison of various analog systems is shown in Figure 13–9. Note that the performance of DSB and SSB is identical on a baseband reference comparison basis, as explained earlier in the chapter. If extended back, the curve would pass through the origin, and there is no threshold effect.

The curve for AM is based on 100% modulation and has the lowest level of any of the curves. For PM, the maximum phase deviation is set at 5, and for FM, the deviation ratio is set at 5. In the case of FM with preemphasis, the ratio $W/f_1 = 7.07$ has been used; this

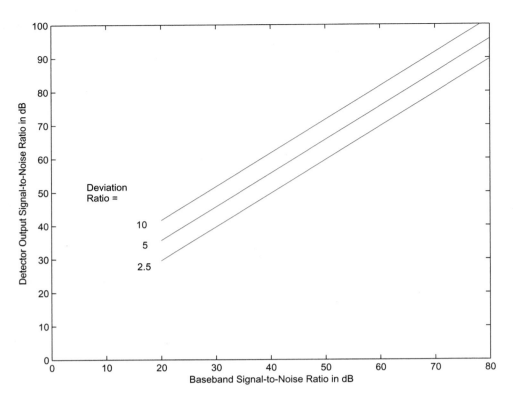

**FIGURE 13–8**
Performance of FM for three different deviation ratios.

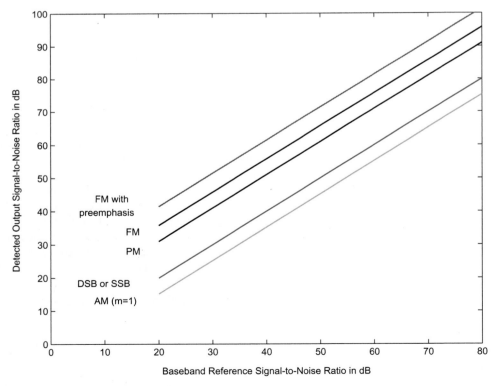

**FIGURE 13–9**
Comparison of analog modulation systems.

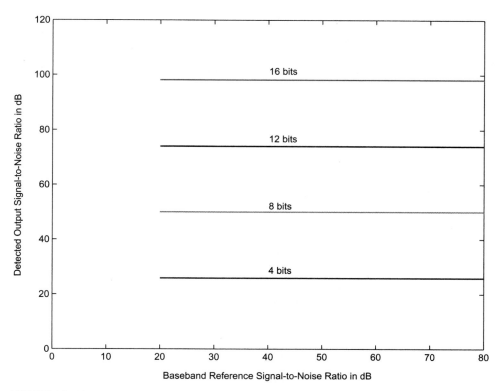

**FIGURE 13–10**
Performance of PCM for different word lengths.

value corresponds to commercial monaural FM broadcasting. By varying different parameters, the AM, PM, and FM curves could be moved up or down, but the trends shown here illustrate the general behavior pattern.

## PCM

As in the case of analog systems, operation above threshold is assumed for PCM, so that the results are applicable to the quantization noise limited region. Curves depicting the detected signal-to-noise ratio of PCM for different word lengths are shown in Figure 13–10. Once the signal level increases beyond the threshold level, the only noise effect is that of quantization noise, and it remains constant with increases in signal-to-noise ratio.

## Comparison of FM with PCM

While many factors must be considered in designing and implementing a communication system, FM tends to be the "best performer" in the analog world, and PCM is the standard for digital encoding and transmission of analog signals. Hence, a comparison between FM and PCM has some interesting features. We will consider FM without preemphasis, but we know that it could be even better with preemphasis.

To provide a common basis for comparison, it is convenient to introduce a *bandwidth expansion factor,* which will be denoted simply as $K$ and defined as

$$K = \frac{B}{W} \tag{13-89}$$

For FM, Carson's rule establishes the bandwidth as

$$B = 2(D + 1)W \tag{13-90}$$

The constant $K$ is then

$$K = \frac{B}{W} = 2(D + 1) \tag{13-91}$$

For PCM, the bandwidth varies with the type of modulation used, but for binary ASK, PSK, and DPSK, the RF bandwidth is equal to the data rate, and is

$$B = R \tag{13-92}$$

Therefore, cases that satisfy this criterion will be used as a basis for comparison. The data rate is the number of bits/second, and each word occupies $N$ bits. Assume that the sampling rate is

$$f_s = 2W \tag{13-93}$$

The data rate must be $N$ times the sampling rate in order that $N$ bits can be transmitted per sample word. Hence,

$$B = R = 2NW \tag{13-94}$$

Therefore,

$$K = \frac{B}{W} = 2N = 2 \log_2 M \tag{13-95}$$

Several curves showing the detected signal-to-noise ratio versus the baseband reference signal-to-noise ratio for different bandwidth expansion factors for both FM and PCM are compared in Figure 13–11.

The results of this comparison are interesting and deserve some discussion. It can be seen that both FM in the analog world and PCM in the digital world have regions in which they are superior to the other, but neither can be regarded as "best." At lower signal-to-noise

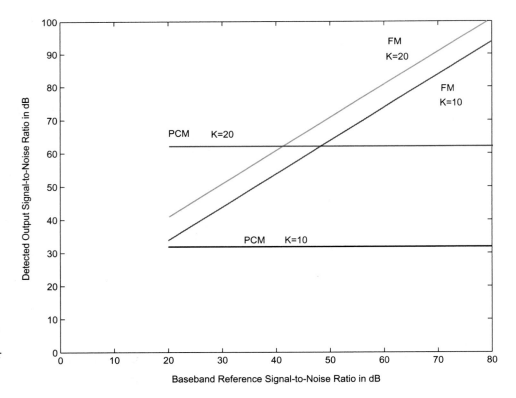

**FIGURE 13–11**

Comparison of FM and PCM for different bandwidth expansion ratios.

ratios (above threshold, of course), PCM appears to be superior, but as the signal-to-noise ratio increases, FM takes the lead. Both systems are capable of exhibiting outstanding performance, but a number of different factors must be considered in deciding which would be best in a given application. Because of the increasing use of digital technology in so many areas of life, it is likely that PCM systems will be the wave of the future.

Because of the superiority of PCM at low signal-to-noise ratios, one can readily surmise why much research has been conducted in the space program to lower the threshold level for PCM, and thus permit operation at very low signal levels from deep space. In this type of operation, PCM would obviously show superiority.

## 13-7  MATLAB® Examples (Optional)

Two M-file programs will be introduced in this section. The first one has the title **analog.m** and can be used in analyzing the performance of analog modulation systems. The second has the title **digital.m** and can be used in analyzing the performance of the binary digital systems considered in this chapter.

Both of the programs involve commands that would sidetrack us from the major emphasis of this text if we stopped to study them in detail. Both utilize **if** statements to allow the user to have the option of the particular modulation system under study. They also utilize print statements that are compatible with C programming commands. A brief summary of each program follows, and the code in each case will be provided for those readers who wish to investigate the structure in more detail.

### analog.m

The code for the program **analog.m** is provided in Figure 13–12. In effect, this program provides virtually all the information of Table 13–1 in computational form. The input and output data are as follows:

Input:  Analog modulation form entered as a code number

Receiver input signal-to-noise ratio in dB

Other information as required for the particular modulation system

Output:  Baseband comparison gain as an absolute value

Baseband comparison gain in dB

Receiver processing gain as an absolute value

Receiver processing gain in dB

Detected output S/N ratio in dB

The program does not have a check on threshold. Therefore, the results are generally valid only if operation above threshold is assumed.

### digital.m

The code for the program **digital.m** is shown in Figure 13–13. The input and output data are as follows:

Input:  Digital modulation form entered as a code number

Number of signal bits in each word

Probability of error for systems other than those in text

Output:  Output S/N ratio as an absolute value

Output S/N ratio in dB

```
%Progam analog.m
fprintf('Enter a number that corresponds to the modulation system.\n\n')
fprintf('For SSB, enter 1.\n')
fprintf('For DSB, enter 2.\n')
fprintf('For AM, enter 3.\n')
fprintf('For PM, enter 4.\n')
fprintf('For FM without preemphasis, enter 5.\n')
fprintf('For FM with preemphasis, enter 6.\n')
fprintf('\n')
n=input('Enter the correct number from the list above. ');
Ridb=input('Enter the receiver input S/N ratio in dB. ');
if n==1
Gb=1;
Gr=1;
elseif n==2
   Gb=1;
   Gr=2;
elseif n==3
   m=input('Enter the modulation factor as a fraction. ');
   Gb=m^2/(2+m^2);
   Gr=2*Gb;
elseif n==4
   delphi=input('Enter the maximum phase deviations in radians. ');
   Gb=0.5*delphi^2;
   Gr=2*Gb*(1+delphi);
elseif n==5
   D=input('Enter the FM deviation ratio. ');
   Gb=1.5*D^2;
   Gr=2*Gb*(1+D);
elseif n==6
      D=input('Enter the FM deviation ratio. ');
      pre_ratio=input('Enter the ratio of 3-dB preemphasis frequency to baseband bandwidth. ');
      factor=(pi/6)*pre_ratio;
      Gb=1.5*factor*D^2;
      Gr=2*Gb*(1+D);
   else
       fprintf('The number entered is not valid.')
   end
fprintf('\n')
Gbdb=10*log10(Gb);
Grdb=10*log10(Gr);
Rodb=Ridb+Grdb;
fprintf('The absolute value of the baseband comparison gain is %g.\n\n',Gb)
fprintf('The decibel value of the baseband comparison gain is %g dB.\n\n',Gbdb)
fprintf('The absolute value of the receiver processing gain is %g.\n\n',Gr)
fprintf('The decibel value of the receiver processing gain is %g dB.\n\n',Grdb)
fprintf('The decibel value of the detected S/N ratio is %g dB.',Rodb)
```

FIGURE 13–12
M-file program analog.m.

```
%Progam digital.m
fprintf('Enter a number that corresponds to the digital system.\n\n')
fprintf('For matched filter BPSK, enter 1.\n')
fprintf('For coherent DPSK, enter 2.\n')
fprintf('For matched filter ASK, enter 3.\n')
fprintf('For envelope detected FSK, enter 4.\n')
fprintf('For other binary systems, enter 5.\n')
fprintf('\n')
n=input('Enter the correct number from the list above. ');
N=input('Enter the number of bits in each word. ');
M=2^N;
rhodb=input('Enter the input bit energy to one-sided noise density in dB. ');
rho=10^(rhodb/10);
if n==1
Pe=0.5*erfc(sqrt(rho));
elseif n==2
  Pe=0.5*exp(-rho);
elseif n==3
  Pe=0.5*erfc(sqrt(rho/2));
elseif n==4
  Pe=0.5*exp(-rho/2);
elseif n==5
  Pe=input('Enter the probability of error. ');
else
    fprintf('The number entered is not valid.')
end
fprintf('\n')
R=(1.5*M^2)/(1+4*M^2*Pe);
Rdb=10*log10(R);
fprintf('The absolute value of the output S/N ratio is %g.\n\n',R)
fprintf('The decibel value of the output S/N ratio is %g dB.\n\n',Rdb)
```

**FIGURE 13–13**
M-file program digital.m.

---

**▌▌ MATLAB EXAMPLE 13-1**     Repeat the analysis of Example 13-3 using the program **analog.m.**

SOLUTION    The program is executed by the command **analog.** The dialogue then begins and data must be entered as required. The results are shown in Figure 13–14. They are in agreement with those of Example 13-3 to the number of decimal places involved in the example.

---

**▌▌ MATLAB EXAMPLE 13-2**     Repeat the analysis of Example 13-6(a) using the program **digital.m.**

SOLUTION    The program is initiated by the command **digital.** The resulting dialogue is shown in Figure 13–15. The program provides an output S/N ratio of about 37.62 dB, which is extremely close to the value of 37.61 dB obtained in Example 13-6(a). Remember that in the text example, the value of probability of error had to be read from a curve, so the difference in the values is negligible. For cases involving a higher probability of error, a greater difference would be likely.    ▌▋

```
>> analog
Enter a number that corresponds to the modulation system.

For SSB, enter 1.
For DSB, enter 2.
For AM, enter 3.
For PM, enter 4.
For FM without preemphasis, enter 5.
For FM with preemphasis, enter 6.

Enter the correct number from the list above. 5
Enter the receiver input S/N ratio in dB. 46.27
Enter the FM deviation ratio. 5

The absolute value of the baseband comparison gain is 37.5.

The decibel value of the baseband comparison gain is 15.7403 dB.

The absolute value of the receiver processing gain is 450.

The decibel value of the receiver processing gain is 26.5321 dB.

The decibel value of the detected S/N ratio is 72.8021 dB.
```

**FIGURE 13–14**
Results for MATLAB Example 13-1.

```
>> digital
Enter a number that corresponds to the digital system.

For matched filter BPSK, enter 1.
For coherent DPSK, enter 2.
For matched filter ASK, enter 3.
For envelope detected FSK, enter 4.
For other binary systems, enter 5.

Enter the correct number from the list above. 1
Enter the number of bits in each word. 6
Enter the input bit energy to one-sided noise density in dB. 10

The absolute value of the output S/N ratio is 5777.47.

The decibel value of the output S/N ratio is 37.6174 dB.
```

**FIGURE 13–15**
Results for MATLAB Example 13-2.

## SystemVue™ Closing Application (Optional)

Insert the text CD in a computer having SystemVue™ installed and activate the program. Open the CD folder entitled **SystemVue Systems** and load the file entitled 13-1.

### Review of System

The system shown on the screen was described in the chapter opening application, and you may wish to review the discussion and system configuration provided there. Recall that the input signal is a 5-Hz sinusoid that frequency modulates a 1-kHz carrier with a deviation of ±50 Hz, and that a PLL detector is used to demodulate the signal. Although there was a noise source in the system, it was set to 0 for the opening application.

### FM Performance with Noise

The desired signal-to-noise ratio (specifically, the carrier-to-noise ratio) to be established at the detector input is 10 dB, which corresponds to an absolute power ratio of 10. For this demonstration, it is simpler to assume a 1-$\Omega$ resistance, and for a carrier amplitude of 1 V, the signal power is 0.5 W. Thus, the noise power required is 0.05 W. Based on a bandwidth of 110 Hz for the detector input filter, show that this would correspond to a hypothetical noise temperature of $32.9 \times 10^{18}$ K.

1. Double click on the noise source, and when the window opens, left-click on **Thermal.** Open the **Parameters** window, set the **Temperature** to **32.9e18,** and the **Resistance** to **1 $\Omega$**.

2. Perform a simulation run and observe the various waveforms. Observe the noise (Token 1) and the FM signal plus noise (Token 2). The latter signal appears to be obscured by the presence of the noise.

3. Observe the output (Token 3). After the settling interval, how does this output compare with the output in the noise-free case? What can you say in this case about the FM detection process as far as the noise is concerned?

## PROBLEMS

**13-1** Calculate the decibel values of (a) baseband comparison gain and (b) receiver processing gain for AM with $m = 0.75$.

**13-2** Calculate the decibel values of (a) baseband comparison gain and (b) receiver processing gain for AM with $m = 0.25$.

**13-3** Calculate the decibel values of (a) baseband comparison gain and (b) receiver processing gain for PM with $\Delta\phi = 10$.

**13-4** Calculate the decibel values of (a) baseband comparison gain and (b) receiver processing gain for PM with $\Delta\phi = 2.5$.

**13-5** Calculate the decibel values of (a) baseband comparison gain and (b) receiver processing gain for FM with $D = 10$.

**13-6** Calculate the decibel values of (a) baseband comparison gain and (b) receiver processing gain for FM with $D = 2.5$.

**13-7** Calculate the decibel values of (a) baseband comparison gain and (b) receiver processing gain for FM with $D = 10$ and preemphasis with $f_1 = 2$ kHz and $W = 10$ kHz.

**13-8** Calculate the decibel values of (a) baseband comparison gain and (b) receiver processing gain for FM with $D = 2.5$ and preemphasis with $f_1 = 1$ kHz and $W = 4$ kHz.

**13-9** The minimum detected signal-to-noise ratio in an FM system with no preemphasis is required to be 80 dB and the deviation ratio is 5. Determine the net minimum signal-to-noise ratio in dB required at the antenna terminals.

**13-10** The minimum detected signal-to-noise ratio in an FM system with no preemphasis is required to be 80 dB and the deviation ratio is 10. Determine the net minimum signal-to-noise ratio in dB required at the antenna terminals.

**13-11** The average signal power at the antenna terminals of a receiver is 20 dBf. The equivalent antenna noise temperature is 100 K and the receiver noise temperature referred to the input is 250 K. For a modulating signal bandwidth of 20 kHz, determine the detected signal-to-noise ratio for FM with $D = 4$. (*Hint:* Don't forget to check for threshold.)

**13-12** The average signal power at the antenna terminals of a receiver is 20 dBf. The equivalent antenna noise temperature is 100 K and the receiver noise temperature referred to the input is 250 K. For a modulating signal bandwidth of 20 kHz, determine the detected signal-to-noise ratio for FM with $D = 8$. (*Hint:* Don't forget to check for threshold.)

**13-13** Some of the PCM systems used in the telephone industry utilize 256 levels for encoding the samples. Determine the theoretical signal-to-noise ratio in decibels. Assume that no decision errors occur.

**13-14** Determine the theoretical signal-to-noise ratio if a signal is encoded into 4-bit words.

**13-15** An audio signal is to be encoded and recorded in digital format. The required signal-to-noise ratio must be at least 80 dB. Determine the minimum number of bits, if it is assumed that there are no decision errors.

**13-16** An audio signal is to be encoded and recorded in digital format. The required signal-to-noise ratio must be at least 60 dB. Determine the minimum number of bits, if it is assumed that there are no decision errors.

**13-17** In a binary digital communication system, the average signal carrier power at the antenna terminals is 75 fW. The equivalent antenna noise temperature is 100 K and the receiver effective noise temperature referred to the input is 356 K. Determine the detected output signal-to-noise ratio for PCM transmission with 6-bit words at a data rate of 1.5 Mbits/s for (a) PSK with matched filter, and (b) DPSK.

**13-18** In a binary digital communication system, the average signal carrier power at the antenna terminals is 75 fW. The equivalent antenna noise temperature is 100 K and the receiver effective noise temperature referred to the input is 356 K. Determine the detected output signal-to-noise ratio for PCM transmission with 6-bit words at a data rate of 1.5 Mbits/s for (a) ASK with matched filter, and (b) envelope detected FSK.

**13-19** For the system of Problem 13-17, assume that the data rate is decreased to 1.06 Mbits/s while all other parameters are unchanged. For PSK with matched filter, determine the detected signal-to-noise ratio.

**13-20** For the system of Problem 13-18, assume that the data rate is decreased to 1.06 Mbits/s while all other parameters are unchanged. For ASK with matched filter, determine the detected signal-to-noise ratio.

**13-21** For the system of Problem 13-17, assume that the number of bits per word is changed to 8 while all other parameters are unchanged. For PSK with matched filter, determine the detected signal-to-noise ratio.

**13-22** For the system of Problem 13-18, assume that the number of bits per word is changed to 8 while all other parameters are unchanged. For ASK with matched filter, determine the detected signal-to-noise ratio.

**13-23** The design specifications in a binary data computer communication system call for the probability of error to not exceed $P_e = 10^{-6}$ at a data rate of 2 Mbits/s. The antenna temperature is 200 K and the effective receiver noise temperature referred to the input is 300 K. Determine the required receiver average carrier power for matched filter PSK.

**13-24** The design specifications in a binary data computer communication system call for the probability of effective receiver noise error to not exceed $P_e = 10^{-6}$ at a data rate of 2 Mbits/s. The antenna temperature is 200 K and the temperature referred to the input is 300 K. Determine the required receiver average carrier power for matched filter ASK.

**13-25** The signal power at the input to a receiver is 200 fW. The antenna noise temperature is 300 K and the effective receiver noise temperature referred to the input is 425 K. The receiver IF bandwidth is 50 kHz. Assuming that the maximum possible baseband bandwidth is employed for each case, determine the baseband bandwidth and detected signal-to-noise ratio in decibels for each of the following modulation systems and conditions: (a) SSB, (b) DSB, (c) AM with 100% modulation, (d) PM with a maximum phase deviation of 5 rad, (e) FM with a deviation ratio of 5 (no preemphasis), and (f) FM with a deviation ratio of 5, preemphasis, and $W/f_1 = 8$.

**13-26** The signal power at the input to a receiver is 150 fW. The antenna noise temperature is 100 K and the noise figure of the receiver is 4 dB. The receiver IF bandwidth is 20 kHz. Assuming that the maximum possible baseband bandwidth is employed for each case, determine the baseband bandwidth and detected signal-to-noise ratio in decibels for each of the following modulation systems and conditions: (a) SSB, (b) DSB, (c) AM with 100% modulation, (d) PM with a maximum phase deviation of 10 rad, (e) FM with a deviation ratio of 10 (no preemphasis), and (f) FM with a deviation ratio of 10, preemphasis, and $W/f_1 = 5$.

**13-27** Consider a communication system having the same operating parameters as in Problem 13-25 except that the IF bandwidth is *not* fixed at 50 kHz. Instead a baseband bandwidth of 10 kHz will be used as the modulating signal and the IF bandwidth of the receiver will be adjusted in each case to match the particular modulation method. Determine the detected output signal-to-noise ratio in decibels for all the methods of Problem 13-25.

**13-28** Consider a communication system having the same operating parameters as in Problem 13-26 except that the IF bandwidth is *not* fixed at 20 kHz. Instead a baseband bandwidth of 5 kHz will be used as the modulating signal and the IF bandwidth of the receiver will be adjusted in each case to match the particular modulation method. Determine the detected output signal-to-noise ratio in decibels for all the methods of Problem 13-26.

**13-29**  The design specifications in a communications system require a detected signal-to-noise ratio of 50 dB. The baseband modulating signal bandwidth is 10 kHz. The antenna noise temperature is 250 K and the effective receiver noise temperature referred to the input is 475 K. Determine the required signal received power for each of the following modulation methods and conditions: (a) SSB, (b) DSB, (c) AM with 100% modulation, (d) PM with a maximum phase deviation of 5, (e) FM with a deviation ratio of 5 (no preemphasis), and (f) FM with a deviation ratio of 5, preemphasis, and $W/f_1 = 10$. It is assumed that the receiver IF bandwidth is adjusted to the optimum value for each case.

**13-30**  The design specifications in a communications system require a detected signal-to-noise ratio of 50 dB. The baseband modulating signal bandwidth is 2 kHz. The antenna noise temperature is 250 K and the noise figure of the receiver is 6 dB. Determine the required signal received power for each of the following modulation methods and conditions: (a) SSB, (b) DSB, (c) AM with 100% modulation, (d) PM with a maximum phase deviation of 10, (e) FM with a deviation ratio of 10 (no preemphasis), and (f) FM with a deviation ratio of 10, preemphasis, and $W/f_1 = 5$. It is assumed that the receiver IF bandwidth is adjusted to the optimum value for each case.

**13-31**  Determine the value of $\Delta\phi$ for PM in which the theoretical performance is the same as at baseband.

**13-32**  Determine the value of $D$ for FM in which the theoretical performance is the same as at baseband.

**13-33**  For a given transmission bandwidth and with no preemphasis, determine the theoretical improvement for FM as compared with PM.

**13-34**  Commercial monaural FM broadcasting uses preemphasis corresponding to a one-pole filter with a time constant of 75 μs. Based on a modulating bandwidth of 15 kHz, determine the theoretical signal-to-noise improvement in decibels by using preemphasis.

**13-35**  Consider a PCM system with 12-bit words. Compute the detected signal-to-noise ratio for the following probabilities of error: (a)$10^{-3}$, (b) $10^{-4}$, (c) $10^{-5}$, (d) $10^{-6}$, (e) $10^{-7}$, (f) $10^{-8}$.

**13-36**  Repeat the analysis of Problem 13-35 for 8-bit words.

# Transmission Lines and Waves  14

## OVERVIEW AND OBJECTIVES

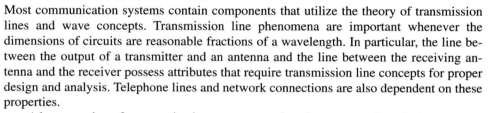

Most communication systems contain components that utilize the theory of transmission lines and wave concepts. Transmission line phenomena are important whenever the dimensions of circuits are reasonable fractions of a wavelength. In particular, the line between the output of a transmitter and an antenna and the line between the receiving antenna and the receiver possess attributes that require transmission line concepts for proper design and analysis. Telephone lines and network connections are also dependent on these properties.

A large number of communication systems employ electromagnetic radiation between transmitting and receiving antennas. The phenomena involve electric and magnetic field concepts along with radiated power. The concept of a plane wave propagating in space is an essential part of the process.

## Objectives

After completing this chapter, the reader should be able to:

1. Discuss the concept of a *distributed parameter circuit* and explain how it differs from a *lumped parameter circuit.*

2. Determine the *delay time* on a transmission line.

3. Estimate the frequency and wavelength conditions for which propagation effects must be considered.

4. Sketch the form of the incremental model of a transmission line, and explain the effects of inductance and capacitance in establishing its behavior.

5. Define *characteristic impedance* for a transmission line.

6. Determine the characteristic impedance for a lossless transmission line in terms of inductance and capacitance per unit length.

7. Discuss *permittivity* and *permeability* as they relate to transmission lines.

8. Determine the velocity of propagation for a lossless transmission line.

9. Determine the various physical and electrical properties of a coaxial cable.

10. Determine the various physical and electrical properties of a parallel-line transmission line.

11. Discuss the requirements for optimum load impedance matching.

12. For a mismatched line, determine the load reflection coefficient and the standing-wave ratio.

13. Define and discuss the *electric field* and the *magnetic field* associated with a *plane wave*.

14. Define and compute *intrinsic impedance* for a lossless medium.

15. Define and compute the *power density* associated with a plane wave.

## 14-1  Propagation Time Effects

Refer to the simple circuit of Figure 14–1. A dc source of value 200 V in series with a 50-Ω resistance is to be connected at $t = 0$ to a cable of length $d$, and a 50-Ω resistance is connected across the opposite end of the cable. The resistances of the cable and connecting wires are considered to be negligible. When the switch is closed, the circuit appears to be a very simple series circuit containing two 50-Ω resistances in series, and by basic circuit theory, the loop current should be $I = 200 \text{ V}/(50 \text{ Ω} + 50 \text{ Ω}) = 2 \text{ A}$.

The current in the loop will actually reach the value of 2 A predicted by basic circuit theory—but it will take some time, however small, before that condition is reached. Following the closing of the switch, a *wave* consisting of both voltage and current will propagate from the source end of the line toward the load. The value of the current will depend on a property of the line called the *characteristic impedance*, whose properties will be introduced shortly. Upon reaching the load, if the value of the resistance is not equal to the characteristic impedance of the line, a second wave of voltage and current will be reflected from the load and will travel toward the source. When it reaches the source, if the source resistance is not equal to the characteristic impedance of the line, another wave will be generated and it will travel back in the direction of the load. This phenomenon is called the *transient response* of the line, and the process will continue until steady-state conditions are reached.

The time required for the circuit to reach the steady-state value of 2 A is based on the length of the line and certain conditions concerning the impedance match between the line and the source and load resistances. In the large body of lumped circuit theory familiar to the reader from dc and ac circuit theory, it is assumed that all propagation effects are negligible and can be ignored. However, when the concepts of a transmission line are introduced, it is necessary to consider *distributed circuit* effects.

We will define $t_\text{p}$ as the *one-way propagation time*, which is the time required for a wave to propagate from one end of a line to the other. Assuming a value of $d$ for the line length and a value of $v$ for the wave propagation velocity, we have

$$t_\text{p} = \frac{d}{v} \tag{14-1}$$

Depending on the physical properties of the line, the velocity of propagation may be equal to the velocity of light, but it may be somewhat less (e.g., in coaxial cables).

Propagation time effects are usually insignificant in most low-frequency applications. However, their possible effects must be considered in either of the following situations:

1. Pulse or digital waveforms are utilized, and the propagation times are within a sizable fraction of a critical time parameter such as a pulse width or, in some cases, a pulse rise time.

2. Modulated waveforms are utilized, and the lengths of cables or circuit connections are within a sizable fraction of the wavelengths of the spectral components of the signal.

**FIGURE 14–1**

Circuit used to illustrate propagation effects.

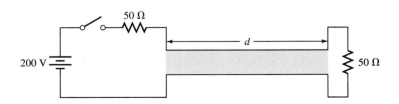

The first effect can be referred to as a *time-domain effect* and the second as a *frequency-domain effect*. Both effects are results of the same phenomenon, but they are different ways of viewing the problem.

Because communication systems generally focus heavily on the frequency-domain representation of the signals, much of our emphasis here will be directed to the second effect.

## Wavelength Review

The relationship between wavelength and frequency was provided in Chapter 1, and will be reviewed here for convenience. For an arbitrary velocity of propagation $v$, the wavelength $\lambda$ is related to the frequency $f$ by the relationship

$$\lambda = \frac{v}{f} \tag{14-2}$$

The velocity of free space is denoted $c$ and its value is $c = 3 \times 10^8$ m/s; whenever that value can be assumed, the relationship becomes

$$\lambda = \frac{c}{f} = \frac{3 \times 10^8}{f} \tag{14-3}$$

It should be stressed that for transmission lines having a dielectric other than air, the velocity of propagation may be smaller than $c$, as will be noted later.

## Frequency-Domain Considerations

In general, propagation effects are significant when the length between two connection points is a reasonable fraction of a wavelength, but what do we mean by a "reasonable fraction"? Actually, it is somewhat arbitrary, and in microwave systems, the value of $\lambda/16$ is sometimes used as a basis. We will choose here to use the simpler value of $0.1\lambda$ as the transition point. In other words, propagation effects must be considered if a line connecting two points is equal to or greater than one-tenth of a wavelength at any frequencies involved. Remember, however, that this is simply a rule of thumb and not an exact formula.

---

**▮▮ EXAMPLE 14-1**

A frequency of 1 MHz is near the center of the commercial AM broadcast band. Using the rule of thumb of $0.1\lambda$ and the free-space velocity, determine the length of line at which transmission line effects must be considered.

**SOLUTION**   The free-space wavelength is

$$\lambda = \frac{3 \times 10^8}{f} = \frac{3 \times 10^8}{1 \times 10^6} = 300 \text{ m} \tag{14-4}$$

Let $l$ represent the length based on the rule of thumb indicated.

$$l = 0.1\lambda = 0.1 \times 300 \text{ m} = 30 \text{ m} \tag{14-5}$$

This is long enough that there is reasonable leeway in dealing with internal circuits at this frequency. However, the distance between a transmitter and the antenna in a commercial AM station is sufficiently great that it must be treated as a transmission line.

Strictly speaking, for the transmission line, its velocity of propagation should be considered as the basis for the computation, and for coaxial cable, it is typically about 2/3 of the velocity of free space. Thus, 20 m could be a more reasonable value for the cable, but it is an estimate in any case.

---

**▮▮ EXAMPLE 14-2**

A frequency of 100 MHz is near the center of the commercial FM broadcast band. Repeat the analysis of Example 14-1 at this frequency.

**SOLUTION**   The free-space wavelength is

$$\lambda = \frac{3 \times 10^8}{f} = \frac{3 \times 10^8}{1 \times 10^8} = 3 \text{ m} \tag{14-6}$$

The value of $l$ is

$$l = 0.1\lambda = 0.1 \times 3 \text{ m} = 0.3 \text{ m} \tag{14-7}$$

This value is about 1 foot, and it clearly illustrates the fact that circuit connection lengths are much more critical in the frequency range employed in commercial FM. The higher-frequency television bands have even shorter wavelengths.

---

**‖ EXAMPLE 14-3**

The frequency of 1 GHz is near the point at which the *microwave region* is often considered to begin. Repeat the analysis of Examples 14-1 and 14-2 at this frequency.

**SOLUTION**   The free-space wavelength is

$$\lambda = \frac{3 \times 10^8}{f} = \frac{3 \times 10^8}{1 \times 10^9} = 0.3 \text{ m} = 30 \text{ cm} \tag{14-8}$$

The value of $l$ is

$$l = 0.1\lambda = 0.1 \times 30 \text{ cm} = 3 \text{ cm} \tag{14-9}$$

This very short length illustrates the difficulty in constructing lumped circuits to operate in the microwave frequency region. Much of the technology above about 1 GHz utilizes wave concepts, as opposed to the low-frequency concepts of voltage and current. Even amplifier circuits typically utilize short sections of microstrip transmission lines to provide optimum coupling between different stages.                                       ‖

---

## 14-2   Transmission Line Electrical Properties

While all transmission lines have losses, the most common initial assumption made in dealing with RF transmission lines of moderate length is to assume that the line is *lossless*. This means that the effects of series resistance in the wires and shunt conductance in the dielectric will be assumed to be negligible. Thus, until stated otherwise, the *lossless* transmission line model will be assumed.

### Distributed Parameter Effects

A *lumped-parameter* circuit is one in which the various parameters are considered to be concentrated in common component forms such as capacitors and inductors. In contrast, a transmission line is a *distributed-parameter circuit,* which means that the effects of various parameters such as capacitance and inductance are distributed over the length on the line. Distributed-parameter effects are more difficult to analyze, and incremental models are usually assumed for a very short distance as the first step in the analysis.

In lumped circuit theory, the symbols $L$ and $C$ are used to represent the inductance and capacitance of lumped models, and the units are henries and farads, respectively. However, in transmission line theory, the same symbols are used to used to represent the incremental values of these parameters per unit length. The basic units in transmission line theory for $L$ are henries/meter (H/m), and the basic units for $C$ are farads/meter (F/m), although these values are usually expressed as nanohenries/meter (nH/m) or microhenries/meter (µH/m) and picofarads/meter (pF/m), respectively, based on their typical sizes. Unless otherwise stated, distributed parameter units will be assumed in this chapter.

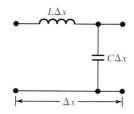

**FIGURE 14–2**
Incremental model used to develop transmission line concept.

An incremental model of a lossless line of length $\Delta x$ is shown in Figure 14–2. The model is considered an *unbalanced* model, because all of the inductance is on one side. It could be adapted to a *balanced* form by dividing the inductance into two equal parts and placing one part in the upper conductor and one part in the lower conductor. An incremental section of the line is assumed to possess a series inductance and a shunt capacitance. The net inductance and capacitance of the section is obtained by multiplying the incremental values by the width of the section. Thus, the net inductance is $L\,\Delta x$ henries, and the net capacitance of the section is $C\,\Delta x$ farads.

### Capacitance

Capacitance results from an electric field existing between two conductors and is a function of the electrical charge. The unit of charge is the coulomb (C), and for a linear circuit, it is directly proportional to the voltage between the conductors. Capacitance is defined as

$$\text{capacitance (F)} = \frac{\text{charge (C)}}{\text{voltage (V)}} \tag{14-10}$$

### Inductance

Inductance results from magnetic flux within a circuit and is a function of the amount of flux enclosed. The unit of magnetic flux is the weber (Wb), and for a linear circuit, it is directly proportional to the current. Inductance is defined as

$$\text{inductance (H)} = \frac{\text{flux (Wb)}}{\text{current (A)}} \tag{14-11}$$

### Characteristic Impedance

Arguably the most important parameter used in specifying transmission line properties is the *characteristic impedance*. Most references use $Z_0$ to represent the characteristic impedance, and the basic unit is the ohm ($\Omega$). It is defined as

$$Z_0 = \frac{\text{voltage for wave traveling in one direction}}{\text{current for wave traveling in same direction as voltage}} \tag{14-12}$$

This means that for a uniform lossless transmission line, the ratio of voltage to current of a single wave traveling in one direction is a constant.

In the most general case, the characteristic impedance is complex, and contains both a resistance and a reactance. For most of the work presented here, the characteristic impedance will be assumed to be a real number. To emphasize this point, we will use the symbol $R_0$ to represent the characteristic impedance when it is real. We define

$$Z_0 = R_0 \quad \text{(all cases where $Z_0$ is real)} \tag{14-13}$$

This is in contrast to many books that use $Z_0$ for both real and complex values. However, this convention should make it clearer to the reader when the characteristic impedance has no imaginary part.

The characteristic impedance of a lossless transmission line is a real number. This is true even though the lossline line contains no resistance. This situation is based on the fact that the characteristic impedance is simply the ratio of a voltage to a current and does not represent dissipation in the usual sense of resistance.

## Characteristic Impedance in Terms of Inductance and Capacitance

Assume a lossless line with incremental parameters $L$ and $C$, with both expressed in their basic units with the same length parameters. The characteristic impedance is given by

$$R_0 = \sqrt{\frac{L}{C}} \tag{14-14}$$

Inductance increases as the spacing between the conductors increases, since the amount of magnetic flux increases. Conversely, capacitance decreases as the spacing between conductors increases, since the electric field intensity decreases. Thus, transmission lines with a wider spacing between conductors will have a larger characteristic impedance than those with a narrower spacing.

---

**▌▌ EXAMPLE 14-4**

A lossless transmission line has $L = 320$ nH/m and $C = 90$ pF/m. Determine the characteristic impedance.

**SOLUTION** The result of Equation 14-14 is employed, but care must be used to ensure that both parameters are expressed in their basic units. We have

$$R_0 = \sqrt{\frac{L}{C}} = \sqrt{\frac{320 \times 10^{-9}}{90 \times 10^{-12}}} = 59.63 \ \Omega \tag{14-15}$$

▌▌

---

## 14-3 Electrical and Magnetic Material Properties

In order to deal with some of the physical properties of transmission lines, and later to deal with electric and magnetic fields, it is necessary to introduce some terms that relate to their properties. These terms will be introduced in this section.

### Dielectric Material

A *dielectric* is an insulating material. In electrical applications, the term usually refers to the materials used to separate the conductors in a capacitor or, for the purpose under consideration, a transmission line. Some examples of dielectric materials are mylar, polyethylene, and free space.

### Permittivity

The concept of *permittivity* is associated with the property of a dielectric medium concerning its effect on capacitance and charge. The exact definition involves some terms that fall outside of the intended coverage here, so we will deal with it in an indirect way. For our purposes, we will say that capacitance between two surfaces is directly proportional to the permittivity of the dielectric medium, and leave it at that.

The basic units of permittivity are farads/meter (F/m). There are three parameters associated with the phenomenon, and they are as follows:

$\varepsilon$ = permittivity of material expressed in farads/meter (F/m)

$\varepsilon_0$ = permittivity of free space = $\dfrac{1}{36\pi} \times 10^{-9}$ F/m = $8.842 \times 10^{-12}$ F/m

$\varepsilon_r$ = relative permittivity (also called the dielectric constant)

The permittivity can be expressed as

$$\varepsilon = \varepsilon_r \varepsilon_0 \tag{14-16}$$

The relative permittivity or dielectric constant is

$$\varepsilon_r = \frac{\varepsilon}{\varepsilon_0} \tag{14-17}$$

The dielectric constant $\varepsilon_r$ is dimensionless, since it is the ratio of two permittivity values. In free space, $\varepsilon = \varepsilon_0$, in which case the dielectric constant is $\varepsilon_r = 1$. For most applications, this value is also assumed for air. Common materials used in transmission lines typically have values well under 10.

### Permeability

The concept of *permeability* is associated with a measure of the magnetic flux density produced by a current. As in the case of permittivity, we will sidestep the exact definition, since some of the terminology is outside the scope of the text. Instead, it will be noted that an increased permeability for magnetic materials within an inductor will result in increased inductance.

The basic units of permeability are henries/meter (H/m). As in the case of permittivity, there are three parameters associated with the phenomenon, as follows:

$\mu$ = permeability of material expressed in henries/meter (H/m)

$\mu_0$ = permeability of free space = $4\pi \times 10^{-7}$ H/m

$\mu_r$ = relative permeability

The permeability can be expressed as

$$\mu = \mu_r \mu_0 \tag{14-18}$$

The relative permeability is

$$\mu_r = \frac{\mu}{\mu_0} \tag{14-19}$$

As in the case of permittivity, the relative permeability is dimensionless, since it is the ratio of two permeability values. In free space, $\mu = \mu_0$, which means that the relative permeability of free space is $\mu_r = 1$.

### Assumptions for Permittivity and Permeability

Many common dielectric materials have values that differ from that of free space. However, the vast majority of common materials have permeability values equal to that of free space. Exceptions are the ferromagnetic materials such as iron, nickel, and cobalt. These materials are never used in transmission line construction. To simplify subsequent references for transmission lines, it will be assumed in most developments that $\mu = \mu_0$. However, the permittivity factors will be assumed as appropriate.

## 14-4  Propagation Velocity

A wave traveling on a transmission line has a velocity that is a function of the line parameters. The velocity $v$ for a *lossless* line is given by

$$v = \frac{1}{\sqrt{LC}} \tag{14-20}$$

The units for the velocity are dependent on the length units for the inductance and capacitance. If the inductance and capacitance are expressed in henries/meter and farads/meter, respectively, the velocity will be expressed in meters/second (m/s).

### Velocity in Terms of Permittivity and Permeability

The velocity of propagation can also be expressed in terms of the permeability and the permittivity of the dielectric material inside the line. For a lossless line, the velocity is given by

$$v = \frac{1}{\sqrt{\mu\varepsilon}} \tag{14-21}$$

where it is understood that $\mu = \mu_0$

## Free-Space Velocity

Assume in Equation 14-21 that $\mu = \mu_0 = 4\pi \times 10^{-7}$ H/m and $\varepsilon = \varepsilon_0 = (1/36\pi) \times 10^{-9}$ F/m. Substituting these values, we obtain

$$v = \frac{1}{\sqrt{4\pi \times 10^{-7} \times (1/36\pi) \times 10^{-9}}} = 3 \times 10^8 \text{ m/s} = c \qquad (14\text{-}22)$$

The result is the velocity of propagation in free space.

## Velocity with Non-Unity Dielectric Constant

It has previously been stated that $\mu = \mu_r\mu_0$ and $\varepsilon = \varepsilon_r\varepsilon_0$. Substitution of these values in Equation 14-21 along with the result of Equation 14-22 leads to

$$v = \frac{c}{\sqrt{\mu_r\varepsilon_r}} = \frac{3 \times 10^8}{\sqrt{\mu_r\varepsilon_r}} \text{ m/s} \qquad (14\text{-}23)$$

As previously stated, it will be assumed that $\mu_r = 1$, which means that the velocity of propagation throughout the remainder of the text can be expressed as

$$v = \frac{c}{\sqrt{\varepsilon_r}} = \frac{3 \times 10^8}{\sqrt{\varepsilon_r}} \text{ m/s} \qquad (14\text{-}24)$$

**EXAMPLE 14-5**    Assume the parameters for the lossless transmission line of Example 14-4 with $L = 320$ nH/m and $C = 90$ pF/m. Determine the velocity of propagation.

**SOLUTION**    The velocity of propagation is given by

$$v = \frac{1}{\sqrt{LC}} = \frac{1}{\sqrt{320 \times 10^{-9} \times 90 \times 10^{-12}}} = 1.863 \times 10^8 \text{ m/s} \qquad (14\text{-}25)$$

**EXAMPLE 14-6**    Determine the dielectric constant for the transmission line of Example 14-5.

**SOLUTION**    From Equation 14-24, the velocity can be expressed in terms of the dielectric constant as

$$1.863 \times 10^8 = \frac{3 \times 10^8}{\sqrt{\varepsilon_r}} \qquad (14\text{-}26)$$

This leads to

$$\varepsilon_r = 2.59 \qquad (14\text{-}27)$$

## 14-5    Properties of Common Transmission Lines

The two most common types of transmission lines are coaxial cables and parallel lines. Their physical and electrical properties will be described in this section. Formulas concerning the inductance per unit length and capacitance per unit length will be presented. These formulas are derived in texts primarily devoted to electric and magnetic field theory. Once these parameters are known, the characteristic impedance for lines assumed to be lossless may be readily determined.

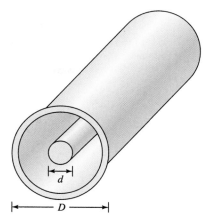

**FIGURE 14–3**

Form of coaxial cable transmission line.

## Coaxial Cable

The most common type of coaxial cable has the geometry shown in Figure 14–3. A center conductor and an enclosing conducting shield realize the two required conductors. A dielectric material separates the shield and conductor. The complete cable is enclosed with an insulating *jacket* such as vinyl.

The common form of coaxial cable that has been described is an example of an *unbalanced line;* that is, the two conductors do not have symmetrical electric and magnetic properties with respect to ground. Instead, the conducting shield is connected to the system ground. Ideally, this should result in an effective shield around the inner conductor and prevent any radiation losses.

In general, coaxial cables are used in a wide number of applications at frequencies extending into the lower microwave region. Because they are unbalanced, coaxial cables may be readily connected to unbalanced sources. If the source or the load is balanced, however, there may be a need for a conversion circuit such as a transformer or a *balun*. The name *balun* represents "balanced-to-unbalanced," and these devices can be constructed from sections of transmission lines. The most significant disadvantages of coaxial cable are the cost and, in some applications, the physical size.

## Capacitance, Inductance, and Characteristic Impedance of Coaxial Cable

Refer to Figure 14–3, and let

$D$ = inner diameter of outer conducting shield

$d$ = diameter of inner conductor

At RF frequencies, the inductance per meter of the line may be expressed as

$$L = \frac{\mu_0}{2\pi} \ln \frac{D}{d} = 2 \times 10^{-7} \ln \frac{D}{d} \text{ H/m} \tag{14-28}$$

where "ln" refers to the natural logarithm.

Under similar conditions, the capacitance per unit length is given by

$$C = \frac{2\pi\varepsilon}{\ln \frac{D}{d}} = \frac{(1/18) \times 10^{-9}\varepsilon_r}{\ln \frac{D}{d}} = \frac{55.56 \times 10^{-12}\varepsilon_r}{\ln \frac{D}{d}} \text{ F/m} \tag{14-29}$$

Assuming a lossless line, the characteristic impedance may then be determined by Equation 14-14 as

$$R_0 = \sqrt{\frac{L}{C}} = \frac{60}{\sqrt{\varepsilon_r}} \ln \frac{D}{d} \text{ } \Omega \tag{14-30}$$

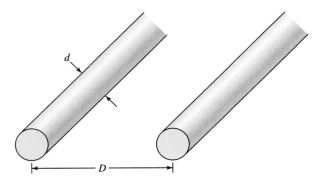

**FIGURE 14–4**
Form of parallel-line
transmission line.

## Parallel Lines

Parallel lines represent the simplest type of geometry, in that the two conductors are of equal size and are spaced apart by a constant separation, as shown in Figure 14–4. The medium between the conductors may be air ("open-wire" lines), or it may be a material such as polyethylene ("twin-lead"). Parallel lines have been used extensively in the telephone industry and for 60-Hz power transmission.

A parallel line is referred to as a *balanced line*. This description means that when the parallel line is operating properly, all electric and magnetic fields are symmetrical with respect to ground, and the impedance of each wire to ground is the same. This balance minimizes the amount of radiation from the line, and other undesirable coupling effects. However, a balanced line should *not* be driven by an unbalanced source (i.e., a source with one side grounded), unless there is a transformer or balun between the source and the line to convert between the forms.

The fields of a parallel line extend some distance from the wires. As the frequency increases, the losses due to unwanted radiation on the line increase. Although parallel lines have been used extensively in the past at TV frequencies for coupling between antenna and receiver, where some losses are acceptable, their use is generally restricted to operation below about 100 MHz or so.

## Capacitance, Inductance, and Characteristic Impedance of Parallel Lines

Refer to Figure 14–4 and let

$D$ = distance between the centers of the two lines

$d$ = diameter of each of the conductors

At RF frequencies, the inductance per meter of the line may be expressed as

$$L = \frac{\mu_0}{\pi} \ln \frac{2D}{d} = 4 \times 10^{-7} \ln \frac{2D}{d} \text{ H/m} \qquad (14\text{-}31)$$

where "ln" refers to the natural logarithm.

Under similar conditions, the capacitance per unit length is given by

$$C = \frac{\pi \varepsilon}{\ln \frac{2D}{d}} = \frac{(1/36) \times 10^{-9} \varepsilon_r}{\ln \frac{2D}{d}} = \frac{27.78 \times 10^{-12} \varepsilon_r}{\ln \frac{2D}{d}} \text{ F/m} \qquad (14\text{-}32)$$

Assuming a lossless line, the characteristic impedance may then be determined by Equation 14-14 as

$$R_0 = \frac{120}{\sqrt{\varepsilon_r}} \ln \frac{2D}{d} \ \Omega \qquad (14\text{-}33)$$

**EXAMPLE 14-6**    A certain coaxial cable has the dimensions that follow.

Diameter of inner conductor $(d) = 0.3$ cm

Inner diameter of outer conductor $(D) = 1.02$ cm

Polyethylene dielectric $(\varepsilon_r = 2.25)$

Determine the following parameters for the cable: (a) inductance per unit length, (b) capacitance per unit length, (c) characteristic impedance, and (d) velocity of propagation.

**SOLUTION**    The quantity $\ln(D/d)$ is required for the first three parameters, and will be calculated first:

$$\ln \frac{D}{d} = \ln \frac{1.02}{0.3} = 1.224 \tag{14-34}$$

(a)  The inductance per unit length can be calculated with Equation 14-28.

$$L = 2 \times 10^{-7} \ln \frac{D}{d} = (2 \times 10^{-7})(1.224) = 244.8 \text{ nH/m} \tag{14-35}$$

(b)  The capacitance per unit length can be calculated with Equation 14-29.

$$C = \frac{55.56 \times 10^{-12} \varepsilon_r}{\ln \frac{D}{d}} = \frac{55.56 \times 10^{-12} \times 2.25}{1.224} = 102.1 \text{ pF/m} \tag{14-36}$$

(c)  The characteristic impedance can be calculated with Equation 14-30.

$$R_0 = \frac{60}{\sqrt{\varepsilon_r}} \ln \frac{D}{d} = \frac{60}{\sqrt{2.25}}(1.224) = 48.96 \ \Omega \tag{14-37}$$

(d)  The velocity of propagation can be calculated with Equation 14-24.

$$v = \frac{c}{\sqrt{\varepsilon_r}} = \frac{3 \times 10^8}{\sqrt{2.25}} = \frac{3 \times 10^8}{1.5} = 2 \times 10^8 \text{ m/s} \tag{14-38}$$

These parameters are in the range of realistic coaxial cables. A "standard" value of characteristic impedance assumed for many coaxial cables is 50 $\Omega$, and the velocity of propagation is $\frac{2}{3}$ of the free space velocity.

## 14-6  Matched Load Impedance

Consider the situation depicted in Figure 14–5. An ac phasor source with internal impedance $\overline{Z}_s$ is connected through a lossless transmission line with characteristic impedance $R_0$ to a load impedance $\overline{Z}_L$. In this case, the load impedance is assumed to be purely resistance, and equal to the characteristic impedance of the line; that is, $\overline{Z}_L = R_L = R_0$. This is an ideal situation as far as the load is concerned. While there will be a delay between the source and the load, *no reflected wave will be generated.* Thus, there will only be a forward wave traveling from the source to the load.

Whether or not maximum power is transferred to the load will depend on the match between the source internal impedance and the line, and that situation is usually easier to

**FIGURE 14–5**

Lossless transmission line with sinusoidal source and resistive load matched to the characteristic impedance.

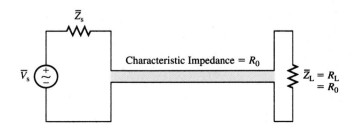

deal with than that of the load. Reactive impedance matching networks are often used to provide an impedance match at the output of a transmitter, and they can be tuned to provide different conditions at different frequencies. It is also possible to provide reactive networks at the input to an antenna to match the impedance, but it is usually more difficult if the frequency is to be varied.

With a matched load, the input impedance $\overline{Z}_{\text{in}}$ at any point along the lossless line is equal to the load impedance, and is

$$\overline{Z}_{\text{in}} = R_L = R_0 \tag{14-39}$$

This means that basic circuit theory may be used to predict the *magnitudes* of the steady-state voltage and current anywhere on the line. (The phase shift will be different at different points due to the time delay, but we will not be concerned with that phenomenon at the moment.)

The rms magnitude $I_{\text{rms}}$ of the current at the input of the line under matched conditions is related to the rms magnitude of the source voltage $V_{\text{rms}}$ by

$$I_{\text{rms}} = \frac{V_{\text{rms}}}{|\overline{Z}_s + R_0|} = \frac{V_{\text{rms}}}{|\overline{Z}_s + R_L|} \tag{14-40}$$

where magnitude bars are used since the source impedance could be complex.

The input power $P_{\text{in}}$ to the line must be the same as the power $P_L$ dissipated in the load, since the line is assumed to be lossless. Hence,

$$P_{\text{in}} = P_L = I_{\text{rms}}^2 R_0 = I_{\text{rms}}^2 R_L \tag{14-41}$$

Assuming that the source has a nonzero value of internal impedance, if there is a match at the source, this resulting power will be the maximum available power that can be obtained from the source.

It should be noted that even under matched conditions, there will be some losses in an actual line, especially if the line is reasonably long. Nevertheless, under matched conditions at the load, the approach used here is one of simplicity, and it can be modified slightly to take care of losses in the line without getting into the more complex conditions of mismatch (to be considered in the next section).

---

**▐▌ EXAMPLE 14-7**

The circuit of Figure 14–6 represents the output of a transmitter connected through a transmission line to an antenna. The input impedance of the antenna is resistive and equal to 50 Ω, and the characteristic impedance of the line is 50 Ω resistive. The nonzero output impedance of the transmitter has been matched to that of the line and is also 50 Ω resistive. The rms value of the source voltage referred to the output of the matching circuit is 400 V. Determine (a) the input impedance of the line, (b) the rms current flowing into the line, (c) the power accepted by the line, (d) the load power if the line is considered lossless.

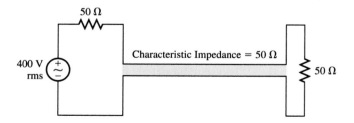

**FIGURE 14–6**
Circuit of Example 14-7.

**SOLUTION**

(a) The line is matched at the load, and the input impedance at any point along the line is simply

$$\overline{Z}_{\text{in}} = R_{\text{in}} = 50 \ \Omega \tag{14-42}$$

(b)  The rms current at the source is determined from basic circuit theory as

$$I_{\text{rms}} = \frac{400 \text{ V}}{50 \ \Omega + 50 \ \Omega} = 4 \text{ A} \qquad (14\text{-}43)$$

(c)  The input power accepted by the line is

$$P_{\text{in}} = I_{\text{rms}}^2 R = (4)^2 \times 50 = 800 \text{ W} \qquad (14\text{-}44)$$

(d)  If the line is lossless, the load power will be equal to the power input to the line, and is

$$P_{\text{L}} = P_{\text{in}} = 800 \text{ W} \qquad (14\text{-}45)$$

If the line is lossless and matched, as assumed here, the magnitudes of both the voltage and current will not vary along the line, although the phase shift will change.

---

**▥ EXAMPLE 14-8**

Assume all the conditions given in Example 14-7, except that the transmission line has a loss of 0.01 dB/m. Determine the load power if the length of the transmission line is 50 m.

**SOLUTION**   Strictly speaking, once the line is assumed to be lossy, the characteristic impedance will become complex and will have both a real and an imaginary part. Hence, the assumption of a perfect match at the load might have to be questioned.

In practice, as long as the length is moderate and the losses not too great, it is reasonable to assume that the matching is essentially perfect and the effect of losses can be added "after the fact." With the length and loss value given here, that should be a reasonable assumption.

The net loss $L_{\text{dB}}$ is given by

$$L_{\text{dB}} = 50 \text{ m} \times 0.01 \text{ dB/m} = 0.5 \text{ dB} \qquad (14\text{-}46)$$

This corresponds to an absolute loss factor $L$ as given by

$$L = 10^{0.5/10} = 1.122 \qquad (14\text{-}47)$$

The actual power reaching the load will then be

$$P_{\text{L}} = \frac{P_{\text{in}}}{L} = \frac{800}{1.122} = 713.0 \text{ W} \qquad (14\text{-}48)$$

This means that $800 - 713 = 87$ W will be dissipated in the line.

The approach taken here should be used with caution if the line is very long or if the loss is very great.  ▥

---

## 14-7  Mismatched Load Impedance

While the matched situation discussed in Section 14-6 is almost always a desired goal in designing a transmission system between a transmitter and an antenna (or between an antenna and a receiver), there are situations in which it is not practical. A good example is an antenna that must be used over a wide frequency range, for which matching is not possible at all frequencies. Another example is that of a portable unit in which the necessary length of a proper antenna is not feasible. Some of the problems associated with a mismatched load will be discussed in this section.

Consider the situation depicted in Figure 14–7. A source with internal impedance $\overline{Z}_{\text{s}}$ is connected through a lossless transmission line with characteristic impedance $R_0$ to an arbitrary load impedance $\overline{Z}_{\text{L}}$. Since we are considering a mismatched situation, we may as well assume that $\overline{Z}_{\text{L}}$ is complex and that it contains both a real and an imaginary part; that is,

$$\overline{Z}_{\text{L}} = R_{\text{L}} + jX_{\text{L}} \qquad (14\text{-}49)$$

where $R_{\text{L}}$ is the resistive part of the load impedance and $X_{\text{L}}$ is the reactive part. Most antennas will exhibit such a complex impedance when they are operated at other than their design frequency range.

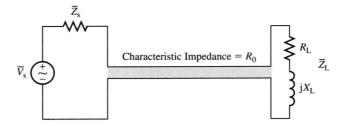

**FIGURE 14–7**
Transmission line with mismatch at load (and possibly at source).

## Reflection Coefficient at Load

We will now introduce a complex parameter $\overline{\Gamma}_L$ that will be denoted the *voltage reflection coefficient at the load*. It is defined as

$$\overline{\Gamma}_L = \frac{\overline{Z}_L - R_0}{\overline{Z}_L + R_0} \tag{14-50}$$

The reflection coefficient is the ratio of the wave reflected from the load to the wave incident to it. We could also define a reflection coefficient at the source, but that is not necessary for our purposes.

## Reflection Coefficient for Matched Load

When $\overline{Z}_L = R_0$, the value of the reflection coefficient is quickly determined as

$$\overline{\Gamma}_L = 0 \quad \text{for a matched load} \tag{14-51}$$

## General Range

In general, the magnitude of the reflection coefficient is bounded by

$$0 \le \Gamma_L \le 1 \tag{14-52}$$

At one extreme, the reflection coefficient is 0, indicating a matched load. Two other conditions are worth noting:

1. When the load is a short circuit; that is, when $\overline{Z}_L = 0$,

$$\overline{\Gamma}_L = -1 \tag{14-53}$$

2. When the load is an open circuit; that is, when $\overline{Z}_L \to \infty$,

$$\overline{\Gamma}_L = 1 \tag{14-54}$$

For either a short circuit or an open circuit, all the power reaching the load is reflected back to the source and the line acts as a reactance. Circuits composed of short circuit or open circuit lines may be used in impedance matching and filter circuits.

## Voltage Standing-Wave Ratio

It has been stated that for a matched load, the magnitude of the voltage or current at any point on a lossless line is the same. However, when there is a mismatch, the voltage and current vary along the line. The greater the mismatch, the greater the variation of either voltage or current. Voltage is easier to measure, so it will be used as the basis for the discussion that follows.

A quantity called the *voltage standing-wave ratio* is widely used in describing the behavior of a transmission line under operating conditions. Some references use VSWR to represent this quantity, but we will use the simpler term $S$. The basic definition is

$$S = \frac{V_{max}}{V_{min}} \qquad (14\text{-}55)$$

Where $V_{max}$ is the maximum voltage on the line and $V_{min}$ is the minimum voltage on the line. The standing-wave ratio is characterized by the inequality

$$1 \leq S \leq \infty \qquad (14\text{-}56)$$

When the line is perfectly matched, $S = 1$, meaning that there is no variation of the voltage along the line. Conversely, when $S = \infty$, the voltage varies between a minimum value of 0 and a finite maximum value, providing an infinite value for the standing-wave ratio.

## Relationship between Reflection Coefficient and Standing-Wave Ratio

It turns out that there is a well-defined relationship between reflection coefficient and standing-wave ratio. The voltage standing-wave ratio may be determined from the magnitude of the reflection coefficient by the following relationship:

$$S = \frac{1 + \Gamma_L}{1 - \Gamma_L} \qquad (14\text{-}57)$$

The inverse relationship can be determined by solving for the magnitude of the reflection coefficient in terms of the standing-wave ratio,

$$\Gamma_L = \frac{S - 1}{S + 1} \qquad (14\text{-}58)$$

Note that only the magnitude of the reflection coefficient may be determined from Equation 14-58. The angle must be determined by some other means.

## Resistive Load and Lossless Line

When the load is purely resistive and the line is lossless, there is a particularly simple relationship for the standing-wave ratio. Thus, let $\overline{Z}_L = R_L$, and continue the assumption that the line is lossless. It can be shown that the standing-wave ratio is given by one of the following two relationships:

$$S = \frac{R_L}{R_0} \qquad (14\text{-}59)$$

or

$$S = \frac{R_0}{R_L} \qquad (14\text{-}60)$$

depending on which value is greater than 1, since $S \geq 1$.

## Reflected Power and Transmitted Power

Let $P_{inc}$ represent the incident power propagating on a lossless transmission toward a load, and assume that it encounters a mismatch at the load. Depending on the extent of the mismatch, some or all of the power will be reflected, and some will be absorbed by the load, with the exception that it is all reflected when there is a short circuit or an open circuit at the load. Let $P_{ref}$ represent the power reflected and let $P_L$ represent the power absorbed by the load. By the conservation of power, the following relationship must be satisfied:

$$P_{inc} = P_{ref} + P_L \qquad (14\text{-}61)$$

It can be shown that the reflected power is related to the incident power by

$$P_{ref} = \Gamma_L^2 P_{inc} \qquad (14\text{-}62)$$

where $\Gamma_L$ is the magnitude of the reflection coefficient at the load. Substituting Equation 14-62 in Equation 14-61, the load power can be expressed as

$$P_L = \left(1 - \Gamma_L^2\right) P_{\text{inc}} \tag{14-63}$$

Practical lines will exhibit some ohmic losses in the wires, but any mismatch at the load will contribute to an additional reduction in load power, as can be seen from Equation 14-63. The *mismatch loss* in decibels can be defined as

$$\text{mismatch loss (dB)} = -10 \log\left(1 - \Gamma_L^2\right) \tag{14-64}$$

The negative sign in Equation 14-64 is necessary in order that the loss be represented as a positive value.

Another quantity sometimes used in characterizing a transmission line is the *return loss* in decibels. It is defined as

$$\text{return loss (dB)} = -10 \log \Gamma_L^2 \tag{14-65}$$

Again, the negative sign results in a positive value for the loss.

## Input Impedance

When a line is mismatched at the load, the input impedance varies alone the line. If the line is lossless, the pattern is periodic and repeats at intervals of one-half wavelength. It can be shown that the input impedance $\overline{Z}_{\text{in}}$ at any distance $d$ from the load for a lossless line is given by

$$\overline{Z}_{\text{in}} = R_0 \frac{(1 + \overline{\Gamma}_L e^{-j4\pi d/\lambda})}{(1 - \overline{\Gamma}_L e^{-j4\pi d/\lambda})} \tag{14-66}$$

For a matched load, this equation reduces to $\overline{Z}_{\text{in}} = R_0$, as previously stated. For a mismatch, however, the input impedance can vary considerably with the length of the line, particularly when there is a high standing-wave ratio. For any given length, the real part of the input impedance can be considered as accepting the input power for a given rms current, and if the line is lossless, this will be the power delivered to the load. Thus, for a high standing-wave ratio, the power delivered to the load may be a very sensitive function of the line length.

---

**▌▌ EXAMPLE 14-9**

A load impedance given by $\overline{Z}_L = 50 + j100 \ \Omega$ is connected to a source through a lossless transmission line with a characteristic impedance of 50 $\Omega$. Determine (a) the load reflection coefficient and (b) the standing-wave ratio.

**SOLUTION**

(a) The reflection coefficient at the load in this case is complex, and is given by

$$\overline{\Gamma}_L = \frac{\overline{Z}_L - R_0}{\overline{Z}_L + R_0} = \frac{50 + j100 - 50}{50 + j100 + 50} = \frac{j100}{100 + j100} = \frac{100\angle 90°}{141.4\angle 45°}$$

$$= 0.7071\angle 45° \tag{14-67}$$

(b) The standing-wave ratio is given by

$$S = \frac{1 + \Gamma}{1 - \Gamma} = \frac{1 + 0.7071}{1 - 0.7071} = 5.828 \tag{14-68}$$

---

**▌▌ EXAMPLE 14-10**

A load impedance given by $\overline{Z}_L = R_L = 100 \ \Omega$ is connected through a lossless transmission line with a characteristic impedance of 300 $\Omega$. Determine (a) the load reflection coefficient and (b) the standing-wave ratio.

**SOLUTION**

(a)  The reflection coefficient at the load in this case is real, and is given by

$$\overline{\Gamma}_L = \frac{R_L - R_0}{R_L + R_0} = \frac{100 - 300}{100 + 300} = -0.5 \tag{14-69}$$

(b)  While Equation 14-57 could be used to determine the standing-wave ratio, it is simpler to use either Equation 14-59 or 14-60, since the both the load impedance and the characteristic impedance are real. The applicable one in this case is Equation 14-60, and we have

$$S = \frac{R_0}{R_L} = \frac{300}{100} = 3 \tag{14-70}$$

---

**▮▮ EXAMPLE 14-11**

For the system of Example 14-10, determine the mismatch loss in dB.

**SOLUTION**    The value is determined from Equation 14-64 as

$$\text{mismatch loss (dB)} = -10\log\left(1 - \Gamma_L^2\right) = -10\log[1 - (0.5)^2]$$
$$= -10\log 0.75 = -10(-0.125) = 1.25 \text{ dB} \tag{14-71}$$

▮▮

---

## 14-8  Electric and Magnetic Fields

In order to deal with electromagnetic radiation, we need to introduce two important field quantities: the electric field and the magnetic field. In a sense, the electric field in space corresponds to voltage in a circuit, and the magnetic field in space corresponds to current in a circuit.

### Electric Field

Strictly speaking, the definition of electric field is the force per unit charge existing between two points or surfaces of electrical charge. For example, a charged capacitor can be thought of as having invisible lines of electric field between the plates, with the lines originating on the positively charged plate and terminating on the negatively charged plate.

Electrical field intensity is denoted by the symbol $E$, and various subscripts and phasor notation may be added as required. The units are volts/meter (V/m).

### Magnetic Field

A magnetic field in the case of static fields arises from current flow. For example, invisible lines of magnetic field will surround and encircle a current-carrying conductor. In contrast to an electric field, magnetic field lines do not have any point sources.

Magnetic field intensity is denoted by the symbol $H$, and various subscripts and phasor notation may be added as required. The units are amperes/meter (A/m).

### Time-Varying Electric and Magnetic Fields

Electromagnetic radiation as utilized in communications is a result of time-varying electric and magnetic fields. It turns out that a time-varying electric field causes a time-varying magnetic field, and vice versa. These concepts serve as the basis for wave propagation.

## 14-9  Plane Wave Propagation

The subject of *plane wave propagation* is very important in the study of communication systems. It is closely related to the preceding development of transmission line phenomena, because of certain analogous properties. At the typical receiving distances from an antenna, the waves may be assumed to be in the form of plane waves. Plane wave propagation may also be assumed for light waves and other forms of electromagnetic radiation.

## Direction of Wave Propagation

The major properties of plane wave propagation may be described by the model of Figure 14–8(a). A plane wave has an electric field in one direction and a magnetic field perpendicular to the electric field. The direction of propagation is perpendicular to the plane containing the electric and magnetic fields. For the model of Figure 14–8(a), the electric field $E_x$ is assumed in the x-direction. The magnetic field $H_y$ is assumed in the y-direction. The direction of propagation for these assumptions is in the z-direction.

The direction of propagation for a plane wave may be determined by the so-called "right-hand" rule. Take the right hand and rotate from $E$ to $H$. The direction of the thumb arranged perpendicular to the plane containing electric and magnetic fields will then point in the direction of propagation. Refer to the diagram of Figure 14–8(b) for further clarification.

## Intrinsic Impedance

Assume next that the medium is lossless, which means that the medium is a perfect dielectric. Based on this assumption, $E_x$ and $H_y$ are related by the equation that follows:

$$\frac{E_x}{H_y} = \eta \tag{14-72}$$

The parameter $\eta$ is defined as *the intrinsic impedance*. Recall that the units of $E_x$ are volts/meter and the units of $H_y$ are amperes/meter. Therefore, the units of $\eta$ are volts/ampere, which are dimensionally equivalent to ohms. This property leads to the interpretation as an impedance. The intrinsic impedance serves for a plane wave the same property as characteristic impedance serves for a single wave of voltage and current on a transmission line.

The intrinsic impedance for a plane wave in a lossless dieelectric medium is a real number expressed as

$$\eta = \sqrt{\frac{\mu}{\varepsilon}} \tag{14-73}$$

For free-space plane wave propagation, $\mu = \mu_0$ and $\varepsilon = \varepsilon_0$. In this case the intrinsic impedance will be represented as $\eta_0$, and when these values are substituted in Equation 14-73, the result is

$$\eta_0 = 120\pi \approx 377 \ \Omega \tag{14-74}$$

We will also assume this value as the intrinsic impedance of air.

Based on the assumption that $\mu = \mu_0$, the intrinsic impedance for any arbitrary dielectric medium is

$$\eta = \frac{\eta_0}{\sqrt{\varepsilon_r}} \approx \frac{377}{\sqrt{\varepsilon_r}} \tag{14-75}$$

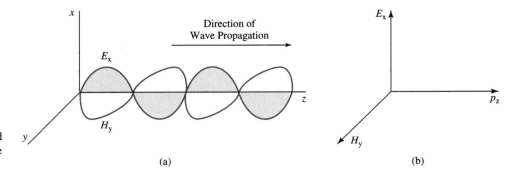

**FIGURE 14–8**
Electric and magnetic fields, and the propagation described by the Poynting vector.

(a)　　　　　　(b)

## Poynting Vector

A plane wave propagates energy in the form of electromagnetic (EM) radiation. The input signal to any receiver antenna is a result of energy "captured" by the antenna from this propagation. Warmth from solar radiation is also a result of plane wave propagation.

Energy propagating from plane wave radiation is usually measured in terms of power, which is the rate of change of the energy transmitted. The concept of *power density* is a very useful parameter for this purpose. The power density can be defined as the power per unit area measured over an area perpendicular (normal) to the direction of propagation. The units of power density are watts/meter$^2$ (W/m$^2$) and, in general, is a vector quantity (having both a magnitude and direction) called the *Poynting vector*.

Based on the directions shown in Figure 14–8, the electric field $E_x$ is aligned with the positive $x$-direction, and the magnetic field $H_y$ is aligned with the positive $y$-direction. This means that the Poynting vector will be aligned with the positive $z$-direction. This power density will be denoted simply as $p_z$ and it can be expressed as

$$p_z = E_x H_y \qquad (14\text{-}76)$$

Observe the units in Equation 14-76. The product of (volts/meter) × (amperes/meter) yields watts/meter$^2$ (W/m$^2$).

## Similarity to Circuit Power

The relationship of Equation 14-76 has the same form as the circuit relationship $P = VI$. The circuit relationship for power in a resistance can also be expressed as $V^2/R$ or $I^2 R$. By analogy, power density can also be expressed in terms of either $E_x$ or $H_y$, and two other expressions for the power density can be readily developed as

$$p_z = \frac{E_x^2}{\eta} \qquad (14\text{-}77)$$

and

$$p_z = H_y^2 \eta \qquad (14\text{-}78)$$

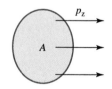

**FIGURE 14–9**

Power propagation at a normal angle through an area.

## Power over a Surface Area

Consider that a Poynting vector is perpendicular to a surface area, as shown in Figure 14–9. Assuming that the power density vector is constant over the area, the total power encompassed by the area is

$$P = p_z A \qquad (14\text{-}79)$$

Integration over the area would be required if the power density is a function of the location on the area.

---

**▌▌ EXAMPLE 14-12**

A certain plane wave in free space has an rms value for the electric field intensity vector of $E_x = 3$ V/m. Assume that the magnetic field intensity vector is in the positive $y$-direction. Determine (a) the magnitude of the magnetic field intensity, (b) the power density of the plane wave, and (c) the total power transmitted through a surface in the $x$-$y$ plane with dimensions of 10 m by 30 m. Assume that the power density is constant over the surface area.

**SOLUTION**

(a) The value of $H_y$ is determined from Equation 14-72:

$$H_y = \frac{E_x}{\eta_0} = \frac{3}{377} = 7.958 \times 10^{-3} \text{ A/m} \qquad (14\text{-}80)$$

(b)   The power density will be determined with Equation 14-77, since it only uses the original data:

$$p_z = \frac{E_x^2}{\eta_0} = \frac{(3)^2}{377} = 23.87 \times 10^{-3} \text{ W/m}^2 = 23.87 \text{ mW/m}^2 \tag{14-81}$$

(c)   The surface area $A$ is

$$A = 10 \text{ m} \times 30 \text{ m} = 300 \text{ m}^2 \tag{14-82}$$

The net power transmitted through this area is

$$P = p_z A = (23.87 \times 10^{-3}) \times 300 = 7.162 \text{ W} \tag{14-83}$$

## 14-10   Multisim® Examples (Optional)

Multisim provides the capability of modeling both lossless and lossy transmission lines. The models are available in the **Miscellaneous** parts bin, which is indicated as **Misc** on the standard component parts toolbar. The three models and their operating parameters are indicated as follows:

### Lossless_Line_Type1

There are two parameters required for this model:

**Nominal Impedance** (This is the characteristic impedance in ohms.)

**Propagation Delay** (This is the one-way delay time in seconds.)

### Lossless_Line_Type 2

There are three parameters required for this model:

**Nominal Impedance** (As in the previous model, this is the characteristic impedance in ohms.)

**Frequency** (This a frequency at which the length satisfies the parameter that follows.)

**Normalized Electrical Length** (This is the length in wavelengths at the preceding frequency.)

Both models can produce identical results, but they represent different ways of providing the specifications. The first type is more convenient for time-domain applications, and the second is more convenient for frequency-domain applications, especially with regard to communication signals.

### Lossy_Transmission_Line

The lossy transmission line requires the entries of all the incremental parameters that follow:

**Length of Transmission Line** expressed in **meters.**

**Resistance per Unit Length** expressed in **ohms/meter.** (This is a *series* parameter.)

**Inductance per Unit Length** expressed in **henries/meter.** (This is a *series* parameter.)

**Capacitance per Unit Length** expressed in **farads/meter.** (This is a *shunt* parameter.)

**Conductance per Unit Length** expressed in **siemens/meter.** (This is a *shunt* parameter.)

It should be noted that if the resistance and conductance values are both entered as **0**, the lossy model reduces back to a lossless model.

**▌▌ MULTISIM EXAMPLE 14-1**    Use the **Lossless_Line_Type2,** along with other components, to model the circuit of Example 14-7 and perform an analysis on the circuit.

**SOLUTION**    The circuit is shown in Figure 14–10. Since the circuit is matched at both ends and the line is considered to be lossless, any length could be used. We will arbitrarily let the **Frequency** and **Normalized electrical length** remain at the default values of **1 MHz** and **0.25** wavelengths, respectively. This means that a frequency of **4 MHz** would represent a full wavelength, and the sinusoidal source will be set at that frequency. However, we wish to stress again that the length is irrelevant because of the assumed matched conditions and the lossless character of the line. The **Nominal Impedance** is set at **50 Ω**. The **RMS** value of the source is **400 V**, which corresponds to an amplitude or peak value of **565.69 V**.

A **Transient Analysis** is performed over 4 cycles of the input signal, which corresponds to a total time duration of **1 μs** at the operating frequency of **4 MHz**. To ensure high accuracy, the **Maximum time step** was set to **1 ns**.

The instantaneous voltage across the input to the transmission line is shown in Figure 14–11. Using a **cursor,** the peak value of the voltage is measured as 282.8444 V. (The perceived accuracy shown here is a bit ludicrous, but it is shown as read.) For all

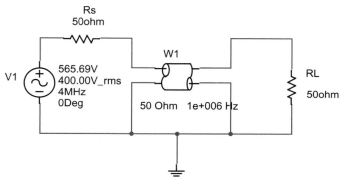

**FIGURE 14–10**
Circuit of Multisim Example 14-1.

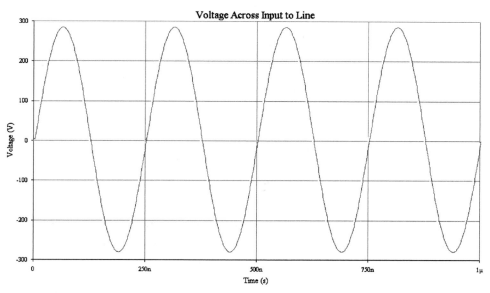

**FIGURE 14–11**
Transmission line input voltage in Multisim Example 14-1.

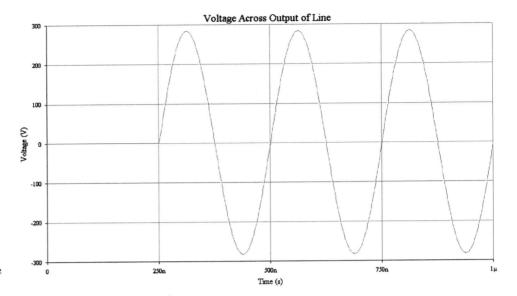

**FIGURE 14–12**

Transmission line output voltage in Multisim Example 14-1.

practical purposes, this value is half of the source peak open-circuit voltage of 565.69 V, as expected. This fact supports the property that the input impedance to the line is 50 $\Omega$.

The instantaneous voltage across the output of the transmission line is shown in Figure 14–12. As expected, there is a delay of **250 ns** before the signal reaches the output (one-quarter wavelength at 1 MHz). Using a **cursor,** the peak value of this voltage is also measured as 282.8444 V. The average power $P$ across the load, which is exactly the same as the input power to the lossless line, is given by

$$P = \frac{V_p^2}{2R_L} = \frac{(282.8444)^2}{2 \times 50} = 800.0 \text{ W} \tag{14-84}$$

This value is exactly the same as obtained in Example 14-7.

**▌▌ MULTISIM EXAMPLE 14-2**  Use the **Lossy_Transmission_Line** model to simulate the conditions of Example 14-8, based on a line of length 50 m and a loss of 0.01 dB/m.

**SOLUTION**    As explained in Example 14-8, some approximations for a low-loss line of moderate length may be reasonable. Let us first "pretend" that the line is lossless and determine the incremental values of $L$ and $C$ necessary to achieve the required parameters. We begin with the following two equations:

$$R_0 = \sqrt{\frac{L}{C}} \tag{14-85}$$

and

$$v = \frac{1}{\sqrt{LC}} \tag{14-86}$$

From the preceding two equations, the values of $L$ and $C$ may be determined in terms of $R_0$ and $v$. Simultaneous solution yields

$$L = \frac{R_0}{v} \tag{14-87}$$

$$C = \frac{1}{R_0 v} \tag{14-88}$$

The value of the characteristic impedance desired is obviously $R_0 = 50\ \Omega$. For the velocity, a common value for cable of this type is $v = 2 \times 10^8$ m/s. Assuming these values, we have

$$L = \frac{R_0}{v} = \frac{50}{2 \times 10^8} = 250\ \text{nH/m} \tag{14-89}$$

$$C = \frac{1}{R_0 v} = \frac{1}{50 \times 2 \times 10^8} = 100\ \text{pF/m} \tag{14-90}$$

Before proceeding further, it should be noted that if we used the preceding two values for $L$ and $C$, while setting $R = 0$ and $G = 0$ in the lossy model, the result would be a lossless model.

Now, how do we set a specified level of losses for the line? Unfortunately, for that purpose we will need to move briefly outside of the scope of this text and use a result developed in texts devoted extensively to transmission lines. Let $\alpha_{dB}$ represent the *attenuation in decibels/ meter (dB/m)*. For lines with small to moderate loss, this value can be approximated as

$$\alpha_{dB} = 8.686 \times \left( \frac{R}{2R_0} + \frac{GR_0}{2} \right) \tag{14-91}$$

where $R$ is the incremental series resistance in ohms/meter ($\Omega$/m) and $G$ is the incremental shunt conductance in siemens/meter (S/m). Thus, we need to find values of $R$ or $G$ that will provide the required value of 0.01 dB/m.

In most practical lines, the effect of the series resistance is usually much greater than that of the shunt conductance. Therefore, for our model, let's make it easy by setting $G = 0$ and letting the series resistance provide all the attenuation. With that choice, we can solve for $R$ from Equation 14-91, and the result is

$$R = \frac{2R_0 \alpha_{dB}}{8.686} = \frac{2 \times 50 \times 0.01}{8.686} = 0.115\ \Omega/\text{m} \tag{14-92}$$

Finally, we are ready to create our model. The circuit with the lossless line replaced by the **Lossy_Transmission_Line** is shown in Figure 14–13. To set the parameters, double left-click on the transmission line and a window with the title **Lossy Transmission Line** will open. The various parameters determined in this section are entered in the slots outlined in the beginning of the section.

A few comments are in order. The length of the line must be entered in **meters.** The other four values must be entered as **basic units per meter.** For this particular problem, the actual numbers entered in order from top to bottom are **50, 0.115, 250e-9, 100e-12,** and **0.** Note that the conductance unit on the window is indicated as **mho,** which is an older unit for conductance.

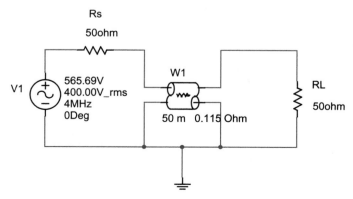

**FIGURE 14–13**
Circuit of Multisim Example 14-2.

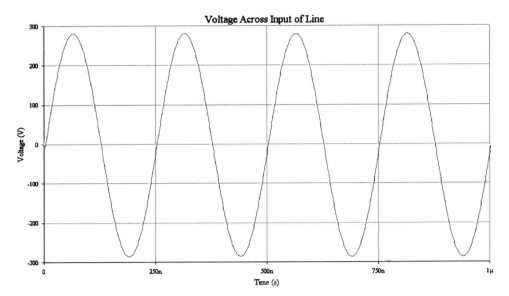

**FIGURE 14–14**
Transmission line input voltage in Multisim Example 14-2.

The voltage across the input to the line is shown in Figure 14–14. Using a cursor, the peak value is measured at 282.8423 V, which for virtually any practical purpose is the same as for the lossless line.

The big difference, which is to be expected, is for the output voltage shown in Figure 14–15. The peak value of this voltage is measured as 267.0674 V (again, all the digits are being shown). The output power is now calculated as

$$P = \frac{V_p^2}{2 \times R_L} = \frac{(267.0674)^2}{2 \times 50} = 713.3 \text{ W} \tag{14-93}$$

The value obtained in Example 14-8 was 713.0, W so the simulation result couldn't be much better!

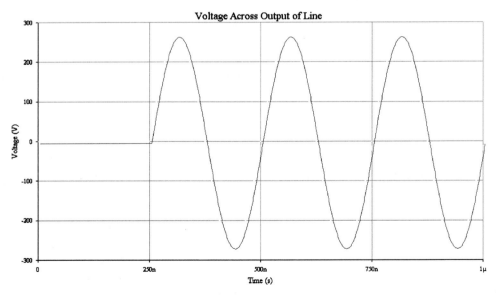

**FIGURE 14–15**
Transmission line output voltage in Multisim Example 14-2.

## 14-11  MATLAB® Examples (Optional)

It was noted in this chapter that when there is a mismatch at the load end of a transmission line, the input impedance to the line will vary with the distance from the load and standing waves will result. Depending on the source impedance, this can result in a variation in load power as the line length is varied.

The variation of the input impedance as a function of the line length was provided in Equation 14-66. Even with a scientific calculator, evaluation of this equation can be a tedious process, which led to the development of the popular Smith Chart, which provides a graphical approach to the evaluation of many properties of a transmission line.

A program has been developed to support the work of this chapter and it has the title **trans_line.m**. The MATLAB code is shown in Figure 14–16. It is based on the assumption of a *lossless* line, but the source and/or load impedances may be complex. We will use the program to support two of the examples within the text.

```
%Title of M-file is trans_line.m
fprintf('\n')
Vs=input('Enter the rms value of the source voltage in volts. ');
R0=input('Enter the characteristic impedance of the line in ohms. ');
Rs=input('Enter the real part of source impedance in ohms. ');
Xs=input('Enter the imaginary part of source impedance in ohms. ');
RL=input('Enter the real part of load impedance in ohms. ');
XL=input('Enter the imaginary part of load impedance in ohms. ');
D=input('Enter the length of line in wavelengths. ');
fprintf('\n')
Zs=Rs+j*Xs;
ZL=RL+j*XL;
Pav=Vs^2/(4*Rs);
Gamma_L=(ZL-R0)/(ZL+R0);
com=Gamma_L*exp(-j*4*pi*D);
Zin=R0*(1+com)/(1-com);
Rin=real(Zin);
Xin=imag(Zin);
Iin=Vs/(Zs+Zin);
PL=(abs(Iin))^2*Rin;
a=abs(Gamma_L);
S=(1+a)/(1-a);
fprintf('The real part of the input impedance is %g ohms.\n',Rin)
fprintf('The imaginary part of the input impedance is %g
ohms.\n',Xin)
fprintf('The standing-wave ratio is %g.\n',S)
fprintf('The maximum available power from the source is %g
watts.\n',Pav)
fprintf('The power delivered to the load is %g watts.\n',PL)
fprintf('The loss due to mismatch is %g dB.\n',10*log10(Pav/PL))
```

**FIGURE 14–16**
M-file program trans_line.m.

The program has the following input and output variables.

Input:    RMS value of source voltage
          Characteristic impedance of line (real value)
          Real part of source impedance
          Imaginary part of source impedance
          Real part of load impedance
          Imaginary part of load impedance
          Length of line in wavelengths

Output:  Real part of input impedance to line
Imaginary part of input impedance to line
Standing-wave ratio
Maximum available power from source
Power delivered to load
Relative power loss in dB due to mismatch

Upon activation of the program, the user will be polled for each of the input variables. For imaginary part entries, they should be entered as *real numbers,* but with the correct sign. Thus, an inductive reactance is entered as a positive real number (no sign) and a capacitive reactance is entered as a negative real number.

---

**▮▮ MATLAB EXAMPLE 14-1**

Apply the program **trans_line** to perform the analysis of Example 14-7.

SOLUTION   The simplicity of this ideal matched situation hardly justifies the use of the program, except as a learning exercise. The dialogue associated with the program is shown in Figure 14–17. The first several values are obvious from the work of Example 14-7.

When polled for the length of line, any value could be used, since this is a matched situation. An arbitrary value of $0.6\lambda$ was selected, so the value entered is 0.6.

The results are listed in the last six lines, and they are in obvious agreement with Example 14-7.

```
>> trans_line

Enter the rms value of the source voltage in volts. 400
Enter the characteristic impedance of the line in ohms. 50
Enter the real part of source impedance in ohms. 50
Enter the imaginary part of source impedance in ohms. 0
Enter the real part of load impedance in ohms. 50
Enter the imaginary part of load impedance in ohms. 0
Enter the length of line in wavelengths. 0.6

The real part of the input impedance is 50 ohms.
The imaginary part of the input impedance is 0 ohms.
The standing-wave ratio is 1.
The maximum available power from the source is 800 watts.
The power delivered to the load is 800 watts.
The loss due to mismatch is 0 dB.
```

**FIGURE 14–17**
Results for MATLAB Example 14-1.

---

**▮▮ MATLAB EXAMPLE 14-2**

Consider the mismatched situation of Example 14-9, and assume that there is a source connected to the input. Assume that the source open-circuit rms voltage is 400 V, and that there is an internal resistive impedance of 50 Ω. Perform the analysis with **trans_line.** Use the same line length as in the preceding example.

SOLUTION   The dialogue associated with this case is shown in Figure 14–18. Note that the reactive part of the load impedance is 100 Ω, representing an inductive reactance. A capacitive reactance would have been entered as a negative number.

For the line length used, the real or resistive part of the input impedance is about 104.2 Ω, and the imaginary or reactive part of the input impedance is about $-133.8$ Ω. Thus, the input impedance $\overline{Z}_{in}$ is about $\overline{Z}_{in} = 104.2 - j133.8$ Ω. As the length of the line

```
>> trans_line

Enter the rms value of the source voltage in volts. 400
Enter the characteristic impedance of the line in ohms. 50
Enter the real part of source impedance in ohms. 50
Enter the imaginary part of source impedance in ohms. 0
Enter the real part of load impedance in ohms. 50
Enter the imaginary part of load impedance in ohms. 100
Enter the length of line in wavelengths. 0.6

The real part of the input impedance is 104.199 ohms.
The imaginary part of the input impedance is -133.799 ohms.
The standing-wave ratio is 5.82843.
The maximum available power from the source is 800 watts.
The power delivered to the load is 400 watts.
The loss due to mismatch is 3.0103 dB.
```

**FIGURE 14–18**
Results for MATLAB Example 14-2.

is varied, the input impedance will assume a variety of values, both inductive and capacitive, and will repeat the pattern at integer multiples of $\lambda/2$. The value of the standing-wave ratio is about 5.828, which agrees with the result of Example 14-9.

While the maximum available power is the same as in the preceding example (800 W), the power reaching the load is reduced to 400 W. The difference in power is due to the serious impedance mismatch at the load, which prevents the maximum available power from being delivered.

There is a very interesting twist to this example that can be verified by the reader with the program. While the input impedance to the line varies considerably with line length, the actual load power remains fixed at 400 W. The reason is that there is an impedance match between the line and the source internal impedance, which results in a reduced power level to the load, but one that is independent of the line length as long as the line is assumed lossless. However, if there is a mismatch at both the source and the load, not only does the input impedance vary with the length, but the power delivered to the load also varies with the length.

## PROBLEMS

**14-1** The Citizen's band is near 27 MHz. Based on the rule of thumb provided in the text, determine the length of line at which transmission line effects should be considered.

**14-2** Based on the rule-of-thumb in the text, determine the length of line at which transmission line effects should be considered for an operating frequency of 300 MHz.

**14-3** An assumed lossless transmission line has $L = 600$ nH/m and $C = 40$ pF/m. Determine the characteristic impedance.

**14-4** An assumed lossless transmission line has $L = 1000$ nH/m and $C = 20$ pF/m. Determine the characteristic impedance.

**14-5** Determine the velocity of propagation of the lossless line of Problem 14-3.

**14-6** Determine the velocity of propagation of the lossless line of Problem 14-4.

**14-7** Determine the dielectric constant for the transmission line of Problems 14-3 and 14-5.

**14-8** Determine the dielectric constant for the transmission line of Problems 14-4 and 14-6.

**14-9** A coaxial cable has physical dimensions as follows:

Inner conductor diameter = 0.12 inches

Inner diameter of outer conductor = 0.48 inches

Dielectric constant of insulating material = 2.7

Determine (a) inductance per unit length, (b) capacitance per unit length, (c) characteristic impedance, and (d) velocity of propagation.

**14-10** A coaxial cable has physical dimensions as follows:

Inner conductor diameter = 5.5 mm

Inner diameter of outer conductor = 33 cm

Dielectric constant of insulating material = 2.3

Determine (a) inductance per unit length, (b) capacitance per unit length, (c) characteristic impedance, and (d) velocity of propagation.

**14-11**  A two-wire balanced transmission line has physical dimensions as follows:

>  Diameter of conductors = 0.25 cm
>
>  Spacing between conductors = 5 cm
>
>  Dielectric constant of insulating material = 2.3

Determine (a) inductance per unit length, (b) capacitance per unit length, (c) characteristic impedance, and (d) velocity of propagation.

**14-12**  An open (air dielectric) two-wire balanced transmission line has physical dimensions as follows:

>  Diameter of conductors = 0.5 inches
>
>  Spacing between the two conductors = 10 inches

Determine (a) inductance per unit length, (b) capacitance per unit length, (c) characteristic impedance, and (d) velocity of propagation.

**14-13**  Some common parameters for several types of coaxial cable are a characteristic impedance of 50 Ω and a velocity of propagation equal to $\frac{2}{3}$ of the velocity of free space. Determine (a) inductance per unit length, (b) capacitance per unit length, and (c) dielectric constant of the insulating material.

**14-14**  Some common parameters for several types of coaxial cable are a characteristic impedance of 72 Ω and a velocity of propagation equal to $\frac{2}{3}$ of the velocity of free space. Determine (a) inductance per unit length, (b) capacitance per unit length, and (c) dielectric constant of the insulating material.

**14-15**  The circuit shown below represents the output of a transmitter connected through a transmission line to an antenna. Determine (a) the input impedance of the line, (b) the rms current flowing into the line, (c) the power accepted by the line, and (d) the load power if the line is considered lossless.

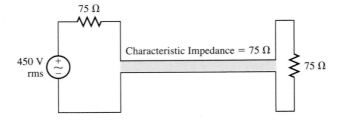

**14-16**  The circuit shown in the following figure represents the output of a transmitter connected through a transmission line to an antenna. Determine (a) the input impedance of the line, (b) the rms current flowing into the line, (c) the power accepted by the line, and (d) the load power if the line is considered lossless.

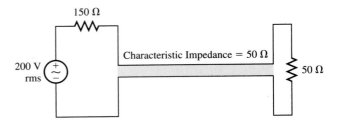

**14-17**  Assume all the conditions in Problem 14-15 except that the transmission line has a loss of 0.005 dB/m. Determine the load power if the length of the transmission line is 80 m.

**14-18**  Assume all the conditions in Problem 14-16 except that the transmission line has a loss of 0.01 dB/m. Determine the load power if the length of the transmission line is 30 m.

**14-19**  A load impedance given by $\overline{Z}_L = 100 - j100$ Ω is connected through a lossless transmission line with a characteristic impedance of 75 Ω. Determine (a) the reflection coefficient at the load, and (b) the standing-wave ratio.

**14-20**  A load impedance given by $\overline{Z}_L = 100 - j100$ Ω is connected through a lossless transmission line with a characteristic impedance of 300 Ω. Determine (a) the reflection coefficient at the load, and (b) the standing-wave ratio.

**14-21**  For the system of Problem 14-19, determine the mismatch loss in dB.

**14-22**  For the system of Problem 14-20, determine the mismatch loss in dB.

**14-23**  For the system of Example 14-9, determine the mismatch loss in dB.

**14-24**  For the system of Example 14-9, determine the return loss in dB.

**14-25**  A certain plane wave in air has an rms value for the magnetic field intensity vector of $H_y = 200$ μA/m. Assume that the electric field intensity vector is in the positive x-direction. Determine (a) the magnitude of the electric field intensity, (b) the power density of the plane wave, and (c) the total power transmitted through a circular surface in the x-y plane with a diameter of 50 m. Assume that the power density is constant over the surface area.

**14-26**  A certain plane wave in sea water has an rms value for the electric field intensity vector of $E_x = 3$ V/m, and the dielectric constant is 80. Assume that the magnetic intensity vector is in the positive y-direction. Determine (a) the magnitude of the magnetic field intensity, (b) the power density of the plane wave, and (c) the total power transmitted through a square surface in the x-y plane with sides of 15 m each. Assume that the power density is constant over the surface area.

**14-27**  The rms values of the electric and magnetic fields in a certain medium are 30 mV/m and 120 μA/m, respectively. Determine (a) the intrinsic impedance, (b) the power density, and (c) the dielectric constant.

**14-28**  The power density of a plane wave propagating in free space is 100 μW/m². Determine (a) the electric field and (b) the magnetic field.

# Introduction to Antennas

## OVERVIEW AND OBJECTIVES

Antennas are very common in today's world—all one need do is ride through any neighborhood to see satellite dish antennas on many houses. Tall vertical antennas are located near radio and television stations, and many repeater antennas are scattered throughout the countryside.

In spite of their prevalence and apparent relative simplicity in construction, antennas are among the most complex electrical devices in terms of their mathematical analysis and design. Antenna specialists have spent large portions of their careers developing the mathematical sophistication and experience required to deal with the electromagnetic phenomena involved.

Fortunately, the communication engineer or technologist can learn enough about the behavior of most common antennas through their external properties, and that will be the focus of this chapter. We will survey a few of the most common antennas, providing somewhat more depth on parabolic dish antennas due to their major application in current and future communication systems.

## Objectives

After completing this chapter, the reader should be able to:

1. Discuss the purpose and general properties of an antenna.
2. Define the gain of an antenna.
3. Determine the distance to the far-field of an antenna pattern.
4. Discuss the general properties of the antenna radiation field and its parameters.
5. Apply the range equation to determine the power density and power received at a given point in space.
6. Define the different types of polarization.
7. Define the radiation resistance of an antenna, and its implications.
8. Discuss the general properties of a dipole antenna.
9. Discuss the general properties of antennas with ground planes.
10. Discuss the general properties of horn antennas.
11. Provide an analysis of the properties of the parabolic reflector antenna, including the gain and beamwidth.

## 15-1   General Properties

An *antenna* is a device that provides the interface between a circuit or a guided wave and a propagating electromagnetic wave, or vice versa. Without an antenna, electromagnetic radiation would consist of the random emission that arises from conductors within a circuit, and any resulting waves would diminish quite rapidly. However, the antenna provides a smooth transition and allows the energy from a transmitter to be converted into an electromagnetic wave. Conversely, an electromagnetic wave impinging on an antenna will be converted to an electrical signal that can be applied to the input of a receiver. Several representative antenna types are shown in Figure 15–1.

### Reciprocity

A concept called the *theorem of reciprocity* states that the properties of the antenna for transmission will be the same for reception. This concept is very useful in studying the various properties of antennas, such as the radiation pattern, and so on. However, this does not necessarily mean that one can simply use a particular receiving antenna for a transmitting antenna and expect the same behavior, since the conductor sizes may not be able to handle the larger currents flowing from a transmitter.

### Isotropic Point Radiator

At the beginning of the study of antennas, the concept of an *isotropic point radiator* is useful. This radiator is a fictitious point source that would radiate power equally well in all directions in a three-dimensional coordinate system. The resulting pattern is called an *omnidirectional pattern.* The isotropic point radiator also serves as a basis for comparison with all types of antennas.

**Radiation Pattern**

The radiation pattern is a plot of the relative intensity of the radiation as a function of the angle in a given plane. The next section is devoted to a development of some of the properties of radiation patterns.

### Antenna Gain and Directivity

The term "antenna gain" can be misleading, since an antenna is usually a passive device and creates no energy. However, all real antennas have electromagnetic properties that permit energy to be focused with more intensity in certain directions as compared with the isotropic point radiator. Thus, energy is "borrowed" from certain directions and added to the energy in other directions, giving the antenna an effective gain in the latter direction or directions.

The three antennas shown in Figure 15–1 represent a wide range in terms of their behavior. The dipole of part (a) has a relatively low gain, while the dish of part (c) has a very high gain. The horn of part (b) has a medium level of gain.

In the mathematical treatment of antenna properties, two gain-related terms are employed: directivity and gain. If the antenna were 100% efficient in transferring the power to an electromagnetic wave, the two terms would be the same. In practice, the term *gain* takes

**FIGURE 15–1**
Representative antenna types.

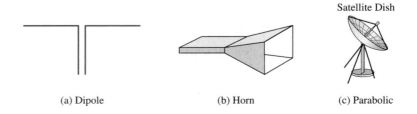

Satellite Dish

(a) Dipole          (b) Horn          (c) Parabolic

into consideration the actual efficiency of the antenna, and is usually the parameter specified in applications. Therefore, we will use the term *antenna gain* exclusively in subsequent developments.

Let $p_d$ represent the power density from an actual antenna at a particular location, and let $p_{di}$ represent the power density at the same point from a hypothetical isotropic point radiator. The antenna gain $G$ in that particular direction can be defined as

$$G = \frac{p_d}{p_{di}} \tag{15-1}$$

The corresponding decibel gain $G_{dB}$ is

$$G_{dB} = 10 \log G \tag{15-2}$$

Normally the gain specified is that corresponding to the maximum value.

### Assumed Reference and Decibel Units

Although dipoles and horns (to be studied later in the chapter) are often used as references in measuring antenna gain, most basic developments utilize gain with respect to the **isotropic radiator.** Many references use the term **dBi** as the abbreviation of the decibel gain with respect to the isotropic radiator. While this practice has many supporters, the authors believe it is slightly misleading, since the practice of adding a letter to a dB value normally implies a reference that has units, but the gain of an isotropic radiator has no units (since it is dimensionless). Hence, *the practice in this text will be to simply use dB for antenna gain, with the understanding that the isotropic radiator is the basis for comparison.*

### Near-Field and Far-Field

The time-varying voltages and currents in an antenna produce electric and magnetic fields that travel away from the antenna at a velocity that is the function of the surrounding medium, which is usually air or free space. Two separate regions of electric and magnetic fields constitute the electromagnetic radiation. The two regions are referred to as the *near-field* and *far-field*. The names are quite descriptive, in that the near-field exists only in the immediate area around the antenna. However, the far-field is responsible for the normal radiation experienced in electromagnetic wave propagation. The far-field is a *transverse* field, which means that the electric and magnetic field intensities are perpendicular to the direction of propagation. This concept was introduced in the last chapter, and is called *plane wave propagation*.

In the near-field region, both transverse and radial fields exist. However, the near-field vanishes fairly quickly as the distance from the antenna increases. The approximate point at which the near-field can be neglected corresponds to a fictitious sphere surrounding the antenna, with radius $R_{ff}$ given by

$$R_{ff} = \frac{2D^2}{\lambda} \tag{15-3}$$

where $D$ is the largest physical linear dimension of the antenna and $\lambda$ is the wavelength. The radiation patterns that display the radiation intensity as a function of angle are normally the far-field patterns.

---

**▌▎ EXAMPLE 15-1**

An ideal half-wave dipole has a theoretical length given by $\lambda/2$. Determine the distance to the boundary between the near-field and the far-field.

**SOLUTION**    The distance to the far-field is determined from Equation 15-3.

$$R_{ff} = \frac{2D^2}{\lambda} = \frac{2(\lambda/2)^2}{\lambda} = \frac{\lambda}{2} \tag{15-4}$$

This interesting result says that the distance to the far-field for an ideal half-wave dipole is the length of the dipole. This result should not be interpreted as applying to any other type of antenna, and is a "mathematical coincidence" rather than a generality.

---

**▋▌ EXAMPLE 15-2**

A high-gain reflector antenna has a diameter of 15 m and operates at 2 GHz. Determine the distance to the boundary between the near-field and the far-field.

**SOLUTION**   The wavelength at 2 GHz is

$$\lambda = \frac{3 \times 10^8}{2 \times 10^9} = 0.15 \text{ m} \tag{15-5}$$

The distance to the far-field in this case is

$$R_{\text{ff}} = \frac{2D^2}{\lambda} = \frac{2(15)^2}{0.15} = 3000 \text{ m} = 3 \text{ km} \tag{15-6}$$

This analysis indicates that the distance to the far-field pattern of a high-gain antenna can be very large, and this makes the measurement of the pattern quite difficult.   ▋▌

---

## 15-2   Antenna Radiation Pattern

The radiation pattern of an antenna is a measure of the relative intensity of the radiation as a function of a three-dimensional spatial orientation. While some three-dimensional plots are useful in visualizing the behavior, most detailed plots are created as two-dimensional plots. This means that two plots in perpendicular planes would be required to fully characterize the behavior. In some cases, one two-dimensional plot may be all that is required, especially if the antenna is omnidirectional in the other plane.

### E-Plane and H-Plane

In general, the electric field and the magnetic field surrounding an antenna in the far-field are perpendicular to each other. One of the two-dimensional plots that can be used to describe the radiation pattern will show the coordinate system emphasizing the electric field, and is referred to as an *E-plane plot*. The other coordinate system will emphasize the magnetic field contribution, and is referred to as an *H-plane* plot.

### Representative Plots

A two-dimensional radiation pattern plot as a function of angle in a rectangular form is shown in Figure 15–2. The same pattern shown on a polar plot is given in Figure 15–3. This latter form is more commonly used in describing an antenna, since the behavior can be more easily visualized in a true spatial sense.

Let $G(\theta)$ represent the actual gain as a function of the angle $\theta$, and let $G_{\text{max}}$ represent the maximum gain. It is common practice to normalize the plots with respect to the maximum gain, and that is the practice here. The normalized gain will be denoted $g(\theta)$ and defined as

$$g(\theta) = \frac{G(\theta)}{G_{\text{max}}} \tag{15-7}$$

The decibel normalized gain will be denoted $g_{\text{dB}}(\theta)$ and is

$$g_{\text{dB}}(\theta) = 10 \log[g(\theta)] = 10 \log \left[ \frac{G(\theta)}{G_{\text{max}}} \right] \tag{15-8}$$

In the normalized form, the maximum gain occurs at the point where the normalized gain is 0 dB.

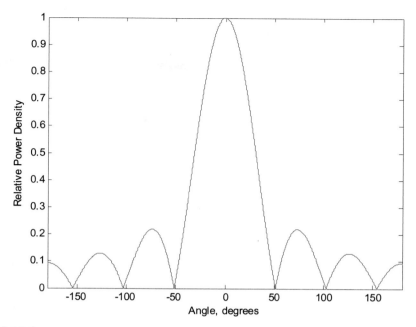

**FIGURE 15–2**
Representative rectangular antenna pattern plot.

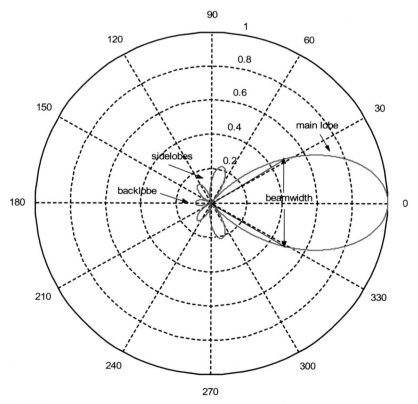

**FIGURE 15–3**
Representative polar antenna pattern plot.

### Beamwidth

The beamwidth of an antenna pattern is similar to the bandwidth of a filter. Normally the beamwidth is the angle between the two points at which the radiation power density is reduced by 3 dB from the maximum power density.

### Gain versus Beamwidth

There is an inverse relationship between the gain and the beamwidth. As the gain increases, the beamwidth decreases, and vice versa.

### Lobes

The various segments of the antenna pattern are referred to as *lobes*. The lobe in the direction of maximum radiation is called the *main lobe,* and the lobes around the side are referred to as *sidelobes*. The null-to-null beamwidth of a typical high-gain antenna is about 2.5 times the 3-dB beamwidth. The lobe opposite to the direction of maximum radiation is referred to as the *backlobe*. For certain antennas where the backlobe is undesired, a parameter called the *front-to-back* ratio is useful.

---

**▍▌ EXAMPLE 15-3**

A parabolic dish antenna has a maximum power gain of $10^5$. Determine the maximum decibel gain.

**SOLUTION**   This simple exercise is designed to familiarize the reader with the order of magnitude of possible antenna gains, and this value can be achieved with a parabolic dish antenna. The decibel maximum gain will be denoted as $G_{\text{max,dB}}$, and is

$$G_{\text{max,dB}} = 10 \log 10^5 = 50 \text{ dB} \qquad (15\text{-}9)$$

▍▌

---

## 15-3   Power Density and Antenna Gain

Consider an isotropic point radiator, and assume that it is located at the center of a sphere of radius $d$. (Normally the symbol $r$ would be used for radius, but eventually $d$ will be used to represent the distance between antennas, and that is the reason for the symbol choice.)

Assume that it is radiating power $P_t$ with equal intensity in all directions. The surface area $A$ of a sphere of radius $d$ is given by

$$A = 4\pi d^2 \qquad (15\text{-}10)$$

Since the power density will be uniform at all points on the surface, it can be determined by dividing the power by the surface area. Let $p_d$ represent the power density, which is given by

$$p_d = \frac{P_t}{A} = \frac{P_t}{4\pi d^2} \qquad (15\text{-}11)$$

A portion of a sphere is illustrated in Figure 15–4. Let $\Delta P$ represent the amount of power passing through an area $\Delta A$. This value is

$$\Delta P = p_d \, \Delta A \qquad (15\text{-}12)$$

If the power density varied over the surface of the sphere it would be necessary to begin with Equation 15-12 and perform an integration, but that is not the case for the isotropic radiator, since the power density is constant everywhere on the surface of the assumed sphere.

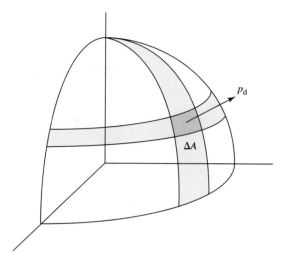

**FIGURE 15–4**

Power density of signal radiating from an antenna.

Next, the antenna gain will be considered. While the maximum gain is usually the value of major interest, to simplify the notation, we will assume an arbitrary transmitting gain $G_t$, which actually makes the equations that follow apply to any angle. Based on the assumed gain, the power density formula of Equation 15-11 can be modified to read

$$p_d = \frac{G_t P_t}{4\pi d^2} \tag{15-13}$$

Consider next the receiving antenna. Any antenna can be described by an *effective capture area* $A_e$, which can be defined in terms of the *received power* $P_r$ and the power density as

$$A_e = \frac{P_r}{p_d} \tag{15-14}$$

Note that the dimensions of the capture area are watts divided by watts/meter$^2$ = meter$^2$, so the units are compatible with area. So what do we mean by the "area of an antenna"? Actually, for certain antennas (such as parabolic dish antennas), it does correspond roughly to a physical area, but for some types of antennas, it is simply a number that acts as if it were an actual physical area.

Solving for the received power from Equation 15-14, we obtain

$$P_r = p_d A_e \tag{15-15}$$

Applying Equation 15-15 to Equation 15-13, we obtain

$$P_r = \frac{A_e G_t P_t}{4\pi d^2} \tag{15-16}$$

There is a fundamental linear relationship between the capture area and the gain of *any* antenna. This relationship is derived in advanced antenna theory texts, and it reads

$$G = \frac{4\pi}{\lambda^2} A_e \tag{15-17}$$

Assume a value of $G_r$ for the receiving antenna gain, and when Equation 15-17 is solved for the capture area in terms of the receiving gain, the received power becomes

$$P_r = \frac{\lambda^2 G_r G_t P_t}{(4\pi)^2 d^2} \tag{15-18}$$

This equation is referred to in different references as the *range equation,* the *link equation,* or the *Friis equation* (in honor of Harold Friis, one of the early pioneers in radio transmission). Where there is any possibility of confusion, we will attach the adjective *one-way* to separate this version from the radar equation, which is based on two-way propagation. The range equation will be used extensively in the communication system analysis that follows in several later chapters.

---

**▌▌ EXAMPLE 15-4**

Determine the power density at a distance of 100 km from a hypothetical isotropic point radiator that is transmitting 100 W.

**SOLUTION**   In later chapters, various equations will be modified to permit distance to be expressed in km, but for the moment, basic units are required. Hence, the distance is $10^5$ m. The power density is then

$$p_d = \frac{P_t}{4\pi d^2} = \frac{100}{4\pi(10^5)^2} = 795.8 \text{ pW/m}^2 \tag{15-19}$$

---

**▌▌ EXAMPLE 15-5**

Assume in the preceding example that the antenna gain in the given direction is 50. Determine the power density at the same point based on the same radiated power.

**SOLUTION**   We could go back and use the form of the equation containing the transmitting antenna gain, but it amounts to simply taking the result of Example 15-4 and multiplying it by 50:

$$p_d = \frac{G_t P_t}{4\pi d^2} = 50 \times 795.8 = 39.79 \text{ nW/m}^2 \tag{15-20}$$

---

**▌▌ EXAMPLE 15-6**

A satellite system operating at 15 GHz has a transmitting antenna gain of $10^4$, a receiving antenna gain of $10^5$, and the distance is 41,000 km. If the transmitted power is 50 W, determine the received power.

**SOLUTION**   To apply the range equation in the form given, we need to determine the wavelength:

$$\lambda = \frac{3 \times 10^8}{15 \times 10^9} = 0.02 \text{ m} \tag{15-21}$$

The distance must be expressed in meters, and the value is $41 \times 10^6$ m. The received power is then

$$P_r = \frac{\lambda^2 G_r G_t P_t}{(4\pi)^2 d^2} = \frac{(0.02)^2 \times 10^4 \times 10^5 \times 50}{(4\pi)^2(41 \times 10^6)^2} = 75.3 \text{ pW} \tag{15-22}$$

The received power is very small, but this value is in the range of satellite received signal power levels.                                                                    ▌▌

---

## 15-4  Polarization

It has been established by convention that the *polarization* of an electromagnetic wave propagating in free space is the direction of the electric field intensity vector in relationship to the surface of the earth. The two basic types of polarization are *linear* and *elliptical*. With *linear polarization,* the electric vector does not change its orientation with respect to an observer. With *elliptical polarization,* the tip of the electric vector traces an ellipse as it moves away from an observer.

### Linear Polarization

Linear polarization can be classified as either *vertical* or *horizontal*. With horizontal polarization, the electric vector is *parallel* to the surface of the earth. With vertical polarization, the electric vector is *perpendicular* to the surface of the earth.

## Circular Polarization

The most widely used form of elliptical polarization is called *circular polarization,* and it is based on the situation in which the rotating electric vector traces a circle as it travels away from the observer. Circular polarization may either be *right-hand circular* or *left-hand circular.* With *right-hand circular polarization,* the electric vector rotates in a *clockwise* direction as the electric vector travels away from the observer. With *left-hand circular polarization,* the electric vector rotates in a *counterclockwise* direction.

Depending on antenna design and orientation, it may transmit horizontal, vertical, right-hand (RH) circular, or left-hand (LH) circular polarization. There is a loss in the neighborhood of 30 dB or more if the transmitting antenna has one form of linear polarization and the receiving antenna has the other linear form. A similar loss also occurs if one antenna is circularly polarized with one sense (RH or LH) and the receiving antenna is circularly polarized with the other sense (LH or RH). These properties may be used to advantage when it is desired to send two channels of information over the same frequency, by sending one channel with one form of polarization and the other channel with the other form.

A linear polarized transmitting antenna can be used in conjunction with a circular polarized receiving antenna, or vice versa, and there will be only a 3-dB loss. This combination is used where the orientation of a linear polarized antenna can change significantly, such as on a missile or spacecraft.

## 15-5 Antenna Impedance and Radiation Resistance

When an antenna accepts power from a source and converts it into radiated energy, an amazing phenomenon occurs. The antenna acts as if it were a lumped impedance to the source. If the antenna is an optimum design for the frequency, the impedance will be real and resistive.

Don't try to measure this resistance with a dc-ohmmeter, since the effort will be futile. Indeed some antennas are a simple open circuit as far as dc is concerned. However, an RF bridge excited at the appropriate frequency can actually measure the impedance.

### Antenna Impedance

The definition of the antenna impedance is the ratio of ac voltage to ac current at the input terminals. In general, the impedance is complex, and will consist of a real part and an imaginary part. As mentioned earlier, the impedance at the proper operating frequency will usually be real and will ideally be matched to the line feeding it. However, the impedances of most antennas are highly frequency dependent, and may have poor characteristics at frequencies different than the design frequency. Some special antenna designs have been developed for operating over a range of frequencies, but it is safe to say that there is some compromise involved.

### Radiation Resistance

Let $P_t$ represent the transmitted or radiated power from an antenna. This power can be assumed to be dissipated in a resistance called the *radiation resistance.* It will be denoted $R_{rad}$, and if the rms current flowing into the terminals of the antenna is $I_{rms}$, the value of the radiation resistance is

$$R_{rad} = \frac{P_t}{I_{rms}^2}$$

(15-23)

If the antenna were 100% efficient, this would be the real part of the input impedance to the antenna. In practice, there are some ohmic losses due to resistance in the conductors of the antenna, and the actual real part of the input impedance is the sum of the radiation resistance and the ohmic resistance.

▌▌ **EXAMPLE 15-7**

The rms current flowing into an antenna is 5 A. The power transmitted is 2 kW. Determine the radiation resistance.

**SOLUTION**    The radiation resistance is

$$R_{rad} = \frac{P_t}{I_{rms}^2} = \frac{2000}{(5)^2} = 80 \ \Omega \qquad (15\text{-}24)$$

▌▌

## 15-6 Dipole Antennas

The *half-wave dipole* is arguably the simplest of all antennas, and while it has very limited gain, it is easy to construct and has a broad area of coverage, making it very popular for a variety of routine applications. The form of construction is shown in Figure 15–5, and it can be viewed as an opening of a two-wire transmission line. The ends of the antenna have voltage maxima, and there is a voltage minimum at the feed point. While the theoretical length of the antenna is $\lambda/2$, in practice the antenna length is reduced by about 5% to compensate for end effects.

### Dipole Input Impedance

At the feed point, there is a voltage minimum and a current maximum. This means that the impedance is a minimum at that point. The actual theoretical value of the input impedance of a half-wavelength dipole is $73 + j42.5 \ \Omega$. By reducing the length by about 5%, the reactive component can be eliminated. Thus, while the name implies a half-wavelength, the actual length should be about 95% of a half-wavelength or approximately 95% of the theoretical length. With the proper length, the half-wave dipole has a resistive impedance of about 73 Ω.

The dipole is a balanced antenna and should be fed by a balanced transmission line. The closest match to the impedance is coaxial cable, which means that a *balun* should be used to properly connect coaxial cable to a dipole.

### Radiation Pattern

The *E*-plane for the half-wave antenna is the plane containing the full length of the dipole and thus the polarization is parallel to the dipole. Refer to Figure 15–6, and assume that the antenna is oriented along the line extending from 0° to 180°, with the feed point at the center of the graph. Maximum radiation occurs in both directions perpendicular to the length as shown and nulls occur at the ends.

In three-dimensional space, the pattern shown would form a sort of "doughnut" shape. The *H*-plane pattern is obtained by looking at the antenna from the ends, and it would appear as a point from that perspective. While not shown, the pattern in that plane would simply be a circle, meaning that the radiation at a given distance from the antenna in that plane is a constant value. We can thus say that the dipole is omnidirectional in the *H*-plane.

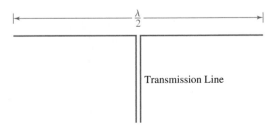

**FIGURE 15–5**
Half-wave dipole fed with balanced transmission line.

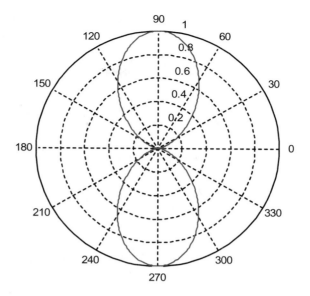

**FIGURE 15–6**
Dipole antenna pattern in
*E*-plane.

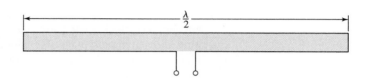

**FIGURE 15–7**
Half-wave folded dipole.

The 3-dB beamwidth of the dipole is 78°. The maximum gain for a lossless half-wavelength dipole is 1.643, corresponding to 2.16 dB.

## Effective Area

The dipole is a good example of an antenna in which the effective area has no relationship to the physical area. Indeed, the only physical area is that of the conductors, which will vary with conductor size. Yet an effective area of a dipole can be determined from the isotropic gain relationship, and is given by

$$A_e = \frac{\lambda^2}{4\pi} G = \frac{\lambda^2}{4\pi}(1.643) = 0.13\lambda^2 \tag{15-25}$$

## Folded Dipole

One form of the half-wave dipole is the *folded dipole* shown in Figure 15–7. A common balanced transmission line is the 300-Ω twin lead, which has been used in both the TV and FM industries. A folded dipole can be constructed from a half wavelength length of 300-Ω twin lead cable. The antenna is fed at the middle of the bottom row. The equivalent resistance seen at the input of the folded dipole is 280 Ω, which provides a reasonably good match to a 300-Ω twin-lead balanced transmission line.

**▐▌ EXAMPLE 15-12**    Determine the effective area of a dipole at a frequency of 100 MHz.

**SOLUTION**    The wavelength corresponding to 100 MHz is

$$\lambda = \frac{c}{f} = \frac{3 \times 10^8}{100 \times 10^6} = 3 \text{ m} \tag{15-26}$$

The capture area is determined from Equation 15-25:

$$A_e = 0.13\lambda^2 = 0.13(3)^2 = 1.17 \text{ m}^2 \qquad (15\text{-}27)$$

## 15-7  Antennas with Ground Planes

Many common antennas, such as those used with commercial AM stations, utilize the earth as an integral part of the antenna. Many mobile antennas utilize the frame of the vehicle as a ground in the same fashion. The general name for a ground utilized with an antenna is that of a *ground plane,* which is a uniform conducting surface beneath an antenna. In commercial AM broadcasting, conductors called *radials* are buried beneath the surface of the earth to enhance the conductivity.

### Reflected Wave from Surface

In a perfect conductor, electromagnetic waves cannot exist, and a wave incident upon the surface of a perfect conductor will be reflected. Figure 15–8 illustrates this situation. To maintain the required value of zero for the electric field at the surface, the reflected wave will undergo a phase shift of 180°.

A technique called *image theory* can be used to determine the properties of an antenna operating above a ground plane. The reflected wave may be thought of as arising from an identical antenna located within the ground plane at the same distance below the surface as the real antenna is above the surface. This concept is illustrated in Figure 15–8. The image antenna may be considered in the same manner as that of a reflected image in a mirror in the optical frequency range.

### Quarter-Wave Vertical Antenna

The *quarter-wave vertical antenna* operating above a ground plane is illustrated in Figure 15–9. This is an unbalanced antenna, and can be connected to a transmitter or receiver

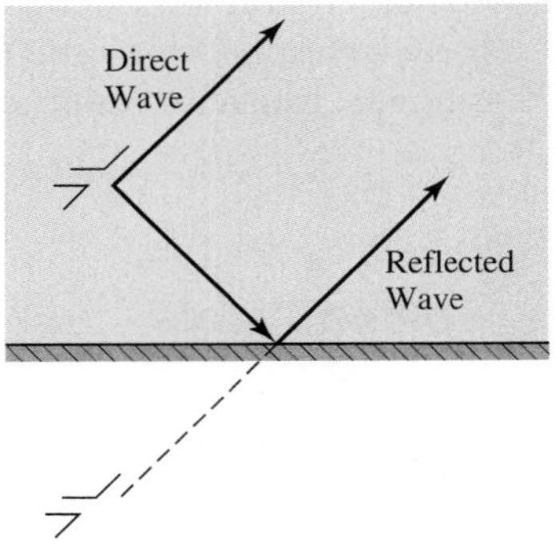

**FIGURE 15–8**
Ground plane below an antenna, and its effect.

**FIGURE 15–9**
Quarter-wave vertical antenna above ground plane, and the effective mirror image below ground.

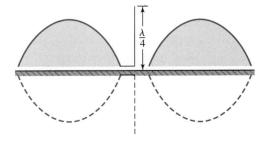

with coaxial cable. It has vertical polarization, and the presence of the ground plane permits optimum operation with half the length of the dipole.

Because of the ground plane, the current flowing into the vertical is twice that of the dipole for a given voltage. Thus, the theoretical radiation resistance of the quarter-wave vertical is 36.5 Ω, and the impedance has a reactive component of $j21$ Ω. The $E$-plane radiation pattern is the same as that of a half-wave dipole above the ground plane, but only half of the pattern exists. The radiation pattern tilts increasingly upward as the ground plane deviates from that of a perfect conductor.

Vertical antennas can be designed with lengths greater than the quarter wavelength. For lengths greater than $5\lambda/8$, there are multiple lobes in the radiation pattern, and the main lobe is located at an angle above the ground plane. For a height of one wavelength, the main lobe is located at an angle of 45° with respect to the ground plane. A major reason for using a longer antenna is to increase the radiation resistance. A large antenna current, commensurate with smaller antenna radiation resistance, results in increased power lost in the ohmic resistance of the antenna wires and transmission line.

## 15-8  Horn Antennas

A microwave waveguide open at the end can transmit or receive electromagnetic radiation and function as an antenna. However, the abrupt change from the waveguide impedance to the intrinsic impedance of free space reflects some of the energy back toward the source. This impedance mismatch can be circumvented by flaring the walls of the waveguide to provide a gradual transition between the waveguide and free-space. The resulting antenna is called a *horn antenna.*

### Types of Horn Antennas

Refer to Figure 15–10 to see illustrations of the three basic types of waveguide horn antennas. They are (a) the $E$-plane horn, (b) the $H$-plane horn, and (c) the pyramidal horn. Based on the orientation shown, the $E$-plane is in the vertical direction and the $H$-plane is in the horizontal direction. The gain of a horn is dependent on the aperture size, frequency, and length of the flare. (Full details are provided in advanced antenna textbooks.) Gains of standard horn antennas vary from about 15 dB at the lowest range of the operating frequency to about 17.5 dB at the upper range.

The beamwidth of a horn antenna depends on the aperture size. As the dimension in the flare direction increases, the 3-dB beamwidth decreases. A sectional horn based on either the $E$-plane or the $H$-plane produces a fan-shaped beam that is narrow in one plane and broad in the other plane. An $E$-plane sectional horn produces a narrow beam in the $E$-plane, and an $H$-plane sectional horn produces a narrow beam in the $H$-plane. A narrow beam in both planes is created by a pyramidal horn.

Horns have high efficiency, and high gains can be achieved by horns with a large aperture. Since the flare must be gradual to provide optimum impedance matching, a high gain requires a long antenna. Therefore, a large, high-gain horn is not too practical, and the parabolic reflector antenna discussed in the next section is more feasible for high-gain applications.

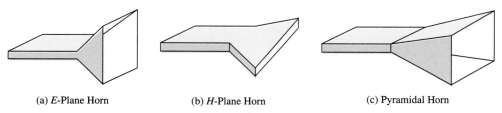

(a) $E$-Plane Horn          (b) $H$-Plane Horn          (c) Pyramidal Horn

**FIGURE 15–10**
Three types of waveguide horns.

## 15-9  Parabolic Reflector Antennas

The construction of the *parabolic reflector antenna* is illustrated in Figure 15–11(a). A parabola rotated about its axis results in a surface called a *paraboloid*. An important parameter for the paraboloid is the *focal point*. The parabolic reflector has a surface created in the form of a paraboloid, and a much smaller primary antenna is located at the *focal point*.

The *focal point* is a point in space such that all plane waves reflected from the surface will converge at that point. Conversely, an isotropic radiator placed at the focal point will radiate a spherical wave, from which the parabolic reflector surface will reflect a plane wave. The plane wave will be present over the area equal to a circle of diameter *D*, as illustrated in Figure 15–11(b).

The primary antenna is used to illuminate the reflector with as close to a uniform electromagnetic wave as possible. The primary antenna should ideally have uniform gain over the angle of the reflector and a gain of zero elsewhere. A good choice for the primary antenna is a pyramidal horn. The horn is usually designed to have a gain at the edge of the paraboloid of about 10 dB below the peak horn gain.

Probably the simplest illustration of the action of a parabolic reflector is the ordinary household flashlight. The actual bulb produces a relative small amount of light, but based on the action of the reflector, a strong beam can be produced.

If the illumination were perfectly uniform, the physical area $A_p$ would be equal to the effective area $A_e$. In practice, the nonuniform nature of the illumination is considered by defining an *illumination factor* $\eta_I$ as

$$\eta_I = \frac{A_e}{A_p} \tag{15-28}$$

This value is usually expressed as a percentage by multiplying by 100. Typical values are in the neighborhood from about 55% to 75%.

### Physical Area

The physical area $A_p$ is the projected area of the aperture in a plane perpendicular to the direction of propagation. It is given by

$$A_p = \frac{\pi D^2}{4} \tag{15-29}$$

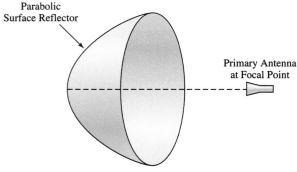

(a) Parabolic Reflector and Primary Antenna

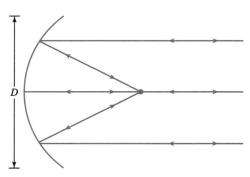

(b) Two-Dimensional Antenna View with Ray Paths

**FIGURE 15–11**
Properties of parabolic reflector antenna.

## Gain

The gain of the parabolic dish antenna can be determined from the basic relationship between gain and aperture area:

$$G = \frac{4\pi A_e}{\lambda^2} = \frac{4\pi}{\lambda^2}\left(\frac{\eta_I \pi D^2}{4}\right) = \eta_I \left(\frac{\pi D}{\lambda}\right)^2 \tag{15-30}$$

## Beamwidth

The approximate 3-dB beamwidth in radians for a parabolic reflector antenna is given by

$$\theta_{3dB} = 1.2\frac{\lambda}{D} \text{ rad} \tag{15-31}$$

The corresponding beamwidth in degrees is approximately

$$\theta_{3dB} = 70\frac{\lambda}{D} \text{ degrees} \tag{15-32}$$

**EXAMPLE 15-11**    The physical diameter of a 10-GHz parabolic reflector is 12 m, and the illumination efficiency is 70%. Determine (a) the effective area, (b) the gain in dB, and (c) the 3-dB beamwidth.

**SOLUTION**    The wavelength corresponding to 10 GHz is

$$\lambda = \frac{3 \times 10^8}{10 \times 10^9} = 0.03 \text{ m} \tag{15-33}$$

(a)  The physical area is determined as

$$A_p = \frac{\pi D^2}{4} = \frac{\pi(12)^2}{4} = 113.1 \text{ m}^2 \tag{15-34}$$

The effective capture area is

$$A_e = \eta_I A_p = 0.7 \times 113.1 = 79.17 \text{ m}^2 \tag{15-35}$$

(b)  The gain is

$$G = \frac{4\pi A_e}{\lambda^2} = \frac{4\pi(79.17)}{(3 \times 10^{-2})^2} = 1.105 \times 10^6 \tag{15-36}$$

Using Equation 15-30 directly could have reduced the number of steps, but the process used should provide some educational value.

The decibel gain is

$$G_{dB} = 10\log 1.106 \times 10^6 = 60.4 \text{ dB} \tag{15-37}$$

(c)  The 3-dB beamwidth in degrees is

$$\theta_{3dB} = 70\frac{\lambda}{D} = 70\left(\frac{0.03}{12}\right) = 0.175 \text{ degrees} \tag{15-38}$$

## 15-10    MATLAB® Example (Optional)

The one example in this section will be devoted to the display of a polar plot of the normalized gain of an antenna.

**MATLAB EXAMPLE 15-1**    Complex patterns can be achieved by utilizing arrays of simple antennas with appropriate spacing between them. By using a combination of phase shifts and spacing, signals traveling in certain directions may reinforce each other, while in other directions, they may cancel.

A relatively simple example is that of two vertical antennas spaced $\lambda/2$ apart. Either antenna alone would exhibit an omnidirectional pattern in the *H*-plane (parallel to the earth's surface). The signals perpendicular to or broadside to the array will reinforce each other, since they are in phase. In the direction aligned with the array, however, a signal from either antenna will have undergone a 180° phase shift when it reaches the other antenna, and the two signals will cancel. In the horizontal plane, the normalized power gain will have the form

$$g(\theta) = \cos^2\left(\frac{\pi}{2}\cos\theta\right) \quad 0 \le \theta \le 2\pi \tag{15-39}$$

where $\theta$ is measured from a line perpendicular to the alignment direction of the two antennas.
Construct a polar plot of the antenna gain pattern.

**SOLUTION**   The angle is generated over the domain from 0 to $2\pi$ by the command

```
>> theta = linspace(0,2*pi,101);
```

The antenna pattern is generated by the command

```
>> g = cos((pi/2)*cos(theta)).^2;
```

Note the period preceding the exponent, which tells the program to square the values on a term-by-term basis.
A polar plot of the normalized gain is then generated by the command

```
>> polar(theta,g)
```

The result, with additional labeling, is shown in Figure 15–12.

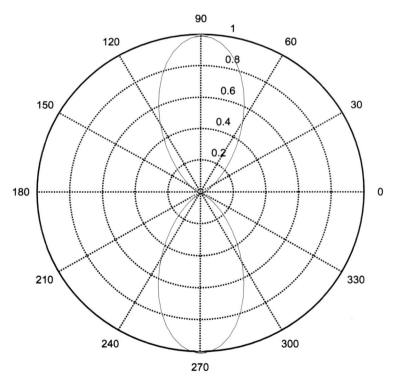

**FIGURE 15–12**
Plot for MATLAB Example 15-1.

## PROBLEMS

**15-1** Determine the distance from a vertical antenna having a length of a quarter wavelength at 1 MHz to the boundary between the near-field and the far-field.

**15-2** A high-gain reflector antenna has a diameter of 12 m and operates at 3 GHz. Determine the distance to the boundary between the near-field and the far-field.

**15-3** A parabolic dish antenna has a maximum absolute power gain of 500,000. Determine the maximum decibel gain.

**15-4** The gain of an antenna is specified as 40 dB. Determine the absolute power gain.

**15-5** Determine the power density at a distance of 50 km from a hypothetical isotropic point radiator that is transmitting 5 W.

**15-6** Determine the power density at a distance of 36,000 km from a hypothetical isotropic point radiator that is transmitting 20 W.

**15-7** Assume in Problem 15-5 that the antenna gain in the given direction is 12 dB. Determine the power density at the same point based on the same radiated power.

**15-8** Assume in Problem 15-6 that the antenna gain in the given direction is 40 dB. Determine the power density at the same point based on the same radiated power.

**15-9** The power density 50 km from a transmitting antenna is 0.5 $\mu$W/m$^2$. If the frequency is 2 GHz and the antenna gain is 30 dB, determine the transmitter power.

**15-10** The power density 10 km from a transmitting antenna is 0.2 $\mu$W/m$^2$. If the frequency is 3 GHz and the antenna gain is 20 dB, determine the transmitter power.

**15-11** An antenna is transmitting 20 W of power. The maximum power density at a distance of 3 km is 5 mW/m$^2$. Determine the gain of the antenna in dB.

**15-12** An antenna is transmitting 300 W of power. The maximum power density at a distance of 1 km is 0.6 mW/m$^2$. Determine the gain of the antenna in dB.

**15-13** An antenna has a gain of 40 dB. Determine the effective area if the frequency is 3 GHz.

**15-14** An antenna has a gain of 6 dB. Determine the effective area if the frequency is 500 MHz.

**15-15** An antenna has an rms current of 6 A flowing into the input, and it is transmitting 3 kW of power. Determine the radiation resistance of the antenna.

**15-16** An antenna has an rms voltage of 160 V across the input terminals, and it is transmitting 500 W of power. Determine the radiation resistance of the antenna.

**15-17** A 6-GHz parabolic reflector antenna has a physical diameter of 12 m. The illumination efficiency is 55%. Determine the following parameters: (a) the effective area, (b) the gain in dB, and (c) the 3-dB beamwidth in degrees.

**15-18** A 5-GHz parabolic reflector antenna has a physical diameter of 4 m. The illumination efficiency is 70%. Determine the following parameters: (a) the effective area, (b) the gain in dB, and (c) the 3-dB beamwidth in degrees.

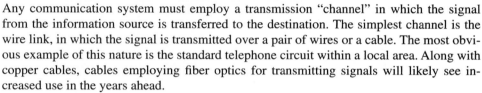

# Communication Link Analysis and Design — 16

Any communication system must employ a transmission "channel" in which the signal from the information source is transferred to the destination. The simplest channel is the wire link, in which the signal is transmitted over a pair of wires or a cable. The most obvious example of this nature is the standard telephone circuit within a local area. Along with copper cables, cables employing fiber optics for transmitting signals will likely see increased use in the years ahead.

While various types of channels have their advantages, the channel concept that has traditionally provided the most capability for information transmission is the use of electromagnetic waves. The concepts of electromagnetic radiation and the use of antennas were introduced in the past two chapters. We are now ready to study the behavior of the link between the transmitting antenna and the receiving antenna. In particular, the study of direct-ray propagation will be heavily emphasized, due to its accelerated use in satellite systems. However, some of the classical methods—such as surface-wave and sky-wave propagation—will also be discussed.

## Objectives

After completing this chapter, the reader should be able to:

1. State the Friis one-way link equation and discuss the various parameters and their significance.
2. Define *effective isotropic radiated power* (EIRP).
3. State and apply the decibel forms of the one-way link equation.
4. Define and calculate *path loss.*
5. Determine the line-of-sight distance between transmitting and receiving antennas.
6. Determine the minimum obstacle clearance required to avoid diffraction effects.
7. State the radar or two-way link equation and discuss the various parameters and their significance.
8. Define *radar backscatter cross-section.*
9. State and apply the decibel form of the two-way link equation.
10. Explain the principle of pulse radar, and how distance is measured.
11. For a pulse radar, determine the *maximum unambiguous range* and the *resolution.*
12. Explain the principle of Doppler radar and how velocity is measured.

13. Determine the Doppler shift for a given velocity, and vice versa.

14. State and apply Snell's law.

15. Define *refraction* and the *index of refraction*.

16. Discuss ground-wave propagation in terms of its properties and applicable frequency range.

17. Discuss sky-wave propagation in terms of its properties and applicable frequency range.

## 16-1 Friis Link Equation

The Friis range or link equation was developed in Chapter 15, and we are now ready to consider how it can be used in practice. This formula is most important in establishing a relationship between transmitted power, received power, antenna gain, and distance between transmitter and receiver. Its use permits a design trade-off between various system link parameters in establishing proper operating levels for a complete RF communication system.

The link equation applies to a system in which there is a transmitting antenna, a receiving antenna, and a direct electromagnetic path without obstacles between the two antennas, as illustrated by the simple diagram of Figure 16–1. It does not apply to the various types of indirect propagation (such as sky-wave bounce and tropospheric scatter). Such indirect propagation techniques have depended on empirical data to determine proper operating conditions, and such results are not always reliable. Consideration of some of the various types of indirect wave propagation will be made later in the chapter.

The trend of the future is definitely toward the use of direct-wave propagation, because of its predictability and reliability. Prior to the advent of communication satellites, direct-ray propagation was limited to a very short range, due to the curvature of the earth and the corresponding difficulty of locating antennas at sufficient heights to utilize direct rays. Following the development of satellite repeater systems, new dimensions for long-range communications became feasible. Direct-ray propagation is now practical between the earth and a satellite, and the signal can then be retransmitted to points around the globe. As we will see with realistic examples later, the amount of power required for direct-ray propagation is relatively modest, utilizing high-gain antennas and microwave frequencies.

### Review of the Link Equation

Let us now review the parameters in the link equation, utilizing basic units at this point.

$\lambda$ = wavelength in meters (m)

$G_t$ = absolute power gain of transmitting antenna (no units)

$G_r$ = absolute power gain of receiving antenna (no units)

$P_t$ = transmitter power in watts (W)

$d$ = distance between transmitting and receiving antennas in meters (m)

The basic form of the equation is then given by

$$P_r = \frac{\lambda^2 G_r G_t P_t}{(4\pi)^2 d^2} \tag{16-1}$$

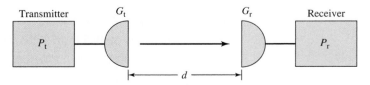

**FIGURE 16–1**
Illustration of direct-ray propagation as described by the Friis range equation.

This formula is also referred to as the *one-way link equation*. Later, we will see that there is a *two-way* version for signals that are transmitted to a destination and are "bounced back," such as in radar.

## General Discussion of the One-Way Link Equation

A number of important deductions can be made from the link equation, some of which are obvious, and some of which are more subtle. Immediately, it is noted that the power at the receiver is directly proportional to the product of the transmitted power, the gain of the transmitter antenna, and the gain of the receiver antenna. Within limits, these three parameters may be adjusted without affecting performance, provided that the product remains the same. For example, a transmitter with a power output of 2 W and an antenna gain of 100 would produce the same received power in the direction of maximum radiation as a transmitter with a power output of 50 W and a gain of 4. As a second example, if a receiving antenna is replaced by one having half the gain, the original received power could be restored by either doubling the transmitted power or by replacing the transmitting antenna with one having twice the gain. From a decibel point of view, doubling either the transmitter power, the transmitter antenna gain, or the receiver antenna gain would increase the received signal level by about 3 dB. Increasing either of the preceding parameters by a factor of 10 would result in an increase of the received signal level by 10 dB.

## Variation with Respect to Distance

The denominator of the link equation indicates that the received power varies inversely with the square of the distance $d$ between transmitter and receiver. Thus, if the distance between transmitter and receiver is doubled, the received power is reduced by a factor of one-fourth.

The decibel variation as a result of the distance variation is interesting. If the distance between the antennas is doubled, the received power level is reduced by about 6 dB. If the distance between the antennas is increased by a factor of 10, the received power level is reduced by 20 dB. The decibel changes in this case are twice as great as for power and antenna changes, due to the square-law relationship for distance.

## Variation with Respect to Wavelength

The variation of received power with respect to wavelength $\lambda$ is more subtle than the other quantities, and requires interpretation to avoid misleading conclusions. From the link equation, it appears that the received power is directly proportional to $\lambda^2$, and this would be true if all other parameters remained the same. The fallacy is that if the wavelength were increased (corresponding to lowering the frequency), the antennas would need to be replaced with new ones having the same gains as the previous ones. As deduced in the previous chapter, lower frequencies require larger antennas for the same gain, so any change in wavelength would mandate a complete redesign of the system. Therefore, accept the $\lambda^2$ as a factor in the equation, but one that cannot easily be changed without a complete system overhaul.

## Losses

In its basic form, the Friis transmission formula is valid under a wide range of operating conditions. In general, however, other factors must be considered in a full system design for worst-case conditions.

Losses resulting from atmospheric absorption are important in certain frequency ranges. Such losses are shown as a function of frequency in Figure 16–2. At frequencies well below 20 GHz, water and oxygen losses are very low. Losses due to water vapor absorption are most significant in the vicinities of 23 and 180 GHz. Similarly, losses due to

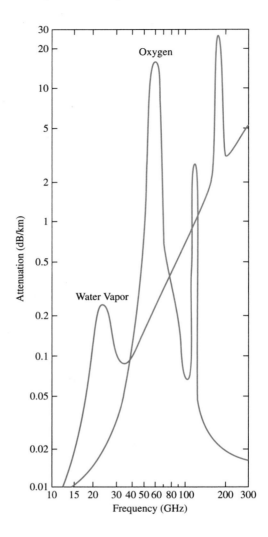

**FIGURE 16–2**
Effects of water vapor and oxygen on atmospheric attenuation, as a function of frequency.

oxygen absorption are most significant in the vicinities of 60 and 120 GHz. It is interesting, however, that there are certain regions in this general frequency range in which these losses are greatly reduced. Such frequencies are called *windows,* and two window locations are 33 and 110 GHz. Operation above about 20 GHz is still in somewhat of a development mode, but such future operation might be forced to utilize these window regions to minimize losses.

Other factors that might have to be considered are antenna pointing errors, bending of waves due to atmospheric refraction, attenuation due to rainfall, and multipath fading. *Multipath fading* results from the successive reinforcement and cancellation of the signal when two or more rays having different path lengths (along with different phase shifts) combine at the receiving antenna.

**▎▌ EXAMPLE 16-1**

A communication system has the following parameters:

$P_t = 5$ W

$G_t(dB) = 13$ dB

$G_r(dB) = 17$ dB

$d = 80$ km

$f = 3$ GHz

Determine the value of the received power.

**SOLUTION**    The basic form of the Friis transmission formula given by Equation 16-1 will be used in this example. We must determine the wavelength, which is

$$\lambda = \frac{3 \times 10^8}{3 \times 10^9} = 0.1 \text{ m} \tag{16-2}$$

The transmitting and receiving antenna gains are given in decibels, and must be converted to absolute form. The values are

$$G_t = 10^{G_t(\text{dB})/10} = 10^{13/10} = 19.95 \tag{16-3}$$

$$G_r = 10^{G_r(\text{dB})/10} = 10^{17/10} = 50.12 \tag{16-4}$$

The units for $d$ must be the same as the units for $\lambda$ (i.e., meters). Hence, $d = 80 \times 10^3$ m. Inserting all the preceding values in the link equation, we have

$$P_r = \frac{(0.1)^2 (19.95)(50.12)(5)}{(4\pi)^2 (80 \times 10^3)^2} = 49.5 \text{ pW} \tag{16-5}$$

Even with the "nice" rounded numbers chosen, the computations are quite messy. At the end of the next section, this example will be worked again using decibel forms throughout the process.

## 16-2  Decibel Forms for the One-Way Link Equation

In the example at the end of the previous section, it was evident that the link equation involves working with a combination of very small numbers and very large numbers. While modern computational aids certainly make this chore easier than in the early days of communications, there has evolved a widespread use of decibel forms to simplify link analysis and interpretation. Anyone who has occasion to search the literature or to review system specifications should be familiar with this approach. In this section, we will show how some of these forms provide a different approach to the analysis.

The process may be initiated from the link equation, which is repeated here for convenience.

$$P_r = \frac{\lambda^2 G_r G_t P_t}{(4\pi)^2 d^2} \tag{16-6}$$

The wavelength may be readily expressed in terms of frequency as $\lambda = 3 \times 10^8 / f$, in which case Equation 16-6 may be expressed as

$$P_r = \frac{569.93 \times 10^{12} G_t G_r P_t}{f^2 d^2} \tag{16-7}$$

where the various constants have been lumped together. Next, a common power reference will be assumed for both transmitted and received power, and the logarithm to the base 10 of both sides will be taken. Both sides can then be multiplied by 10 to give the results in decibel form. This process leads to

$$P_r(\text{dBx}) = P_t(\text{dBx}) + G_t(\text{dB}) + G_r(\text{dB}) + 147.56 - 20 \log f - 20 \log d \tag{16-8}$$

where dBx represents decibels with respect to any standard reference (e.g., dBW, dBf, etc.). However, it is essential that both $P_r$ and $P_t$ *have the same reference.*

Note that "20" appears in front of the logarithms of frequency and distance, a result of the squares for those variables in Equation 16-7. In the form of Equation 16-8, frequency is measured in **hertz** and distance is measured in **meters.** It is more convenient to work with frequency in either MHz or GHz and to work with distance in kilometers. The equations may be converted to those forms, resulting in certain additional constants. One form, after some manipulation, is

$$P_r(\text{dBx}) = P_t(\text{dBx}) + G_t(\text{dB}) + G_r(\text{dB}) - 32.44$$
$$- 20 \log f(\text{MHz}) - 20 \log d(\text{km}) \tag{16-9}$$

Note that in this form, frequency is expressed in **MHz** and distance is expressed in **km.**

### Effective Isotropic Radiated Power

Returning momentarily to the basic form of Equation 16-1, a term called the *effective isotropic radiated power* (EIRP) is defined as

$$EIRP = G_t P_t \tag{16-10}$$

The EIRP is the effective power in the direction of radiation resulting from the antenna gain in that direction. In many cases, this will be the maximum gain. In decibel form, the first two terms of Equation 16-9 are the terms that produce this effect, and we can express the decibel form of the EIRP as

$$EIRP(\text{dBx}) = P_t(\text{dBx}) + G_t(\text{dB}) \tag{16-11}$$

This is the *effective decibel power level* in the direction of radiation resulting from the action of the antenna. For our immediate purposes, we will leave it expressed by the product on the right-hand side of Equation 16-10, but the reader should be aware of the EIRP term since it is used in some forms of the link equation and is often specified as a design parameter.

### Path Loss

The last three terms on the right-hand side of Equation 16-9 have negative signs in front of them. This means that they can be interpreted as *loss terms*. Leaving off the negative signs, we will denote the *one-way path loss* as $\alpha_1(\text{dB})$, which can be defined as

$$\alpha_1(\text{dB}) = 20 \log f(\text{MHz}) + 20 \log d(\text{km}) + 32.44 \tag{16-12}$$

where frequency is expressed in **MHz** and distance is expressed in **km**.

An alternate form of the path loss is

$$\alpha_1(\text{dB}) = 20 \log f(\text{GHz}) + 20 \log d(\text{km}) + 92.44 \tag{16-13}$$

where frequency is expressed in **GHz** and distance is expressed in **km**.

### One-Way Link Equation Using Path Loss Definition

The form of the one-way link equation using the various preceding definitions is then

$$P_r(\text{dBx}) = P_t(\text{dBx}) + G_t(\text{dB}) + G_r(\text{dB}) - \alpha_1(\text{dB}) \tag{16-14}$$

or

$$P_r(\text{dBx}) = EIRP(\text{dBx}) + G_r(\text{dB}) - \alpha_1(\text{dB}) \tag{16-15}$$

where $EIRP(\text{dBx})$ was defined in Equation 16-12.

While some of the preceding steps may seem like going around in circles, these final forms put everything in nice little decibel-form "packages" that simplify the process of communication system analysis and design, especially when investigating the various trade-offs.

---

**▌▌ EXAMPLE 16-2**

Rework Example 16-1 using the decibel approach developed in this section.

**SOLUTION**  First, the transmitted power of 5 W must be converted to a decibel reference form. We will choose to use units of dBW as the basis since the power is expressed directly in watts. Hence,

$$P_t(\text{dBW}) = 10 \log \frac{P_t(\text{W})}{1 \text{ W}} = 10 \log 5 = 6.99 \text{ dBW} \tag{16-16}$$

Next, the path loss must be determined. Since the frequency is given in GHz, the form of Equation 16-13 will be used. We have

$$\alpha_1(\text{dB}) = 20 \log f(\text{GHz}) + 20 \log d(\text{km}) + 92.44$$
$$= 20 \log 3 + 20 \log 80 + 92.44$$
$$= 9.54 + 38.06 + 92.44$$
$$= 140.04 \text{ dB} \tag{16-17}$$

Since the transmitted power in expressed in dBW, the link equation will be expressed in that form, and it is

$$P_r(\text{dBW}) = P_t(\text{dBW}) + G_t(\text{dB}) + G_r(\text{dB}) - \alpha_1(\text{dB})$$
$$= 6.99 + 13 + 17 - 140.04$$
$$= -103.05 \text{ dBW} \tag{16-18}$$

Conversion of this value back to watts is accomplished as follows:

$$P_r(W) = 10^{P_r(\text{dBW})/10} = 10^{-103.05/10} = 49.5 \text{ pW} \tag{16-19}$$

which agrees with the result of Example 16-1.

Was this problem any easier to work in decibel form than the basic analysis of Example 16-1? It probably was not, in this case. In fact, it may have been even more unwieldy to work it out this way.

The true beauty of this approach is best seen when an overall system containing many parts must be designed or analyzed. In such cases, the various trade-offs can be more easily visualized and their effects weighted and compared. Since many system component specifications are given directly in decibel form, they can be easily adapted to this form in working with a complete system. Additional losses not considered here (such as antenna pointing error, rain attenuation, etc.) may be more easily entered into the analysis when the decibel form is used. In fact, many system designers utilize a so-called *link budget,* in which all of these terms in decibel form may be considered almost in the same way as a financial budget.

Irrespective of any discussions concerning advantages or disadvantages of working with decibel forms, the fact is that the entire communications industry utilizes this approach, so one must learn to work with such forms in order to deal with system specifications, design, and analysis.

---

**▐▌ EXAMPLE 16-3**

The distance from the earth to the moon is approximately 240,000 miles. Determine the path loss at (a) 100 MHz, (b) 1 GHz, and (c) 10 GHz. (*Note:* 1 mile = 1.609 km.)

**SOLUTION**  First, the distance in miles must be converted to kilometers to use the formulas developed.

$$d(\text{km}) = 240,000 \text{ miles} \times 1.609 \text{ km/mile} = 386.2 \times 10^3 \text{ km} \tag{16-20}$$

(a)  At 100 MHz, the form of Equation 16-12 will be used.

$$\alpha_1(\text{dB}) = 20 \log f(\text{MHz}) + 20 \log d(\text{km}) + 32.44$$
$$= 20 \log 100 + 20 \log 386.2 \times 10^3 + 32.44$$
$$= 40 + 111.7 + 32.44$$
$$= 184.1 \text{ dB} \tag{16-21}$$

(b)  At 1 GHz, the form of Equation 16-13 is slightly easier to use.

$$\alpha_1(\text{dB}) = 20 \log f(\text{GHz}) + 20 \log d(\text{km}) + 92.44$$
$$= 20 \log 1 + 20 \log 386.2 \times 10^3 + 92.44$$
$$= 0 + 111.7 + 92.44$$
$$= 204.1 \text{ dB} \tag{16-22}$$

(c) At 10 GHz, the form of Equation 16-13 will be used again.

$$\alpha_1(\text{dB}) = 20 \log f(\text{GHz}) + 20 \log d(\text{km}) + 92.44$$
$$= 20 \log 10 + 20 \log 386.2 \times 10^3 + 92.44$$
$$= 20 + 111.7 + 92.44$$
$$= 224.1 \text{ dB} \tag{16-23}$$

From these results, it can be readily deduced that the path loss increases by 20 dB for each increase in the frequency by a factor of 10. However, within limits, it is easier to build antennas having a higher gain as the frequency increases, which can offset the increasing path losses.

---

**▌▌ EXAMPLE 16-4**

At a frequency of 1 GHz, determine the path loss at the following distances: (a) 1 km, (b) 10 km, and (c) 100 km.

**SOLUTION**  Since the frequency is given in GHz, the form of Equation 16-13 is easier to use.

(a) At a distance of 1 km, the path loss is

$$\alpha_1(\text{dB}) = 20 \log f(\text{GHz}) + 20 \log d(\text{km}) + 92.44$$
$$= 20 \log 1 + 20 \log 1 + 92.44$$
$$= 0 + 0 + 92.44$$
$$= 92.44 \text{ dB} \tag{16-24}$$

(b) At a distance of 10 km, the path loss is

$$\alpha_1(\text{dB}) = 20 \log f(\text{GHz}) + 20 \log d(\text{km}) + 92.44$$
$$= 20 \log 1 + 20 \log 10 + 92.44$$
$$= 0 + 20 + 92.44$$
$$= 112.4 \text{ dB} \tag{16-25}$$

(c) Finally, at a distance of 100 km, the path loss is

$$\alpha_1(\text{dB}) = 20 \log f(\text{GHz}) + 20 \log d(\text{km}) + 92.44$$
$$= 20 \log 1 + 20 \log 100 + 92.44$$
$$= 0 + 40 + 92.44$$
$$= 132.4 \text{ dB} \tag{16-26}$$

These results indicate that for each increase in distance by a factor of 10, the path loss increases by 20 dB.

---

**▌▌ EXAMPLE 16-5**

An analog system requires an antenna signal power of 50 pW to meet the required detected signal-to-noise ratio. Other system parameters are given as follows:

$$G_t(\text{dB}) = 3 \text{ dB}$$
$$G_r(\text{dB}) = 4 \text{ dB}$$
$$f = 500 \text{ MHz}$$
$$d = 80 \text{ km}$$

Assuming direct-ray propagation, determine the minimum value of the transmitted power required.

**SOLUTION**  Once again, we will choose to work in the units of dBW. The value 50 pW corresponds to a level in dBW of

$$P_r(\text{dBW}) = 10 \log \frac{P_r(\text{W})}{1 \text{ W}} = 10 \log 50 \times 10^{-12} = -103.01 \text{ dBW} \tag{16-27}$$

The path loss is determined from Equation 16-12 as

$$\alpha_1(\text{dB}) = 20 \log f(\text{MHz}) + 20 \log d(\text{km}) + 32.44$$
$$= 20 \log 500 + 20 \log 80 + 32.44$$
$$= 53.98 + 38.06 + 32.44$$
$$= 124.48 \text{ dB} \tag{16-28}$$

The link equation is then rearranged to determine the transmitter power as

$$P_t(\text{dBW}) = P_r(\text{dBW}) + \alpha_1(\text{dB}) - G_t(\text{dB}) - G_r(\text{dB})$$
$$= -103.01 + 124.48 - 3 - 4$$
$$= 14.47 \text{ dBW} \tag{16-29}$$

The power level in watts is then

$$P_t = 10^{P_t(\text{dBW})/10} = 10^{14.47/10} = 10^{1.447}$$
$$= 28.0 \text{ W} \tag{16-30}$$

---

**⫼ EXAMPLE 16-6**

In a binary digital communication system, the average signal carrier power at the receiver terminals for the specified probability of error is required to be 200 f W. For the link portion of the system, assume the following parameters:

$G_t(\text{dB}) = 30 \text{ dB}$

$G_r(\text{dB}) = 20 \text{ dB}$

$f = 4 \text{ GHz}$

$d = 40{,}000 \text{ km}$

Assuming direct-ray propagation, determine the required transmitter power.

**SOLUTION**   To illustrate a slightly different approach in this problem, the received power will be determined in units of dBf. We have

$$P_r(\text{dBf}) = 10 \log \frac{P_r(\text{fW})}{1 \text{ fW}} = 10 \log 200 = 23.01 \text{ dBf} \tag{16-31}$$

Since the frequency is given in GHz, it is simpler to employ the path loss form of Equation 16-13, and the value is

$$\alpha_1(\text{dB}) = 20 \log f(\text{GHz}) + 20 \log d(\text{km}) + 92.44$$
$$= 20 \log 4 + 20 \log 40{,}000 + 92.44$$
$$= 12.04 + 92.04 + 92.44$$
$$= 196.52 \text{ dB} \tag{16-32}$$

The required transmitted power is

$$P_t(\text{dBf}) = P_r(\text{dBf}) + \alpha_1(\text{dB}) - G_t(\text{dB}) - G_r(\text{dB})$$
$$= 23.01 + 196.52 - 30 - 20$$
$$= 169.53 \text{ dBf}$$
$$= 19.53 \text{ dBW} \tag{16-33}$$

Notice in the last step that the power level in dBf was converted to dBW by subtracting 150 dB.

The power level in watts is then

$$P_t(\text{W}) = 10^{P_t(\text{dBW})/10} = 10^{19.53/10}$$
$$= 89.74 \text{ W} \tag{16-34}$$

In Example 16-5, we worked with dBW, while in Example 16-6, we worked with dBf until the last step. It doesn't matter as long as consistency is maintained in the process, and in the final analysis, the results must be interpreted in the correct units.   ⫼

## 16-3   Line-of-Sight Propagation

In the frequency range in which direct-wave propagation is feasible, there are numerous relatively short-range applications whose transmission can be achieved without the benefit of satellite relay systems. This includes microwave repeater stations, cell-phone transmission, public service and utility two-way systems, communication to aircraft, military applications, and many others. In general, this type of transmission is limited by two major factors: curvature of the earth's surface, and natural obstacles such as buildings and mountains. The first effect can be circumvented partially by locating transmitting and receiving antennas as high above the earth as possible. The second effect can only be alleviated if the path between transmitter and receiver can be located above potential blocking objects.

### Line-of-Sight Distance along Smooth Earth

Because the natural properties of a perfect sphere can be described mathematically, it is possible to precisely predict the possible distance between a transmitting antenna and a receiving antenna in terms of their heights above the earth's surface. However, the situation is complicated by the fact that the mathematical equations derived by the perfect sphere model actually underestimate the distance somewhat. Apparently, early experimenters discovered that some bending of the radio waves occurs, and that the earth acts as if it had a radius about $\frac{4}{3}$ of its actual radius (in terms of the line-of-sight prediction). In fact, this phenomenon has been so well researched that special graph paper based on an assumption of $\frac{4}{3}$ of the earth's actual radius is available to assist designers of microwave terrestrial systems.

Refer to Figure 16–3 for the discussion that follows. Based on the assumption of a reasonably smooth earth, and considering the $\frac{4}{3}$ factor, the distance $d(\text{km})$ in kilometers over which direct line-of-sight transmission is feasible can be approximated as

$$d(\text{km}) = \sqrt{17h_T(\text{m})} + \sqrt{17h_R(\text{m})} \qquad (16\text{-}35)$$

where

$$h_T(\text{m}) = \text{transmitter antenna height in meters}$$
$$h_R(\text{m}) = \text{receiver antenna height in meters}$$

Note that there are mixed units in the equation; that is, the antenna heights are in meters and the distance is in kilometers. Various constants in the development have taken care of the mixed units.

An alternate version, in which the distance is measured in miles and the antenna heights are measured in feet, is the following:

$$d(\text{miles}) = \sqrt{2h_T(\text{feet})} + \sqrt{2h_R(\text{feet})} \qquad (16\text{-}36)$$

where

$$h_T(\text{feet}) = \text{transmitter antenna height in feet}$$
$$h_R(\text{feet}) = \text{receiver antenna height in feet}$$

Both of these equations should be interpreted as reasonable approximations at best. In fact, the constants have been rounded somewhat, so if you calculate the distance from either formula and convert the units to the other, slightly different results are obtained. Again,

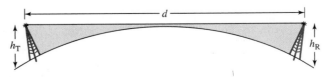

**FIGURE 16–3**
Direct-ray propagation between two antennas on the earth's surface.

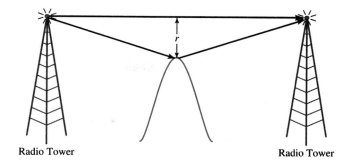

**FIGURE 16–4**
Effect of diffracted signal combining with primary signal.

Radio Tower                                    Radio Tower

these are based on a reasonably smooth earth, and would not be very good approximations in a mountainous region or in a city (in the latter case, they could be useful if the antennas were located on the tops of some of the tallest buildings).

## Diffraction

A relatively complex subject in the study of wave theory is that of *diffraction*. Without attempting to get into the mathematical form of the subject, it is a phenomenon that results in the scattering of electromagnetic energy when a portion of a wave encounters an object. It has been determined that the effects of diffraction are minimized if the electromagnetic path clears any object by a distance equal to 60% of the distance to the *first Fresnel zone*. Fresnel zones are a result of a theory known as the Huygens–Fresnel principle, which states that an object that diffracts waves acts as if it were a second source for the waves. The direct and refracted waves combine, but they either reinforce each other or partially cancel each other. This phenomenon is illustrated in Figure 16–4.

## Clearance Required

The distance in meters $r\,(\mathrm{m})$ to about 60% of the first Fresnel zone (i.e., the distance above which the direct ray between transmitter and receiver should pass) is approximated by the following formula:

$$r(\mathrm{m}) = 10.4\sqrt{\frac{d_1(\mathrm{km})d_2(\mathrm{km})}{f(\mathrm{GHz})\,[d_1(\mathrm{km}) + d_2(\mathrm{km})]}} \tag{16-37}$$

where

$d_1(\mathrm{km})$ = distance between object and transmitter in km

$d_2(\mathrm{km})$ = distance between object and receiver in km

$f(\mathrm{GHz})$ = frequency in GHz

---

**▮▮ EXAMPLE 16-7**

A utility company has its dispatching antenna located at a height of 50 m. The trucks have antennas located at heights of 5 m. Based on reasonably smooth terrain, determine the approximate range for transmission.

**SOLUTION**   The applicable formula is Equation 16-35. We have

$$d(\mathrm{km}) = \sqrt{17h_{\mathrm{T}}(\mathrm{m})} + \sqrt{17h_{\mathrm{R}}(\mathrm{m})}$$
$$= \sqrt{17 \times 50} + \sqrt{17 \times 5}$$
$$= 29.15 + 9.22$$
$$= 38.37\ \mathrm{km} \tag{16-38}$$

▮▮

## 16-4    Radar Link Equation

The term **radar** is a contraction for r̲adio d̲etection a̲nd r̲anging. In this section, a development and discussion of the radar link transmission formula will be presented. This form of the link equation is a "two-way" version, in that power is transmitted along a direct path in one direction, but the ultimate power level of interest is the amount that is returned along the same path in the opposite direction.

### Model for Analysis

The model for developing the radar equation is shown in Figure 16–5. The radar receiver is located at essentially the same position as the radar transmitter and may share some of the same circuits. In the figure, the signal is transmitted to the right, and a portion of the energy that encounters the target is backscattered to the left toward the radar receiver. The reader may be tempted to use the term "reflection" to describe this process, and the term is commonly used in casual conversation. Strictly speaking, however, a reflected wave has a precise meaning in terms of a coherent relationship with the incident wave. The term *backscattered wave* represents energy that is a combination of both coherent and noncoherent components, which is the case with most radar signals. Consequently, the term *backscatter* will be used as the basis for discussing the radar return phenomena. The term *echo* will also be used.

The signal returning to the source location is processed by the radar receiver. Some means must be provided to avoid overloading the low-level receiver by leakage energy from the transmitter. In the simplest classical radar system, the transmitter is turned off during the "listen" interval, but other means of isolation are available. In this development, we will assume that the same antenna, by means of appropriate switching or isolation, is used for both transmitting and receiving.

### Power Density at Target

The first part of the development parallels that of the one-way link equation. From the work of Chapter 15, the power density $p_d$ of the transmitted signal in the vicinity of the target can be expressed as

$$p_d = \frac{G P_t}{4\pi d^2} \tag{16-39}$$

where $G$ is the common antenna gain. (No subscript is required, since there is only one antenna.)

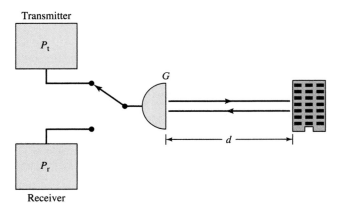

**FIGURE 16–5**
Model used in developing the radar link equation.

## Backscatter Cross Section

We now introduce a term $\sigma$, called the *radar backscatter cross section,* which is defined as

$$\sigma = \frac{\text{power backscattered in direction of source (W)}}{\text{incident power density (W/m}^2)} \tag{16-40}$$

Observe that $\sigma$ has the units of m$^2$, so it is an "area" parameter. The backscattered power $P_{\text{bs}}$ is then determined as

$$P_{\text{bs}} = \sigma p_{\text{d}} = \frac{\sigma G P_{\text{t}}}{4\pi d^2} \tag{16-41}$$

In some references, the backscatter cross section is denoted **RCS**.

At the receiver, the power density $p_{\text{d1}}$ at the antenna is

$$p_{\text{d1}} = \frac{P_{\text{bs}}}{4\pi d^2} = \frac{\sigma G P_{\text{t}}}{(4\pi)^2 d^4} \tag{16-42}$$

The total power $P_{\text{r}}$ captured by the antenna is

$$P_{\text{r}} = p_{\text{d}} A_{\text{e}} = \frac{\sigma G P_{\text{t}} A_{\text{e}}}{(4\pi)^2 d^4} \tag{16-43}$$

We may then express $A_{\text{e}}$ as

$$A_{\text{e}} = \frac{\lambda^2}{4\pi} G \tag{16-44}$$

Substitution of Equation 16-44 in Equation 16-43 yields

$$P_{\text{r}} = \frac{\sigma \lambda^2 G^2 P_{\text{t}}}{(4\pi)^3 d^4} \tag{16-45}$$

This result is the *radar transmission equation,* and it provides an insight into the trade-offs possible with radar systems. Observe that the power is proportional to the square of the power gain of the common antenna, and is inversely proportional to the fourth power of the distance between the source and the target.

The most nebulous quantity in the radar equation is the radar cross section $\sigma$. While this quantity has the dimensions of area and is treated as such, it is actually a complex function of the surface roughness, surface composition, angle of incidence, and many other factors. Much research has been conducted through the years to establish reasonable estimates of the radar cross section for many different types of surfaces, and several books devoted to this topic have been written. Anyone using the radar equation for system design would likely need to utilize such information to assist in estimating power return levels.

In military applications, *stealth* airplanes have utilized designs that minimize the cross-section area and maximize the absorption of the wave. This minimizes the probability of detection by enemy radar systems.

## Decibel Form of Radar Equation

In the same spirit as for the one-way link equation, decibel forms for the two-way radar equation can be readily developed. Without going through all the details in this case, the *two-way path loss in decibels* will be denoted $\alpha_2(\text{dB})$, and may be expressed as

$$\alpha_2(\text{dB}) = 20 \log f(\text{GHz}) + 40 \log d(\text{km}) + 163.43 - 10 \log \sigma(\text{m}^2) \tag{16-46}$$

The two-way link equation is then

$$P_{\text{r}}(\text{dBx}) = P_{\text{t}}(\text{dBx}) + 2G_{\text{t}}(\text{dB}) - \alpha_2(\text{dB}) \tag{16-47}$$

Note in Equation 16-46 that 40 (rather than 20, as in the one-way link equation) is the multiplier in the decibel distance term in the two-way radar equation, since it is based on an inverse fourth-power variation. Note also that the common antenna gain in decibels is multiplied by 2 in the radar case, since it is used in both directions.

**▐▌ EXAMPLE 16-8**    A radar system observing a target is characterized by the following parameters:

transmitted power $= 10$ kW

antenna gain $= 25$ dB

frequency $= 3$ GHz

distance to target $= 50$ km

radar cross section $= 20$ m$^2$

Determine the received power.

**SOLUTION**    First, the two-way path loss will be computed using Equation 16-46.

$$\alpha_2(\text{dB}) = 20 \log f(\text{GHz}) + 40 \log d(\text{km}) + 163.43 - 10 \log \sigma(\text{m}^2)$$
$$= 20 \log 3 + 40 \log 50 + 163.43 - 10 \log 20$$
$$= 9.54 + 67.96 + 163.43 - 13.01$$
$$= 227.92 \text{ dB} \tag{16-48}$$

The transmitted power will be expressed in dBW.

$$P_t(\text{dBW}) = 10 \log \frac{P_t(W)}{1 \text{ W}} = 10 \log 10^4 = 40 \text{ dBW} \tag{16-49}$$

The received power is then determined as

$$P_r(\text{dBW}) = P_t(\text{dBW}) + 2G_t(\text{dB}) - \alpha_2(\text{dB})$$
$$= 40 + 2(25) - 227.92$$
$$= -137.92 \text{ dBW} \tag{16-50}$$

The absolute power is then

$$P_r = 10^{P_r(\text{dBW})/10} = 10^{-137.92/10}$$
$$= 16.1 \times 10^{-15} \text{ W} = 16.1 \text{ fW} \tag{16-51}$$

▐▌

## 16-5  Pulse Radar

The oldest and most basic form of radar is called *pulse radar,* which is used primarily for the measurement of distance. The power is transmitted in short bursts, and the time delay is measured for the return energy to be detected back at the source location. From a measurement of the time delay $T_d$, the distance $d$ to the object may be computed as

$$d = \frac{cT_d}{2} = 1.5 \times 10^8 T_d \tag{16-52}$$

The factor of 2 in the denominator of the equation results from the fact that the actual round-trip distance of the signal is *2d*, and this net distance must be divided by 2 to yield the actual distance to the object. In radar terminology, the object is often referred to as the "target", but the designation does not necessarily imply any military significance.

There are several important parameters that are significant in determining the performance of pulse radar. Consider the periodic waveform shown in Figure 16–6. This baseband waveform may be considered the modulating waveform for a high-frequency carrier, usually in the microwave region. Since it would be impossible to show more than a few cycles in the figure, we will skip the RF form altogether and deal only with this gating signal. Just remember that a high frequency RF transmitter is turned on during the time the pulse is on, and is off for the remainder of the period.

Let $\tau$ represent the width of the RF pulse, and let $T$ represent the period. Thus, the RF transmitter is turned on for a relatively short time $\tau$ and is turned off for an interval $T - \tau$.

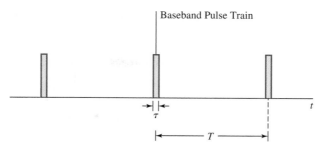

**FIGURE 16–6**
Periodic baseband gating function for pulse radar.

During this latter time interval, the receiver "listens" for an echo, in which the time can be used to directly determine the distance according to Equation 16-52. In any modern system, the various waveforms will be used with calibrated instrumentation to produce an output reading directly in distance units. Moreover, the antenna may be changing its orientation at a rate that would permit the observation of a wide field of view.

## Pulse Repetition Frequency

Let $f_p$ represent the frequency of the baseband pulse train. It is given by

$$f_p = \frac{1}{T} \tag{16-53}$$

Like most specialty areas, radar has its own terminology. This frequency is often referred to as the *pulse repetition frequency* (prf) or the *pulse repetition rate* (prr).

## Maximum Unambiguous Range

Any echo observed at the receiver may have originated from the pulse transmitted at the beginning of the particular cycle or it may have originated from a pulse transmitted during a previous cycle. The *maximum unambiguous range* $d_{max}$ is the greatest distance away from the radar that can be measured within one cycle. This value is

$$d_{max} = \frac{cT}{2} = \frac{c}{2f_p} \tag{16-54}$$

The design and expected performance of the radar must work around this potential ambiguity. Note that the maximum unambiguous range increases with the period of the baseband gating waveform.

## Resolution

Another parameter is the smallest target that can be measured based on the pulse width. If the pulse is too wide, it will fail to distinguish two targets that are very close together, and will give a distorted view of the size of the target. This value can also represent the closest distance to a target that can be measured, and will be denoted as $d_{min}$. This value is

$$d_{min} = \frac{c\tau}{2} \tag{16-55}$$

This value is directly proportional to the width $\tau$ of the gating pulse.

## Trade-Offs

As with most engineering systems, there are trade-offs that must be considered. A greater unambiguous range calls for an increasing value of the period. Likewise, finer resolution calls for a shorter pulse width. However, decreasing the pulse width means that the transmission bandwidth must be increased, which would decrease the signal-to-noise ratio. Moreover, increasing the period or decreasing the pulse width will decrease the average transmitted power, which will degrade the signal processing. Therefore, compromises are always made, depending on the constraints that are most significant in a particular application.

### Clutter

Clutter is any superfluous return that is not part of the desired target being measured. There are sophisticated techniques for removing or minimizing clutter while maximizing the desired signal return.

---

**█▌ EXAMPLE 16-9**

A pulse radar system operating at 10 GHz measures an echo 400 μs after the pulse is transmitted. Determine the distance to the target.

**SOLUTION**   The basic relationship of Equation 16-52 is applicable. We have

$$d = \frac{cT_d}{2} = 1.5 \times 10^8 T_d = 1.5 \times 10^8 \times 400 \times 10^{-6} = 60 \text{ km} \qquad (16\text{-}56)$$

---

**█▌ EXAMPLE 16-10**

A pulse radar system operates at a frequency of 10 GHz with a pulse repetition frequency of 2 kHz and a pulse width of 6 μs. Determine (a) the maximum unambiguous range, and (b) the resolution or minimum range.

**SOLUTION**

(a)  The maximum unambiguous range is determined from Equation 16-54.

$$d_{\max} = \frac{cT}{2} = \frac{c}{2f_p} = \frac{3 \times 10^8}{2 \times 2 \times 10^3} = 75 \text{ km} \qquad (16\text{-}57)$$

(b)  The resolution is determined from Equation 16-55.

$$d_{\min} = \frac{c\tau}{2} = \frac{3 \times 10^8 \times 6 \times 10^{-6}}{2} = 900 \text{ m} \qquad (16\text{-}58)$$

---

## 16-6   Doppler Radar

Another important type of radar is *Doppler radar*. While pulse radar is primarily used to measure distance to a target, Doppler radar can be used to measure the speed of a moving object, a capability that may have provided unpleasant experiences for readers who have been caught speeding. The concept will be developed in some detail in this section.

Consider the airplane in Figure 16–7, which is moving at a velocity $v$ in the direction toward the receiver on the right. (We will assume, of course, that the plane is far enough away that it poses no danger of collision with the receiver!) Assume that the plane is transmitting a sinusoidal signal of cyclic frequency $f_c$ and radian frequency $\omega_c = 2\pi f_c$. If the plane were not moving, the voltage signal received by the observer would have the form

$$v(t) = A \sin \omega_c t = A \sin 2\pi f_c t \qquad (16\text{-}59)$$

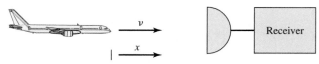

**FIGURE 16–7**
Illustration of one-way doppler shift.

(Don't confuse the symbol $v$ representing velocity with the symbol $v(t)$ representing voltage.) The plane is actually moving with velocity $v$ toward the receiver, and this causes the phase of the transmitted sinusoidal function to advance at a greater rate than if it were stationary. Indeed, in the distance of one wavelength $\lambda$, the phase shift due to the plane movement alone will have advanced by $2\pi$ radians. This means that the signal may actually be expressed as

$$v(t) = A \cos\left(2\pi f_c t + \frac{2\pi}{\lambda}x\right) \tag{16-60}$$

where $x$ is the distance traveled from some arbitrary beginning reference point.

## Instantaneous Frequency

Let us momentarily return to the concept of instantaneous frequency, as considered in Chapter 7 on FM. The instantaneous radian frequency $\omega_i(t)$ associated with the argument $\theta_i(t)$ of the cosine function in Equation 16-60 can be expressed as

$$\omega_i(t) = \frac{d\theta_i(t)}{dt} = 2\pi f_c + \frac{2\pi}{\lambda}\frac{dx}{dt} = 2\pi f_c + \frac{2\pi}{\lambda}v \tag{16-61}$$

where $v = dx/dt$ is the velocity of the airplane.

The instantaneous cyclic frequency is

$$f_i(t) = \frac{\omega_i(t)}{2\pi} = f_c + \frac{v}{\lambda} = f_c + f_D \tag{16-62}$$

where $f_D$ is the Doppler shift. In effect, the frequency appears at the receiver to be higher than the transmitted frequency by the value $f_D$. Of course, if the plane were moving away from the observer, the Doppler shift would be considered negative and the net effect would be to decrease the received frequency.

In the form of Equation 16-62, the Doppler shift is $f_D = v/\lambda$. A more convenient form can be obtained by setting $\lambda = c/f_c$ and substituting that value in the expression for the Doppler shift. The result for the **one-way Doppler** shift then becomes

$$f_D = \frac{v}{c}f_c \tag{16-63}$$

This form clearly shows that the Doppler shift is directly proportional to the velocity of the moving object and the frequency being transmitted. Note, however, that this is a **one-way Doppler shift** since the signal being transmitted is received at a stationary point and is not returned to the transmitter.

## Two-Way Doppler Shift

Now consider the situation depicted in Figure 16–8. This situation differs from the previous one in that the signal is transmitted from the Doppler radar on the right and is backscattered from the plane on the left back to the Doppler radar receiver. The frequency is measured at the receiver on the right. Since the phase is being advanced on both the signal being transmitted and the return echo from the target, the result is that the Doppler shift is *twice as great* as before. Thus, the **two-way Doppler shift** is given by

$$f_D = \frac{2v}{c}f_c \tag{16-64}$$

**FIGURE 16–8**

Illustration of two-way doppler shift along direct path to radar.

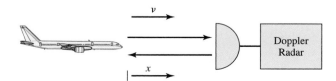

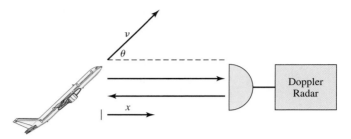

**FIGURE 16–9**
Illustration of two-way doppler shift at an angle.

This is the form that would be measured at a stationary Doppler transmitter-receiver on the right based on the airplane moving in a straight line toward the Doppler radar. If both the radar and the target were moving, the appropriate velocity in Equation 16-64 would be the *relative velocity* between them. If they are moving toward each other, the Doppler shift is positive; whereas, if they are moving away from each other, the Doppler shift is negative.

## Two-Way Doppler Shift at an Angle

Next, consider the situation shown in Figure 16–9. The target is moving at a velocity $v$, but it is moving at an angle $\theta$ with respect to a straight line between the plane and the radar. The component of velocity in the direction of the radar is now $v \cos\theta$. The Doppler shift is then given by

$$f_{\mathrm{D}} = \frac{2v}{c} f_{\mathrm{c}} \cos\theta \qquad (16\text{-}65)$$

If $\theta = 0°$, Equation 16-65 reduces to Equation 16-64 and the Doppler shift is maximum. Conversely, if $\theta = 90°$, there is no Doppler shift.

---

**▌▌ EXAMPLE 16-11**

A Doppler radar operating at 15 GHz is viewing a target moving directly toward it at a speed of 25 m/s. Determine the Doppler shift.

**SOLUTION**

$$f_{\mathrm{D}} = \frac{2v}{c} f_{\mathrm{c}} = \frac{2 \times 25}{3 \times 10^{8}} \times 15 \times 10^{9} = 2500 \text{ Hz} = 2.5 \text{ kHz} \qquad (16\text{-}66)$$

The frequency shift is typically determined by multiplying in a mixer the return signal with a signal proportional to the signal being transmitted. One of the output components of the mixer is directly proportional to the difference frequency and, after filtering, can be calibrated directly in terms of the velocity of the target.

---

**▌▌ EXAMPLE 16-12**

A Doppler radar operating at 10 GHz is being used to measure the speed of an automobile moving directly toward it. The frequency shift is 2 kHz. Determine the speed of the automobile in miles per hour.

**SOLUTION**    We start with Equation 16-64 and solve for velocity. We have

$$v = \frac{cf_{\mathrm{D}}}{2f_{\mathrm{c}}} \qquad (16\text{-}67)$$

We could take several different paths at this point. One way would be to express the speed of light in meters/second, determine the velocity in the same units, and then convert to miles per hour. An alternate approach is to express the speed of light as 186,000 miles/second, which would make the result appear in miles per second. By an additional multiplication of 3600 seconds/hour, the result will be in miles per hour. We will take the second approach. First, the velocity in miles/second is

$$v = \frac{186{,}000 \times 2 \times 10^3}{2 \times 10^{10}} = 18.6 \times 10^{-3} \text{ mi/s} \tag{16-68}$$

Conversion to miles/hour yields

$$v = 18.6 \times 10^{-3} \ \frac{\text{mi}}{\text{s}} \times 3600 \ \frac{\text{s}}{\text{hr}} = 67.0 \ \text{mi/hr} \tag{16-69}$$

Watch out for a traffic ticket!

## 16-7 Reflection and Refraction

When an electromagnetic wave encounters a boundary between two media, some of the energy is reflected by the boundary while the remaining portion is transmitted into the second medium. This situation is similar to the transmission line case, in which the traveling waves on the transmission line arrive at an impedance mismatch. A portion of the wave is reflected by the mismatch, and the remaining portion is transmitted past the impedance mismatch.

The direction of propagation for the plane wave can approach the boundary at any angle. The direction of propagation for the reflected plane wave is a function of the angle of the incident wave and the smoothness of the surface of the boundary between the two media, in terms of the wavelength of the electromagnetic propagation.

### Perfect Conductor Reflection

The reflection from a conductor with a smooth surface is illustrated in Figure 16–10. An *incident* electromagnetic wave in medium 1 is represented by $p_i$. The angle between the incident wave and the normal line from the surface is called the *angle of incidence* and is denoted $\theta_i$. The smooth conducting surface acts like a mirror, and the reflected wave is denoted $p_r$. The angle between the reflected wave and the normal line from the surface is called the *angle of reflection* and is denoted $\theta_r$. For the perfect conductor with a smooth surface, *the angle of reflection is equal to the angle of incidence;* that is,

$$\theta_r = \theta_i \tag{16-70}$$

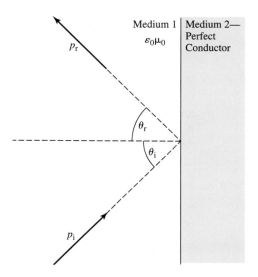

**FIGURE 16–10**

Reflection at the boundary of a specular surface.

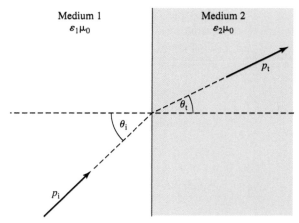

**FIGURE 16–11**
Refraction at the boundary between two dielectric media.

If medium 2 had been a dielectric, some of the energy in the incident wave would have been transmitted into medium 2, and the magnitude of the reflected wave would have been reduced. The angle of reflection, however, would remain the same for a smooth boundary surface.

## Refraction

Consider now the situation depicted in Figure 16–11, containing a smooth interface between two different dielectric media. Medium 1 contains an incident electromagnetic wave indicated by $p_i$. Some of the energy will be reflected, and the angle of reflection will equal the angle of incidence as indicated earlier, but that component is not shown here. Instead, the emphasis in this figure will be directed toward the component that enters medium 2, which is called the *refracted wave*.

As has been the case all along, assume that the permeability in medium 2 is the same as that of medium 1, and that both have the value of the free-space permeability $\mu_0$. However, assume that the permittivity in medium 1 is $\varepsilon_1$ and the permittivity in medium 2 is $\varepsilon_2$. This means that the respective velocities of propagation in the two media are

$$v_1 = \frac{1}{\sqrt{\mu_0 \varepsilon_1}} \tag{16-71}$$

and

$$v_2 = \frac{1}{\sqrt{\mu_0 \varepsilon_2}} \tag{16-72}$$

It can be shown with geometric optics that the wave will change directions as it passes through the boundary. Let $\theta_t$ represent the *angle of refraction* as illustrated in Figure 16–11. (Think of the subscript "t" as representing "*transmitted*" wave.) If the velocity in medium 2 is less than the velocity in medium 1, the wave will bend in the direction of the normal, and that is the situation depicted in the figure.

## Snell's Law

Snell's law provides the relationship between the angle of incidence $\theta_i$ and the angle of refraction (or transmission) $\theta_t$. It can be stated as

$$\frac{\sin \theta_t}{\sin \theta_i} = \frac{v_2}{v_1} \tag{16-73}$$

Substituting expressions for the velocities, Equation 16-73 can be expressed as

$$\frac{\sin \theta_t}{\sin \theta_i} = \sqrt{\frac{\varepsilon_1}{\varepsilon_2}} = \sqrt{\frac{\varepsilon_{r1}}{\varepsilon_{r2}}} \tag{16-74}$$

where the permittivity values have been expressed in terms of the dielectric constant values, and the common value of $\varepsilon_0$ has been cancelled in the last form.

### Index of Refraction

A term used extensively in optics is the *index of refraction*. The index of refraction $n$ is defined as *the ratio of the velocity of propagation in a vacuum to the velocity of propagation in a particular medium*. The values in this case are given by

$$n_1 = \sqrt{\varepsilon_{r1}} \tag{16-75}$$

$$n_2 = \sqrt{\varepsilon_{r2}} \tag{16-76}$$

Snell's law can then be expressed as

$$\frac{\sin \theta_t}{\sin \theta_i} = \frac{n_1}{n_2} \tag{16-77}$$

Because of these various forms, it is easy to get mixed up in interpreting the formulas. A worthwhile point to remember is that *the side having the highest velocity will have the largest angle measured from the normal*. The equations can then be arranged in the appropriate form.

The bending of electromagnetic waves plays a major role in sky-wave radio propagation, as will be discussed in Section 16-9. It is also important in fiber-optic transmission systems.

---

**❚❚ EXAMPLE 16-13**

An electromagnetic wave propagating in air encounters a boundary with a material having a dielectric constant of 4. The angle of incidence in air is 50°. Determine the angle of refraction.

**SOLUTION**   We will assume that the index of refraction in air is $n_1 = 1$. The index of refraction in the second medium is

$$n_2 = \sqrt{\varepsilon_2} = \sqrt{4} = 2 \tag{16-78}$$

The angle of refraction is determined as follows:

$$\frac{\sin \theta_t}{\sin \theta_i} = \frac{n_1}{n_2} = \frac{1}{2} \tag{16-79}$$

This leads to

$$\sin \theta_t = \frac{1}{2} \sin \theta_i = \frac{1}{2} \times \sin 60° = 0.4330 \tag{16-80}$$

The angle of refraction is then

$$\theta_t = \sin^{-1} 0.4330 = 25.66° \tag{16-81}$$

Note that the wave bends toward the normal in medium 2, a result of the higher index of refraction and lower velocity in that medium.   ❚❚❚

---

## 16-8   Ground-Wave Propagation

Direct-wave propagation as predicted by the Friis equation is very reliable at the higher frequencies used for space and satellite communications. However, at lower frequencies, and especially with vertical polarization, direct-wave propagation is limited due to ground effects.

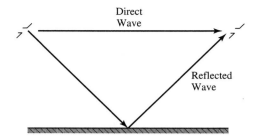

**FIGURE 16–12**
Space wave consisting of direct wave and reflected wave.

In the discussion that follows, the **direct wave** as predicted by the Friis formula will be considered as one of the components of a **space wave,** due to common terminology and usage.

## Ground Wave

The **ground wave** is commonly referred to as the electromagnetic propagation that travels from a transmitting antenna to a receiving antenna along or near the surface of the earth, without leaving the lower portion of the earth's atmosphere. The ground wave can be divided into a **surface wave** and a **space wave,** which travels in the lower portion of the earth's atmosphere near the surface. The **space wave** consists of two components: the **direct wave** and the **indirect** or **ground-reflected wave.** The two waves are illustrated in Figure 16–12.

## Surface Wave

The surface wave is used primarily for vertical polarization propagation below 3 MHz. The horizontally polarized signal is quickly attenuated, since the electric field vector is parallel to the earth's surface. The vertical polarized surface wave is the propagation mode for the standard AM broadcast band. It can provide coverage up to about 100 miles in the AM band during the daytime. At 300 kHz, vertically polarized surface waves can provide coverage to about 500 miles over good conductivity earth and to 1000 miles over ocean water. The use of the surface wave above 3 MHz is very limited due to its restricted coverage. At 30 MHz, the typical range is 10 miles or less.

## Space Wave

The space wave is very poor for communications below 3 MHz. The reflected wave undergoes a 180° phase shift when it is reflected from the earth's surface. For antennas located near the earth's surface and operating at long wavelengths, the distances that the direct and indirect waves travel are essentially the same. Therefore, when arriving at the receiving site, they will be 180° out of phase and will cancel. The vertically polarized signal is reflected better than the horizontally polarized signal. Also, as the frequency is increased, the path difference in terms of wavelength is increased, and the cancellation between the direct and indirect waves is less significant.

## 16-9   Sky-Wave Propagation

In addition to the ground wave discussed in the previous section, an antenna with a broad beam or one having an orientation toward space will radiate a *sky wave.* As the sky wave travels through the earth's atmosphere, it encounters a region known as the *ionosphere.* The ionosphere consists of several layers of ionized gases that are located from 40 to 300 km above the earth's surface. The electromagnetic radiation is both refracted and attenuated by the ionosphere, depending on the frequency of the electromagnetic radiation, the radiation angle from the transmitting antenna, and the degree of ionization of the atmospheric gases at that level in the ionosphere. If the electromagnetic radiation is within a certain frequency range, typically 3 to 30 MHz, the signal will be refracted—that is, bent back toward the earth's surface and received by a receiving site on the earth. If the frequency is below a certain frequency, typically 3 MHz, it will be totally absorbed by the ionosphere. If the frequency is above a certain frequency, typically 30 MHz, it will pass through the ionosphere and will continue into outer space.

It should be noted that the ionosphere is between the earth's surface and the orbits of communication satellites. Therefore, satellite communication systems must use frequencies sufficiently high enough to be unaffected by the passage through the ionosphere. Satellite systems operate in the microwave region at frequencies typically above 3 GHz.

## Skip Effect

Waves transmitted at too great an angle will not be bent by a sufficient amount to return to the earth. However, there is *a critical angle* at which waves in the appropriate frequency range can be bent by a sufficient amount to return to the earth. The *skip distance* is the distance along the earth's surface from the transmitting site to the point on the surface where a wave returns. The range between the ground wave and the skip distance is the *quiet zone,* where no transmission can be received.

## The Ionosphere and Its Layers

The ionosphere is a region where the pressures are so low that the constituents of the atmosphere are ionized; that is, a molecule loses an electron and becomes a positively charged ion. The region, therefore, consists of both free electrons and ions. With the low pressure, the density of molecules is such that it takes a longer time for recombination to occur. The major source of energy for the ionization process comes from solar ultraviolet radiation, although solar X rays and meteor radiation also play a role. The major ionization process begins right after sunrise, peaks around local noon, and decays after sunset.

During the daytime, the ionization is in four distinct layers, due to the ionization requirements of the different constituents of the atmosphere. The ionization peaks at about the midrange of each layer and tapers off both above and below the altitude of each layer's maximum ionization. The lowest ionized layer is the D layer, which lies between 60 and 92 km (about 37 to 57 miles) above the earth's surface. The next ionized layer is the E layer, which lies between 100 and 115 km (about 62 to 71 miles) above the surface. The next ionized layer is the F layer, which lies between 160 and more than 500 km (about 100 to over 310 miles) above the surface.

During daytime hours, the F layer breaks into two distinct layers: the F1 layer and the F2 layer. After the sun sets, the ionization in the D and E layers ceases due to the fairly rapid recombination of electrons and ions, and these layers disappear. The F1 layer, which is lower and weaker than the F2 layer, also disappears at night, and the F2 layer drops in altitude. The F2 layer is slow to recombine and, therefore, lasts throughout the night, reaching a minimum sometime after midnight.

The use of the ionosphere to "bounce" electromagnetic radiation back to the earth is a technique that requires a great deal of information about a large number of factors. These factors include location on the earth of both the transmitting and receiving stations, time of day, season of the year, and degree of solar activity.

## Maximum Usable Frequency

The proper choice of the frequency of the electromagnetic radiation is important to the successful completion of a communication link. There is a frequency that is defined as the *maximum usable frequency* (MUF). If a variable-frequency transmitter is used to radiate an electromagnetic wave straight up (at a 90° angle), the highest frequency that is reflected back to the earth for that layer in the ionosphere is the *vertical incidence critical frequency*. The MUF is the vertical incidence critical frequency for that layer. The MUF is a variable that changes significantly as a function of the time of day, solar activity, propagation path distance, and propagation path between the transmitting and receiving stations. For example, the MUF between the eastern United States and Europe has varied from 7 MHz to 70 MHz because of these variables. Predictions for the MUF as a function of the propagation path, time of year, and time of day are published periodically by several agencies worldwide.

The degree of refraction in the ionosphere is a function of the frequency. Therefore, since the transmitting antenna typically radiates at a constant radiation angle relative to the surface, the skip distance is a function of frequency. The shorter the distance between the transmitter and receiver, the lower the MUF will be for that propagation path.

### Lowest Usable Frequency

A second important frequency is the *lowest usable frequency* (LUF). If the frequency of the electromagnetic radiation is lowered from the MUF, the losses due to absorption in the atmosphere will increase as the frequency decreases. The frequency at which these losses cause the signal strength to fall below the background noise is the LUF. Therefore, it is desirable to operate as close to the MUF as the available frequency allocation will allow. Since each type of communication system is restricted to operate in specified frequency allocation bands, sometimes the frequency range from the LUF to the MUF is so narrow that there will be no allocated frequency band for successful communications.

## 16-10  MATLAB® Examples (Optional)

The emphasis in this section will be on the use of two programs to assist in link analysis. The first program **link1.m** involves the computation of parameters in a typical one-way communication system from transmitter to receiver. The second program **link2.m** is based on a radar link analysis, with transmission from a source to a target and the return signal along the same path. A brief description of each program follows.

### link1.m

This program is based on the one-way link analysis of Section 16-3; the MATLAB code is shown in Figure 16–13.

| | |
|---|---|
| Input: | Transmitter output power |
| | Transmitting antenna gain |
| | Receiver antenna gain |
| | Frequency |
| | Distance between transmitter and receiver |
| Output: | One-way path loss |
| | Received power |

**FIGURE 16–13**
M-file program link1.m.

```
%Program link1.m
fprintf('\n')
Pt=input('Enter the transmitter power in watts. ');
Gtdb=input('Enter the transmitter antenna gain in dB. ');
Grdb=input('Enter the receiver antenna gain in dB. ');
f=input('Enter the frequency in GHz. ');
d=input('Enter the distance in km. ');
alpha1=20*log10(f)+20*log10(d)+92.44;
Ptdbw=10*log10(Pt);
Prdbw=Ptdbw+Gtdb+Grdb-alpha1;
Prdbf=Prdbw+150;
fprintf('\n')
fprintf('The path loss is %g dB.\n',alpha1)
fprintf('The received power is %g dBW.\n',Prdbw)
fprintf('Alternately, the received power is %g dBf.\n\n',Prdbf)
```

```
%Program link2.m
fprintf('\n')
Pt=input('Enter the transmitter power in watts. ');
Gtdb=input('Enter the transmitter antenna gain in dB. ');
f=input('Enter the frequency in GHz. ');
d=input('Enter the distance in km. ');
sigma=input('Enter the radar cross-section in square meters. ');
alpha2=20*log10(f)+40*log10(d)+163.43-10*log10(sigma);
Ptdbw=10*log10(Pt);
Prdbw=Ptdbw+2*Gtdb-alpha2;
Prdbf=Prdbw+150;
fprintf('\n')
fprintf('The path loss is %g dB.\n',alpha2)
fprintf('The received power is %g dBW.\n',Prdbw)
fprintf('Alternately, the received power is %g dBf.\n\n',Prdbf)
```

**FIGURE 16–14**
M-file program link2.m.

## link2.m

This program is based on the two-way radar link analysis of Section 16-5; the MATLAB code is shown in Figure 16–14.

Input:   Transmitter output power
         Gain of common antenna
         Frequency
         Distance between transmitter and target
         Radar cross-section area

Output:  Two-way path loss
         Received power

▌▌ **MATLAB EXAMPLE 16-1**

Apply the program **link1.m** to perform some of the computations of Examples 16-1 and 16-2.

**SOLUTION**    The various values are requested upon initiating the program. The dialogue and the resulting output data are shown in Figure 16–15. The results are in agreement with those of Examples 16-1 and 16-2.

```
>> link1

Enter the transmitter power in watts. 5
Enter the transmitter antenna gain in dB. 13
Enter the receiver antenna gain in dB. 17
Enter the frequency in GHz. 3
Enter the distance in km. 80

The path loss is 140.044 dB.
The received power is -103.055 dBW.
Alternately, the received power is 46.9455 dBf.
```

**FIGURE 16–15**
Results for MATLAB
Example 16-1.

▌▌ **MATLAB EXAMPLE 16-2**

Apply the program **link2.m** to perform some of the computations of Example 16-8.

**SOLUTION**    The syntax of the program is very similar to the one-way analysis, except that there is only one gain required, and a value of the radar cross-section must be entered. The dialogue and the resulting output data are shown in Figure 16–16. The results are in agreement with those of Example 16-8.

>> link2

Enter the transmitter power in watts. 1e4
Enter the transmitter antenna gain in dB. 25
Enter the frequency in GHz. 3
Enter the distance in km. 50
Enter the radar cross-section in square meters. 20

The path loss is 227.921 dB.
The received power is -137.921 dBW.
Alternately, the received power is 12.0791 dBf.

**FIGURE 16–16**
Results for MATLAB
Example 16-2.

## PROBLEMS

**16-1** A communication system has the following parameters:

$$P_t = 20 \text{ W}$$
$$G_t(\text{dB}) = 15 \text{ dB}$$
$$G_r(\text{dB}) = 22 \text{ dB}$$
$$d = 50 \text{ km}$$
$$f = 10 \text{ GHz}$$

Determine the received power using the basic Friis transmission formula (not the decibel form).

**16-2** A communication system has the following parameters:

$$P_t = 120 \text{ W}$$
$$G_t(\text{dB}) = 30 \text{ dB}$$
$$G_r(\text{dB}) = 24 \text{ dB}$$
$$d = 40,000 \text{ km}$$
$$f = 15 \text{ GHz}$$

Determine the received power using the basic Friis transmission formula (not the decibel form).

**16-3** Repeat the analysis of Problem 16-1 using the decibel form of the one-way link equation.

**16-4** Repeat the analysis of Problem 16-2 using the decibel form of the one-way link equation.

**16-5** The closest distance between Earth and Mars is about $54.5 \times 10^6$ km. Determine the decibel path loss between the two planets at (a) 1 GHz and (b) 10 GHz.

**16-6** The closest distance between Earth and Jupiter is about $600 \times 10^6$ km. Determine the decibel path loss between the two planets at (a) 1 GHz and (b) 10 GHz.

**16-7** Assume that a radio signal is being sent from Earth to Mars based on the distance given in Problem 16-5. If there was an immediate response when it arrived, how long would it be from the time that you sent the message until you received an answer?

**16-8** Assume that a radio signal is being sent from Earth to Jupiter based on the distance given in Problem 16-6. If there was an immediate response when it arrived, how long would it be from the time that you sent the message until you received an answer?

**16-9** At a frequency of 15 GHz, determine the path loss in decibels at the following distances: (a) 1 km, (b) 100 km, and (c) 10 megameters (Mm).

**16-10** At a frequency of 6 GHz, determine the path loss in decibels at the following distances: (a) 5 km, (b) 50 km, and (c) 500 km.

**16-11** An analog system requires an antenna signal power of 100 pW to meet the required detected signal-to-noise ratio. Other system parameters are given as follows:

$$G_t(\text{dB}) = 4 \text{ dB}$$
$$G_r(\text{dB}) = 6 \text{ dB}$$
$$f = 900 \text{ MHz}$$
$$d = 50 \text{ km}$$

Assuming direct-ray propagation, determine the minimum value of the transmitter power required.

**16-12** An analog system requires an antenna signal power of 50 dBf to meet the required detected signal-to-noise ratio. Other system parameters are given as follows:

$$G_t(\text{dB}) = 12 \text{ dB}$$
$$G_r(\text{dB}) = 20 \text{ dB}$$
$$f = 5 \text{ GHz}$$
$$d = 100 \text{ km}$$

Assuming direct-ray propagation, determine the minimum value of the transmitter power required.

**16-13** In a binary digital communications system, the bit energy required at the receiver input terminals is $50 \times 10^{-18}$ J. Other system parameters are as follows:

$$G_t(\text{dB}) = 24 \text{ dB}$$
$$G_r(\text{dB}) = 26 \text{ dB}$$
$$f = 10 \text{ GHz}$$
$$d = 40,000 \text{ km}$$

Assuming direct-ray propagation, determine the minimum value of the transmitter power required if the data rate is 5 kbits/s.

**16-14** For the system of Problem 16-13, determine the minimum value of the transmitter power required if the data rate is increased to 40 kbits/s.

**16-15** Based on a height of 200 m for both a transmitting and a receiving antenna, determine the approximate range in kilometers for transmission based on reasonably smooth terrain.

**16-16** A company has its base antenna at a height of 60 ft. All vehicles have their antennas at heights of 8 ft. Based on reasonably smooth terrain, determine the approximate range in miles for transmission.

**16-17** A radar system observing a target is characterized by the following parameters:

transmitted power = 5 kW

antenna gain = 30 dB

frequency = 6 GHz

distance to target = 50 km

radar cross section = 30 m$^2$

Determine the received power.

**16-18** A radar system observing a target is characterized by the following parameters:

transmitted power = 40 dBW

antenna gain = 28 dB

frequency = 10 GHz

distance to target = 100 km

radar cross section = 50 m$^2$

Determine the received power.

**16-19** Measurements are performed to determine the cross section of a target, and the following parameters are measured:

transmitted power = 25 dBW

antenna gain = 26 dB

frequency = 5 GHz

distance to target = 2 km

received power = 60 dBf

Determine the radar cross section in m$^2$.

**16-20** Measurements are performed to determine the cross section of a target, and the following parameters are measured:

transmitted power = 50 dBm

antenna gain = 30 dB

frequency = 10 GHz

distance to target = 5 km

received power = 30 dBf

Determine the radar cross section in m$^2$.

**16-21** A pulse radar system operates at a frequency of 12 GHz with a pulse repetition period of 2 ms and a pulse width

of 3 μs. Determine (a) the maximum unambiguous range, and (b) the resolution.

**16-22** A target is located at a distance of 12 km from the radar. Determine the time from the beginning of the transmitted pulse to the received echo.

**16-23** A Doppler radar operating at 9 GHz is viewing a target moving directly away from it at a speed of 40 m/s. Determine the doppler shift.

**16-24** A Doppler radar operating at 12 GHz is viewing a target moving directly toward it at a speed of 72 km/hour. Determine the doppler shift.

**16-25** A Doppler radar operating at 10 GHz is located on a police car moving at 50 mph and is pointed backwards toward a car overtaking it. The frequency shift is 500 Hz. Determine the speed of the approaching vehicle.

**16-26** Consider the situation depicted in Problem 16-25 with the police car moving at 50 mph. If the Doppler shift is −300 Hz, determine the speed of the vehicle behind the police car.

**16-27** Consider the situation shown in the following figure, with the Doppler radar operating at 12 GHz. If the frequency shift is 1 kHz, determine the speed of the car in miles per hour.

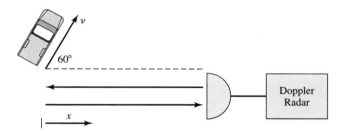

**16-28** Consider the situation shown in the following figure, with the Doppler radar operating at 10 GHz. Determine the Doppler shift if the car is moving at 50 mph.

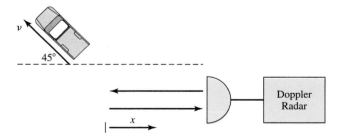

**16-29** An electromagnetic wave propagating in a medium with a dielectric constant of 3 encounters a boundary with air. The angle of incidence is 30°. Determine the angle of refraction.

**16-30** An electromagnetic wave propagating in a medium with a dielectric constant of 3 encounters a medium with an dielectric constant of 6. The angle of incidence is 60°. Determine the angle of refraction.

# Satellite Communications

## OVERVIEW AND OBJECTIVES

Arthur C. Clarke, a British Royal Air Force electronics officer and author of *2001: A Space Odyssey,* is credited with proposing the concept of satellite communications. In 1945 he wrote an article in *Wireless World* suggesting that a satellite placed in an equatorial orbit with a radius of about 42,242 km would have the same rotation rate as that of the earth and would appear to be fixed with respect to any position on the earth. An orbit of this type is called a *geostationary orbit.* A satellite placed in this orbit would be able to provide relay communications between any two points that would be electromagnetically visible; that is, any two points from which direct-ray propagation would be possible. Three satellites spaced 120° apart could then conceivably communicate between any two points on the earth and between themselves.

The first satellite communication systems were implemented in the 1950s, and development has continued to the present. Many commercial and military satellites presently occupy orbits around the earth. Satellite communications are much more reliable for long-range communication than the earlier propagation methods. Not only are satellite systems used for large-scale commercial and military purposes, but many households utilize these services for entertainment purposes. This trend is expected to continue in the years ahead.

### Objectives

After completing this chapter, the reader should be able to:

1. Describe briefly the purposes and advantages of satellite communications.
2. Define the primary forces acting on a satellite, and describe how a stable orbit can be achieved.
3. Describe the general operation of a satellite communication system.
4. Define the difference between the *uplink* and the *downlink.*
5. Describe frequency allocations and types of satellite communication services.
6. Discuss the types of antennas used on communication satellites for uplinks and downlinks.
7. Discuss the types of antennas used by earth stations for uplinks and the downlinks.
8. Determine the range from any earth station to a particular communication satellite in a geostationary orbit.
9. Define the $G/T$ ratio and its application to satellite communication receivers.
10. Determine the $C/N$ ratio for both the uplink and the downlink in a satellite communication system.

## 17-1   Orbital Mechanics Primer

The authors make no claim toward being experts on orbital mechanics, so if there are any readers with an extensive background in that subject, they are requested to excuse any oversimplifications that follow. However, the authors feel that electrical and electronic specialists can benefit from understanding some of the basic concepts of orbital mechanics that make satellite communication feasible.

### Forces

Newton's first law states than any object will remain in its existing state unless acted on by external forces. This is obvious for an object at rest, but many people initially have trouble accepting the notion that a body in motion would continue to move at the same speed forever unless acted on by external forces. The notion is hard to accept because it is impossible near the earth to create motion without some external forces. However, in outer space there are objects that keep moving for centuries unless they are influenced by an external force.

There are two primary types of forces that are crucial in the operation of a satellite. They are **centrifugal force** and **gravitational force.** Refer to Figure 17–1 for the discussion that follows. The larger object is assumed to be the earth and the smaller is the satellite. The quantity $r$ is the distance from the satellite to the center of the earth.

### Centrifugal Force

If an object is constantly changing directions, a force is generated that attempts to keep it moving in the original direction. The primary case for consideration is when an object is moving in a circular or near-circular path around a reference center. The resulting force is called the *centrifrigal force* and is denoted $f_c$ in Figure 17–1. Using SI units, the force can be measured in newtons (N). If the object is moving at a velocity $v$ measured in meters/second, the centrifugal force is given by

$$f_c = \frac{mv^2}{r} \tag{17-1}$$

where $m$ is the mass in kilograms and $r$ is measured in the basic unit of meters. (Later, we will change the radius in some formulas to kilometers.) A simple example of centrifugal force is an object at the end of a rope. If the rope is twirled in a circle at a sufficiently high speed, it will cause an outward pull and will assume a circular pattern.

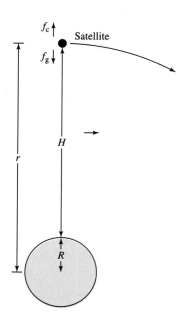

**FIGURE 17–1**
Satellite in motion and external forces.

## Gravitational Force

The force $f_g$ in Figure 17–1 is the *gravitational force,* which is directed toward the earth. The value of this force is given by

$$f_g = mg \tag{17-2}$$

The quantity $g$ is the gravitational constant measured in m/s$^2$. You may remember from basic physics that the value of $g$ near the earth's surface is approximately 9.8 m/s$^2$. In fact, most of the problems that you worked in physics probably assumed that you were on or near the earth's surface, in which case the constant value can be assumed. The added problem with orbital mechanics is that the value of $g$ varies inversely with the square of the distance from the center of the earth. We will get to that point shortly, but first let us consider the basic force balance equations.

## Balance between Centrifugal and Gravitational Forces

Assume conditions are such that the centrifugal and gravitational forces exactly balance each other. This means that we can set $f_c$ in Equation 17-1 equal to $f_g$ in Equation 17-2, in which case we have

$$\frac{mv^2}{r} = mg \tag{17-3}$$

This leads to

$$v = \sqrt{gr} \tag{17-4}$$

Note that the mass cancels, indicating that the velocity is independent of the actual mass of the satellite. This result is the velocity at a given radius such that the centrifugal and gravitational forces exactly balance. We next need to consider how the gravitational constant varies with the distance above the earth.

## Gravitational Variation with Altitude

In the development that follows, a minor amount of rounding will be made in order to make the results as easy to use as feasible. More exact values may be found in various handbooks if required. (In fact, some of the universal constants are still being revised for accuracy to this day.)

The value of the gravitational constant $g$ as a function of the radius to the center of the earth is given approximately by

$$g = \frac{400 \times 10^{12}}{r^2} \tag{17-5}$$

with $g$ measured in m/s$^2$ and $r$ measured in meters.

It is more desirable to measure the distance to the satellite from the earth's surface than from the center. The radius of the earth will be denoted $R$, and can be approximated as $R \approx 6.4 \times 10^6$ m. Let $H$ represent the distance between the earth's surface and the satellite, on a straight line between the center of the earth and the satellite; that is, the closest distance between the earth and the satellite. We can then express $r$ in meters as

$$r = H + R = H + 6.4 \times 10^6 \tag{17-6}$$

(Recall that we also use $R$ to represent data rate in bits per second, but there should be very little chance of confusing the dual use of the symbol.)

Substituting Equation 17-6 in Equation 17-5, the value of $g$ is closely approximated as

$$g = \frac{400 \times 10^{12}}{(H + 6.4 \times 10^6)^2} \tag{17-7}$$

We will leave as an exercise for the reader to show that the value obtained at the earth's surface from this approximation is not too bad.

When the expression for $g$ from Equation 17-7 and the expression for $r$ from Equation 17-6 are substituted in Equation 17-4, the following expression for velocity is obtained:

$$v = \frac{20 \times 10^6}{\sqrt{H + 6.4 \times 10^6}} \tag{17-8}$$

Just in case we might be getting confused, the expression for velocity in Equation 17-8 is the value in meters/second for a satellite at a height $H$ in meters above the earth's surface that would result in a stable orbit. For this purpose, we will keep the quantities involved in their basic units.

## Geostationary Orbit

Through the years, as the technology has progressed, satellites have been placed at various orbits, and the expression of Equation 17-8 permits the determination of the required velocity that would theoretically cause equilibrium of the external forces. However, our major interest is that of a **geostationary orbit** (as discussed earlier in the chapter), and most satellites operate in that mode. Assume that an orbit is determined such that the stable velocity is exactly equal to the rotational velocity of the earth. From the perspective of an observer on the earth, the satellite would then appear to be stationary.

Let's see if we can determine the approximate height above the earth's surface for this orbit. Let $H_s$ represent the height or altitude above the earth's surface at the geostationary orbit. The *period $T$* of the rotation is the time required to travel through one complete circular orbit. We know that the earth rotates once every 24 hours. The period in seconds is given by

$$T = 24 \text{ hours} \times 60 \text{ minutes/hour} \times 60 \text{ seconds/minute}$$
$$= 86{,}400 \text{ seconds} \tag{17-9}$$

The total distance traveled in this period is the circumference $C$ around the earth at the particular orbit, given by

$$C = \text{velocity} \times \text{time} = 86{,}400v \tag{17-10}$$

However, the circumference must also satisfy the basic conditions associated with a circle; that is,

$$C = 2\pi r = 2\pi(H_s + 6.4 \times 10^6) \tag{17-11}$$

where $H = H_s$ has been used in the equation.

Setting Equation 17-10 equal to Equation 17-11, we obtain a relationship between velocity and altitude, which reads

$$H_s + 6.4 \times 10^6 = 13{,}751v \tag{17-12}$$

Next, the result of Equation 17-12 can be substituted directly in Equation 17-8, and after some simplification, we obtain

$$v^{1.5} = 170.55 \times 10^3 \tag{17-13}$$

This equation can be solved by raising both sides of the equation to the power of $\frac{2}{3}$. The result is

$$v = 3075.2 \text{ m/s} \tag{17-14}$$

The height above the surface of the earth is then determined from Equation 17-12 as

$$H_s = 35.887 \times 10^6 \text{ m} = 35{,}887 \text{ km} \tag{17-15}$$

This value agrees quite closely with one published value of 35,786 km when the satellite is directly above the reference point on the earth's surface. It should be noted that there are slight variations in the literature concerning the exact value. We will use the slightly rounded value of 36,000 km in many subsequent computations.

### Geostationary versus Geosynchronous

The term **geosynchronous** is often used to mean the same thing as **geostationary.** Strictly speaking, the first term could refer either to rotation with the earth (geostationary) or rotation in a direction opposite to that of the earth. Since the latter situation is of no practical interest, we will consider the two terms as synonymous for our purposes.

**▐▌ EXAMPLE 17-1**

A satellite is to be placed in an orbit 1000 km above the earth's surface. Determine (a) the required velocity, (b) the circumference of the rotation, and (c) the period of the rotation.

**SOLUTION**

(a)  The velocity is determined from Equation 17-8 by substituting $H = 10^6$ m.

$$v = \frac{20 \times 10^6}{\sqrt{H + 6.4 \times 10^6}} = \frac{20 \times 10^6}{\sqrt{10^6 + 6.4 \times 10^6}} = 7352 \text{ m/s} \qquad (17\text{-}16)$$

(b)  The circumference is

$$C = 2\pi(H + R) = 2\pi(10^6 + 6.4 \times 10^6) = 46.50 \times 10^6 \text{ m} \qquad (17\text{-}17)$$

(c)  The period is

$$T = \frac{C}{v} = \frac{46.5 \times 10^6}{7352} = 6325 \text{ s} = 105.4 \text{ minutes} \qquad (17\text{-}18)$$

▐▌

## 17-2   Satellite Alignment

The alignment of a satellite for geostationary tracking purposes is an interesting exercise in geometry and will be pursued briefly in this section. Refer to Figure 17–2 for the development that follows. The satellite antenna is assumed to be in the northern hemisphere, and is rotating in an equatorial orbit. It is obvious that the size of the antenna and mount are **highly exaggerated** with respect to the earth in order to show the appropriate angles!

### Angle of Elevation

Let $R$ represent the radius of the earth, which will be closely approximated as $R = 6400$ km. Since we will be considering only geostationary orbits in the remainder of the chapter, we will drop the subscript from $H_s$ and set $H = 36,000$ km as the approximate normal distance above the equator for a geostationary orbit. Let $L$ represent the **latitude,** which is measured as an angle from 0° to 90° along the northern hemisphere. The value 0°

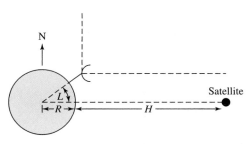

(a) Polar Axis Set at Elevation Angle

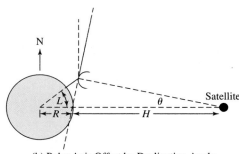

(b) Polar Axis Offset by Declination Angle

**FIGURE 17–2**
Geometry of satellite alignment.

corresponds to the equator and 90° N corresponds to the North Pole. For the purpose of alignment, the latitude is equal to the **angle of elevation.** In Figure 17–2(a), the polar axis of the antenna is set at the angle of elevation, which is a desirable reference orientation for space astronomy. However, for satellite purposes, it would be looking into deep space and would miss the satellite.

### Polar Axis Offset

In order to align with the satellite, the polar axis of the antenna should be offset by an additional angle. This angle will be denoted $\delta$ and called the **declination offset angle.** The situation is shown in Figure 17–2(b). By a careful analysis of the trigonometric situation depicted, it can be shown that the declination offset angle is given by

$$\delta = \tan^{-1}\left[\frac{R \sin L}{H + R(1 - \cos L)}\right] = \tan^{-1}\left[\frac{6400 \sin L}{36000 + 6400(1 - \cos L)}\right] \qquad (17\text{-}19)$$

where the distances appearing in the formula are expressed in kilometers. In actual practice, it has been determined that slight variations in the value predicted by the formula permit better tracking over a wider arc. This slight adjustment has been one of the reasons why a slight amount of rounding has been justified in developing the formulas given here.

---

**▌▌ EXAMPLE 17-2**

Norfolk, Virginia is located at an approximate latitude of 37°. Determine the declination offset angle.

**SOLUTION**    The angle is determined from Equation 17-19.

$$\delta = \tan^{-1}\left[\frac{6400 \sin 37°}{36000 + 6400(1 - \cos 37°)}\right] = \tan^{-1}\left[\frac{6400(0.6018)}{36000 + 6400(1 - 0.7986)}\right]$$

$$= \tan^{-1}\left(\frac{3852}{37.29 \times 10^3}\right) = \tan^{-1}(0.1033) = 5.90° \qquad (17\text{-}20)$$

▌▌

---

## 17-3 Spacecraft Communication Systems

A common form of a satellite communication system, obviously not drawn to scale, is illustrated in Figure 17–3. The *uplink* on the left consists of a *transmitting earth station,* which transmits information to a satellite located in a geostationary orbit. The antenna on the satellite receives the uplink signal and transfers it to a *transponder.* The transponder amplifies the signal and translates the modulation to a new carrier frequency. In the case of digital

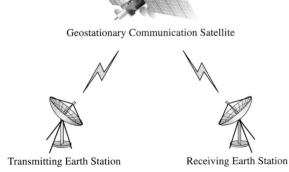

Geostationary Communication Satellite

Transmitting Earth Station          Receiving Earth Station

**FIGURE 17–3**
Basic satellite communication system.

data, the signal can also be regenerated to reduce noise and distortion. The new signal is then applied to the transmitter aboard the satellite and the transmitting antenna radiates the signal back to the earth. This latter portion of the overall path is called the *downlink.*

While the system of Figure 17–3 suggests information flow in one direction only, it is possible for each ground station to operate as either a transmitting or a receiving station by the use of different uplink and downlink frequencies. A major reason for separate uplink and downlink frequencies is to minimize the likelihood of the high power of the downlink transmitter affecting the weak signal at the uplink receiver input.

### C-Band

Some of the early communication satellites utilized frequencies in the C-band microwave range. The uplink frequencies typically ranged from about 5925 to 6425 MHz. The downlink frequencies typically ranged from about 3700 to 4200 MHz. This provided about 500 MHz of bandwidth in each direction. Satellite channels are typically spaced 20 MHz apart. The even channels operate with one antenna polarization, and the odd channels utilize orthogonal polarization. Polarization may be either (1) linear polarization utilizing vertical and horizontal polarization, or (2) circular polarization utilizing right-handed and left-handed polarization. This process is called *polarization diversity*. Thus, even though the channels are 20 MHz apart, the use of alternate channel polarization provides each user 40 MHz. The actual bandwidth utilized is 36 MHz, which results in 4 MHz of guardband.

In the earlier development phase, C-band frequencies were almost ideal considering external noise and atmospheric loss. Assuming that the exact satellite location is known, a narrow beamwidth can be employed so that a satellite 2° away in geostationary orbit will not experience interference. This means that a high-gain, large-aperture antenna will be required. Conversely, the satellite may be required to transmit to many stations located in different positions on the earth, which means that the satellite antenna must have a wider beamwidth. The wider beamwidth corresponds to a lower gain.

A receiving antenna in the C-band region looking toward outer space can have a relatively low level of externally received noise for certain *look angles*. The *look angle* is the angle measured from a line perpendicular to the earth's surface. Received noise as a function of look angle and frequency is shown in Figure 17–4. Galactic noise decreases in the microwave frequency region, and is below atmospheric noise above 3 GHz. As can be seen, the noise level decreases with decreasing look angle, but is relatively constant for a given look angle over a range of frequencies from about 1 to 10 GHz. The peak at 22 GHz results from absorption due to water vapor in the atmosphere, and the peak at 60 GHz results from absorption by oxygen in the atmosphere. The presence of rain results in additional losses and noise. Losses due to rain increase with increasing frequency and are at a minimum in the C-band region. Thus, C-band from about 3.7 GHz to about 6.425 GHz is a region of minimum atmospheric noise.

One significant problem with the C-band is that many terrestrial microwave radio communication systems operate in the same band. A large number of commercial telephone communication systems operating in this band existed prior to the advent of satellite systems. To circumvent the potential of interference, the level of radiated power from the satellite is restricted. Moreover, earth-based uplink stations have been restricted to locations that minimize potential interference.

### Ku-Band

In the 1970s, the Ku-band was established for exclusive use in satellite communication systems, which would eliminate the problem of interference with terrestrial systems. The uplink frequencies in this band are from about 14.0 to 14.5 GHz, and the downlink frequencies are from about 11.7 to 12.7 GHz. The use of the Ku-band permits increased EIRP levels from the satellite transmitter. Another advantange is that smaller earth-based antennas can be used to achieve the same gain and beamwidth as the much larger C-band antennas.

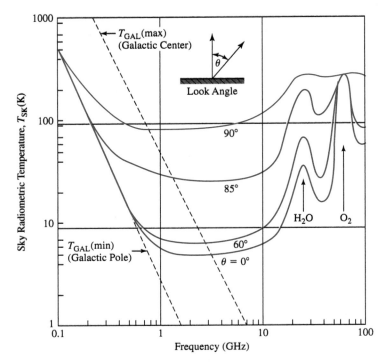

**FIGURE 17–4**
External atmospheric noise. F. T. Ulaby, R. K. Moore, and A. K. Fung, *Microwave Remote Sensing: Active and Passive, Vol. 1, Microwave Remote Sensing Fundamentals and Radiometry* (Reading, MA: Addison-Wesley, 1981), p. 287, Fig. 5.9(b). Used with permission.

Moreover, mobile Ku-band uplinks are practical and have made possible live television coverage of news and sports events.

One disadvantage of the Ku-band is a small increase in rain attenuation and atmospheric noise. In situations in which outages are likely, the use of *space diversity* can circumvent this problem. *Space diversity* is the process of employing multiple uplink earth stations at different locations, with the goal of minimizing the probability of rain outages at all stations.

### K-Band

A fairly recent development is the K-band frequency region from about 27.5 to 30 GHz for uplinks and from about 17.7 to 20.2 GHz for downlinks. The National Aeronautics and Space Administration (NASA) has investigated this frequency region as a possibility for future satellite communication services.

### Frequency Allocations and Services

The allocation and assignment of frequency utilization is a task performed by the International Telecommunication Union (ITU), an agency of the United Nations. More than 150 nations are members of the ITU. Representatives of member nations meet at the World Administrative Radio Conference (WARC) at periodic intervals to review and update the International Table of Frequency Allocations and the International Radio Regulations. WARC deals with frequency allocations from 10 kHz to 300 GHz and with three types of satellite systems: (1) Fixed Satellite Service (FSS), (2) Mobile Satellite Service, and (3) Broadcast Satellite Service. Some of the typical frequency allocations for nongovernmental usage in North and South America are listed in Table 17–1.

Table 17–1    Satellite Frequency Allocations

| Fixed Satellite Service (FSS) | |
|---|---|
| *Downlink* | *Uplink* |
| 3.7 to 4.2 GHz | 5.85 to 7.075 GHz |
| 4.5 to 4.8 GHz | 12.7 to 13.25 GHz |
| 10.7 to 12.2 GHz | 14.0 to 14.5 GHz |
| 17.7 to 20.2 GHz | 27.5 to 30.00 GHz |
| **Broadcast Satellite Service (BSS)** | |
| *Downlink* | *Uplink* |
| 2.5 to 2.655 GHz | 2.655 to 2.690 GHz |
| 12.2 to 12.7 GHz | 17.3 to 17.80 GHz |
| 22.5 to 23.0 GHz | |
| **Mobile Satellite Service (MSS)** | |
| *Downlink* | *Uplink* |
| 1.530 to 1.559 GHz | 1.6265 to 1.6605 GHz |
| 19.2 to 20.2 GHz | 29.5 to 30.0 GHz |

## 17-4    Antennas aboard Satellites

The two antennas located on a satellite are the uplink receiving antenna and the downlink transmitting antenna. They can be classified in three ways according to the desired coverage. The beamwidths may be chosen for (1) global coverage, (2) hemispheric coverage, or (3) spot coverage.

Global coverage has the broadest beamwidth, and the intent is to cover the electromagnetically visible portion of the earth's surface. Hemispheric coverage is designed to cover either the lower or upper hemisphere. The most narrow beamwidths are those chosen for spot coverage based on a narrow "footprint."

### Beamwidth and Area Coverage

The 3-dB beamwidth required for global coverage is about 18° from a geostationary orbit. The 3-dB beamwidth required for hemispheric coverage of the continental United States is about 8°. The 3-dB beamwidths of spot beams depend on the desired area of coverage but can be as narrow as 0.5° to 1°. In some cases, multiple spot beams are employed, such as the coverage intended for a single country. A diameter of approximately 630 km can be covered with a 1° beamwidth.

The 3-dB beamwidth of a parabolic antenna of diameter $D$ was given in Chapter 15 and is

$$\theta_{3\text{dB}} = 70\frac{\lambda}{D} \text{ degrees} \qquad (17\text{-}21)$$

where $\lambda$ is the wavelength expressed in the same units as $D$.

The gain of a parabolic reflector antenna relative to an isotropic radiator diameter was also given in Chapter 15 and is

$$G = \eta_{\text{I}}\left(\frac{\pi D}{\lambda}\right)^2 \qquad (17\text{-}22)$$

where $\eta_{\text{I}}$ is the illumination efficiency factor.

### Relationship between Gain and Beamwidth

An interesting relationship directly connecting the 3-dB bandwidth and the gain can be deduced from the preceding two equations. Assume that the ratio $D/\lambda$ is determined from

Equation 17-21 in terms of the 3-dB beamwidth. If this ratio is substituted in Equation 17-22, after some rearrangement and combining of constants, the following equation is obtained:

$$G = \frac{(70\pi)^2 \eta_I}{\theta_{3dB}^2} \approx \frac{48,000\eta_I}{\theta_{3dB}^2} \qquad (17\text{-}23)$$

The constant has been rounded slightly for simplification, since it is an approximation and the angle is expressed in degrees. Equation 17-23 indicates that the absolute gain is inversely proportional to the square of the 3-dB beamwidth.

---

**▮▮ EXAMPLE 17-3**

The lowest downlink frequency for C-band is 3.7 GHz. Consider a satellite transmitter operating at this frequency with the goal of providing coverage of the continental United States, which requires a 3-dB beamwidth of about 8°. Assuming an illumination efficiency of 60%, determine the diameter and the gain of the downlink antenna.

**SOLUTION**   The wavelength corresponding to the downlink frequency is

$$\lambda = \frac{c}{f} = \frac{3 \times 10^8}{3.7 \times 10^9} = 8.108 \text{ cm} \qquad (17\text{-}24)$$

The diameter may be determined from Equation 17-21.

$$D = \frac{70\lambda}{\theta_{3dB}} = \frac{70 \times .08108}{8} = 0.7095 \text{ m} \qquad (17\text{-}25)$$

The gain may be determined from Equation 17-22.

$$G = \eta_I \left(\frac{\pi D}{\lambda}\right)^2 = 0.6 \times \left(\frac{0.7095\pi}{0.08108}\right)^2 = 453.4 \qquad (17\text{-}26)$$

The corresponding decibel gain is

$$G_{dB} = 10 \log 453.4 = 26.57 \text{ dB} \qquad (17\text{-}27)$$

▮▮

---

## 17-5    Earth Station Antennas

There are two design considerations for earth station antennas, for both the uplink and the downlink: (1) the antenna gain must be sufficiently large to achieve the required signal-to-noise ratio, and (2) since satellites are located at 2° intervals in the geostationary orbit, the 3-dB beamwidth for the uplink must be sufficiently narrow that it does not cause interference with adjacent satellites. Likewise, the downlink earth station 3-dB beamwidth must be sufficiently narrow that it does not receive signals from two or more adjacent satellites.

### Beamwidth Constraint

The angle between the closest nulls on each side of the maximum gain in a parabolic reflector antenna is usually about 2.5 times the 3-dB beamwidth. If the maximum gain of the earth station is oriented directly toward the desired satellite position, the two adjacent satellites would be located approximately at the first nulls. The nulls would be about 4° apart, resulting in a 3-dB beamwidth of about 1.6°. This situation is illustrated in Figure 17–5. An earth station antenna meeting the preceding criteria would have a diameter of

$$D = 70 \frac{\lambda}{\theta_{3dB}} = \left(\frac{70}{1.6}\right) \lambda = 43.75\lambda \qquad (17\text{-}28)$$

The isotropic gain of a lossless antenna may then be determined from Equation 17-23.

$$G = \frac{(70\pi)^2 \eta_I}{\theta_{3dB}^2} \approx \frac{48,000\eta_I}{\theta_{3dB}^2} \qquad (17\text{-}29)$$

The angle in this equation must be expressed in degrees.

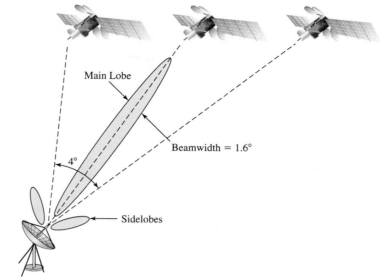

**FIGURE 17–5**
Earth station antenna with a beamwidth of 1.6 degrees.

Note that the gain required for a fixed beamwidth is independent of the frequency for an assumed illumination efficiency. However, understand that the size of the antenna for a given gain will be dependent on the frequency.

## Maximum Gain Estimates

Example 17-4 shows that the gain corresponding to a 3-dB beamwidth of 1.6° and an efficiency of 60% is 40.5 dB. If more gain is required, the diameter of the antenna can be increased. However, the 3-dB beamwidth decreases as the gain is increased. Due to problems of pointing (pointing losses and reflector surface accuracy), the smallest practical beamwidth is about 0.3°. Example 17-5 shows that the upper limit of gain is about 55 dB.

## Ku-Band

The development of the Ku-band satellite communication frequency bands provided a significant reduction in the size of earth station antennas. The use of Ku-band instead of C-band for a given gain reduces the antenna diameters by a factor of approximately 3, thereby reducing the area by a factor of 9. Therefore, the antennas are not only less expensive, they are also more portable and easier to handle. The future use of Ku-band reduces the antenna diameter by a factor of approximately 5 over that of C-band, and it reduces the area by a factor close to 25. This is a significant reduction in area, and will enhance the use of very small mobile earth stations.

---

**▌▌ EXAMPLE 17-4**

For an illumination efficiency of 60%, determine the gain required to achieve a 3-dB beamwidth of 1.6°.

**SOLUTION**  The absolute gain value may be determined directly from Equation 17-29 by substituting the values of efficiency and beamwidth:

$$G = \frac{0.6 \times 48,000}{(1.6)^2} = 11,250 \tag{17-30}$$

The decibel gain is then

$$G_{dB} = 10 \log G = 10 \log 11,250 = 40.5 \text{ dB} \tag{17-31}$$

**EXAMPLE 17-5**

For an illumination efficiency of 60% and a minimum practical beamwidth of about 0.3°, determine the gain required.

**SOLUTION**    The absolute gain is determined from Equation 17-29 and is

$$G = \frac{0.6 \times 48,000}{(0.3)^2} = 320,000 \tag{17-32}$$

The decibel gain is

$$G_{dB} = 10 \log G = 10 \log 320,000 = 55.1 \text{ dB} \tag{17-33}$$

**EXAMPLE 17-6**

The lowest uplink frequency for C-band is 5.925 GHz. Consider a ground station transmitter operating at this frequency with a desired beamwidth of 1.6°. Assuming an illumination efficiency of 60%, determine the diameter of the ground station uplink antenna.

**SOLUTION**    In Example 17-4, it was determined that the gain required for a beamwidth of 1.6° with an illumination efficiency of 60% was 40.5 dB, or an absolute gain of 11,250. The wavelength at 5.925 GHz is

$$\lambda = \frac{3 \times 10^8}{5.925 \times 10^9} = 5.063 \text{ cm} \tag{17-34}$$

Equation 17-21 provides a relationship between the diameter and wavelength, and it can be used to determine the diameter. We have

$$D = 70 \frac{\lambda}{\theta_{3dB}} = \left(\frac{70}{1.6}\right)\lambda = 43.75\lambda = 43.75 \times 0.05063 = 2.215 \text{ m} \tag{17-35}$$

## 17-6  Satellite Link Analysis

As previously discussed, there are two communication links involved in most satellite systems. The **uplink** transmission involves communication from the earth to the satellite. The **downlink** involves communication from the satellite back to the earth. Both of these links involve direct-ray propagation, and the Friis transmission formula may be used. As a quick review, the received power $P_r$ is related to the transmitted power $P_t$ by the following relationship:

$$P_R = \frac{\lambda^2 G_t G_r P_t}{(4\pi)^2 d^2} \tag{17-36}$$

where $G_t$ and $G_r$ are the absolute gains of the transmitting and receiving antennas, respectively, and $d$ is the distance between the antennas in meters. Several alternate forms utilizing decibel measures were developed in earlier chapters, and will be utilized later.

### System Noise

The noise temperature and noise figure concepts were developed in some detail in Chapter 12. It was shown that the effect of a lossy network such as a transmission line could increase the effective noise temperature at the input to the receiver.

The overall system noise temperature at the input of a communication receiver can be divided into two components:

1. **Internal Receiver Noise.** This is the noise generated in the receiver, and represented by an equivalent noise temperature $T_e$ referred to the input.

2. **Input Source Noise.** The input source noise consists of external atmospheric noise, and any additional component introduced by losses in the transmission line between the antenna and receiver input. The external noise includes atmospheric noise, galactic noise, and human-made noise. It will be denoted $T_i$ as referred to the receiver input.

The net system noise temperature $T_{\text{sys}}$ referred to the input of the receiver is given by

$$T_{\text{sys}} = T_e + T_i \tag{17-37}$$

The system noise power $N_{\text{sys}}$ referred to the receiver input is then given by

$$N_{\text{sys}} = kT_{\text{sys}}B \tag{17-38}$$

## Carrier-to-Noise Ratio

For optimum performance, the received carrier power must exceed the received noise power level by an amount based on the performance criteria. The criteria could be the bit error rate in a digital system, or the detected signal-to-noise ratio for an analog system. The requirements will often be dictated by a minimum carrier power to noise power ratio $(C/N)$ at the receiver input terminals. The value is usually specified in decibels, and typically varies from about 7 to 25 dB or more.

The carrier power to system noise power ratio can be determined as

$$C/N = \frac{P_r}{N_{\text{sys}}} = \frac{\lambda^2 G_r G_t P_t}{(4\pi d)^2 kT_{\text{sys}}B} \tag{17-39}$$

This equation can be factored in the following form:

$$C/N = (G_t P_t)\left(\frac{\lambda}{4\pi d}\right)^2 \left(\frac{G_r}{T_{\text{sys}}}\right)\left(\frac{1}{kB}\right) \tag{17-40}$$

Referring to the right-hand side of Equation 17-40, the first factor is the *effective isotropic radiated power (EIRP)*. The second factor represents those quantities that produce the *one-way path loss* $\alpha_1$ (dB) when the equation is converted to decibel form. The third term is referred to casually as the $G/T$ ratio when expressed in decibel form, and is a figure of merit for a satellite receiver. The last factor has no special name but contributes a significant amount.

## Decibel Form

The decibel form of the one-way link equation was introduced in Chapter 16 and is applicable to satellite link analysis. However, it is customary in satellite systems to carry the analysis one step further, to provide a decibel form for the carrier-to-noise ratio that includes the $G/T$ ratio and other constants. This can be accomplished by taking the logarithms to the base 10 of both sides of Equation 17-40 and expanding. An additional factor can be added to take care of miscellaneous losses. Without showing all the steps, the $C/N$ ratio in decibels can be expressed as

$$(C/N)_{\text{dB}} = EIRP(\text{dBW}) - \alpha_1(\text{dB}) + (G/T)_{\text{dB/K}} + 228.6 - 10\log B - L_{\text{misc}}(\text{dB}) \tag{17-41}$$

The various terms in the equation are defined as follows:

$$EIRP(\text{dBW}) = 10\log P(\text{W}) + G_t(\text{dB}) \tag{17-42}$$

$$\alpha_1(\text{dB}) = \text{one-way path loss} = 20\log f(\text{GHz}) + 20\log d(\text{km}) + 92.44 \tag{17-43}$$

$$(G/T)_{\text{dB/K}} = 10\log\left(\frac{G_r}{T_{\text{sys}}}\right) = G_r(\text{dB}) - 10\log T_{\text{sys}} \tag{17-44}$$

The quantity $L_{\text{misc}}(\text{dB})$ represents any additional losses, such as transmission line losses, pointing errors, rain attenuation, and so on.

The value of 228.6 arises from $10\log k = 10\log 1.38 \times 10^{-23} = -228.6$. However, since $k$ is in the denominator of Equation 17-39, the negative of the value is taken and the result becomes a positive value in Equation 17-41.

Note that the units for the $G/T$ parameter are indicated as decibels/kelvin (dB/K).

## Distance between Earth and Satellite

If a satellite receiver were located on the equator at the same longitude as the transmitter, the distance used in performing an analysis of the path loss between a geostationary satellite and the earth would be approximately 36,000 km. However, at all other points on the earth, the distance is a function of the latitude and longitude. The formula that follows can be used to determine the distance between a satellite and any point on the earth's surface. The values listed have been slightly "refined" from the approximate values considered earlier, and are more exact. The distance $d$ in kilometers from any point on the Earth to a satellite in a geostationary orbit above the equator is given by

$$d(\text{km}) = \sqrt{H^2 + 2R(H + R)(1 - \cos L \cos l)} \qquad (17\text{-}45)$$

where

$H =$ satellite altitude $= 35{,}786$ km

$R =$ radius of earth $= 6378$ km

$L =$ latitude of the earth station in degrees

$l =$ difference between longitude of the earth station and the satellite in degrees

When the values of $H$ and $R$ are substituted in Equation 17-45, the equation may be reduced to

$$d(\text{km}) = 35.786 \times 10^3 \sqrt{1 + 0.42(1 - \cos L \cos l)} \qquad (17\text{-}46)$$

---

**▐▌ EXAMPLE 17-7**

Determine the distance from an earth station located in New York City to a satellite located at 127° west longitude. The latitude of New York is 40.5° N and the longitude is 70.2° W.

**SOLUTION**   The difference in longitude is

$$l = 127.0° - 70.2° = 56.8° \qquad (17\text{-}47)$$

Substitution of this value and $L = 40.5°$ into Equation 17-46 results in

$$d(\text{km}) = 35.786 \times 10^3 \sqrt{1 + 0.42(1 - \cos L \cos l)}$$
$$= 35.786 \times 10^3 \sqrt{1 + 0.42(1 - \cos 40.5° \cos 56.8°)}$$
$$= 39.932 \times 10^3 \text{ km} \qquad (17\text{-}48)$$

---

**▐▌ EXAMPLE 17-8**

A C-band communications satellite is located at 91° W longitude. Various specifications for the **uplink** are as follows:

uplink earth station location: New York City, 40.5° N latitude, 70.2° W longitude.

uplink frequency $= 6.125$ GHz

uplink transmitter output $= 100$ W

uplink earth station antenna gain $= 55$ dB

bandwidth $= 36$ MHz

satellite receiver antenna gain $= 27$ dB (501.2)

satellite receiver noise figure $= 3$ dB

net input noise temperature (including losses) $= 300$ K

atmospheric loss at 6.125 GHz $= 1.6$ dB

Determine the $C/N$ ratio at the receiver for the uplink.

SOLUTION   The EIRP for the uplink earth station is

$$EIRP\,(\text{dBW}) = P_t(\text{dBW}) + G_t(\text{dB}) = 20 + 55 = 75\ \text{dBW} \tag{17-49}$$

where the transmitter power is $10 \log P_t(W) = 10 \log 100 = 20\ \text{dBW}$.

The distance for the uplink from New York City to the satellite is

$$
\begin{aligned}
d(\text{km}) &= 35.786 \times 10^3 \sqrt{1 + 0.42(1 - \cos L \cos l)} \\
&= 35.786 \times 10^3 \sqrt{1 + 0.42(1 - \cos 40.5^\circ \cos 20.8^\circ)} \\
&= 37.897 \times 10^3\ \text{km}
\end{aligned} \tag{17-50}
$$

where the difference in longitudes is $l = 91^\circ - 71.2^\circ = 20.8^\circ$.

The path loss is

$$
\begin{aligned}
\alpha_1(\text{dB}) &= 20 \log f(\text{GHz}) + 20 \log d(\text{km}) + 92.44 \\
&= 20 \log 6.125 + 20 \log 37.897 \times 10^3 + 92.44 \\
&= 199.75\ \text{dB}
\end{aligned} \tag{17-51}
$$

The noise temperature of the satellite receiver is

$$T_e = (F - 1) \times 290 = 290\ \text{K} \tag{17-52}$$

The system temperature at the satellite receiver is

$$T_{\text{sys}} = T_i + T_e = 300 + 290 = 590\ \text{K} \tag{17-53}$$

The $G/T$ ratio for the satellite receiver in dB is

$$(G/T)_{\text{dB/K}} = G(\text{dB}) - 10 \log T_{\text{sys}} = 27 - 10 \log 590 = -0.71\ \text{dB/K} \tag{17-54}$$

The carrier power to noise power ratio at the satellite receiver is

$$
\begin{aligned}
(C/N)_{\text{dB}} &= EIRP(\text{dBW}) - \alpha_1(\text{dB}) + (G/T)_{\text{dB/K}} + 228.6 - 10 \log B - L_{\text{misc}}(\text{dB}) \\
&= 75 - 199.75 - 0.71 + 228.6 - 10 \log 36 \times 10^6 - 1.6 \\
&= 75 - 199.75 - 0.71 + 228.6 - 75.56 - 1.6 \\
&= 25.98\ \text{dB}
\end{aligned} \tag{17-55}
$$

---

**▌▍ EXAMPLE 17-9**

For the system whose uplink was analyzed in Example 17-8, assume that the signal is being received in Atlanta, Georgia, and that the downlink parameters are as follows:

downlink receiver: Atlanta, 32° N latitude, 90° W longitude.

downlink frequency = 3.9 GHz

downlink EIRP = 47.8 dBW

bandwidth = 36 MHz (same as uplink)

user receiver antenna gain = 42 dB ($15.85 \times 10^3$)

user receiver noise figure = 2.5 dB (1.778)

user receiver external input noise temperature (including losses) = 150 K

atmospheric loss at 6.125 GHz = 1.0 dB

Determine the $C/N$ ratio at the receiver for the downlink.

SOLUTION   Note that the EIRP in dBW is given for the satellite transmitter, so we do not need to know the transmitter antenna gain.

The distance for the uplink from Atlanta to the satellite can be determined as

$$
\begin{aligned}
d(\text{km}) &= 35.786 \times 10^3 \sqrt{1 + 0.42(1 - \cos L \cos l)} \\
&= 35.786 \times 10^3 \sqrt{1 + 0.42(1 - \cos 32^\circ \cos 1^\circ)} \\
&= 36.911 \times 10^3\ \text{km}
\end{aligned} \tag{17-56}
$$

where the difference in longitudes was $l = 91^\circ - 90^\circ = 1^\circ$.

The path loss is

$$
\begin{aligned}
\alpha_1(\text{dB}) &= 20 \log f(\text{GHz}) + 20 \log d(\text{km}) + 92.44 \\
&= 20 \log 3.9 + 20 \log 36.911 \times 10^3 + 92.44 \\
&= 195.60\ \text{dB}
\end{aligned}
\tag{17-57}
$$

The noise temperature of the satellite receiver is

$$
T_e = (F - 1) \times 290 = 225.6\ \text{K}
\tag{17-58}
$$

The system temperature at the satellite receiver is

$$
T_{\text{sys}} = T_i + T_e = 150 + 225.6 = 375.6\ \text{K}
\tag{17-59}
$$

The $G/T$ ratio for the satellite receiver in dB is

$$
(G/T)_{\text{dB/K}} = G(\text{dB}) - 10 \log T_{\text{sys}} = 42 - 10 \log 375.6 = 16.25\ \text{dB/K}
\tag{17-60}
$$

The carrier power to noise power ratio at the satellite receiver can now be determined as

$$
\begin{aligned}
(C/N)_{\text{dB}} &= EIRP(\text{dBW}) - \alpha_1(\text{dB}) + (G/T)_{\text{dB/K}} + 228.6 - 10 \log B - L_{\text{misc}}(\text{dB}) \\
&= 47.8 - 195.60 + 16.25 + 228.6 - 10 \log 36 \times 10^6 - 1 \\
&= 47.8 - 195.60 + 16.25 + 228.6 - 75.56 - 1 \\
&= 20.5\ \text{dB}
\end{aligned}
\tag{17-61}
$$

## 17-7   MATLAB® Examples (Optional)

A program for performing a satellite link analysis will be presented in this section. The program has the title **satellite.m**, and it can be used for either an uplink or a downlink analysis. Two examples will be used to illustrate the program.

The M-file is shown in Figure 17–6. The input and output data are as follows:

Input:   Transmitter output power
Transmitter antenna gain
Receiver antenna gain
Frequency
Bandwidth
Receiver noise temperature
Antenna external noise temperature
Miscellaneous losses
Ground station latitude
Difference in satellite and ground station longitudes

Output:  Range between satellite and ground station
Transmitter EIRP
Path loss
System noise temperature
Receiver input power
G/T ratio at the receiver
C/N ratio at the receiver

Since specifications are sometimes given in alternate forms, some comments are in order. If the EIRP of the transmitter is given in watts, enter it as the requested value for transmitter power and enter the transmitter decibel gain as 0 dB. If the EIRP of the transmitter is given in decibel form, first convert to dBW if necessary. Then enter the transmitter power as 1 W and enter the EIRP as if it were the antenna gain in dB. For either case, the result will be correct, since the EIRP deals with the effects of both the transmitter power and the transmitter antenna gain.

```
%Program satellite.m
fprintf('\n')
Pt=input('Enter the transmitter power in watts. ');
Gtdb=input('Enter the transmitter antenna gain in dB. ');
Grdb=input('Enter the receiver antenna gain in dB. ');
f=input('Enter the frequency in GHz. ');
B=input('Enter the bandwidth in Hz. ');
Te=input('Enter the receiver noise temperature in kelvin. ');
Text=input('Enter the external noise temperature in kelvin. ');
Lmisc=input('Enter the value of miscellaneous losses in dB. ');
L=input('Enter the ground station latitude in degrees. ');
l=input('Enter the difference in longitude in degrees. ');
Lrad=(pi/180)*L;
lrad=(pi/180)*l;
rkm=35.786e3*sqrt(1+0.42*(1-cos(Lrad)*cos(lrad)));
alpha1=20*log10(f)+20*log10(rkm)+92.44;
EIRPdbw=10*log10(Pt)+Gtdb;
Tsys=Te+Text;
GoverT=Grdb-10*log10(Tsys);
Prdbw=EIRPdbw-alpha1+Grdb;
Prdbf=Prdbw+150;
CoverNdb=EIRPdbw-alpha1+GoverT+228.6-10*log10(B)-Lmisc;
fprintf('\n')
fprintf('The range is %g km.\n',rkm)
fprintf('The EIRP is %g dBW.\n',EIRPdbw)
fprintf('The path loss is %g dB.\n',alpha1)
fprintf('The system noise temperature is %g K.\n',Tsys)
fprintf('The received power is %g dBf.\n',Prdbf)
fprintf('The G/T ratio at the receiver is %g dB/K.\n',GoverT)
fprintf('The C/N ratio at the receiver is %g dB.\n\n',CoverNdb)
```

**FIGURE 17–6**
M-file program satellite.m.

If the G/T ratio in dB/K is given for the receiver, it will be necessary to convert it to an absolute value and separate the gain factor from the temperature for the purpose of using the program. If the gain is known, the temperature may be arbitrarily separated into the three separate components for entering the data into the program.

**▌▐ MATLAB EXAMPLE 17-1**    Rework the uplink analysis of Example 17-8 using the program **satellite.m**.

**SOLUTION**    Refer back to Example 17-8 for a summary of all the data involved. The noise figure of the receiver must be manually converted to an equivalent noise temperature, although it would be very easy to modify the program to accept noise figure instead of noise temperature.

The dialogue for this example is shown in Figure 17–7. Except for insignificant round-off and different numbers of decimal places, the results agree with those of Example 17-8.

**▌▐ MATLAB EXAMPLE 17-2**    Rework the downlink analysis of Example 17-9 using the program **satellite.m**.

**SOLUTION**    The solution dialogue is given in Figure 17–8. Since the EIRP is given as 47.8 dBW, a "trick" approach is utilized by entering the power as if it were 1 W and then the antenna gain is entered as if it were 47.8 dB. As in the preceding example, the results are in agreement except for minor roundoff and different numbers of decimal places.

>> satellite

Enter the transmitter power in watts. 100
Enter the transmitter antenna gain in dB. 55
Enter the receiver antenna gain in dB. 27
Enter the frequency in GHz. 6.125
Enter the bandwidth in Hz. 36e6
Enter the receiver noise temperature in kelvin. 290
Enter the external noise temperature in kelvin. 300
Enter the value of miscellaneous losses in dB. 1.6
Enter the ground station latitude in degrees. 40.5
Enter the difference in longitude in degrees. 20.8

The range is 37896.8 km.
The EIRP is 75 dBW.
The path loss is 199.754 dB.
The system noise temperature is 590 K.
The received power is 52.2458 dBf.
The G/T ratio at the receiver is -0.70852 dB/K.
The C/N ratio at the receiver is 25.9743 dB.

**FIGURE 17–7**
Results for MATLAB
Example 17-1.

>> satellite

Enter the transmitter power in watts. 1
Enter the transmitter antenna gain in dB. 47.8
Enter the receiver antenna gain in dB. 42
Enter the frequency in GHz. 3.9
Enter the bandwidth in Hz. 36e6
Enter the receiver noise temperature in kelvin. 225.6
Enter the external noise temperature in kelvin. 150
Enter the value of miscellaneous losses in dB. 1
Enter the ground station latitude in degrees. 32
Enter the difference in longitude in degrees. 1

The range is 36911.2 km.
The EIRP is 47.8 dBW.
The path loss is 195.604 dB.
The system noise temperature is 375.6 K.
The received power is 44.1955 dBf.
The G/T ratio at the receiver is 16.2527 dB/K.
The C/N ratio at the receiver is 20.4853 dB.

**FIGURE 17–8**
Results for MATLAB
Example 17-2.

## PROBLEMS

**17-1**  The lowest downlink frequency for the Ku-band Direct Broadcast Satellite service in America is 12.2 GHz. Consider a satellite transmitter operating at this frequency with the goal of providing coverage of the continental United States, which requires a 3-dB beamwidth of 8°. Assuming an illumination efficiency of 70%, determine the diameter and the gain of the downlink antenna.

**17-2**  Assume a downlink frequency of 18 GHz. Consider a satellite transmitter operating at this frequency with the goal of providing coverage of the continental

United States, which requires a 3-dB beamwidth of 8°. Assuming an illumination efficiency of 70%, determine the diameter and the gain of the downlink antenna.

**17-3**  The lowest downlink frequency for the C-band is 3.7 GHz. Consider a satellite transmitter operating at this frequency with the goal of providing global coverage of the earth, which requires a 3-dB beamwidth of 18°. Assuming an illumination efficiency of 70%, determine the diameter and the gain of the downlink antenna.

**17-4** Assume a downlink frequency of 18 GHz. Consider a satellite transmitter operating at this frequency with the goal of providing global coverage of the earth, which requires a 3-dB beamwidth of 18°. Assuming an illumination efficiency of 70%, determine the diameter and the gain of the downlink antenna.

**17-5** The lowest uplink frequency for the Ku-band Direct Broadcast Satellite service in America is 17.3 GHz. Consider a ground station transmitter operating at this frequency with a desired beamwidth of 1.6°. Assuming an illumination efficiency of 70%, determine the diameter and the gain of the ground station uplink antenna.

**17-6** Assume an uplink frequency of 31 GHz. Consider a ground station transmitter operating at this frequency with a desired beamwidth of 1.6°. Assuming an illumination efficiency of 70%, determine the diameter and the gain of the ground station uplink antenna.

**17-7** The approximate latitude and longitude of New York City are 40.5° N and 70.2° W. Determine the distance from an earth station located there to a communications satellite located at 105° W longitude.

**17-8** Repeat the analysis of Problem 17-7 for a communications satellite located at 85° W.

**17-9** A certain Ku-band communications satellite is located at 99° W longitude. Various specifications for the **uplink** are as follows:

> uplink earth station location: New York City, 40.5° N latitude, 70.2° W longitude.

> uplink frequency = 14.02 GHz

> uplink transmitter output = 500 W

> uplink earth station antenna gain = 55 dB

> bandwidth = 43 MHz

> satellite receiver antenna gain = 30 dB

> satellite receiver noise figure = 4.8 dB

> satellite receiver external noise temperature = 300 K

> atmospheric loss at 14.02 GHz = 2.2 dB

Determine the $C/N$ ratio at the receiver for the uplink.

**17-10** For the system whose uplink was given in Problem 17-9, assume that the signal is being received in Atlanta, Georgia, and that the downlink parameters are as follows:

> downlink receiver: Atlanta, 32° N latitude, 90° W longitude.

> downlink frequency = 11.72 GHz

> downlink EIRP = 50.4 dBW

> bandwidth = 43 MHz (same as uplink)

> user receiver antenna gain = 46 dB ($39.81 \times 10^3$)

> user receiver noise figure = 3.5 dB (2.239)

> user receiver external noise temperature = 200 K

> atmospheric loss at 11.72 GHz = 1.5 dB

Determine the $C/N$ ratio at the receiver for the uplink.

# Data Network Communications Basics

# 18

## OVERVIEW AND OBJECTIVES

In the next two chapters, we will consider network communications in the context of a typical office environment. This chapter deals with general networking concepts and wired networks. Chapter 19 deals with wireless networks.

Some office computers are dedicated to providing specialized services, such as storing files, processing e-mail, or printing. Other computers are used by workers to access those services from their desktop. However, the same network communications principles and techniques can be used in other environments, such as the automation of a manufacturing plant, or the control of the processes and machinery in an office building.

In a stand-alone computer, operations are relatively simple. An operating system (OS) such as Windows XP Professional enables the user to communicate with the computer. The user indicates the application to be loaded. The OS locates the application on the hard drive or other storage media. It then loads the application in memory and transfers control to the application. The user is then able to access and manipulate files, access other services on the computer, or use printers or other devices. The user communicates with the application and the application communicates with the OS. In most cases, it is the OS that actually performs the requested operations. All of this is relatively simple, because all data communications takes place within the computer itself.

However, in a network environment, the files accessed may be on a remote computer, not the user's computer. Even the application may be remotely located. Not only must the user be able to communicate with processes in the local computer, the user must also be able to communicate with processes in the remote computer. A connection is established between the local computer and the remote computer, and information passes between the computers. Generally, we call the computer that initiates the connection the *client* and the computer to which the connection is made the *server*. In some contexts, a computer that provides services to other computers is called the *server* and computers using those servers are called *clients,* but these two uses of the terms client and server are generally compatible. If services are required, it is generally the computer requiring the services that initiates the connection.

In this chapter and in Chapter 19 we will examine the methods and equipment that makes this remote access to network resources possible.

As in all areas of technology, many new terms will be encountered in these chapters, as well as familiar terms used in unfamiliar ways. Definitions will be provided as each term is encountered for the first time. In addition, networking has its own set of acronyms, and a list of many of the more common ones is included at the end of the chapter.

## Objectives

After completing this chapter, the reader should be able to:

1. Describe, in general terms, how data communications is accomplished in a network.
2. Describe the IEEE 802 project.
3. Describe the OSI seven-layer model, and explain why and how it is used.
4. Describe the Ethernet, token ring, frame relay, and ATM networking technologies.
5. Describe the NetBEUI, IPX/SPX, and TCP/IP protocols.
6. Understand IP addressing and its use in the Internet.
7. Understand the operation of typical networking applications, such as Internet browsing and Voice over IP.
8. Describe the operation of hubs, switches, bridges, routers, and other devices.

## SystemVue™ Opening Application (Optional)

Insert the text CD in a computer having SystemVue™ installed and activate the program. Open the CD folder entitled **SystemVue Systems** and open the file entitled 18-1.

### Sink Tokens

| Number | Name | Token Monitored |
|--------|--------|-----------------|
| 0 | Input | 3 |
| 1 | Clock | 4 |
| 2 | Output | 5 |

### Operational Tokens

| Number | Name | Function |
|--------|------|----------|
| 3 | Input Signal | generates ASCII code for WOW |
| 4 | Clock Generator | generates clock to Manchester code input signal |
| 5 | Output XOR | encodes input signal |

The system shown on the screen consists of two *source* tokens (Tokens 3 and 4), an exclusive-OR (Token 5), and three *sink* tokens (Tokens 0, 1 and 2). Token 3 reads the contents of a text file (**Manchester_WOW.txt**) containing the pulse representing the ASCII codes for the letters WOW, with the most significant bit first and a parity bit for odd parity. The first character consists of a binary 0 as a parity bit and the ASCII code for "W" (1010111). The second character consists of a binary 0 as a parity bit and the ASCII code for "O" (1001111). The third character, like the first, consists of a binary 0 as a parity bit and the ASCII code for "W" (1010111). This pulse train is monitored by Token 0, which displays the pulse train on the screen and also in the Analysis screen. The file **Manchester_ WOW.txt** contains ten 1s for each positive pulse and ten −1s for each negative pulse. Since the sample rate is 100 Hz, this represents an input signal at a rate of 10 bps, which is monitored by Sink 0.

Token 4, monitored by Sink 1, generates a square wave at 10 Hz that is used to encode the input signal with the Manchester code. Manchester encoding represents each bit with a transition at the center of the bit. The transition is HIGH to LOW for a 0 and LOW to HIGH for a 1. Applying the input signal and the clock to an XOR gate produces that encoding. Assuming that the clock phase produces a HIGH during the first half-cycle and a LOW during the second half-cycle, if the input is LOW, the XOR output will be HIGH during the first half-cycle and LOW during the second half-cycle, producing the HIGH-to-LOW transition at the midpoint of the pulse. Conversely, if the input is a HIGH, the XOR output will be LOW during the first half-cycle and HIGH during the second half-cycle, producing the LOW-to-HIGH transition at the midpoint of the pulse. Sink 2 monitors the output of the XOR gate, Token 5.

## Checking the Settings

You will not need to modify any settings in this exercise, but you can observe the values of the parameters by placing the cursor over a token. A window will appear containing the parameters of that token.

## Activating the Simulation

Left-click the **Run System** button or press F5 to start the simulation. A moving blue line at the bottom of the screen indicates that the simulation is running. Note that when the line has reached the right end of its excursion, the waveforms appear in the box on the screen to the right of each sink. Left-click the **Analysis Window** button to open the Analysis window.

In the Analysis window, W0 shows the input signal, W1 shows the clock, and W2 shows the output. You can expand each of these windows to show a single pulse by placing the cursor at the 0 at the top of a display, and dragging the cursor to the right and down so that it creates a rectangle around the waveform from the left end of the waveform (0 sec) to the fourth grid mark (0.8 sec, the grid mark to the left of 1 on the horizontal scale). When you do this, the display expands so that it shows just the first eight pulses. Note that the first input symbol is 01010111. The first 0 is the parity bit, and 1010111 binary or 57 hex is the ASCII code for the uppercase W. If you expand W1 and W2 similarly, you can see the clock signal in W1, which has one cycle during each pulse period of the input signal, and the output in W2. Note that the output has a transition at the middle of each pulse period, and that the transitions are HIGH to LOW when the input pulse is LOW and LOW to HIGH when the input pulse is HIGH. This can be seen most clearly during the period 500e-3 to 800e-3, when there are three HIGH pulses in a row.

Now, scroll the waveform in W0 so that the 800e-3 grid mark is at the left end of the display. This can be done easily by holding down the shift key while left-clicking on the waveform in W0, and then moving the mouse to the right. The further to the right you move the mouse, the faster the waveform scrolls. If it scrolls to far, move the mouse to the left while holding down the shift key. Scroll the W1 and W2 waveforms the same distance. The 1.2-second marker should be at the center of each window. Note that the input waveform in W0 is 01001111. Again, the first 0 is the parity bit for odd parity and the remaining bits are the ASCII code for the upper case letter O, 1001111 binary or 4F hexadecimal.

There is a limit to how far you can scroll the waveform. To show the third symbol, right-click each waveform and select **Rescale** from the menu. That will return each waveform to its original scale. Then drag a rectangle from 1.6 sec to 2.4 sec on each waveform. You will see the third byte of data expanded to fill the window. Note that it is an ASCII W like the first byte.

The text file entitled **Manchester_WOW.txt** contains the input for Token 3. This is simply a table of values of the output of that token at each sample time. At the time of the first sample period, the program reads the first value, −1, interprets it as −1 V, and outputs it. At the time of the second sample period, the program reads the second value, −1, interprets it as −1 V, and outputs it. It does this at each sample time. The Word file entitled **18-1 initial settings.doc** contains views of the system and of the time settings. The sample frequency is 100 Hz and the file contains 10 entries for each pulse, so this represents data sent at a rate of 10 bps. After you have completed the chapter, you will see the effect of changing the phase of the clock.

## 18-1    Network Operations and Operating Systems

### Objectives of Networking

Soon after scientists, engineers, and others started using computers to manipulate and store information, some of them recognized the benefit of exchanging data between computers. If several people view data stored in the same file located on a shared server, they can be confident that they all view the same version of the file. Loading an application on a local computer from a single shared source makes it easier to ensure that all users have access to the most current version of the application. Sharing an expensive peripheral across a network makes it available to more users and therefore more cost-effective. Sharing information across a network makes that information more available to everyone who needs it.

### Network Operations

Today, we are all familiar with the operation of our personal computer. If we want to run a program (also called an *application*) that program must be available on the computer's hard drive or other portable device (the *storage media*). When we click on the program's icon or type the program name at a prompt, the program is loaded into memory and run (*executed*). The computer's *operating system* is a set of dedicated programs that enables us to communicate with the computer and carries out our desires within its capability. Among other things that the OS does for us is to locate the program that we want to run, load it into memory, and start its execution.

Network operations mirror those that occur on our personal computer. If an application must be loaded into memory, it must be located on a storage media. If the storage media is attached to the local computer, the computer's OS locates the application. If the storage media is on a remote computer, the local computer must communicate with the remote computer to obtain the file containing the application. The local OS transfers control to a *network operating system* (NOS). The NOS enables the local computer to tell the remote computer what file is needed, and to transfer the file from the remote computer to the local computer.

We classify network operations as either *peer-to-peer* or *client/server*. In a peer-to-peer network, illustrated by Figure 18–1, a local OS (such as Windows XP Professional) is designed primarily to support local operations. However, the OS contains components to enable the user to share files on the local hard drive, or printers attached to the local computer, with another computer. All that is needed is a network connection between the computers, and an OS on the other computer that also has the components necessary for file and printer sharing. No computer in a peer-to-peer network is dedicated to providing services to other computers in the network. What is shared and with whom it is shared is at the discretion of the operator of the local computer (the *user*). Most often, a user is operating the computer from the *console* (keyboard and monitor), even when files or printers are being shared. Generally, the OS on each computer in a peer-to-peer network is the same, but this is not necessary as long as they are compatible.

In a client/server network, selected computers, the *servers,* are dedicated to providing services to other computers on the network. Those other computers, the *clients,* enable users to access the servers. Figure 18–2 illustrates a client/server network. Servers have different functions: *File servers* provide central storage for files needed by multiple users

PC          PC                    Printer

**FIGURE 18–1**
Peer-to-peer network.

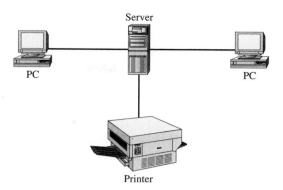

**FIGURE 18–2**
Client/server network.

across the network. *Application servers* contain programs that multiple users need to run, such as database programs, or that provide services to many users, such as e-mail programs. *Print servers* enable multiple users to print documents on centralized printers. Often a network server may perform a combination of these services. Server operating systems (such as Windows 2003 Server) are optimized for their function of providing services, and often for the services that they provide. Thus, in general, the server OS is different from the client OS. Furthermore, there is generally no user operating the server from the console while it is providing services to the network.

Sharing of printers or other *peripherals* (devices attached to the computer to provide ancillary functions) between computers is not necessarily networking. For example, prior to Windows, the OS of a personal computer did not generally enable easy sharing of files or printers. To make printer sharing possible, the printer outputs of two computers (their *parallel ports*) could be connected to a switch which was in turn connected to a printer. By throwing the switch, one or the other of the computers was connected directly to the printer. No networking components in the computer OS are required to share a printer in this way.

---

**▐▌ EXAMPLE 18-1**

A student and his roommate each has a laptop computer that is used for coursework. Both computers run Windows XP Professional as an OS. The two students are enrolled in the same curriculum and frequently work on joint projects. They decide that it would be advantageous if they could share files between their computers. In addition, they decide that they could purchase a better printer if they pooled their money. Therefore they each buy a *network adapter* which enables their computer to be connected to a network, install it in their computer, and connect the two computers together with a network cable. Then they connect the printer to one of the computers and share it over their network. Is this a peer-to-peer network or a client/server network? Why?

**SOLUTION**   Theirs is a peer-to-peer network, exactly as shown in Figure 18–1. First, neither computer is dedicated to providing services to other computers rather than being using as a local computer. The primary function of each computer is to be used by its owner to do coursework. Sharing files and the printer is a secondary use of the computer. Second, both laptops are running Windows XP Professional, an OS that is essentially a local OS.                                                          ▐▌

---

## Network Operating Systems

NOSs have evolved from very basic to very sophisticated systems. Initially, only text-based files could be shared across a network. Today it is possible to share graphics, audio, and video information. It is possible for people in different parts of the world to participate in a conference in which they see and hear each other, and each make inputs to a document or graphic file located on the computer of one of the participants (see Section 18-6).

While there have been a number of NOS's developed through the years, there are three that currently dominate the market: Windows, NetWare, and Unix. Although the three were originally very different, many of those differences have disappeared in recent versions. All three support *single logon,* a feature that allows a user to obtain access to all authorized resources in the network by logging on at a single client computer. All three support the organization of network objects into a hierarchy in which the location of a specific object can be found using a directory service. All three have utilities that make the management of files and printers, performance monitoring, troubleshooting, and remote administration easier. What is different is how each performs these operations. Moreover, the data communications methods used by each are generally the same, and will be discussed in more detail later in the chapter.

### LANs, MANs, and WANs

For convenience, a network can be classified as a local area network (LAN), metropolitan area network (MAN), or wide area network (WAN). This classification is based upon the reach or span of the network, and its complexity. In this classification scheme, two computers in the same room linked by an Ethernet cable clearly constitute a LAN and the Internet clearly constitutes a WAN. In between these two extremes, there is a spectrum of networks of various sizes and complexities that can rather arbitrarily be classified as LAN, MAN, or WAN. There are some guidelines that can be used, however.

A MAN is a group of interconnected LANs in a metropolitan area or a small city. The distance between the LANs in a MAN generally requires that they be interconnected using fiber-optic cable or wireless media. Although there are no hard and fast rules that differentiate a LAN from a MAN, the outer limit of the individual LAN is often taken to be the place where the media changes from copper to fiber-optics or wireless. This is not an absolute criterion, because both fiber-optic cable and wireless media can be used to interconnect devices within a LAN. The maximum extent of a MAN is often considered to be on the order of 30 miles. For example, a university might have a main campus and a satellite campus located in the same city but several miles away. The networks in the buildings in the main campus or in the satellite campus each might be classified as a LAN, and the whole university's network might be classified as a MAN. The elements of each LAN (and, often, the cabling that connects the LANs) is usually the property of the organization to which the MAN belongs. Sometimes, however, the cabling that connects the LANs belongs to some other entity, such as a telephone company.

A WAN is a group of interconnected LANs that span a distance greater than that of a MAN. Thus, an organization that has networked computers in New York and in San Francisco, configured so that computers in one city can access resources in the other city, has a WAN. The links between the cities are referred to as *WAN links.* To extend the example of the preceding paragraph, a university might have a main campus in one city and a branch in another city. Each of these locations could consist of one or more LANs. The two locations could be joined by a WAN link so that users in the branch could access resources on the main campus, and vice versa. Although the components of the LANs in each location would normally be the property of the university, generally the components of the WAN link will belong to another entity, such as a telephone company.

## 18-2   Standards and the IEEE 802 Project

### The Need for Standards

The techniques used in early network devices were largely proprietary, belonging to the manufacturer of the device. Although a network interface card (NIC) from Company A performed the same functions as a NIC from Company B, the voltage levels used or the sequence of signals sent were not necessarily the same. Therefore, a Company A NIC could not be used with a Company B hub. The result was that networks from different vendors were not compatible, and the cost of networking components tended to be high. If all network devices are manufactured in accordance with a standard set of specifications, components manufactured by different vendors are interchangeable and can interoperate. This can also lead to competition between vendors, leading to lower prices. Also, conforming with standards tends to make new devices backward compatible with older devices, protecting a user's investment in technology. This encourages users to make that investment, because they can have some confidence that they will not have to replace everything when technology advances occur.

## Standards Organizations

Many organizations have been established to create standards for various aspects of computing and networking. Some of these organizations are made up of representatives of the manufacturers of the devices used. Others are sponsored by government agencies. Still others are sponsored by professional organizations. Two of most important standards organizations for network data communications are the Institute of Electrical and Electronics Engineers (IEEE) and the Internet Engineering Task Force (IETF). The IEEE initiated the IEEE 802 project, discussed below, which established standards for local area and metropolitan area networks (LANs and MANs). The IETF is responsible for establishing standards for the Internet.

## The IEEE 802 Project

In February of 1980 (thus the designation 802), the first meeting of an IEEE subcommittee to establish networking standards was held. As a result of this committee's work, a number of working groups were created, as indicated in Table 18–1. The groups that are marked "Active" in the table are still working on standards, although some have already published standards. Others, marked "Hibernating," have published their standards and are currently not active. Still others, marked "Disbanded," did not publish a standard.

We will look at some of the published IEEE 802 standards in this and the following chapter. In particular, IEEE 802.3, *CSMA/CD Access Method,* provides the standard for the

**Table 18–1**   IEEE Working Groups

| Standard | Title | Status |
|----------|-------|--------|
| 802.1 | High Level Interface (HILI) | Active |
| 802.2 | Logical Link Control (LLC) | Hibernating |
| 802.3 | CSMA/CD | Active |
| 802.4 | Token Bus | Hibernating |
| 802.5 | Token Ring | Hibernating |
| 802.6 | Metropolitan Area Network (MAN) | Hibernating |
| 802.7 | Broadband | Hibernating |
| 802.8 | Fiber-Optics | Disbanded |
| 802.9 | Integrated Services LAN Interface (ISLAN) | Hibernating |
| 802.10 | Interoperable LAN Security (SILS) | Hibernating |
| 802.11 | Wireless LAN (WLAN) | Active |
| 802.12 | Demand Priority | Hibernating |
| 802.14 | Cable-TV Broadband Communication Network | Disbanded |
| 802.15 | Wireless Personal Area Network (WPAN) | Active |
| 802.16 | Broadband Wireless Access (BBWA) | Active |
| 802.17 | Resilient Packet Ring (RPR) | Active |
| 802.18 | Radio Regulatory | Active |
| 802.19 | Coexistence | Active |
| 802.20 | Mobile Wireless Access | Active |

Ethernet technology which we will consider in more detail in Section 18-8. IEEE 802.11, *Wireless,* which is actually several standards, is the subject of Chapter 19.

## 18-3 The OSI Seven-Layer Model

Regardless of the NOS involved, there are certain functions that must be performed on a client to prepare data and control signals for transmission across a network to a server, and on a server to process the data and control signals and to respond to the client. Several models have been developed to assist in understanding and describing NOS's, but the most widely used is the open systems interconnect (OSI) model developed by the International Organization for Standardization. As shown in Figure 18–3, this model divides the functions that must be performed into seven categories or layers. Programs in the top layer, the application layer, perform the functions that are necessary to interface with the local OS. Programs and devices in the bottom layer, the physical layer, perform the functions necessary to actually transmit the data to another computer. Intermediate layers perform other functions required for successful networking, and will be described below. The protocols (software modules) in each layer receive data and provide services to the layers above and below. We will consider the functions performed in each layer in order, from bottom to top, and then will look at how these layers work together to accomplish the transfer of information across a network.

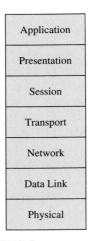

**FIGURE 18–3**
OSI seven-layer model.

### Overview

In network data communications, a *protocol* is a set of rules that specify how functions will be performed to achieve a certain result. For communication to succeed between two computers, a common protocol must be employed in both computers. For example, a protocol specifies the coding that will be used to convert logical 1s and 0s into electrical pulses on the cabling between the computers. If the sending computer and the receiving computer use different protocols, the received pulses cannot be properly converted into data that can be used in the receiving computer. Rather than define a single protocol that encompasses every function that must be performed to move data from an application on one computer to an application on another computer, the functions involved are usually divided into a set of protocols. The resultant set of protocols is often called a *protocol stack* or a *protocol suite.* The OSI seven-layer model provides a way to systematically describe such protocol stacks. Furthermore, different sets of protocols can be used to accomplish the same final result. Therefore, there are different protocol suites available, all of which can be described using the OSI model. The most common suites are TCP/IP, IPX/SPX, and NetBEUI. TCP/IP, the protocol suite used on the Internet, is also described in more detail in Section 18-4. NetBEUI and IPX/SPX are described in Section 18-5.

Protocols in one layer of the OSI model provide services to the protocols in the layers above and below. When sending data, an application-layer protocol accepts information from an application running on the computer, processes that information in accordance with its rules, and then passes the processed information to a protocol in the layer below. Ideally, that next layer is the presentation layer. However, actual protocols do not always conform perfectly to the OSI model, and may span two or more layers. Thus, in the TCP/IP protocol suite, http, the protocol that processes data from Web browsers, performs all of the functions included in the application, presentation, and session layers of the OSI model.

Figure 18–4 represents the flow of data between an application on a client and an application on a server. Data from the client application is passed to an application-layer protocol, where it is *encapsulated* into a protocol data unit (PDU) and passed to the presentation layer. Generally, *encapsulation* involves adding a header to the PDU received from the layer above to create a new PDU and passing that PDU to the layer below, as illustrated in Figure 18–5. It may also involve making significant changes to the data before adding the header. The header attached by a protocol at each layer provides the information needed by the corresponding protocol in the remote computer to restore the data to the state that it was in before encapsulation. Each succeeding layer processes and encapsulates what

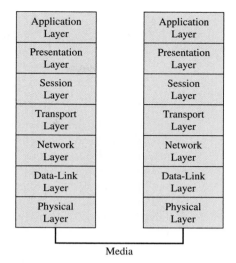

**FIGURE 18–4**
OSI between two computers.

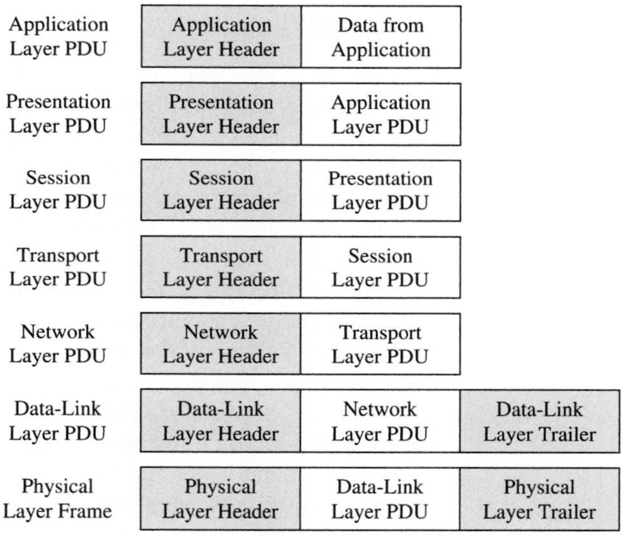

| | | | |
|---|---|---|---|
| Application Layer PDU | Application Layer Header | Data from Application | |
| Presentation Layer PDU | Presentation Layer Header | Application Layer PDU | |
| Session Layer PDU | Session Layer Header | Presentation Layer PDU | |
| Transport Layer PDU | Transport Layer Header | Session Layer PDU | |
| Network Layer PDU | Network Layer Header | Transport Layer PDU | |
| Data-Link Layer PDU | Data-Link Layer Header | Network Layer PDU | Data-Link Layer Trailer |
| Physical Layer Frame | Physical Layer Header | Data-Link Layer PDU | Physical Layer Trailer |

**FIGURE 18–5**
OSI frames.

it received from the protocol in the layer above, and passes the result to a protocol in the layer below. At the physical layer, the logical data from the data-link layer is encoded into a frame and placed on the media for transmission to the server. The process is reversed in the server. At each layer, the header is stripped off, information in the header is used to process the data, and the result is passed to a protocol in the layer above. The information in the header, for example, might identify the protocol to which the package should be passed.

## Physical Layer

The physical layer consists of the media across which the information travels from one computer to another, and the devices that make this possible. This includes the devices that accomplish this transmission of data, and the software and devices that are responsible for getting the information out of and into the computers involved. The protocols in the physical layer receive control signals and data from the data-link layer in the form of packets of

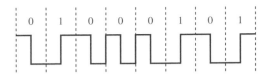

**FIGURE 18–6**
Manchester encoding.

logical 1s and 0s, and convert them, in conjunction with physical layer hardware, into the appropriate physical signals for the media. For example, if the computers are connected with coaxial cable, a network interface card and driver software in the sending computer will convert the logical 1s and 0s of the data into electrical pulses for transmission in the cable. Figure 18–6 shows Manchester encoding, a form of encoding used with Ethernet, which is described later in this chapter. Note that this is similar to biphase-L encoding studied in Section 9-4. However, the implementation specified by IEEE 802.3 utilizes a low-to-high transition at the center of a bit to represent a "1" and a high-to-low transition at the center of a bit to represent a "0". A NIC and driver software in the receiving computer will convert the pulses back into logical 1s and 0s for delivery to a data-link layer protocol. If fiber optic cable were used between the computers, the conversion at the physical layer would be between the logical 1s and 0s and light pulses.

## Data-Link Layer

Data-link layer protocols are responsible for packaging the data to be transmitted to the remote computer into appropriate size packets or frames for the transmission technology being used. These protocols also control access to the media and ensure that the packets are delivered to the destination on the local network without error. How this is accomplished depends upon the transmission technology used. For example, if Ethernet (the most popular transmission technology for local networks) is used, each Ethernet NIC has a unique 12-digit hexadecimal address assigned to it called the MAC or hardware address. After a frame of data is created, the data-link layer attaches a header and a trailer to the frame. The header contains the MAC addresses of the source and destination computer or device, along with control information required by the Ethernet standard. The trailer contains a checksum calculated from the contents of the frame, used to determine if any errors occur during transmission. Media access control for Ethernet utilizing a process called Carrier Sense Multiple Access with Collision Detection (CSMMA/CD) is discussed later in this chapter, but is exercised at the data-link layer. The physical layer of each computer or device on the network segment converts the signal from the media back to a logical frame, and passes it to the data link layer, where the header and trailer are removed and examined. The destination MAC address in the header is compared with the local MAC address. If they are not the same, the frame is discarded. If they are the same, a checksum is calculated from the contents of the frame and compared to that included in the trailer. If these do not match, an error has occurred in transmission and the frame is discarded. If they match, a frame acknowledging receipt of the data is returned to the sender, using the sender's MAC address from the header. The data itself, after removal of the header and trailer, is passed to the appropriate network-layer protocol. The sender of the packet waits for a designated period of time for an acknowledgement. If no acknowledgment is received, the sender retransmits the packet. Other transmission technologies perform similar functions, although the format of the frames transmitted, the contents of the header and trailer, the method used for error detection and control, and the media access control method will differ.

## Network Layer

The network layer is responsible for addressing frames of data so that they can reach their ultimate destination, even if that destination is not on the local network. The network layer works with logical addresses able to identify destinations on remote network segments as well as on local network segments. In a complex network, devices called routers are used to join network segments. If the ultimate destination of a packet is on a remote network segment, a network layer protocol must determine the path to that destination and identify the address of the first router to which the frame must be sent. A network layer header is attached that includes the logical address (rather than the MAC address) of the source and destination. Different addressing schemes are used, depending upon the protocol used, such as TCP/IP or IPX/SPX. These protocols and their addressing are discussed in more detail in sections 18-4 and 18-7. After the network layer header and trailer are appended, the packet is passed to the data-link layer. If the ultimate destination is local, the MAC address of the ultimate destination is used to deliver the packet. If the ultimate destination is remote, the MAC address of the router on the local network segment is used to deliver the packet to the router. The router processes the data-link layer header as discussed above. When it discovers that the packet is addressed to itself, it passes the packet to the router's network-layer protocol. That protocol strips off the network-layer header and trailer and uses their contents to determine where the packet should be sent next. If the ultimate destination is a network segment to which the router has direct access, it attaches a header and trailer containing appropriate address and control information and passes the packet to its data-link layer for delivery to the destination. If the ultimate destination is more remote, it determines the address of a router that can move the packet along its path to that ultimate destination, attaches a header and trailer containing address and control information for the next router, and passes the packet to its data-link layer for delivery to that router.

The network layer of a sending computer is also is responsible for segmenting data that is too large for inclusion in a single frame into multiple frames. The network layer of the receiving computer is then responsible for reassembling those frames, based upon control information received from the sending computer.

## Transport Layer

Protocols in the transport layer are responsible for ensuring that data is transferred from sender to receiver reliably and error free. These protocols can employ either connection-oriented or connectionless communication. An example of a connection-oriented protocol is the Transport Control Protocol, discussed in more detail in Section 18-4. A connection-oriented protocol establishes a connection between two computers and then transfers data in the context of that connection. The transport-layer protocol in the receiving computer acknowledges receipt of all packets. If the sending computer's transport-layer protocol does not receive an acknowledgment for a packet in a timely manner, the packet is retransmitted. A connection-oriented protocol is generally used when the volume of data transferred requires the transmission of multiple packets.

A connectionless protocol, on the other hand, simply transmits a packet and forgets it. If the packet does not reach its destination, it is the responsibility of a protocol in a higher layer to detect that and take corrective action. A connectionless protocol is generally used for control information, and when the data to be transferred can be accomplished in a single packet. Since the overhead of setting up and breaking down a connection is not required, connectionless communication is faster than connection-oriented communication.

## Session Layer

Session-layer protocols are responsible for defining how a communication session is set up, maintained, and terminated between two computers. This may involve establishing the identity of the computers involved, defining how much data is to be transferred, acknowledging the receipt of the data, and determining when data transmission is complete.

## Presentation Layer

Presentation-layer protocols are responsible for formatting data for transmission across the network in the sending computer, and formatting data for use by the application in the receiving computer. This may include the conversion of character sets, and the encryption and compression of data.

## Application Layer

Application-layer protocols are responsible for communicating with applications on the host computer, and are not the applications themselves. A word processor running on a computer would use an application-layer protocol to access a file on a remote computer, or to print a document on a remote printer. The application-layer protocols provide the interface between the applications running on a computer and the other protocols in the protocol stack. Error recovery may also be supported at the application layer.

---

**▐▌ EXAMPLE 18-2**

A certain protocol in a network stack takes input from the user application, formats that input for transmission across the network, encrypts the data, incorporates provisions to determine if errors in transmission have occurred and corrects any such errors, sets up communication sessions with servers to which the user wants to connect, and determines when the communication session is complete. To what layer or layers of the OSI seven-layer model does this protocol belong? Why?

**SOLUTION**    Accepting input from a user application and error recovery are application-layer functions. Encrypting data is a presentation-layer function. Setting up communications sessions and determining when the communication session is complete are session-layer functions. Therefore this protocol spans the application, presentation, and session layers.

▐▌

---

## 18-4    TCP/IP and IP Addressing

TCP/IP, the networking protocol of the Internet, is the most common networking protocol used, even in LANs. It consists of the protocols shown in Figure 18–7, plus others not shown. TCP/IP was initially developed before the OSI model was created. Therefore, the upper layer protocols, such as HTTP and FTP, perform the functions of the application,

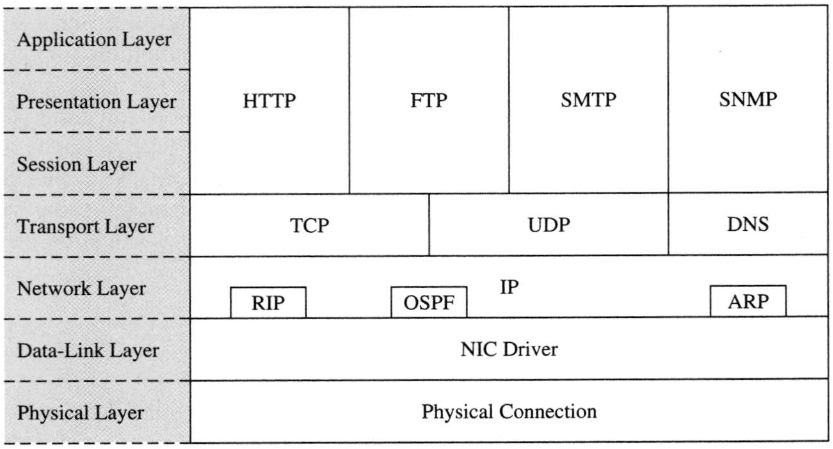

| Application Layer | | | | |
|---|---|---|---|---|
| Presentation Layer | HTTP | FTP | SMTP | SNMP |
| Session Layer | | | | |
| Transport Layer | TCP | | UDP | DNS |
| Network Layer | RIP | OSPF | IP | ARP |
| Data-Link Layer | NIC Driver | | | |
| Physical Layer | Physical Connection | | | |

**FIGURE 18–7**
TCP/IP protocols.

presentation, and session layers. The name of the protocol suite is derived from the principal transport and network layer protocols, Transport Control Protocol (TCP) and Internet Protocol (IP).

## TCP/IP Operation

The HyperText Transport Protocol (HTTP) enables applications to access Web services. Web browsers are the most common HTTP-compatible applications, but late versions of many other applications, such as Microsoft Word, are now able to use HTTP to access resources on remote servers. A protocol, such as HTTP, contains commands that an application uses to access the services of the protocol. Along with the commands, the application passes data to the protocol, which either instructs the protocol how to service the application or is sent by the protocol to an application on the remote computer. The File Transfer Protocol (FTP) performs the functions necessary to transfer files between a local computer and a remote computer. An application written for FTP, for example, might show directory listings of files on both the local and the remote computer, and initiate the transfer of a file from one computer to the other by highlighting a file and clicking on an arrow. Applications written to interact with the Simple Mail Transfer Protocol (SMTP) can be used to write e-mail messages or to move e-mail messages from one computer to another. Applications written for the Simple Network Management Protocol are used to configure and monitor network computers and other devices.

There are two transport protocols used in TCP/IP, the Transport Control Protocol (TCP) and the User Datagram Protocol (UDP). TCP is the connection-oriented protocol for this suite. It guarantees delivery of packets by requiring that receipt of all packets sent be acknowledged by the remote computer. Section 18-5 provides an example of the use of TCP to browse the Internet. UDP, on the other hand, is a connectionless protocol. It sends a packet and forgets it, relying on a higher-layer protocol to ensure delivery, if that is necessary. The advantage of UDP is that it is fast and has very little overhead. This makes it ideal for applications like Voice over IP, discussed in Section 18-6.

Usually, the application identifies the remote computer by a name, such as *www.mycompany.com*. However, to route packets to the remote host, an IP address is needed. The domain name service (DNS) resolves names to IP addresses, by querying databases on network computers known as DNS servers.

Once the IP address of the remote computer is known, the Internet Protocol (IP), is used to determine the route through the network to reach the remote computer. The form of IP addresses and how they are used will be discussed in more detail below. Each IP address consists of two parts, a network address and a host address. In TCP/IP standards, known as requests for comments (RFCs), individual computers are referred to as *hosts*. The network address identifies a unique network segment or group of hosts. This may be the segment of which the sending computer is part, a local network segment, or a segment that is separated from the local network segment by many other network segments. Routers, which will be discussed in Section 18-10, are used to pass a packet from one network segment to another. IP maintains a table of all known network addresses and the path through the network to reach them. IP uses routing protocols, such as Routing Information Protocol (RIP) or open shortest path first (OSPF), to maintain the routing table.

Actual delivery of packets on a local network segment is accomplished using the physical or media access control (MAC) address, which is normally burned into a NIC when it is manufactured. Before IP can pass a packet to a data-link protocol for delivery, it uses the Address Resolution Protocol (ARP) to obtain the MAC address that corresponds to the logical or IP address. ARP maintains a table of logical-to-physical address conversions. If the address needed is not in the table, ARP broadcasts a request to all computers on the local network segment, asking the computer with the specified logical address to respond with its physical address.

If the destination computer is on the local network, IP provides the MAC address of the destination computer to the data-link layer. If the destination computer is on a remote network segment, IP provides the MAC address of a router on the local network that is in the path to the remote network. The data-link layer then sends the packet through the network.

## IP Addresses

Every device, or host, in a TCP/IP network is assigned a unique address, called an IP address. This address (which can be resolved into two components, a network address and a host address) makes it possible to connect to remote computers, even across the Internet. When a person types the name of a Web site into a Web browser, a TCP/IP protocol called the domain name service (DNS) resolves that name to an IP address. The person's computer then determines the address of the network on which the remote computer is located, and initiates a connection with that computer. In this section we will examine the structure of IP addresses and how they are used to locate hosts on remote networks. There are two forms of IP address, IPv4, which uses 32-bit binary numbers as addresses, and IPv6, which uses 128-bit binary numbers as addresses. IPv4 is discussed below.

An IPv4 address is a 32-bit binary number. If only computers had to use IP addresses, they could be left in binary number form. However, it is frequently necessary for people to speak, write, and type IP addresses. 32-bit binary numbers are difficult for people to use, so the dotted decimal notation was developed. The 32-bit number is divided into four octets or groups of 8 bits. Each bit in an octet is weighted. The least significant bit (LSb) has a weight of 1. Thus, if that bit is a 1, its decimal value is also 1. Of course, if the bit is a 0, its decimal value is 0. The next more significant bit has a weight of 2 so that if that bit is a 1, its decimal value is 2. Moving to the left, each bit has a value that is twice that of the bit to its right until the most significant bit (MSb) is reached. Its weight is 128. Thus, each octet has a decimal value between 0 and 255: if all bits are 0s, the decimal value is 0 and if all bits are 1s, the decimal value is 255 ($128 + 64 + 32 + 16 + 8 + 4 + 2 + 1$). More generally, a binary 01101101 has a decimal value of $64 + 32 + 8 + 4 + 1$ or 109. The dotted decimal notation is obtained by dividing the 32-bit IP address into four octets, expressing each octet of the IP address by its decimal equivalent, and separating the decimal equivalents with dots. Thus, the dotted decimal notation for an IP address that consisted of 32 1s would be 255.255.255.255.

---

**▐▌ EXAMPLE 18-3**

Express in the dotted decimal notation an IP address whose binary value is 10110110011011000001110110000001.

**SOLUTION**   First, divide the binary number into its four octets. The first octet is 10110110, the second is 01101100, the third is 00011101, and the fourth is 10000001.

Next, find the decimal equivalent of the first octet, 10110110. Each bit that is a 0 has a value of 0 times its weight, or 0. Each bit that is a 1 has a value that is 1 times its weight. Therefore the most significant bit has a value of $1 \times 128 = 128$. The third and fourth bits have values of $1 \times 32 = 32$ and $1 \times 16 = 16$. The sixth and seventh bits have values of $1 \times 4 = 4$ and $1 \times 2 = 2$. Therefore, the decimal value of the number is $128 + 32 + 16 + 4 + 2 = 182$.

Similarly, the second octet, 01101100, has a decimal value of $64 + 32 + 8 + 4 = 108$. The third octet, 00011101, has a decimal value of $16 + 8 + 4 + 1 = 29$. The fourth octet, 10000001, has a decimal value of $128 + 1 = 129$. Putting these four numbers together and separating them with dots gives a dotted decimal value of 182.108.29.129.   ▐▌

---

## Class A, B, and C Networks

When the IP addressing scheme was first set up, several classes of networks were identified. The most important of these were designated Class A, Class B, and Class C. Class A networks use the first octet (first 8 bits) to identify the network address (or network ID) and the remaining three octets to identify the host address (or host ID). Class B networks use the first two octets (first 16 bits) to identify the network address and the last two octets to identify the host address. Class C networks use the first three octets to identify the network address and the last octet to identify the host address.

### Class A Networks

A Class A network has a zero for the left-hand bit of its first octet. Thus, the value of the first octet can range from 00000000 to 01111111. However, there are restrictions on addressing. According to these restrictions, a first octet that is all 0s cannot be used for a Class A network because it is used to indicate a host on a local network. Furthermore, the Class A network whose first octet decimal value is 127 is reserved for loopback addresses. Therefore, Class A networks have IP addresses ranging from 1.0.0.0 to 126.255.255.255. In addition, some of the addresses in that range are reserved for special purposes, such as broadcasts and private addresses. The remaining addresses identify nodes (also called hosts) on the network. The first octet of the Class A IP address, combined with 0s in the remaining octets (e.g., 10.0.0.0), identifies the network and is called the *network ID*. The remaining octets in the IP address identify nodes on that network and are called the *host ID*. A host ID of x.0.0.0 is the network ID, where x represents the first octet of the Class A network ID. x.255.255.255 is reserved for the network broadcast address. Therefore, the range of host IDs in each Class A network is x.0.0.1 to x.255.255.254.

---

**▌▌ EXAMPLE 18-4**

A computer is assigned an IP address of 10.1.103.235. What is the network ID and the host ID of that computer?

**SOLUTION**  An IP address with a first octet of 10 is a Class A network. Therefore, the network ID is 10.0.0.0. The last three octets of the IP address represent the host on the network. Therefore the host ID is 0.1.103.235.

---

**▌▌ EXAMPLE 18-5**

As noted above, there are 126 possible Class A networks. How many possible nodes can exist on each Class A network?

**SOLUTION**  The nodes are represented by the last three octets of the IP address. Since there are 24 bits in the last three octets, it is possible to have $2^{24} = 16,777,216$ different combinations of 0s and 1s. However, if all 24 bits are 0, that is used to represent the network ID. A Class A IP address with the last 24 bits all set to 1 is used as the broadcast address for that network. Therefore there are $2^{24} - 2 = 16,777,214$ possible nodes on a Class A network.                                                                                      ▌▌

---

### Class B Networks

There is a 1 in the MSb of the first octet of Class B IP addresses, followed by a 0 in the next bit. Therefore, the network IDs of Class B addresses range from 128.0.0.0 to 191.255.0.0. The first two octets of the IP address identify the network ID. Therefore, there are $2^{14}$ or 16,384 possible Class B networks. The exponent, 14, is because there are 14 bits in the first two octets after the 10 in the first octet is removed. Each Class B network can contain $2^{16} - 2 = 65,534$ hosts. We subtract 2 because, in each network, x.y.0.0 and x.y.255.255 (where x.y identifies the network) are reserved for the network ID and the network broadcast address, respectively.

### Class C Networks

The first three bits of the first octet of a Class C IP address are 110. Therefore, Class C network IDs range from 192.0.0.0 to 223.255.255.0. There can be $2^{21} = 2,097,152$ possible Class C networks, each with $2^8 - 2 = 254$ possible hosts. We subtract 2 because, in each network, x.y.z.0 and x.y.z.255 (where x.y.z identifies the network) are reserved for the network ID and the network broadcast address, respectively.

Table 18–2    Subnet Masks

| Network Class | Subnet Mask (Binary) | Subnet Mask (Dotted Decimal) |
|---|---|---|
| Class A | 11111111  00000000  00000000  00000000 | 255.0.0.0 |
| Class B | 11111111  11111111  00000000  00000000 | 255.255.0.0 |
| Class C | 11111111  11111111  11111111  00000000 | 255.255.255.0 |

## Subnet Mask

Although a computer could determine the network ID and the host ID by examining the first few bits on the first octet of the IP address, another method was developed. This method involves the use of a 32-bit binary number called the subnet mask. When the subnet mask is logically ANDed with the IP address it produces the network ID. The subnet mask for a Class A network must contain all 1s in the first octet and all 0s in the remaining octets. Thus, expressed in dotted decimal notation, the subnet mask for a Class A network is 255.0.0.0. The result of performing the AND operation between a Class A address and the Class A subnet mask is the network ID. For example, for an IP address of 10.25.21.2, if each of the last 24 bits of the IP address is ANDed with one of the last 24 bits of the subnet mask, the result is a 0 for each bit. Table 18–2 lists the subnet masks for Class A, B, and C networks.

## 18-5    Internet Browsing

To access a Web site across the Internet, a user starts a Web browser (such as Netscape Navigator or Internet Explorer). The address of the target Web site (its URL) is typed into the browser and the enter key is struck. The browser passes the URL (along with other information, including the type and version of the browser) to the HyperText Transfer Protocol (http)—see Figure 18–8. This protocol performs all application, presentation, and session layer functions for connecting to a Web site. These functions include ensuring that all information needed by the Web site to communicate with the client have been provided, and that the information is in the proper format. Http is also responsible for establishing, maintaining, and terminating the connection to the Web server when the exchange of information is complete. The session that is established may consist of many transmissions in both directions, and may last for a protracted length of time.

Http will initiate a communication with the Web site by passing a PDU to the Transmission Control Protocol (TCP), which identifies the name of the Web server. TCP will

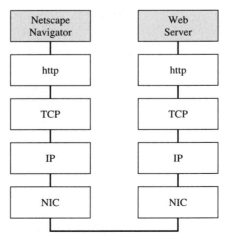

**FIGURE 18–8**
Internet browsing example.

assign a *port number* to the session, so that it knows to what http process to send replies. The PDU, along with the TCP port number, is then passed to the Internet Protocol (IP). IP must convert the Web server name to an address that can be used to locate the computer hosting the Web site on the Internet (an IP address), using another protocol called the domain name service (DNS). IP must also determine how the packet is to be routed to the Web server by analyzing the IP address. IP attaches a header to the PDU containing the source and destination IP addresses and the source and destination TCP port numbers. The PDU is passed to a protocol at the data-link layer that adds a header containing the source and destination MAC addresses (which identify the local NIC and the NIC to which the resulting frame will be sent). The NIC driver then causes the NIC to convert the data to a series of pulses on the media.

At the destination Web server the process is reversed. The information will be passed from the NIC driver through the IP and TCP protocols to http on the server. The server's http protocol will negotiate with the client's http protocol to establish the communications session between the two computers. Once the session is established, information is transferred between the client's browser and the server's Web service. When all communication is complete, the http protocols in the two computers will terminate the session.

It should be noted that http on the local computer is being used by an application on the local computer to provide communications with applications on the remote computer. On the remote computer, http provides access to applications that perform the necessary functions. For example, the purpose of the communication may be to query a database on the remote computer. A client application on the local computer formats a query and passes it to the local http. The local http passes the query to the remote http. The remote http passes it to an application on the remote computer that can query the database and return the desired information. The response is formatted by the remote application and passed to http which transmits it to the local http. The local http passes the response to the application which initially prepared the query.

During this session, http, TCP, and IP have very different responsibilities. When IP passes a packet to a data-link layer protocol its responsibility is ended. IP simply sends a packet and forgets it. TCP, on the other hand must remember the packet long enough to ensure that the packet reaches its destination. Once TCP on the receiving computer has acknowledged receipt of the packet, TCP can forget it. TCP is only responsible for one transmission at a time. Http, on the other hand, is responsible for all of the transmissions necessary to complete a transfer of information and must ensure that the requirements of the local application are satisfied.

## 18-6    Voice over IP

Originally, TCP/IP was used just to transfer textual information between computers, mainly because transmission speeds were very slow. As transmission speeds increased and computer power became greater, methods were developed to transfer larger files, such as graphics. Finally, data communication speeds and computer power increased enough to make the transfer of real-time audio and video information feasible. With the right software and hardware installed, it is possible for a computer user with a high-speed Internet connection to connect across the Internet to any other similarly equipped computer user, and exchange audio and video signals in real time.

There are three phases to such an interchange of signals. First, an appropriate connection must be established between the two computers, so we have an *initialization phase*. Second, the actual signals must be exchanged, so we have an *execution phase*. Finally, the connection between the two computers must be disconnected, so we have a *disestablishment phase*. On each of the endpoint computers, client software must be operating to accomplish each action that is required in each phase. Just as we used a combination of protocols to browse the Internet, a combination of protocols is used to accomplish this real-time audio and video exchange. In this section we will examine some of the protocols that can be used for this interchange of signals.

Voice over IP (VoIP) is often referred to as a substitute for placing telephone calls across the public switched telephone network. If I can use my microphone and speaker equipped computer to connect to a friend's similarly equipped computer across the country, I can talk to that friend without having to pay long distance telephone fees. There are service providers who provide such Internet voice services and vendors who manufacture and sell telephones that are intended for such Internet communications. However, we are concerned with how such interchanges of signals can be accomplished, not what services are available.

## The Initialization Phase

Some aspects of initialization are common to all Internet connectivity, and will not be considered in any detail here. For example, to connect to another computer across the Internet, the IP address of the other computer must be known. We will assume that the IP address is known and it is accessible from the Internet. A protocol or combination of protocols is needed to properly configure the connection between the two endpoint computers to support the transfer of audio and video signals. The Internet Engineering Task Force has developed a protocol called the Session Initialization Protocol (SIP), documented in RFC 3261, which provides the necessary request/response transaction model for establishing the connection. In Figure 18–9, adapted from RFC 3261, Alice in Atlanta wants to talk to Bob in Biloxi. In this example, Alice and Bob are using a service that provides each with a SIP address, similar to an e-mail address, and the connection is made with the aid of the Altanta.com proxy server and the Biloxi.com proxy server, which are part of that service.

Since Alice doesn't know Bob's IP address or the address of the Biloxi.com proxy server, Alice's computer sends an INVITE packet to her local (Atlanta.com) proxy server, transaction [1], using Bob's SIP address, Bob@Biloxi.com. The Atlanta.com proxy server sends an INVITE packet to the Biloxi.com proxy server, transaction [2], and also responds to Alice's computer with a 100 (trying) packet, transaction [3]. The Biloxi.com proxy server, which does know Bob's IP address, sends an INVITE packet to Bob's computer, transaction [4], and responds to the Atlanta.com proxy server by sending a 100 (trying) packet, transaction [5]. Bob's computer will respond with a 180 (ringing) packet, transaction [6], which is passed on by both proxy servers to Alice's computer, transactions [7] and [8]. If Bob is available and willing to talk to Alice, his computer sends a 200 (OK) packet to his local proxy server, transaction [9], which is relayed to Alice's computer, transactions [10] and [11]. Alice's computer then sends an ACK packet directly to Bob's computer to complete the initialization.

The SIP protocol does not limit how the actual transfer of signals between the computers takes place, and therefore does not include all of the information necessary to establish the connection in elements of its protocol. What it does is include the details of the

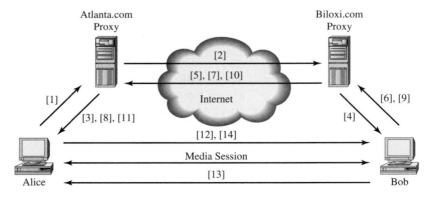

**FIGURE 18–9**
Voice over IP example.

session, such as the type of media or the sampling rate, formatted by another protocol (such as the Session Description Protocol (SDP) of RFC 2327). The SDP message containing the details of the session is, in effect, an attachment to the SIP INVITE packet.

## Execution Phase

Both audio and video involve analog signals, so the first step in transmitting them is to convert them to digital using an analog-to-digital converter. To reduce the bandwidth required, compression using μ-law encoding is often used, as described in Section 9-3. The resulting data stream is broken up into segments and encoded into packets for transmission. These data stream segments are typically about 20 ms in length for voice, which is a compromise between excessive overhead and the disruptive effect of lost packets. Overhead involves adding fields to a packet for control and routing purposes. Those additional fields are the same length regardless of the length of the audio data in the packet. Therefore, half as much overhead is involved if 20 ms audio segments are transmitted as if 10 ms audio segments are transmitted. On the other hand, if a packet doesn't reach its destination, loss of a 10 ms audio segment is less disruptive than loss of a 20 ms audio segment. However, 20 ms is short enough that loss of an occasional packet is not likely to be noticed.

Once the connection is established between Alice's computer and Bob's computer, other protocols control the transfer of the signals between the computers. Commonly this is a combination of the Real-Time Transport Protocol (RTP), RFC 3550, with UDP and IP. RTP provides services such as payload type identification, sequence numbering, timestamping, and delivery monitoring, needed for delivery of real-time data such as audio and video. The main feature of real-time data is that it must be delivered with the right timing and in sequence in order to be intelligible. Ideally, that means delivering the packet to its destination immediately. There are a number of factors in packet delivery over the Internet that make this difficult to accomplish. Packets can be lost or corrupted in transit. Congestion on the network can cause packets to be delayed. Even without losses and delays, the length of time it takes a packet to reach its destination is going to be measured in milliseconds. That transit time, known as *latency,* can be tolerated as long as all packets take the same length of time to reach their destination.

If packets take different lengths of time to reach their destination the audio can be broken up or distorted. This difference in the transit time of the packets is called *jitter*. Assume: (1) packet A contains the first 20 ms of Alice's comments to Bob and takes 50 ms to reach Bob's computer; (2) packet B contains the next 20 ms of Alice's comments and takes 120 ms to reach Bob's computer; (3) packet B contains the third 20 ms segment and takes 40 ms to reach Bob's computer. Therefore, the order of arrival of these packets at Bob's computer is A, C, and B, which results in the audio being scrambled. The sequence numbering provided by RTP can enable Bob's computer to put the packets back in the right sequence. The timestamps provided by RTP can enable Bob's computer to determine how bad the jitter is and how much buffering is required to faithfully reproduce the audio.

To mitigate the problem created by jitter, packets are placed in a buffer when received, instead of being played back immediately. While in the buffer, they can be placed in the right sequence and the playback delayed enough that all packets can be played back in the right sequence and with the correct spacing. A determination of the amount of jitter helps determine how large the buffer must be, and how much delay should be induced before playback.

UDP is used as the transport protocol because it is fast and involves minimum overhead. Although there is no guarantee that every packet will reach its destination (as there is with TCP), VoIP can tolerate the occasional lost packet. In fact, if TCP were used instead, more buffering would be required because of the delays resulting from the TCP error-recovery process. With TCP, if a packet is lost or corrupted in transit it must be retransmitted. This could increase the jitter significantly.

The VoIP process can thus be summarized as follows (although we use audio in the summary, transmission of video is accomplished the same way). The application software on the sender's computer applies the output of the microphone or telephone to an analog-to-digital converter. The resulting digital data stream is broken up into short segments by this application, and transferred to the RTP program module on the sender's computer. RTP adds a header to the segment, which contains the information needed to properly sequence the segment on the receiver's computer and to monitor delivery, and delivers the resulting PDU to the UDP program module on the sender's computer. UDP adds a header containing the information needed by UDP on the receiver's computer to get the segment to RTP on that remote computer, and delivers the PDU to IP. IP adds a header containing the necessary routing information, and delivers the packet to the network driver and interface card that will convert the logical 1s and 0s of the packet into electrical (or optical) signals for transmission through the network. At the receiving computer the order of processing is reversed, and the packet is passed up the protocol stack through IP, UDP, and RTP. At each protocol, the header is stripped off and its contents used to process the packet prior to passing up to the next protocol in the stack. Finally, RTP buffers the segments, puts them in the right sequence, and then they are delivered to the sound card or to a telephone for playback.

## 18-7   NetBEUI and IPX/SPX

### NetBEUI

The NetBIOS Extended User Interface (NetBEUI) is based upon the Network Basic Input/Output System (NetBIOS) that was developed for IBM's PC-Net, a basic network product marketed by IBM in the 1980s. Later it became the standard network protocol for the early Microsoft networks. It is a small, fast protocol that is still useful for small networks. Computers on the network are accessed by their MAC addresses; the protocol has no provision for network addresses and therefore is not routable. It is this characteristic that restricts its use to small networks.

NetBEUI just consists of the two protocols, NetBIOS at the session layer and NetBEUI at the transport and data-link layers. In its typical Microsoft Windows implementation, it relies on Microsoft OS components to perform application- and presentation-layer functions. Additional data-link and physical-layer functions are performed by a NIC and its driver. There are no network-layer functions performed, although NetBEUI is sometimes represented as being a transport, network, and data-link layer protocol.

Although NetBEUI is used only with NetBIOS, the opposite is not true. In addition to NetBEUI, NetBIOS can use TCP/IP or IPX/SPX to establish, maintain, and terminate its network connections. Computers on a NetBIOS network are identified by a 15-character name. Computers on the network periodically broadcast their NetBIOS names and MAC addresses. Each computer on the network maintains a table of these names and MAC addresses for use in communicating with other computers.

### IPX/SPX

IPX/SPX, the protocol suite developed originally for use with Novell's NetWare NOS, consists of a number of protocols, shown in Figure 18–10. Like TCP/IP, its principal upper-layer protocols perform the functions of the application, presentation, and session layers. The name of the protocol suite is derived from the two transport layer protocols, Internetwork Packet Exchange (IPX) and Sequenced Packet Exchange (SPX). IPX provides connectionless communications and SPX provides connection-oriented communications.

Most client/server functions in this suite are performed by the NetWare Core Protocol (NCP). Originally, NetWare provided primarily file sharing and printing services, and NCP provides these services to applications. The Service Advertising Protocol (SAP) is used by file and print servers to advertise their services to other computers on the network. Periodically, SAP packets are broadcast to identify the services available and the address of the computer.

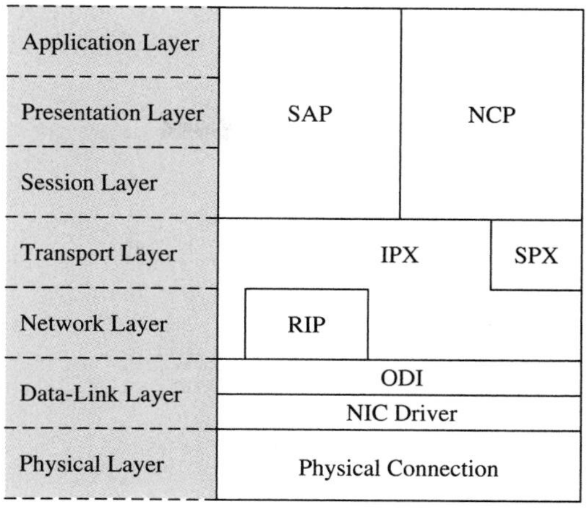

| Application Layer | | |
| Presentation Layer | SAP | NCP |
| Session Layer | | |
| Transport Layer | IPX | SPX |
| Network Layer | RIP | |
| Data-Link Layer | ODI | |
| | NIC Driver | |
| Physical Layer | Physical Connection | |

**FIGURE 18–10**
IPX/SPX protocols.

IPX is the principal transport and network layer protocol used to address and route most IPX/SPX traffic on the network. It provides connectionless communications and is therefore fast but unreliable. IPX relies upon the upper-layer protocol to ensure that reliable communications occur. If greater reliability is required at the transport and network layers, SPX works with IPX to provide connection-oriented communications at the expense of speed.

The Open Data-Link Interface (ODI) provides an interface between IPX and the NIC driver that enables a single network interface card to support multiple protocol suites. For example, it is possible to install both TCP/IP and IPX/SPX on a computer and have both protocol suites access the network through a single NIC. In addition, ODI enables a single protocol suite to access the network through multiple NICs. For example, a server might be providing services to two separate network segments. A separate NIC and its driver is used to access each network segment. ODI enables one or more protocol suites on the computer to connect or bind to either or both NICs.

When IPX/SPX is bound to a NIC, the NIC is configured with an internal network number and an external network number. These correspond to the host address and network address of TCP/IP. It is thus possible to uniquely identify a network segment in an IPX/SPX network as well as a computer on that segment. Network segments can be connected by routers and paths between remote network segments can be identified.

## 18-8    Network Architectures

We can classify a network as a local area network (LAN), a metropolitan area network (MAN), or a wide area network (WAN), based upon the geographical extent of the network. LANs are small, typically contained in a single building. MANs encompass a larger area, perhaps as large as a campus or a city. WANs encompass large geographic areas. The architectures required to provide effective data communications in each of these networks are different. In this section we will examine four of these architectures. Token ring and Ethernet are more appropriate for use on a LAN. FDDI might be used on a MAN, and ATM might be used on a WAN. There are other architectures that could be used in any of these types of network. Also, MANs and WANs will usually use a combination of architectures in different parts of the network. Even in a LAN it is possible to have token ring used on some network segments and Ethernet used on others.

### Token Ring

A major difference between network architectures relates to how a computer gains access to the network. In token ring, access to the network media is obtained by possessing the

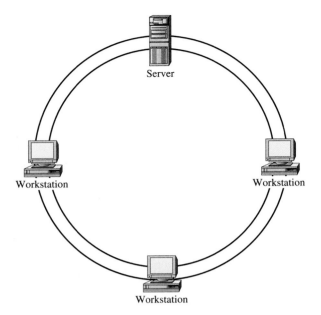

**FIGURE 18–11**
Token ring network.

*token.* When the network first starts up, a special packet called a token is created. The topology of the network, its physical layout, is in the form of a ring, as shown in Figure 18–11. The computers are connected to the ring. The first computer that starts on the network creates the token and passes it to the computer next to it. If that computer has no traffic for the network it passes the token to the next computer along the ring. On the other hand, if the computer receiving the token does have traffic for another computer on the network it sends the traffic along with the token to the destination. The destination computer sets a flag in the token to indicate that the traffic was received and returns the token to the source computer, enabling it to send more traffic. When that computer is finished sending its traffic it forwards the token to the next computer on the ring. The IEEE 802.5 standard defines the operation of this architecture. This was a popular architecture earlier in the history of networking, but Ethernet has largely replaced it as the major LAN architecture.

## Ethernet

Ethernet uses a *bus topology,* as shown in Figure 18–12, although most modern implementations give the appearance of a star, because of a central hub or switch to which all

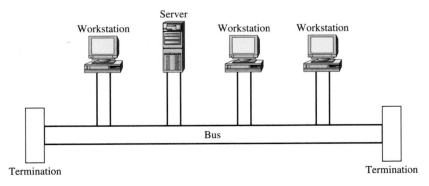

**FIGURE 18–12**
Bus-based network.

computers are connected. A packet inserted into the bus at any point will travel in both directions toward the end of the bus. Terminators at each end of the bus prevent the packet from being reflected back along the bus when it reaches the end. Access to the media is accomplished using Carrier Sense Multiple Access with Collision Detection (CSMA/CD), as defined in the IEEE 802.3 standard. If a computer has data to send, it listens to the line to see if there is already traffic on the bus. If it doesn't sense any traffic on the line it sends its data. If another computer is using the bus, the computer with data to send waits a random period of time and listens again. If two computers send packets at the same time, a collision occurs. Computers monitor the channel while data is being sent, and both computers will detect the collision because the signal strength on the line more than doubles when a collision occurs. Both computers will wait a random period of time after a collision before trying to send their data again.

Originally, Ethernet operated at 10 Mbps (Megabits per second), and there are four different implementations of 10 Mbps Ethernet, depending upon the media used. *10Base5* uses thicknet coaxial cable, *10Base2* uses thinnet coaxial cable, *10BaseT* uses unshielded twisted-pair (UTP) cable, and *10BaseF* uses fiber-optic cable. In these designations, the first number represents the speed of transmission in Mbps, "Base" means that baseband transmission is used, and the last part indicates the media type. While "F" for fiber-optic and "T" for twisted-pair are evident, 5 for thicknet and 2 for thinnet are not. Thicknet cable supports a maximum transmission distance of 500 meters and thinnet supports a maximum transmission distance of 200 meters. More details about coaxial cables are included in Section 18-9.

Currently there are Ethernet standards for transmission speeds of 10 Mbps, 100 Mbps, and 1000 Mbps. The 100 Mbps and 1000 Mbps standards support only twisted-pair and fiber-optic media types.

Ethernet places data on a network in one of four possible frame types. The frame type used normally depends upon the protocol suite involved. Older versions of NetWare over IPX/SPX use the Ethernet 802.3 frame. Later versions of NetWare that still use IPX/SPX normally use the Ethernet 802.2 frame. EtherTalk (for Apple computers) and mainframes normally use the Ethernet SNAP frame. TCP/IP uses the Ethernet II frame. Although the information contained in the frames is largely the same, the frames are not compatible; two computers using Ethernet must use the same frame type to communicate.

Figure 18–13 shows the structure of the Ethernet II frame. The preamble identifies the start of the frame and enables the receiving computer to synchronize with the transmission. The last byte of the preamble is the Start of Frame Delimiter (SFD), a single byte containing 10101011. The destination and source addresses are the respective MAC addresses. The type field identifies the type of data that is in the packet (for example, TCP). The data field contains the upper-layer headers and the actual data being transmitted. Finally, the CRC field contains the results of applying the CRC-32 error-checking algorithm to the data, and is used for integrity verification.

For transmission, the logical 1s and 0s are encoded using Manchester encoding. Two conflicting definitions of Manchester encoding exist. One defines Manchester encoding as identical to biphase-L, shown in Figure 9–7, with a high-to-low transition at the center of the bit representing a 1 and a low-to-high representing a 0. The other, shown in Figure 18–6, which is used in the IEEE 802.3 standard for the CSMA/CD access method, specifies a 1 as having a low-to-high transition at the center of the bit and a 0 as having a high-to-low transition. The presence of a transition at the center of every bit time makes Manchester encoding self-synchronizing. The receive clock simply needs to lock on to those transitions to remain synchronized.

| Preamble 8 Bytes | Destination Address 6 Bytes | Source Address 6 Bytes | Type 2 Bytes | Data (Variable) 46–1500 Bytes | CRC 4 Bytes |
|---|---|---|---|---|---|

**FIGURE 18–13**
Ethernet frame.

## FDDI

The fiber distributed data interface (FDDI) uses a pair of counter-rotating rings and the token-passing media access method with fiber-optic cabling to provide 100 Mbps transmission speeds over distances up to 100 kilometers (60 miles). The two rings transmit data in opposite directions, and one, the *primary ring,* is normally used for data transmission. The *secondary ring* is intended as a backup in case the primary ring fails.

Several factors improve the transmission speed of FDDI. Unlike token ring, when a station on an FDDI network receives the token it does not send the token with each data packet and then wait for the token to circumnavigate the ring and return to it before it can send additional packets. FDDI uses an algorithm to determine when it is safe to send the next packet without causing interference, so multiple packets can be on the ring simultaneously. When a computer finishes sending data, it does not have to wait for the token to return to it around the ring; it can pass the token to its downstream neighbor immediately. Also, it is possible to use both rings to transmit data, effectively doubling the transmission speed.

Individual computers or other devices in the FDDI network connect directly to one or both rings without the use of hubs or other central components. Fault tolerance is built into the system. A computer should receive periodic packets from its upstream neighbor within a certain timeframe. If communication with the upstream neighbor fails on the primary ring, the computer switches communication to the secondary ring.

FDDI is normally used as a backbone for interconnecting networks that use other architectures, such as Ethernet or token ring. At the time that FDDI was developed, most current architectures operated at speeds of 10 Mbps or less. With the advent of high-speed Ethernet and other high-speed architectures, FDDI is less important than it was.

## ATM

Most network architectures use variable-length frames to transmit data. The asynchronous transfer mode (ATM) uses small, fixed-size, 53-byte cells, as shown in Figure 18–14, to transfer data over dedicated circuits. Each cell contains a 5-byte header, shown in Figure 18–15, that identifies the circuit, and 48 bytes of data. At the beginning of an ATM session, a dedicated circuit is established between two end systems. The physical network in which this dedicated circuit is established is that of some large communication company, such as AT&T.

Its fixed cell size enables ATM to rely heavily upon hardware (switches) rather than computer logic to route the cells over the appropriate circuit. As a result, ATM can achieve transmission speeds appropriate for the efficient transfer of real-time data such as audio, video, multimedia, teleconferencing, and so forth.

| Header 5 Bytes | Payload 48 Bytes |
|---|---|

**FIGURE 18–14**
ATM cell.

| GFC 4 Bits | VPI 8 Bits | VCI 16 Bits | PTI 3 Bits | CLP 1 Bit | HEC 8 Bits |
|---|---|---|---|---|---|

**FIGURE 18–15**
ATM header.

## 18-9   Media

### Coaxial Cable

Several networking architectures have standards for coaxial cable, but the same cable and connector types cannot always be used for more than one architecture. For example, Ethernet uses 50-ohm coaxial cable, and ARCnet, an older, slower technology, uses either 93-ohm or 75-ohm coaxial cable. Only the Ethernet standards will be considered here.

As mentioned earlier, two Ethernet standards for transmission of data over coaxial cable exist, 10Base2 and 10Base5, also known as thinnet and thicknet. There are no Ethernet specifications for speeds greater than 10 Mbps using coaxial cable. The characteristics of 10Base2 and 10Base5 are compared in Table 18–3. Figure 18–16 shows a segment of a typical thinnet (10Base2) network. The coaxial cable forms a linear bus, which must be terminated at each end with a 50-ohm load to prevent reflections when a signal reaches the end of the cable. Connections to the computers are made with BNC T connectors. As shown in Figure 18–16, two lengths of coaxial cable are connected to the ends of the T connector to provide continuity of the bus. The third part of the T connector is connected to a male BNC connector on the computer's network interface card. Thus, the T connector is connected directly to the computer. This means that lengths of coaxial cable run from the back of one computer to the backs of two other computers on the network. Although it is possible to run the coaxial cable through the walls and ceilings of a building to wall plates with BNC connectors, and then run lengths of coax from the wall plate to a computer, this is an awkward way to network computers in an office environment. Thicknet uses a different method of connecting computers to the coaxial cable, but its method is also awkward.

**Table 18–3**   Comparison of Thinnet and Thicknet

| Characteristic | 10Base2 | 10Base5 |
|---|---|---|
| Maximum cable length | 185 meters | 500 meters |
| Bandwidth | 10 Mbps | 10 Mbps |
| Topology | Linear bus | Linear bus |
| Maximum number of segments | 5 | 5 |
| Maximum number of repeaters | 4 | 4 |
| Maximum number of populated segments | 3 | 3 |
| Maximum number of devices per segment | 30 | 100 |
| Maximum number of devices per network | 1024 | 1024 |
| Connector type | BNC | BNC |
| Bend radius | 360 deg/ft | 30 deg/ft |
| Termination | Required | Required |
| Cable type | RG-58 A/U | RG-8 |
| Cable diameter | ~0.25 inch | ~.4 inch |
| Immunity to electrical interference | Very good | Excellent |

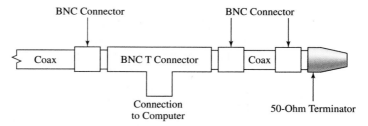

**FIGURE 18–16**
Segment of 10Base2 network.

A disadvantage of either implementation using coaxial cable is that if a break in the cable occurs, the entire network is down. For example, if someone disconnects the coax from the T connector on the back of a computer to move the computer, the entire network is disabled. If the T connector is removed from the back of the computer but left connected to the coax, the computer can be moved without disrupting network operations.

The maximum cable lengths for thinnet and thicknet listed in Table 18–3 result from attenuation of the signal as it travels along the coaxial cable. As the signal is propagated along the cable, each NIC that it passes extracts a small amount of energy from the signal in order to read the MAC address of the packet. In addition to this loss of energy, additional energy is dissipated in the resistance of the cable. The result is that after a signal has traveled far enough along the cable it is not strong enough to be reliably read by a NIC. The maximum cable lengths specified are selected to ensure that signal strengths are still strong enough for reliable communications.

It is possible to extend the length of a network beyond that dictated by the maximum cable length. If a repeater is used to regenerate the signal before it is too weak for reliable communication, it can be propagated another maximum cable length. There is a limit, however. This limit is defined in the 5-4-3 Rule, which states that there can be no more than 5 network segments, separated by no more than 4 repeaters, and no more than 3 of these segments can be populated by computers. This rule applies to both thinnet and thicknet, and to combinations of thinnet and thicknet.

---

**▌▌ EXAMPLE 18-3**

Thinnet is to be used to network the computers in each of three computer laboratories in a university building. There will be eighteen computers in each laboratory in three rows of six each. The computers are five feet apart in each row and there is four feet from the back of the computer table in each row to the front of the computer table behind it. The computer tables are thirty inches deep. In addition to networking the computers in each laboratory, the three laboratories are to be connected into a single network. The closest point on any two laboratories is one thousand feet apart. How can this be accomplished?

**SOLUTION**   Thinnet has a maximum cable length of 185 meters or approximately 600 feet. This is more than enough to network the computers in each computer lab, but not enough to join the labs. Also, the eighteen computers to be networked in each lab is less than the limit of thirty per network segment for thinnet. Therefore, each computer lab can be networked, individually, as a network segment, using thinnet. Thicknet has a maximum cable length of 500 meters, or approximately 1640 feet, which is enough to connect two labs. A length of thicknet can be run from Lab 1 to Lab 2. Repeaters can connect the thinnet networks of the two labs to the thicknet cable. A second length of thicknet cable can be used to connect Lab 2 to Lab 3. The 5-4-3 Rule is satisfied: there are five segments, three of thinnet and two of thicknet; four repeaters are used; and only three of the segments are populated with computers.                                                                     ▌▌

---

## Twisted-Pair Cable

Twisted-pair cable is just two pieces of copper wire twisted together, like ordinary telephone cable. The twisted pair may have a foil shield surrounding it (STP) or it may be left unshielded (UTP). Typically, in computer network applications, cables are made up of 2, 3, or 4 pairs of wires. As is the case for coaxial cable, different networking architectures have different specifications for twisted-pair cable. For example, token ring uses STP and Ethernet uses UTP. Unlike coaxial cable, specifications do exist for twisted-pair cable at speeds greater than 10 Mbps.

The 100BaseTX specification has largely replaced the 10BaseT specification as the most common Ethernet cabling method. Some characteristics of the 10BaseT and the 100BaseTX specifications are shown in Table 18–4. Twisted-pair cable is classified by the Category

Table 18–4   Comparison of 10BaseT and 100BaseT Ethernet Specifications

| Characteristic | 10BaseT | 100BaseTX |
|---|---|---|
| Cable type | Category 3, 4 or 5 UTP | Category 5 UTP |
| Maximum cable segment length | 100 meters | 100 meters |
| Maximum number of segments | 1024 | 1023 |
| Maximum number of devices per segment | 2 | 1 |
| Maximum number of devices per network | 1024 | 1024 |
| Transmission speed | 10 Mbps | 100 Mbps |

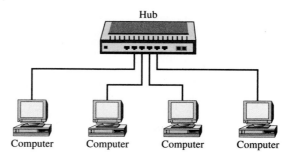

**FIGURE 18–17**
10BaseT network.

designation. Categories 3, 4, and 5 are used for Ethernet, with Category 5 being the highest quality. The principal difference between categories is the number of twists per inch, and Category 5 has the highest number of twists per inch.

Twisted-pair cabling is popular because it is inexpensive and easy to install. In a typical Ethernet installation, the cables are run from wall plates in the rooms where networked computers will be used to a central location. A hub is installed in the central location to which all of the cables are attached (see Figure 18–17). Thus, the network topology is physically that of a star. However, the ports in the hub are connected to form a logical bus, which is the standard arrangement for Ethernet. A single cable is run from a connector in a wall plate to a connector in a NIC in a computer. The twisted-pair cables can be disconnected from the hub, or from the wall plate, or from the computer, without disrupting network operations.

The hubs used in Ethernet networks normally are powered, and have repeaters between each pair of ports. Therefore, the maximum segment length shown in Table 18–4 is normally the maximum distance between the hub and a computer on the network. Hubs in the network can be connected with twisted-pair cable if the maximum segment length is not exceeded. If the distance between hubs exceeds 100 meters, coaxial cable can be used to connect hubs in 10BaseT networks. If the distance between hubs exceeds the maximum segment length for coaxial cable, fiber-optic cable can be used at any of the Ethernet speeds.

## Fiber-Optic Cable

Specifications exist for fiber-optic cable for Ethernet as well as for FDDI and other network architectures. Some characteristics of common Ethernet fiber-optic standards are listed in Table 18–5. Although 10BaseF could be used to connect computers to a hub, until recently twisted-pair cabling has been much less expensive, and it is easier to install. Therefore, fiber-optic cables have been mainly used to interconnect remote parts of a network rather than to connect to individual computers. However, with optical fiber becoming less expensive, and with Gigabit Ethernet becoming more common, that is changing.

**Table 18–5**   Comparison of 10BaseF and 1000BaseLX Ethernet Specifications

| Characteristic | 10BaseF | 1000BaseLX |
|---|---|---|
| Cable type | Fiber-optic | Fiber-optic |
| Maximum cable segment length | 2000 meters | Up to 5000 meters |
| Maximum number of segments | 1023 | 1023 |
| Maximum number of devices per segment | 2 | 2 |
| Maximum number of devices per network | 1024 | 1024 |
| Transmission speed | 10 Mbps | 1000 Mbps |

### Wireless

It is possible to interconnect network devices without the use of cables. Two technologies are commonly used, both of which will be the subject of Chapter 19. Infrared technology has the advantage of being inherently more secure, but provides a more limited range, and transmissions are blocked by solid objects. The use of radio transmissions has become more prevalent, and several IEEE standards exist for this technology.

## 18-10   Other Devices

### Physical Layer Devices

The physical layer of the OSI model is also known as *Layer 1*. Therefore, the devices discussed in this section can be referred to as physical layer devices or as Layer 1 devices.

*Network interface cards* provide the circuitry and connectors necessary to translate the logical 1s and 0s used in the protocols within the computer to the electrical or light pulses that travel through the media connecting the network devices. The primary component of a NIC is a *transceiver,* which is a device that does the actual conversion of the data.

*Repeaters* are used to amplify signals that have become attenuated and distorted as they travel along a cable. As a signal travels along a cable, eventually it is so degraded that transceivers are no longer able to reliably recognize the data represented by the signal. In analog circuits, amplifiers are used to restore the signal strength of attenuated signals. Unfortunately, amplifiers are unable to distinguish between the signal and the noise that has been added to it as it propagates. In computer networks, *repeaters* perform the function of restoring digital signals. Unlike an amplifier, which simply increases the amplitude of the input, a repeater interprets the input signal and generates a new signal at the output that duplicates the information content of the input, without the attenuation and distortion that was present at the input.

Repeaters have many limitations. Because they simply output the same pulse pattern that they receive, the LAN technologies on both sides of the repeater must be the same. A repeater cannot convert a token ring signal to an Ethernet signal. However, repeaters can convert similar signals from one media type to another. For example, a repeater can be used to convert a 10BaseT signal to a 10Base5 signal for interconnecting two hubs with a coaxial cable. These limitations arise because repeaters operate at the physical layer of the OSI model. The repeater has no ability to identify what is in the signals; all it can do is replicate the shape of the signal. However, once a signal has passed through a repeater, the cable segment on the output of the repeater has the full maximum segment length for that media. For example, when a thicknet coaxial cable is used to join two Ethernet hubs, the hubs can be located 500 meters apart.

Another limitation of the repeater is that it introduces a certain amount of delay between the input and output signals. For example, if a 0 was represented by 0 volts and a 1 was represented by $+5$ volts, the standard might say that anything less than 1 volt is a 0 and anything greater than $+3$ volts is a 1. This provides some latitude for noise and for attenuation. As long as the noise doesn't exceed 3 volts, it won't cause a 0 to be mistaken for a 1 or a 1 to be mistaken for a 0. However, the voltage on a cable cannot change value instantaneously. When changing from a 0 to a 1, it takes time for the voltage to rise to +3 volts, so the output isn't going to start rising until some time after the input starts rising. Each repeater in the network adds to the *propagation delay* of the system. This is the reason that the 5-4-3 Rule allows only four repeaters.

As mentioned earlier, *hubs* are used to interconnect computers and other devices in a network. Hubs can be classified as *active* or *passive,* and the two forms are often used together. An example of a *passive* hub is a punchdown block. This serves as the termination of the UTP cables that go from a central location to the connectors in wall plates where computers are to be installed. The cables connect to the back of the punchdown block, and there is an RJ-45 connector, a port, on the front of the punchdown block, that is then connected internally to each cable. No regeneration takes place in the punchdown block, so no electrical power is applied. Each port on the punchdown block is then connected to a port on an active hub, such as an Ethernet hub. This Ethernet hub has a repeater between each pair of ports, and therefore must have power applied. Furthermore, the circuitry in the Ethernet hub provides the bus structure that is required for Ethernet. Inside the hub, the signal travels from port to port and therefore from computer to computer, just as it would if the media were thinnet or thicknet. So, although the network topology physically is a star, logically it is a linear bus.

In effect, active hubs are multiport repeaters, and are therefore physical layer devices. Although the ports are usually all the same type, that is not necessary. A hub can have a number of ports for UTP connectors, usually 8, 12 or 24, and in addition have one or more ports for BNC or fiber-optic connectors.

## Data-Link Layer Devices

The data-link layer of the OSI model is also referred to as *Layer 2* of the model. Therefore the devices discussed below are often called Layer 2 devices.

*Bridges* can also be used to connect segments of a network, instead of using repeaters. However, bridges operate at the data-link layer or Layer 2 of the OSI model, in addition to the physical layer. They have the ability to read the data-link layer header on the packets that they receive, which contains the source and destination MAC addresses. Most bridges have the ability to build a *bridging table,* a list of the MAC addresses of the computers on each segment attached to the bridge. When a packet is received, the bridge compares the source and destination MAC addresses. If they are on the same segment, the bridge discards the packet. Delivery can be completed without any action on the part of the bridge. If the source and destination MAC addresses are not the same, the bridge attempts to find the destination address and the corresponding segment in its bridging table. If it succeeds, it forwards the packet on the appropriate segment. If it cannot find the destination address in its bridging table, it sends the packet to all attached segments, except the one from which it received the packet.

Instead of addressing packets to specific computers, broadcasts are often used. For example, if a computer in a TCP/IP network doesn't know the MAC address of a destination computer, it sends an ARP packet addressed to all of the computers on the network. This packet identifies the IP address of the destination computer and asks that computer to reply with its MAC address. In a NetBEUI network, NetBEUI sends a broadcast giving the NetBIOS name of the destination computer, which requests the computer with that name to reply with its MAC address. When a bridge receives a broadcast packet, it delivers the packet to all network segments to which it is connected.

The use of bridges in networks makes networking more efficient. Instead of having all network traffic on a single segment on which only one computer can transmit at a time, bridges allow the network to be separated into several segments. One computer on each segment can be transmitting at the same time. By properly designing the network so that computers that communicate with each other frequently are on the same segment, network efficiency can be enhanced. For example, if a particular client frequently accesses files on a specific server, the client and the server should be on the same segment. That way the traffic between them is isolated to that segment and doesn't interfere with traffic between computers on other segments.

Although most bridges are used to join segments using the same network architecture, such as Ethernet, it is possible to use bridges to translate between network architectures. Thus, a *translation bridge* could be used to connect an Ethernet network to a FDDI network, or a FDDI network to a token ring network. The FDDI network might be a backbone joining several Ethernet or token ring networks. This ability relies upon data-link layer protocols in the bridge. However, because the bridge also operates at the physical layer, it can link dissimilar media, just like a repeater.

*Switches* are similar to hubs but more intelligent. A hub connects an incoming signal to all ports, so that all attached computers receive the signal, whether or not the signal is meant for them. A switch is able to identify the destination of the signal, and connect the port of the source computer to the port of the destination computer. Using a hub, the available bandwidth of the network is shared by all of the computers on the network. Using a switch, the entire available bandwidth can be made available to the two computers involved in the communication session. Of course, when it is appropriate, a switch can apply the incoming signal to all of the ports, just as a hub does, sharing the bandwidth among all connected computers.

Switches can also be configured to route transmissions to selected groups of computers. Each group of computers, called a *virtual LAN (VLAN),* operates as though it were a separate local area network.

## Network Layer Devices

Network layer devices are also known as *Layer 3* devices, because the network layer is the third layer of the OSI model.

The most common network layer devices are *routers.* Protocols at the network layer are able to work with the logical addresses of computers on remote networks in order to provide access to resources on more complex internetworks. The ultimate example of this is the Internet.

Figure 18–18 shows an internetwork made up of many networks connected by routers. Routing protocols on each of the routers, operating at the network layer of the OSI model, develop *routing tables* in each router that list all of the networks and the possible paths to the networks from the router. For example, a routing table would list four paths to Network 5 from Network 1: through router A, router B, and router D; through router A, router C, and router E; through router F and router E; and through router G. The table would also indicate how many routers a packet would have to go through to get from Network 1 to Network 5. The number of routers that must be traversed is known as the *cost* and is measured in *hops.* Therefore, the path through Router G would have a cost of one hop and would be the most efficient path. The path through router F and router E would have a cost of two hops and be the next most efficient path. The other two paths would each have a cost of three hops. Routing protocols select the path with the least cost.

The IP address can be separated into a network address and a host address. For example, 100.218.119.41 might represent a network address of 100.218.0.0 and a host address of 0.0.119.41. If a router received a packet destined for that IP address, the packet would be passed up the protocol stack to the network layer protocol of the router. There, the destination network address in the network layer header would be read, and an attempt would be made to locate the network address in the routing table. The least-cost path to the destination would be found. One entry in the table for that path is the address of the next router in the path. The packet would be forwarded to the next router. The network layer header of the forwarded packet would contain the original source and destination IP addresses, but the data-link header would contain the MAC addresses of the sending router and the next

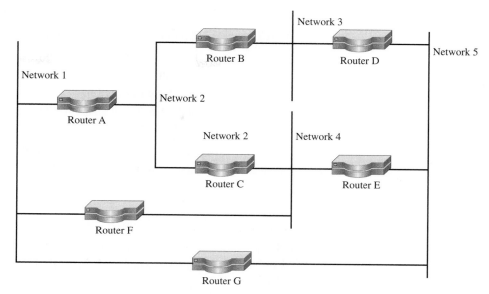

**FIGURE 18–18**
Routed network.

router in the path. The packet would be sent on the network segment that is common to the two routers. For example, in Figure 18–18, if router C was sending a packet to a computer on Network 5, the packet would be placed on Network 4, using the MAC address of router E.

If the router in the above example received a packet with a network address it could not locate in its address table, it would discard the packet rather than forward it. This is one of the differences between how a router handles traffic and how a bridge handles traffic. Another difference involves how broadcasts are handled. Bridges pass broadcasts but routers discard them. This makes sense if you consider the two common uses for broadcasts mentioned in the discussion of bridges. In a routed network, MAC addresses are only useful in the network segment in which the computer with the MAC address exists. Routers do not use MAC addresses to determine where to send packets. Therefore, to forward a broadcast requesting a MAC address from one network to another in a routed network simply adds noise to the traffic on the network.

*Brouters* are special devices that combine the capabilities of bridges and routers. If a brouter receives a packet of a routable protocol, it selects the best path for it to its destination as a router would. On the other hand, when a brouter receives a packet of a nonroutable protocol, it forwards it based upon the destination MAC address as a bridge would. Therefore, brouters maintain both a routing table and a bridging table. A brouter might be used in a network using both TCP/IP and NetBEUI.

## 18-11 Computer Network Acronyms

| | |
|---|---|
| ADSL | Asymmetric Digital Subscriber Line |
| ARP | Address Resolution Protocol |
| ATM | Asynchronous Transmission Mode |
| BIND | Berkley Internet Name Domain |
| BIOS | Basic Input/Output System |
| BRI | Basic-Rate Interface |
| CSMA/CD | Carrier Sense Multiple Access with Collision Detection |

| | |
|---|---|
| CSU/DSU | Channel Services Unit/Data Service Unit |
| DACL | Discretionary Access Control List |
| DLL | Data Link Layer (also Dynamic Link Library) |
| DNS | Domain Name Service |
| DSL | Digital Subscriber Line |
| FDDI | Fiber Distributed Data Interface |
| HTTP | Hypertext Transfer Protocol |
| ICANN | Internet Corporation for Assigned Names and Numbers |
| IEEE | Institute of Electrical and Electronics Engineers |
| IETF | Internet Engineering Task Force |
| IP | Internet Protocol |
| IPX/SPX | International Packet Exchange/Sequence Packet Exchange |
| ISDN | Integrated Services Digital Network |
| ISO | International Organization for Standardization |
| IXC | Interchange Carrier |
| LAN | Local Area Network |
| LATA | Local Access and Transport Area |
| MAC | Media Access Control |
| MAN | Metropolitan Area Network |
| NAT | Network Address Translation |
| NCP | Network Core Protocol |
| NetBEUI | NetBIOS Extended Users Interface |
| NetBIOS | Network Basic Input/Output System |
| NIC | Network Interface Card |
| ODI | Open Data-Link Interface |
| OSI | Open System Interconnect |
| OSPF | Open Shortest Path First |
| PIN | Personal Identification Number |
| POP | Point of Presence (also Post Office Protocol) |
| PRI | Primary-Rate Interface |
| PSTN | Public Switched Telephone Network |
| RFC | Request for Comments |
| RIP | Routing Information Protocol |
| SAP | Service Advertising Protocol |
| SDSL | Single-Line Digital Subscriber Line (also Symmetric Digital Subscriber Line) |
| SMTP | Simple Mail Transfer Protocol |
| STP | Shielded Twisted Pair |
| TCP/IP | Transmission Control Protocol/Internet Protocol |
| TLD | Top-Level Domain |
| URL | Uniform Resource Locator |
| UTP | Unshielded Twisted Pair |
| WAN | Wide Area Network |
| WWW | Worldwide Web |

## SystemVue™ Closing Application (Optional)

Insert the text CD in a computer having SystemVue™ installed and activate the program. Open the CD folder entitled **SystemVue Systems** and load the file entitled 18-1.

### Description of System

The system shown on the screen is the same one that was used in the chapter opening application. In this exercise, however, you will change the phase of the clock to see what effect that has on the waveforms.

First, run the simulation by pressing F5 or left-clicking the **Run System** button. Note that the clock pulse in W1 starts HIGH and then goes LOW. Then, right-click Token 4 and select **Edit Parameters** from the menu. The **Pulse Train (Token 4)** window will open. Change the entry in the **Phase (deg)** text box from 0 to 180 and click on OK. Then run the simulation by pressing F5 or left-clicking the **Run System** button.

When you open the Analysis window, if the **Load New Sink Data** (Ctrl-N) button is blinking at the left end of the tool bar, you will need to left-click that button to refresh the display. With the display current you should see that the clock has changed from starting from HIGH to LOW to starting from LOW to HIGH. You can also see in W2 that the output waveform has changed. A LOW input is now encoded with a LOW-to-HIGH transition at the midpoint of the pulse, and a HIGH input is now encoded with a HIGH-to-LOW transition at the midpoint of the pulse.

## PROBLEMS

**18-1** Two college roommates decide to purchase a printer together and connect it to their computers. They buy an AB switch that consists of three parallel connectors and a switch. Two of the connectors, designated port A and port B, are connected to the parallel ports of their respective computers by a suitable cable. The third connector on the AB switch is connected to their printer with a suitable cable. The switch connects either port A or port B to the printer. Is this a peer-to-peer network, a client/server network, or neither? Why?

**18-2** Two roommates each have a laptop computer that they use for coursework. They decide that it would be advantageous if they could share files and share a printer. However, each often takes her laptop to class at times when the other is working in their room. They decide to purchase a third computer to host the shared files and the shared printer. They also purchase a hub, network adapters, and the necessary cabling to connect the three computers. Is this a peer-to-peer network, a client/server network, or neither? Why?

**18-3** Many of the employees in your department create documents that must be printed on a color printer. Currently, this is accomplished by copying the documents to a CD-ROM and taking them off site to be printed. Your boss has identified a high-capacity, color laser printer with a duplexer attachment that he proposes sharing on the department network. He has asked you to help him provide justification to his boss for its purchase, based upon sharing the printer on the network. What do you suggest as justification?

**18-4** A simple encryption method is to substitute one symbol for another, such as the substitution of E for A, F for B, G for C, and so on, in a text message. If this method of encryption were used in a presentation-layer protocol in a server, what information would have to be included in the presentation-layer header to enable the client's presentation protocol to decrypt the data?

**18-5** The physical-layer protocols of a certain NOS can only transmit frames that have a maximum size of 2,000 bytes. Data is passed down through the protocol stack in much larger frames. What layer contains the protocol that is responsible for segmenting each frame of the data into multiple frames?

**18-6** An application provides data to the protocol stack in a character set that cannot be utilized on the server to which the data is being sent. What layer in the protocol stack contains the module that is responsible for converting the character set to one that can be used on the server?

**18-7** What are the characteristics of a connection-oriented transport-layer protocol?

**18-8** What are the characteristics of a connectionless transport-layer protocol?

**18-9** Describe the process by which protocols at the session layer of a client computer communicate with corresponding protocols at the session layer of a server.

**18-10** Describe the process by which protocols at the network layer of a client computer communicate with the corresponding protocols at the network layer of a server.

**18-11** A host has been assigned the IP address 17.42.207.21. What class of address is this, and what is the network ID?

**18-12** A host has been assigned the IP address 199.23.103.198. What class of address is this, and what is the network ID?

**18-13** A host has been assigned the IP address 207.55.28.172. What class of address is this, and what is the network ID?

**18-14** What protocol does IP use to determine the IP address of a host when the host name is known?

**18-15** What protocol does IP use to determine the MAC address of a host, and how does that protocol work?

**18-16** RFCs are available on the Internet, and can be located using Google. Obtain a copy of RFC 3261, *SIP: Session Initiation Protocol,* and list the five facets of establishing and terminating multimedia communications that are supported by SIP.

**18-17** RFCs are available on the Internet and can be located using Google. Obtain a copy of RFC 3550, *RTP: A Transport Protocol for Real-Time Applications.* List the two closely linked parts that make up RTP, and describe the function of each.

**18-18** You are part of a design team that is planning the installation of a network for an organization that is moving into a new building. You have been tasked with determining what type of cabling should be used, and how the installation should be accomplished. The floor that will contain the organization's offices is 1800 feet long and 600 feet wide. One of the members of the team has suggested that Category 5 twisted-pair cable should be used to wire the offices, and that the switches should be placed in a wiring hub in the northeast corner of the building. What problems will be encountered if this plan is followed?

**18-19** You are part of a design team that is planning the installation of a network for an organization that is moving into a new building. You have been tasked with determining what type of cabling should be used, and how the installation should be accomplished. The floor that will contain the organization's offices is 1800 feet long and 600 feet wide. One of the members of the team has suggested that Category 5 twisted-pair cable should be used to wire the offices, and that the switches should be placed in a wiring hub in the center of the floor. What problems will be encountered if this plan is followed?

# Wireless Network Communication

# 19

## OVERVIEW AND OBJECTIVES

Wired networks—those that use coaxial, twisted-pair, or fiber-optic cabling—provide access to shared resources and enable collaborative capabilities, but lack mobility. The computers in a wired network must be relatively fixed in position, because of the constraints of physical connections. Efforts to break these physical connections have involved using either infrared light or radio frequency transmissions as the medium. In this chapter, these techniques will be described, compared, and evaluated.

### Objectives

After completing this chapter, the reader should be able to:

1. Describe the salient features of the major wireless communications techniques.
2. Compare the capabilities of IEEE 802.11a, 802.11b, and 802.11g wireless network communications.
3. Discuss the advantages and disadvantages of different wireless network communications methods.
4. Indicate situations in which each of the wireless network communications methods are most appropriate.

## SystemVue™ Opening Application (Optional)

Insert the text CD in a computer having SystemVue™ installed and activate the program. Open the CD folder entitled **SystemVue Systems** and open the file entitled 19-1.

### Sink Tokens

| Number | Name | Token Monitored |
|--------|------|-----------------|
| 0 | Input | 4 |
| 1 | Barker Code | 5 |
| 2 | DSSS Signal | 6 |
| 3 | Output | 8 |

597

## Operational Tokens

| Number | Name | Function |
|--------|------|----------|
| 4 | Input | generates input bit train |
| 5 | Barker Code | provides code for chipping |
| 6 | XOR | modulates bit train with code |
| 7 | Barker Code | provides code for demodulation |
| 8 | XOR | reconstructs signal |

The system shown on the screen has two *source* tokens on the far left. Token 4 produces a 1-MHz random pulse sequence that is the digital input to the system. Token 5 produces a Barker code sequence, an 11-chip sequence at 11 MHz, 10110111000. It is used as part of the *direct sequence spread spectrum* (DSSS) modulation method for an IEEE 802.11b wireless LAN. Token 6 is an exclusive-OR (XOR) gate used to combine the input and the Barker code. This 11-MHz pulse train is used to modulate a carrier and produces a $\sin x/x$ waveform with zero crossings 11 MHz on either side of the carrier; this is the transmitted signal. Token 7 also produces the Barker code at 11 MHz, and is used in conjunction with another XOR gate (Token 8) to recover the original input pulse train from the received and demodulated transmitted signal. The output of Token 8 is the reproduced input signal.

Sinks 0, 1, 2, and 3 enable us to view the waveforms. Sink 0 shows the input waveform. Sink 1 shows the Barker code waveform from one of the two Barker code generators. The output of Token 7 is identical to the output from Token 5. Sink 2 shows the output of the first XOR gate after the Barker code is applied to the input. Sink 3 shows the output of the second XOR gate after the Barker code is used to extract the input signal.

## Checking the Settings

In this exercise you will not need to modify any settings, but it is instructive to observe the values of the settings. The parameters of each token can be observed by placing the cursor over the token. A window will appear containing the parameters of that token.

## Activating the Simulation

Left-click the **Run System** button or press F5 to start the simulation. A moving blue line at the bottom of the screen indicates that the simulation is running. Left-click the **Analysis Window** button to open the Analysis window. If the windows for the four sinks are minimized, left-click the **Restore Up** button to open the windows, but do not maximize them. With the windows open but not maximized, left-click the **Tile Horizontal** button.

## What You See

You should see four waveforms. W0, the window for Sink 0, displays the input waveform. W1, the window for Sink 1, displays the Barker code. W2, the window for Sink 2, displays the output of the XOR after the input and the Barker chips are combined. Finally, W3, the window for Sink 3, displays the recovered input signal. Since the input is random in nature, it will be different each time the simulation is run. However, the signals in W0 and in W3 should be the same.

If you examine the waveforms of W1 and W2 closely, you should observe that they are identical during the time that the input is low, and complements of each other when the input is high. This phase relationship between the Barker code and the modulating waveform is how the intelligence is incorporated into the transmitted signal, and is why an XOR gate can be used to recover the input pulse train. When the Barker code and the modulating waveform are in phase, the output of an XOR gate will be low, or a "0"; when the two waveforms are out of phase, the output of an XOR gate will be high, or a "1."

After you have completed study of this chapter, you will modify this system to see the effect of using different 11-MHz pulse trains. You will see that the use of the sequence 10110111000 is not critical, as long as the same sequence is used as inputs to both XOR gates. However, you will also see that if the sequences are not the same, the process does not work.

## 19-1 Origins of Wireless Networking

Early attempts at networking computers without using coaxial cable or other wires utilized infrared (IR) transmissions. An IR transceiver consists of a light-emitting diode (LED) optimized for the IR region of the spectrum as the transmitter, and a photodiode or phototransistor as the receiver. A lens is used to concentrate or disburse the IR beam, as desired. A narrow beam width can be used as a link between two devices, or a wide beam width can be used that can be detected by multiple devices. Several problems limit the effectiveness of IR as a medium. At best, the range is very limited, and the transmissions cannot pass through the walls of a room. In fact, any opaque object can block the transmissions. Furthermore, the maximum speed that can be attained with IR is much less than is acceptable for most networking applications, on the order of 100 Kbps; 100 Mbps is the standard speed in wired networks.

In spite of these limitations, IR is used to provide wireless links between devices. Many laptop computers and personal data assistants (PDAs) contain IR ports. These ports are usually compatible with a standard developed by the Infrared Data Association (IrDA), a nonprofit trade association created to define infrared standards. These IrDA ports are used to make wireless connections between computers and print devices, between computers and PDAs for the purpose of backing up PDA data, or between computers to transfer files. Such connections are one-to-one, as illustrated in Figure 19–1, because of the narrow beamwidth involved in the IrDA standard.

Wide IR beams can be used to provide connectivity between multiple computers or devices, as shown in Figure 19–2. One technique that has been used to interconnect multiple devices involves broad beams that are reflected off the ceiling to provide a relatively wide area of coverage. However, these techniques have not experienced much commercial interest.

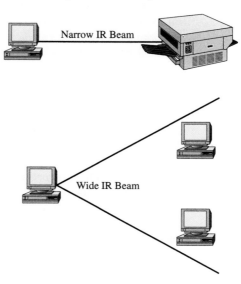

**FIGURE 19–1**
Narrow-beam IR.

Narrow IR Beam

**FIGURE 19–2**
Wide-beam IR.

Wide IR Beam

Attempts were also made to utilize RF transmissions to link network components. Although the technology was successful, each vendor who developed wireless devices used proprietary techniques. Therefore, a network could only be assembled using components from the same manufacturer. The use of RF as the medium was encouraging, however, because the area covered by an RF-based network is larger than that covered by an IR network. Furthermore, walls and other physical barriers do not usually block the transmissions. It was recognized that standards were needed to enable the products of different manufacturers to interoperate, and to lead to reduced costs of components. The Institute of Electrical and Electronic Engineers (IEEE), through its 802 committee, has been instrumental in the development of standards for wireless networks.

▌▌ EXAMPLE 19-1

The IrDA describes IrDA DATA, its standard for data transmission, as "recommended for high-speed short-range, line-of-sight, point-to-point cordless data transfer—suitable for HPCs, digital cameras, handheld data collection devices, etc." Describe how IrDA could be used in a warehouse inventory system.

SOLUTION    One possibility is to have each bin in the warehouse labeled with a bar code. A handheld computer or PDA with an attached bar code reader and appropriate software can be used as a recording device. It reads the bar code on the bin; then the operator types in the number of items in that bin, and presses a key to enter the results into a database in the recording device. Periodically the operator points the IrDA port of the recording device at an IrDA port on a network computer, and transfers files to synchronize the inventory database on a network server with that on the recording device.    ▌▌

## 19-2  IEEE 802.11

The IEEE 802.11 standard, as first approved in 1997, defines three physical (PHY) layers for wireless networking. Other standards in the 802.11 series, such as 802.11a, 802.11b, and 802.11g, add additional PHY layers to supplement this basic framework but do not replace it. In this section we will examine the structure of wireless networking as defined in 802.11. We will consider the provisions of 802.11a, 802.11b, and 802.11g in later sections.

In developing 802.11, an objective of the committee was to make this standard compatible with other standards in the 802 series. The IEEE 802 standards divide the data link of the OSI seven-layer model into two sublayers, logical link control (LLC) and media access control (MAC). They also define different physical layers for various media types. Wireless networking provisions are all included in the MAC sublayer and the PHY layer. By incorporating all wireless network provisions into these two elements of the seven-layer model, compatibility with software and devices based on protocols in the higher layers is assured.

Any device that can participate in a wireless network is called a *station*. This will usually be a computer, or a PDA, containing a wireless network adapter and appropriate software. Two or more stations that communicate wirelessly constitute a basic service set (BSS). If all of the stations in a BSS communicate directly with each other, the BSS is said to be an independent BSS (IBSS), as shown in Figure 19–3. A device that is used to interface wireless devices with a wired network is called an *access point* (AP). An AP can also facilitate communication between other stations. That is, station A can communicate with station B through the AP rather than directly. A BSS that includes an AP is called an infrastructure BSS (there is no acronym for this). Figure 19–4 shows an infrastructure BSS.

The range of stations in a BSS is limited, for two reasons. First, the FCC and other regulatory bodies impose a low limit on the power used by the transmitters in wireless adapters. Second, the antennas used by these stations are usually omnidirectional. Generally speaking, however, an infrastructure BSS will cover a wider area than an IBSS, because the AP acts as a

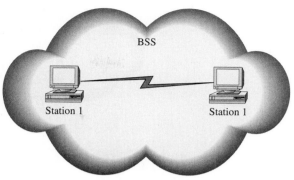

**FIGURE 19–3**
An IBSS wireless network.

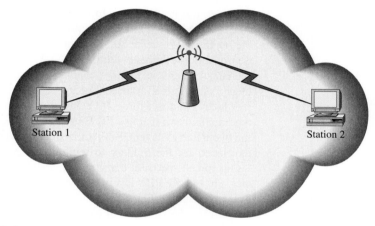

**FIGURE 19–4**
Infrastructure BSS.

**FIGURE 19–5**
Two stations communicating through an AP.

repeater and can enable connections between two stations that can each access the AP, but are too far from each other to be able to connect directly. Figure 19–5 illustrates this concept.

When an AP is connected to a wired network, it forms a bridge between the wireless LAN and the wired LAN. If the wired LAN has other APs connected to it, the wired LAN functions as a *distribution system* (DS) to enable the APs to communicate with, and to forward traffic between, each other. When this occurs, the individual BSSs that exist around each AP become part of an extended service set (ESS). Figure 19–6 illustrates multiple BSSs connected by a DS.

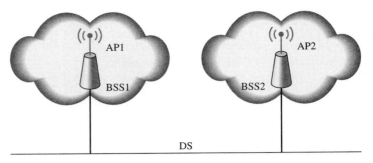

**FIGURE 19–6**
Two BSSs connected to a DS.

Although a wireless LAN based upon the 802.11 standard has many similarities to Ethernet, it is not Ethernet. Ethernet controls access to media using Carrier Sense Multiple Access with Collision Detection (CSMA/CD) to minimize collisions between packets on the medium. The IEEE 802.11 standard requires the use of a variation of this called Carrier Sense Multiple Access with Collision Avoidance (CSMA/CA). In addition to listening on a channel to ensure that the channel is free, CSMA/CA uses request to send (RTS) and clear to send (CTS) frames to control access.

802.11 defines three PHY layers for wireless networking, one that uses IR as the medium and two that use RF. Emphasis on the development of wireless devices centers on the two RF PHY layers. Both utilize the spread spectrum transmission method that was first developed for the military to minimize interference and provide secure communications. Using spread spectrum transmission, a single transmission utilizes multiple carrier frequencies. The two RF PHY layers use two different spread spectrum methods, Frequency Hopping Spread Spectrum (FHSS) and Direct Sequence Spread Spectrum (DSSS). Each is discussed below. First, however, let us examine some of the goals that the IEEE 802.11 subcommittee had in defining its standard.

## 19-3 The Goals, the Challenges, and the Solutions

### The Goals

Although there are good reasons for using wireless networks with fixed stations, the primary goal of wireless networking is *mobility,* to allow portable computer users to access network resources while free from the restrictions of a network cable. A laptop or notebook computer that must be tethered to a hub by a length of twisted-pair cable does not exploit the portability of the computer. In addition to limiting how far the user can roam by its length, the cable itself is underfoot, and hence a hindrance. A wireless connection to the network should enable the user to roam at will, at least in a designated area.

A second goal of wireless networking is *interoperability.* Wireless networking components from different manufacturers need to be able to work together. If components of a wireless network can only work with other components from the same vendor, many of the potential uses of wireless networking will be restricted, if not impossible. Without interoperability, a wireless user who accessed his company network with a laptop that was also used to access a home network for Internet access would have to ensure that the components used at home were from the same vendor as the components used by his company. Reliable wireless access in airports, hotels, cybercafes, and so on, would not be available without interoperability.

A third goal of wireless networking is *usability,* or ease of use. It should be easy for a portable computer user to connect to a wireless network. Ideally, the portable computer of an authorized user should recognize the existence of the wireless network, determine the network parameters, and connect to the network, without excessive additional configuration.

A fourth goal of wireless networking is *compatibility*. The operating system and applications on a computer should be able to work the same way, and as well, with a wireless connection as with a wired connection. Ideally, the only change that should be necessary to switch from a wired to a wireless connection is to replace the network interface card (NIC) for the wired connection with a wireless NIC, and load the proper drivers.

A fifth goal of wireless networking is *speed*. Data flow through a wireless connection should not be appreciably slower than through a wired connection.

## The Challenges

The two media that have been used for wireless networking are IR and RF transmissions. Both have severe limitations. The major limitations of IR were stated in Section 19-1. IR is used for wireless connections of devices such as printers to computers, but not for general wireless networking. The range of RF transmission is limited by transmitter power and antenna size. The very nature of portable computers, their restricted size and weight and ability to operate from batteries, dictates that transmitter power will be low and antenna size small, both of which are factors that will limit the distance that can separate components of a wireless network. In addition, the FCC and other regulatory bodies restrict the maximum power that can be used in the frequency bands that are utilized for wireless networking.

Wireless communications must function efficiently in the presence of noise and other forms of interference, such as microwave ovens, narrowband transmissions, and other RF sources. Furthermore, multipath fading can prevent frames from being reliably received by a station. Thus, connections between wireless stations must function over a medium that is much less reliable than wired LANs.

Furthermore, complexity is introduced into the wireless network by the methods that are used to manage traffic between stations and APs. The higher the speed used in transmissions, the more difficult it is for the communication to be error free. Whatever the current state of the technology, if speed is increased enough, accuracy will decrease to the point that further attempts to increase speed will actually cause a decrease in speed, because of the number of retransmissions that are necessary to correct errors. This tends to make wireless networking slower than wired networking.

The flow of data in wired networks is restricted to the wires themselves. There is no such restriction with wireless networks. When RF is used, any appropriate receiver within the range of the antenna of the transmitter can receive the data. Thus, wireless networking is inherently less secure than wired networking. It is possible to drive down a street with a properly configured computer and detect the presence of wireless networks in the buildings and houses that are passed.

## The Solutions

Several methods are used to extend the limited range of wireless LANs. An AP can extend the range of wireless stations, as shown in the ESS of Figure 19–6. Instead of two wireless components connecting to each other, each connects to a fixed component, which relays the data between the two components. The IEEE 802.11 standard specifically requires that the protocols used in wireless networking conform as closely as possible to the Ethernet standard. The use of a DS to connect multiple APs can further extend the range. However, the DS must include provisions for managing the flow of data between APs, for identifying which AP a particular station communicates with, and for maintaining connections when a station moves from one BSS to another.

Unlike Ethernet, in which it is assumed that a transmitted frame will be successfully received, the IEEE 802.11 standard requires that all transmitted frames be acknowledged. This ensures that frames that are not successfully received because of interference or other factors are retransmitted, and helps overcome the unreliability that can occur in wireless LANs because of interference.

The 802.11 standard defines two transmission speeds, 1 Mbps (million bits per second) and 2 Mbps. This was slow, when compared to the 10 Mbps that was standard in wired Ethernet LANs at the time that the 802.11 standard was adopted. Efforts to increase that speed led to 802.11b, which supplements 802.11 by adding a new PHY layer to the standard that supports speeds up to 11 Mbps. (The 802.11b standard will be discussed in more detail in a later section.) The 802.11a standard was also approved, which supports speeds up to 54 Mbps, but utilizes the 5.7 GHz Industrial, Scientific and Medical (ISM) frequency band instead of the 2.4 GHz ISM band used by 802.11 and 802.11b. In June of 2003, the 802.11g standard was approved, which also supports speeds up to 54 Mbps but utilizes the 2.4 GHz ISM band. Furthermore, 802.11g provides compatibility with 802.11a.

The IEEE 802.11 standard incorporates encryption of data in accordance with Wired Equivalent Privacy (WEP). Encryption using WEP, which is optional, is intended to protect data while it is being transmitted between station and AP. However, serious flaws in WEP have been discovered, and it is generally recommended that it not be relied upon to protect sensitive data. Developing security for wireless LANs is an ongoing effort.

## 19-4   Spread Spectrum

The radio transmission methods that we know best involve concentrating the RF into as narrow a bandwidth as possible, consistent with the information to be transmitted. Interference between signals is avoided by not transmitting two signals on the same carrier frequency. AM, FM, and TV stations have frequency allocations, based on geographical locations, that are designed to minimize interference between stations using the same carrier frequency.

Spread spectrum techniques were first developed for the military as a means of communicating in the presence of jamming. Spread spectrum transmissions minimize interference by transmitting on different frequencies in a predetermined pattern. Figure 19–7 illustrates this concept. F1, F2, F3, and so on, are bands of frequency of a specified bandwidth. T1, T2, T3, and so on, are time intervals during which the data is to be transmitted. During T1, data is transmitted in F4. During T2, data is transmitted in F7. During T3, data is transmitted in F1. The same sequence of transmissions is known to the receiver, so that during T1 it listens on F4, during T2 it listens on F7, and during T3 it listens on F1. As long as two transmitters don't use the same sequence of frequencies and thus don't both transmit on the same frequency at the same time, they will not interfere with each other. Of course, there are some technical difficulties involved in this process. There must be some way of synchronizing the transmitter and receiver both in time and in the sequence of frequencies to be used. The FCC requires the use of spread spectrum technology for unlicensed transmitters in the ISM bands using low power.

### Frequency Hopping Spread Spectrum

In FHSS, transmissions are made at the different frequencies in a pseudorandom pattern. Both the transmitter and receiver have to have the same pattern, and timing between them is critical. Multiple FHSS devices can share the same spectrum, however, as long as they use different, noninterfering patterns. Even narrowband devices can operate in the same

**FIGURE 19–7**

Example of Frequency Hopping Spread Spectrum.

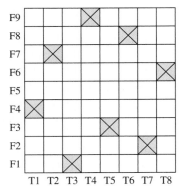

frequency range with minimum interference. The spread spectrum devices use low power, so that a high-powered narrowband device that is using a frequency that is momentarily being used by the FHSS device will be able to overpower the FHSS device. The data transmitted at that frequency by the FHSS device will be lost, but that is only one of many FHSS transmissions, and that data will be retransmitted when the frame is not acknowledged.

When the 802.11 standard was ratified in 1997, FHSS devices were the first to find commercial use. FHSS is relatively easy and inexpensive to implement. The standard divides the 2.4 GHz ISM band into 1-MHz channels. Channel numbers start with channel 0, with a center frequency of 2.400 GHz, and are incremented in 1-MHz steps up to channel 95, at a center frequency of 2.495 GHz. In the United States, Canada, and most of Europe, channels 2 through 79, which cover the range from 2.402 to 2.479 GHz, are allowed for wireless networking. Modulation of data bits in the 802.11standard involves frequency shifts from the center frequency.

The term "pseudorandom pattern" is used to describe the hop pattern because an FHSS transmission involves a brief transmission on one channel, then transmission on another channel, and then another channel, and so on, in a pattern that ensures that the transmitted energy is widely distributed over the entire 79 MHz of bandwidth. In the United States, the FCC regulates how the channels are used in part 15 of the FCC rules. These rules require that there be at least 75 hopping channels in the band, that each channel must be no more than 1 MHz wide, and that no more than 0.4 seconds can be spent in any one channel in a 30-second period. 802.11 requires that the minimum hop rate be 2.5 hops/second and that the minimum hop distance be 6 MHz. In order for a receiver to detect the transmissions, the receiver has to know the sequence of channels, and must be synchronized in time with the transmitter. Therefore, standard hop patterns are specified in 802.11 in accordance with the following formula:

$$f_x(i) = [b(i) + x] \bmod(79) + 2 \qquad (19\text{-}1)$$

where $f_x(i)$ is the $i$th channel in hopping sequence $x$, and the $b(i)$ are the base hopping sequences specified in Table 42 of the standard. Table 42 lists values of $i$ from 1 to 79, with the corresponding values of $b(i)$. There are 78 hopping patterns in the United States, Canada, and most of Europe, divided into three sets, numbered 1, 2, and 3. As specified in the 802.11 standard, set 1 consists of hopping sequences or patterns 0, 3, 6, 9, through 75; set 2 consists of hopping patterns 1, 4, 7, 10, through 76; and set 3 consists of hopping patterns 2, 5, 8, 11, through 77. Synchronization of transmitter and receiver is accomplished by having the transmitter periodically send a beacon frame. Among other fields, the beacon frame contains a time stamp and the FH parameter set. The FH parameter set includes the hop pattern number and a hop index. The hop index indicates what next value of $i$ in Table 42 the transmitter will use at the next hop following the time indicated in the time stamp.

---

**|||  EXAMPLE 19-2**

The first 10 values of $b(i)$ in Table 42 are 0, 23, 62, 8, 43, 16, 71, 47, 19, 61. That is, $b(1) = 0$, $b(2) = 23$, $b(3) = 62$, and so on. What are the first 10 channels in hop sequence 9?

**SOLUTION**

For $i = 1$, $b(i) = 0$. Therefore, $f_9(1) = [0 + 9]\bmod(79) + 2 = 9 + 2 = 11$.

For $i = 2$, $b(i) = 23$. Therefore, $f_9(2) = [23 + 9]\bmod(79) + 2 = 32 + 2 = 34$.

For $i = 3$, $b(i) = 62$. Therefore, $f_9(3) = [62 + 9]\bmod(79) + 2 = 71 + 2 = 73$.

For $i = 4$, $b(i) = 8$. Therefore, $f_9(4) = [8 + 9]\bmod(79) + 2 = 17 + 2 = 19$.

For $i = 5$, $b(i) = 43$. Therefore, $f_9(5) = [43 + 9]\bmod(79) + 2 = 52 + 2 = 54$.

For $i = 6$, $b(i) = 16$. Therefore, $f_9(6) = [16 + 9]\bmod(79) + 2 = 25 + 2 = 27$.

For $i = 7$, $b(i) = 71$. Therefore, $f_9(7) = [71 + 9]\bmod(79) + 2 = 1 + 2 = 3$.

For $i = 8$, $b(i) = 47$. Therefore, $f_9(8) = [47 + 9]\bmod(79) + 2 = 56 + 2 = 58$.

For $i = 9$, $b(i) = 19$. Therefore, $f_9(9) = [19 + 9]\bmod(79) + 2 = 28 + 2 = 30$.

For $i = 10$, $b(i) = 61$. Therefore, $f_9(10) = [61 + 9]\bmod(79) + 2 = 70 + 2 = 72$.

**▌▌ EXAMPLE 19-3**

The first 10 values of $b(i)$ in Table 42 are 0, 23, 62, 8, 43, 16, 71, 47, 19, 61. That is, $b(1) = 0$, $b(2) = 23$, $b(3) = 62$, and so on. A beacon frame from this transmitter indicates that the transmitter is using hop sequence 1 and that the hop index is 7. What channel did the transmitter use after the next hop following the time of the time stamp?

**SOLUTION**    From the sequence shown, $b(7)$ is 71 and the hop sequence $(x)$ is 1. Substituting into equation 19-1,

$$f_1(7) = [71 + 1] \bmod(79) + 2$$

But $[71 + 1] \bmod(79)$ is just the remainder when $71 + 1$ or 72 is divided by 79. This remainder is 72. Adding $72 + 2$ gives 74 as the seventh channel in hopping pattern 1. Table B.2 in the 802.11 standard lists the patterns in hopping sequence set 2. A review of that table verifies that the seventh hop in hopping pattern 1 is channel 74. Thus, a receiver that has received this beacon frame will know not only what channel the transmitter will hop to next, but also what channel the transmitter will hop to in each future hop.    ▌▌

**Table 19–1**  Symbol Encoding for 2-Level Gaussian Frequency-Shift Keying

| 2GFSK, 1 Mbps | |
|---|---|
| **Symbol** | **Frequency** |
| 1 | $F_c + f_d$ |
| 0 | $F_c - f_d$ |

**Table 19–2**  Symbol Encoding for 4-Level Gaussian Frequency-Shift Keying

| 4GFSK, 2 Mbps | |
|---|---|
| **Symbol** | **Frequency** |
| 10 | $F_c + f_{d1}$ |
| 11 | $F_c + f_{d2}$ |
| 01 | $F_c - f_{d2}$ |
| 00 | $F_c - f_{d1}$ |

## FHSS Modulation Methods

Either 2-level gaussian frequency-shift keying (2GFSK) or 4-level GFSK (4GFSK) is used to modulate FHSS data. In 2GFSK, a "1" is transmitted by shifting the transmitted frequency higher by a certain deviation, $f_d$. Conversely, a "0" is transmitted by shifting the frequency down by $f_d$, as shown in Table 19–1. Figure 19–8 illustrates 2GFSK. The symbol rate is 1 Mbps, which means that one bit is transmitted each microsecond. Since the carrier frequency is approximately 2.4 GHz, hundreds of cycles of the RF signal are transmitted for each bit. This is necessary in order to be able to accurately measure the transmitted frequency, and to allow slow transitions between bits. If the frequency is changed too rapidly, a larger bandwidth is required to accommodate the signal (see Chapter 3). The frequency is measured at approximately the center of the symbol period to determine whether a "1" or a "0" was sent. Because one bit is sent each microsecond, the speed is 1 Mbps.

If 4GFSK is used, each level corresponds to a pair of symbols. Two deviations are defined, $f_{d1}$ and $f_{d2}$, resulting in four possible frequencies in each channel, $F_c + f_{d1}$, $F_c + f_{d2}$, $F_c - f_{d2}$, and $F_c - f_{d1}$ where $F_c$ is the center frequency of the channel and $f_{d1}$ is the larger of the two frequency deviations. Table 19–2 shows the symbols involved with each transmitted frequency. The symbol rate is still 1 Mbps, but since each symbol includes two characters, the effective data rate is 2 Mbps. Figure 19–9 illustrates 4GFSK.

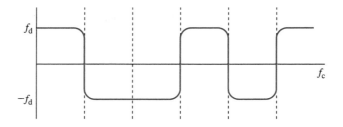

**FIGURE 19–8**
Example of 2GFSK modulation.

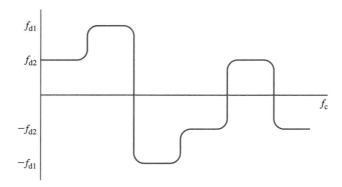

**FIGURE 19–9**
Example of 4GFSK modulation.

**EXAMPLE 19-4**

Part of an FHSS frame is being transmitted on 2.4 GHz ISM channel 11 using 2GFSK. Assume that $f_d = 160$ kHz. (a) What is the frequency corresponding to a 0? (b) What is the frequency corresponding to a 1?

**SOLUTION**   Channel 0 has a center frequency of 2.400 GHz. The channel center frequencies are separated by 1 MHz. Therefore, the center frequency of channel 11 is 2.411 GHz or 2411 MHz.

(a) Since $f_d = 160$ kHz or 0.160 MHz, a 0 is represented by $F_c - f_d = 2411 - 0.16 = 2410.84$ MHz.

(b) A 1 is represented by $F_c + f_d = 2411 + 0.16 = 2411.16$ MHz.

**EXAMPLE 19-5**

Part of an FHSS frame is being transmitted on 2.4 GHz ISM channel 25 using 4GFSK. Assume that $f_{d1} = 216$ kHz and $f_{d2} = 72$ kHz. (a) What is the frequency corresponding to a 00 bit pair? (b) What is the frequency corresponding to a 01 bit pair? (c) What is the frequency corresponding to a 10 bit pair? (d) What is the frequency corresponding to a 11 bit pair?

**SOLUTION**   Channel 25 has a center frequency of $2.400 + .025 = 2.425$ GHz or 2425 MHz.

(a) Since $f_{d1} = 216$ kHz or 0.216 MHz, a 00 is represented by $F_c - f_{d1} = 2425 - 0.216 = 2424.984$ MHz.

(b) A 01 is represented by $F_c - f_{d2} = 2425 - 0.072 = 2424.928$ MHz.

(c) A 10 is represented by $F_c + f_{d1} = 2425 + 0.216 = 2425.216$ MHz.

(d) A 11 is represented by $F_c + f_{d2} = 2425 + 0.072 = 2425.072$ MHz.

## Direct Sequence Spread Spectrum

In DSSS, instead of spreading the power over the 2.4 GHz ISM band by transmitting at one frequency for a period and then hopping to a different frequency for the next period, the power is spread over a bandwidth of 22 MHz by applying concepts studied in Chapter 3. The symbol rate for the data is 1 Mbps, as in FHSS. An XOR operation is performed between each bit of the data and an 11-chip "Barker" code. The bits of the Barker code are called *chips* to create a distinction between them and the bits of the data, and the duration of each chip is $\frac{1}{11}$ of a microsecond. The effect of the XOR operation is such that the output of the operation is the unmodified Barker code if the data bit is a 1, and the complement of the Barker code if the data bit is a 0. However, since each chip is $\frac{1}{11}$ of a microsecond, the resulting function of frequency is a $\sin x/x$ function with zero crossings every 11 MHz. The main lobe is centered around the center frequency, and has a width of 22 MHz. The lobes between the first and second zero crossing on either side of the center frequency are filtered so that they are at least 30 dB below the main lobe. The lobes between the second and third zero crossings are filtered to be at least 50 dB below the main lobe. This filtering ensures that those lobes do not interfere with adjacent channels. The DSSS channel width is 5 MHz instead of the 1-MHz width used in FHSS. Because the bandwidth of a transmission is 22 MHz, two transmitters in the same operating area must be transmitting at least five channels apart in order not to interfere with each other.

The DSSS Barker code is 10110111000, and this is the pattern that is transmitted if the data bit is a 1; 01001000111 is transmitted if the data bit is a 0. A correlator is used in the receiver to reverse the spreading process of the transmitter. The effect of this is to concentrate the components of the transmitted waveform. Then, at the receiver, it is only necessary to count the number of "1s" received during the 1-μs symbol interval. If there are six "1s" the data bit is a 1. If there are five "1s" the data bit is a 0. If interference causes an ambiguity, the receiver is able to analyze the received pattern to resolve the ambiguity and determine the data bit.

### Differential Binary Phase-Shift Keying

DSSS uses differential phase-shift keying (DPSK) to modulate the signal. Like FHSS, there are two forms of DPSK used, differential binary phase-shift keying (DBPSK) to achieve a speed of 1 Mbps and differential quadrature phase-shift keying (DQPSK) to achieve a speed of 2 Mbps. Instead of increasing or decreasing frequency by some deviation, DPSK maintains the frequency constant and changes the phase. In DBPSK, a 0 is represented by zero phase shift since the last symbol time, and a 1 is represented by a 180° phase shift since the last symbol time.

---

**▌▌ EXAMPLE 19-6**

Sketch a waveform of DBPSK used to encode 010110.

**SOLUTION**  Each bit is encoded by either a 0° or a 180° phase shift from the previous bit. Therefore, there should be no phase shift for the first symbol, a 180° phase shift between the first and second symbol, no phase shift between the second and third symbol, a 180° phase shift between the third and fourth symbol, a 180° phase shift between the fourth and fifth symbol, and no phase shift between the fifth and sixth symbol. Figure 19–10 is a sketch of DBPSK used to encode 010110, using two cycles per symbol instead of the hundreds that would normally appear.

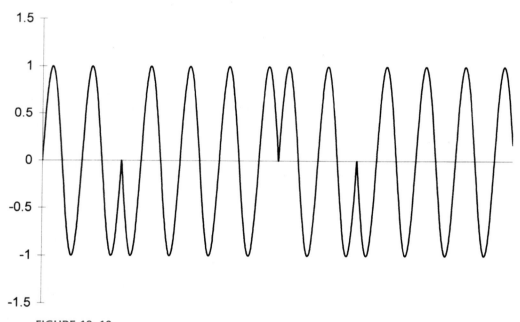

**FIGURE 19–10**
DBPSK used to encode 010110.

---

### Differential Quadrature Phase-Shift Keying

Like 2GFSK, DBPSK only encodes one bit per symbol. This results in a 1 Mbps speed. In DQPSK, four phases are used, so that two bits can be encoded per symbol. A 0° phase shift since the last symbol represents 00, a 90° phase shift represents 01, a 180° phase shift represents 11, and a 270° phase shift represents 10. Tables 19–3 and 19–4 list these phase shifts.

Table 19–3   Symbol Encoding for Differential Binary Phase-Shift Keying

| 2GFSK, 1 Mbps | |
| --- | --- |
| Symbol | Phase Shift |
| 0 | 0° |
| 1 | 180° |

Table 19–4   Symbol Encoding for Differential Quadrature Phase-Shift Keying

| 4GFSK, 2 Mbps | |
| --- | --- |
| Symbol | Phase Shift |
| 00 | 0° |
| 01 | 90° |
| 11 | 180° |
| 10 | 270° |

▌▌ EXAMPLE 19-7

Sketch a waveform of DQPSK used to encode 001110000001.

**SOLUTION**   Each pair of bits is encoded by shifting the phase 0°, 90°, 180°, or 270° after the last pair of bits. Since the first two bits are 00, there is no phase shift before the symbol. The second pair of bits is 11, so there is a 180° phase shift between the first and second symbol. The third pair of bits is 10, so there is a 270° phase shift between the second and third symbol. The fourth and fifth pair of bits are both 00, so there is no phase shift between the third and fourth symbol or between the fourth and fifth symbol. The last pair of bits is 01, so there is a 90° phase shift between the fifth and sixth symbol. Figure 19–11 shows DQPSK used to encode 001110000001, with two cycles per symbol instead of hundreds of cycles.

As noted above, both the FHSS PHY and the DSSS PHY only provide data rates of 1 Mbps and 2 Mbps, both of which are very slow for modern networks. In 1999, two supplements to the 802.11 standard were ratified, IEEE 802.11a and 802.11b. Each defined a new PHY for 802.11 that supported higher data rates. Because it was easier to implement, the 802.11b standard found commercial application first, but commercialization of both

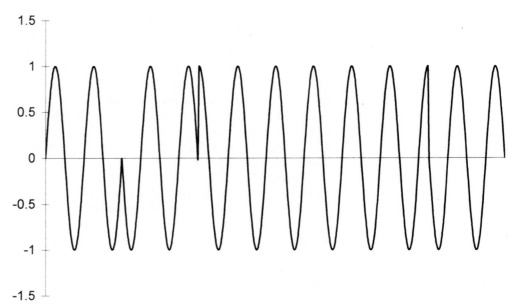

**FIGURE 19–11**
DQPSK used to encode 001110000001.

standards did occur. In 2003, the IEEE 802.11g supplement was ratified, which also supports higher data rates. Each is discussed below. Each of these standards is summarized in Table 19–5 and described below.

**Table 19–5** Comparison of 802.11 Technologies

| Standard | Frequency Band | Data Rates | Modulation Methods | Architecture |
|---|---|---|---|---|
| 802.11 | 2.4 GHz | 1 Mbps | 2GFSK | FHSS |
| | | 2 Mbps | 4GFSK | FHSS |
| | | 1 Mbps | DBPSK | DSSS |
| | | 2 Mbps | DQPSK | DSSS |
| | IR | 1 Mbps | 16-PPM | |
| | | 2 Mbps | 4-PPM | |
| 802.11a | 5 GHz | 6 and 9 Mbps | BPSK | OFDM |
| | | 12 and 18 Mbps | QPSK | |
| | | 24 and 36 Mbps | 16-QAM | |
| | | 48 and 54 Mbps | 64-QAM | |
| 802.11b | 2.4 GHz | 1 Mbps | DBPSK | HR/DSSS |
| | | 2 Mbps | DQPSK | |
| | | 5.5 Mbps | DBPSK | |
| | | 11 Mbps | DQPSK | |
| 802.11g | 2.4 GHz | 1 Mbps | DBPSK | ERP-DSSS |
| | | 2 Mbps | DQPSK | |
| | | 5.5 Mbps | DBPSK | |
| | | 11 Mbps | DQPSK | |
| | | 6 and 9 Mbps | BPSK | |
| | | 12 and 18 Mbps | QPSK | |
| | | 24 and 36 Mbps | 16-QAM | |
| | | 48 and 54 Mbps | 64-QAM | |

## 19-5  IEEE 802.11b

The maximum speed in both FHSS and DSSS is 2 Mbps, and both will slow down to 1 Mbps if interference is significant. These speeds are not adequate for modern networking, so efforts have continued to improve the speed of wireless networking. The first approach to achieve commercial success is a variation on the DSSS of 802.11, known as High Rate, Direct Sequence Spread Spectrum (HR/DSSS). Instead of using DPSK to encode the bits, the HR/DSSS PHY uses an encoding method called complementary code keying (CCK). In CCK, DQPSK is used in conjunction with certain mathematical transforms to transmit 4- or 8-bit code symbols at a symbol rate of 1.375 million code symbols per second. Thus, speeds of 5.5 Mbps and 11 Mbps per second can be achieved, depending upon whether 4-bit or 8-bit code symbols are sent.

For 5.5 Mbps, data is divided into 4-bit symbols. These symbols are further divided into 2-bit segments. The first 2-bit segment is used to select the phase shift for DQPSK encoding. For odd symbols, a 00 bit pattern produces a 0 radian phase shift, a 01 bit pattern produces a $\pi/2$ radian phase shift, a 11 bit pattern produces a $\pi$ radian phase shift, and a 10 bit pattern produces a $3\pi/2$ radian phase shift. For odd symbols, 00 corresponds to a $\pi$ radian phase shift, 01 corresponds to a $3\pi/2$ radian phase shift, 11 corresponds to 0 radian phase shift, and 10 corresponds to a $\pi/2$ radian phase shift. The second 2-bit segment is used to select one of four code words that function like the Barker code to spread the spectrum.

For 11 Mbps, the operation is similar, except that the data is divided into 8-bit symbols. Each symbol is further divided into a 2-bit segment and a 6-bit segment. The 2-bit segment is used to select the phase shift for DQPSK encoding. The 6-bit segment is grouped into three pairs, and each pair is used to determine a code word used to spread the spectrum. Thus, instead of using static Barker codes to provide the spreading, code words that contain information are used to perform the spreading. In a 5.5 Mbps transmission, four data bits are encoded in each symbol. Two of these bits result from using DQPSK and the other two from the use of CCK. In an 11-Mbps transmission, eight data bits are encoded in each symbol. Two data bits result from using DQPSK and the other six result from the use of CCK. HR/DSSS equipment is backward compatible with DSSS, so that it is possible for HR/DSSS equipment to also operate at rates of 1 Mbps and 2 Mbps. The 802.11 and 802.11b standards both use the 2.4 GHz ISM band, and 802.11b equipment has been very successful commercially.

In addition to CCK modulation at 5.5 Mbps and 11 Mbps, IEEE 802.11b also defines Packet Binary Convolutional Coding (PBCC). Unlike CCK, 8 chip codes are used for both 5.5 Mbps and 11 Mbps, but DBPSK is used to provide the 5.5 Mbps rate and DQPSK is used for the 11 Mbps rate.

In order for a device to be IEEE 802.11b compatible, it must support Barker code modulation at 1 Mbps and 2 Mbps, and CCK at 5.5 Mbps and 11 Mbps. PBCC modulation at 5.5 Mbps and 11 Mbps is optional.

## 19-6  IEEE 802.11a

The 2.4 GHz ISM band is very crowded. As a result, the 802.11a standard, which defines a "High-Speed Physical Layer in the 5 GHz Band," was developed. This PHY uses the 5.7 GHz ISM band, and commercial products based on this standard are available. The 5.7 GHz ISM band has the advantage of greater available bandwidth. However, path losses are greater at 5.7 GHz than at 2.4 GHz, and higher-frequency equipment typically requires more power to achieve comparable results than lower-frequency equipment. IEEE 802.11a uses orthogonal frequency-division multiplexing (OFDM). This process uses multiple subcarriers to encode a single transmission. This is a variation on standard frequency-division multiplexing, in which a number of signals share a channel by separating the subcarrier frequencies enough to provide a guard band between adjacent signals. In OFDM, the subcarriers are selected so that the peak of one subcarrier occurs at the zero crossings of the adjacent subcarriers, which eliminates the need for guard bands. In OFDM, the inverse fast Fourier transform (IFFT) is used to combine all of the subchannels into a composite waveform. The receiver then uses the fast Fourier transform (FFT) to separate the subchannels from the received signal. The IEEE 802.11a standard supports twelve nonoverlapping channels.

Throughput with OFDM is a function of the bandwidth used; the greater the bandwidth, the greater the throughput. However, as the width of the channel is increased, fewer channels can be operated in the available spectrum. The 802.11 working group compromised on the use of 20 MHz operating channels, which allows speeds as high as 54 Mbps per channel. There are 52 subcarriers in each 20 MHz channel, 48 of which transmit data. A symbol rate of 250,000 symbols per second is used, and different speeds are available that depend upon the type of encoding used. Speeds of 6 and 9 Mbps are achieved using BPSK with one bit per subchannel or 48 bits per symbol. That would seem to provide a speed of 12 Mbps (250,000 symbols per second times 48 bits per symbol). However, either one-half or one-quarter of these bits are used for error correction, so that actual speeds achieved are 6 and 9 Mbps.

QPSK is used to achieve speeds of 12 or 18 Mbps by encoding 2 bits per symbol. If one-half of the bits are used for error correction, the speed is 12 Mbps. If only one-quarter of the bits are used for error correction, the speed is 18 Mbps. Quadrature amplitude modulation (QAM) is used to achieve speeds greater than 18 Mbps. Speeds of 24 and 36 Mbps are achieved using 16-QAM, which encodes 4 bits using 16 symbols. Speeds of 48 and 54 Mbps are achieved using 64-QAM, which encodes 6 bits using 64 symbols.

## 19-7 IEEE 802.11g

When the 802.11 standard was developed, the FCC did not allow the use of OFDM in the 2.4 GHz ISM band. That regulation was changed, and 802.11g achieves speeds up to 54 Mbps by using OFDM (like IEEE 802.11a) but operates in the 2.4 GHz ISM band. This makes backward compatibility with 802.11b possible.

There are a number of different rates defined in the IEEE 802.11g standard, utilizing a number of modulation methods. Some of these rates are mandatory for a device compatible with the standard, but most are optional. The mandatory rates include 1 and 2 Mbps using the Barker code; 5.5 and 11 Mbps using CCK; and 6, 12, and 24 Mbps using OFDM. Optional rates using PBCC are 5.5, 11, 22, and 33 Mbps. Optional rates using OFDM are 9, 18, 36, 48, and 54 Mbps. Optional rates using DSSS-OFDM are 6, 9, 12, 18, 24, 36, 48, and 54 Mbps.

### SystemVue™ Closing Application (Optional)

Insert the text CD in a computer having SystemVue™ installed and activate the program. Open the CD folder entitled **SystemVue Systems** and load the file entitled 19-1.

#### Description of System

The system shown on the screen is the same one that was used in the chapter opening application. In this exercise, however, you will modify the outputs of Tokens 5 and 7 to see what effect having different types of 11-MHz pulse trains will have. Tokens 5 and 7 currently are producing a pulse train defined by a text file, **Barker.txt,** which is available on the CD. Examining that file, you will see that it consists of two columns of numbers. The left column shows time in $\frac{1}{11}$-μs increments. The right column shows the corresponding pulse amplitude. This file is used to define the pulse train from both Token 5 and Token 7, so their outputs are identical.

Double-click Token 5. A window entitled **SystemVue Source Library** will open. Currently, **Import** is selected as the group to which Token 5 belongs. Left-click **Noise/PN** to open that group. Left-click the button labeled **PN Seq** to turn Token 5 into a random pulse generator. Then left-click the button labeled **Parameters.** In the text box under **Rate (Hz),** type in **11e+6.** Then left-click **OK** to return to the previous menu. Repeat the same procedure for Token 7 and change it to a **PN Seq** token with the same rate as Token 5.

Left-click the **Run System** button or press F5 to run the simulation. When you examine the **Analysis** window, you should see a display in which the input signal is not reconstructed. The reason for this is that the input from Token 7 is no longer identical to the input from Token 5.

Now, break the connection between Token 7 and Token 8. Make a connection between Token 5 and Token 8, so that Token 5 is providing an input to both Tokens 6 and 8. It may look like the connection from Token 5 to Token 8 is going through Token 6, but it is not. When you run the simulation, a **SystemVue Connection Analysis** window will open that tells you that Token 7 has no output connection. Left-click the button in that window labeled **Run System** to run the simulation without Token 7. When you open the **Analysis** window, you should see waveforms which indicate that even though the "Barker" code is not 10110111000, as long as the same sequence is used in both the transmitter and the receiver, the input pulse train can be recovered.

## PROBLEMS

**19-1** The IrDA describes IrDA DATA, its standard for data transmission, as "recommended for high-speed short-range, line-of-sight, point-to-point cordless data transfer—suitable for HPCs, digital cameras, handheld data collection devices, etc." Describe how IrDA could be used to connect a print device to a computer.

**19-2** The IrDA describes IrDA DATA, its standard for data transmission, as "recommended for high-speed short-range, line-of-sight, point-to-point cordless data transfer—suitable for HPCs, digital cameras, handheld data collection devices, etc." Describe how IrDA could be used to connect a personal data assistant (PDA) to a computer.

**19-3** The first 10 values of $b(i)$ in Table 42 of the IEEE 802.11 standard are 0, 23, 62, 8, 43, 16, 71, 47, 19, 61. That is, $b(1) = 0$, $b(2) = 23$, $b(3) = 62$, and so on. What are the first 10 channels in hop sequence 28?

**19-4** The first 10 values of $b(i)$ in Table 42 of the IEEE 802.11 standard are 0, 23, 62, 8, 43, 16, 71, 47, 19, 61. That is, $b(1) = 0$, $b(2) = 23$, $b(3) = 62$, and so on. What are the first 10 channels in hop sequence 77?

**19-5** The first 10 values of $b(i)$ in Table 42 of the IEEE 802.11 standard are 0, 23, 62, 8, 42, 16, 71, 47, 19, 61. That is, $b(1) = 0$, $b(2) = 23$, $b(3) = 62$, and so on. A beacon frame from this transmitter indicates that the transmitter is using hop sequence 25 and that the hop index is 1. What channel did the transmitter use after the next hop following the time of the time stamp?

**19-6** The first 10 values of $b(i)$ in Table 42 of the IEEE 802.11 standard are 0, 23, 62, 8, 42, 16, 71, 47, 19, 61. That is, $b(1) = 0$, $b(2) = 23$, $b(3) = 62$, and so on. A beacon frame from this transmitter indicates that the transmitter is using hop sequence 8 and that the hop

index is 7. What channel did the transmitter use after the next hop following the time of the time stamp?

**19-7** Part of an FHSS frame is being transmitted on 2.4 GHz ISM channel 59 using 2GFSK. Assume that $f_d = 160$ kHz. (a) What is the frequency corresponding to a 0? (b) What is the frequency corresponding to a 1?

**19-8** Part of an FHSS frame is being transmitted on 2.4 GHz ISM channel 43 using 2GFSK. Assume that $f_d = 160$ kHz. (a) What is the frequency corresponding to a 0? (b) What is the frequency corresponding to a 1?

**19-9** Part of an FHSS frame is being transmitted on 2.4 GHz ISM channel 17 using 4GFSK. Assume that $f_{d1} = 216$ kHz and $f_{d2} = 72$ kHz.

(a) What is the frequency corresponding to a 00 bit pair?
(b) What is the frequency corresponding to a 01 bit pair?
(c) What is the frequency corresponding to a 10 bit pair?
(d) What is the frequency corresponding to a 11 bit pair?

**19-10** Part of an FHSS frame is being transmitted on 2.4 GHz ISM channel 1 using 4GFSK. Assume that $f_{d1} = 216$ kHz and $f_{d2} = 72$ kHz.

(a) What is the frequency corresponding to a 00 bit pair?
(b) What is the frequency corresponding to a 01 bit pair?
(c) What is the frequency corresponding to a 10 bit pair?
(d) What is the frequency corresponding to a 11 bit pair?

**19-11** Make a sketch of DBPSK used to encode 011011. Use two cycles per symbol.

**19-12** Make a sketch of DBPSK used to encode 001100. Use two cycles per symbol.

**19-13** Make a sketch of DQPSK used to encode 000110000011. Use two cycles per symbol.

**19-14** Make a sketch of DQPSK used to encode 001010111100. Use two cycles per symbol.

# Optical Communications

## OVERVIEW AND OBJECTIVES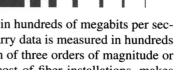

The capacity of copper to carry data is measured, at best, in hundreds of megabits per second (Mbps). The capacity of fiber-optic cable (fiber) to carry data is measured in hundreds of gigabits per second (Gbps). This increase in bandwidth of three orders of magnitude or more, in conjunction with significant reductions in the cost of fiber installations, makes optical fiber the medium of choice for the future. Fiber has other advantages as well. Transmission over optical networks is inherently far more secure than other media. Major bandwidth improvements are being made by improvements to the components that inject the signals into the fiber and detect the signals at the far end, so installed fiber will not soon become obsolete and need to be replaced by improved cable.

Not long ago, the cost of fiber and its installation made it suitable only for high-speed backbone and long-distance transmission networks. Now fiber is competitive with twisted-pair for installation to the desktop, and has the potential to provide virtually unlimited bandwidth in LANs as well as in MANs and WANs.

In this chapter we will examine the components of a fiber-optics (FO) communications system, learn how optical signals are transmitted through fiber, consider the factors that limit the performance of these systems, and perform a link analysis of FO systems similar to that performed for RF communications systems in Chapter 16. The measure of performance of a FO communications system is generally the bit error rate (BER), as was the case with systems studied in previous chapters.

## Objectives

After completing this chapter, the reader should be able to:

1. Understand how signals are transmitted over optical fibers.
2. Describe the difference between *single-mode* and *multimode* operation.
3. Describe the operation and limitations of LEDs and lasers as light sources.
4. Describe the operation and limitations of PIN and avalanche photodiodes as light detectors.
5. Enumerate and describe the major loss and dispersion mechanisms in fiber.
6. Enumerate and describe the factors outside the fiber that limit system performance.
7. Explain how *wavelength-division multiplexing* works.
8. Perform a link analysis of a FO communications system.

## 20-1  Overview of Optical Communications

### Optical Signals

Signaling with light is an ancient practice, which has evolved into a modern, high-speed, secure data communications method. The lighting of a signal fire on a headland to announce the sighting of an approaching enemy fleet is a form of optical signaling. The use of flashing light and Morse code to communicate between ships is a form of optical signaling that is faster and more versatile than the signal fire. An infrared remote control used to change channels on a television is a form of optical signaling, as is the transfer of data between a calculator or portable computer and a printer using IR ports. However, the optical signals that are the subject of this chapter are those produced by an LED or laser and injected into a fiber-optic cable for transmission.

The transmission signals that we have considered in earlier chapters have ranged in frequency from a few kHz to a few GHz, perhaps as high as $20 \times 10^9$ Hz. The frequencies used for optical fiber communications are in the infrared (IR) range at approximately $2 \times 10^{14}$ Hz or 200 THz (1 THz = $1 \times 10^{12}$ Hz). In fact, the full range of the frequencies used for optical-fiber communications is only from approximately 167 THz to approximately 400 THz, a difference of 233 THz. Until recently, these frequencies have been the realm of scientists, not communications engineers, and scientists developed the practice of referring to these frequencies by their wavelengths. 167 THz corresponds to a wavelength of 1800 nm (nanometers) or $1800 \times 10^{-9}$ m; 400 THz corresponds to a wavelength of 750 nm. We will generally express wavelength in nanometers, although it is also appropriate to express it in micrometers ($\mu$m). Thus, 1800 nm and 1.8 $\mu$m are the same wavelength.

---

**◼◼ EXAMPLE 20-1**

An IR laser operates at a wavelength of 1300 nm. What is the frequency of the laser?

**SOLUTION**    Recall that the relationship between frequency and wavelength is

$$f = \frac{300 \times 10^6}{\lambda} \tag{20-1}$$

Therefore,

$$f = \frac{300 \times 10^6}{1300 \times 10^{-9}} = 2.31 \times 10^{14} = 231 \text{ THz} \tag{20-2}$$

Although the signals transmitted over fiber can be either analog or digital, digital signals predominate. These digital optical signals are binary. A pulse of light can represent a binary 1, and the absence of light for a pulse period can represent a binary 0. Thus, an optical signal can be created by turning a light on and off at a rate determined by a stream of electrical pulses. This means that digital optical communications is inherently serial in nature. The optical signal is injected into an optical fiber for transmission. Although visible light could be used for this transmission, available fibers have large losses in the visible light spectrum (approximately 400 nm to 700 nm). Figure 20–1 shows the approximate typical loss in dB/km

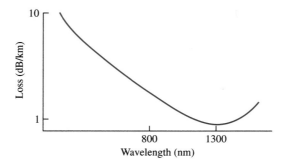

**FIGURE 20–1**

Fiber loss as a function of wavelength.

for modern optical fibers as a function of wavelength. Note that the range of minimum loss is in the IR range between about 1200 nm and 1800 nm, and in this range the loss is less than 1 dB/km. As we shall see, this loss is not the only factor that limits how far an optical signal can be propagated through fiber without requiring amplification or regeneration. In practice, there are three regions of the IR spectrum that are commonly used for optical communications, designated the 850 nm band, the 1310 nm band, and the 1550 nm band.

### What Is an Optical Fiber?

An optical fiber is a strand of glass or plastic, thinner than a human hair, consisting of a core, cladding, and a protective buffer coating, as shown in Figure 20–2. Plastic fiber has much greater loss than glass fiber, so only glass fiber will be considered in this chapter. Under the right circumstances, the difference in index of refraction between the core and the cladding causes the fiber to act as a waveguide for light waves, containing the optical signal within the core and propagating it down the fiber. (The details of this process will be presented in the next section.)

Glass fiber cables can be divided into two general types, single mode and multimode. Light travels through a fiber in electromagnetic modes, just as RF energy travels through a waveguide. If the core of the fiber is sufficiently large, multiple modes (e.g., $HE_{11}$, $HE_{21}$, $TM_{01}$, $TE_{01}$, etc.) can exist simultaneously in the fiber. A single-mode cable has a very narrow core, typically 8 to 10 microns (micrometers), and supports only one mode of transmission. A multimode cable has a much wider core, typically 50, 62.5, or 100 microns, and supports the transmission of multiple modes simultaneously. Both types typically have a cladding diameter of 125 microns and outer coating diameter of 245 microns. The operation of each type is presented in the next section.

### What Is an Optical Fiber Communications System?

As shown in Figure 20–3, a FO communications system consists of an electrical signal, circuitry to drive an optical transmitter from this signal, the optical fiber, an optical receiver to demodulate the light and convert the output back to an electrical signal, and output circuitry

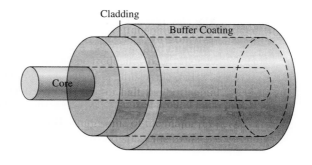

**FIGURE 20–2**
Fiber structure.

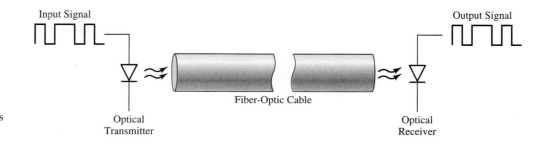

**FIGURE 20–3**
Fiber-optic communications system.

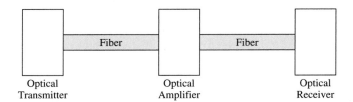

**FIGURE 20–4**
Optical amplifier for regeneration of optical signal.

to deliver the electrical output to its destination. Depending upon the length of the optical-fiber cable run, there may also be amplifiers (as shown in Figure 20–4) to regenerate the signal periodically as it travels from source to destination.

## Optical Fiber Advantages and Disadvantages

Optical fiber has many advantages over wire as a medium for data communications. Modern fiber has extremely low losses, on the order of 1 dB/km or less, so the signal can be propagated much greater distances in fiber without requiring regeneration. Fiber is capable of supporting bandwidths that are thousands of times greater than can be supported by copper. Power requirements are low, because low-power signals are transported by fiber. Because the signal is carried by light waves, fiber is immune to electromagnetic interference. Fiber is not an electrical conductor, so it can be used safely in the vicinity of high-voltage equipment. Fiber-optic cables are lighter than copper cables.

On the other hand, optical fiber has some disadvantages. Optical fibers are difficult to splice, so more skilled personnel are required to install fiber-optic cables. The diameter of the signal-carrying core is so small that an optical signal can be blocked by a dust particle, making it necessary to ensure extreme cleanliness when making connections. The equipment used to inject signals into optical fibers and to detect the signals transported by optical fibers is more complex and therefore more expensive. Optical fibers cannot carry electricity. Therefore, if cables must have one or more amplifiers in inaccessible places to regenerate the signal, such as in undersea cables, copper wires must be included to provide the electrical power.

## 20-2   Optical Fibers

### Snell's Law

The speed of light traveling through a transparent object depends upon the density of the object. For example, light travels faster through air than it does through water or through glass. Figure 20–5 shows a ray of light passing through a boundary between air and glass. If the angle the ray makes with the boundary is 90°, like ray A, the direction of travel is not changed. However, if the angle the ray makes with the boundary is not 90°, like ray B, the direction of travel changes. This change in direction is called refraction. In Figure 20–5, $\Theta_i$, the angle that the ray makes with the normal to the boundary in the first medium, is

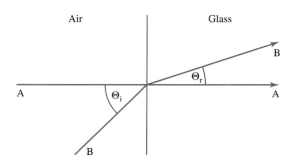

**FIGURE 20–5**
Refraction of light.

called the angle of incidence, and $\Theta_r$, the angle that the ray makes with the normal to the boundary in the second medium, is called the angle of refraction. Because light travels faster in air than in glass, $\Theta_i$ is larger than $\Theta_r$. If the ray had passed through the boundary from glass to air, $\Theta_i$ would be smaller than $\Theta_r$.

If we measured the angle of incidence and the resulting angle of refraction for a large number of angles of incidence, and then plotted $\sin \Theta_i$ versus $\sin \Theta_r$, the result would be a straight line. The slope of that line is the ratio of a property of the two media called the index of refraction ($n$). The index of refraction for air is 1, and the index of refraction for crown glass is 1.52. Snell's law states that

$$n_i \times \sin \Theta_i = n_r \times \sin \Theta_r \tag{20-3}$$

where $n_i$ is the index of refraction for the incident medium, $\Theta_i$ is the angle of incidence, $n_r$ is the index of refraction for the refractive medium, and $\Theta_r$ is the angle of refraction.

---

**▮▮ EXAMPLE 20-2**

A beam of light is traveling through air, and strikes the surface of a block of crown glass at an angle of 30° from the normal to the surface, as shown in Figure 20–6. What is the angle of refraction?

**SOLUTION**    In Figure 20–6, $\Theta_i = 30°$, $n_i = 1.00$, and $n_r = 1.52$. Therefore, $\Theta_r$ can be calculated as follows:

$$n_i \times \sin \Theta_i = n_r \times \sin \Theta_r \tag{20-4}$$

$$\sin \Theta_r = \frac{n_i}{n_r} \times \sin \Theta_i \tag{20-5}$$

$$\Theta_r = \sin^{-1}\left(\frac{n_i}{n_r} \times \sin \Theta_i\right) \tag{20-6}$$

$$\Theta_r = \sin^{-1}\left(\frac{1.00}{1.52} \times \sin 30°\right) \tag{20-7}$$

$$\Theta_r = \sin^{-1}(0.66 \times 0.5) = \sin^{-1}(0.33) = 19.27° \tag{20-8}$$

Note that the angle of refraction is less than the angle of incidence. This occurs because the light is passing from a less dense into a denser medium. When light travels from a medium having a lower index of refraction to one having a higher index of refraction, it can never be reflected by the boundary.

A phenomenon called *total internal reflection,* in which all of the incident light reflects off the boundary and remains in the incident material, occurs when the incident material is the more dense material, the light reaches a boundary with a less dense material, and the angle of incidence is greater than an angle called the *critical angle.* The critical angle is the angle of incidence that causes the angle of refraction to be 90°. Figure 20–7 illustrates the critical angle.

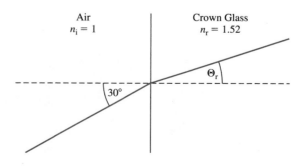

**FIGURE 20–6**
Refraction of light through boundary between air and crown glass.

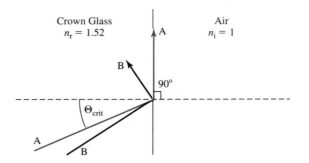

**FIGURE 20–7**
Critical angle of refraction.

**EXAMPLE 20-3**

What is the critical angle if a ray of light passing through crown glass (which has an index of infraction of 1.52) reaches a boundary with air (which has an index of refraction of 1.00)?

**SOLUTION** $\Theta_r = 90°$, $n_i = 1.52$, and $n_r = 1.00$. Therefore,

$$n_i \times \sin \Theta_i = n_r \times \sin \Theta_r \tag{20-9}$$

$$\sin \Theta_i = \frac{n_r}{n_i} \times \sin \Theta_r \tag{20-10}$$

$$\Theta_i = \sin^{-1} \left( \frac{n_r}{n_i} \times \sin \Theta_r \right) \tag{20-11}$$

$$\Theta_r = \sin^{-1} \left( \frac{1.00}{1.52} \times \sin 90° \right) \tag{20-12}$$

$$\Theta_r = \sin^{-1}(0.66 \times 1) = \sin^{-1}(0.66) = 41.30° \tag{20-13}$$

Therefore, light passing through the crown glass and striking the boundary between the glass and air with an angle of incidence greater than of 41.30° will be completely reflected into the glass, as shown by ray B in Figure 20–7.

## Cone of Acceptance

Light enters the fiber through the end of the core, which must be carefully cut and polished to provide a suitable surface for the light to enter. Once inside the core, the boundary that determines whether total internal reflection will occur is that between the core and the cladding. As stated earlier, the core has a greater index of refraction than the cladding. Therefore, light striking the boundary between the core and the cladding at an angle of incidence greater than the critical angle will remain in the core. Light striking the boundary at an angle of incidence less than the critical angle will leave the core, and represents a loss of signal. This is illustrated in Figure 20–8. $\Theta_A$ is less than $\Theta_{crit}$, so ray A is able to escape

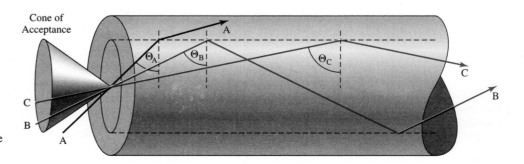

**FIGURE 20–8**
Total internal reflection and cone of acceptance.

from the fiber core. $\Theta_B$ and $\Theta_C$ are greater than $\Theta_{crit}$, so rays B and C are reflected from the boundary between the core and the cladding and remain in the core. Rays B and C will continue to be reflected from the boundary as they travel through the fiber. Note from Figure 20–8 that the rays that enter the fiber with angles greater than $\Theta_{crit}$ must enter the core with angles contained in the cone labeled *cone of acceptance*.

## Optical Fiber Structure

Pure silica, the basic ingredient in optical fibers, has an index of infraction of 1.45. This can be changed by adding impurities to the silica to raise or lower its index of infraction. Adding germanium, for example, will raise the index of infraction, while adding fluorine will lower the index of infraction. Manufacturing of optical fibers involves drawing the strand of fiber from molten silica in a process that forms the core and the cladding at the same time, by adding one dopant to raise the index of refraction of the core and another to lower the index of refraction of the cladding. As the fiber is drawn, a protective coating is also applied to it before it is wound onto a take-up spool. A strand of fiber 12 km long can be stored on a standard eight foot diameter spool. However, FO cables generally contain multiple fiber strands, each capable of acting as an optical waveguide.

## Transmission and Refraction in Optical Fibers

Transmission of light through an optical fiber is dependent upon achieving total internal reflection, as described above. The light is inserted into the core, which has a higher index of refraction than the cladding, at an angle greater than the critical angle. The fiber acts like a waveguide, containing the light wave and transporting it to the far end. Like waveguides, fibers can have different transmission modes, depending upon the diameter of the core. Small diameter cores (8–10 microns) have only one strong propagation mode, and are known as single-mode fibers. Single-mode fibers are used for long distance communication because their losses are very low: signals can travel hundreds of km through single-mode fibers without requiring regeneration. Larger diameter cores (50–100 microns) support multiple propagation modes simultaneously, and are called multimode fibers. Without regeneration, transmission distance in multimode fibers is limited (by losses in the fiber) to a fraction of that of single mode fiber.

A single wavelength in a single-mode fiber can carry data rates of many GHz, but wavelength-division multiplexing can be used to transmit many wavelengths through a single fiber. Thus, the aggregate bandwidth in a single fiber can be measured in THz. Since fiber-optic cables typically contain tens or hundreds of fibers, a single cable can handle extremely large amounts of data.

## 20-3   Optical Transmitters

An optical transmitter performs several critical functions in a FO communications system. First, it must provide light at a frequency that minimizes attenuation and distortion as it propagates through the FO waveguide. Second, it must modulate the light coupled into the fiber with the binary signal to be transmitted at the required data rate. Third, the power in the light it provides must be sufficient to be detected reliably by the optical receiver at the far end of the fiber. Fourth, it must couple the light it produces efficiently into the fiber within the cone of acceptance, so that the light will be transmitted through the fiber rather than be refracted in the cladding.

Optical transmitters are commercially available modules that typically utilize either a light-emitting diode (LED) or a laser diode (LD) as the light source, along with the electronics necessary to couple in the electrical signal, the mechanism necessary to modulate the light, and the mechanism necessary to couple the light to the fiber. In selecting an appropriate transmitter for a system, there are several factors to consider. The transmitter must provide sufficient power to meet the system's BER requirement after the losses in the system are taken into consideration. The transmitter must be the correct size to couple the light into

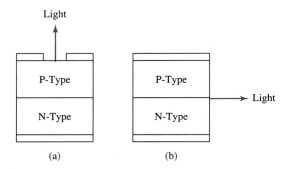

**FIGURE 20–9**
(a) Surface-emitting LED; (b) edge-emitting LED.

the type of fiber being used, and be capable of coupling the light into the cable efficiently. The transmitter must be capable of modulating the light at the data rates required by the system, and must be sufficiently linear to prevent the production of harmonics and distortion. The transmitter must also meet cost, size, weight, and reliability constraints.

## LED

An LED is basically a forward-biased PN junction diode, in which recombination of electrons and holes in the depletion region generates photons of light. The structure is designed to allow some light photons to escape, as shown in Figure 20–9. If the photons are allowed to escape in a direction normal to the junction, the diode is called a surface-emitting LED, as shown in Figure 20–9a. If the photons escape in a direction parallel to the junction, the diode is called an edge-emitting LED, as shown in Figure 20–9b. Edge-emitting LEDs tend to have a narrower beam of emitted light than surface-emitting LEDs. This makes it easier to couple more of the light power into the fiber.

## LD

Laser diodes are more complex than LEDs, although the basic mechanism is still that of a forward-biased PN junction diode. There are several different types of LDs, including multilongitudinal mode (MLM), single longitudinal mode (SLM), single longitudinal mode with distributed feedback (DFB), and vertical-cavity surface-emitting laser (VCSEL). In an LD, the PN junction is contained in a cavity that has a reflective surface at one end and a partially reflective surface at the other. Below a certain threshold current, the LD acts like an LED and produces noncoherent light. At currents above that threshold, laser action occurs. The light reflecting back and forth between the reflective surfaces is reinforced at a wavelength determined by the physical size of the cavity, and canceled at other wavelengths. Some of this light escapes through the partially reflective surface, and forms the output of the LD. In many LDs the partially reflective end is normal to the junction, so the LD acts as an edge-emitting source, as shown in Figure 20–10a. The VCSEL, however, has the partially reflective surface parallel to the junction, so it acts as a surface-emitting source, as shown in Figure 20–10b.

## LED versus LD

Both the LED and the LD can satisfy the requirements for the light source in an optical transmitter for many applications. Each can be modulated with a digital electrical signal, and can generate an optical beam that is compatible with fiber-optic cables. The LED tends to be more reliable, have better linearity, and cost less than the LD, while the LD generally can be modulated at higher data rates, can produce higher optical power, and couples energy to the fiber-optic cable more efficiently. Several factors are involved in these differences.

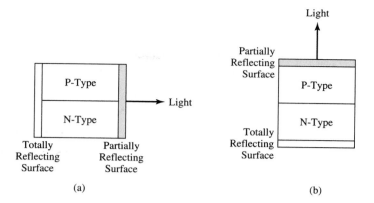

**FIGURE 20–10**

(a) Laser diode; (b) vertical-cavity surface-emitting laser.

The light produced by an LD is coherent; that produced by an LED is not. The light emission events in a laser are triggered by a common stimulus, so that the photons in the light beam produced have a specific phase relationship with each other. The result is a very narrow beam, both spatially and spectrally. Spatially, the beam is focused into a very narrow beamwidth that can be directed into the core of the fiber-optic cable. Spectrally, an LD typical has a bandwidth of about 1 nm at a wavelength of 850 nm, compared to a bandwidth of about 40 nm at 850 nm for an LED. At a wavelength of 1310 nm, the bandwidth figures increase to about 3 nm for the LD and 80 nm for the LED. In general, the narrower the bandwidth of the beam, the greater the distance it can travel in the fiber without regeneration.

Both the LED and the LD can be modulated by varying the current through the diode. This can be accomplished either by having the input signal turn the diode completely on or off, or by having the input signal change the current through the diode between two set values. In the first case, the presence of light can represent a 1 and the absence of light can represent a 0. In the later case, a light of high intensity can represent a 1 and a light of low intensity can represent a 0. For a given diode, the latter method may support higher data rates; however, such direct modulation causes some shift in the frequency of the light. This shift is called *chirp*. Chirp can cause the light pulses to broaden, which can degrade the system's BER.

An alternative to direct modulation (i.e., turning the source on and off) is the use of an optical modulator. An optical modulator is placed between a continuous light source and the fiber, and acts like a shutter, blocking or passing the light. A solid-state optical modulator can be integrated into the same chip as the LED or LD.

Because the LD can provide higher power and a more concentrated beam of light (both spatially and spectrally), it is more suitable than the LED for coupling light into single-mode fibers, and is used when long-distance transmission without regeneration is required. The LED is typically used for short-distance transmissions.

A lens can be used to help concentrate the output of either an LED or a LD for transmission into the fiber, as shown in Figure 20–11. This can significantly improve the efficiency of coupling the light into the fiber.

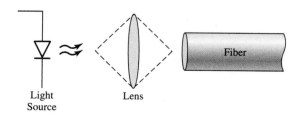

**FIGURE 20–11**

Lens used to concentrate light into fiber.

## 20-4 Optical Receivers

An optical receiver or light detector has one primary function, to convert the light received from the fiber to an electrical signal. An important parameter of optical receivers is their *responsivity* (R) or sensitivity. This is defined as the ratio of the electrical output current to the incident light power, and is measured in amperes per watt (A/W). Responsivity is directly proportional to the wavelength of the light, and to the quantum efficiency of the photodiode. Quantum efficiency ($\eta$) is a measure of the number of electrons produced by each incident photon of light. The relationship between responsivity, quantum efficiency, and wavelength is given by

$$R = \frac{\eta\lambda}{1234} \text{ A/W} \tag{20-14}$$

where $\lambda$ is the wavelength in nanometers.

The PIN photodiode and the avalanche photodiode (APD) are the two types of light detector most commonly used in FO communication systems. The characteristics of each will be considered below.

### PIN Photodiode

A PIN photodiode is a PN junction diode with an intrinsic region between the P- and N-type materials. PIN photodiodes for use in fiber-optic communications systems are usually formed from silicon, or from an alloy of indium arsenide (InAs) and gallium arsenide (GaAs) called InGaAs. Physically, the diode provides a window through which light can reach the depletion region of the junction. The diode is reverse-biased, and photons reaching the depletion region break covalent bonds to generate electron-hole pairs. This results in a current in the external circuit that is proportional to the light intensity. PIN photodiodes can be operated at the 5–15 volt supply voltage levels normally associated with integrated circuits and printed circuit boards.

Silicon PIN photodiodes are most suitable for applications in the 850 nm band. Their responsivity peaks near 850 nm and drops off rapidly above 900 nm. These photodiodes typically have a quantum efficiency of about 80%, and responsivity of about 0.55 A/W at a wavelength of 850 nm.

---

**▌▌ EXAMPLE 20-4**

A silicon PIN photodiode with a quantum efficiency of 80% is used to detect light with a wavelength of 850 nm. What is the responsivity of the photodiode?

**SOLUTION** Substituting the quantum efficiency and the wavelength into Equation 20-14,

$$R = \frac{\eta\lambda}{1234} \text{ A/W} = \frac{(0.8)(850)}{1234} = 0.55 \text{ A/W} \tag{20-15}$$

An InGaAs PIN photodiode, on the other hand, has a responsivity that tends to be constant over the wavelength range of 1300 nm to 1600 nm, and drops off rapidly above about 1650 nm. At the two longer wavelength bands, 1310 nm and 1550 nm, an InGaAs PIN photodiode typically has a quantum efficiency between 60% and 70% and a responsivity between 0.6 A/W and 0.9 A/W.

---

**▌▌ EXAMPLE 20-5**

An InGaAs PIN photodiode with a quantum efficiency of 60% is used to detect light with a wavelength of 1310 nm. What is the responsivity of the photodiode?

**SOLUTION** Substituting the quantum efficiency and the wavelength into Equation 20-14,

$$R = \frac{\eta\lambda}{1234} \text{ A/W} = \frac{(0.6)(1310)}{1234} = 0.64 \text{ A/W} \tag{20-16}$$

▌▌

### Avalanche Photodiode (APD)

An APD is a PIN photodiode that utilizes a high electric field in the intrinsic region to produce gain by avalanche multiplication. This high electric field requires a much higher supply voltage than that required for ordinary PIN photodiodes. Silicon APDs have a gain of at least 100, and may be as high as 1000, and are usable over a wavelength range of 300 nm to 1100 nm. Germanium APDs and InGaAs APDs have a much lower gain, typically 30 to 40. Germanium APDs are usable over a wavelength range of 800 nm to 1600 nm, while InGaAs APDs are usable from 900 nm to 1700 nm.

### PIN Photodiode versus APD

If system requirements (which generally means the BER) can be met with a PIN photodiode, that is generally preferred. PIN photodiodes are more reliable, and may have a mean time between failures that is 10 times that of APDs. PIN photodiodes are less temperature-dependent than APDs, and are much less expensive. As noted above, they operate at power supply levels consistent with most integrated circuits, whereas APDs require much higher power supply voltages. However, when fiber links are long, it may be necessary to use an APD in order to meet BER requirements.

## 20-5 Fiber-Optic Communications System Performance

The performance of a fiber-optic communications system depends upon a number of factors, including the type and length of fiber-optic cable used, the type of light source and detector, the power of the optical signal, post-detection amplification, the efficiency of the connections or splices when lengths of cable are connected, the efficiency of the coupling between the light source and detector and the cable, and environmental conditions (such as temperature). We will consider these factors in this section, and perform a link analysis for optical communications systems similar to that performed for radio-frequency communications systems in Chapter 16.

### Fiber Losses

Losses in the fiber-optic waveguide are expressed in dB/km. Losses in fiber generally result either from absorption of energy by molecules in the fiber or from radiation. Absorption of energy occurs at certain wavelengths. For example, water trapped in the fiber in the form of OH ions absorbs strongly at 1390 nm. The molecules of silica itself absorb some of the energy in the transmitted light, although this is minimal compared to other losses, and will be less than 0.03 dB/km. Imperfections in the fiber (such as fluctuations in the density, bubbles, cracks and irregularities in the boundary between the core and the cladding) can cause losses. Fiber can be fabricated from plastic or glass. That fabricated from plastic is less expensive, but has much greater losses than that fabricated from glass. Therefore, glass cables are preferred for most communications applications, and are the only type considered here.

Another factor that causes loss in fiber is bending of the cable. As shown in Figure 20–12, when a cable is bent too much, it is possible for some of the light to pass through the core-cladding boundary and escape from the optical waveguide. As long as the radius of the bend does not exceed the manufacturer's recommendation, this loss should be minimal. Also, during manufacture of the cable, imperfections called *microbends* can occur, which also cause losses.

### Dispersion

Light propagates through an optical waveguide in a manner similar to the way RF energy propagates through a microwave waveguide, as a transverse electric field, transverse magnetic field, or as a hybrid field that is a combination of the two. As in a microwave

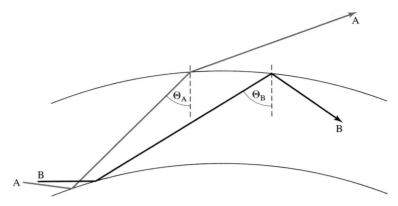

**FIGURE 20–12**
Bend in fiber causes light loss.

waveguide, it is possible for multiple waves or modes to exist simultaneously in the optical waveguide if the waveguide is physically large enough to support them. The two basic types of fiber, multimode with a core that is 50 to 100 microns in diameter, and single mode with a core that is 8 to 10 microns in diameter, present very different problems to the optical system designer. Because of the larger core diameter, it is easier to couple light from the source into the multimode cable. In addition, the larger diameter of a multimode cable allows it to support multiple modes simultaneously, while the smaller diameter of a single mode cable only allows it to support propagation of a single mode—but this advantage comes with its own cost.

The significance of the ability to support multiple modes is that the modes follow different paths through the core as they travel down the waveguide, as illustrated in Figure 20–13. In this figure, the mode with the shortest path travels straight down the core. The mode with the longest path is reflected frequently from the boundary between the core and the cladding. Because these path lengths are different, a pulse of light tends to spread out as it travels further from the source. The light traveling the longer path arrives at any point in the optical waveguide later than the light traveling the shorter path. This spreading of the pulses limits the distance from the source that the transmitted signal can be successfully detected. This is illustrated in Figure 20–14, which shows pulses at different positions along the fiber. On the

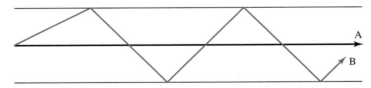

**FIGURE 20–13**
Dispersion.

Distance along Fiber

**FIGURE 20–14**
Pulse spreading caused by dispersion.

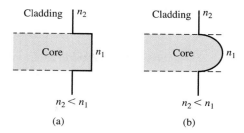

**FIGURE 20–15**
Index of refraction variation in (a) step-index fiber, and (b) graded-index fiber.

left, the pulses have just been injected into the fiber. In the middle, dispersion has caused the pulses to spread, but it is still possible to discriminate between a 1 and a 0. On the right, the spreading is so extensive that the individual pulses can no longer be identified.

This spreading out of the pulse as it propagates through the optical waveguide is called *intermodal dispersion,* and is a greater problem at higher data rates and for longer path lengths. Single-mode cable does not have this intermodal dispersion problem, so an optical signal of a given intensity at the source can be successfully detected at a much greater distance from the source in single-mode than in multimode cable.

Although intermodal dispersion is not a problem in single-mode cables, a different dispersion phenomenon, called *chromatic dispersion,* is present in all fiber-optic cables. Even within the same mode, different wavelengths of light travel at different speeds. This is because the index of refraction of a particular material varies inversely with wavelength. Thus the speed of higher-wavelength light is less than that of lower-wavelength light. Since the light in the core of the optical wavelength generally consists of a range of wavelengths rather than a single wavelength, the result is that pulses spread and are attenuated as they travel away from the source. Since the loss due to chromatic dispersion is much less than that due to intermodal dispersion, single-mode cables are used for long cable runs.

As stated in Section 20-2, the mechanism that causes the light to be completely contained in the core of the cable is that the core has a higher index of refraction than the cladding. This makes it possible for reflections from the boundary between the core and the cladding to cause the light to be completely contained in the core. Originally, all fibers had an abrupt transition between the core index of refraction ($n_1$) and the cladding index of refraction ($n_2$), as shown in Figure 20–15a. Such fiber-optic cables are called *stepped-index* cables. It is also possible to fabricate fiber-optic cables in such a way that the index of refraction varies gradually, as shown in Figure 20–15b. These *graded-index* cables are more expensive to manufacture, but dispersion is reduced, because the modes with the longer paths travel at a greater speed.

---

**▌▌ EXAMPLE 20-6**

A FO cable will be used to connect two buildings that are 800 meters apart. The insertion loss of a single mode FO cable is 0.4 dB/km at the wavelength to be used. How much loss will using that cable introduce?

**SOLUTION**   Let $L$ (dB) be the loss.

$$L\ (\text{dB}) = 0.4\ \text{dB/km} \times 0.8\ \text{km} = 0.32\ \text{dB} \tag{20-17}$$

Therefore using this cable will introduce 0.32 dB of loss.

---

**▌▌ EXAMPLE 20-7**

Instead of using the single-mode cable of Example 20-6, multimode cable having an insertion loss of 2.7 dB/km is to be used. What loss will the multimode cable introduce?

**SOLUTION**

$$L \text{ (dB)} = 2.7 \text{ dB/km} \times 0.8 \text{ km} = 2.16 \text{ dB} \tag{20-18}$$

Therefore the multimode cable will introduce significantly more loss than the single-mode cable.

## Coupling and Connector Losses

Connectors are used with fiber-optic cables to couple the cables to components in the system. The attachment of those connectors to the cables introduces losses, which can be very high if great care is not taken. Furthermore, failure to ensure cleanliness when connecting the cable to a component can introduce large losses. A speck of dust can be larger than the diameter of a single-mode cable; if introduced into the connection, it can block transmission of light. In general, connectors applied to the fiber by the manufacturer have less insertion loss than those applied in the field, on the order of 0.5 dB per connector as compared to 1 dB or more.

When two lengths of fiber-optic cable are joined, either by splicing or with connectors, the mating surfaces must be carefully cut and polished to minimize loss through that connection. Surfaces that are not sufficiently smooth, as shown in Figure 20–16, introduce losses. Also, connections in which the cores of the connected cables are not perfectly aligned, as shown in Figure 20–17, will introduce losses because the light is not coupled completely from one fiber to the other. Even carefully installed connectors and skillfully prepared connections will introduce losses, but these can be minimized.

## Link Analysis

FO communications systems will be either loss-limited or dispersion-limited. A system is loss-limited if the loss of signal power over the distance between optical transmitter and optical receiver reduces the BER below an acceptable value before dispersion makes the signal undetectable. A system is dispersion-limited if the power at the optical receiver is adequate but the spreading of the pulse by dispersion reduces the BER below an acceptable value. If the system is loss-limited, the effect of dispersion is accounted for by adding a dispersion penalty, as illustrated in Example 20-8.

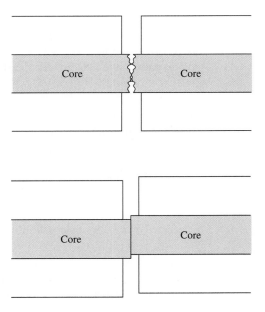

**FIGURE 20–16**
Example of poor surface preparation before connection.

**FIGURE 20–17**
Example of poor core alignment during connection.

**EXAMPLE 20-8**

Figure 20–18 is a sketch of a 100-km fiber-optic link. The light source is an LD having an output of 0 dBm. Although the sketch shows only a single splice, there will be a splice in the fiber every 2 km, and it is estimated that the insertion loss at each splice is 0.03 dB. Single-mode cable with a loss of 0.25 dB/km will be used. The insertion losses of the connectors at the transmitter and receiver ends of the cable are estimated to be 0.5 dB each. Assume a dispersion penalty of 1 dB. A PIN photodiode receiver that requires an input of −30 dBm to achieve a BER of $1 \times 10^{-10}$ (the required BER) has been proposed as the detector. Can this system perform adequately?

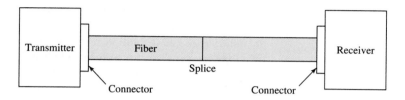

**FIGURE 20–18**
Fiber-optic communications system.

**SOLUTION**  The output of the source is 0 dBm. 2-km lengths of fiber are to be spliced to form the 100-km fiber link. There will be 50 lengths of fiber, so 49 splices will be required. Total insertion loss for the splices is 1.47 dB ($49 \times 0.03$ dB). The connector at each end introduces 0.5 dB of loss, so total insertion loss for connectors is 1 dB. The fiber itself introduces a loss of 25 dB ($100 \times 0.25$ kB/km). The dispersion penalty is 1 dB. Therefore, the total loss is $1.47 + 1 + 25 + 1 = 28.47$ dB. The signal reaching the detector will be 0 dBm $-28.47$ dB $= -28.47$ dBm.

Since this is greater than the −30 dBm threshold required for a BER of $1 \times 10^{-10}$, the system should perform adequately when new. However, deterioration over the lifetime of the system will increase the loss. Adding 3 dB for end-of-life deterioration is reasonable. If this is added, the total loss is 31.47 dB.

The new figure for the signal reaching the detector after consideration of end-of-life deterioration is −31.47 dB, which is less than the threshold for the required BER.

In addition, we have not allowed any margin for such things as slack in the cable, and losses that exceed what was estimated. Therefore, it would be prudent to select either a source with a higher output or a detector with a lower threshold.

## 20-6 Multiplexing Concepts and Components

When fiber was first used as a medium for transmission of signals, each strand of fiber supported the transmission of only one wavelength. Since the bandwidths of most signals are at most a few hundred MHz, the inherent high bandwidth of fiber was not utilized. Furthermore, demands for bandwidth were and are increasing at a high rate, and the cost of installing new fiber plants is high. It would be advantageous if greater bandwidth could be obtained from existing fiber plants. But commercially feasible methods of multiplexing and demultiplexing optical signals did not exist, so frequency-division multiplex .g (called wavelength-division multiplexing or wave-division multiplexing in the optical world) could not be utilized. However, technology prevailed, and optical multiplexers and demultiplexers were developed and are now available as "off the shelf" equipment.

### Wavelength-Division Multiplexing

Wavelength-division multiplexing (WDM) enables signals from different sources to be coupled into the same optical fiber. In Figure 20–19, signals from four different sources, each having a distinct wavelength, are combined in the multiplexer. The combined signal

**FIGURE 20–19**
Wavelength-division multiplexing.

is then coupled into the fiber for transmission to the demultiplexer. At the demultiplexer the four signals are separated, and each is coupled to a different detector. In this diagram, the multiplexer and demultiplexer are both labeled MUX, because in optical systems both functions can generally be performed by the same unit.

Commercially available optical multiplexers incorporate the light sources for each channel and the optical components for combining the wavelengths into a single unit. The inputs to the multiplexer are electrical signals, and the output is an optical signal containing multiple modulated wavelengths, with each electrical signal modulating a different wavelength. These multiplexers also typically act as demultiplexers. If the fiber is used as the input to the multiplexer, it separates an optic signal containing multiple wavelengths into separate electrical signals, each corresponding to a different optical wavelength, as shown in Figure 20–20.

The channel separation in WDM systems must be great enough to prevent interference between adjacent channels. As shown in Figure 20–21, each channel should have a defined passband that contains the signal transmitted on that channel, and there should be a guard band between channels.

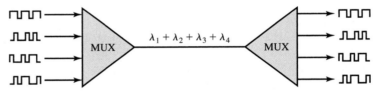

**FIGURE 20–20**
WDM with electrical signal inputs and outputs.

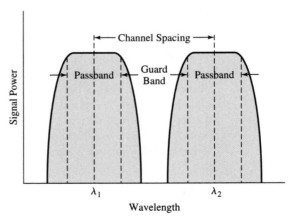

**FIGURE 20–21**
Adjacent channels and guard band in WDM.

## Dense Wavelength-Division Multiplexing

When the WDM channels are close together, the term *dense wavelength-division multiplexing (DWDM)* is used. The International Telecommunications Union (ITU) has defined standard channels in the 1550 nm band for 50- and 100-GHz channel spacing. Channels that are this densely packed require multiplexers that contain very sharp optical filters to provide adequate isolation between wavelengths.

## PROBLEMS

**20-1**  Determine the frequency of an optical signal having a wavelength of 850 nm.

**20-2**  Determine the frequency of an optical signal having a wavelength of 1550 nm.

**20-3**  Two optical signals are 100 GHz apart. What is their separation in wavelength?

**20-4**  Two optical signals are 50 GHz apart. What is their separation in wavelength?

**20-5**  The index of refraction of crown glass is 1.52, and that of air is 1. A ray of light in the crown glass strikes the boundary with an angle of incidence of 25°. What is its angle of refraction?

**20-6**  The index of refraction of air is 1, and that of water is 1.33. A ray of light in the air strikes the boundary with an angle of incidence of 50°. What is its angle of refraction?

**20-7**  What is the critical angle beyond which a ray of light striking the boundary between crown glass and air will be completely reflected back into the crown glass? The index of refraction of crown glass is 1.52, and that of air is 1.

**20-8**  What is the critical angle beyond which a ray of light striking the boundary between water and air will be completely reflected back into the water? The index of refraction of water is 1.33, and that of air is 1.

**20-9**  A silicon PIN photodiode with a quantum efficiency of 70% is used to detect light with a wavelength of 820 nm. What is the responsivity of the photodiode?

**20-10**  An InGaAs PIN photodiode with a quantum efficiency of 60% is used to detect light with a wavelength of 1550 nm. What is the responsivity of the photodiode?

**20-11**  Single-mode fiber with a loss of 0.3 dB/km will be used to span 140 km. How much loss will that cable introduce?

**20-12**  Multimode fiber with a loss of 2.3 dB/km will be used to span 900 m. How much loss will that cable introduce?

**20-13**  An LD having an output of 2 dBm will be used as the light source for a FO communications link that will be established between two locations that are 140 km apart. Single-mode fiber with an insertion loss of 0.35 dB/km will be used. There will be a splice in the fiber every 2 km, and it is estimated that the insertion loss at each splice is 0.02 dB. The insertion loss of the connectors at the transmitter and receiver ends of the cable is estimated to be 0.6 dB each. Assume a dispersion penalty of 1 dB and a margin of at least 5 dB. What threshold must the optical receiver have (i.e., what minimum input signal to the receiver must produce the required BER)?

**20-14**  A FO communications link will be established between two locations that are 1500 m apart. Multimode fiber with an insertion loss of 2.2 dB/km will be used. The connectors at the ends of the fiber have a loss of 1 dB each. A PIN photodiode receiver that requires an input of $-28$ dBm to achieve a BER of $1 \times 10^{-10}$ will be used. Assume a dispersion penalty of 1 dB and a margin of at least 5 dB. What is the minimum output required of the light source?

# Consumer Communication Systems

# 21

## OVERVIEW AND OBJECTIVES

The general principles of communication systems have been developed throughout the text and many practical system concepts have been introduced. This last text chapter will be devoted to discussion of the operation of some common commercial systems that consumers use on a daily basis. Throughout this development, you will be able to see many applications of the theory considered earlier in the text.

It should be stressed that the various parameters and assumptions provided in this chapter could change as new standards and technological developments are introduced. It should also be stressed that there are variations within the literature concerning some of the current standards. These variations may reflect different manufacturers' interpretations and applications of the standards. However, this chapter should impart to the reader a definite appreciation of the progress made in the communication field, and most of the systems discussed here are widely available and very visible to most consumers.

## Objectives

After completing this chapter, the reader should be able to:

1. Discuss the general principles of operation of stereo frequency modulation (FM) systems.
2. Show how the left and right channels are separated in a stereo FM signal.
3. Sketch the baseband spectral form of an FM signal and identify the components.
4. Describe the scanning process in a television (TV) system.
5. Estimate the required bandwidth to process a video signal.
6. Sketch the spectral form of a TV channel and identify the signals involved.
7. Describe the operation of a monochrome TV transmitter.
8. Describe the operation of a monochrome TV receiver.
9. Identify the primary colors and discuss the *RGB* video color system.
10. Describe the process by which color information is inserted into a TV channel.
11. Describe the encoding process for generating a color signal at the transmitter.
12. Describe the decoding process for extracting a color signal at the receiver.
13. Discuss some of the properties of high-definition color television.
14. Describe the basic operation of a telephone circuit.

## 21-1   Stereo Frequency Modulation

We will begin the development with a consideration of the process by which frequency modulation broadcasting systems were modified to provide stereo, hereafter referred to as *stereo FM*. The *monaural* (or one-channel) FM system had been in operation for many years when the concept of stereo FM was first considered.

All broadcasting systems in the United States of America are regulated by the Federal Communications Commission (FCC), an agency of the United States government. The FCC assigns operating frequencies, power levels, and many other standards to which radio and television stations are required to adhere. One might say that the FCC is the "guardian" of the frequency spectrum, a resource that must be managed in a manner similar to land and other natural resources.

As a general rule, the FCC requires that any new communication technology must not immediately force obsolescence of existing technology. This philosophy was applied to stereo FM and, as we will see later, it was also applied to color television. Thus, when stereo FM was being developed, it was essential that the millions of existing monaural receivers should still be able to receive monaural versions of the signals with minimal or no alteration. This type of boundary condition creates some interesting challenges, but it also provides an opportunity to apply communication theory in rather innovative ways.

### Stereo FM Signal Processing at the Transmitter

The process for simultaneously generating monaural and stereo FM is illustrated by the block diagram of Figure 21–1. The two signals for stereo FM are denoted $L$ (for left) and $R$ (for right), referring to the orientation of the sound sources as well as the intended orientation of the speakers at the receiver. The two signals could represent the outputs of two separate microphones, or the two channels at the output of a sound recording medium such as a compact disc (CD). Preemphasis is applied to the two channels in accordance with the performance discussions of Chapter 13. Following deemphasis at the receiver, the result should be an improvement in the signal-to-noise ratio, as discussed in that chapter.

Along the upper path, the two signals are added together, and the result is the composite monaural signal, which is the only signal of interest as far as monaural transmission is concerned. Although the original $L$ and $R$ signals have been modified by preemphasis, to simplify the development, the original notation will be retained and the sum will be denoted simply as $L + R$.

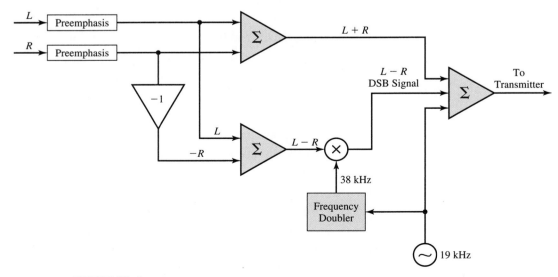

**FIGURE 21–1**
Processing scheme for baseband stereo FM signals.

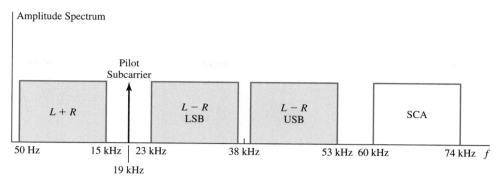

**FIGURE 21–2**
Baseband spectrum of commercial FM stereo signal.

The signal $R$ is inverted with an amplifier having a gain of $-1$ to yield $-R$. This signal is added to $L$ in the lower summer, and the output is denoted $L - R$. The spectral width of $L - R$ is essentially the same as $L + R$; that is, about 15 kHz, so it must be processed in a manner that does not conflict with $L + R$.

As we progress through the remainder of the discussion, the reader may wish to refer to Figure 21–2 along with Figure 21–1. Figure 21–2 shows the spectral allocation for the signals being developed prior to RF modulation, and this layout will support the work of Figure 21–1.

To provide a unique location for $L - R$, this signal is applied as the input to a balanced modulator in which the carrier frequency is 38 kHz. The result is a DSB signal centered at 38 kHz, but with a spectrum ranging from about $38 - 15 = 23$ kHz to about $38 + 15 = 53$ kHz.

Note from Figure 21–1 that the 38-kHz subcarrier was derived from a 19-kHz subcarrier by doubling the frequency. The 19-kHz pilot subcarrier is then added to the baseband signal $L + R$ and the DSB version of $L - R$. The reason for using a 19-kHz pilot subcarrier for transmission instead of a 38-kHz subcarrier is that the latter would be difficult to extract within the narrow region between the two sidebands. However, there is "clear territory" around 19 kHz, and this simplifies the signal processing greatly.

Carrier synchronization for detection is not a big problem with this system, since there is phase coherency between the modulation subcarrier at the transmitter and the detection subcarrier at the receiver. In practice, an additional time delay is required in the $L + R$ path to compensate for the additional time delay for the more complex processing in the $L - R$ path, but that is not shown in Figure 21–1.

In addition to the various signals already considered, the FCC provides a *subsidiary carrier authorization* (SCA), which is shown from 60 kHz to 74 kHz. This spectral allocation can be used for other purposes, including certain types of background music, and so on. Clearly, commercial FM demonstrates a common application of frequency-division multiplexing (FDM).

## Stereo FM Processing at the Receiver

The block diagram of Figure 21–3 illustrates the processing for both monaural FM and stereo FM. The system shown begins at the output of the FM detector. For monaural processing, the signal is passed through a deemphasis circuit to compensate for the preemphasis at the transmitter, and then to an audio amplifier and speaker as illustrated along the top row.

The more complex processing is that of a stereo receiver. Along the top row of the stereo receiver diagram, the $L + R$ signal is recovered by passing the composite signal through a low-pass filter whose cutoff frequency is just above 15 kHz. Some additional delay (not shown) may be required to match the additional delay of the more complex $L - R$ component as it is extracted.

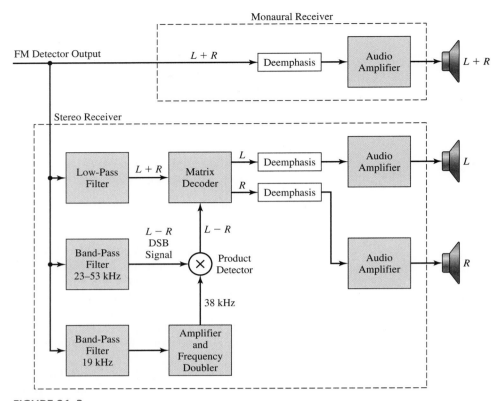

**FIGURE 21–3**
Block diagram showing monaural and stereo receiver signal processing.

The $L - R$ component initially is a DSB signal occupying a spectral range from about 23 kHz to 53 kHz, so it must be separated from the composite signal by a band-pass filter as shown.

The 19-kHz subcarrier must be extracted by either a very narrow band-pass filter or by a phase-locked loop (PLL). This signal is amplified to a suitable level, and the frequency is doubled to 38 kHz.

The output of the product detector is $L - R$ after eliminating the superfluous component in the product demodulation process centered at twice the subcarrier frequency. (The low-pass filter that removes that component is not shown.)

For the development that follows, we will define two variables as

$$x = L + R \tag{21-1}$$

$$y = L - R \tag{21-2}$$

The matrix decoder has $x$ and $y$ as its two inputs. From a simple mathematical point of view, we have two simultaneous equations with two unknowns, so we should be able to determine $L$ and $R$ from $x$ and $y$.

Solving the equations for $L$ and $R$ in terms of $x$ and $y$, we obtain

$$L = 0.5x + 0.5y \tag{21-3}$$

and

$$R = 0.5x - 0.5y \tag{21-4}$$

Therefore, $L$ and $R$ are readily determined by the linear processing indicated, and this is performed in the matrix decoder.

Once $L$ and $R$ are determined, deemphasis can be applied to each, and the resulting signals are amplified and applied to separate speakers.

**EXAMPLE 21-1**

A rather crude, but relatively simple, approach to obtaining a rough estimate of the difference in signal-to-noise performance between commercial stereo FM and monaural FM will be illustrated in this example. The FCC allocates each FM station a bandwidth of 200 kHz. The highest modulating frequency of monaural FM is 15 kHz, and it was shown in Chapter 7 that the resulting bandwidth according to Carson's rule was about 180 kHz, based on a deviation ratio of 5. With stereo FM, the highest modulating frequency (assuming no SCA service) is about 53 kHz. The deviation ratio must obviously be reduced to permit the wider bandwidth of stereo FM. Based on this assumption, and with the processing gain parameter from Chapter 13, determine the difference in performance between stereo FM and monaural FM based on the same signal-to-noise input. For this purpose, disregard preemphasis, and assume that the full 200-kHz channel allocation is used for stereo.

**SOLUTION**   Let $D_1 = 5$ represent the monaural value. From the work of Chapter 13, the receiver processing gain $G_{R1}$ is given by

$$G_{R1} = 3D_1^2(1 + D_1) = 3(5)^2(6) = 450 \qquad (21\text{-}5)$$

Let $D_2$ represent the deviation ratio based on a baseband bandwidth of 53 kHz and utilization of the full channel bandwidth of 200 kHz. This deviation ratio is determined from Carson's rule as follows:

$$B_T = 200 \times 10^3 = 2(1 + D_2)W = 2(1 + D_2) \times 53 \times 10^3 \qquad (21\text{-}6)$$

$$D_2 = \frac{200 \times 10^3}{2 \times 53 \times 10^3} - 1 = 0.887 \qquad (21\text{-}7)$$

Let $G_{R2}$ represent the processing gain for stereo FM. It will be estimated as

$$G_{R2} = 3D_2^2(1 + D_2) = 3(0.887)^2(1.887) = 4.454 \qquad (21\text{-}8)$$

The ratio of the two processing gains is

$$\frac{G_{R2}}{G_{R1}} = \frac{4.454}{450} = 0.0099 \qquad (21\text{-}9)$$

The decibel difference is

$$\text{dB difference} = 10\log\frac{G_{R2}}{G_{R1}} = 10\log 9.9 \times 10^{-3} \approx -20\,\text{dB} \qquad (21\text{-}10)$$

Accept this as a rough estimate, for which the validity of the procedure could be questioned. However, it is close to the actual value that has been determined by more sophisticated approaches. The result means that the price paid for stereo FM is about a 20-dB degradation in signal-to-noise performance. Fortunately, FM provides enough leeway that the process is quite successful if the input signal-to-noise level is sufficient.

## 21-2  Monochrome Television

Commercial television (TV) broadcasting stands out as one of the most complex communication systems encountered in the electronic industry. It is taken for granted by hundreds of millions of people on a daily basis, yet it utilizes some of the most sophisticated signal processing schemes. In the next several sections, we will explore some of the features of this ubiquitous medium from the standpoint of the principles covered throughout the book.

The parameters discussed here are those of the National Television Systems Committee (NTSC). These standards are utilized in the United States and a number of other countries. Different standards are employed in other parts of the world, but the general principles apply; only the values of the operating parameters are different.

Since black-and-white (b&w) or *monochrome* TV is considerably simpler in concept than color TV, this section will be primarily oriented to a discussion of the original principles and standards employed in that medium. Monochrome operation is so intertwined with color operation, however, that it is nearly impossible to separate the discussions of the two media. Thus, much of the information in this section applies to color TV as well.

## Scanning

We begin with a discussion of how the picture is generated on a TV screen by the scanning process. Based on the averaging effect of our eyes, we might think that an entire picture appears instantaneously across the screen, but that is an illusion. The picture is actually created by lines moving across the screen at a very high rate. If all the lines actually transmitted were visible on the screen, the picture would be composed of more than 500 lines. In practice, a few of the lines are not visible, for reasons that will be discussed later.

Some video devices (such as certain DVD players and VCRs) employ **progressive scanning,** which means that an entire picture or **frame** is created by one complete scanning process, in which all the lines are generated in sequence. To reduce flicker based on the original standards, standard commercial TV utilizes a different technique called **interlaced scanning.** With interlaced scanning, a **frame** consists of **two fields.** In the first field, the odd-numbered lines (1, 3, 5, etc.) are scanned on the screen, and in the second field, the even-numbered lines (2, 4, 6, etc.) appear. Thus, it takes two fields to complete a frame.

The scanning process, with considerable exaggeration for clarity, is illustrated in Figure 21–4. Field 1 begins in the upper left-hand corner and sweeps across the screen from left to right, as shown. The overall time for one line sweep, including the time for the beam to return, is about 63.5 μs. Note that the line is also moving vertically as it sweeps across the screen, although the angle shown is highly exaggerated for clarity. When the line reaches the right-hand side of the screen, the picture is **blanked** for a short interval of time, and the beam sweeps back to the left-hand side of the screen. This process is called **retrace,** and it has been eliminated from the figure for clarity. The retrace interval is short compared to the trace time, and the next line starts at a vertical level only slightly lower than the level reached at the right-hand side prior to retrace.

This process continues until the 263rd line begins on the left. This line reaches the bottom at the midpoint, as shown. The **vertical blanking interval** now occurs, in which the beam moves back to the top and begins at approximately the same horizontal position that

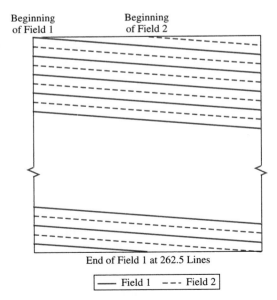

**FIGURE 21–4**
Interlaced scanning systems
for television picture.

*Note:* Retrace lines are not shown.

was reached at the bottom. This point begins the scanning process for Field 2 (the lines for this field are shown as dashed). Since Field 2 begins at the horizontal midpoint, its path does not conflict with that of Field 1. In fact, the lines for field 2 are **interlaced** with those of Field 1. Field 2 terminates at the lower right-hand corner as shown, and another vertical retrace occurs during the next blanking interval. Thus, a new Field 1 begins in the upper left-hand corner, and the entire process is repeated.

### Frame Parameters

The original parameters for the monochrome process are as follows:

Frame rate = 30 frames per second

Field rate = 60 fields per second

Lines per frame = 525

Lines per field = 262.5

The original horizontal sweep rate $f_H$ is determined as

$$f_H = 525 \text{lines/frame} \times 30 \text{ frames/s} = 15{,}750 \text{ lines/s} = 15.75 \text{ kHz} \qquad (21\text{-}11)$$

This means that the period $T_H$ associated with the assumed horizontal sweep rate is

$$T_H = \frac{1}{f_H} = \frac{1}{15{,}750} = 63.5 \ \mu\text{s} \qquad (21\text{-}12)$$

These results indicate that the original horizontal sweep oscillator for black-and-white TV operates at a frequency of 15,750 Hz, meaning that there are 15,750 lines generated each second. We will see later that this value is very slightly modified for color TV.

While the horizontal sweep process is generating many lines per second, the vertical sweep must also operate to successively generate new fields. The vertical sweep rate $f_V$ for black-and-white TV is the same as the field rate, and is given by

$$f_V = 60 \text{ Hz} \qquad (21\text{-}13)$$

The net vertical period $T_V$ is

$$T_V = \frac{1}{f_V} = \frac{1}{60} = 16.67 \text{ ms} \qquad (21\text{-}14)$$

### Horizontal Period

The form of the video voltage signal over the duration of a little more than one line interval is illustrated in Figure 21–5. Increasing brightness corresponds to an upward vertical direction, and increasing darkness (or decreasing brightness) corresponds to a downward

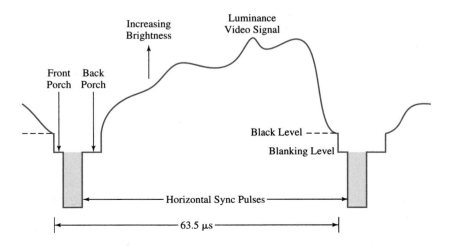

**FIGURE 21–5**
Format of horizontal sweep over one line, plus a portion of the next line.

vertical direction. There is a particular level, shown by the dashed line, that represents full screen blackness, and below it is a **blanking level.** A signal at or below this level will completely blank out the screen. This region is sometimes referred to as the "blacker than black" region.

The dual functions of the horizontal sync pulses are (1) to blank out the screen during the retrace intervals, and (2) to synchronize the horizontal oscillator in the receiver so that it is in alignment with that of the transmitter. Shortly, we will see that vertical synchronization must also be provided. Readers may have seen a malfunctioning TV receiver in which the picture is moving either horizontally, vertically, or in both directions, and this phenomenon is usually caused by a problem with the sync circuits.

Another property of the horizontal sweep interval is that of the flat levels just before and just after the sync pulses. These levels have the unusual but descriptive names of "front porch" and "back porch," respectively. Later we will see that the latter interval plays a key role in the color process, but the present illustration does not show that action.

Between the blanking sync pulses, the video information is provided. The signal shown there is the black-and-white video, and it is called the **Luminance Signal** (often called the **luma** for short). The signal intensity and variation during a given interval between sync pulses represents the varying intensity and variation of the picture information for that particular horizontal line. Note again that the brighter the spot, the higher the level of this time-varying video signal voltage.

A point that will come out later (in dealing with color TV) is the somewhat subtle property that the video is actually being sampled; as such, it assumes some of the character of a sampled signal as developed much earlier in the book. In fact, the signal is in effect being sampled at a rate of 15.75 kHz by virtue of the periodic interruptions of the signal at that frequency. We will return to that important point later.

## Vertical Period

In addition to the thousands of times per second that the screen is blanked for horizontal retrace, the screen must also be blanked 60 times per second to allow the vertical retrace to take place. For this purpose, vertical sync pulses are required, and they must assume a different character in order for the receiver to distinguish them from the horizontal sync pulses. As in the case of horizontal sync, the vertical sync pulses must blank the screen during the vertical retrace interval as well as synchronize the vertical sweep oscillator. The vertical blanking interval is much longer than the horizontal blanking interval, and it is much more complex. There are actually two vertical blanking intervals per frame, one for the first field and one for the second field. There are slight differences in the two blanking intervals, based on the difference in the timing sequence.

Refer to Figure 21–6 in the discussion that follows. Part a deals with Field 1 and part b deals with Field 2. For this purpose, Field 1 is defined as the field in which scanning begins in the upper left-hand corner and ends at the middle of the bottom, while Field 2 is assumed to begin in the top middle and end at the far right of the bottom.

The vertical blanking interval begins with six equalizing pulses. The width between successive pulse beginnings is $T_H/2$, so the net width of this interval is $3T_H$. The vertical sync pulse may also be considered as having a width of $3T_H$, and if it were unnecessary to consider the simultaneous horizontal effect, it could be a single pulse of that width. However, it is necessary to maintain horizontal sync during the time of the vertical sync signal. Therefore, what would otherwise be a wide negative pulse is *serrated* or sliced as shown, so that horizontal synchronization can be maintained during that interval by the narrow transitions.

The vertical sync pulse is followed by another sequence of six equalizing pulses, and the width of that interval is also $3T_H$. Finally, an additional interval of about $12T_H$, consisting of a regular train of additional horizontal pulses, appears at the end of the vertical blanking interval to allow sufficient time for the oscillators to realign and begin a new field.

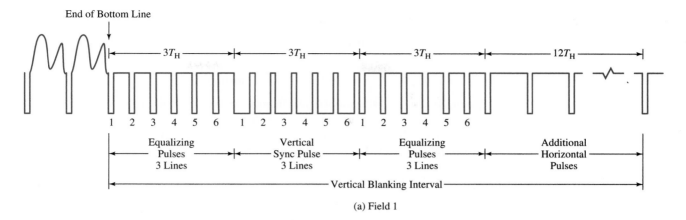

(a) Field 1

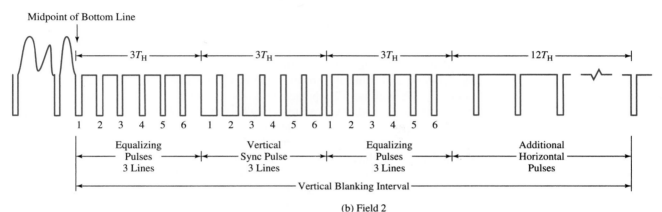

(b) Field 2

**FIGURE 21–6**

Formats for the vertical blanking intervals for the two fields.

Field 2 follows essentially the same format except that it begins at the midpoint of a horizontal sweep interval, as shown in Figure 21–6b. Once this time shift is considered, it becomes clearer why the different forms of horizontal synchronizing pulses must occur at intervals of $T_H/2$ during the vertical blanking interval. Otherwise, the initial shift of a half-cycle of the horizontal sweep would be misaligned during the vertical interval. A detailed examination of both fields will reveal that the horizontal synchronization is maintained throughout the process.

The net time $T_{VB}$ for the vertical blanking interval is the sum of the various time intervals shown on the figure:

$$T_{VB} = 3T_H + 3T_H + 3T_H + 12T_H = 21T_H = 21 \times 63.5 \ \mu s \approx 1.33 \ ms \qquad (21\text{-}15)$$

This means that about 21 lines are blanked out in each field, or about 42 lines of a complete frame are lost. Fortunately, this loss does not cause any serious problem with the visual information. However, although we initially say that there are 525 lines of information in a black-and-white frame, the actual number appearing on the screen is about $525 - 42 = 483$ lines.

Incidentally, due to the length of the vertical blanking period relative to the horizontal sweep interval, the electron beam would actually undergo a zig-zag pattern on the screen during the vertical blanking interval if it were visible. However, it becomes visible again at the correct position at the top of the screen for a given field, based on the timing sequence provided in alternate fields.

## Aspect Ratio

The *aspect ratio* is defined as follows:

$$\text{aspect ratio} = \frac{\text{picture width}}{\text{picture height}} \tag{21-16}$$

The original aspect ratio was 4/3 and there are millions of TV sets in existence having that ratio (including those of the authors at the time of this writing). However, the newer standard, which is being used in many new TVs (as well as for many programs) is 16/9. For the computations that follow, the original aspect ratio of 4/3 will be used.

## Video Bandwidth

An important parameter that had to be considered in the early days of television is that of the required video bandwidth. It is a baseband bandwidth, in that its lowest frequency is dc. Indeed, if the TV signal did not have a dc component, it would not be possible to see any extended interval on the screen having the same shade or color. However, the frequency range of video extends far beyond the range required for audio.

Different references estimate this bandwidth in slightly different ways, and the results vary somewhat. Ultimately, however, the FCC limitation comes into play and serves as a boundary. We will use an approach here that the authors find user friendly, but don't be puzzled if you find other approaches elsewhere that result in slight differences.

The bandwidth estimation reduces to an observation of the number of possible changes that could be made on the screen per unit of time, which makes the process somewhat similar to that of encoding binary digital data. First, consider the finite lines that make up the scanning process. We have seen that the process begins with 525 lines, although some are not visible. Assume in a hypothetical situation that each line must take on a different shade or video level. The result would be 525 possible changes for a frame in the vertical direction at a given horizontal position. If the screen geometry is assumed to be symmetrical, the number of possible changes in a horizontal direction at a given vertical level will be about $(4/3) \times 525 = 700$ possible changes, based on an aspect ratio of 4/3. These numbers also relate to the concept of *picture elements* or **pixels.** While the actual number of pixels on a screen does not necessarily agree with the numbers provided here, the possible number of changes that could theoretically occur in a frame could be as great as the following:

$$\text{maximum possible changes in a frame} = 525 \times 700 = 367{,}500 \tag{21-17}$$

However, there are 30 complete frames per second, so the maximum total number $N$ of possible changes that could occur per second could theoretically be as great as

$$N = 30 \times 525 \times 700 \approx 11 \times 10^6 \text{ changes/second} \tag{21-18}$$

Next, we will return to a concept that was introduced very early in the book and used in dealing with pulse and digital transmission, namely the approximate bandwidth required to reproduce a pulse. The time $\tau$ for a given picture element is the reciprocal of the value of $N$, which is

$$\tau = \frac{1}{11 \times 10^6} = 90.9 \text{ ns} \tag{21-19}$$

The approximate bandwidth $B_\text{T}$ could then be estimated as

$$B_\text{T} \approx \frac{0.5}{\tau} = 0.5 \times 11 \times 10^6 \approx 5.5 \text{ MHz} \tag{21-20}$$

In practice, this value can be relaxed somewhat due to several factors. First, the requirement for every pixel on the screen to assume a different shade or vertical signal level for each frame is extremely unlikely, and would never occur in a practical sense. Second, imperfections in the scanning process reduce the resolution effectiveness somewhat, so that some smearing across pixels does occur and is tolerable. Finally, the effective number of pixels in a typical standard television screen is somewhat less than the number of changes predicted by Equation 21-17. In practice, the actual video bandwidth has been established at about 4.2 MHz; this value, coupled with additional processing to be discussed later, has proven to be adequate based on standards utilized for many years. Nevertheless, the need for the signal to respond to rapid changes encountered in video action scenes does necessitate a very large bandwidth. In particular, the newer technology associated with high-definition television standards has necessitated the development of tubes with a much higher pixel density.

The major point of this development is that the bandwidth is directly proportional to the number of possible horizontal changes times the number of vertical changes times the number of frames per unit time. If any of those parameters changes, the bandwidth will change in the same proportion.

## TV Channel Spectrum

We begin the discussion of a TV channel by considering the black-and-white components in the spectral layout; we will add the color components in the next section. Each commercial TV channel in the United States has a width of 6 MHz, allocated by the FCC. Consider the *idealized* layout shown in Figure 21–7. Block-like spectral forms have been used, to delineate the boundaries more clearly, but the actual spectral forms have some realistic rounding and rolloff on the sides. The vertical levels are not necessarily shown to scale.

The process that follows may seem a bit odd, but it is convenient to assign $f = 0$ to the location of the video carrier and perform frequency measurements with respect to that point. The video carrier is located 1.25 MHz above the lower end of the TV channel, so the horizontal coordinate scale shown varies from $f = -1.25$ MHz to $f = 4.75$ MHz, thus constituting a bandwidth of 6 MHz. Of course, the actual RF frequencies involved vary considerably according to the particular channel, but the measurements provide a reference for a given channel.

A large portion of the channel must necessarily be devoted to the video information, since it has been stated earlier that a realistic bandwidth is about 4.2 MHz. What type of modulation should be used? Clearly, FM is out of the picture (no pun intended), since the bandwidth for any reasonable deviation ratio would be considerably greater than 6 MHz. Even conventional AM or DSB would require about 8.4 MHz and that is too much.

What about SSB, which could be argued as requiring only about 4.2 MHz? The problem is that video requires response all the way down to dc, as previously noted, and SSB is

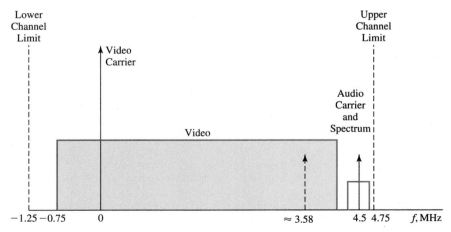

**FIGURE 21–7**
Idealized layout of a standard TV channel with color subcarrier added for later reference.

not suited for transmitting a dc component. Recall that with SSB, it is necessary to pass one sideband completely while rejecting the other, and that requires some guard band around dc.

The amazing answer to this dilemma was the development of **vestigial sideband plus carrier (VSB + C)**. This process is based on transmitting one sideband without alteration, a carrier, and a portion or *vestige* of the other sideband. The resulting bandwidth is greater than that of SSB but considerably less than that of conventional AM or DSB. The theory behind VSB is somewhat more difficult than the other forms of AM, but it has been made to work well for commercial TV.

Referring back to Figure 21–7, the video carrier is located at the reference value of $f = 0$, which is 1.25 MHz above the lower edge of the channel. The upper sideband is transmitted without change, but only about 0.75 MHz of the lower sideband is transmitted, as can be perceived from the sketch. The upper limit of the video is about 4.2 MHz above the carrier, or about 5.45 MHz above the lower edge of the band. As we will see later, a special type of filter is required in the receiver to assure that the process works.

To eliminate the need to show the spectral layout again in the next section, the location of the color subcarrier is shown on this figure. It is located about 3.58 MHz above the video carrier. (The exact value will be given in the next section.)

## Audio

Compared with the video component of the signal, the audio requirements are much more modest. The audio is transmitted as an FM signal with a deviation of $\pm 25$ kHz. The FM audio subcarrier is located exactly 4.5 MHz above the video carrier, or at a frequency of 5.75 MHz above the left-hand end of the channel. This provides adequate bandwidth to allow the lower end of the audio spectrum to exceed the highest video frequency. The difference of 4.5 MHz between the video and audio carriers is an important parameter in dealing with TV receiver processing, and is the basis of the *intercarrier system,* to be discussed later.

The TV audio can be transmitted as a stereo signal in the same fashion as for commercial FM. The details are not shown here, but since it is readily available, the horizontal sweep frequency is used for the reference stereo subcarrier in this case.

---

**⫼ EXAMPLE 21-2**

What percentage of the available scan lines for TV actually appear on the screen?

**SOLUTION**    A frame consists of 525 lines, of which about 483 actually appear on the screen. Thus the percentage is

$$\frac{483}{525} \times 100\% = 92\% \tag{21-21}$$

---

**⫼ EXAMPLE 21-3**

For the FM audio contained in a TV channel, determine the deviation ratio, assuming monaural sound transmission based on a 15-kHz modulating signal.

**SOLUTION**    The maximum deviation is $\pm 25$ kHz, and the deviation ratio $D$ is

$$D = \frac{25\,\text{kHz}}{15\,\text{kHz}} = 1.667 \tag{21-22}$$

---

**⫼ EXAMPLE 21-4**

Determine the transmission bandwidth for a monaural television FM signal.

**SOLUTION**    Based on Carson's rule, the bandwidth $B_\text{T}$ is

$$B_\text{T} = 2(\Delta f + W) = 2(25 + 15) = 80\,\text{kHz} \tag{21-23}$$

---

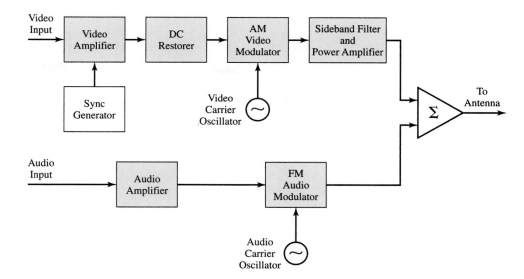

**FIGURE 21–8**
Block diagram of black-and-white (monochrome) TV transmitter.

## Monochrome Transmitter

A block diagram displaying the primary signal processing functions for a black-and-white or monochrome TV transmitter is shown in Figure 21–8. The left-hand side shows a video input, which could be from any applicable source such as a TV camera or a video recording. Since color is not being considered at this point, the signal will be assumed to be in luminance form.

A video amplifier first amplifies the signal. Simultaneously, the sync signals are inserted on the video by a sync generator. A so-called dc restorer circuit is used to reestablish the proper dc level to the video and the sync components. This composite signal is then applied to an AM video modulator, which produces an AM output for the video. This signal is then filtered by a sideband filter to eliminate a significant portion of the lower sideband, and that signal is amplified by a power amplifier.

The lower path shows the processing for the audio. The audio signal is first amplified by an audio amplifier and the resulting frequency modulates the audio carrier. The video and audio are then combined in a *diplexer* and applied to the antenna. A *diplexer* is a special circuit that combines two signals in an additive sense, but which provides filtering and isolation to avoid the output power from one source being diverted into the output circuit of the other source.

It turns out that the video signal is reversed in vertical level for modulating the carrier. The result is illustrated in Figure 21–9 for the AM process. Observe that the negative

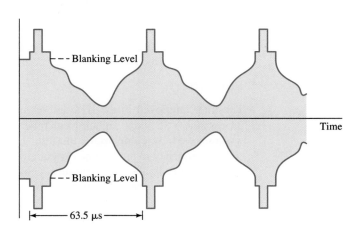

**FIGURE 21–9**
Video modulated signal form.

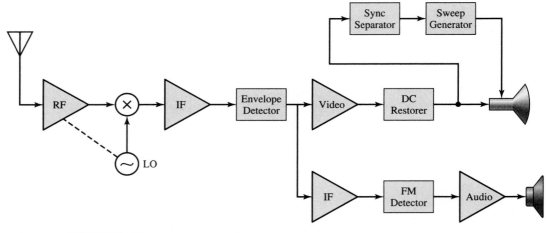

**FIGURE 21–10**
Block diagram of monochrome TV receiver.

blanking pulses considered earlier now appear as positive pulses on the positive envelope of the modulated signal. The signal will, of course, be reversed again at the receiver after demodulation.

There are several reasons for this modulation reversal. First, the average transmitted power is less than if the modulation were applied in the original direction, so the transmission efficiency is improved. Second, the blanking pulses occupy the highest level, so in the event of signal fading or reduction in strength, there is a greater probability of synchronization and signal stability being retained than if the opposite form were used.

## Monochrome Receiver

The basic form of a monochrome TV receiver is shown in Figure 21–10. As is common with most receivers, it is a superheterodyne form and contains an RF amplifier, a mixer, and a local oscillator. The video carrier position of the IF amplifier following the mixer is typically located at 45.75 MHz, and it must have a bandwidth of 6 MHz.

The composite video signal is demodulated by an envelope detector. This transfers the composite signal back to baseband. Zero frequency or dc corresponds to the location of the original video carrier in the bandpass spectrum. Therefore, the audio carrier is now at a location of 4.5 MHz, and an IF amplifier centered at that frequency is used to amplify and filter the audio. In effect, the audio IF occurred as a natural process, and this approach is called an *intercarrier* system.

On the video track, the signal is amplifed by a video amplifier, and a dc restorer may be used to establish the proper dc level. A sync separator circuit removes the sync signals, and they are used to synchronize the sweep generator circuits. The same sync signals also blank the screen in the video amplifier path.

## VSB Filter

The filter employed in the IF amplifier must have a special shape in order to optimally process the VSB video signal. Remember that a significant portion of the lower sideband was removed. To compensate, the filter must display a form of odd symmetry about the carrier frequency to approximately equalize the degradation of one of the sidebands.

If the LO is above the frequency of the incoming signal, which is the usual case, the sense of the spectrum of the difference component is reversed (as was shown in Chapter 5). However, to simplify the discussion that follows, the original spectral sense will be assumed to have been retained. In effect, this would be based on an LO located

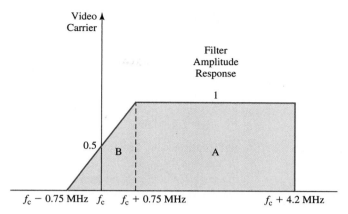

**FIGURE 21-11**

Ideal response of VSB filter in receiver to compensate for asymmetry in video sidebands.

below the incoming frequency. If the LO is above the incoming frequency, the only change in the strategy of the discussion would be to reverse the direction of the spectral diagram.

Consider the idealized filter characteristic shown in Figure 21-11. To simplify the discussion, the nomalized level of the flat region of the amplitude response is assumed to be 1. In the frequency range from 0.75 MHz below the carrier to 0.75 MHz above the carrier, the ideal filter characteristic should have odd symmetry about the normalized level of 0.5. With frequencies expressed in MHz, the amplitude response $A(f)$ can be described by the equations that follow.

$$A(f) = 1 \qquad\qquad \text{for region A}$$
$$= \frac{0.5(f - f_c)}{0.75} + 0.5 \quad \text{for region B} \qquad (21\text{-}24)$$

A spectral component in the original baseband spectrum above 0.75 MHz will have only one component in the spectrum, and will experience a relative weight of 1.

A spectral component in the original baseband spectrum below 0.75 MHz will have two components in the spectrum. Let $f_1 = f_c - \Delta f$ represent a frequency on the left-hand side of the carrier, and $f_2 = f_c + \Delta f$ represent the corresponding frequency on the right-hand side in region B. Note that both $f_1$ and $f_2$ correspond to the same baseband frequency. The relative weight of the component on the left-hand side of the carrier frequency is

$$A(f_1) = \frac{0.5(-\Delta f)}{0.75} + 0.5 = 0.5 - 0.67\Delta f \qquad (21\text{-}25)$$

The relative weight of the component on the right-hand side of the carrier frequency is

$$A(f_2) = \frac{0.5(\Delta f)}{0.75} + 0.5 = 0.5 + 0.67\Delta f \qquad (21\text{-}26)$$

It would now be an easy step to add the two preceding components, in which case the $\Delta f$ components would cancel and the net sum be 1, which is the same as for a component in region A. Indeed, we have chosen this as an intuitive way to explain the concept.

Actually, it is somewhat more complicated, because of the mathematical relationship between the carrier and sidebands for a signal having a carrier and asymmetrical sidebands. To analyze this completely, we would need to introduce modulated phasors or analytic signals, topics that have not been considered in this text. In a complete mathematical analysis, the results deduced are approximate and apply with high accuracy only when the modulation index is fairly low. However, the success of commercial TV certainly indicates that the process does work.

## 21-3   Color Television

We have discussed the basic principles of black-and-white or monchrome television. We now wish to continue with the evolution of the color system. The widely employed standards in the United States were actually developed in the early part of the 1950s, but it took some time before the process was fully implemented.

### Primary Colors

Before discussing how color TV works, a little background information on color is appropriate. The three *primary colors* are **red (R)**, **green (G)**, and **blue (B)**. All colors, including black and white, can be represented as a combination of these three colors. A process known as **RGB video** is used with computer monitors and certain high-end video systems. This process utilizes the three primary colors as separate signals to create combinations of different colors. However, the direct use of the three separate signals for standard commercial TV is not practical, as will be seen shortly.

### Where Does the Color Signal Fit?

As discussed earlier for commercial FM, the FCC has traditionally required backward compatibility for new systems. Therefore, when color TV was being developed, the original black-and-white signal had to be maintained in order that the millions of black-and-white TVs of that era would not suddenly become obsolete. The television standards in existence did not have sufficient bandwidth to permit the addition of three separate primary color signals on top of the already existing luminance or black-and-white signal. Something had to be done and, fortunately, engineering ingenuity provided a way.

First, consider the luminance or black-and-white signal that appears between the sync pulses, which is denoted in the industry as the $Y$ signal (not to be confused with yellow). Since all colors can be represented in terms of the three primary colors, it turns out that the luminance signal can be expressed very closely by the following equation:

$$Y = 0.299R + 0.587G + 0.114B \qquad (21\text{-}27)$$

The $Y$ signal is thus the ordinary black-and-white video occupying most of the transmitted video signal, but it may be considered as a linear combination of the three primary colors.

Now think back for the moment to stereo FM. In order for the left and right signals to be separated, there was a need to transmit two separate combinations, meaning in a mathematical sense that to solve for two variables, there was a need for two independent equations. The same concept can be extended to color TV. In order to transmit the three primary colors $R$, $G$, and $B$, there must be three linearly independent signals. One, of course, is the $Y$ signal, but there is a need for two more.

Unless one served on the committee that established the current color standards or carefully researched all of the documents produced, it is impossible to know exactly how the standards evolved, but it is likely that a lot of trial and error went into the process. Remember a boundary condition was that the existing black-and-white standards had to be maintained or at least subjected to unnoticeable changes, and the channel bandwidth had to be maintained at 6 MHz.

Two considerations that influenced the strategies can be formulated as follows:

1. The color information should be located within a channel as far away as practicable from the luminance information, to minimize interference with it.

2. Simultaneously, and in contrast, the color information couldn't be too close to the audio information or a different type of interference would result.

Serious studies of video spectral forms were likely made, and an interesting conclusion was deduced. It turns out that the luminance information at the higher frequencies tends to be clustered around the spectral lines representing the harmonics of the horizontal sweep rate. These lines are spaced apart by a frequency of 15.75 kHz, based on the original sweep

**Table 21–1**    NTSC Color Television Standards

| Parameter | Value or Form |
|---|---|
| Horizontal scan frequency | 15,734.26 Hz |
| Vertical scan frequency | 59.94 Hz |
| Frames per second | 29.97 |
| Fields per frame | 2 |
| Lines per frame | 525 |
| Lines per field | 262.5 |
| Color subcarrier frequency | 3.579545 MHz |
| Lines per field | 262.5 |
| Video bandwidth | 4.2 MHz |
| Aspect ratio | 4/3 |
| Video signal | AM modulation, vestigial sideband |
| Video modulation | Negative |
| Audio signal | FM modulation |
| RF channel bandwidth | 6 MHz |

rate. In view of this property, it was decided to place the color carrier at a point midway between two horizontal spectral lines, which amounts to an odd integer multiple of half the horizontal sweep rate.

One other problem apparently surfaced, and that was the possible intermodulation between the color carrier and the audio carrier. The difference between these two frequencies should also be located at a midpoint between horizontal spectral lines, so that it does not interfere with video signal.

A slight amount of "twiddling" was performed during the process, none of which has led to any noticeable problems with compatibility. The actual NTSC values that resulted from these changes are listed in Table 21–1.

Most of the parameters have already been introduced and are unchanged, but a few are worth discussing. Most noticeable is that the horizontal sweep frequency was changed from about 15,750 Hz to 15,734.26 Hz. The color frequency is 3.579545 MHz above the video carrier, and this frequency with respect to the video carrier turns out to be exactly the 455th harmonic of half of the sweep frequency, which is certainly an odd integer multiple of half the sweep frequency. This places the color carrier at a position of 227.5 times the horizontal frequency, meaning that it is midway between two harmonics of the horizontal frequency. The vertical scan frequency was changed to 59.94 MHz. These minor alterations permit all required frequencies at the transmitter to be derived from coherently locked signal sources.

Next, consider the difference frequency between the color carrier and the audio carrier. As measured from the video carrier, this is $4.5\,\text{MHz} - 3.579545\,\text{MHz} = 920.455\,\text{kHz}$. This is exactly the 117th harmonic of half the horizontal sweep frequency, which would place this beat frequency at a null point.

A spectral diagram illustrating some of the major properties is shown in Figure 21–12. The scale has been broken in order to display the forms of both the horizontal spectral lines

**FIGURE 21–12**

Illustration of chrominance spectral lines interleaved with luminance spectral lines.

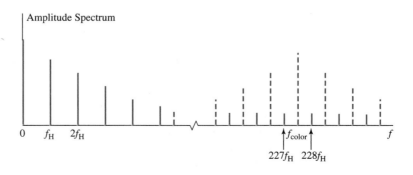

(solid lines) and the color spectral lines (dashed lines). This figure shows only the line spectral components, and does not show either the luminance or the color information that appears between the lines.

A carrier frequency has been selected, but it is necessary to send two more signals—so how is this achieved? The answer is to use quadrature modulation, which was introduced in conjunction with digital communication. One signal modulates an in-phase ($I$) subcarrier and the other modulates a quadrature ($Q$) subcarrier at the same frequency. The two signals are referred to quite naturally as the $I$ signal and the $Q$ signal, respectively. So that all three signals can be shown together for convenience, the $Y$, $I$, and $Q$ signals are as follows:

$$Y = 0.299R + 0.587G + 0.114B \tag{21-28}$$

$$I = 0.596R - 0.275G - 0.321B \tag{21-29}$$

$$Q = 0.212R - 0.523G + 0.311B \tag{21-30}$$

As previously noted, the $Y$ signal is called the *luminance* or *luma* signal. The $I$ and $Q$ signals constitute the *chrominance* or color components and they are often referred to by the shortened name *chroma*.

Apparently, the human eye is more forgiving on color bandwidth than on the basic luminance bandwidth. It turns out that the minimum bandwidth required for the spectral components of the $I$ signal is about 1.5 MHz, and for the $Q$ signal about 0.5 MHz. Of course, later work with high-definition TV resulted in significant changes, but based on the other limitations inherent in standard color TV, the bandwidths used have been adequate. The $I$ signal utilizes VSB and has a spectrum encompassing about 1.5 MHz on the lower side of the color subcarrier and about 0.5 MHz on the upper side. The $Q$ signal utilizes about 0.5 MHz.

## Color Burst Information

In order to provide a reference for color synchronous detection and for adjusting certain color properties, several cycles of a standard level sinusoid at the color subcarrier frequency are inserted once per horizontal line. The color burst signal is inserted during each horizontal blanking interval. The signal is actually added to the back porch of the horizontal pulse, as illustrated in Figure 21–13 . The level is kept below the blanking level so that no signal appears on the screen during that interval. However, the information is extracted by the TV set and used in the detection and color adjustment processes.

## Color Transmitter

Some of the major color signal processing functions that must be performed in the TV transmitter are shown in Figure 21–14. The three color signals $R$, $G$, and $B$ are first converted to the components $Y$, $I$, and $Q$ in accordance with the equations considered earlier. The $Y$ component is processed by a low-pass filter (LPF) to limit the bandwidth to 4.2 MHz,

**FIGURE 21–13**
Color burst signal added to back porch of horizontal blanking pulse.

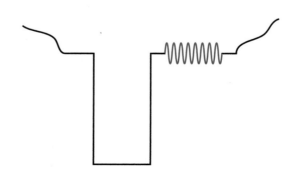

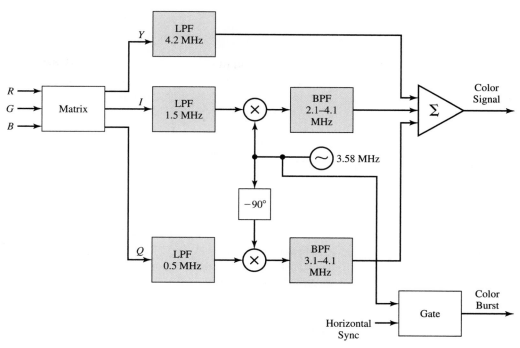

**FIGURE 21–14**
Modulation system for color subcarrier.

but it needs no further processing at the transmitter end. The *I* component is processed by a low-pass filter that limits its bandwidth to 1.5 MHz. It modulates the color subcarrier at 3.58 MHz. The band-pass filter (BPF) permits a full lower sideband and a vestige of the upper sideband.

The quadrature carrier is obtained by a phase shift of −90° of the carrier, and it is modulated by the *Q* information. Both sidebands of that component are transmitted, and the filter shown limits the spectrum from about 3.1 MHz to 4.1 MHz. The color subcarrier oscillator also supplies the color burst signal to the back porch of the horizontal sync pulse once per horizontal cycle, as shown.

## Color Receiver

A block diagram illustrating the demodulation process for a color receiver is shown in Figure 21–15. The operations shown here are those that follow the detection of the video signal. The top row shows the luminance or black-and-white signal, whose bandwidth is limited to 4.2 MHz. Because of the presence of the 3.58 MHz color subcarrier, it is necessary to eliminate that component from the spectrum. This is achieved by a "trap" or band-rejection filter.

The middle path of the diagram illustrates the processing required for the chrominance components. A band-pass filter ensures that only the chrominance components will be processed. A phase-locked loop (PLL) extracts a 3.58 MHz component from the back porch of the horizontal signal, which is used to synchronously demodulate the *I* and *Q* components of the signal. It also turns out that an adjustment of the phase of the carrier reference can be used to adjust the phase.

## Color Matrix

The three simultaneous equations that determine *Y*, *I*, and *Q* in terms of *R*, *G*, and *B* can be reversed to determine *R*, *G*, and *B* in terms of *Y*, *I*, and *Q*. The reader is invited to carry out

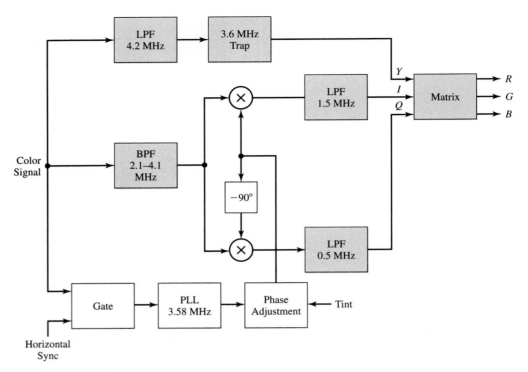

**FIGURE 21–15**
Demodulation system for color information.

this exercise; the results, with numbers rounded to three decimal places, are

$$R = Y + 0.956I + 0.620Q \tag{21-31}$$

$$G = Y - 0.272I - 0.647Q \tag{21-32}$$

$$B = Y - 1.108I + 1.705Q \tag{21-33}$$

Note that the coefficient of $Y$ in each case is the constant 1. The transformation is performed in the matrix operation shown in Figure 21–15. The outputs are the three primary colors, and they are applied to the three respective control terminals of the color picture tube.

It will be stated by the authors without reservation that this was a brilliant piece of engineering, and the proof of the pudding is that it works!

## High-Definition Television

At the time this book is being prepared, *high-definition television* (HDTV) is definitely on the scene and poised to be a major step in the continuing evolution of the consumer electronics industry. In general, HDTV refers to any standard that will provide a significant increase in the quality of the picture while simultaneously providing better sound.

The evolution of HDTV standards has been a complex process, and a number of different approaches have been investigated. Though HDTV is already available to those consumers who can afford it and wish to pursue the highest excellence in TV quality, the standards are still evolving. The treatment provided here is intended to provide an insight to the reader as to the issues involved and the approach to the technology.

We have seen that the both the monochrome and color TV standards utilize 525 lines of resolution. With that resolution, we have also seen that the fit within a 6-MHz bandwidth is still rather tight. How, then, can the resolution be increased with the same bandwidth? Alternately, is it possible to change the system so that more channels can be used for an HDTV signal? These are only two of the types of issues that were seriously considered in developing the process.

### Digital versus High-Definition Television

The terminology surrounding HDTV can be confusing because the term *digital television* (DTV) is also frequently used by various organizations, especially by cable television companies. Strictly speaking, DTV means that the signal has been converted to digital format, transmitted in that form, and possibly converted back to analog at the receiver. It does not necessarily mean that there is any special enhancement in the resolution or other properties, although it does enjoy some of the attributes of digital transmission. However, high definition television, which may also be transmitted with digital processing, refers to the enhancement resulting from greater resolution and other factors.

HDTV is combined with Dolby Digital surround sound, denoted 5.1 surround sound or AC-3. There are approximately eighteen different DTV formats, of which six are HDTV formats. Three of the formats used in HDTV are as follows:

720p    $1280 \times 720$ pixels with progressive scanning

1080i    $1920 \times 1080$ pixels with interlaced scanning

1080p    $1920 \times 1080$ pixels with progressing scanning

The secret to obtaining a picture with much greater resolution in a 6-MHz bandwidth is to utilize data compression. A standard for achieving this is MPEG-2. The acronym MPEG stands for "Motion Pictures Expert Group." The process is based on the fact that not every element of a given frame changes before the next frame. The software records only the changes in the image, and other elements remain the same as in the previous frame. This process reduces the amount of data to the order of about 2% of what would be required if every element of every frame were transmitted. MPEG-2 is also used with DVD discs. This process permits a possible direct interaction between the TV and a computer.

A typical standard color picture tube has about 210,000 pixels. The highest resolution HDTV standards call for about 2 million pixels, resulting in close to a tenfold increase in resolution. In contrast to the ratio of 4/3 for traditional TV, the aspect ratio for HDTV is 16/9. Most new TV sets have this ratio, whether they are actually HDTV or designated as "HDTV ready." The latter sets will require a converter box to accept HDTV signals. The technology required for HDTV depends very heavily on data compression. It employs advanced encoding techniques, and will utilize data transmission at the rate of 19.38 Mbits/s.

## 21-4    The Commercial Telephone System

Because of its common presence in all facets of life, a short description of the commercial telephone system will be provided in this section. We will consider the effects as they relate to wired connections in homes and businesses.

### Local Loop

A simplified connection between a station and a user is illustrated in Figure 21–16. This circuit is referred to as a *local loop*. The hook switches located on the right are open when the telephone is at rest in its cradle. This is called the *on-hook* condition, and there is a path to the voice circuits on the extreme right. The voltage $V_{battery}$ is commonly supplied by a battery, so that the telephone will work during a power outage. Of course, the battery is recharged on a regular basis as long as power is available.

The battery voltage is typically about 48 V, and this is the approximate voltage at the telephone when it is in the *on-hook condition* previously mentioned. The positive terminal of the battery is grounded; this is referred to in telephone jargon as the *tip*, referring to the tip of a telephone plug used in earlier exchanges. Normally, the color of this wire is green. The negative terminal is referred to as the *ring*, and it is normally red. Usually, two other wires, black and yellow, appear in the wiring of a house. The second set of wires permits two separate telephone lines to be used in the house.

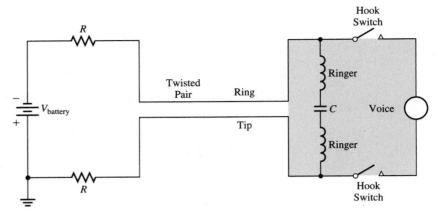

**FIGURE 21–16**
Simplified telephone circuit.

Assume that the user wishes to make a call. This process is achieved by lifting the handset, which closes the hook switches. The resulting condition is called *off hook,* and the voltage across the handset drops considerably due to the dc current that flows through the resistance in the twisted-pair line. The dc current is typically in the range from 20 to 80 mA.

### Dialing

Modern telephones utilize touch-tone dialing, a process characterized by different tone frequencies representing different numbers. The tones are received at the station and decoded, and they used to route the connection to the appropriate destination.

When a call is received, the proper path is through the series connection of the ringer coils and the capacitor. The ringing is accomplished by superimposing a frequency of 20 Hz at a level of about 100 V on the dc.

### Cell Phones

Walk down any street or through any public place in today's world and you will see many people conducting business or simply chatting with someone on a cell phone. A cell phone is actually a miniature *transceiver* (*trans*mitter plus *receiver*) that is connected between users through a network of repeater stations, and interfaced with the wired telephone system. Typical frequencies are in the 900-MHz, 2-GHz, and 5-GHz range. The term *cell* refers to the fact that hexagonal areas, denoted as *cells,* have been established for common use in different areas.

## PROBLEMS

**21-1**   The maximum frequency deviation for the audio carrier in a TV channel is $\pm 25$ kHz. Based on a modulating frequency of 15 kHz, determine the receiver processing gain in decibels.

**21-2**   For the audio subcarrier described in Problem 21-1, determine the baseband comparison gain.

**21-3**   Commercial TV channel 3 occupies the range from 60 MHz to 66 MHz. Based on the channel allocation process described in the text, determine the frequencies of (a) the video carrier, (b) the audio carrier, and (c) the color carrier.

**21-4**   Commercial TV channel 7 occupies the range from 174 MHz to 180 MHz. Based on the channel allocation process described in the text, determine the frequencies of (a) the video carrier, (b) the audio carrier, and (c) the color carrier.

**21-5**   Most TV receivers utilize high-side LO injection; that is, the LO frequency is higher than the incoming signal. The typical IF frequency of 45.75 MHz is based on the video carrier location. For channel 3 as considered in Problem 21-3, determine (a) the LO frequency and (b) the image frequency.

**21-6**  Perform the analysis of Problem 21-5 for channel 7 as described in Problem 21-4.

**21-7**  Based on the analysis of Problem 21-5, what are the frequencies in the IF amplifier range for (a) the video carrier, (b) the audio carrier, and (c) the color carrier?

**21-8**  The *Phase Alternating Line* (PAL) system is a format used in some countries other than the United States. The system utilizes 625 lines, a field rate of 50 Hz, a frame rate of 25 Hz, and an aspect ratio of 4/3. Based on the analysis used in the text for estimating an upper bound for the video bandwidth, determine a similar value for the PAL system. (As in the United States, the actual bandwidth assigned is somewhat less.) How would you expect the resolution of a picture generated with the PAL system to compare with one generated by the NTSC system?

**21-9**  Assume a color frequency with respect to the video carrier of 3.579545 MHz. With your calculator set to at least five decimal places, show that this is an odd integer multiple of half the horizontal sweep frequency of 15,734.26 kHz, and determine the integer.

**21-10**  Using the numbers provided in Problem 21-9, show that the difference between the audio carrier and the color frequency is an odd integer multiple of half the horizontal sweep frequency, and determine the integer.

# APPENDIX A

# MATLAB® FFT Programs for Spectral Analysis

This appendix describes three programs available on the supporting CD for determining the frequency spectrum of a time signal. The three programs and their intended applications are as follows:

| | |
|---|---|
| **fourier_series_1** | Determines magnitudes of $C_n$ coefficients in one-sided amplitude-phase form and plots them as a line spectrum for positive frequencies (including dc). |
| **fourier_series_2** | Determines magnitudes of $\overline{V}_n$ coefficients in two-sided exponential form and plots them as a line spectrum for positive frequencies (including dc). |
| **fourier_transform** | Determines magnitude of $\overline{V}(f)$ and plots it as a continuous function of positive frequency. |

All of these operations utilize the fast Fourier transform (FFT) function of MATLAB and must, therefore, be considered reasonable approximations to the actual spectral functions. However, the results are quite accurate if interpreted properly, and may be considered as the spectral representations for most practical purposes. The MATLAB M-files for the three programs are provided in Figures A–1, A–2, and A–3.

## Time Function

Prior to executing any of the programs, it is necessary to define the time function, which will be denoted **v**. The time function should be defined as a row vector at equally spaced values of time, and the operation usually works best when the number of points is *even*.

At one extreme, the function could be defined on a point-by-point basis, although this could be quite laborious if there are many points. More commonly, it will be defined by a series of equations. For example, suppose the function **v** is to be defined over a 100-point interval with the value 2 for the first 10 points and the value 0 for the last 90 points. A command to accomplish this is

```
>> v = [2*ones(1,10) zeros(1,90)];
```

In defining the function, one must decide what particular interpretation is required. If the function is considered periodic, either **fourier_series_1** or **fourier_series_2** should be used, depending on whether a one-sided or a two-sided amplitude spectrum is desired. In this case, the total time interval is considered *one cycle*. Note that while the second program is based on the two-sided form, it only provides a magnitude plot for the positive frequency range. The negative frequency range is simply the mirror image about dc.

If the function is nonperiodic, **fourier_transform** should be used. In the actual sense of the FFT, the function is still considered periodic and the total time interval is the period.

```
%Title of M-File is fourier_series_1.m
%Program to Determine Amplitude Spectrum in One-Sided Form Using FFT
fprintf('\n')
fprintf('The function v must already be in memory.\n')
fprintf('It should be a row vector defined over one cycle.\n\n')
N=size(v,2);
fprintf('The number of points in one cycle is %g.\n\n',N)
delt=input('Enter the time step between points in seconds. ');
fprintf('\n')
fs=N*delt;
f1=1/fs;
fprintf('The fundamental frequency is %g Hz.\n\n',f1)
fprintf('The highest unambiguous frequency is %g Hz.\n\n',0.5/delt)
fprintf('Enter integers requested as multiples of the fundamental.\n')
fprintf('For example, dc would be entered as 0.\n')
fprintf('The fundamental would be entered as 1.\n')
fprintf('The 2nd harmonic would be entered as 2, etc.\n')
fprintf('The highest integer that should be entered is %g. \n\n',N/2-1)
N1=input('Enter the lowest integer for plotting. ');
fprintf('\n')
N2=input('Enter the highest integer for plotting. ');
fa=N1*f1;
fb=N2*f1;
V1=abs(fft(v))/N;
V2(1)=V1(1);
V2(2:N/2)=2*V1(2:N/2);
C=V2(N1+1:N2+1);
f=fa:f1:fb;
stem(f,C)
xlabel('Frequency, Hz')
ylabel('Amplitude')
title('Amplitude Spectrum Based on One-Sided Form Using FFT')
grid
```

**FIGURE A–1**
MATLAB M-file for
fourier_series_1.m.

However, if the actual function constitutes only a short portion of the total time interval, it is reasonable in most cases to consider the function nonperiodic. In some cases, it may be desirable to add a long trail of zeros to the function so that it appears more like a nonperiodic function.

For the nonperiodic assumption, the reciprocal of the total time duration becomes the frequency increment between successive spectral components. Moreover, a continuous curve is extrapolated between successive spectral components, providing an approximation to the required continuous nature of the Fourier transform.

To illustrate this concept further, consider the function **v** defined earlier in this discussion. If this function is considered periodic, then the period is based on 100 points, and the function is considered a periodic pulse train with a duty cycle of 0.1. On the other hand, the function could be considered nonperiodic for the purpose of approximating the spectrum of a nonperiodic pulse. In this case, the reciprocal of the total time duration becomes the frequency step between successive components.

## Initiating the Program

Once the time function **v** has been defined, any of the three commands indicated earlier may be typed in the **Command Window** to initiate the program. A dialogue appears, requesting specific values to be entered. As the values are entered, various quantities are calculated and listed. Eventually, the amplitude spectrum will be plotted as a function of frequency. For the first two programs, the result will be a line spectrum; for the last program, the result will be approximated by a continuous curve. Examples of the applications of these programs are provided throughout the text.

FIGURE A–2
MATLAB M-file for
fourier_series_2.m.

```
%Title of M-File is fourier_series_2.m
%Program to Determine Amplitude Spectrum in Two-Sided Exponential Form Using FFT
fprintf('\n')
fprintf('The function v must already be in memory.\n')
fprintf('It should be a row vector defined over one cycle.\n\n')
N=size(v,2);
fprintf('The number of points in one cycle is %g.\n\n',N)
delt=input('Enter the time step between points in seconds. ');
fs=N*delt;
f1=1/fs;
fprintf('The fundamental frequency is %g Hz.\n\n',f1)
fprintf('The highest unambiguous frequency is %g Hz.\n\n',0.5/delt)
fprintf('Enter integers requested as multiples of the fundamental.\n')
fprintf('For example, dc would be entered as 0.\n')
fprintf('The fundamental would be entered as 1.\n')
fprintf('The 2nd harmonic would be entered as 2, etc.\n')
fprintf('The highest integer that should be entered is %g \n\n',N/2-1)
N1=input('Enter the lowest integer for plotting. ');
fprintf('\n')
N2=input('Enter the highest integer for plotting. ');
fa=N1*f1;
fb=N2*f1;
V1=abs(fft(v))/N;
V2(1:N/2)=V1(1:N/2);
C=V2(N1+1:N2+1);
f=fa:f1:fb;
stem(f,C)
xlabel('Frequency, Hz')
ylabel('Amplitude')
title('Amplitude Spectrum Based on Two-Sided Form Using FFT')
grid
```

FIGURE A–2
MATLAB M-file for
fourier_series_2.m.

```
%Title of M-File is fourier_transform.m
%Program to Determine Amplitude Spectrum of Fourier Transform Using FFT
fprintf('\n')
fprintf('The function v must already be in memory.\n')
fprintf('It should be a row vector defined over the time interval.\n\n')
N=size(v,2);
fprintf('The total number of points is %g.\n\n',N)
delt=input('Enter the time step between points in seconds. ');
fprintf('\n')
fs=N*delt;
f1=1/fs;
fprintf('The frequency increment between points is %g Hz. \n\n',f1)
fprintf('The highest unambiguous frequency is %g Hz. \n\n',0.5/delt)
fprintf('Enter integers as multiples of the frequency increment.\n')
fprintf('For example, dc would be entered as 0.\n')
fprintf('The lowest nonzero frequency would be entered as 1.\n')
fprintf('The next frequency would be entered as 2, etc.\n')
fprintf('The highest integer that should be entered is %g.\n\n',N/2-1)
N1=input('Enter the lowest integer for plotting. ');
N2=input('Enter the highest integer for plotting. ');
fa=N1*f1;
fb=N2*f1;
V1=delt*abs(fft(v));
V2(1:N/2)=V1(1:N/2);
V=V2(N1+1:N2+1);
f=fa:f1:fb;
plot(f,V)
xlabel('Frequency, Hz')
ylabel('Amplitude')
title('Amplitude Spectrum of Fourier Transform Using FFT')
grid
```

FIGURE A–3
MATLAB M-file for
fourier_transform.m.

# Introduction to Multisim®

## B-1 General Discussion

Multisim is a comprehensive circuit analysis program that permits the modeling and simulation of a wide variety of electrical and electronic circuits. It offers a very large component database, schematic entry, analog/digital circuit simulation, and many other features, including seamless transfer to printed circuit board (PCB) layout packages. The program is a product of Interactive Image Technologies Ltd.* It has evolved from the company's earlier Electronics Workbench (EWB) program. The circuit simulation portion of the program is based on the popular SPICE program (Simulation Program with Integrated Circuit Emphasis). SPICE was developed at the University of California at Berkeley and was a batch-oriented program. However, Multisim is interactive and offers many user-friendly features. The company is now owned by National Instruments, Inc.

A major feature of Multisim is that the schematic diagram is created on the screen using a mouse and various window options. The type of analysis desired is then applied to the circuit, and the results can be viewed in a number of ways.

One of the most valuable features of Multisim is that the source excitation and instrumentation functions closely parallel those of a basic electronics laboratory, and the procedures that are used in obtaining data are very similar to those of the "real world." Hence, it closely approaches the concept of an ideal "virtual laboratory." For example, the test and measurement models contain voltmeters, ammeters, a multimeter, a function generator with several output waveforms, a two-channel oscilloscope, and other instruments. These units must be wired into the circuit in essentially the same fashion as in an actual laboratory. Thus, good laboratory skills can be taught very easily using a computer and software.

The treatment provided in this appendix is primarily concerned with general familiarization and the creation of circuit schematics on the screen. (The various applicable types of analysis will be discussed within the text as necessary.) It is not intended as a complete treatment of Multisim, but the combination of the appendix and the various examples in the text should provide the reader with sufficient proficiency to use the program in support of communication circuits and systems.

The major instructions in this appendix are based on Version 8, which was released just before this book went into production. Along with Version 8 circuits, Version 7 and Version 6 (called 2001) circuits are also included on the accompanying disk.

Like most software, there will likely be revisions and updates that appear in Multisim after this text is printed. Therefore, there might be some minor changes in some of the instructions and/or windows in future updates. When in doubt, use the **Help** file.

---

* Interactive Image Technologies, Ltd., 111 Peter Street, Suite 801, Toronto, Ontario, Canada M5V 2H1.

## B-2  Overview

Upon activating Multisim Version 8, the initial screen will probably have a form similar to that shown in Figure B–1. However, there are numerous options for displaying and moving different utilities on the screen, and the previous settings may have been left in a different form than those in the figure. Therefore, don't panic if the screen is a little different, but as we discuss various options, you will see some of the possible ways that the display can be adjusted.

The initial blank space occupying most of the screen is called the **Circuit Window**, and the initial name at the top is **Circuit1**. If this space does not appear, left-click on **File** and then left-click on **New**. If this space has been minimized, left-click on the **Maximize** button in the upper right-hand area.

### Toolbars

Various rows of options called **Toolbars** appear at the top and sides of the screen. The various toolbars may be activated or deactivated on the screen. Left-click on **View** and then hold the cursor over **Toolbars**. A window will open containing the names of all the toolbars, with a check mark beside each one that is visible. Left-click on a given toolbar name to toggle the check mark.

Some of the menu choices and the options within menus are those encountered in most Windows-based applications, and some are peculiar to Multisim. When the mouse arrow is moved over most of the buttons, their names appear on the screen. If in doubt about the function of a given operation, refer to the **Help** file. You do not need to understand all of the options available to get started with Multisim. Experience will lead to familiarity with those features most appropriate to your needs. We will concentrate here on the ones that are essential to get started, and as experience with the program is acquired, more options will become apparent.

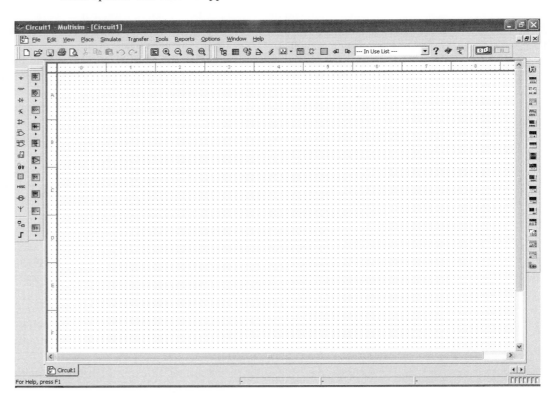

**FIGURE B–1**
Initial screen of Multisim Version 8.

Some options to initially locate are the command buttons to perform an analysis. Search for a button that looks similar to a triangular waveform. When the cursor is brought over it, the name **Grapher/Analysis List** will appear. The square button is used to activate the grapher, which is the screen for displaying a curve of some circuit variable. The small rectangular button beside it can be used to set the various forms of analysis used to simulate circuits. The latter operation can also be performed by a left-click on the **Simulate** button followed by a left-click on **Analyses**. The particular analysis desired can then be selected.

After you eventually create a circuit, you can save it under a different name by left-clicking on **File** and then left-clicking on **Save as**. You can then provide an appropriate name for the file. You can navigate through your file system according to your personal preference for file storage. All circuit files in Version 8 will automatically be assigned the extension **ms8** when you save them. Circuits in Version 7 have the extension **ms7**, and circuits in Version 6 (2001) have the extension **msm**.

The various buttons arranged in two columns on the left-hand side of the screen are the **Component Toolbars**. When the arrow is placed above a given bin, the name assigned to the particular parts bin appears on the screen. Each of these toolbars may be moved around on the screen for convenience if desired. To move a bar, left-click on the area at the top, hold down the left mouse button, drag it to any desired location, and then release the mouse button.

The various buttons arranged in a column on the right-hand side of the screen form the **Instruments Toolbar**. When the arrow is placed above a given button, the name of the instrument appears on the screen.

In the large blank area on the screen, the **Circuit Window**, you will create the circuit schematic. A schematic of a typical circuit containing instrumentation is shown in Figure B–2. This particular circuit was used as an example on the Multisim software disk and has the title **ActiveBandPassFilter.ms8**.

## B-3  Creating a Circuit Diagram

The process of creating a circuit diagram or schematic consists of *selecting* and *dragging* the components from a parts bin and *connecting* the components using *wire*. In some cases, it may be desirable to drag all the components out to the circuit window first and then wire them together. However, the procedures may be interchanged at any time, and parts may be moved around to make room for other parts as necessary.

### Using a Grid for Layout

The default option in Version 8 is to have a grid appear on the screen and it is very helpful in laying out the components. However, the grid can be toggled on and off if desired from the **Show Grid** option reached with the **View** button.

### Node Numbers

All nodes in the circuit will eventually have a node number or name, whether assigned by Multisim or by the user. The default condition for Multisim 8 initially displays this information on the screen. Changes in labeling options may be performed through the **Options** button and the subsequent menu choices.

### Parts Bin Toolbars

As previously noted, the **Component Toolbars** are represented by the columns of buttons to the left of the circuit window. Starting with the battery symbol at the top, each button identifies a particular parts bin based on a common grouping. As you move the arrow across each button, the name is identified.

## Active BandPass Filter

See the Circuit Description Box for instructions.

(Turn the view of the description on and off by selecting View/Circuit Description Box.)

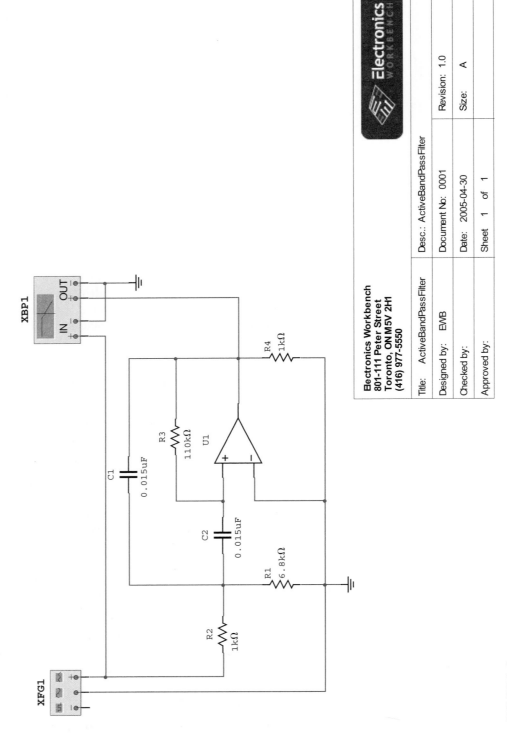

| Electronics Workbench 801-111 Peter Street Toronto, ON M5V 2H1 (416) 977-5550 | Desc.:: ActiveBandPassFilter | |
|---|---|---|
| Title: ActiveBandPassFilter | Document No: 0001 | Revision: 1.0 |
| Designed by: EWB | Date: 2005-04-30 | Size: A |
| Checked by: | | |
| Approved by: | Sheet 1 of 1 | |

FIGURE B–2

A schematic of a typical Multisim circuit included with the program.

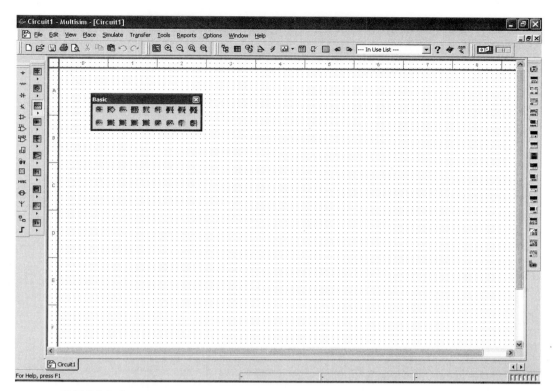

**FIGURE B–3**
Multisim Version 8 screen with the Basic Family parts bin open.

To open a given parts bin, left-click on or below its button. Figure B–3 shows the **Basic Family** parts bin, containing passive components (*R*, *L*, and *C*) and a number of other parts. A parts bin may be closed by left-clicking on the **X** in the upper right-hand corner.

As you look through various bins, don't worry if at first there seems to be an overwhelming number of parts. You will quickly learn to find those that you use often, but it is unlikely that you will ever need more than a small number of the available parts.

## Selecting and Dragging a Part from a Bin

To obtain a part for the circuit being created, first left-click on the bin button containing the part. For certain parts, you may need to use a little trial and error before finding the right bin. After the part is identified in the appropriate bin window, move the arrow to the correct button and left-click on it. You can then release the mouse button, since that component has now been "locked" to the arrow. With many parts, an additional **Components Browser** window will open with a title such as **Select a Component**. You can then select the value or parts number of the component. When the choice is made, click **OK**. The arrow will appear with a marker attached to it, and it may be moved to the position at which it will be placed. Don't be overly concerned about its exact location at this time since it can be changed later. When you reach the approximate location at which the component is to be placed, left-click the mouse and the component will be released at that point on the screen.

## Selecting and Moving Components

It is frequently desirable to select and move a component that is already in the schematic workplace. This can be achieved by moving the arrow to the component position and placing it in contact with the component. At the point where contact is established, left-click once and hold the mouse button down. Four small rectangles now appear as if to create a box around the component. While holding the left mouse button down, you can then drag the component to a new location on the screen. The labels will move with the part. When the proper location is reached, release the left button and the component will be detached. The rectangles may be eliminated by left-clicking outside of the area of the component.

The selection process just described may also be used to move the label or value of the component or the node number or name. This is accomplished by bringing the arrow in contact with the label or value instead of the component itself, and the remainder of the procedure is the same.

## Rotating a Component

Components always have a fixed orientation when they come out of the parts bin, and this may or may not be the desired final orientation. All components may be rotated by integer multiples of 90°. First, the component must be selected by left-clicking on it.

Rotation can be achieved most easily by means of a left-click on the **Edit** menu followed by a left-click on **Orientation**. The options are **Flip Horizontal, Flip Vertical, 90 Clockwise,** and **90 CounterCW** (counter clockwise). The titles are somewhat self-explanatory, but the reader may wish to experiment with these operations when an actual circuit is constructed. Two successive steps of either one of the flip operations or four successive steps of either one of the rotate operations will restore the original orientation.

## Deleting a Component

To delete a component, first select it. The simplest way to delete it is to press the **Delete** key on the keyboard, although there is a **Delete** option under the **Edit** menu. If the component is a two-terminal device to which external wires have been connected, the component will be replaced by a short between the terminals.

## Ground

All circuits must have at least one ground point. The quickest way to obtain a ground symbol is to use the **Power Source Family Bar**, which is identified by the battery symbol. You can also obtain a ground from the **Sources** bin. The ground establishes the point from which all voltages are measured. To eliminate extra wiring, ground points may be used at various points in the circuit. Again, however, there must be at least one ground symbol, and it is automatically established within the program as node **0**.

## Wiring the Components Together

At some point, you will be ready to start wiring the components together. You may have all the parts that you need in the circuit window, or you may prefer to wire them as you go along. For this purpose, you will need *wire*.

A basic rule to follow is that *all terminals that are to be connected must have a section of wire between them.* There is a temptation in some cases to simply bring component terminals together, in which case they may appear to be connected. Unless there is a section of wire between them, however, they will not connect. Think of the wire as having solder at each end; without it, there will not be a connection. This rule applies to the ground symbol as well as all components.

The program automatically establishes node numbers from **1** upward in the order in which you wire components together, although all ground symbols assume node **0**. Therefore, if you wish to establish any kind of numbering system, it is suggested that you connect the nodes in the desired numerical order. However, this is not absolutely necessary, since you can perform some renumbering later.

To start a wire from a given terminal, move the arrow as close as possible to the tip of the terminal. At the point where a connection is possible, the arrow will then change to a set of crosshairs. Left-click at this point, and the process creates a piece of wire that can be moved to some other terminal. The wire will initially appear as a dashed line until the final connection is made. It is recommended that you move in straight lines parallel to either the horizontal or vertical sides of the screen. You can make one 90° turn without any further clicking but if you make more than one turn, you will need to left-click at the point of the second and subsequent turns. When you reach the desired connection point, left-click at the terminal. The dashed line will now appear as a solid line, meaning that you have wired the two terminals together. A node number will now appear somewhere along the wire. It will take a little practice to become fully proficient in this process.

To delete a section of wire, first select it by left-clicking on it. Then press the **Delete** key.

## Opening a Component Properties Window

In addition to their initial part numbers and values, many components will require further parameter values and properties to be established prior to running a circuit simulation. Moreover, you may desire to change some of the parameters without having to go back to the parts bin. This process can be performed at any time after the component is obtained from the parts bin. One of the great virtues of circuit simulation is the ease with which parameter values may be changed throughout a simulation study.

Parameter values and properties are specified in the **Component Properties** window. To open such a window, proceed as if you were planning to select the given component and place the arrow on the component. However, perform a double left-click on the component, and the window should appear. The name of the window depends on the particular type of component. A wide variety of properties may be specified. For the simplest types of component, only one or two values need to be specified, while others require a multitude of entries. Within the text examples, many of these windows are discussed in detail. After you have entered all the data required to specify the component, left-click **OK**. If for any reason you do not change any values, or if you wish to return to the previous state, left-click **Cancel**. The component may now be deselected by moving the arrow away from the component and left-clicking.

Since a section of wire is treated essentially the same as a component, the preceding process may be used to open a properties window for changing its characteristics. This concept will be discussed in the next paragraph as it relates to node numbering.

## Node Numbering and Grid Options

The node numbers will appear on the schematic in the same order as they were connected. If you wish to change a node number, first double left-click on the wire section. In this case, the window that opens has the title **Net**. Next, type the desired node number or name in the **Net name** slot and left-click **OK**.

If you are renumbering, the program will not allow you to use a number that is already active. If you have a strong desire to use a fixed node numbering order, you should try to connect the nodes in the order planned. However, you can always temporarily label nodes with numbers outside of the range so that you can get access to a desired number. You can also label nodes with names such as **IN** and **OUT**.

## Analysis and Simulation

At a very basic level, where laboratory skills are being emphasized, the use of virtual instruments and sources is very desirable. Whenever the virtual instruments are used, the circuit is activated by either the **Run/stop simulation button** or by the switch displaying a 0 and a 1. Some of the circuits in Chapters 9 and 10 require this procedure.

For most of the analysis and simulation examples within the text, the **Analyses** menu yields results that are easier to use, and virtual instruments are not required for this purpose. Instead, left-click on the **Analyses** button. A window will then open providing a number of options for analysis. Many of these will be used within the text examples.

# APPENDIX C

# MATLAB® Primer

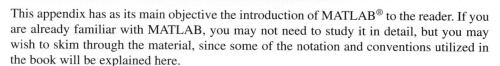

## OVERVIEW AND OBJECTIVES

This appendix has as its main objective the introduction of MATLAB® to the reader. If you are already familiar with MATLAB, you may not need to study it in detail, but you may wish to skim through the material, since some of the notation and conventions utilized in the book will be explained here.

For readers desiring a more detailed reference, we recommend *Technical Analysis and Applications with MATLAB* by William D. Stanley (Thomson Delmar Learning, 2005, Clifton Park, NY). Portions of this appendix were extracted from that source.

MATLAB is a general software package for technical computing, mathematical analysis, and system simulation. It is a product of The MathWorks, Inc., and is widely employed in industry, government, and education. In addition to the basic MATLAB program, there are numerous supplements called *toolboxes* that provide software applications for specific specialty areas.

The MathWorks, Inc. has an excellent record of maintaining relatively seamless compatibility as new versions are released. Depending on the age of this book and its ultimate market endurance, there may be some changes that occur in the MATLAB environment while the book is still in use, but it can be reasonably expected that the company will continue to maintain the fine record established thus far.

## Objectives

After completing this appendix, the reader should be able to perform the following operations with MATLAB:

1. Describe the **Desktop Layout** and the various windows associated with it.

2. Enter scalar values in the **Command Window** and perform operations such as clearing the screen (**clc**), clearing values in memory (**clear**), and determining variables using the **who** and **whos** commands.

3. Perform addition (+) and subtraction (-).

4. Perform multiplication (*) and division (/).

5. Perform exponentiation (^).

6. Perform the square root operation (**sqrt**).

---

® MATLAB is a registered trademark of The MathWorks, Inc., 3 Apple Hill Drive, Natick, MA 01760-2098 USA. Tel: 508-647-7000, fax: 508-647-7001, e-mail: info@mathworks.com, Web: www.mathworks.com.

7. Explain the hierarchy of arithmetic operations.

8. Explain *nesting* and apply it to arithmetic operations.

9. Write an *M-file program* that can be saved and used again for different input data.

## C-1 Desktop Layout

Very little will be said about installing or activating the program, since that process could vary somewhat from one computer to another. Anyone using this program is assumed to know how to turn on a computer, click on an icon, or go to a program menu to activate a program. Upon activating the program, the **Desktop Layout** appears on the screen.

The discussion that follows and the computer-generated examples in the text are based on Version 7, Release 14 and the optional Signal Processing Toolbox. The Student Version does not contain that particular toolbox, but it may be obtained by students from the company for a modest additional cost. The Student Version also contains the program Simulink, which provides a block diagram approach for simulating systems. However, the latter program is not required to support this book.

### Windows on Desktop Layout

The default screen based on the version employed is shown in Figure C–1. Brief descriptions of the windows will be provided in the next few paragraphs.

The upper left area toggles between **Workspace** and **Current Directory** windows. The **Workspace** provides a list of the variables used in the current work session. The **Current Directory** provides a list of the MATLAB programs available in the given directory.

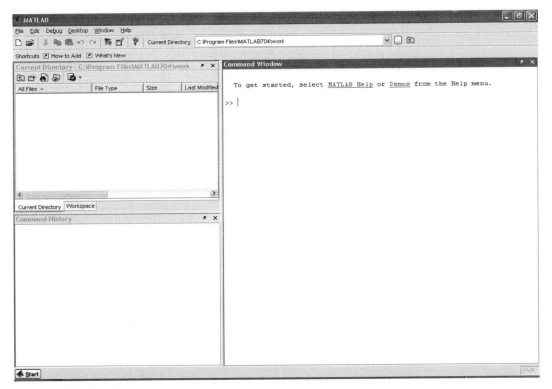

**FIGURE C–1**
Initial screen of MATLAB Version 7.

The **Command History** on the lower left shows a continuous record of the commands used in the analysis.

In the Professional Version, the window on the right is called the **Command Window**, and we will use that term in all subsequent references. In the Student Version, it is called the **Student Version Command Window**. This is the area in which all commands are entered for using MATLAB on an interactive basis. Commands for executing **M-file** programs may also be entered in that window.

The different windows may be restructured if desired, and such actions are generated with the **View** option on the upper toolbar. Once you become familiar with the program, you may wish to experiment with different options. To restore the original layout, first left-click on **View**. Next, left-click on **Layout** and then left-click on **Default.**

## Prompt

MATLAB automatically inserts the prompt **>>** at the beginning of each line in the **Command Window** of the Professional Version. The corresponding prompt for the Student Version is **EDU>>**. We will assume the shorter prompt of the Professional Version throughout the text, and the presence of the prompt on a line will alert the reader to the fact that a MATLAB command and possible results will follow.

## Command Window

The **Command Window** may be used on an interactive basis by simply typing in the commands after **>>** for the Professional Version or after **EDU>>** for the Student Version, on a line-by-line basis. Depress **Enter** on the keyboard after each line is entered, and any results generated by the command will immediately appear on the screen, unless the addition of a semicolon has suppressed the display (to be discussed later), or the command is part of a sequence.

If you desire to have the **Command Window** stand alone, move the mouse arrow to the small arrow on the upper right-hand side of the **Command Window**. You will see the statement **Undock from Desktop**. Next, left-click on the arrow, and the window will now separate from the **Desktop Layout**. It can now be maximized to occupy the entire screen if desired.

## C-2  Getting Started with the Command Window

Let's begin by assuming that MATLAB has been activated and that the **Desktop Layout** appears on the screen. The focus at this point will be directed toward the **Command Window**. If desired, you could maximize the **Command Window** and let it take over the screen, but it's probably better for the moment to leave it as it is. There are numerous menu options for changing type formats, number of digits appearing on the screen, and other things, but again it's better to wait until you have experience with the simple things first.

The main identifier for the Professional Version is that each instruction line in the **Command Window** begins with the prompt **>>**, which is automatically inserted by MATLAB. To clearly illustrate that point, the text will show all entries in the **Command Window** beginning with that identifier. They will also be left-justified as in the **Command Window**. Following the typing of a command, it is activated by depressing **Enter**.

## Notational Conventions

In the MATLAB examples, we will usually display commands and results exactly as they appear on the screen. This means that the type style will be different from the standard mathematical style in the remainder of the book, but the symbols should be perfectly clear.

To illustrate this concept, the mathematical format of a common variable in the text is $x_1$, with the italicized form of $x$ and a subscript for the 1. In a MATLAB command, the variable will probably be indicated as x1, with a nonitalicized x and a 1 that is not sub-scripted. We don't believe this change in format will cause any heartburn, and we want to keep the MATLAB notation as close as possible to the way it appears on a computer screen in the default mode. Occasionally, there may be a variable that could be stated in either form, in which case an arbitrary choice will be made. Be alert to these different styles of notation, but remember that it is not a result of sloppy editing.

## Command Window versus M-files

The **Command Window** will allow you to work in a manner similar to that of a calculator. You can enter a command and see the numerical results immediately. The important thing is that it is interactive. This is in contrast to the writing of **M-files**, which is performed in a different window. An **M-file** is in reality a computer program that can be written, saved, and activated as many times as desired. We will introduce the process of developing an **M-file** in Section C-4.

As indicated earlier, for all commands written on the screen following either the $>>$ or **EDU** $>>$ prompt, the command is activated by **Enter**. This will be assumed in all cases, whether stated or not.

## Clear Commands

At any time that you wish to clear the screen, the following command can be used:

```
>> clc
```

The variables previously defined remain in memory, but you will have a clear screen.

If you wish to delete all the variables in memory, the following command can be used:

```
>> clear
```

Suppose you wish to delete a variable x but retain all others in memory. This can be done by the command

```
>> clear x
```

## C-3  Basic Arithmetic Operations in the Command Window

In this section, we will take the reader back to some of the most basic arithmetic opera-tions learned in elementary and junior high school. Please don't be insulted by this step. To use MATLAB properly, you need to understand how the most elementary operations are performed, and what better way is there to learn such operations than with simple arithmetic?

### Spacing

As a general rule, blank spaces between distinct sections of a command line may be used to avoid a crowded appearance, *provided* that they don't disrupt a particular command function or number. Thus, x=5, x = 5, and x  =  5 are all acceptable. However, if the num-ber is 51, you will get an error message if you try to put a space between the 5 and the 1. Likewise a variable x1 cannot be expressed as x 1. The same holds for various functions that will be introduced throughout the text.

Before inserting a space, ask yourself whether adding a space will either change the meaning or create an ambiguity in the interpretation. If in doubt, it is probably better to avoid the blank space and have the symbols run consecutively, as long as you can read the expression.

To make it easier for the reader, we will frequently use spaces in the examples, and that should provide some guidance as to what is acceptable.

## Entering Values

Numbers can be entered directly in the **Command Window** using the keyboard. For example, suppose we wish to enter the value 5. We simply type 5 and then depress **Enter**. The screen appears as follows:

```
>> 5

ans =

    5
```

If one does not assign a specific name to a quantity, MATLAB gives it the name **ans** (for answer). Suppose we desire to assign the name x to the value of 5. In that case, we type $x = 5$ and depress Enter. The screen appears as follows:

```
>> x = 5

x =

    5
```

As long as we don't delete or redefine x, any subsequent references to x will be treated as the value 5. At any point, if we wish to review the value of the quantity, we simply type x and depress **Enter**. The screen appears as follows:

```
>> x

x =

    5
```

In anticipation of subsequent operations, let us define y as 3. The operation follows.

```
>> y = 3

y =

    3
```

## Addition

Now suppose we wish to add the preceding two values to form z1 as the sum. We could write

```
>> z1 = 5 + 3

z1 =

    8
```

In this case we typed out the actual values again since they were simple and easily remembered. However, a more systematic approach, based on the utilization of variables defined earlier, is

```
>> z1 = x + y

z1 =

    8
```

Thus, variables in memory may be manipulated in arithmetic expressions by using their names.

## Subtraction

Next, suppose we wish to form z2 as y - x. We have

```
>> z2 = y - x

z2 =

    -2
```

The value of z2 is a negative number as expected. To enter a negative number, simply place a hyphen (-) in the space preceding the positive number. Thus negative 2 would be entered as -2.

Note that addition and subtraction utilize the same symbols encountered in ordinary arithmetic. As will be seen shortly, this is not the case for multiplication and division.

## Checking on the Variables

Suppose now that we wish to pause to see what quantities that have been defined or computed up to this point are in memory. The command **whos** will provide a list as follows:

```
>> whos

Name    Size      Bytes Class

x       1x1          8 double array

y       1x1          8 double array

z1      1x1          8 double array

z2      1x1          8 double array
```

Grand total is 4 elements using 32 bytes.

Because MATLAB is matrix oriented, ordinary single-valued variables are considered arrays with 1 row and 1 column, that is, 1x1 arrays as shown. Such quantities are called *scalars*. Each scalar value is stored as a double array number with 8 bytes. An alternate command to **whos** is **who**. The latter command will list the variables, but not their dimensions.

The information provided by the last two commands may also be found in the **Current Directory** window.

## Multiplication

Next, we will consider multiplication. Multiplication of scalars is denoted by placing an asterisk (*) between the numbers. Thus, if z3 is the product of x and y defined earlier, we could write

```
>> z3 = 5*3

z3 =

    15
```

However, making use of the names of the variables, a better approach is

```
>> z3 = x*y

z3 =

    15
```

## Division

Division with scalars can be achieved by using the forward slash /. Assume that z4 is determined by dividing x by y. We have

```
>> z4 = x/y

z4 =

    1.6667
```

The actual value in memory is more accurate than the value displayed on the screen. Options for displaying numbers in a variety of fashions are available, but will not be pursued at this point.

## Alternate Division Form

In the previous division form, the order is numerator, forward slash, and denominator. An alternate MATLAB form is denominator, reverse slash, and numerator. Thus, the preceding division could also be expressed as

```
>> z4 = y\x

z4 =

    1.6667
```

While this latter form may seem awkward, it turns out to be useful in dealing with matrix forms.

## Exponentiation

Exponentiation of scalars is achieved by utilizing the *caret* or *circumflex* symbol ^. Suppose z5 is defined as the square of x. We have

```
>> z5 = x^2

z5 =

    25
```

## Suppressing the Listing of Computational Results

Suppose we desire to perform a computation but prefer not to have the value or values displayed on the screen. This is often the case when the results are intermediate and not the final values being sought. This situation also arises with dimensioned variables when long lists would move the data on the screen out of the immediate area of interest. Suppression of the screen value for any operation can be achieved by placing a semicolon ( ; ) at the end of the command. For example, let z6 represent y raised to a power of x. If we prefer not to have the value immediately shown on the screen, the operation would be

```
>> z6 = y^x;
```

where the preceding result was followed with **Enter**. Nothing appeared on the screen, but the value of z6 was computed and is in memory. At some later time, we can determine the value simply by typing z6 and depressing **Enter.**

```
>> z6

z6 =

    243
```

## Entering in Exponential Form

The numbers entered thus far have all been simple values, but when very small or very large values are to be entered, the use of exponential forms can be employed. For example, a microwave frequency $f = 15\,\text{GHz}(1\,\text{GHz} = 10^9\,\text{Hz})$ can be entered as

```
>> f = 15e9

f =

   1.5000e+010
```

Note that MATLAB expresses the value according to the rules of standard scientific notation.

A constant that appears in noise analysis is Boltzmann's constant k. This value is $1.38 \times 10^{-23}$ joules/kelvin (J/K). It can be entered as

```
>> k = 1.38e-23

k =

   1.3800e-023
```

## Square Root

The square root of a scalar quantity x is denoted by **sqrt(x)**. Let z7 represent this value. We can type and enter

```
>> z7 = sqrt(x)

z7 =

   2.2361
```

## Hierarchy

There is a hierarchy or order associated with expressions that are to be evaluated with computer software. The default hierarchy is usually in the following order: *parentheses, exponentiation, multiplication, division, addition,* and *subtraction,* although the order for multiplication and division can sometimes be confusing in software equations. When there is ambiguity in a sequence of multiplication and division operations, the order is from left to right.

In most standard mathematical equations, this chain of operations is usually evident from the manner in which the equation is written. For example, consider the algebraic equation

$$y = 5x^3 + 4$$

It should be evident that the first operation required is that of raising $x$ to the third power and then multiplying that value by 5. The result is then added to 4 to form $y$.

If $x$ is a single value and this equation were written in MATLAB, it would appear as follows:

```
>> y = 5*x^3 + 4
```

where it is assumed that x is in memory. The hierarchy remains the same as in the algebraic equation.

Consider next the equation

$$y = (5x)^3 + 4$$

This equation is quite different from the first one. In the latter equation, we are to multiply $x$ by 5 and then take the third power of the result. This value is then added to 4.

The corresponding MATLAB equation is

```
>> y=(5*x)^3 + 4
```

Thus, parentheses may be used to alter the order of computation. When in doubt about the order, parentheses are recommended to ensure that the desired order is obtained.

## Nesting

To create complex orders of operation, we can place parentheses within parentheses, provided that we ultimately have the same number of left parentheses as right parentheses. The order of computation is from the innermost to the outermost. The process of putting parentheses within parentheses is called *nesting*.

To illustrate this point, let us consider a MATLAB algorithm in which we start with the value x defined earlier, and proceed to determine a desired result y, as follows:

1. Add 3 to x.
2. Square the result of step 1.
3. Multiply the result of step 2 by 6.
4. Add 8 to the result of step 3.
5. Take the square root of the result of step 4 and the value obtained is y.

One way to accomplish the chore would be to introduce four intermediate values, which we will denote u1, u2, u3, and u4. The number following u in each case corresponds to the step number. The MATLAB commands follow.

```
>> u1 = 3+x;

>> u2 = u1^2;

>> u3 = 6*u2;

>> u4 = 8+u3:

>> y = sqrt(u4)

y =

   19.7990
```

Note that we suppressed the printing of the intermediate variables since their values were not of prime interest.

Now let's see how this operation can be performed in one step.

```
>> y = sqrt(8+(6*((3+x)^2)))

y =

   19.7990
```

Note that the operation starts with the innermost set of parentheses and proceeds toward the outside. Note also that the number of left parentheses is equal to the number of right parentheses (four in each case).

We have put more parentheses in this example than actually necessary in order to make a point: When in doubt, adding parentheses to ensure that the computations are performed in the desired order is fine. Can the reader see how one or more sets of parentheses in this expression could be eliminated?

Either approach in the preceding example is acceptable. You can choose to break up a complex expression into a series of simpler operations by introducing intermediate variables, or you can choose to nest some or all of them together, as you see fit. The important thing is to use the form that is the easiest for you to understand, and easiest to check in the event of an error.

## Some Constants in MATLAB

The constant $\pi$ appears so often in scientific analysis that MATLAB has built it into the library, and it is denoted simply as **pi.** The value can be expressed as

```
>> pi

ans =

    3.1416
```

Again, we emphasize that the value in memory is far more accurate than the result displayed on the screen in the default numerical format.

Another constant that is built into MATLAB is **eps** (for *epsilon*). It is a very small value that can sometimes serve to avoid indeterminate forms such as 0/0. The value can be seen as

```
>> eps

ans =

    2.2204e-016
```

This is not a "standard" definition but one employed by MATLAB based on the program's numerical precision.

## Imaginary Numbers

Consider the square root of -1.

```
>> sqrt(-1)

ans =

    0 + 1.0000i
```

The quantity $\sqrt{-1}$ is denoted in complex number theory as i, and the result from MATLAB agrees with that, although it carries the answer out to the default number of decimal places. The quantity 0 in front means that there is no "real part" to the particular value. Later, we will see that a *complex number* in general has both a *real part* and an *imaginary part*.

In electrical/electronics engineering and technology, the quantity j is normally used to represent $\sqrt{-1}$ since *i* is used for current flow. MATLAB will actually accept either i or j, but it always expresses the result on the screen as **i.**

If it is desired to enter a purely imaginary number, there are four ways it can be entered. Consider for example the number 5*i* or *i*5 (both forms are acceptable in mathematical equations). The four commands that follow illustrate different ways of entering this value:

```
>> 5i

ans =

    0 + 5.0000i

>> 5j

ans =

    0 + 5.0000i

>> i*5

ans =

    0 + 5.0000i

>> j*5

ans =

    0 + 5.0000i
```

Note that when the **i** or the **j** is placed after the number, the asterisk representing multiplication is not required. However, when the **i** or the **j** is placed in front, the asterisk is required. Moreover, the printing on the screen always appears as an **i**, and always follows the number.

## Division by Zero

Division by zero will yield either **Inf** ("infinity") or **NaN** ("not a number"), depending on the form. As examples, if the command **1/0** is entered, the result is **Inf**, and if the command **0/0** is entered, the result is **NaN**.

## Help File

One of the most valuable resources available in the MATLAB program is the **Help** file. It is virtually impossible in any single book to provide documentation on all the possible situations that might be encountered in using the many operations. Therefore, get in the habit of going to the **Help** file for assistance when required.

The **Help** file may be opened by left-clicking on the **Help** button on the upper toolbar. A small window will open, providing access to a number of features. The primary help documentation is opened by left-clicking on the **MATLAB Help** option. The next window that opens will display a number of different ways to access information, and the reader may wish to experiment with them. The author tends to use the **index** option more frequently, but that is a personal preference. In that option, one or more key words may be typed and entered, and if suitable documentation is available, it will be provided. Certain words may also lead to similar terms that might be applicable in a given case.

## Numerical Formats

All computations in MATLAB are performed in double precision, so most values obtained usually have a much greater precision than the values of physical parameters being studied. However, the results may be displayed in various formats. In fact, the **Help** file lists about a dozen different formats, some of which are of interest only in specialized applications; for example, the hexadecimal format a base of 16 using consisting of the numbers 0 through 9 plus the letters a through f.

We believe it would be more confusing than helpful to introduce all of these formats at this point in the text. Rather, by typing **format** in the **index** slot previously referred to in the **Help** file, the various formats may be investigated whenever the need arises. However, a few of the most common ones will be discussed here.

The default format is denoted **short**, and it is the one that has been employed so far in the text. Moreover, it will be used more than any other format throughout the remainder of the text, since it is the default format and is convenient to use. It utilizes a scaled fixed-point format with 5 digits.

In general, any of the formats may be activated by the command

```
>> format type
```

where *type* represents a particular code name for the format, examples of which will be illustrated in the steps that follow. At any time, the default short format may be restored by either of the following commands:

```
>> format short
```

or simply

```
>> format
```

Now we will illustrate a few of the formats by using the constant **pi**. First, we have the default short format.

```
>> pi

ans =

    3.1416
```

Next, consider the **long** format

```
>> format long

>> pi

ans =

    3.14159265358979
```

Wow! That's certainly as much or more than we could ask for. Now consider the **short e** format.

```
>> format short e

>> pi

ans =

    3.1416e+000
```

This latter format is useful for very large and very small values for which scientific notation is desired. In fact, the short format may temporarily change to this format if required.

If you are using MATLAB as you follow along with the text, it is appropriate now to switch back to the default **short** format by the command

```
>> format
```

## C-4  Programming with M-Files

Thus far, all of our work has been performed within the **Command Window** on an interactive basis. This pattern will continue throughout the text, since most commands will be introduced on an individual basis, and the results may be readily observed. In general, the **Command Window** is usually best for fast results, when the procedure may not need to be repeated and where only a few lines of code are involved.

Consider next the situation where a computational procedure may need to be repeated for more than one set of input data, or the code consists of a relatively long list of commands. In this case, it is usually more convenient to develop an **M-file**, which may be saved and used as often as desired. An **M-file** of this type (or **M-file script**, as it is often called) is a MATLAB-based computer program for performing a particular analysis. Thus, MATLAB may be considered as a valid form of programming language. In general, any of the commands that may be applied in the **Command Window** may be combined into an **M-file** program. Instructions for accepting input data for different cases may be added to the programs.

### Creating an M-file

In general, any "text only" or ASCII editor can be used to create an **M-file**. However, the simplest procedure is to use the built-in editor within MATLAB. To use the MALAB editor, left-click on **File** and then left-click on **New**. Next, left-click on the **M-file** option. A window will open for the purpose of preparing the program. Comments may be freely added in the program by placing the percentage character % at the beginning of each comment line. Type all the commands required and enter each one individually. None of the MATLAB commands will be executed during the entry process, and you may correct errors in typing. Basically, you are creating a program that will be executed later, and MATLAB will not perform any computations at this time. The prompt symbols will *not* appear at the beginning of each line of code, and they should *not* be added. Place only the appropriate command on each executable line, and place a % at the beginning of each comment line.

## Saving a File

After a program has been completed, it should be saved. Left-click on **File** and then left-click on **Save As**. If the computer belongs to someone else or to an institution (e.g., a college computer), you may wish to save the file on a floppy disk. If it is your own computer, you may wish to save it in the MATLAB default **work** folder. In general, the name *must begin with a letter,* but it may contain numbers within the name. However, the name may *not* contain any symbols that could be confused with an arithmetic operation such as +, -, *, /, and so on. The underscore symbol _ *may* be used to separate portions of the name. Finally, the name must be continuous; that is, spaces are not permitted in the name. If the MATLAB editor is used, it will automatically assign an extension of **.m** to the name as it is saved. If a different editor is used to create the file, the extension **.m** should be added to the end of the name.

## Setting a Path

Assume next that you have saved the program and desire to run it. Return to the **Command Window** for this purpose. (The next procedure may be unnecessary if you saved the file in the **work** folder.)

   If you saved the file to a folder outside of the MATLAB directory or to a floppy disc, you may need to set a path from the MATLAB directory to that particular location. If in doubt, try to run the program by the procedure of the next paragraph and see if it runs. If the program is not recognized, you will need to set a path. First left-click on **File**. When the menu opens, left-click on **Set Path** and then left-click on **Add Folder**. Search through the file structure on the right to identify the path required to reach the appropriate folder. Then left-click on **OK**. If necessary, use the **Help** file to assist in this process.

## Running a Program

Within the **Command Window**, type the name of the program after >> or **EDU** >> at the beginning of the first line, and press **Enter**. You should *not* add the extension **.m** since MATLAB assumes it by default.

   The program will either run as it should on the first trial, or, more likely, an error statement will appear, telling you that something is wrong with one or more lines of code. If there is an error, you will need to open the file and attempt to identify and correct the problem.

   MATLAB does not necessarily catch all possible errors on the first attempt. In long programs, it may take several attempts to catch all the possible errors.

## Functions

A second type of M-file is that of a **function**. MATLAB contains a large number of built-in functions, many of which will be considered throughout the text. However, special functions may be created and added to the user's library for convenience.

## Extending a Line

Most commands used in M-files within the text will easily fit on one line. However, if it is necessary to use more than one line for a command, place three periods in succession **...** at the end of the first line and any other lines that are to be continued. This combination of periods is referred to as an *ellipsis*.

## Examples

The reader desiring to develop M-files will find various examples throughout the text that can serve as guidelines.

## MATLAB DRILL EXERCISES

All of the exercises should be performed within the MATLAB **Command Window**. As was the case within the chapter, some may seem trivial, but they are presented for the same purpose as much of this chapter; that is, practice and familiarization.

As further practice in converting from standard mathematical notation to MATLAB forms, all values and operations will be given in the standard form. You must then enter them in the MATLAB format. First, enter the values of $x$ and $y$, and then determine all of the operations in problems that follow. At any point that you need to clear the screen, use the **clc** command, but if you should clear the memory with **clear**, you will need to reenter the values of $x$ and $y$.

$$x = 5$$
$$y = 8$$

**A-1**  $z_1 = x + y$

**A-2**  $z_2 = x - y$

**A-3**  $z_3 = xy$

**A-4**  $z_4 = \dfrac{x}{y}$

**A-5**  $z_5 = x^3$

**A-6**  $z_6 = 2^5$

**A-7**  $z_7 = (2^5)^2$

**A-8**  $z_8 = 2x + 3y$

**A-9**  $z_9 = 2x^2 + 3y^2$

**A-10**  $z_{10} = (2x + 3y)^2$

**A-11**  $z_{11} = 2\pi$

**A-12**  $z_{12} = \pi^2$

**A-13**  $z_{13} = 2\pi^2$

**A-14**  $z_{14} = (2\pi)^2$

**A-15**  $z_{15} = -1.6 \times 10^{-19} \times \pi^3$

**A-16**  $z_{16} = 3 \times 10^8 \times 5 \times 10^{-6}$

**A-17**  $z_{17} = \sqrt{x^2 - 4}$

**A-18**  $z_{18} = \sqrt{4 - x^2}$

**A-19**  The following algorithm is used to generate a variable $z_{19}$, starting with $x$.

1. Square $x$.

2. Subtract 12 from the result of 1.

3. Take the square root of the result of 2.

4. Add 9 to the result of 3.

5. Square the result of 4 and the answer is $z_{19}$.

Determine the solution using four intermediate variables on a step-by-step basis.

**A-20**  The following algorithm is used to generate a variable $z_{20}$, starting with $y$.

1. Take the square root of $y$.

2. Add 4 to the result of 1.

3. Form the square of the result of 2.

4. Subtract 8 from the result of 3.

5. Multiply the result of 4 by $\pi$ and the answer is $z_{20}$.

Determine the solution using four intermediate variables on a step-by-step basis.

**A-21**  Solve Problem A-19 with one equation, using nesting.

**A-22**  Solve Problem A-20 with one equation, using nesting.

# Answers to Selected Odd-Numbered Problems

**Chapter 1**

**1-1**   555.6 m

**1-3**   3.409 m

**1-5**   15 MHz

**1-7**   28.75 dB

**1-9**   39.81

**1-11**   13.98 dB

**1-13**   $251.2 \times 10^{18}$

**1-15**   89.13 to 112.2

**1-17**   44.08 dB

**1-19**   **(a)** 49.29 dBm    **(b)** 19.29 dBW    **(c)** 169.29 dBf

**1-21**   **(a)** 60 dBm    **(b)** 30 dBm    **(c)** 0 dBm
        **(d)** −30 dBm    **(e)** −60 dBm

**1-23**   **(a)** 100    **(b)** 20 dB    **(c)** 20 dB

**1-25**   40 dBm out

**1-27**   **(a)** 2.236 V    **(b)** 22.36 V

**1-29**   **(b)** 32.81 dB

**1-31**   **(a)** 33 dBf    **(b)** 25 dB

**Chapter 2**

**2-1**   **(a)** 160 V    **(b)** $1000\pi$ rad/s
        **(c)** 500 Hz    **(d)** 2 ms

**2-3**   **(a)** 0.01 A    **(b)** 1 Mrad/s
        **(c)** 159.2 kHz    **(d)** 6.283 μs

**2-5**   **(a)** $13\cos(200t + 22.62°)$
        **(b)** $13\sin(200t + 112.62°)$

**2-7**   $10.39\cos 3000t - 6\sin 3000t$

**2-9**   dc, 250, 500, 750, 1000 Hz

**2-11**   **(a)** dc, 100, 200 Hz

**2-13**   **(a)** 22.49 V    **(b)** 10.12 W    **(c)** 2.88, 4, 3.24 W

**2-15**   **(a)** dc component    **(b)** cosine terms only
        **(c)** both odd and even numbers    **(d)** −6 dB/octave
        **(e)** 125 Hz

**2-17**   **(a)** dc component
        **(b)** cosine terms only
        **(c)** both odd and even numbers
        **(d)** −12 dB/octave
        **(e)** 83.33 kHz

**2-19**   **(a)** no dc component
        **(b)** both cosine and sine terms
        **(c)** odd numbers
        **(d)** −12 dB/octave
        **(e)** 100 Hz

**2-21**   **(a)** 1 kHz, 16.21 V, 2.628 W
        3 kHz, 1.801 V, 32.4 mW
        5 kHz, 0.6485 V, 4.20 mW

**2-23**   **(a)** dc, 6.366 A, 2026.4 W
        500 Hz, 10 A, 2500 W
        1000 Hz, 4.244 A, 450.3 W
        2000 Hz, 0.849 A, 18.0 W

**2-25**   **(a)** dc, 12 V, 2.88 W
        500 Hz, 15.279 V, 2.334 W
        1.5 kHz, 5.093 V, 0.259 W
        2.5 kHz, 3.056 V, 93.4 mW

**2-27**   **(a)** dc, 3 V, 180 mW
        1.25 kHz, 3.820 V, 145.9 mW
        3.75 kHz, 1.273 V, 16.2 mW
        6.25 kHz, 0.764 V, 5.8 mW

**2-29**   99.92%

**2-31**   96.64%

**Chapter 3**

**3-1**   **(a)** $20\dfrac{\sin 0.2\pi f}{0.2\pi f}$
        **(b)** 20, 15.14, 4.667, −3.118
        **(c)** 0 dB, −2.420 dB, −12.62 dB, −16.14 dB

**3-3** (a) $20\left(\dfrac{\sin 0.2\pi f}{0.2\pi f}\right)^2$

(b) 20, 11.46, 1.094, 0.4862

(c) 0 dB, −4.840 dB, −25.24 dB, −32.28 dB

**3-5** zero crossings at 5 Hz, 10 Hz, 15 Hz, etc.

**3-7** lines at dc, 1.25 Hz, 2.5 Hz, 3.75 Hz, 6.25 Hz, 7.5 Hz, 8.75 Hz, 11.25 Hz, etc.

zero crossings at 5 Hz, 10 Hz, etc.

**3-9** lines at dc, 1 MHz, 2 MHz, 3 MHz, 4 MHz; first zero crossing at 5 MHz

**3-11** center at 250 kHz; first zero crossings at 125 kHz and 375 kHz

**3-13** center at 1 MHz; lines spaced 250 kHz apart; first zero crossings at 500 kHz and 1.5 MHz

**3-15** center at 10 MHz; first zero crossings at 9 MHz and 11 MHz

**3-17** center at 10 MHz; lines spaced 200 kHz apart; first zero crossings at 9 MHz and 11 MHz

**3-19** (a) 2 kHz    (b) 0.125 ms    (c) 0.25

**3-21** (a) 2 MHz    (b) 5 kHz    (c) 50 s

**3-23** (a) O-6    (b) 20 minutes

## Chapter 4

**4-1** $\dfrac{R}{R + j\omega L}$

**4-3** $\dfrac{j\omega RC}{1 + j\omega RC}$

**4-5** (a) $\dfrac{R}{\sqrt{R^2 + (\omega L)^2}}$

(b) $-\tan^{-1}\left(\dfrac{\omega L}{R}\right)$

(c) $20\log\left(\dfrac{R}{\sqrt{R^2 + (\omega L)^2}}\right)$

**4-7** (a) $\dfrac{\omega RC}{\sqrt{1 + (\omega RC)^2}}$

(b) $\dfrac{\pi}{2} - \tan^{-1}\omega RC$

(c) $20\log\left(\dfrac{\omega RC}{\sqrt{1 + (\omega RC)^2}}\right)$

**4-9** (a) $\dfrac{1}{1 - \omega^2 + j\sqrt{2}\omega}$

(b) $\dfrac{1}{\sqrt{1 + \omega^4}}$    $-\tan^{-1}\left(\dfrac{\sqrt{2}\omega}{1 - \omega^2}\right)$

**4-11** (a) $\dfrac{1}{\omega}\tan^{-1}\left(\dfrac{\sqrt{2}\omega}{1 - \omega^2}\right)$

(b) $\dfrac{\sqrt{2}(1 + \omega^2)}{1 + \omega^4}$

**4-13** (a) 1 MHz    (b) 12.5 MHz

**4-15** (a) 2 MHz    (b) 25 MHz

**4-17** 2 Mbits/s

**4-19** 500 MHz

## Chapter 5

**5-1** (a) 3.559 MHz    (b) −19

**5-3** 50.66 pF

**5-5** (a) 4.003 MHz    (b) 3.995 MHz

**5-7** (a) 4.5 MHz    (b) 500 kHz

**5-9** 2, 3

**5-11** 32, 48 kHz

**5-13** 115, 117, 123, 125 kHz

**5-15** $V_2$: 5 to 7 and 155 to 157 MHz, $V_3$: 5 to 7 MHz, same spectral sense

**5-17** $V_2$: 5 to 7 and 167 to 169 MHz, $V_3$: 5 to 7 MHz, reversed spectral sense

**5-19** 155 to 157 MHz

**5-21** 167 to 169 MHz

**5-23** (a) 130 MHz

**5-25** (a) 370 MHz

**5-27** LO: 4.3 to 5.8 MHz, image: 5.1 to 6.6 MHz

**5-29** LO: 2.7 to 4.2 MHz, image: 1.9 to 3.4 MHz

**5-31** first LO: 21 MHz, second LO: 420 MHz

## Chapter 6

**6-1** (a) 1194 to 1206 kHz    (b) 12 kHz

**6-3** (a) 1196.5, 1198.5, 1199.5, 1200.5, 1201.5, 1203.5 kHz

(b) 7 kHz

**6-5** (a) 1194 to 1200 kHz    (b) 1200 to 1206 kHz

(c) 6 kHz

**6-7** (a) 1196.5, 1198.5, 1199.5 kHz

(b) 1200.5, 1201.5, 1203.5 kHz

(c) 3 kHz

**6-9** (a) 3.5, 1.5, 0.5, 2396.5, 2398.5, 2399.5 kHz

(b) 3.5, 1.5, 0.5 kHz

**6-11** (a) 0.5, 1.5, 3.5, 2400.5, 2401.5, 2403.5 kHz

(b) 0.5, 1.5, 3.5 kHz

**6-13** (a) 0.3, 1.3, 3.3, 2400.7, 2401.7, 2403.7 kHz

(b) 0.3, 1.3, 3.3 kHz

**6-15** (a) all components of Problem 6-3(a) and 1.2 MHz

(b) 7 kHz

**6-17** (a) 250, 150 V    (b) 300, 100 V    (c) 400, 0 V

**6-19** 50%

**6-21** (a) 15 V    (b) 30 V    (c) 60 V

**6-23** (a) 98, 102 kHz    (b) 98, 100, 102 kHz

(c) 98 kHz    (d) 102 kHz

**6-25** (a) 300 W    (b) 600 W

**6-27** (a) 50 W    (b) 100 W

**6-29** 1581 V and 31.62 A

**6-31** (a) 56.25, 75 kW    (b) 112.5, 200 kW

      (c) 6.25, 25 kW

**6-33** (a) 1677 V, 33.54 A    (b) 1936 V, 38.73 A

## Chapter 7

**7-1** (a) $40\pi(1 - e^{-2t})$ rad

      (b) $80\pi e^{-2t}$ rad/s    $40e^{-2t}$ Hz

**7-3** (a) 95.49 MHz    (b) 500 Hz    (c) 40

      (d) 20 kHz    (e) 4 W

**7-5** 40 rad

**7-7** $20\cos(6 \times 10^8 t + 80\sin 500\pi t)$

**7-9** $20\cos(6 \times 10^8 t + 40\sin 500\pi t)$

**7-11** $8\cos(2\pi \times 90 \times 10^6 t + 25\sin 4\pi \times 10^3 t)$

**7-13** 7 kHz

**7-15** 24 kHz

**7-17** 38 kHz

**7-19** 30 kHz

**7-21** (a) 51.2 MHz    (b) 384 kHz    (c) 38.4

**7-23** 2 doublers and 4 triplers

**7-25** –5 kHz/V

**7-27** 0.120 V/kHz

## Chapter 8

**8-1** (a) 16 kHz    (b) 62.5 μs

**8-3** (a) 20 kHz    (b) 50 μs

**8-5** 22 kHz

**8-7** 22.05 kHz

**8-9** (a) 75,000    (b) 800 μs

**8-11** 3 kHz

**8-13** in kHz: 2, 14, 18, 30, 34, 46, 50

**8-15** in kHz: 1, 2, 8, 9, 11, 12, 18, 19, 21, 22, 28, 29, 31, 32

**8-17** (a) 15 kHz    (b) 30 kHz

**8-19** 1.25 MHz

## Chapter 9

**9-1** (a) 64    (b) 1024

**9-3** 9 bits

**9-5** 0000, 0010, 0011, 0110, 1001, 1100, 1111

**9-7** 1110, 1100, 1100, 1000, 0101, 0010, 0000

**9-9** (a) $15.26 \times 10^{-6}$    (b) 305.2 μV    (c) 0.99998

      (d) 19.9997 V    (e) $7.629 \times 10^{-6}$    (f) 152.6 μV

**9-11** (a) $30.52 \times 10^{-6}$    (b) 305.2 μV    (c) 0.99997

      (d) 9.9997 V    (e) $15.26 \times 10^{-6}$    (f) 152.6 μV

**9-13** (a) 3.760 V    (b) 4.379 V    (c) 5.000

**9-17** Use a two-input AND gate with NRZ-L signal as one input, and a square-wave as the other input.

**9-19** (a) 1.544 Mb/s    (b) 772 kHz

**9-21** (a) 28    (b) 1 channel at 5 kHz, 19 channels at 1 kHz, 4 channels at 200 Hz, 4 channels at 50 Hz

**9-23** 32 kHz

**9-25** 57 kHz

**9-27** 32 kHz

**9-29** self-checking

## Chapter 10

**10-1** (a) 2 MHz    (b) 6 MHz    (c) 12 MHz

**10-3** (a) 5.025 Mb/s    (b) 8.304 Mb/s    (c) 11.62 Mb/s

**10-5** 5.025 Mb/s

**10-7** 6 MHz

**10-9** (a) 8    (b) 17.99 dB

**10-11** 100 kHz

**10-13** 5

## Chapter 11

**11-1** 0011 0101 0110

**11-3** 0110 1000 1001

**11-5** 1000100

**11-7** 11000100

**11-9** 2112 Mbps

**11-11** 32 Mbps

**11-13** The LSB is LOW. The next six bits are LOW, HIGH, LOW, LOW, LOW, HIGH. The parity bit is HIGH.

**11-15** Y

**11-17** I L U V U

**11-19** station A: SNRM U frame; station B: Unnumbered Acknowledgement (UA) U frame

**11-21** RR U frame with N(R)=5

**11-23** 010

**11-25** computer: IN token; scanner: ACK handshake and DATA0 and DATA1 packets

## Chapter 12

**12-1** (a) 6.72 μV    (b) 21.2 μV    (c) 67.2 μV

      (d) 212 μV    (e) 672 μV    (f) 694 μV

**12-3** 6.72 V

**12-5** 4 μV

**12-7** 20 fW

**12-9**   2 nW

**12-11**   1.10 mV

**12-13**   **(a)** $4 \times 10^{-21}$ W/Hz      **(b)** $4 \times 10^{-16}$ W/Hz

**12-15**   $3.623 \times 10^6$ K

**12-17**   82.8 pW

**12-19**   **(a)** 5 dB      **(b)** 3.162

**12-21**   627 K

**12-23**   **(a)** 0 K      **(b)** 290 K      **(c)** 1540 K

**12-25**   **(a)** 1, 0 dB      **(b)** 1.5, 1.76 dB      **(c)** 2, 3.01 dB

**12-27**   **(a)** 261 K      **(b)** 1.90

**12-29**   1.90

**12-31**   **(a)** 870 K      **(b)** 6.02 dB

**Chapter 13**

**13-1**   **(a)** −6.585 dB      **(b)** −3.575 dB

**13-3**   **(a)** 16.99 dB      **(b)** 30.41 dB

**13-5**   **(a)** 21.76 dB      **(b)** 35.19 dB

**13-7**   **(a)** 25.94 dB      **(b)** 39.36 dB

**13-9**   53.47 dB

**13-11**   43.95 dB

**13-13**   49.92 dB

**13-15**   13 bits

**13-17**   **(a)** 36.0 dB      **(b)** 31.9 dB

**13-19**   37.8 dB

**13-21**   40.0 dB

**13-23**   154.8 fW

**13-25**   **(a)** 50 kHz, 26.0 dB

 **(b)** 25 kHz, 29.0 dB

 **(c)** 25 kHz, 24.3 dB

 **(d)** 4.17 kHz, 47.8 dB

 **(e)** 4.17 kHz, 52.6 dB

 **(f)** 4.17 kHz, 58.8 dB

**13-27**   **(a)** 33.0 dB      **(b)** 33.0 dB      **(c)** 28.2 dB

 **(d)** 44 dB      **(e)** 48.7 dB      **(f)** 55.0 dB

**13-29**   **(a)** 10.0 pW      **(b)** 10.0 pW      **(c)** 30.0 pW

 **(d)** 0.8 pW      **(e)** 267 fW      **(f)** 51.0 fW

**13-31**   1.414

**13-33**   4.77 dB

**13-35**   **(a)** 25.7 dB      **(b)** 35.7 dB      **(c)** 45.7 dB

 **(d)** 55.7 dB      **(e)** 65.1 dB      **(f)** 71.8 dB

**Chapter 14**

**14-1**   1.11 m

**14-3**   122.5 Ω

**14-5**   $2.04 \times 10^8$ m/s

**14-7**   2.16

**14-9**   **(a)** 277.3 nH/m      **(b)** 108.2 pF/m      **(c)** 50.62 Ω

 **(d)** $182.6 \times 10^6$ m/s

**14-11**   **(a)** 1.476 μH/m      **(b)** 17.32 pF/m      **(c)** 291.9 Ω

 **(d)** $197.8 \times 10^6$ m/s

**14-13**   **(a)** 250 nH/m      **(b)** 100 pF/m      **(c)** 2.25

**14-15**   **(a)** 75 Ω      **(b)** 3 A      **(c)** 675 W

 **(d)** 675 W

**14-17**   615.6 W

**14-19**   **(a)** 0.5114 ∠−46.22°      **(b)** 3.093

**14-21**   1.317 dB

**14-23**   3.01 dB

**14-25**   **(a)** 75.4 mV/m      **(b)** 15.08 μW/m²      **(c)** 29.61 mW

**14-27**   **(a)** 250 Ω      **(b)** 3.6 μW/m²      **(c)** 2.274

**Chapter 15**

**15-1**   37.5 m

**15-3**   56.99 dB

**15-5**   159.2 pW/m²

**15-7**   2.522 nW/m²

**15-9**   15.71 W

**15-11**   44.51 dB

**15-13**   7.958 m²

**15-15**   83.33 Ω

**15-17**   **(a)** 62.20 m²      **(b)** 54.95 dB      **(c)** 0.292°

**Chapter 16**

**16-1**   229 pW

**16-3**   −96.41 dBW or 229 pW

**16-5**   **(a)** 247.2 dB      **(b)** 267.2 dB

**16-7**   **(a)** 363.3 s or 6.06 min

**16-9**   **(a)** 115.96 dB      **(b)** 155. 96 dB      **(c)** 195.96 dB

**16-11**   35.5 W

**16-13**   701 W

**16-15**   116.6 km

**16-17**   30.3 fW

**16-19**   176 m²

**16-21**   **(a)** 300 km      **(b)** 450 m

**16-23**   −2.4 kHz

**16-25**   66.8 mi/hr

**16-27**   55.9 mi/hr

**16-29**   60°

**Chapter 17**

**17-1**   0.2152 m   27.53 dB

**17-3**   0.3153 m   20.49 dB

**17-5**   0.7587 m   40.54 dB

**17-7**   $38.5 \times 10^3$ km

**17-9**   25.6 dB

## Chapter 18

**18-1** Neither peer-to-peer nor client/server. Neither computer has access to resources on the other computer. When the AB switch is in either position, one computer is connected directly to the printer and the other is disconnected.

**18-3** Existence of a network printer will allow print jobs to be printed without the delay required to copy files to a CD-ROM and take them to an off-site printer. This also saves labor required to carry the files to the off-site printer.

**18-5** network layer

**18-7** guaranteed delivery of packets

**18-9** Session layer adds header to PDU received from presentation-layer protocol. Header is removed and read by session layer in remote computer.

**18-11** Class A; 17.0.0.0

**18-13** Class C; 207.55.28.0

**18-15** ARP broadcasts IP address. Host with that IP address replies with its MAC address.

**18-19** Each segment of TP wiring has a maximum length of 100 meters. Even if repeaters were used, the maximum distance from the wiring closet that can be reached with TP segments is 500 meters. Thus, the entire floor cannot be reached from a corner wiring closet with TP wiring.

## Chapter 19

**19-1** Print device and computer each need IR port. Computer needs appropriate driver. Computer and print device placed with IR ports facing each other so data can be transferred from computer to print device.

**19-3** 30, 53, 13, 38, 73, 46, 22, 77, 49, 12

**19-5** 27

**19-7** (a) 2458.84 MHz, (b) 2459.16 MHz.

**19-9** (a) 2416.784 MHz, (b) 2416.928 MHz, (c) 2417.216 MHz, (d) 2417.072 MHz

## Chapter 20

**20-1** 352.9 THz

**20-2** 193.6 THz

**20-3** 0.7596 nm at $\lambda = 1500$ nm

**20-4** 0.3749 nm at $\lambda = 1500$ nm

**20-5** 40°

**20-6** 35.2°

**20-7** 41.1°

**20-8** 48.8°

**20-9** 0.465 A/W

**20-10** 0.754 A/W

**20-11** 42 dB

**20-12** 2.07 dB

**20-13** −55.6 dBm

**20-14** −16.7 dBm

## Chapter 21

**21-1** 13.47 dB

**21-3** (a) 61.25 MHz    (b) 65.75 MHz    (c) ≈64.83 MHz

**21-5** (a) 107 MHz    (b) 152.75 MHz

**21-7** (a) 45.75 MHz    (b) 41.25 MHz    (c) ≈42.17 MHz

**21-9** 455

# Index